T0235259

Lecture Notes in Computer Science 10311

Commenced Publication in 1973
Founding and Former Series Editors:
Gerhard Goos, Juris Hartmanis, and Jan van Leeuwen

Editorial Board

David Hutchison
  Lancaster University, Lancaster, UK
Takeo Kanade
  Carnegie Mellon University, Pittsburgh, PA, USA
Josef Kittler
  University of Surrey, Guildford, UK
Jon M. Kleinberg
  Cornell University, Ithaca, NY, USA
Friedemann Mattern
  ETH Zurich, Zurich, Switzerland
John C. Mitchell
  Stanford University, Stanford, CA, USA
Moni Naor
  Weizmann Institute of Science, Rehovot, Israel
C. Pandu Rangan
  Indian Institute of Technology, Madras, India
Bernhard Steffen
  TU Dortmund University, Dortmund, Germany
Demetri Terzopoulos
  University of California, Los Angeles, CA, USA
Doug Tygar
  University of California, Berkeley, CA, USA
Gerhard Weikum
  Max Planck Institute for Informatics, Saarbrücken, Germany

More information about this series at http://www.springer.com/series/7410

Raphael C.-W. Phan · Moti Yung (Eds.)

Paradigms in Cryptology – Mycrypt 2016

Malicious and Exploratory Cryptology

Second International Conference, Mycrypt 2016
Kuala Lumpur, Malaysia, December 1–2, 2016
Revised Selected Papers

 Springer

Editors
Raphael C.-W. Phan
Multimedia University, MMU
Cyberjaya
Malaysia

Moti Yung
Snap and Columbia University
New York, NY
USA

ISSN 0302-9743   ISSN 1611-3349 (electronic)
Lecture Notes in Computer Science
ISBN 978-3-319-61272-0   ISBN 978-3-319-61273-7 (eBook)
DOI 10.1007/978-3-319-61273-7

Library of Congress Control Number: 2017944355

LNCS Sublibrary: SL4 – Security and Cryptology

© Springer International Publishing AG 2017
This work is subject to copyright. All rights are reserved by the Publisher, whether the whole or part of the material is concerned, specifically the rights of translation, reprinting, reuse of illustrations, recitation, broadcasting, reproduction on microfilms or in any other physical way, and transmission or information storage and retrieval, electronic adaptation, computer software, or by similar or dissimilar methodology now known or hereafter developed.
The use of general descriptive names, registered names, trademarks, service marks, etc. in this publication does not imply, even in the absence of a specific statement, that such names are exempt from the relevant protective laws and regulations and therefore free for general use.
The publisher, the authors and the editors are safe to assume that the advice and information in this book are believed to be true and accurate at the date of publication. Neither the publisher nor the authors or the editors give a warranty, express or implied, with respect to the material contained herein or for any errors or omissions that may have been made. The publisher remains neutral with regard to jurisdictional claims in published maps and institutional affiliations.

Printed on acid-free paper

This Springer imprint is published by Springer Nature
The registered company is Springer International Publishing AG
The registered company address is: Gewerbestrasse 11, 6330 Cham, Switzerland

Preface

The first and only Mycrypt up to 2015 was Mycrypt 2005, wherein an unconventional cryptography session featured papers with ideas that were beyond the norm. Since Mycrypt 2005, substantial breakthroughs have been made in crypto and security in the past decade, including notions pertaining to malicious security, i.e., where security is no longer simply against bad guys but where good guys who are conventionally viewed as mostly defensive can equally be adversarial. These align well with recent trends wherein trusted parties need not necessarily be trustworthy, and where insiders can potentially be malicious.

Mycrypt 2016 was a rejuvenation of the Mycrypt series with a particular focus on paradigm-shifting crypto research and thinking outside the current box. Jointly organized by the Multimedia University and Cyber Security Malaysia in cooperation with the International Association for Cryptologic Research (IACR), it was held in Kuala Lumpur, Malaysia, during December 1–2, 2016, at the Pullman Kuala Lumpur City Centre Hotel, just before Asiacrypt 2016 took place in Hanoi.

The technical Program Committee (PC) comprised 41 experts from 19 countries with an additional 51 external reviewers. We adopted a hybrid journal-like style where there were two separate and independent calls each with its own review process. Each submission was put through two stages, where after stage one some papers without major issues were accepted or those without scientific merit rejected, while others were put through to stage two for a further review after obtaining rebuttals from authors. Every worthwhile submission was reviewed by at least two reviewers, and up to six reviewers for some submissions requiring substantial deliberations. The server handling the submissions, reviews, and discussions was hosted by Microsoft via its Conference Management Toolkit (CMT). We thank all the PC members and external reviewers for their passion and commitment to see this through. Note that in defying convention, we have listed the PC and external reviewers in alphabetical order based on first name first! This is not a new paradigm, but is somewhat unusual in the West.

A total of 51 complete submissions were processed through the review phases after filtering checks, from which 21 papers were finally accepted for inclusion in the technical program, with authors spanning 14 countries. Of these submissions, three paradigm-revisiting papers that garnered a significant number of positive comments and interest from PC members and external reviewers are listed in these proceedings under the "Revisiting Tradition" category. Papers under "Different Paradigms" touch on alternative or new perspectives on doing cryptography, while the "Cryptofication" papers aim to bridge the gap between the physical world and the cryptographic world. "Malicious Crypto" papers deal with the issues of backdoors, malware, and leakages, while papers under "Advances in Cryptanalysis" aim to revisit existing cryptanalytic techniques toward new formulations or measures. The "Primitives and Features" papers are those that propose new variants of cryptographic primitives or new features. Finally, the "Cryptanalysis Correspondence" section presents two concise papers that had the unique feature of being able to obtain rebuttals from the authors of the attacked schemes as part of an extended review process.

After much deliberation and considering the points made by PC members and external reviewers, the paper "Another Look at Tightness II: Practical Issues in Cryptography" by Sanjit Chatterjee, Neal Koblitz, Alfred Menezes and Palash Sarkar was awarded Best Paper.

Mycrypt 2016 had the pleasure of two keynote talks, namely, Xavier Boyen of QUT speaking on human primacy in crypto and Neal Koblitz of the University of Washington discussing paradigm shifts in our disciplinary culture. There was also an IEICE invited talk sponsored by the IEICE Malaysia Section, by Goichiro Hanaoka on user-friendly crypto.

As an additional twist to an otherwise conventionally structured program, several Insight Papers were solicited for presentation at Mycrypt 2016 and/or inclusion in the proceedings. Insight Papers were invited from researchers on their celebrated or recent breakthrough results: these were notably on multi-prover interactive proofs, human encryption, algebraic cryptanalysis, watermarking programs, polytopic cryptanalysis, and the division property. Authors of such papers had the option of presenting at Mycrypt 2016 physically or pre-recording a video clip to be shown during Mycrypt 2016 to the audience.

We thank the general chairs, Wei-Chuen Yau and Geong-Sen Poh, as well as the local secretariat led by Cyber Security Malaysia for supporting the venue logistics and outreach. We are also grateful to Tourism Malaysia for sponsoring the cultural dance performance during the banquet dinner.

Starting with Mycrypt 2016, the Mycrypt series will continue to focus on beyond-norm, paradigm-shifting, unconventional cryptography as we firmly believe that a field can only advance if its research community revisits conventional paradigms, rocks the crypto boat, questions the status quo, and raises controversial issues. Mycrypt will thus henceforth be known as the International Conference on *M*alicious and Exploratory *Crypt*ology, and be co-located with major crypto/security conferences to maximize the impact of its frontier-stretching theme.

April 2017   Raphaël C.-W. Phan
                                                      Moti Yung

Organization

Program Chairs

Raphaël C.-W. Phan   MMU, Malaysia
Moti Yung   Snap and Columbia University, USA

General Chairs

Wei-Chuen Yau   Xiamen University, Malaysia
Geong-Sen Poh   MIMOS, Malaysia

Program Committee

Adam Young   Cryptovirology Labs, USA
Aggelos Kiayas   National and Kapodistrian University of Athens,
                              Greece
Ahmad-Reza Sadeghi   Technische Universität (TU) Darmstadt, Germany
Alfred Menezes   University of Waterloo, Canada
Andrew Odlyzko   University of Minnesota, USA
Angela Sasse   University College London (UCL), UK
Arjen Lenstra   EPFL, Switzerland
Bart Preneel   University of Leuven (KU Leuven), Belgium
Catherine Meadows   Naval Research Lab, USA
Chris Mitchell   Royal Holloway, University of London (RHUL), UK
David Naccache   Université Paris II, France
Ed Dawson   Queensland University of Technology (QUT),
                              Australia
Elisa Bertino   Purdue University, USA
Graham Steel   Inria, France
Gregory Neven   IBM Research–Zurich, Switzerland
Helena Handschuh   Cryptography Research Inc., USA
Ivan Visconti   Università degli Studi di Salerno, Italy
Jean-Philippe Aumasson   Kudelski Security, Switzerland
Jonathan Katz   University of Maryland, USA
Josef Pieprzyk   QUT, Australia
Kaoru Kurosawa   Ibaraki University, Japan
Kristian Gjøsteen   NTNU, Norway
Lars Knudsen   Danmarks Tekniske Universitet (DTU), Denmark
Marine Minier   LORIA, France
Markus Dürmuth   Ruhr-Universität Bochum (RUB), Germany
Markus Jakobsson   Agari and ZapFraud, USA
Moti Yung (Chair)   Snap and Columbia University, USA

Nasir Memon	New York University, USA
Neal Koblitz	University of Washington, USA
Orr Dunkelman	University of Haifa, Israel
Peter Y.A. Ryan	Université du Luxembourg
Phillip Rogaway	University of California (UC) Davis, USA
Raphaël C.-W. Phan (Chair)	Multimedia University (MMU), Malaysia
Reihaneh Safavi-Naini	University of Calgary, Canada
Ronald Cramer	CWI and Universiteit Leiden, The Netherlands
San Ling	Nanyang Technological University (NTU), Singapore
Sherman S.M. Chow	Chinese University of Hong Kong
Shiho Moriai	NICT, Japan
Tatsuaki Okamoto	Nippon Telegraph and Telephone (NTT), Japan
Vincent Rijmen	KU Leuven and iMinds, Belgium
Yvo Desmedt	University of Texas at Dallas, USA and UCL, UK

External Reviewers

Alexander Koch	Jens Zumbrägel	Rafael Misoczki
Alfredo Rial Duran	John Chan	Robert Granger
Anton Stiglic	Jonathan Bootle	Russell W.F. Lai
Benjamin Wesolowski	Jong Hwan Park	Scott Ruoti
Cristina Onete	Kazuki Yoneyama	Sébastien Canard
Daniel Masny	Keita Emura	Shiwei Zhang
Duane Wilson	Khoa Nguyen	Sigurd Eskeland
Edoardo Persichetti	Koutarou Suzuki	Subhamoy Maitra
Frédérique Elise Oggier	Laura Luzzi	Sylvain Ruhault
Gäetan Leurent	Luisa Siniscalchi	Tom Ristenpart
Guomin Yang	Mark Marson	Vanishree Rao
Hossein Siadati	Michele Ciampi	Viet Tung Hoang
Huaxiong Wang	Nicolas Sendrier	Wakaha Ogata
Hyung Tae Lee	Palash Sarkar	Yannick Seurin
Irene Giacomelli	Peng Jiang	Yu Sasaki
Jae Hong Seo	Peter B. Rønne	Yu Yu
Jean-Pierre Tillich	Pierre-Louis Cayrel	Yusi James Zhang

In Cooperation with

The International Association for Cryptologic Research (IACR)

Sponsoring Institutions

Cyber Security Malaysia
IEICE Malaysia Section
Multimedia University (MMU)
Tourism Malaysia

Contents

Keynotes

The Case For Human Primacy in Cryptography (Summary from
the Keynote Lecture)...................................... 3
   X. Boyen

Time for a Paradigm Shift in Our Disciplinary Culture?................ 11
   Neal Koblitz

Revisiting Tradition

Another Look at Tightness II: Practical Issues in Cryptography 21
   Sanjit Chatterjee, Neal Koblitz, Alfred Menezes, and Palash Sarkar

Another Look at Anonymous Communication: Security and Modular
Constructions 56
   Russell W.F. Lai, Henry K.F. Cheung, Sherman S.M. Chow,
   and Anthony Man-Cho So

Challenges with Assessing the Impact of NFS Advances on the Security
of Pairing-Based Cryptography............................... 83
   Alfred Menezes, Palash Sarkar, and Shashank Singh

Different Paradigms

Key Recovery: Inert and Public 111
   Colin Boyd, Xavier Boyen, Christopher Carr, and Thomas Haines

Honey Encryption for Language: Robbing Shannon to Pay Turing?........ 127
   Marc Beunardeau, Houda Ferradi, Rémi Géraud, and David Naccache

Randomized Stopping Times and Provably Secure Pseudorandom
Permutation Generators 145
   Michal Kulis, Pawel Lorek, and Filip Zagorski

Cryptofication

A Virtual Wiretap Channel for Secure Message Transmission............ 171
   Setareh Sharifian, Reihaneh Safavi-Naini, and Fuchun Lin

Necessary and Sufficient Numbers of Cards for Securely Computing
Two-Bit Output Functions   193
  Danny Francis, Syarifah Ruqayyah Aljunid, Takuya Nishida,
  Yu-ichi Hayashi, Takaaki Mizuki, and Hideaki Sone

Malicious Cryptography

Controlled Randomness – A Defense Against Backdoors
in Cryptographic Devices.   215
  Lucjan Hanzlik, Kamil Kluczniak, and Mirosław Kutyłowski

Malware, Encryption, and Rerandomization – Everything Is Under Attack . . .   233
  Herman Galteland and Kristian Gjøsteen

Protecting Electronic Signatures in Case of Key Leakage   252
  Mirosław Kutyłowski, Jacek Cichoń, Lucjan Hanzlik, Kamil Kluczniak,
  Xiaofeng Chen, and Jianfeng Wang

Advances in Cryptanalysis

A New Test Statistic for Key Recovery Attacks Using Multiple Linear
Approximations   277
  Subhabrata Samajder and Palash Sarkar

Tuple Cryptanalysis: Slicing and Fusing Multisets.   294
  Marine Minier and Raphaël C.-W. Phan

Improvements of Attacks on Various Feistel Schemes   321
  Emmanuel Volte, Valérie Nachef, and Nicolas Marrière

Primitives and Features

Updatable Functional Encryption.   347
  Afonso Arriaga, Vincenzo Iovino, and Qiang Tang

Linking-Based Revocation for Group Signatures: A Pragmatic Approach
for Efficient Revocation Checks   364
  Daniel Slamanig, Raphael Spreitzer, and Thomas Unterluggauer

CARIBE: Cascaded IBE for Maximum Flexibility and User-Side Control . . .   389
  Britta Hale, Christopher Carr, and Danilo Gligoroski

Multi-authority Distributed Attribute-Based Encryption with Application
to Searchable Encryption on Lattices   409
  Veronika Kuchta and Olivier Markowitch

One-Round Exposure-Resilient Identity-Based Authenticated Key
Agreement with Multiple Private Key Generators   436
    Atsushi Fujioka

Cryptanalysis Correspondence

Attacks on the Basic cMix Design: On the Necessity of Commitments
and Randomized Partial Checking.   463
    Herman Galteland, Stig F. Mjølsnes, and Ruxandra F. Olimid

Cryptanalysis of an Identity-Based Convertible Undeniable
Signature Scheme   474
    Rouzbeh Behnia, Syh-Yuan Tan, and Swee-Huay Heng

Invited and Insight Papers

Towards User-Friendly Cryptography   481
    Goichiro Hanaoka

Multi-prover Interactive Proofs: Unsound Foundations.   485
    Claude Crépeau and Nan Yang

Human Public-Key Encryption   494
    Houda Ferradi, Rémi Géraud, and David Naccache

Two Philosophies for Solving Non-linear Equations
in Algebraic Cryptanalysis   506
    Nicolas T. Courtois

Watermarking Cryptographic Programs   521
    Ryo Nishimaki

From Higher-Order Differentials to Polytopic Cryptyanalysis   544
    Tyge Tiessen

Division Property: Efficient Method to Estimate Upper Bound
of Algebraic Degree   553
    Yosuke Todo

Author Index   573

Keynotes

The Case For Human Primacy in Cryptography
(Summary from the Keynote Lecture)

X. Boyen[(✉)]

QUT, Brisbane, Australia

Abstract. In the history of "adversarial" technologies, cryptology has the unique distinction of offering an exponential advantage to the defence over the offence, as long as it is implemented correctly and, crucially, the defence platform is not otherwise compromised. Sadly, the ubiquitous devices wherein people increasingly confide their life stories and secrets, have dubious allegiances, and appear hardly worthy of any trust.

This lecture is concerned with the "clouds" that gather over the integrity of our global computing landscape, from an individual, not institutional, perspective. We argue on the basis of the ongoing disappearance of user-accessible honest computing, that the cryptographic research community should strive for security and privacy models that give absolute primacy to human principals, rather than their electronic surrogates.

We advocate a human-centric agenda as the core of a broader research programme to build trustworthy systems on top of untrusted platforms. A promising approach is through (strong) human-powered cryptography.

It is customary when designing or analysing a security protocol, to consider and even name human protagonists, such as Alice and Bob, as if they were truly in charge of the protocol's execution: dutifully discharging all of the specified cryptographic operations whereby some designated security goal can be achieved (e.g., key exchange to create secure communication over an unsafe channel, etc.).

In this merry Wonderland, Alice is a model modular exponentiation expert. She races through the twists of serpentine elliptic curves, pedal-to-the-metal, while snacking on a side of AES by the mega-bite. Her cryptic dance abides by a flawless choreography...But in reality, outside of the Rabbit Hole, the truth is that Alice is quite helpless without Hal, the proxy computing device that does all of the heavy lifting, for her and on her behalf (or so she thinks).

Typically unacknowledged in this picture, is that Hal may be unreliable or even treacherous. Bugs happen, and they tend to be the norm rather than the exception, even in the stern and buttoned-down security industry. In a not-too-distant future, as machine learning takes over many traditional areas of computer programming, perhaps Hal might one day have its own unpredictable agenda? More prosaically, but with far greater contemporary relevance, what if computing tools and other "smart devices" sold to the unsuspecting public were already less than honest, being beholden to other masters? As we shall see, such questions

© Springer International Publishing AG 2017
R.C.-W. Phan and M. Yung (Eds.): LNCS 10311, Mycrypt 2016, pp. 3–10, 2017.
DOI: 10.1007/978-3-319-61273-7_1

underscore the reality that security cannot be credibly outsourced; it needs to be rooted in the very entity for which it is sought, without intermediary, which for human security/privacy means the human itself being their own root of trust.

1   Clouded Computing

The explosive allegations by whistle-blowers Binney and Snowden of (then-) unlawful and anti-constitutional surveillance that their national security agency had engaged on its own citizens (among others), undoubtedly mark a defining turn in the storied history of personal computing.

It was alleged that tracking and monitoring of ordinary people on a huge scale was being conducted, through highly sophisticated electronic means, e.g., by hooking into *private* interactions that *private* individuals had with *personal* devices *privately* owned and paid for. Such massive and indiscriminate electronic mass surveillance on a scale commensurate to an entire country, required implicit or explicit industry complicity, in the hardware, software, telecommunication, and internet sectors, which were relentlessly sought, subverted, and exploited.

Of course, surveillance of citizenry by their government is nothing new, but its execution through *consumer* electronics with industry involvement is a novel phenomenon that is seemingly affecting people's attitudes. Not only are much of the deeds exposed by Binney and Snowden being banalised by retroactive legislation (whose constitutionality many are doubting), the general apathy from the media appears to have emboldened the private sector to follow suit.

The past couple of years in particular have seen a veritable recrudescence of documented cases of deceptive "clouded" computing, marketed at the public. The seeds have been planted for decades, but it is only recently that the industry has drastically ramped up the invasiveness of their own for-profit "user data collection" programmes, while being defiantly unapologetic about it. Part of the problem is also a pattern of legal quasi-immunity afforded to the software industry in the relevant country, while the media distracts with outside threats.

Indeed, much of the press, albeit with highly laudable exceptions, takes little issue with either government or corporate mass surveillance. On the contrary, it likes to focus on "hackers", especially foreign ones, for the "foreign" political bogeyman *du jour*. Of course, hacking is a real threat, but only one side of the equation. The other is the sorry state of commercial software stewing a first-to-market rush culture of quantity over quality, even ignoring the basic question of trust when that culture self-servingly propandises the death of privacy.

The "Internet-of-Things" fad epitomises both attitudes, infamous for *nanny* home appliances morphing from Matrix minders to botnets of zombies. In both cases the user is being watched; only the watcher is changing. The actual harm is thankfully limited by the paucity of stealable data from the typical IoT fridge (but beware those "personal assistant" camera/microphone-laden Trojan horses that people are lured to invite in their homes). Hacks directed at phones and actual computers are potentially much more serious, with consequences ranging from data destruction (via ransomware) to identity fraud (via spyware).

At the extreme end of the spectrum, hacking can be state-sponsored and wielded as a weapon (remember Stuxnet, with its alleged collateral damage). The lines are blurred between those and the new *civil* cyber-warfare of the sort that Binney and Snowden were blowing the whistle about, who, lest we forget, ultimately depended on a *military* chain of command.

2   Rain Men

We now turn to the role played by certain corporations in undermining the personal sanctity of computing, though we really cannot give justice to all such astonishing case studies of electronic voyeurism funded by their own victims.

It all starts with a business model that can be said to have arisen about two decades ago: the business of stalking unsuspecting members of the public on the then-nascent World-Wide Web, to uncover, record, and monetise their every digital footprints.

Around that time was founded a certain advertising-through-technology company whose bold research and development ambitions revolved (and still revolve) on the collection and monetisation of every bit of datum about every last man, woman, and possibly child, starting with the very patrons of its marquee internet search engine. (That is not to say that it remains the king of online indiscreetness; that crown now likely belongs to internet social networks, whose model it is to entice people to gossip on-the-record about others, tracking their relations and interests, while also offering an illusion of private messaging for confidences.)

Relevant to our matter, the advertiser has now become a dominant software integrator for mobile devices, which happen to make extraordinarily piercing and intrusive ogling platforms. Indeed, smart phones with their multiple radios and redundant network gateways provide an unsquelchable path for the exfiltration of a trove of personal data, collected in part through the vast array of sensors that the hardware device manufacturers are all but seemingly mandated to provide. This collection and exfiltration is secured and obscured by the convenient excuse of regulatory compliance for radio transmitters, invoked as the reason for locking both the hardware and software against modifications (this contrasts with the personal computer market, where, until recently, proprietary software could be substituted for open-source alternatives of the computer owner's choosing).

Evidently, such "data oriented" business models must be successful, for privacy-abiding personal devices are hardly ever to be found any more.

On the traditional personal-computer front, also, does invasive data collection appear sharply on the rise. Perhaps the most blatant example of such, performed by a corporation on a massive scale, is to be found in the current version of a dominant legacy personal computer operating system, used by millions all over the world. Said OS has reportedly been laden with a euphemistically called array of "telemetry" activated by default, including a full key logger transmitting key strokes to the corporate watcher. Now ponder: keyboard and pointer are the primary and sometimes only conduits by which humans enter information into personal computers. Clearly, no private message, business secret, nor *user password* could possibly stay safe for very long on such a platform, no matter how strong the cryptography of any security software installed on top of it.

As an indicator of this vendor's motivations, and harbinger of its controlling intentions, many paying customers of prior versions have complained to have had the update forced upon them against their consent. Reportedly, lawsuits are presently unfolding (despite the uphill battle of overcoming a jurisprudence that often seems to excuse software vendors from warranty and liability).

3   Umbrellas

Despite what looks (or is portrayed by the press) as public apathy, the foregoing manoeuvres by governments and corporations alike are not going on without stern opposition. Communities of advocates, from ethicists to computer and legal professionals, have been and continue to be highly vocal to denounce new cases of mass surveillance and "user data collection" as they are being exposed, and work to steer this ship in the courts of law and public opinion.

On the technical side, a massive and increasingly growing counter-current of top-quality "free and open-source software" (FOSS), e.g., in the Linux and BSD ecosystems, is gained mainstream exposure as a suitable alternative to many commercial offerings. As part and parcel of treating patrons with respect, *software libre* provides unique benefits, such as the freedom to implement and distribute improvements, and a long-term viability owing to immunity to vendor lock-in. Open software also more naturally espouses basic security principles and honours people's rights to privacy, enforced partly by reputation and partly by the technical difficulty of hiding spyware and malware in fully open source code.

Regarding open-source, the movement far predates the rise of electronic surveillance, but it is fair to say that it is boosted by it, as a predictable "immune system" defence reaction. Similar justifications have been advanced to explain the rise of cryptocurrencies and their appeal with a segment of the population. Cryptography itself is enjoying a spike of interest in the public discourse, as it did twenty-five years ago during the first War on (civilian) Cryptography.

As a testament to the resilience of high-profile open-source projects, a number of attempts by suspected state-sponsored saboteurs to plant malware into the Linux kernel have seemingly been caught and repelled over the years. A new controversial *system daemon* is now aggressively pushed into Linux distributions; if one were to attribute to malice its overarching anti-features, one might grin it could betray a shift to a softer target, likely as it is to be shunned by experts who would otherwise hunt for bugs and backdoors as they do in the kernel itself.

Whether this turns out to be benign or malign, the point is that large-scale subversion attempts appear harder to pull off in an open-source ecosystem.

4   Perfect Storm

Unfortunately, despite the indisputable success of open-source software, neither end-users nor hobbyist communities have been able yet to eschew closed-course hardware with the obvious associated risks. (Although, just as affordable 3D printers are currently revolutionising hobbyist desktop manufacturing, printable integrated circuits might one day allow people to build their own open-source trusted devices at home from the ground up, but we are not there yet.)

On the war against secure personal computing, OEM manufacturers appear mostly neutral at the moment. Yet, there are unmistakable signs of their being subjected to substantial pressure to alter that stance, and not in the favour of the individual customer's rights. Much of that pressure seems to come from the aforementioned dominant legacy PC OS vendor. OEMs are made to implement an ostensibly "secure" boot mechanism for all desktop and laptop retail PCs running that OS. By all discernable accounts, this initiative has little to do with actual security (the interminable litany of incidents incurred by said OS would quickly disabuse anyone of that notion) but it effectively cements the proprietary OS by locking out bespoke open-source alternatives (lip service is being paid by permitting a few "open-but-unmodifiable-source" pre-compiled kernels).

The push for such locked bootloaders in (not-so)-general-purpose personal computers mimics the *bold trail* "blazed" by the mobile phone industry, but here without the ostensible excuse of radio transmitters for locking it all down. The exclusion mechanism involves digital signatures; this is one example of institutional use of cryptography *against* the end-user and customer. As we have seen earlier, the particular OS that such use of cryptography is meant to cement, is also the one whereupon the deployment of secure cryptography, outside of the vendor's purview, might just be a fool's errand.

5   Hurricane's Eye

Equally alarming, but much less obvious, is the threat of nearly incontrovertible surveillance and "private cryptography minders", *in hardware*, that loom on the horizon. At the moment, we can only speculate as to what form this might take; the following scenarios are at best speculative educated guesswork.

Our first hint comes from reports, by security and industry observers, that for the past decade a premier maker of branded CPUs for PCs has laden its chips with a sophisticated self-contained secondary computer subsystem.[1] Such "management engine" embeds a powerful CPU (allegedly a competitor's ARM core, in delightful irony) that runs opaque firmware that has resisted all reverse-engineering attempts disclosed so far. The subsystem is autonomous and totally invisible (hence useless) to the paying customer. It nevertheless has full hardware access, including to memory, input devices, and wired/wireless networks. The subsystem is being acknowledged as "enterprise-class"; nevertheless, it has reportedly been present across the board in all of the maker's CPUs, including crippled models destined for the entry-level consumer market, for the last decade. Recall that high-volume chip lithography costs increase super-linearly with area. That the designer would opt to devote precious die space for such complex yet

---

[1] Cf. eg.: Joanna Rutkowska (2015) https://blog.invisiblethings.org/papers/2015/x86_harmful.pdf.

ostensibly pointless subsystem (outside of the "enterprise-class" market), makes so little *economic sense* to us that it begs for a *hidden agenda* explanation.

Our second clue comes from the very technical documents leaked by Edward Snowden in 2013, which, *inter alia*, mention technically credible methods for the exfiltration of data by virtually undetectable semi-passive radio, based on a modicum of software and some *integratable* tiny hardware on the target machine.

Our third hunch comes from the CPU architecture's instruction set itself, which leads us to imagine how hypothetical *hardware moles* speculatively buried deep within an off-the-shelf computer, and operating only at a very low level (e.g., at the machine level itself, merely tapping the seemingly meaningless stream of "text" and "data" words being fed to the main CPU for execution) might acquire sensitive information worthy of and suitable for exfiltration, such as keys.

Further connecting the dots, one might arrive at the rather sobering epiphany that there is no imperative reason, other than a quaint sense of obligation of decency, why virtually all laptop and desktop computers in the hands of the public, would not already be *irremediably bugged*, in hardware, *today*.

To speak in the abstract, it would not be a stretch to ascribe to some nations' security agencies the desire to plant undetectable hardware moles *at manufacture* in general-purpose computers sold to the public, to serve as beach heads for mass surveillance or targetted spying, or to bypass unsanctioned cryptography. Were this ever to happen, hypothetically speaking of course, the personal threats to security would be nearly impossible to counter merely by software substitution, notwithstanding the heroics of the FOSS community.

6   Looking for Sunshine

As cryptography and security researchers, it is our duty not only to respond to present dangers, but also to anticipate and parry future threats, even as far fetched as the ones imagined in the previous sections. The picture being dressed up is quite gloomy indeed, as it concerns the denial of *individual* command of computing and of *personal* privacy and security for the rest of us, mere humans. (This does not concern institutional computing, as large entities undoubtfully have the clout to procure solutions that meet their specifications.)

Trustworthy computing over untrusted platforms is the name of the game, and the house has the upper hand. How should one go about to play it?

Open-source auditable software will certainly continue to play a key role, but it is highly unlikely that software by itself could defeat hardware subversion.

Open-source desktop micro-manufacturing of bespoke integrated electronics may one day bring some bright sunshine to clear up the field, but the technology is in its infancy, and there is no doubt that vested interests will unleash tricks, laws, and regulations to stymie or delay its popularisation.

Until such day, or as part of a multi-pronged strategy, cryptography may provide a credible approach to *creating trust out of mistrust*. The difficulty here is that *trust* and *mistrust* are to live very close to each other: the former to be gained by a human interacting with a device only worthy of the latter. Carefully bridging this gap will require models that give *primacy to humans* (and primacy specifically to the designated first-person human principal beneficiary) while viewing any proxy device as a separate entity and potential adversary.

Bringing such models to fruition will involve the bootstrapping of entirely new cryptographic protocols, in whose execution humans must take an active and essential role, complete with (humanly feasible) cryptographic calculations and (humanly memorable) secrets. It is clear that with obvious human cognitive limitations, achieving strong security under *human primacy* models will not be an easy task. Nevertheless, we believe that it is feasible: At the lecture was demonstrated a candidate *public-key* cryptosystem allowing *human decryption* of short messages, without any device or even a scratch pad.[2]

Whichever path is taken, personal security will necessitate personal effort.

7 Conclusion

Peddlers of tied systems would like us to believe in *Privacy's untimely demise.* Governments all too often seem to condone or even foster surveillance-enabling behaviours, rather than stand for their constituents' security and privacy rights. These actions seriously undermine the computer and data security that matters: not *theirs*, but *ours*—private individuals' held captive to such devices.

The take-home message is that *trustworthiness* is increasingly scarce in the computer world. If safety in computing is ever to obtain, it has to be rooted in inherently trustworthy foundations, i.e., self-referential or very close thereto. Top-to-bottom design or vetting of private computing devices is one possible approach, for those with the technical wherewithal to do such things. Another path is through the potent trust-leveraging tools offered by cryptography, but the necessary computations are keeping its benefits out of direct human reach.

Regardless of the direction taken, there is *no free lunch* with data security and privacy. For those whose seek them, one's direct involvement in trust assurance or establishment is critical (or at the very least, the ability to rely on trusted experts for the same). This will be true in all the proposed solutions, whether trust arises from the hand-assembly of a clean hardware and software stack, or from the direct human execution of sound cryptographic algorithms.

Cryptography indeed provides very potent weapons in the war *for* privacy, drawing on immaterial mathematical truths resistant to censorship and embargo. Cryptography is unfortunately not very human-friendly, which has impeded its development for direct human benefit. If true human cryptography is ever to become a reality, for human privacy or security, it will have to involve new formal models and protocols constructed with *and for* humans as executors and beneficiaries. The feasibility of such a research programme is still largely open-ended, but there are promising leads, and the implications could be profound.

---

[2] An upcoming paper focuses on the technical details, which go far beyond the scope of this piece.

Acknowledgements. The author thanks the Programme Chairs and Programme Committee for the kind invitation to speak at MᴏCʀʏᴘᴛ "Paradigm-Shifting Cryptography" 2016. The author is supported by ARC Future Fellowship grant FT140101145.

Disclaimer

All claims made in this piece are opinions or speculations of the author only, intended to be thought provoking, editorialised from past and present public records and testimonials too numerous and disparate to be enumerated.

Time for a Paradigm Shift in Our Disciplinary Culture?

Neal Koblitz$^{(\boxtimes)}$

Department of Mathematics, University of Washington, Seattle, USA
koblitz@uw.edu

1 Introduction

The well-known KISS principle of engineering — *Keep It Simple, Stupid!* — is also of value in cryptography. In certain subfields, such as lattice-based crypto and indistinguishability obfuscation, the proposed constructions pay little heed to the KISS principle. Even the descriptions of the proper functioning of the protocols are frightfully complicated (by comparison with RSA or ECC, for example), and the security analyses and guidelines for parameter selection are even more problematic.

But even something as wonderful as the KISS principle can be taken too far, as I learned to my chagrin during my early years of work in cryptography. In the late 1980s, when I wrote or spoke about ECC, I wanted to use the simplest possible examples, and these were the supersingular curves. For instance, just take the equation $y^2 = x^3 - x$ over a field of p elements with $p \equiv 3 \pmod 4$ or else $y^2 = x^3 - 1$ with $p \equiv 2 \pmod 3$, where p is chosen so that respectively $(p+1)/4$ or $(p+1)/6$ is prime. As long as $\log_2 p \geq 163$ we would have 80 bits of security, which at the time was enough.

A few years later, Menezes et al. [14] showed that the discrete log problem (DLP) on such a curve can be reduced to the DLP in the finite field of p^2 elements, and even in the early 1990s this was insecure.

A short time later, when I proposed Hyperelliptic Curve Cryptography (HCC), I again made a very erroneous judgment about parameter selection. My favorite example was a genus-191 hyperelliptic curve over the field of 2 elements whose jacobian has group order divisible by a prime of more than 160 bits. So I was confident in its security. In fact, I thought that a high-genus curve would be more secure than a low-genus one. I couldn't have been more wrong! Although the genus is a measure of the topological complexity of the curve, it was a rookie mistake for me to confuse that with the computational complexity of the corresponding DLP. In 1994, Adleman et al. [1] found a subexponential-time algorithm for the DLP on high-genus curves. As a result, my genus-191 example was totally insecure, even in 1994. At present, based on work of Gaudry, Diem, and others, we believe that HCC with $g \geq 3$ is less secure than ECC (which is the $g = 1$ case of HCC); the case $g = 2$ is the only one that seems to be as secure as ECC.

© Springer International Publishing AG 2017
R.C.-W. Phan and M. Yung (Eds.): LNCS 10311, Mycrypt 2016, pp. 11–18, 2017.
DOI: 10.1007/978-3-319-61273-7_2

In this way I learned early on how easy it is to make serious mistakes in cryptography. Making mistakes is not necessarily a bad thing if we learn from them, and one lesson to be learned is to exercise caution in giving assurances to each other and to the general public. We should present our results with care and humility. The pressures on us — characteristic of a modern capitalist, consumer society — to act like salespeople, advertisers, and hypesters, should be resisted.

One expects to encounter giant egos among entrepreneurs, professional athletes, media celebrities, and (especially in the U.S.) politicians. In contrast, a long historical tradition in the intellectual professions, including science and mathematics, has been to discourage

- extreme competitiveness;
- boastfulness and self-promotion;
- aggressive marketing of one's own work;
- angry, intemperate responses to criticism;
- arrogance.

However, the disciplinary culture of our community has become remarkably tolerant of aberrant behavior by researchers — and the worst conduct is sometimes by very prominent people. We have strayed far from the standards that one would expect from scientists. With alarming frequency we see

- abstracts and introductions to papers that exaggerate the authors' contributions with misleading and inaccurate claims;
- shameless self-promotion in invited talks;
- arrogance toward those who belong to different subdisciplines or social groups;
- anger and retaliation in response to criticism, or simply ignoring work that questions the prevailing notions.

2   Important Work Gets Dismissed or Ignored

I will give two examples. First, in the late 1990s, Blake-Wilson and Menezes [4] showed that certain standardized signature protocols (but not all) are vulnerable to an attack that they named the Duplicate Signature Key Selection (DSKS) attack. To illustrate how such an attack works, let's take the example of an online lottery. Alice chooses her number N and sends it in with her signature $s_{\text{Alice}}(N)$. If her number wins, she claims her winnings by showing her certified public key to the officials, who verify that the signature on the winning number was hers.

But before Alice gets around to doing this, a thief named Bob computes a key pair that satisfies the condition that the same signature on N verifies as his, that is, $s_{\text{Bob}}(N) = s_{\text{Alice}}(N)$. He quickly gets it certified and claims the money before Alice shows up.

DSKS attacks have never attracted the interest of people working in provable security. They've been mentioned briefly by Canetti and Dodis, but only in order to dismiss them as unimportant. It is true that DSKS does not violate security of signature schemes under the standard Goldwasser–Micali–Rivest (GMR) security model [6]. But the GMR definition was formulated in 1984, way before anyone was thinking of online lotteries as an application of a signature scheme. It seems a little imprudent to rigidly adhere to a 1/3-century old definition even after it has been shown to be inadequate in certain settings. For more details, see [10].

A second example of work that deserves to be much better known than it is concerns hybrid encryption schemes. Five years ago Zaverucha posted an important analysis [16] of the security of such schemes in the multi-user setting, which, obviously, is where they are normally deployed. He found practical attacks on certain implementations that had been developed by H. Krawczyk and others and had been "proved secure" by them — but of course only in the single-user setting. Zaverucha's work has been all but ignored by provable security researchers. For more details see [16] and Appendix B to the article [5] in this volume.

3 Exaggerated Advertising of One's Own Work

I'll give two examples, both from leading members of our profession. First, here's an excerpt from the introduction to a widely-cited Micali and Reyzin paper [15] on leakage resilience:

> We focus on the strongest possible adversary, so as to capture what is cryptographically possible in the worst possible, physically observable setting. In particular, we
> • consider an adversary that has full (and indeed adaptive) access to any leaked information; . . .
> • construct pseudorandom generators that are provably secure against all physical-observation attacks.
> Our model makes it easy to meaningfully restrict the power of our general physically observing adversary.

Reading these paragraphs could give the practical cryptographer false hopes. Despite the extravagant promises, the paper contains no concrete construction of any pseudorandom generator, let alone one that resists side-channel attacks. Nor do the authors give any techniques that "make it easy to meaningfully restrict the power" of the side-channel attacker.

A second example is from Hugo Krawczyk's invited talk at Asiacrypt 2010. In it Krawczyk described how his work (along with Bellare and Canetti) in developing HMAC in the 1990s had to satisfy both the engineers and theoreticians and achieve a balance between practicality and theoretical soundness. His slide concluded: "Balance regained, and the rest is history."

This was the first time I saw the expression "the rest is history" used about the speaker's own work! Normally one uses that expression about what someone else did. For instance, one can say, "In the 17th century Leibniz and Newton invented calculus, and the rest is history." Or, "In 1939 Einstein signed a letter to U.S. President Roosevelt urging him to start a nuclear weapons program, and the rest is history."

In the mathematical world, I could not imagine someone saying, "In the 1990s I did such-and-such, and the rest is history." That level of boastfulness would be considered socially unacceptable among mathematicians and, I believe, scientists in most fields. But in our community no one bats an eyelash.

And HMAC is not beyond controversy, although the words "the rest is history" would imply that it is.

In 2012 Menezes and I showed that the main concrete security guarantee for the pseudorandom-function property of HMAC that was claimed by Bellare [2] at Crypto 2006 was based on flawed reasoning. It now seems that no practically meaningful prf-property of the sort in Bellare's paper can be proved for HMAC if it is implemented with MD5 or SHA-1.

In addition, in a recent paper on 1-key nested MACs [11] Menezes and I showed that the same security theorems (the tight reduction for the secure-MAC property and the nontight reduction for the prf-property) can be proved for a broad class of MACs, and there's no good reason to believe that the HMAC construction is the best in this class for either security or efficiency.

4   Responses to Criticism

One way to distinguish a healthy disciplinary culture from an unhealthy one is to examine the way leaders of the field respond to criticism. How did Bellare react to our critique [9] of his Crypto 2006 paper, which we posted in February 2012? He wrote me [3]:

> I find your current manuscript insulting to me personally and also wrongful in the way it represents the field. I had in the past been supportive of the goals of the Another Look series, [unlike] most cryptographers I know, who have reacted violently to every paper in the series. . .

This is amazing — "reacted violently to every paper in the series"! Is this a rational response?

Bellare has never responded publicly to the specific criticism of the fallacy in his concrete security analysis for HMAC. Rather, he has simply accused Menezes and me of being ignorant of the basics of modern cryptography. His failure to deal seriously with this issue is especially regrettable because the security of HMAC is not just a theoretical question — HMAC is one of the most popular and widely-used message authentication codes.

My second example concerns a prominent researcher's reaction to criticism of a flaw in his Crypto 2005 paper [12]. In that paper Krawczyk described his modified version of the Menezes–Qu–Vanstone key agreement protocols, which he called HMQV. He claimed that by supplying a proof of security that did not require a certain step of the original MQV protocol, he could increase efficiency at the same time as he proved security. Astoundingly, the Crypto program committee had accepted the paper after a superficial reading without asking any of the designers of MQV for comment.

The omitted MQV step was a public key validation that had been introduced to prevent known attacks. When Menezes finally got to see the paper, he immediately saw that some of Krawczyk's protocols fell victim to the same attacks. How could this be, if they're "provably secure" without the public key validation?

Menezes started reading the proof carefully, and soon found a glaring flaw (see [13] for details). Krawczyk — and the Program Committee — had been so mesmerized by the (fallacious) proof that they had failed to see the vulnerability.

How did Krawczyk react to this embarrassing setback to HMQV? He denied that it was of any importance, and responded angrily when I described this episode in a 2007 article [8] in the *Notices of the Amer. Math. Society*. In a letter to the *Notices* he wrote:

> Contrary to what Koblitz claims, the HMQV work represents a prime example of the success of theoretical cryptography, not only in laying rigorous mathematical foundations for cryptography at large, but also in its ability to guide us in the design of truly practical solutions to real-world problems.

Part of what provable security researchers mean by the last phrase (as explained, for example, in [7]) is that they can improve efficiency by dropping unnecessary steps, where "unnecessary" means that the proof of security doesn't require them. Never mind that HMQV as published in Crypto 2005 was insecure, because of the omitted step. Never mind that its "rigorous mathematical" proof of security was fallacious.

Why do prominent researchers react so angrily to criticism? Why do they expect everyone to think that they are perfect and never make mistakes? Is it because of their personalities? Do they behave this way in the non-crypto world with their families and friends? My guess is they don't. They probably have pleasant, normal personalities in the outside world, and the reason for their behavior in the crypto world is that our disciplinary culture tolerates and even encourages bad behavior.

5   Harmful Effects

Have Menezes and I suffered reprisals for our "Another Look" series of papers? For example, have Bellare's colleagues who "reacted violently to every paper in the series" sometimes blocked (or attempted to block) publication of our work? Of course. But that does us no harm really, because we can easily find ways around it — and in any case we're established old guys with secure jobs in university math departments. So there's not much they can do except wallow in impotent rage.

The harmful effects of the aggressive behavior and angry reaction to criticism are felt most of all in the younger generation of recent PhDs who are starting out and have no job security. They are pressured to conform, are socialized into a sycophantic attitude toward the old guys, and are less likely to challenge reigning paradigms and go in radically new directions. The message to young people is a new KISS principle: Kiss up to the establishment, flatter your elders, do what they say, and *never, never* criticize them. This is not the way that science progresses.

There's another harmful effect of our disciplinary culture that concerns me: it discourages most women. In almost all cultures of the world, women from a young age are socialized into thinking that aggressive competitiveness and excessive boasting are improper behavior for women. Even when they see men behaving that way, they grow up with the understanding that such male behavior is wrong for them. Plenty of men also find a hyper-competitive, egotistical disciplinary culture to be unpleasant. But on average women are even more likely to find it disagreeable.

In the U.S., this is reflected in the statistics about female participation in computer science (from which we inherited our disciplinary culture) versus mathematics (which has a less competitive disciplinary culture). In the 1970s and 1980s, when the young field of computer science had a very different disciplinary culture from what it has now, there was a higher percentage of women studying at an advanced level in computer science than in mathematics. However, at present roughly 30% of math PhDs go to women, whereas women get only about 20% of computer science PhDs. Although no one intended to discriminate against women, our disciplinary culture in effect does precisely that.

6   Conclusion

In our time — especially since January 2017 — irrational and vindictive behavior, unrestrained boastfulness, and rejection of well-established norms of scientific inquiry are being more and more associated with American national character. When certain prominent U.S.-based cryptographers react angrily to technical criticisms of their work, retaliate against their critics, and continue to hype their work without mentioning the flaws or weaknesses that have been discovered, they are conforming to this unfortunate stereotype of Americans.

It is time for our community to return to the scholarly values that were articulated in the ancient world, for example, by the great leader of Islam of the 7th century, Ali Ibn Abi Talib, who said

The most harmful disaster for the intellect is arrogance.

And also:

When proven wrong, the wise man will correct himself and the ignorant will keep arguing.

We need to make a paradigm shift in our disciplinary culture before we can claim that our field deserves to be called a "science." We should agree to use words like "scientist" and "scholar" only for those who

- present their results modestly and honestly, highlighting the limitations and never overstating their accomplishments; and
- respond to criticism by thanking the critics and withdrawing any claims that are shown to be fallacious or questionable.

We should use words like "marketer" and "hypester" for those who

- make exaggerated or misleading claims in their abstract or introduction;
- engage in aggressive self-promotion; or
- respond to critics with anger and retaliation, rather than carefully addressing the technical issues that the critics have raised.

References

1. Adleman, L.M., DeMarrais, J., Huang, M.-D.: A subexponential algorithm for discrete logarithms over the rational subgroup of the Jacobians of large genus hyperelliptic curves over finite fields. In: Adleman, L.M., Huang, M.-D. (eds.) ANTS 1994. LNCS, vol. 877, pp. 28–40. Springer, Heidelberg (1994). doi:10.1007/3-540-58691-1_39
2. Bellare, M.: New proofs for NMAC and HMAC: security without collision-resistance. In: Dwork, C. (ed.) CRYPTO 2006. LNCS, vol. 4117, pp. 602–619. Springer, Heidelberg (2006). doi:10.1007/11818175_36
3. Bellare, M.: Email to Koblitz, N., 24 February 2012
4. Blake-Wilson, S., Menezes, A.: Unknown key-share attacks on the station-to-station (STS) protocol. In: Imai, H., Zheng, Y. (eds.) PKC 1999. LNCS, vol. 1560, pp. 154–170. Springer, Heidelberg (1999). doi:10.1007/3-540-49162-7_12
5. Chatterjee, S., Koblitz, N., Menezes, A., Sarkar, P.: Another look at tightness II. In: Phan, R.C.-W., Yung, M. (eds.) Mycrypt 2016. LNCS, vol. 10311, pp. 21–55. Springer, Cham (2017)
6. Goldwasser, S., Micali, S., Rivest, R.: A "paradoxical" solution to the signature problem. In: Proceedings of the 25th Annual IEEE Symposium on the Foundations of Computer Science, pp. 441–448 (1984)
7. Katz, J., Lindell, Y.: Introduction to Modern Cryptography. Chapman and Hall/CRC, London (2007)
8. Koblitz, N.: The uneasy relationship between mathematics and cryptography. Not. Amer. Math. Soc. **54**, 972–979 (2007)
9. Koblitz, N., Menezes, A.: Another look at HMAC. J. Math. Cryptol. **7**, 225–251 (2013)
10. Koblitz, N., Menezes, A.: Another look at security definitions. Adv. Math. Commun. **7**, 1–38 (2013)
11. Koblitz, N., Menezes, A.: Another look at security theorems for 1-key nested MACs. In: Koç, Ç. (ed.) Open Problems in Mathematics and Computational Science, pp. 69–89. Springer, Cham (2014)

12. Krawczyk, H.: HMQV: a high-performance secure Diffie-Hellman protocol. In: Shoup, V. (ed.) CRYPTO 2005. LNCS, vol. 3621, pp. 546–566. Springer, Heidelberg (2005). doi:10.1007/11535218_33
13. Menezes, A.: Another look at HMQV. J. Math. Cryptol. **1**, 47–64 (2007)
14. Menezes, A., Okamoto, T., Vanstone, S.: Reducing elliptic curve logarithms to logarithms in a finite field. IEEE Trans. Inf. Theory **39**, 1639–1646 (1993)
15. Micali, S., Reyzin, L.: Physically observable cryptography. In: Naor, M. (ed.) TCC 2004. LNCS, vol. 2951, pp. 278–296. Springer, Heidelberg (2004). doi:10.1007/978-3-540-24638-1_16
16. Zaverucha, G.M.: Hybrid encryption in the multi-user setting. http://eprint.iacr.org/2012/159.pdf

Revisiting Tradition

Another Look at Tightness II:
Practical Issues in Cryptography

Sanjit Chatterjee[1], Neal Koblitz[2], Alfred Menezes[3(✉)], and Palash Sarkar[4]

[1] Department of Computer Science and Automation,
Indian Institute of Science, Bengaluru, India
`sanjit@csa.iisc.ernet.in`
[2] Department of Mathematics, University of Washington, Seattle, USA
`koblitz@uw.edu`
[3] Department of Combinatorics and Optimization, University of Waterloo,
Waterloo, Canada
`ajmeneze@uwaterloo.ca`
[4] Applied Statistics Unit, Indian Statistical Institute, Kolkata, India
`palash@isical.ac.in`

Abstract. How to deal with large tightness gaps in security proofs is a vexing issue in cryptography. Even when analyzing protocols that are of practical importance, leading researchers often fail to treat this question with the seriousness that it deserves. We discuss nontightness in connection with complexity leveraging, HMAC, lattice-based cryptography, identity-based encryption, and hybrid encryption.

1 Introduction

The purpose of this paper is to address practicality issues in cryptography that are related to nontight security reductions. A typical security reduction (often called a "proof of security") for a protocol has the following form: A certain mathematical task $\mathcal{P}$ reduces to the task $\mathcal{Q}$ of successfully mounting a certain class of attacks on the protocol—that is, of being a successful adversary in a certain security model. More precisely, the security reduction is an algorithm $\mathcal{R}$ for solving the mathematical problem $\mathcal{P}$ that has access to a hypothetical oracle for $\mathcal{Q}$. If the oracle takes time at most T and is successful with probability at least ϵ (here T and ϵ are functions of the security parameter k), then $\mathcal{R}$ solves $\mathcal{P}$ in time at most T' with probability at least ϵ' (where again T' and ϵ' are functions of k). We call $(T'\epsilon)/(T\epsilon')$ the *tightness gap*. The reduction $\mathcal{R}$ is said to be *tight* if the tightness gap is 1 (or is small); otherwise it is *nontight*. Usually $T' \approx T$ and $\epsilon' \approx \epsilon$ in a tight reduction.

A tight security reduction is often very useful in establishing confidence in a protocol. As long as one is not worried about attacks that lie outside the security model (such as side-channel attacks, duplicate-signature key selection attacks, or multi-user attacks [56]), one is guaranteed that the adversary's task is at least as hard as solving a certain well-studied mathematical problem (such as integer

© Springer International Publishing AG 2017
R.C.-W. Phan and M. Yung (Eds.): LNCS 10311, Mycrypt 2016, pp. 21–55, 2017.
DOI: 10.1007/978-3-319-61273-7_3

factorization) or finding a better-than-random way to predict output bits from a standardized primitive (such as AES).

The usefulness of a nontight security reduction is more controversial. If, for example, the tightness gap is 2^{40}, then one is guaranteed that the adversary's task is at least 2^{-40} times as hard as solving the mathematical problem or compromising AES. Opinions about whether nontightness is a cause of concern depend on how much importance one attaches to quantitative guarantees. In his paper [11] explaining practice-oriented provable security, Bellare writes:

> Practice-oriented provable security attempts to explicitly capture the inherently *quantitative* nature of security, via a *concrete* or *exact* treatment of security.... This enables a protocol designer to know exactly how much security he/she gets. (emphasis in original)

In contrast, some researchers minimize the importance of quantitative security and object strongly when someone criticizes a practice-oriented provable security result for giving a useless concrete security bound. For example, an anonymous reviewer of [54] defended the nonuniform proof in [12], acknowledging that its nonuniformity "reduces the quantitative guarantees" but then stating:

> Many proofs do not yield tight bounds, but they still are powerful qualitative indicators of security.

This reviewer characterized the use of the word "flaw" in [54] in reference to a fallacious analysis and erroneous statement of quantitative guarantees as "misleading" and "offensive," presumably because the "qualitative indicators" in [12] were still valid.

What makes the nontightness question particularly sensitive is that cryptographers are supposed to be cautious and conservative in their recommendations, and sources of uncertainty and vulnerability are not supposed to be swept under the rug. In particular, one should always keep in mind the possibility of what Menezes in [64] calls the *nightmare scenario*—that there actually is an attack on the protocol that is reflected in the tightness gap.

In [27] the authors presented attacks on MAC schemes in the multi-user setting—attacks that are possible because the natural security reduction relating the multi-user setting to the single-user setting is nontight. Similar attacks on protocols in the multi-user setting were given for a network authentication protocol, aggregate MAC schemes, authenticated encryption schemes, disk encryption schemes, and stream ciphers.

In Appendix B we describe the attacks of Zaverucha [77] on hybrid encryption in the multi-user setting. In Sect. 4 we describe another situation where the tightness gap reflects the fact that there's an actual attack, in this case due to Pietrzak [40,70].

A practical issue that is closely related to the nontightness question is the matter of safety margins. There are at least two kinds of safety margins: (1) parameter sizes that give significantly more bits of security than are currently needed, and (2) "optional" features in a protocol that are believed (sometimes because of tradition and "instinct" rather than any rigorous security argument) to help prevent new attacks or attacks that are outside the commonly used security models.

At present it is widely agreed that it is prudent to have at least 128 bits of security.[1] Why not 96? In the near future it is unlikely that anyone (even the N.S.A.) will expend 2^{96} operations to break a protocol. The reason for insisting on 128 bits of security is that one should anticipate incremental improvements in cryptanalytic attacks on the underlying mathematical problem that will knock several bits off the security level. If nontightness has already reduced the security assurance provided by the proof from 128 to 96 bits (and if the parameter sizes have not been increased so as to restore 128 bits of security), then even relatively small advances in attacking the mathematical problem will bring the security assurance further down to a level where a successful attack on the protocol is feasible in principle.

A common explanation of the value of security proofs is that features that are not needed in the proof can be dropped from the protocol. For instance, Katz and Lindell make this point in the introduction to [49]. However, in Appendix B (see also Sect. 5 of [56]) we shall find that optional features included in protocols often thwart attacks that would otherwise reduce the true security level considerably.

On the one hand, there is widespread agreement that tight proofs are preferable to nontight ones, many authors have worked hard to replace nontight proofs with tighter proofs when possible, and most published security reductions duly inform the reader when there is a large tightness gap. On the other hand, authors of papers that analyze protocols that are of practical importance almost never suggest larger parameters that compensate for the tightness gap. Presumably the reason is that they would have to sacrifice efficiency. As Bellare says [11],

> A weak reduction means that to get the same level of security in our protocol we must use larger keys for the underlying atomic primitive, and this means slower protocols.

Indeed, many standardized protocols were chosen in part because of security "proofs" involving highly nontight security reductions. Nevertheless, we are not aware of a single protocol that has been standardized or deployed with larger parameters that properly account for the tightness gaps. Thus, acknowledgment of the nontightness problem remains on the level of lip service.

In Sects. 2–6 we discuss nontightness in connection with complexity leveraging, HMAC, lattice-based cryptography, and identity-based encryption; in Appendix B we discuss Zaverucha's results on nontightness in security proofs for hybrid encryption in the multi-user setting. In the case of HMAC, in view of the recent work [40,54] on the huge tightness gaps in pseudorandomness results, in Sect. 4 we recommend that standards bodies reexamine the security of HMAC when used for non-MAC purposes (such as key derivation or passwords) or with MD5 or SHA1.

---

[1] By "k bits of security" we mean that there is good reason to believe that, if a successful attack (of a specified type) takes time T and has success probability ϵ, then $T/\epsilon > 2^k$.

2   Complexity Leveraging

"Complexity leveraging" is a general technique for proving that a cryptographic protocol that has been shown to be *selectively secure* is also *adaptively secure*. Here "selectively secure" means that the adversary has to select its target before it is presented with its inputs (e.g., public keys, signing oracles, etc.), whereas "adaptive security" means that the adversary is free to select its target at any time during its attack. The second type of adversary is in general much stronger than the first type. Thus, selective security is in principle a much weaker result than adaptive security, and so is not usually relevant to practice. Because selective security is often easier to prove than adaptive security, researchers devised the method of complexity leveraging to convert any selective security theorem into an adaptive security theorem.

Complexity leveraging has been used to prove the adaptive security of many kinds of cryptographic protocols including identity-based encryption [23], functional encryption [39], constrained pseudorandom functions [25], and constrained verifiable random functions [35]. In Sect. 2.1 we illustrate the problems with complexity leveraging in the context of signature schemes. In Sect. 2.2 we consider the case of identity-based encryption.

2.1   Signature Schemes

The most widely accepted definition of security of a signature scheme is against an *existential forger under chosen-message attack*. This means that the forger is given a user's public key and is allowed $\leq q$ queries, in response to which she is given a valid signature on each queried message. The forger is successful if she then forges a signature for any message M other than one that was queried.

A much weaker property is security against a *selective forger*. In that case the adversary is required to choose the message M for which she will forge a signature before she even knows the user's public key. She cannot modify M in response to the public key or the signature queries, and to be successful she must forge a signature on the original M. Selective security is obviously much weaker than existential security. A theorem that gives only selective security is not generally regarded as satisfactory for practice.

Complexity leveraging works by converting an arbitrary existential forger into a selective forger, as follows. The selective forger Cynthia guesses a message M, which she desperately hopes will be the message on which the existential forger eventually forges a signature. She then runs the existential forger. She is successful if the message forged is M; otherwise she simply tries again with a different guess. Her probability of success in each run is $\epsilon = 2^{-m}$, where m is the allowed bitlength of messages. The bound m on the message length could be large, such as one gigabyte.

Fortunately for Cynthia, in practice messages are normally hashed, say by SHA256, and it is the hash value that is signed. Thus, Cynthia needs to guess the 256-bit hash value of the message on which the existential forger forges a signature, not the message itself. Her probability of success is then 2^{-256}, and so the tightness gap in going from selective to existential security is 2^{256}.

Suppose, for example, that we have an integer-factorization-based signature protocol for which selective security has been shown to be tightly equivalent to factoring. How large does the modulus N have to be so that the corresponding existential security theorem gives us a guarantee of 128 bits of security? If only 3072-bit N is used, then the protocol will have 128 bits of selective security, but complexity leveraging gives us no existential security, because of the 2^{256} tightness gap. In order to have 128 bits of existential security, we need to have $128 + 256 = 384$ bits of security against factoring N, and this means roughly 40,000-bit N. Even though this is what we must do if we want complexity leveraging to give us the desired security, no one would ever seriously recommend deploying 40,000-bit moduli. Thus, from a practical standpoint complexity leveraging gives us nothing useful here.

2.2   Identity-Based Encryption

Boneh and Boyen [23] used bilinear pairings on elliptic curves to design an identity-based encryption scheme. They proved that their scheme is selectively secure in the sense that the adversary has to select the target before she gets the public parameters and access to the appropriate oracles (see Sect. 6 for background on identity-based encryption). The highlight of the proof is that it does not invoke the random oracle assumption.

Boneh and Boyen [23, Theorem 7.1] then used complexity leveraging to prove that a generic identity-based encryption scheme that is selectively secure is also adaptively secure. The proof has a tightness gap of $2^{2\ell}$, where ℓ is the desired security level and 2ℓ is the output length of a collision-resistant hash function (the hash function is applied to the identifiers of parties). Boneh and Boyen remarked that the reductionist proof is "somewhat inefficient" and explained that the desired level of security can be attained by increasing the parameters of the underlying pairing.

Suppose now that one desires 128 bits of security. Suppose also that the proof of selective security for the identity-based encryption scheme is tight. Then one can achieve 128 bits of selective security by using an (asymmetric) bilinear pairing $e : \mathbb{G}_1 \times \mathbb{G}_2 \to \mathbb{G}_T$ derived from a prime-order Barreto-Naehrig (BN) elliptic curve E over a finite field $\mathbb{F}_p$ [10]. Here, p is a 256-bit prime, $\mathbb{G}_1 = E(\mathbb{F}_p), \mathbb{G}_2$ is a certain order-n subgroup of $E(\mathbb{F}_{p^{12}})$, and $\mathbb{G}_T$ is the order-n subgroup of $\mathbb{F}_{p^{12}}^*$, where $n = \#E(\mathbb{F}_p)$. This pairing is ideally suited for the 128-bit security level since the fastest attacks known on the discrete logarithm problems in $\mathbb{G}_1, \mathbb{G}_2$ and $\mathbb{G}_T$ all take time approximately 2^{128}.[2] If resistance to adaptive attacks is desired, then to account for the tightness gap of 2^{256} a pairing $e : \mathbb{G}_1 \times \mathbb{G}_2 \to \mathbb{G}_T$ should be selected so that the fastest attacks known on the discrete logarithm problems

---

[2] We are not accounting for recent progress by Kim and Barbulescu [51] in algorithms for computing discrete logarithms in $\mathbb{G}_T$. This will lead to working with even larger parameters.

in $\mathbb{G}_1, \mathbb{G}_2$ and $\mathbb{G}_T$ take time at least 2^{384}. If the protocol is implemented using BN curves, then one now needs $p^{12} \approx 2^{40000}$ and thus $p \approx 2^{3300}$. Consequently, computations in $\mathbb{G}_1$ and $\mathbb{G}_T$ will be over 3300- and 40000-bit fields, instead of 256- and 3072-bit fields had the reduction been tight. Hence, the tightness gap that arises from complexity leveraging has a very large impact on efficiency.

3   Nonuniformity to Achieve Better Tightness

Informally speaking, the difference between a *nonuniform* algorithm to solve a problem $\mathcal{P}$ and the more familiar notion (due to Turing) of a uniform algorithm is that the former is given an "advice string," depending on the input length (and usually assumed to be of polynomial size in the input length). In general, a nonuniform algorithm is more powerful than a uniform one because the advice string may be very helpful in solving $\mathcal{P}$. Several prominent researchers have repeatedly claimed that security theorems that are proved in the nonuniform model of computation are stronger than theorems proved in the uniform model, because they provide assurances against successful attacks by nonuniform as well as uniform adversaries. In their lecture notes for their 2008 course at MIT [42], Bellare and Goldwasser state:

> Clearly, the nonuniform adversary is stronger than the uniform one. Thus to prove that "something" is "secure" even in presence of a nonuniform adversary is a better result than only proving it is secure in presence of a uniform adversary. (p. 254)

In an email explaining why his paper [12] did not inform the reader that the security reduction was being given in the nonuniform model, Bellare wrote [13]:

> I had no idea my paper would be read by anyone not familiar with the fact that concrete security is nonuniform.

What these researchers are failing to take into account is that the use of the nonuniform model makes *the hypothesis as well as the conclusion* of the theorem stronger. Thus, the theorem's assumption that a certain mathematical task is hard or that a certain compression function cannot be distinguished from a random function has to allow nonuniform algorithms. It is usually very difficult to get any idea of the strength of the commonly-used primitives against nonuniform attacks, and in practice they are not designed to withstand such attacks. See [55] for a discussion of the history of confusion about this issue in the literature and a detailed rebuttal of the arguments in favor of the nonuniform model in cryptography.

Whether or not nonuniform algorithms for a problem $\mathcal{P}$ are known that are much faster than uniform ones depends very much on the problem $\mathcal{P}$.

Example 1 (No known difference between uniform and nonuniform). There is no known nonuniform algorithm for the general integer factorization problem that is faster than the fastest known uniform algorithms.

In the next two examples, let $\mathcal{H}_k$ be a fixed family of hash functions, one for each security level k. In both examples, suppose that the input is k written in unary (this is a trick used to allow the input length to be different for different k).

Example 2 (Trivial in the nonuniform model). For a well-constructed family $\mathcal{H}_k$, by definition one knows no efficient uniform algorithm for finding a collision. In contrast, one has a trivial nonuniform algorithm, since the advice string can consist of two messages whose hash values are equal.

Example 3 (Between these two extremes). Consider the problem of distinguishing a hash function $\mathcal{H}_k$ in a family of keyed hash functions from a random function; a function for which this cannot be done with non-negligible success probability is said to have the pseudorandom function property (PRF). More precisely, an attack on the PRF property is an algorithm that queries an oracle that with equal probability is either the hash function with hidden key or else a random function and, based on the responses, can determine which it is with probability $\epsilon + 1/2$ of being correct, where the *advantage* ϵ is significantly greater than 0. For a well-constructed hash function no uniform algorithm is known that is faster than simply guessing the key, and this has advantage roughly $T/2^\ell$, where ℓ is the key-length and T is the time (here we are assuming that each query takes unit time). However, there is a simple nonuniform algorithm that runs in unit time and distinguishes a hash function with hidden key from a random function with advantage roughly $2^{-\ell/2}$—an advantage that would take the uniform algorithm time $T \approx 2^{\ell/2}$ to achieve. Our advice string is a message M that has a very special property with respect to $\mathcal{H}_k$ when averaged over all possible keys. For example, let M be a message that maximizes the probability that the 29th output bit is 1 rather than 0. The nonuniform algorithm then queries M to the oracle; if the oracle's response has 29th bit equal to 1, it guesses that the oracle is the hash function with hidden key, but if the 29th bit is 0, it guesses that the oracle is a random function. It follows by an easy argument from the theory of random walks that the expected advantage of this nonuniform algorithm is roughly $2^{-\ell/2}$.

As pointed out in [55], almost all security proofs in the literature are valid in the uniform model of complexity, and only a few use what's sometimes called *coin-fixing* to get a proof that is valid only in the nonuniform model. As far as we are aware, none of the nonuniform theorems in the literature have hypotheses of the sort in Examples 1 and 2; all are like Example 3, that is, the task whose hardness is being assumed is easier in the nonuniform model, but not trivial. The authors' main purpose in using coin-fixing in these cases is to achieve a tighter security reduction than they could have achieved in the uniform model.

Unfortunately, it is easy to get tripped up if one attempts to use coin-fixing to get a stronger result—authors fool themselves (and others) into thinking that their result is much stronger than it actually is. The most important example of a researcher who was led astray by his belief in the nonuniform model is Bellare in his Crypto 2006 paper [12] on HMAC. We will summarize this story and carry it up to the present by discussing some errors in his revised version [14], which was recently published in the *Journal of Cryptology*.

4   The HMAC Saga

HMAC [17,19] is a popular hash-function-based message authentication code (MAC). The controversy about nonuniform reductions concerns security proofs of the PRF property (see Example 3 of Sect. 3) of NMAC, which is a MAC that is closely related to HMAC. We shall discuss NMAC rather than HMAC, because the extension of results from NMAC to HMAC has generated relatively little controversy (see [57] for an analysis of 1-key variants of NMAC).

By a compression function we mean a function $z = f(x, y)$, where $y \in \{0, 1\}^b$ and $x, z \in \{0, 1\}^c$; typically $b = 512$ and c is equal to either 128 (for MD5), 160 (for SHA1), or 256 (for SHA256).

Given a compression function f, to construct an iterated hash function $\mathcal{H}$ one starts with an *initialization vector* IV, which is a publicly known bitstring of length c that is fixed once and for all. Suppose that $M = (M_1, \ldots, M_m)$ is a message consisting of $m \leq \mathfrak{m}$ b-bit blocks (where $\mathfrak{m}$ is the bound on the block-length of messages; for simplicity we suppose that all message lengths are multiples of b). Then we set $x_0 = $ IV, and for $i = 1, \ldots, m$ we recursively set $x_i = f(x_{i-1}, M_i)$; finally, we set $\mathcal{H}(M) = \mathcal{H}_{\mathrm{IV}}(M) = x_m$, which is the c-bit hash value of M.[3]

Suppose that Alice shares two secret c-bit keys K_1 and K_2 with Bob, and wants to create an NMAC-tag of a message M so that Bob can verify that the message came from Alice. She first uses K_1 as the IV and computes $\mathcal{H}_{K_1}(M)$. She pads this with $b - c$ zeros (denoted by a 0-superscript) and sets her tag $t(M)$ equal to $\mathcal{H}_{K_2}(\mathcal{H}_{K_1}(M)^0)$.

The purpose of finding a security reduction for NMAC is to show that if one has confidence that the compression function f enjoys a certain security property, then one can be sure that NMAC has the same property. Two decades ago HMAC was first proposed by Bellare et al. [17,19]. In [17] they proved (assuming weak collision-resistance of $\mathcal{H}$) that if f has the secure-MAC property, then so does NMAC. (The secure-MAC property is analogous to existential security of signatures, see Sect. 2.) The proof in [17] was tight. It was also short and well-written; anyone who was considering using HMAC could readily verify that the proof was tight and correct.

In 2006 Bellare [12] published a different security reduction for NMAC. First, he dispensed with the collision-resistance assumption on $\mathcal{H}$, which is a relatively strong assumption that has turned out to be incorrect for some real-world iterated hash functions. Second, he replaced the secure-MAC property with the stronger PRF property, that is, he showed that if $f(x, y)$ (with x serving as the hidden key) has the PRF property, then so does NMAC. This was

---

[3] In iterated hash functions one also appends a "length block" to the message M before hashing. We are omitting the length block for simplicity.

important in order to justify the use of HMAC for purposes other than message authentication—in applications where the PRF property is desired, such as key-derivation protocols [34,45,60] and password-systems [66].

Remark 1. A third advantage (not mentioned in [12,14]) of assuming the PRF property rather than collision-resistance arises if one derives a concrete security assurance using the best known generic attacks on the property that the compression function is assumed to have. As far as we know the best generic attack on the PRF property using classical (i.e., uniform and non-quantum) algorithms has running time $\approx 2^c$ (it amounts to guessing the hidden key), whereas the birthday-paradox attack on collision-resistance only takes time $\approx 2^{c/2}$. Other things being equal, one expects that c must be twice as great if one is assuming collision-resistance than if one is assuming the PRF property.

However, in 2012 Koblitz and Menezes found a flaw in [12]. For Bellare, who along with Rogaway popularized the concept of "practice-oriented provable security" [11], his theorem was not merely a theoretical result, but rather was intended to provide some concrete assurance to practitioners. Thus, it was important for him to determine in real-world terms what guarantee his theorem provided. To do this, Bellare's approach was to take the fastest known generic attack on the PRF property of a compression function, and evaluate what his theorem then implied for the security of NMAC. In his analysis he took the key-guessing attack (see Example 3 of Sect. 3) as the best generic attack on f, and concluded that NMAC is secure "up to roughly $2^{c/2}/\mathfrak{m}$ queries." For instance, for a bound of $\mathfrak{m} = 2^{20}$ on the block-length of messages Bellare was claiming that NMAC-MD5 is secure up to 2^{44} queries and NMAC-SHA1 up to 2^{60} queries. (In 2006, MD5 and SHA1 were common choices for hash functions.)

Bellare failed to account for the fact that, because of his "coin-fixing," i.e., nonuniform security reduction, he was logically required to examine security of f against *nonuniform* attacks, not just uniform attacks. As we saw in Sect. 3, there are simple generic nonuniform attacks on the PRF property that have a much higher success probability than the key-guessing attack. If one repeats Bellare's analysis using the nonuniform attack described in Sect. 3, one finds that NMAC's security is guaranteed only up to at most $2^{c/4}/\sqrt{\mathfrak{m}}$ queries, that is, 2^{22} for NMAC-MD5 and 2^{30} for NMAC-SHA1. That level of security is of little value in practice.

When we say that Bellare's paper had a basic flaw, we have in mind the definition of the f-word that was given by Stern et al. [76], who said:

> The use of provable security is more subtle than it appears, and flaws in security proofs themselves might have a devastating effect on the trustworthiness of cryptography. By flaws, we do not mean plain mathematical errors but rather ambiguities or misconceptions in the security model.

Now let us bring this story up to the present. In an effort to determine what can be said about the relation between the PRF property of the compression function f and the PRF property of NMAC, Koblitz and Menezes [54] gave a uniform security reduction that had tightness gap $\mathfrak{m} \cdot \max(2, \; q^2/(2^c\epsilon))$, where ϵ is a measure of the PRF-security of f and q is a bound on the number of queries. They had to use a stronger version of the PRF property of f (a version that's similar to the property used in [18]); a corollary of their theorem then gave a tightness gap of $2\mathfrak{m}q$ if one assumes only standard PRF-security of f.[4]

The interpretation in [54] of the authors' Theorem 10.1 and Corollary 10.3 on NMAC security is pessimistic. Those results assume the single-user setting and strong properties of f; moreover, they have large tightness gaps. The authors conclude:

> We would not want to go out on a limb and say that our Theorem 10.1 is totally worthless. However, its value as a source of assurance about the real-world security of HMAC is questionable at best.

Specifically, they caution that "In our opinion none of the provable security theorems for HMAC with MD5 or SHA1 [...] by themselves provide a useful guarantee of security." For instance, suppose that the query bound q is 2^{30}, the block-length bound $\mathfrak{m}$ is 2^{25}, and the number of users n is 2^{25}. (As we shall see in Appendix B, the step from single-user to multi-user setting introduces an additional factor of n in the tightness gap.) Then the number of bits of security drops by $30 + 25 + 25 = 80$ due to these tightness gaps. In other words, the guarantees drop to 48 bits and 80 bits in the case of MD5 and SHA1, respectively.

Remark 2. If SHA256 is used in order to have at least 128 bits of HMAC security, then there is such a huge safety margin that even these tightness gaps do not lower the security to an undesirable level, at least if one assumes that there is no attack on the PRF property of the SHA256 compression function that is faster than the generic key-guessing one. This is because key-guessing takes time $\approx 2^{256}$, leaving a safety margin of 128 bits. One reason SHA256 might be used for HMAC even if only 128 bits of security are required is that the user might need SHA256 for other protocols that require collision-resistance and so she cannot allow fewer than 256 bits of hash-output; in the interest of simplicity she might decide to use a single hash function everywhere rather than switching to SHA1 for HMAC.

Remark 3. The above comment about a huge safety margin when SHA256 is used in HMAC applies only if a 256-bit key and 256-bit message tags are used. Not all standards specify this. For example, the NIST standard [33] recommends 128-bit HMAC keys for 128 bits of security and allows 64-bit tags. The recommendations in [33] are supported by an *ad hoc* analysis, but are not supported by any provable security theorem.

---

[4] The early posted versions of [54] contained a serious error that was pointed out to the authors by Pietrzak, namely, the theorem is given assuming only the PRF property rather than the strong PRF property that is needed in the proof. This error was explained and corrected in the posted versions and the published version. Soon after the corrected version was posted, Pietrzak posted a paper [70] containing a different proof of essentially the same result as in Corollary 10.3 of Theorem 10.1 of [54] (see also [40]).

Aside from the issue of the tightness gaps, there is another fundamental reason why the theorems in [12,14,54] about security of NMAC and HMAC under the PRF assumption offer little practical assurance. To the best of our knowledge, the PRF assumption has never been seriously studied for the compression functions used in MD5, SHA1, or SHA256; in fact, we are not aware of a single paper that treats this question. Moreover, when those compression functions were constructed, the PRF property was not regarded as something that had to be satisfied – rather, they were constructed for the purpose of collision-resistance and pre-image resistance. Thus, in the case of the concrete hash functions used in practice, we have no evidence that could rule out attacks on the PRF property that are much better than the generic ones. It would be very worthwhile for people to study how resistant the concrete compression functions are to attacks on the PRF property; in the meantime it would be prudent not to rely heavily on theorems that make the PRF assumption.

Remark 4. The situation was quite different for AES, since a longstanding criterion for a good block cipher has been to have the pseudorandom permutation (PRP) property with respect to the secret (hidden) key. That is, an adversary should not be able to distinguish between the output of a block cipher with hidden key and that of a random permutation. The PRF property is close to the PRP property as formalized by the PRP/PRF switching lemma (see Sect. 5 of [73]), and so it is reasonable to assume that AES has the PRF property. On the other hand, the criteria for judging hash constructions have been very different from those for judging encryption.

Remark 5. In [15] the authors prove security of a MAC scheme called AMAC, which is a prefix-MAC in which the output of the hash function is truncated so as to thwart the extension attacks to which prefix-MACs are susceptible. As in the case of the HMAC papers discussed above, the authors of [15] assume that the underlying compression function is a PRF. Their proof has the remarkable feature that it does not lose tightness in the multi-user setting. On the other hand, the tightness gap in the single-user setting is much larger than in the above security reductions for HMAC—namely, roughly q^2m^2. With, for instance, $q \approx 2^{30}$ and $m \approx 2^{25}$ one has a tightness gap of 110 bits. The paper [15] does recommend the use of SHA512, and if one assumes 512 bits of PRF-security for its compression function, then we have such a large safety margin that a 2^{110} tightness gap is not worrisome. Nevertheless, it should be stressed that the PRF assumption is a very strong one that, to the best of our knowledge, has never been studied or tested for the SHA512 compression function.

Remark 6. In [43], Goldwasser and Kalai propose a notion of what it means for a complexity assumption to be reasonable in the context of reductionist security proofs. Among other things, the assumption should be falsifiable and non-interactive. Since the assumption that the compression function in a hash function such as MD5, SHA1, SHA256 or SHA512 has the PRF property is an interactive one, it does not meet the Goldwasser-Kalai standard for a reasonable cryptographic assumption. Rather, in the words of Goldwasser and Kalai, such an assumption "can be harmful to the credibility of our field."

Returning to our narrative, in 2015 Bellare [14] published a revised version of [12] in *J. Cryptology* that, regrettably, just muddied the waters because of errors and unclarities in his abstract and introduction that could easily mislead practitioners. First of all, the first sentence of the abstract states that the 1996 paper [17] proved "HMAC... to be a PRF assuming that (1) the underlying compression function is a PRF, and (2) the iterated hash function is weakly collision resistant." In fact, only the secure-MAC property, not the PRF property, was proved in [17].[5]

In the second place, in the concluding paragraph of the introduction of [14] Bellare gives the impression that Pietrzak in [70] proved tight bounds for the PRF-security of NMAC:[6] "Tightness estimates [in the present paper] are now based on the blackbox version of our reductions and indicate that our bounds are not as tight as we had thought. The gap has been filled by Pietrzak [70], who gives blackbox reduction proofs for NMAC that he shows via matching attack to be tight."[7] A practitioner who reads the abstract and introduction of [14] but not the technical sections would probably go away believing that PRF-security of NMAC has been proved to be tightly related to PRF-security of the compression function. This is false. In fact, it is the opposite of what Pietrzak proved.

What Pietrzak showed in [40,70] was that the mq tightness gap cannot be reduced in the general case (although the possibility that better tightness might conceivably be achieved for a special class of compression functions wasn't ruled out). He found a simple attack on NMAC that shows this. This is far from reassuring—it's what Menezes in [64] called the "nightmare scenario." To put it another way, Pietrzak's attack shows a huge separation in PRF-security between the compression function and NMAC. The desired interpretation of a security reduction of the sort in [14,54] or [40] is that it should tell you that the study of a certain security property of a complicated protocol is unnecessary if one studies the corresponding property of a standard primitive. In this case the tightness gap along with Pietrzak's attack show that this is *not* the case.

It is unfortunate that neither of Bellare's papers [12,14] discuss the practical implications of the large tightness gap. It would be interesting to know

---

[5] The abstract to [40] also erroneously states that "NMAC was introduced by Bellare, Canetti and Krawczyk [Crypto96], who proved it to be a secure pseudorandom function (PRF), and thus also a MAC, assuming that (1) f is a PRF and (2) the function we get when cascading f is weakly collision-resistant".

[6] In this quotation Bellare uses the word "blackbox" in a non-standard way. Later in his paper he defines a "blackbox" reduction to be one that is constructible and a "non-blackbox" reduction to be one that is non-constructible. However, when comparing a proof as in [12] that uses "coin-fixing" with more recent proofs that do not, the standard terms are nonuniform/uniform rather than non-blackbox/blackbox.

[7] The section "Our Contributions" in [40] starts out: "Our first contribution is a simpler, uniform, and as we will show, basically tight proof for the PRF-security of NMAC^f assuming only that f is a PRF." The authors apparently meant to say that their tightness gap is best possible, i.e., cannot be improved. Their proof is not tight, however—far from it. Their tightness gap is nq, essentially the same as in Corollary 10.3 of [54].

why he disagrees with the conclusion of Koblitz–Menezes that the tightness gaps and other weaknesses render the security reductions (proved by them in Theorems 10.1 and Corollary 10.3 of [54]) "questionable at best" as a source of real-world assurance. In view of Pietrzak's recent work, which shows that the tightness gap cannot be removed and reflects an actual attack, it is particularly puzzling that even the revised paper [14] has nothing to say about the practical implications of this weakness in the security reductions for HMAC.

We conclude this section with a recommendation. *Standards bodies should reexamine—taking into account tightness gaps—the security of all standardized protocols that use HMAC for non-MAC purposes such as key derivation or passwords. The same should be done for HMAC-protocols using hash functions such as MD5 or SHA1 that are not believed to have weak collision-resistance in the sense of* [17].

In some cases adjustments should be made, such as mandating a feature that is currently optional (such as a nonce or a randomization) in order to prevent known attacks; in other cases the recommended parameters or choices of hash function may need to be changed in order to account for the tightness gaps. Protocols that use HMAC as a MAC and use a collision-resistant hash function do not have to be reexamined, because in that case [17] has a tight security reduction. (However, in view of the multi-user attacks discussed in Appendix B, the standards for any protocol that is used in a setting with a large number of users should be modified if necessary to account for the multi-user/single-user tightness gap.)

5 Lattice-Based Quantum-Safe Crypto

The reason for intense interest in lattice-based cryptography can be traced back to the early years of public key, when Merkle–Hellman proposed the knapsack public-key encryption system. It aroused a lot of interest both because of its superior efficiency (compared to RSA) and the supposedly high level of confidence in its security, since it was based on an NP-hard problem. Within a few years Shamir, Brickell and others completely broke both the original knapsack and modified versions of it. It turned out that the knapsack was based on an easy subproblem of the NP-hard subset sum problem, not on hard instances. This was a traumatic experience for researchers in the nascent field of public-key cryptography. The lesson learned was that it would be good to base systems on hardness of a problem for which the average case is provably equivalent to the hardest case (possibly of a different problem).

There was a lot of excitement (even in the popular press) when Ajtai–Dwork announced a lattice-based encryption scheme based on such a problem [2,3]. Since that time much of the motivation for working on lattice-based systems (especially now that standards bodies are looking for quantum-safe cryptographic protocols that have provable security guarantees) is that many of them can be proved to have worst-case/average-case equivalence. (For a comprehensive overview of research on lattice-based cryptography in the last ten years, see [69].)

In this section we shall look at the worst-case to average case reductions from the standpoint of tightness.

First, though, it is important to recognize that equivalence between average and worst cases is not the Holy Grail for cryptographers that some might think. As Dan Bernstein has noted (quoted in [43]), long before Ajtai-Dwork we had discrete-log cryptosystems over characteristic-two fields. For each k the Discrete Log Problem (DLP) in the group $\mathbb{F}_{2^k}^*$ is random self-reducible, meaning that instances can be randomized. This gives a tight equivalence between hardest instances and average instances. However, the DLP in those groups has long been known to be weaker than the DLP in the multiplicative group of prime-order fields [30], and recently it was completely broken [9].

Meanwhile the general DLP in the multiplicative group of prime fields $\mathbb{F}_p^*$ does not have this nice self-reducibility property, since for a given bitlength of p one has vastly different levels of difficulty of the DLP. Yet as far as we know these groups are secure for suitably chosen p of bitlength >1024.

5.1 Lattices

A (full rank) *lattice* L in $\mathbb{R}^n$ is the set of all integer linear combinations of n linearly independent vectors $B = \{v_1, v_2, \ldots, v_n\}$. The set B is called a *basis* of L, and the dimension of L is n. If the v_i are in $\mathbb{Z}^n$, then L is said to be an integer lattice; all lattices in this section are integer lattices. The *length* of a vector is its Euclidean norm. For each $1 \le i \le n$, the *ith successive minimum* $\lambda_i(L)$ is the smallest real number r such that L has i linearly independent vectors the longest of which has length r. Thus, $\lambda_1(L)$ is the length of a shortest nonzero vector in L.

5.2 Lattice Problems

Let L be an n-dimensional lattice. When we say that we are "given a lattice" L, we mean that we are given some arbitrary basis for L.

A well-studied lattice problem is the Shortest Vector Problem (SVP): Given L, find a lattice vector of length $\lambda_1(L)$. The SVP problem is NP-hard. The fastest classical algorithms known for solving it have provable running time $2^{n+o(n)}$ [1] and heuristic running time $2^{0.337n+o(n)}$ [61]. The fastest quantum algorithm known for solving SVP has heuristic running time $2^{0.286n+o(n)}$ [61]. More generally, one can consider the Approximate-SVP Problem (SVP$_\gamma$), which is the problem of finding a nonzero lattice vector of length at most $\gamma \cdot \lambda_1(L)$. If $\gamma > \sqrt{n}$, then SVP$_\gamma$ is unlikely to be NP-hard [41]. In fact, if $\gamma > 2^{n \log \log n / \log n}$, then SVP$_\gamma$ can be solved in polynomial time using the LLL algorithm. For $\gamma = 2^k$, the fastest algorithm known for SVP$_\gamma$ has running time $2^{\tilde{\Theta}(n/k)}$, where the $\tilde{\Theta}$ term hides a constant factor and a factor of a power of $\log n$ (see [69]).

A related problem to SVP$_\gamma$ is the Approximate Shortest Independent Vectors Problem (SIVP$_\gamma$): Given L, find n linearly independent lattice vectors all of which have length at most $\gamma \cdot \lambda_n(L)$. The hardness of SIVP$_\gamma$ is similar to that of SVP$_\gamma$ [21]; in fact, SIVP$_{\sqrt{n}\gamma}$ polynomial-time reduces to SVP$_\gamma$ [65].

5.3   Learning with Errors

The Learning With Errors (LWE) problem was introduced by Regev in 2005 [71]. The LWE problem and the related R-LWE problem (see [63]) have been extensively used to design many cryptographic protocols including public-key encryption, identity-based encryption, and fully homomorphic encryption. Public-key encryption schemes based on LWE (and R-LWE) are also attractive because no quantum algorithms for solving LWE are known that perform better than the fastest known classical algorithms. Thus, LWE-based public-key encryption schemes are viable candidates for post-quantum cryptography.

Let $q = q(n)$ and $m = m(n)$ be integers, and let $\alpha = \alpha(n) \in (0,1)$ be such that $\alpha q > 2\sqrt{n}$. Let χ be the probability distribution on $\mathbb{Z}_q$ obtained by sampling from a Gaussian distribution with mean 0 and variance $\alpha^2/2\pi$, and then multiplying by q and rounding to the closest integer modulo q; for more details see [71]. Then the (search version of the) LWE problem is the following: Let s be a secret vector selected uniformly at random from $\mathbb{Z}_q^n$. Given m samples $(a_i, a_i \cdot s + e_i)$, where each a_i is selected independently and uniformly at random from $\mathbb{Z}_q^n$, and where each e_i is selected independently from $\mathbb{Z}_q^n$ according to χ, determine s. Intuitively, in LWE you are asked to solve a linear system modulo q, except that the constants on the right of the system are given to you with random errors according to a Gaussian distribution.

The decisional version of LWE, called DLWE, asks us to determine whether we have been given m LWE samples $(a_i, a_i \cdot s + e_i)$ or m random samples (a_i, u_i), where each u_i is selected independently and uniformly at random from $\mathbb{Z}_q$.

5.4   Regev's Reduction

Regev [71] proved the following remarkable result[8].

Theorem 1. *If there exists an efficient algorithm that solves DLWE (in the average case), then there exists an efficient quantum algorithm that solves SIVP$_\gamma$ in the worst case where $\gamma = \tilde{O}(n/\alpha)$.*

Suppose now that a lattice-based cryptosystem has been designed with a reductionist security proof with respect to the hardness of average-case DLWE. By Theorem 1, this cryptosystem also has a reductionist security proof with respect to the hardness of SIVP$_\gamma$ in the *worst case*. This is widely interpreted as providing ironclad assurance for the security of the cryptosystem since there is compelling evidence that the well-studied SIVP$_\gamma$ problem is hard in the worst case when γ is small.

However, Regev's theorem and similar results are asymptotic. Although results of this type are interesting from a qualitative point of view, it is surprising that in the literature there are virtually no attempts to determine the

---

[8] Regev's theorem can also be stated with the GapSVP$_\gamma$ problem instead of SIVP$_\gamma$. Given an n-dimensional lattice L and a number $r > 0$, GapSVP$_\gamma$ requires that one output "yes" if $\lambda_1(L) \le r$ and "no" if $\lambda_1(L) > \gamma r$ (either "yes" or "no" is allowed if $r < \lambda_1 \le \gamma r$).

concrete security assurances that worst-case to average-case results such as Theorem 1 provide for lattice-based cryptosystems. That is, in the lattice-based cryptography literature concerning worst-case/average-case results, practice-oriented provable security in the sense of Bellare-Rogaway (as explained in the quote from [11] in the Introduction) is conspicuous by its absence.

Remark 7. Suppose that one has a polynomial-time reduction of a well-studied worst-case problem Π_1 to an average-case problem Π_2. Then, if one assumes that the worst-case instances of Π_1 are not polytime solvable, then the reduction provides the assurance that no polynomial-time algorithm can solve Π_2 on average. This asymptotic assurance is viewed by some as ruling out "structural weaknesses" in Π_2; for example, see Sect. 5.1 of [62]. However, in the absence of a concrete analysis, the reduction by itself does not guarantee the hardness of fixed-sized instances of Π_2.

A closer examination of Theorem 1 reveals several obstacles to using it to obtain concrete security assurances for DLWE-based cryptosystems. We list five such difficulties. Whereas the first and second are widely acknowledged in the literature, there is scant mention of the remaining three difficulties.

1. One needs to assess the hardness of SIVP$_\gamma$ under *quantum* attacks and not just under attacks on classical computers.
2. For parameters n, q and α that arise in DLWE-based cryptosystems, the SIVP$_\gamma$ problem is likely *not* NP-hard. Thus, the evidence for worst-case hardness of SIVP$_\gamma$ instances that arise in lattice-based cryptography is not as compelling as the evidence for the worst-case hardness of an NP-hard problem.
3. Very little work has been done on concretely assessing the hardness of SIVP$_\gamma$. As mentioned in Sect. 5.2, the fastest attack on SIVP$_\gamma$ where $\gamma = 2^k$ has running time $2^{\tilde{\Theta}(n/k)}$; however this expression for the running time is far from concrete.
4. The statement of Theorem 1 uses "efficient" to mean "polynomial time in n". However, the exact tightness gap in the reduction of worst-case SIVP$_\gamma$ to average-case DLWE has to the best of our knowledge never been stated.
5. A more precise formulation of DLWE involves several parameters including the number of available samples and the adversary's advantage in distinguishing between LWE and random samples. In practice, these parameters have to be chosen based on the security needs of the DLWE-based cryptosystem. However, there is little discussion in the literature of concrete values for these parameters in the context of specific protocols. All the reductionist security claims that we examined for DLWE-based cryptosystems are stated in asymptotic terms and make liberal use of the phrases "polynomial time," "polynomial number," and "non-negligible."

Section 5.5 elaborates on (4) and (5).

5.5   Analysis of Regev's Reduction

A careful examination of Regev's proof of Theorem 1 (see Appendix A for details) reveals the following refined statement. For concreteness, we will take $q = n^2$

and $\alpha = 1/(\sqrt{n}\log^2 n)$, whence $\gamma = \tilde{O}(n^{1.5})$; these are the parameters proposed by Regev for his DLWE-based public-key encryption scheme [71]. Suppose that there is an algorithm W_1 that, given $m = n^c$ samples, solves DLWE for a fraction $1/n^{d_1}$ of all $s \in \mathbb{Z}_q^n$ with advantage at least $1/n^{d_2}$. Then there is a polynomial-time algorithm W_2 for solving SIVP$_\gamma$ that calls the W_1 oracle a total of

$$O(n^{11+c+d_1+2d_2}) \tag{1}$$

times. The tightness gap is thus $O(n^{11+c+d_1+2d_2})$. While this is polynomial in n, it can be massive for concrete values of n, c, d_1 and d_2.

Suppose, for example, that one takes $n = 1024$ ($n = 1024$ is used in [5,26] for implementations of an R-LWE based cryptosystem). In a DLWE-based encryption scheme such as Regev's [71], the public key is a collection of $m = n^{1+\epsilon}$ LWE samples and the secret key is s; for simplicity we take $m = n$ whence $c = 1$. The encryption scheme is considered to be insecure if an attacker can distinguish between encryptions of 0 and 1 with advantage at least $1/n^d$ for some $d > 0$ depending on the security parameter. This advantage is assessed over choices of public-private key pairs and the randomness in the encryption algorithm. Regev showed that such an adversary can be used to solve DLWE for a fraction $1/4n^d$ of all $s \in \mathbb{Z}_q^n$ with advantage at least $1/8n^d$; thus $d_1 \approx d$ and $d_2 \approx d$. If one is aiming for the 128-bit security level, then a reasonable choice for d might be 12.8. Then, ignoring the hidden constant in the expression (1), the tightness gap is $n^{50.4} \approx 2^{504}$. Thus, if average-case DLWE can be solved in time T, then Theorem 1 shows that SIVP$_\gamma$ can be solved by a quantum algorithm in time $2^{504}T$. As mentioned above, the fastest quantum algorithm known for solving SVP has running time $2^{0.286n+o(n)}$. If we assume that this is also the fastest quantum algorithm for solving SIVP$_\gamma$ and ignore the $o(n)$ term in the exponent, then the algorithm has running time approximately $2^{293} \ll 2^{504}T$. Thus, for our choice of parameters Theorem 1 provides no assurances whatsoever for the hardness of average-case DLWE or for the security of the encryption scheme. In other words, even though Theorem 1 is viewed by many as providing "powerful qualitative indicators of security" (in the words of the anonymous reviewer quoted in Sect. 1), the quantitative security assurance it provides is vacuous.

Remark 8. The condition $\alpha q > 2\sqrt{n}$ is needed for Regev's proof of Theorem 1 to go through. It was later discovered that this condition is indeed necessary for security. In 2011, Arora and Ge [7] showed that if $\alpha q = n^t$, where $t < 1/2$ is a constant and $q \gg n^{2t}\log^2 n$, then there is a subexponential $2^{\tilde{O}(n^{2t})}$ algorithm that solves LWE. This attack is touted as a demonstration of the importance of security proofs—Theorem 1 anticipated the Arora-Ge attack which was discovered 6 years after Theorem 1 was proven. In the same vein, one can wonder about the implications of the large tightness gap in Theorem 1 for the concrete hardness of DLWE. One needs to ask: Is the tightness gap anticipating yet-to-be-discovered algorithms for solving DLWE that are considerably faster than the fastest algorithms for solving SIVP$_{n^{1.5}}$? The answer to this question has major consequences for the security of DLWE-based protocols.

On the other hand, if one were to select a larger value for n while still targeting the 128-bit security level, then the large tightness gap in (1) might not be a concern if there is a very large safety margin—large enough so that the fastest quantum algorithm for solving the corresponding SIVP$_\gamma$ is believed to have running time 2^k for $k \gg 128$. While this necessitates selecting a larger value of n, the impact on the cryptosystem's performance might not be too large. Thus, there remains the possibility that Theorem 1 can indeed provide meaningful security assurances for DLWE-based cryptosystems in practice. In order for this to occur, the following problems should be further investigated:

1. Determine concrete lower bounds for the worst-case quantum hardness of SIVP$_\gamma$ (or GapSVP$_\gamma$) in terms of n and γ.
2. Determine whether the tightness gap in Regev's worst-case to average-case reduction (see the estimate (1)) can be improved. Such improvements might be achieved either through a closer analysis of Regev's reduction, or else by formulating new reductions.
3. Determine appropriate values of c, d_1 and d_2.
4. Assess the tightness gap in the reductionist security proof for the cryptosystem (with respect to average-case DLWE).

Similarly, it would be very worthwhile to assess whether the analogue of Theorem 1 for the R-LWE problem provides any meaningful assurances for cryptosystems based on R-LWE using parameters that have been proposed in recent work [4,5,26,67]. We note that the worst-case to average-case reduction for R-LWE [63] is with respect to SVP$_\gamma$ in so-called ideal lattices (that is, lattices that come from ideals in rings). Deriving concrete bounds on the hardness of SVP$_\gamma$ for these lattices is more challenging than deriving concrete bounds on the hardness of SIVP$_\gamma$ for arbitrary lattices.

Remark 9. In preparation for the possible advent of large-scale quantum computers, standards organizations have begun examining candidates for public-key cryptosystems that withstand attacks by quantum computers (see [58]). Public-key cryptosystems based on R-LWE are considered to be one of the leading candidates for these quantum-safe standards. Initial deployment of quantum-safe cryptosystems will likely be for the protection of highly sensitive data whose confidentiality needs to be assured for several decades. For these applications, long-term security guarantees will be more important than short-term concerns of efficiency. Thus, it would be prudent to select parameters for R-LWE cryptosystems in such a way that the worst-case to average-case reductions provide meaningful concrete security guarantees. As mentioned above, the degradation in performance that results from larger lattice parameters might not be of great concern for high-security applications.

Remark 10. NTRU is a lattice-based public-key encryption scheme that was first presented in 1996 (see [46,47]) and has been standardized by several accredited organizations including ANSI [6] and IEEE [48]. NTRU uses lattices that arise from certain polynomial rings. The algebraic structure of these lattices facilitate implementations that are significantly faster than public-key encryption schemes based on LWE and R-LWE. Despite its longevity, NTRU is routinely disparaged in the theoretical cryptography literature because, unlike the case of public-key encryption schemes based on LWE or R-LWE (including some variants of NTRU that were proposed more recently [75]), there are no worst-case to average-case reductions to support the security of its underlying lattice problems. However, as we have noted, whether or not these asymptotic worst-case to average-case reductions provide meaningful concrete security assurances is far from being understood. Thus, the claim that, because of worst-case/average-case reductions, the more recent lattice-based encryption schemes have better security than classical NTRU rests on a flimsy scientific foundation.

In [68] Peikert describes asymptotic analyses of the security of lattice-based systems, and concludes:

> ...worst-case reductions give a hard-and-fast guarantee that the cryptosystem is at least as hard to break as the *hardest* instances of some underlying problem. This gives a true lower bound on security, and prevents the kind of unexpected weaknesses that have so often been exposed in schemes that lack such reductions.

This would be true in a meaningful sense if the reductions were tight and if the underlying problem were SIVP$_\gamma$ for a small γ (small enough so that SIVP$_\gamma$ is NP-hard or so that there is reason to have confidence that there are no efficient algorithms for SIVP$_\gamma$). However, neither is the case. When discussing asymptotic results and writing for a broad readership interested in practical cryptography, the use of such terms as "hard-and-fast guarantee" and "true lower bound on security" is inappropriate and misleading, because in real-world cryptography the normal interpretation of these terms is that one has concrete practical security assurances.

6   Tightness in Identity-Based Encryption

By way of counterpoint to the main theme of this paper—the potential dangers in ignoring tightness gaps in security reductions—we now discuss the case of Boneh-Franklin Identity-Based Encryption (IBE), where a large tightness gap is, we believe, of no concern. The evidence for this belief is that an informal (but convincing) argument allows one to reduce to the case where the adversary is not allowed any key-extraction queries.

An identity-based encryption scheme offers the flexibility of using any string — in particular, the identity of an individual or entity—as a public key. There is an authority called the Private Key Generator which publishes its own public parameters, including a public key, and maintains a master secret key. To obtain a decryption key corresponding to her identity, a user in the system applies to the Private Key Generator, which performs appropriate checks (possibly including physical checks) to ascertain the identity. Then the Private Key Generator uses its public parameters and master secret key to generate the decryption key corresponding to the identity. This decryption key is transmitted to the user through a secure channel. Anybody who wishes to securely send a message uses the identity of the recipient and the public parameters to perform the encryption. The recipient can decrypt using her decryption key.

Security of an IBE scheme is modeled using a game between a simulator and an adversary [24]. The game models security against an attack by a set of colluding users attempting to decrypt a ciphertext intended for a user outside the set.

In the initial phase, the simulator sets up an instance of the scheme based on the security parameter. The simulator generates the public parameters, which are given to the adversary, and the master secret key. The adversary is allowed to adaptively make key-extraction queries to the simulator, who must provide the decryption keys corresponding to identities of the adversary's choosing. At some point, the adversary provides the simulator with an identity id^* (called the target identity) and two messages M_0 and M_1 of equal length. The simulator randomly chooses a bit b and provides the adversary with C^*, which is an encryption of M_b for the identity id^*. The adversary continues making key-extraction queries in an adaptive manner. Finally, the adversary outputs its guess b'; its advantage in winning the game is defined to be $|\Pr[b = b'] - 1/2|$. The adversary may not make more than one key-extraction query for the same id; and of course it must not have queried the simulator for the decryption key of id^*, as otherwise the game becomes trivial to win. The adversary's resources are measured by the time that it takes and the number of key-extraction queries that it makes.

The model that we have described provides what is called IND-ID-CPA security (indistinguishability for ID-based encryption under key-extraction[9] attack). This model does not allow the adversary to make decryption queries. The model where such queries are also allowed is said to provide IND-ID-CCA (chosen ciphertext) security.

The first efficient IBE construction is due to Boneh and Franklin [24]. Their scheme—and in fact all subsequent efficient IBE constructions—uses bilinear pairings. A (symmetric) bilinear pairing is a map $e : \mathbb{G} \times \mathbb{G} \rightarrow \mathbb{G}_T$, where $\mathbb{G} = \langle P \rangle$ and $\mathbb{G}_T$ are groups of some prime order p, that satisfies the following conditions: $e(aP, bP) = e(P, P)^{ab}, e(P, P) \neq 1$, and e is efficiently computable. Practical bilinear pairings are obtained from elliptic curves where $\mathbb{G}$ is a subgroup of points on an appropriately chosen elliptic curve and $\mathbb{G}_T$ is a subgroup of the multiplicative group of a finite field.

Identity-based encryption schemes are proved secure under various computational hardness assumptions. We mention the basic bilinear Diffie-Hellman (BDH) assumption and two of its derivatives. The bilinear Diffie-Hellman (BDH) assumption is that computing $e(P, P)^{abc}$ given (P, aP, bP, cP) is infeasible. The decisional bilinear Diffie-Hellman (DBDH) assumption is that distinguishing between the distributions $(P, aP, bP, cP, e(P, P)^{abc})$ and $(P, aP, bP, cP, e(P, P)^z)$, where a, b, c and z are independent and uniform random choices from $\mathbb{Z}_p$, is

---

[9] In the IBE setting "CP" does not stand for *chosen plaintext* but rather for *clave pedida*, which means "requested key" in Spanish.

infeasible. The gap bilinear Diffie-Hellman (GBDH) assumption is that computing $e(P, P)^{abc}$ given (P, aP, bP, cP) and access to a DBDH oracle is infeasible.

We now briefly describe the basic Boneh-Franklin IBE scheme. The Private Key Generator sets up the scheme by selecting a generator P of the group $\mathbb{G}$; choosing a random s from $\mathbb{Z}_p$ and setting $Q = sP$; and selecting two hash functions $H_1 : \{0,1\}^* \to \mathbb{G}$, $H_2 : \mathbb{G}_T \to \{0,1\}^n$. The public parameters are (P, Q, H_1, H_2) while the master secret key is s. Given an identity id $\in \{0,1\}^*$, let $Q_{id} = H_1(\text{id})$; the decryption key is defined to be $d_{id} = sQ_{id}$. Encryption of an n-bit message M for the user with identity id is done by first choosing a random r in $\mathbb{Z}_p$ and then computing the ciphertext $(C_1, C_2) = (rP, M \oplus H_2(e(Q, Q_{id})^r))$. Decryption is made possible from the relation $e(Q, Q_{id})^r = e(rP, d_{id})$.

Note that the basic Boneh-Franklin scheme does not provide chosen-ciphertext security, because the message occurs in the ciphertext only in the last XOR step. This means that a plaintext M can be determined from its ciphertext (C_1, C_2) by asking for the decryption of the ciphertext (C_1, C_2'), where C_2' is C_2 with the first bit flipped. One can, however, obtain IND-ID-CPA security results for the basic Boneh-Franklin scheme under the assumption that H_1 and H_2 are random oracles.

Using the Fujisaki-Okamoto transformation [36], the basic Boneh-Franklin IBE scheme can be converted into a scheme, called FullIdent (see [24]), that provides IND-ID-CCA security. To get FullIdent the basic scheme is modified as follows. First, a random $\rho \in \{0,1\}^n$ is chosen and r is set equal to $H_3(\rho, M)$, where H_3 is a hash function that maps bitstrings to integers mod p; we then define $C_1 = rP$ as before. The second component C_2 of the ciphertext is defined by $C_2 = \rho \oplus H_2(e(Q, Q_{id})^r)$ (that is, the hash value is XORed with ρ rather than with M), and we also need a third component C_3 defined by $C_3 = M \oplus H_4(\rho)$, where H_4 is a hash function that maps $\{0,1\}^n$ to $\{0,1\}^n$. The decryption proceeds by first computing $\rho = C_2 \oplus H_2(e(C_1, d_{id}))$ and then $M = C_3 \oplus H_4(\rho)$. But the decryption rejects the ciphertext unless it is validated by checking that $H_3(\rho, M)P = C_1$. This last check is very important, since it prevents an adversary from generating a valid ciphertext for an unknown message M.

Boneh and Franklin [24] argued for the IND-ID-CCA security of their construction using a three stage reduction based on BDH; the reduction turned out to be flawed. Galindo [38] provided a corrected reduction which resulted in a tightness gap of q_H^3, where q_H is the maximum number of queries made to any of the random oracles H_1, H_2, H_3 or H_4. Zhang and Imai [78] provided a direct reduction based on the same BDH assumption with a tightness gap of $q_D \cdot q_E \cdot q_H$, where q_D bounds the number of decryption queries and q_E bounds the number of key-extraction queries made by the adversary.[10] The tightness gap can be reduced to $q_E \cdot q_H$ by making the following change to the simulation of the H_3 random oracle in the proof of Theorem 1 in [78]: when the simulator

---

[10] In Table 1 of [78], Zhang and Imai claim that their security reduction has a tightness gap of $q_E \cdot q_H$; this assertion is repeated in Table 4 of [8]. However, they neglected to account for the tightness gap arising from the running times in Theorem 1 of their paper.

responds to a query (σ_i, M_i) with r_i, it stores g^{r_i} in addition to (σ_i, M_i, r_i) in its "H_3-list" (here we're using the notation of the proof in [78] rather than our own notation, in which σ would be ρ and g^r would be rP). With this change, the simulator can respond to all q_D decryption queries in time q_D instead of $q_D \cdot q_H$ (we are ignoring the time to sort and search the H_3-list). As a result, the lower bound for the BDH-time now has order equal to the sum of the query bounds $q_D + q_{H_2} + q_{H_4} + q_E$, which is essentially the adversary's running time. In other words, in this way we can remove the tightness gap in the running times, and we're left with the tightness gap $q_E \cdot q_{H_2}$ that comes from the success probabilities in Theorem 1 of [78].

As noted in [8], the tightness gap reduces further to q_E if one is willing to base the security on the presumably stronger DBDH or GBDH assumptions. In practice, the hash functions in the IBE constructions are publicly known functions. Thus, the number of queries made to these functions by the adversary can be quite high—q_H could be 2^{64} or even 2^{80} for powerful adversaries. The number of key-extraction queries q_E, on the other hand, will be lower.

An informal argument can be used to show why the tightness gaps in the reductions for Boneh-Franklin IBE are inconsequential for real-world security. Namely, we claim that key-extraction queries give no useful information to the adversary, and so without loss of generality we may take $q_E = 0$; in that case, as mentioned above, there is a tight reduction based on the DBDH or GBDH assumption. Recall that in response to a queried id, the Private Key Generator returns $Q_{id} = H_1(id)$, where H_1 is a random oracle, and $d_{id} = sQ_{id}$. This can be simulated by the adversary itself, who chooses $k \bmod p$ at random and sets $Q_{id} = kP$ and $d_{id} = kQ$. Note that this does not give a valid formal reduction from the case when $q_E > 0$ to the case when $q_E = 0$, because the adversary does not get the "true" key pair of the user, whose public point is produced by the random oracle H_1. However, it is hard to conceive of any difference this could possibly make in the adversary's effectiveness against the IND-ID-CCA security of FullIdent.

Remark 11. In Sect. 3.1 of [53] Koblitz and Menezes made an analogous informal argument in order to conclude that the tightness gap in the security reduction for RSA Full Domain Hash should not be a cause of concern. These examples show, as remarked in [52], that "whether or not a cryptographic protocol lends itself to a tight security reduction argument is not necessarily related to the true security of the protocol... the question of how to interpret a nontight reductionist security argument has no easy answer."

7   Conclusions

Reductionist arguments can contribute to our understanding of the real-world security of a protocol by providing an ironclad guarantee that certain types of attacks are infeasible as long as certain hardness assumptions remain valid. However, even this limited kind of assurance may, as we have seen, turn out to be meaningless in practice if the reduction is nontight and the parameters have not been increased to account for the tightness gap. In order to properly evaluate provable security claims, one needs to study the tightness issue. In this paper we have given examples of the type of analysis of tightness that should be performed, but much work remains to be done. Among the open problems are the following:

1. Examine all uses of complexity leveraging to see whether or not the concrete adaptive security results are meaningful.
2. Evaluate the effect on the required parameter sizes of nontightness in security proofs for HMAC and adjust standards accordingly, particularly in applications that require the pseudorandom function property; also study whether or not the commonly used hash compression functions are likely to satisfy the PRF assumption.
3. Carefully evaluate all lattice-based protocols that have worst-case-to-average-case reductions to see what meaningful concrete bounds, if any, follow from these reductions.
4. For protocols whose security reductions lose tightness in the multi-user setting or the multi-challenge setting (or both), determine how parameter sizes should be increased to account for this.

Acknowledgments. We wish to thank Greg Zaverucha for extensive help with Appendix B as well as useful comments on the other sections, Michael Naehrig for reviewing and commenting on Sect. 5, Somindu C. Ramanna for providing helpful comments on an earlier draft of Sect. 6, Ann Hibner Koblitz for editorial suggestions, and Ian Blake, Eike Kiltz, and Chris Peikert for helpful feedback and suggestions. Of course, none of them is responsible for any of the opinions expressed in this article.

A   Concrete Analysis of Regev's Worst-Case/ Average-Case Reduction

Let $q = q(n)$ and $m = m(n)$ be integers, and let $\alpha = \alpha(n) \in (0,1)$ be such that $\alpha q > 2\sqrt{n}$. Let χ be the probability distribution on $\mathbb{Z}_q$ obtained by sampling from a Gaussian distribution with mean 0 and variance $\alpha^2/2\pi$, and then multiplying by q and rounding to the closest integer modulo q. Then the (search version of the) LWE problem is the following: Let s be a secret vector selected uniformly at random from $\mathbb{Z}_q^n$. Given m samples $(a_i, a_i \cdot s + e_i)$, where each a_i is selected independently and uniformly at random from $\mathbb{Z}_q^n$, and where each e_i is selected independently from $\mathbb{Z}_q$ according to χ, determine s. The decisional version of LWE, called DLWE, asks us to determine whether we have been given m LWE samples $(a_i, a_i \cdot s + e_i)$ or m random samples (a_i, u_i), where each u_i is selected independently and uniformly at random from $\mathbb{Z}_q$.

Regev [71] proved that the existence of an efficient algorithm that solves DLWE in the average case implies the existence of an efficient quantum algorithm that solves SIVP_γ in the worst case where $\gamma = \tilde{O}(n/\alpha)$. In the remainder of this section we provide justification for the following refinement of Regev's theorem:

Claim. Let $q = n^2$ and $\alpha = 1/(\sqrt{n} \log^2 n)$, whence $\gamma = \tilde{O}(n^{1.5})$. Suppose that there is an algorithm W that, given $m = n^c$ samples, solves DLWE for a fraction $1/n^{d_1}$ of all $s \in \mathbb{Z}_q^n$ with advantage at least $1/n^{d_2}$. Then there is a polynomial-time algorithm W' for solving SIVP$_\gamma$ that calls the W oracle a total of $O(n^{11+c+d_1+2d_2})$ times.

A.1   Gaussian Distributions

Recall that the *Gaussian distribution* with mean 0 and variance σ^2 is the distribution on $\mathbb{R}$ given by the probability density function

$$\frac{1}{\sqrt{2\pi} \cdot \sigma} \exp\left(\frac{-x^2}{2\sigma^2}\right).$$

For $x \in \mathbb{R}^n$ and $s > 0$, define the Gaussian function scaled by s:

$$\rho_s(x) = \exp\left(\frac{-\pi \|x\|^2}{s^2}\right).$$

The *Gaussian distribution* D_s *of parameter* s over $\mathbb{R}^n$ is given by the probability density function

$$D_s(x) = \frac{\rho_s(x)}{s^n}.$$

Note that D_s is indeed a probability distribution since $\int_{x \in \mathbb{R}^n} \rho_s(x) \, dx = s^n$.

If L is a lattice, we can define

$$\rho_s(L) = \sum_{x \in L} \rho_s(x).$$

Then the *discrete Gaussian probability distribution* $D_{L,s}$ of width s for $x \in L$ is

$$D_{L,s}(x) = \frac{\rho_s(x)}{\rho_s(L)}.$$

Let L be a (full-rank integer) lattice of dimension n.

A.2   Concrete Analysis

In this section the tightness gap of a reduction algorithm from problem A to problem B is the number of calls to the oracle for B that are made by the reduction algorithm.

Regev's worst-case/average-case reduction has two main components:

1. The reduction of (search-)LWE to average-case DLWE (denoted DLWE$_{ac}$).
2. The reduction of worst-case SIVP$_\gamma$ to LWE.

Reduction of LWE to DLWE$_{ac}$. This reduction has three parts.

Part I. Worst-Case to Average-Case. DLWE$_{wc}$ denotes the worst-case DLWE problem. Lemma 4.1 in [71] shows that an algorithm W_1 that solves DLWE$_{ac}$

for a fraction $\frac{1}{n^{d_1}}$ of all $s \in \mathbb{Z}_q^n$ with acceptance probabilities differing by at least $\frac{1}{n^{d_2}}$ can be used to construct an algorithm W_2 that solves DLWE_{wc} with probability essentially 1 for *all* $s \in \mathbb{Z}_q^n$. The algorithm W_2 invokes W_1 a total of $O(n^{d_1+2d_2+2})$ times.

Part II. Search to Decision. Lemma 4.2 in [71] shows that an algorithm W_2 which solves DLWE_{wc} for all $s \in \mathbb{Z}_q^n$ with probability essentially 1 can be used to construct an algorithm W_3 that solves (search-)LWE for *all* $s \in \mathbb{Z}_q^n$ with probability essentially 1. Algorithm W_3 invokes W_2 a total of nq times, so this reduction has a tightness gap of nq.

Part III. Continuous to Discrete. Lemma 4.3 in [71] shows that an algorithm W_3 that solves LWE can be used to construct an algorithm W_5 that solves $\text{LWE}_{q,\Psi_\alpha}$. (See [71] for the definition of the $\text{LWE}_{q,\Psi_\alpha}$ problem.) This reduction is tight.

Reduction of SIVP$_\gamma$ to LWE. This reduction has tightness gap $6n^{6+c}$. The reduction has two parts.

Part I. DGS to LWE. Let $\epsilon = \epsilon(n)$ be some negligible function of n. Theorem 3.1 of [71] shows that an algorithm W_4 that solves $\text{LWE}_{q,\Psi_\alpha}$ given m samples can be used to construct a *quantum* algorithm W_9 for $DGS_{\sqrt{2n}\cdot\eta_\epsilon(L)/\alpha}$. Here, $\eta_\epsilon(L)$ is the "smoothing parameter with accuracy ϵ", and $DGS_{r'}$ (discrete Gaussian sampling problem) is the problem of computing a sample from the discrete Gaussian probability distribution $D_{L,r'}$ where $r' \geq \sqrt{2n} \cdot \eta_\epsilon(L)/\alpha$.

Let $r = \sqrt{2n} \cdot \eta_\epsilon(L)/\alpha$. Let $r_i = r \cdot (\alpha q/\sqrt{n})^i$ for $i \in [0,3n]$. Algorithm W_9 begins by producing n^c samples from $D_{L,r_{3n}}$ (Lemma 3.2 in [71]); the W_4 oracle is not used in this step. Next, by repeatedly applying the 'iterative step,' it uses the n^c samples from D_{L,r_i} to produce n^c samples from $D_{L,r_{i-1}}$ for $i = 3n, 3n - 1, \ldots, 1$. Since $r_0 = r$, the last step produces the desired sample from $D_{L,r}$.

The iterative step (Lemma 3.3 in [71]) uses n^c samples from D_{L,r_i} to produce one sample from $D_{L,r_{i-1}}$; this step is then repeated to produce n^c samples from $D_{L,r_{i-1}}$. Thus, the iterative step is executed a total of $3n \cdot n^c = 3n^{1+c}$ times.

Each iterative step has two parts.

1. The first part invokes W_4 a total of n^2 times:
    - Lemma 3.7 in [71] uses W_4 to construct an algorithm W_5 that solves $\text{LWE}_{q,\Psi_\beta}$; W_5 invokes W_4 n times.
    - Lemma 3.11 in [71] uses W_5 and the n^c samples from D_{L,r_i} to construct an algorithm W_6 that solves the $\text{CVP}^{(q)}_{L^*,\alpha q/\sqrt{2}r_i}$ problem. The reduction is tight.
    - Lemma 3.5 in [71] uses W_6 to construct an algorithm W_7 that solves the $\text{CVP}_{L^*,\alpha q/\sqrt{2}r_i}$ problem. Algorithm W_7 invokes W_6 n times.
2. The second part (Lemma 3.14 in [71]) uses W_7 to construct a *quantum* algorithm W_8 that produces a sample from $D_{L,r_{i-1}}$. This reduction is tight.

Since each iterative step has tightness gap n^2, the total tightness gap for the reduction of DGS to LWE is $3n^{3+c}$.

Part II. SIVP$_\gamma$ *to DGS* Lemma 3.17 in [71] uses W_9 to construct an algorithm W_{10} that solves SIVP$_{2\sqrt{2}n\eta_\epsilon(L)/\alpha}$. Algorithm W_{10} invokes W_9 $2n^3$ times.

Lemma 2.12 in [71] states that $\eta_\epsilon(L) \leq \sqrt{\omega(\log n)} \cdot \lambda_n(L)$ for some negligible function $\epsilon(n)$. Thus

$$\gamma = \frac{2\sqrt{2}n\eta_\epsilon(L)}{\alpha \cdot \lambda_n(L)} = \frac{2\sqrt{2}n\sqrt{\omega(\log n)}}{\alpha} = \tilde{O}\left(\frac{n}{\alpha}\right) = \tilde{O}(n^{1.5}).$$

Summary. Regev's reduction of SIVP$_\gamma$ to DLWE$_{ac}$ has tightness gap

$$n^{d_1+2d_2+2} \cdot nq \cdot 3n^{3+c} \cdot 2n^3 = 6n^{11+c+d_1+2d_2}.$$

B   Nontightness and Multi-user Attacks

In an important paper that has been all but ignored by the cryptographic research community, Zaverucha [77] showed that "provably secure" hybrid encryption, as described in several standards, is insecure in the multi-user setting if certain permitted (and even recommended) choices are made in the implementation. Because this work should be much better known than it is, we shall devote this section to explaining and summarizing [77]. We shall focus on hybrid encryption schemes in the comprehensive ISO/IEC 18033-2 standard [74].

We first recall the definition in [16] of IND-CCA security (Indistinguishability under Chosen-Ciphertext Attack) of encryption in the multi-user setting. Suppose there are n users. The adversary is given n public keys, a decryption oracle for each public key, and an LR (left-or-right encryption) oracle for each public key. The adversary can query each decryption oracle up to q_D times and each LR oracle up to q_{LR} times. A decryption query simply asks for a chosen ciphertext to be decrypted under the corresponding public key. An LR query works differently. The n LR-oracles all have a hidden random bit b in common. The adversary chooses two equal-length messages M_0 and M_1 to query to one of the LR-oracles, which then returns an encryption C^* of M_b. The adversary is not permitted to query C^* to the decryption oracle for the same public key. The adversary's task is to guess b with success probability significantly greater than $1/2$.

Remark 12. This "multi-challenge" security model (that is, $q_{LR} > 1$) can also be used in the single-user setting, but almost never is ([22] is a rare exception); in the standard IND-CCA security model $q_{LR} = 1$. We shall later give a simple attack that shows that the standard IND-CCA is deficient and should be replaced by the multi-challenge model.

Remark 13. In [16] the authors give a generic reduction with tightness gap $n \cdot q_{LR}$ between the multi-user and single-user settings. In the full version of [16] they also give a construction that shows that this tightness bound is optimal; that is, they describe a protocol that can be attacked with $n \cdot q_{LR}$ times the advantage in the multi-user setting than in the single-challenge single-user setting. Their construction is contrived and impractical; later we shall describe a simple attack on hybrid encryption that shows that in practice as well as in theory the generic tightness bound in [16] is best possible. That is, the attack described below reduces security by a factor equal to n times the number of messages sent to each user (see Remark 16). (In specific cases tighter reductions are sometimes possible—for example, the paper [16] contains a reduction with tightness gap q_{LR} in the case of the Cramer–Shoup public-key encryption scheme [31].)

We now recall the setup and terminology of hybrid encryption. The encryption has two stages: a key-encapsulation mechanism (KEM) using a public-key cryptosystem (with the recipient's public/secret key pair denoted PK/SK), and a data-encapsulation mechanism (DEM) using a symmetric-key cryptosystem that encrypts the data by means of the shared key K that is produced by the KEM. The KEM takes PK as input and produces both the key material K by means of a key-derivation function (KDF) and also a ciphertext C_1 that will enable the recipient to compute K; the DEM takes K and the message M as input and produces a ciphertext C_2. The recipient decrypts by first using C_1 and SK to find K and then using C_2 and the symmetric key K to find M.

Among the public-key systems commonly used for KEM are Cramer-Shoup [31] and ECIES (ElGamal encryption using elliptic curves, see [74]); symmetric-key systems commonly used for DEM are AES in cipher block chaining (CBC) mode and XOR-Encrypt using a hash function with a counter. (We will describe this in more detail shortly.) The KDF is a publicly known way to produce key material of a desired length L from a shared secret that's computed using the public-key system.

Suppose, following [74], that we use 128-bit AES in CBC-mode with zero initialization vector for DEM. Let MAC denote a message authentication code that depends on a 128-bit key. Our KDF produces two 128-bit keys $K = (k_1, k_2)$. To send a 128m-bit message M, we set C_2 equal to a pair (C', t), where C' is the 128m-bit ciphertext computed below and $t = \text{MAC}_{k_2}(C')$ is its tag. The ciphertext $C' = (C'_1, \ldots, C'_m)$ is given by: $C'_1 = \text{AES}_{k_1}(M_1)$, $C'_i = \text{AES}_{k_1}(C'_{i-1} \oplus M_i)$ for $i = 2, \ldots, m$.

After receiving $(C_1, C_2) = (C_1, C', t)$, the recipient first uses C_1, SK, and the KDF to find (k_1, k_2), and then uses the shared key k_2 to verify that t is in fact the tag of C'; otherwise she rejects the message. Then she decrypts using k_1.

Alternatively, for DEM we could use XOR-Encrypt with a hash function $\mathcal{H}$ as follows. To send a message M consisting of m 256-bit blocks, we have the KDF generate a 256m-bit key $k_1 = (k_{1,1}, \ldots, k_{1,m})$ by setting $k_{1,i} = \mathcal{H}(z_0 \| i)$, where z_0 is a shared secret produced by KEM, and also a MAC-ing key k_2. The MAC works as before, but now C' is determined by setting $C'_i = M_i \oplus k_{1,i}$. This is the hash function with counter (CTR) mode mentioned above.

In [32] Cramer and Shoup gave a tight proof that hybrid encryption has IND-CCA security under quite weak assumptions. The MAC-scheme need only be "one-time secure" (because it receives a new key k_2 for each message), and the symmetric encryption function need only be one-time secure against passive adversaries—in particular, there is no need for randomization (again the reason is that it gets a new key k_1 for each message). In accordance with the general principle that standards should not require extra features that are not needed in the security reductions, the standards for hybrid encryption [74] do not require randomization in the symmetric encryption; nor do they impose very stringent conditions on the KDF. In addition, in [74] Shoup comments that if KEM is implemented using the Cramer–Shoup construction [31], which has a security proof without random oracles, and if DEM is implemented using AES-CBC, then it is possible to prove a tight security reduction for the hybrid encryption scheme without the random oracle assumption. Thus, anyone who mistrusts random oracle proofs should use AES-CBC rather than XOR-Encrypt. All of these security proofs are given in the single-user setting.

B.1   Attacks in the Multi-user Setting

We now describe some of the attacks of Zaverucha [77] in the multi-user setting, which of course is the most common setting in practice. Let $n = 2^a$ be the number of users. First suppose that the DEM is implemented using AES128 in CBC-mode. Suppose that Bob sends all of the users messages that all have the same first two blocks (M_1, M_2) (that is, they start with the same 256-bit header). The rest of the message blocks may be the same (i.e., broadcast encryption), or they may be different. The adversary Cynthia's goal is to read at least one of the 2^a messages. She guesses a key k that she hopes is the k_1-key for one of the messages. She computes $C_1'' =\mathrm{AES}_k(M_1)$ and $C_2'' = \mathrm{AES}_k(C_1'' \oplus M_2)$ and compares the pair (C_1'', C_2'') with the first two blocks of ciphertext sent to the different users.[11] If there's a match, then it is almost certain that she has guessed the key $k_1 = k$ for the corresponding message. That is because there are 2^{128} possible keys k_1 and 2^{256} possible pairs (C_1', C_2'), so it is highly unlikely that distinct keys would give the same (C_1', C_2'). Once Cynthia knows k_1—each guess has a $2^{-(128-a)}$ chance of producing a match—she can quickly compute the rest of the plaintext. This means that even though the hybrid encryption scheme might have a tight security reduction in the single-user setting that proves 128 bits of security, in the multi-user setting it has only $128 - a$ bits of security. Commenting on how dropping randomization in DEM made his attack possible, Zaverucha [77] calls this "an example of a provable security analysis leading to decreased practical security."

Remark 14. In modern cryptography—ever since the seminal Goldwasser-Micali paper [44]—it has been assumed that encryption must always be probabilistic. In [74] this principle is violated in the interest of greater efficiency because the security proof in [32] does not require randomization. This decision was bold, but also rash, as Zaverucha's attack shows.

---

[11] We can suppose that the 2^a ciphertexts are sorted according to their first two blocks (or perhaps stored using a conventional hash function). Then one iteration of the attack takes essentially unit time, since it just requires computing (C_1'', C_2'') and looking for it in the sorted table. Since the expected number of iterations is 2^{128-a}, the running time of the attack is $T = 2^{128-a}$ (and the success probability is essentially 1).

Remark 15. A time–memory–data tradeoff can be applied to speed up the online portion of the attack; see Remark 7 in [27]. Namely, at a cost of precomputation time 2^{128-a} and storage size $2^{2(128-a)/3}$, the secret key k of one of the 2^a users can be determined in time $2^{2(128-a)/3}$.

Remark 16. The above attack can also be carried out in the single-user setting if we suppose that Bob is sending Alice $2^{a'}$ different messages that all have the same header (M_1, M_2). Since different keys are generated for different messages (even to the same user), there is no need for the recipients of the messages to be different. This gives a reduction of the number of bits of security by a'. This attack shows the need for the multi-challenge security model even in the single-user setting. Thus, even in the single-user setting the standard security model for encryption is deficient because it fails to account for the very realistic possibility that Bob uses hybrid encryption as standardized in [74] to send Alice many messages that have the same header.

Remark 17. Note that if the $2^{a'}$ messages are broadcast to 2^a users, then obviously the reduction in security is by $a' + a$ bits. In some circumstances $a' + a$ could be large enough to reduce the security well below acceptable levels. For example, if $a' + a > 32$, it follows that what was thought to have 128 bits of security now has fewer than 96, which, as remarked in Sect. 1, is not enough. It should be emphasized that the security is reduced because of actual practical attacks, not because of a tightness gap that could conceivably be removed if one finds a different proof.

We note that the above attack does not in general work if DEM is implemented using XOR-Encrypt. (Of course, someone who does not trust security proofs that use random oracles would not be using XOR-Encrypt, and so would be vulnerable.) But Zaverucha has a different attack on hybrid encryption with XOR-Encrypt that works for certain KDF constructions.

B.2   Attacks on Extract-then-Expand with XOR-Encrypt

The most commonly used KDF takes the shared secret z_0 produced in KEM and derives a key of the desired length by concatenating $\mathcal{H}(z_0\|i)$ for $i = 1, \dots$. However, at Crypto 2010, as Zaverucha [77] explains,

> Krawczyk argues that cryptographic applications should move to a single, well-studied, rigorously analyzed family of KDFs. To this end, he formally defines security for KDFs, presents a general construction that uses any keyed pseudorandom function (PRF), and proves the security of his construction in the new model. The approach espoused by the construction is called *extract-then-expand*. [...] The HKDF scheme is a concrete instantiation of this general construction when HMAC is used for both extraction and expansion.

The Extract-then-Expand key derivation mechanism was soon standardized [28, 29,59]. In particular, RFC 5869 describes HKDF, which instantiates the Extract-then-Expand mechanism with HMAC, and states that HKDF is intended for use in a variety of KDF applications including hybrid encryption.

Extract-then-Expand works in hybrid encryption as follows. Suppose that z_0 is the shared secret produced in KEM. The Extract phase produces a bitstring $z_1 = \text{Extract}(z_0)$, perhaps of only 128 bits, which is much shorter than z_0. (The Extract phase may also depend on a "salt," but this is optional, and we shall omit it.) Then the key material K is obtained by a function that expands z_1, i.e., $K = \text{Expand}(z_1, L)$, where L as before is the bitlength of K. (There is also the option of putting some contextual information inside the Expand-function, but we shall not do this.)

We now describe Zaverucha's attack on hybrid encryption when Extract-then-Expand with 128-bit z_1 values is used as the KDF and XOR-Encrypt is used for message encryption. Suppose that Bob sends messages to 2^a users that all have the same header (M_1, M_2) and the same bitlength L. Cynthia's goal is to recover at least one of the plaintexts. Rather than guessing a key, she now guesses the bitstring z_1. For each guess she computes $K = \text{Expand}(z_1, L)$ and $C_i'' = M_i \oplus k_{1,i}, i = 1, 2$. When she gets a match with (C_1', C_2') for one of the users, she can then recover the rest of the plaintext sent to that user: $M_i = C_i' \oplus k_{1,i}, i > 2$.

Note that this attack does not work for XOR-Encrypt with the KDF using $\mathcal{H}(z_0 \| i)$ described above. Once again the "provably secure" choice of Extract-then-Expand turns out to be vulnerable, whereas the traditional choice of KDF is not. Zaverucha comments that "In this example, replacing a commonly used KDF in favor of a provably secure one causes a decrease in practical security."

As discussed in [77], Zaverucha's attacks can be avoided in practice by putting in features that are not required in the standard single-user single-challenge security proofs. It would be worthwhile to give proofs of this.

Open Problem. Give a tight security reduction for hybrid encryption in the multi-user multi-challenge security model (random oracles are permitted) if DEM uses either: (1) randomized encryption rather than one-time-secure encryption (for example, AES-CBC with random IV that is different for each message and each recipient), (2) XOR-Encrypt using $\mathcal{H}(z_0 \| i)$ for the KDF, (3) XOR-Encrypt using HKDF with a recipient- and message-dependent salt in the Extract phase and/or recipient- and message-dependent contextual information in the Expand phase.

We conclude this section by noting a curious irony. As we remarked in Sect. 1, it is very rare for a standards body to pay much attention to tightness gaps in the security reductions that are used to support a proposed standard or to whether those security reductions were proved in the multi-user or single-user setting. However, recently the IETF decided that the standard for Schnorr signatures [72] should require that the public key be included in the hash function. The reason was that Bernstein [20] had found a flaw in the tight reduction from an adversary in the single-user setting to an adversary in the multi-user setting that had been given by Galbraith et al. [37], and he had proved that a tight security reduction could be restored if the public key is included in the hash function. (Later Kiltz et al. [50] gave a tight security reduction *without* needing to include the public key in the hash function; however, their assumptions are stronger than in [37], and it is not yet clear whether their result will cause the IETF to go back to dropping the public key from the hash input.)

The peculiar thing is that the tightness gap between single-user and multi-user settings is only a small part of the tightness problem for Schnorr signatures.

Lemma 5.7 in [50] gives a security proof in the random oracle model for the Schnorr signature scheme in the single-user setting. The proof has a tightness gap equal to the number of random oracle queries, which can be very large—in particular, much larger than the number of users in the multi-user setting. Even a tight single-user/multi-user equivalence leaves untouched the large tightness gap between Schnorr security and hardness of the underlying Discrete Log Problem. It should also be noted that the IETF was responding to the error Bernstein found in a proof, not to any actual attack that exploited the tightness gap (we now know that such an attack is probably impossible, because of the recent proof in [50] that under a certain reasonable assumption there is no single-user/multi-user tightness gap).

In the meantime, standards bodies have done nothing to address Zaverucha's critique of the standardized version [74] of hybrid encryption, which allows implementations that have far less security than previously thought, as shown by actual attacks.

References

1. Aggarwal, D., Dadush, D., Regev, O., Stephens-Davidowitz, N.: Solving the shortest vector problem in 2^n time via discrete Gaussian sampling. In: Proceedings of the 47th Annual Symposium Foundations of Computer Science, pp. 733–742 (2015)
2. Ajtai, M.: Generating hard instances of lattice problems. In: Proceedings of the 28th Annual ACM Symposium on Theory of Computing, pp. 99–108. ACM (1996)
3. Ajtai, M., Dwork, C.: A public-key cryptosystem with worst-case/average-case equivalence. In: Proceedings of the 29th Annual ACM Symposium on Theory of Computing, pp. 284–293. ACM (1997)
4. Albrecht, M., Player, R., Scott, S.: On the concrete hardness of learning with errors. J. Math. Cryptol. **9**, 169–203 (2015)
5. Alkim, E., Ducas, L., Pöppelmann, T., Schwabe, P.: Post-quantum key exchange - a new hope. In: Proceeding of the 25th USENIX Security Symposium, pp. 327–343 (2016)
6. ANSI X9.98: Lattice-Based Polynomial Public Key Establishment Algorithm for the Financial Services Industry, Part 1: Key Establishment, Part 2: Data Encryption (2010)
7. Arora, S., Ge, R.: New algorithms for learning in presence of errors. In: Aceto, L., Henzinger, M., Sgall, J. (eds.) ICALP 2011. LNCS, vol. 6755, pp. 403–415. Springer, Heidelberg (2011). doi:10.1007/978-3-642-22006-7_34

8. Attrapadung, N., Furukawa, J., Gomi, T., Hanaoka, G., Imai, H., Zhang, R.: Efficient identity-based encryption with tight security reduction. In: Pointcheval, D., Mu, Y., Chen, K. (eds.) CANS 2006. LNCS, vol. 4301, pp. 19–36. Springer, Heidelberg (2006). doi:10.1007/11935070_2

9. Barbulescu, R., Gaudry, P., Joux, A., Thomé, E.: A heuristic quasi-polynomial algorithm for discrete logarithm in finite fields of small characteristic. In: Nguyen, P.Q., Oswald, E. (eds.) EUROCRYPT 2014. LNCS, vol. 8441, pp. 1–16. Springer, Heidelberg (2014). doi:10.1007/978-3-642-55220-5_1

10. Barreto, P.S.L.M., Naehrig, M.: Pairing-friendly elliptic curves of prime order. In: Preneel, B., Tavares, S. (eds.) SAC 2005. LNCS, vol. 3897, pp. 319–331. Springer, Heidelberg (2006). doi:10.1007/11693383_22

11. Bellare, M.: Practice-oriented provable-security. In: Damgård, I.B. (ed.) EEF School 1998. LNCS, vol. 1561, pp. 1–15. Springer, Heidelberg (1999). doi:10.1007/3-540-48969-X_1

12. Bellare, M.: New proofs for NMAC and HMAC: security without collision-resistance. In: Dwork, C. (ed.) CRYPTO 2006. LNCS, vol. 4117, pp. 602–619. Springer, Heidelberg (2006). doi:10.1007/11818175_36

13. Bellare, M.: email to N. Koblitz, 24 February 2012

14. Bellare, M.: New proofs for NMAC and HMAC: security without collision-resistance. J. Cryptol. **28**, 844–878 (2015)

15. Bellare, M., Bernstein, D.J., Tessaro, S.: Hash-function based PRFs: AMAC and its multi-user security. In: Fischlin, M., Coron, J.-S. (eds.) EUROCRYPT 2016. LNCS, vol. 9665, pp. 566–595. Springer, Heidelberg (2016). doi:10.1007/978-3-662-49890-3_22

16. Bellare, M., Boldyreva, A., Micali, S.: Public-key encryption in a multi-user setting: security proofs and improvements. In: Preneel, B. (ed.) EUROCRYPT 2000. LNCS, vol. 1807, pp. 259–274. Springer, Heidelberg (2000). doi:10.1007/3-540-45539-6_18. https://cseweb.ucsd.edu/ mihir/papers/musu.html

17. Bellare, M., Canetti, R., Krawczyk, H.: Keying hash functions for message authentication. In: Koblitz, N. (ed.) CRYPTO 1996. LNCS, vol. 1109, pp. 1–15. Springer, Heidelberg (1996). doi:10.1007/3-540-68697-5_1

18. Bellare, M., Canetti, R., Krawczyk, H.: Pseudorandom functions revisited: the cascade construction and its concrete security. In: Proceedings of the 37th Annual Symposium Foundations of Computer Science, pp. 514–523 (1996). http://cseweb.ucsd.edu/users/mihir/papers/cascade.pdf

19. Bellare, M., Canetti, R., Krawczyk, H.: HMAC: keyed-hashing for message authentication, Internet RFC 2104 (1997)

20. Bernstein, D.: Multi-user Schnorr security, revisited. http://eprint.iacr.org/2015/996.pdf

21. Blömer, J., Seifert, J.: On the complexity of computing short linearly independent vectors and short bases in a lattice. In: Proceedings of the 31st Annual ACM Symposium on Theory of Computing, pp. 711–720. ACM (1999)

22. Boldyreva, A.: Strengthening security of RSA-OAEP. In: Fischlin, M. (ed.) CT-RSA 2009. LNCS, vol. 5473, pp. 399–413. Springer, Heidelberg (2009). doi:10.1007/978-3-642-00862-7_27

23. Boneh, D., Boyen, X.: Efficient selective-ID secure identity based encryption without random oracles. http://eprint.iacr.org/2004/172.pdf

24. Boneh, D., Franklin, M.: Identity-based encryption from the Weil pairing. SIAM J. Comput. **32**, 586–615 (2003)

25. Boneh, D., Waters, B.: Constrained pseudorandom functions and their applications. In: Sako, K., Sarkar, P. (eds.) ASIACRYPT 2013. LNCS, vol. 8270, pp. 280–300. Springer, Heidelberg (2013). doi:10.1007/978-3-642-42045-0_15

26. Bos, J., Costello, C., Naehrig, M., Stebila, D.: Post-quantum key exchange for the TLS protocol from the ring learning with errors problem. In: Proceedings of the 2015 IEEE Symposium on Security and Privacy, pp. 553–570 (2015)

27. Chatterjee, S., Menezes, A., Sarkar, P.: Another look at tightness. In: Miri, A., Vaudenay, S. (eds.) SAC 2011. LNCS, vol. 7118, pp. 293–319. Springer, Heidelberg (2012). doi:10.1007/978-3-642-28496-0_18

28. Chen, L.: Recommendation for key derivation using pseudorandom functions (revised), NIST SP 800–108 (2009)

29. Chen, L.: Recommendation for key derivation through extraction-then-expansion, NIST SP 800–56C (2011)

30. Coppersmith, D.: Fast evaluation of logarithms in fields of characteristic two. IEEE Trans. Inf. Theory **30**, 587–594 (1984)

31. Cramer, R., Shoup, V.: A practical public key cryptosystem provably secure against adaptive chosen ciphertext attack. In: Krawczyk, H. (ed.) CRYPTO 1998. LNCS, vol. 1462, pp. 13–25. Springer, Heidelberg (1998). doi:10.1007/BFb0055717

32. Cramer, R., Shoup, V.: Design and analysis of practical public-key encryption schemes secure against adaptive chosen ciphertext attack. SIAM J. Comput. **33**, 167–226 (2003)

33. Dang, Q.: Recommendation for applications using approved hash algorithms, NIST SP 800–107 (2012)

34. Dierks, T., Allen, C.: The TLS protocol, Internet RFC 2246 (1999)

35. Fuchsbauer, G.: Constrained verifiable random functions. In: Abdalla, M., Prisco, R. (eds.) SCN 2014. LNCS, vol. 8642, pp. 95–114. Springer, Cham (2014). doi:10.1007/978-3-319-10879-7_7

36. Fujisaki, E., Okamoto, T.: Secure integration of asymmetric and symmetric encryption schemes. In: Wiener, M. (ed.) CRYPTO 1999. LNCS, vol. 1666, pp. 537–554. Springer, Heidelberg (1999). doi:10.1007/3-540-48405-1_34

37. Galbraith, S., Malone-Lee, J., Smart, N.: Public key signatures in the multi-user setting. Inf. Process. Lett. **83**, 263–266 (2002)

38. Galindo, D.: Boneh-Franklin identity based encryption revisited. In: Caires, L., Italiano, G.F., Monteiro, L., Palamidessi, C., Yung, M. (eds.) ICALP 2005. LNCS, vol. 3580, pp. 791–802. Springer, Heidelberg (2005). doi:10.1007/11523468_64

39. Garg, S., Gentry, C., Halevi, S., Raykova, M., Sahai, A., Waters, B.: Candidate indistinguishability obfuscation and functional encryption for all circuits. http://eprint.iacr.org/2013/451.pdf

40. Gaži, P., Pietrzak, K., Rybár, M.: The exact PRF-security of NMAC and HMAC. In: Garay, J.A., Gennaro, R. (eds.) CRYPTO 2014. LNCS, vol. 8616, pp. 113–130. Springer, Heidelberg (2014). doi:10.1007/978-3-662-44371-2_7

41. Goldreich, O., Goldwasser, S.: On the limits of nonapproximability of lattice problems. J. Comput. Syst. Sci. **60**, 540–563 (2000)

42. Goldwasser, S., Bellare, M.: Lecture Notes on Cryptography, July 2008. http://cseweb.ucsd.edu/mihir/papers/gb.pdf

43. Goldwasser, S., Kalai, Y.: Cryptographic assumptions: a position paper. http://eprint.iacr.org/2015/907.pdf

44. Goldwasser, S., Micali, S.: Probabilistic encryption. J. Comput. Syst. Sci. **28**, 270–299 (1984)

45. Harkins, D., Carrel, D.: The internet key exchange (IKE), Internet RFC 2409 (1998)

46. Hoffstein, J., Howgrave-Graham, N., Pipher, J., Whyte, W.: Practical lattice-based cryptography: NTRUEncrypt and NTRUSign. In: Vallée, B., Nguyen, P.Q. (eds.) The LLL Algorithm, pp. 349–390. Springer, Heidelberg (2010). doi:10.1007/978-3-642-02295-1_11

47. Hoffstein, J., Pipher, J., Silverman, J.H.: NTRU: a ring-based public key cryptosystem. In: Buhler, J.P. (ed.) ANTS 1998. LNCS, vol. 1423, pp. 267–288. Springer, Heidelberg (1998). doi:10.1007/BFb0054868

48. IEEE 1363.1: Standard Specification for Public Key Cryptographic Techniques Based on Hard Problems over Lattices (2008)

49. Katz, J., Lindell, Y.: Introduction to Modern Cryptography. Chapman and Hall/CRC, London (2007)

50. Kiltz, E., Masny, D., Pan, J.: Optimal security proofs for signatures from identification schemes. In: Robshaw, M., Katz, J. (eds.) CRYPTO 2016. LNCS, vol. 9815, pp. 33–61. Springer, Heidelberg (2016). doi:10.1007/978-3-662-53008-5_2

51. Kim, T., Barbulescu, R.: Extended tower number field sieve: a new complexity for the medium prime case. In: Robshaw, M., Katz, J. (eds.) CRYPTO 2016. LNCS, vol. 9814, pp. 543–571. Springer, Heidelberg (2016). doi:10.1007/978-3-662-53018-4_20

52. Koblitz, N., Menezes, A.: Another look at "provable security". II. In: Barua, R., Lange, T. (eds.) INDOCRYPT 2006. LNCS, vol. 4329, pp. 148–175. Springer, Heidelberg (2006). doi:10.1007/11941378_12

53. Koblitz, N., Menezes, A.: Another look at 'provable security'. J. Cryptol. **20**, 3–37 (2007)

54. Koblitz, N., Menezes, A.: Another look at HMAC. J. Math. Cryptol. **7**, 225–251 (2013)

55. Koblitz, N., Menezes, A.: Another look at non-uniformity. Groups Complex. Cryptol. **5**, 117–139 (2013)

56. Koblitz, N., Menezes, A.: Another look at security definitions. Adv. Math. Commun. **7**, 1–38 (2013)

57. Koblitz, N., Menezes, A.: Another look at security theorems for 1-key nested MACs. In: Koç, Ç.K. (ed.) Open Problems in Mathematics and Computational Science, pp. 69–89. Springer, Cham (2014). doi:10.1007/978-3-319-10683-0_4

58. Koblitz, N., Menezes, A.: A riddle wrapped in an enigma. IEEE Secur. Priv. **14**, 34–42 (2016)

59. Krawczyk, H., Eronen, P.: HMAC-based extract-and-expand key derivation function (HKDF), Internet RFC 5869 (2010)

60. Krawczyk, H.: Cryptographic extraction and key derivation: the HKDF scheme. In: Rabin, T. (ed.) CRYPTO 2010. LNCS, vol. 6223, pp. 631–648. Springer, Heidelberg (2010). doi:10.1007/978-3-642-14623-7_34

61. Laarhoven, T., Mosca, M., van de Pol, J.: Finding shortest lattice vectors faster using quantum search. Des. Codes Crypt. **77**, 375–400 (2015)

62. Lyubashevsky, V., Micciancio, D., Peikert, C., Rosen, A.: SWIFFT: a modest proposal for FFT hashing. In: Nyberg, K. (ed.) FSE 2008. LNCS, vol. 5086, pp. 54–72. Springer, Heidelberg (2008). doi:10.1007/978-3-540-71039-4_4

63. Lyubashevsky, V., Peikert, C., Regev, O.: On ideal lattices, learning with errors over rings. J. ACM **60**, 43:1–43:35 (2013)

64. Menezes, A.: Another look at provable security, Invited talk at Eurocrypt 2012. http://www.cs.bris.ac.uk/eurocrypt2012/Program/Weds/Menezes.pdf

65. Micciancio, D., Goldwasser, S.: Complexity of Lattice Problems: A Cryptographic Perspective. Springer, New York (2002). doi:10.1007/978-1-4615-0897-7

66. M'Raihi, D., Bellare, M., Hoornaert, F., Naccache, D., Ranen, O.: HOTP: an HMAC-based one time password algorithm, Internet RFC 4226 (2005)
67. Peikert, C.: Lattice cryptography for the internet. In: Mosca, M. (ed.) PQCrypto 2014. LNCS, vol. 8772, pp. 197–219. Springer, Cham (2014). doi:10.1007/978-3-319-11659-4_12
68. Peikert, C.: 19 February 2015 blog posting. http://web.eecs.umich.edu/~cpeikert/soliloquy.html
69. Peikert, C.: A decade of lattice cryptography. http://eprint.iacr.org/2015/939
70. Pietrzak, K.: A closer look at HMAC. http://eprint.iacr.org/2013/212.pdf
71. Regev, O.: On lattices, learning with errors, random linear codes, cryptography. J. ACM **56**, 34:1–34:40 (2009)
72. Schnorr, C.P.: Efficient identification and signatures for smart cards. In: Quisquater, J.-J., Vandewalle, J. (eds.) EUROCRYPT 1989. LNCS, vol. 434, pp. 688–689. Springer, Heidelberg (1990). doi:10.1007/3-540-46885-4_68
73. Shoup, V.: Sequences of games: a tool for taming complexity in security proofs. http://eprint.iacr.org/2004/332.pdf
74. Shoup, V.: ISO/IEC 18033-2:2006, Information Technology – Security Techniques – Encryption Algorithms – Part 2: Asymmetric Ciphers (2006). http://www.shoup.net/iso/std6.pdf
75. Stehlé, D., Steinfeld, R.: Making NTRU as secure as worst-case problems over ideal lattices. In: Paterson, K.G. (ed.) EUROCRYPT 2011. LNCS, vol. 6632, pp. 27–47. Springer, Heidelberg (2011). doi:10.1007/978-3-642-20465-4_4
76. Stern, J., Pointcheval, D., Malone-Lee, J., Smart, N.P.: Flaws in applying proof methodologies to signature schemes. In: Yung, M. (ed.) CRYPTO 2002. LNCS, vol. 2442, pp. 93–110. Springer, Heidelberg (2002). doi:10.1007/3-540-45708-9_7
77. Zaverucha, G.M.: Hybrid encryption in the multi-user setting. http://eprint.iacr.org/2012/159.pdf
78. Zhang, R., Imai, H.: Improvements on security proofs of some identity based encryption schemes. In: Feng, D., Lin, D., Yung, M. (eds.) CISC 2005. LNCS, vol. 3822, pp. 28–41. Springer, Heidelberg (2005). doi:10.1007/11599548_3

Another Look at Anonymous Communication
Security and Modular Constructions

Russell W.F. Lai[1], Henry K.F. Cheung[2], Sherman S.M. Chow[1(✉)],
and Anthony Man-Cho So[2]

[1] Department of Information Engineering, The Chinese University of Hong Kong,
Sha Tin, N.T., Hong Kong
{wflai,sherman}@ie.cuhk.edu.hk
[2] Department of Systems Engineering and Engineering Management,
The Chinese University of Hong Kong, Sha Tin, N.T., Hong Kong
{kfcheung,manchoso}@se.cuhk.edu.hk

Abstract. Anonymous communication is desirable for personal, financial, and political reasons. Despite the abundance of frameworks and constructions, anonymity definitions are usually either not well-defined or too complicated to use. In between are ad-hoc definitions for specific protocols which sometimes only provide weakened anonymity guarantees. This paper addresses this situation from the perspectives of syntax, security definition, and construction. We propose simple yet expressive syntax and security definition for anonymous communication. Our syntax covers protocols with different operational characteristics. We give a hierarchy of anonymity definitions, starting from the strongest possible to several relaxations. We also propose a modular construction from any key-private public-key encryption scheme, and a new primitive – oblivious forwarding protocols, of which we give two constructions. The first is a generic construction from any random walk over graphs, while the second is optimized for the probability of successful delivery, with experimental validation for our optimization. Anonymity is guaranteed even when the adversary can observe and control all traffic in the network and corrupt most nodes, in contrast to some efficient yet not-so-anonymous protocols. We hope this work suggests an easier way to design and analyze efficient anonymous communication protocols in the future.

Keywords: Anonymous communication · Key-privacy · Oblivious forwarding · Global adversary

1 Introduction

Since the seminal work of Chaum [9], the notion of anonymous communication has been extensively studied in the past decades. The goal of anonymous communication is to hide the correspondence between senders and receivers of messages.

S.S.M. Chow is supported by the Early Career Scheme and the Early Career Award of the Research Grants Council, Hong Kong SAR (CUHK 439713).

ⓒ Springer International Publishing AG 2017
R.C.-W. Phan and M. Yung (Eds.): LNCS 10311, Mycrypt 2016, pp. 56–82, 2017.
DOI: 10.1007/978-3-319-61273-7_4

In a stricter sense, the identities of the senders and/or receivers may also need to be hidden. There are plentiful reasons for having anonymous communication, such as to act against censorship and mass surveillance, to protect the privacy of personal preferences, and to express minority opinions. The use of anonymous communication has become increasingly popular among the general public, as indicated by the success of the Tor network [14].

1.1 Anonymity Against a Global Adversary

Very often, research on anonymous communication focuses on achieving low latency, while the anonymity guarantee is not well defined. Pfitzmann and Hansen [24] consolidated informally a collection of terminologies (*e.g.*, unlinkability, anonymity, unobservability) which are commonly used in the literature. Hevia and Micciancio [17] formally gave indistinguishability-based definitions of many of these terminologies, and showed that unobservability is the strongest notion against passive eavesdroppers, yet all the definitions are actually equivalent under efficient transformations. Gelernter and Herzberg [16] extended the work of Hevia and Micciancio [17] to the setting with adaptive adversaries including malicious receivers. In particular, sender anonymity against malicious receivers is considered the strongest anonymity possible in this setting. Unfortunately, not many of the recent works used these formal definitions: They are too complicated, as admitted by Gelernter and Herzberg [16], or not that well-known to the practical community. It is desirable to have a more accessible security definition, as *simple* as the indistinguishability definition (IND-CPA/CCA) for public-key encryption, yet *expressive* enough to capture the security properties desired by anonymous communication protocols.

A particular class of anonymous communication systems aims to provide provable anonymity (under corresponding ad-hoc definitions) with the presence of adversaries which globally observe all traffic of the network. Perhaps the most basic protocol within this class is the buses [5], which circulates a large array of ciphertexts (the bus) along a fixed route covering all nodes in the network. The reduced-seats buses [18] and the taxis [19] have improved efficiency upon the buses by reducing the size of the ciphertext carrier. At its extreme, Young and Yung [29] recently proposed the *Drunk Motorcyclist* (DM) where each ciphertext carrier (the motorcycle) only carries a single ciphertext. The ciphertext only travels to a random neighboring node upon arriving each node, hence the name Drunk Motorcyclist. Young and Yung [29] also fixed a flaw in the previous buses, reduced-seats buses, and taxis protocols by pointing out that key-private public-key encryption schemes should be used instead of ordinary ones. In a nutshell, this class of protocols initiated by the buses protocol [5] works by routing packets in a way that is independent to the intended receivers. Note that whether this routing strategy is deterministic (*e.g.*, buses) or probabilistic (*e.g.*, DM) does not matter in terms of anonymity.

For simplicity, we consider a communication network as a strongly connected (*i.e.*, each node is reachable from any other node) directed graph with N nodes, where packets can only travel along the edges. For other graphs, we can always consider the subgraphs containing the nodes connected from each sender node[1].

1.2   Our Results

In view of the existing complicated definitions of anonymity, we make mainly theoretical but also technical contributions. Theoretically, we present a simple algorithmic syntax which aims to capture a wide class of anonymous communication protocols. We also propose a simple indistinguishability-based definition which captures the strongest possible anonymity known in the literature, namely, unobservability and sender anonymity against malicious receivers [16], simultaneously. The simple formulation can hopefully make analyzing anonymous communication protocols an easier job. We further provide several relaxations of the anonymity notion so that the level of anonymity is still reasonably strong, yet finding efficient constructions is plausible.

Next, we show that the confidentiality of messages and the routing mechanisms can be decoupled, formalizing the idea of Young and Yung [29]. Specifically, we construct anonymous communication (AC) protocols generically from a key-private public-key encryption scheme and a new primitive called oblivious forwarding (OF) protocol. With this generic approach, we can now focus on constructing the conceptually simpler building block, namely, oblivious forwarding protocols. We then propose a generic construction of oblivious forwarding protocols from any random walk algorithm over graphs.

Our main technical contribution lies in our second construction of oblivious forwarding protocols, which is specially designed for optimizing the probability of successful delivery (p_{success}). This construction ensures that the most "unfortunate" nodes, *i.e.*, those located in the most isolated areas of the network, receive packets intended for it with at least a fair probability.

We evaluate the optimality by testing our constructions over randomly generated strongly connected graphs. Our test records p_{success} and the number of hops traveled for each sender-receiver pair. The test results show that our optimized protocol performs much better in terms of p_{success} in realistic networks. Due to page constraint, we refer the readers to the full version for the experiment results.

1.3   Technical Overview

We briefly introduce the design of our second construction. Consider a network represented by a strongly connected directed graph, such that packets are routed deterministically according to the routing table stored in each node. Each sender node in the network may not have the complete view of the network. Specifically, it only knows a partial list of intermediate nodes between itself and each receiver node. These partial paths form a tree rooted at the sender node.

---

[1] Also see Sect. 7 for a discussion on the network environment and deploying our protocols on the internet.

Suppose that node i has a packet for node j. Instead of the non-anonymous approach of always sending this packet to the real receiver j directly, it picks a dummy receiver j' according to a distribution independent of the real receiver, and forwards the packet to the dummy according to the routing table. The hope is that the real receiver j is located along the path to the node j'.

The question is then what the distribution of the dummy receivers should be. Intuitively, the nodes which are the least likely to receive the packet are those located at the leaves of the tree. Thus, it is natural to assign the uniform distribution over the set of leaf nodes. Indeed, we show that this distribution is in some sense optimal using standard arguments in linear programming.

2 Related Work

2.1 Anonymous Communication Protocols

Anonymous communication protocols can be classified roughly into two categories. One class (which is also our focus) provides strong anonymity. Another class features low latency and scalability, which often rely on trusted participants, servers, or other third parties. Examples in this class include the classical Crowds [26], mix networks (Mixnets) [9], and onion routing [25] (*e.g.*, Tor [14]).

Crowds provides sender anonymity by having the sender randomly forward requests to crowd members until the request eventually reaches the receiver. Hordes [21] replaces the reply mechanism of Crowds and onion routing by multicast to improve efficiency.

Mixnets collect packets from different sources, shuffle and forward them to the next hop in a random order. Multiple layers of encryption are used. The message is in the inner-most layer.

The idea of layered encryption is also applied in onion-routing, where a sender randomly selects a path of "somewhat trusted" routers and encrypts its packet to the routers along this path in layers, so that the last router can send the inner-most content to the intended receiver. The Tor network [14] is the most widely deployed anonymous communication network which uses onion routing as its underlying routing mechanism.

A common problem in most of these schemes (in the case of Mixnets, we consider the efficient variants which do not use zero-knowledge proofs) is that, anonymity is not guaranteed against adversaries which can observe or even control the traffic. In particular, the sender and receiver will be known to the first and last relay respectively for Tor.

While our work focuses on a highly distributed setting where each user in the network sends and forwards packets individually, some recent work utilize cooperation among users to achieve strong anonymity and efficiency at the same time (*e.g.*, Dissent [12,28] and Riposte [11]).

Dissent introduces semi-trusted servers to make dining cryptographers networks (DC-nets) [10] practical in a decent scale. It provides anonymity if at least one of the servers is honest. Dissent provides also accountability which is not considered in most anonymous communication protocols. However, Dissent ideally assumes that all members remain connected and send correct signed messages during one round. It takes a long $(O(N))$ time to exclude a single disruptor.

Riposte works on a slightly different setting where a huge number of users wish to post on a shared bulletin board anonymously. Compared to Dissent, Riposte provides similar privacy guarantee, and is able to identify malicious users faster.

2.2   Frameworks for Anonymity Analysis

While we focus on giving a simple definition for the strongest possible anonymity, where leakage of anonymity is negligible, Backes *et al.* [3,4] formulated a framework AnoA to qualitatively and quantitatively analyze abstract anonymous communication protocols. Their definition is similar to ours in the sense that they also consider a security game played between a powerful adversary and a challenger. Similar to our relaxations, they also introduce adversary classes as wrappers of the powerful adversary to capture realistic attacks. There are several differences:

First, instead of an indistinguishability-style definition which quantifies anonymity leakage additively, they give a differential-privacy-style definition, which quantifies anonymity leakage both multiplicatively and additively. This is helpful for analyzing imperfect anonymous communication protocols which provide weak anonymity. However, we remark that we can easily add the multiplicative factor to our definition as well if desired.

Second, the anonymous communication protocols in AnoA are modeled abstractly as general interactive Turing machines, whereas in our work we give a simple syntax to capture a wide range of anonymous protocols. We think this makes our anonymity definition easier to use.

Lastly, AnoA considers static corruption, *i.e.*, the set of corrupted entities is chosen at the beginning of the anonymity game, while we consider adaptive corruption.

Besides the general framework, much effort has been made over the decades to analyze various anonymous communication protocols under different anonymity definitions and adversarial capabilities. Recent examples include a probabilistic analysis to onion routing [15], the analysis of Tor in the UC framework [2, 15], fingerprinting attacks on onion routing [23], and accountable anonymous communication [1,12,13,27,28].

3   Preliminary

3.1   Notations

Let λ be the security parameter. All algorithms take 1^λ as input implicitly. Let ϕ be the empty set. Let $[N]$ be the set $\{1, 2, \ldots, N\}$. $P = (p_{ij})_{i,j=1}^{N}$ denotes an N-by-N matrix with the (i,j)-th entry given by p_{ij}. $x = (x_i)_{i=1}^{N}$ denotes an N-dimensional (column) vector with the i-th entry given by x_i. Let A be a probabilistic algorithm. $x \leftarrow \mathsf{A}(\cdot)$ denotes the computation of x output from A. Let S be a set and $X, Y \sim S$ be distributions over S. $x \leftarrow S$ denotes the sampling of a uniformly random $x \in S$, and $x \leftarrow X$ denotes the sampling of $x \in S$ according to the distribution X. We denote by $X \approx Y$ that the distributions are identical. $x := y$ denotes assigning the value of y to the variable x. "$\oplus$" denotes the XOR operation.

3.2 Key-Private Public-Key Encryption

Syntax. A public-key encryption scheme is a tuple of PPT algorithms $\mathsf{PKE} =$ (Setup, KGen, Enc, Dec) defined below.

$\mathsf{pp} \leftarrow \mathsf{Setup}(1^\lambda)$ is a probabilistic algorithm which inputs the security parameter λ, and outputs a public parameter pp.

$(\mathsf{pk}, \mathsf{sk}) \leftarrow \mathsf{KGen}(\mathsf{pp})$ is a probabilistic algorithm which inputs the public parameter pp, and outputs a public key pk and a secret key sk.

$c \leftarrow \mathsf{Enc}(\mathsf{pk}, m)$ is a probabilistic algorithm which inputs the public key pk and a message m, and outputs a ciphertext c.

$m \leftarrow \mathsf{Dec}(\mathsf{sk}, c)$ is a deterministic algorithm which inputs the secret key sk and a ciphertext c, and outputs a message m.

Correctness. We say PKE is correct if it holds that

$$\Pr[m' = m : \mathsf{pp} \leftarrow \mathsf{KGen}(1^\lambda); (\mathsf{pk}, \mathsf{sk}) \leftarrow \mathsf{KGen}(\mathsf{pp});$$
$$c \leftarrow \mathsf{Enc}(\mathsf{pk}, m); m' \leftarrow \mathsf{Dec}(\mathsf{sk}, c)] \geq 1 - \mathsf{negl}\,(\lambda).$$

Key-Privacy. Key-privacy of PKE is introduced by Bellare *et al.* [6], which requires that no PPT adversary can distinguish between ciphertexts produced from two public keys chosen by the challenger.

In this work, we consider a slightly modified definition, where the adversary is given access to a corruption oracle which returns the secret key of the requested party. PKE is key-private under chosen ciphertext attack (IK-CCA2) if, for any PPT adversary $\mathcal{A}$,

$$|2 \Pr[\text{IK-CCA2}^{\mathcal{A}}_{\mathsf{PKE}}(1^\lambda) = 1] - 1| \leq \mathsf{negl}\,(\lambda)$$

where the probability is taken over the random coins of the adversary and the experiment IK-CCA2$^{\mathcal{A}}_{\mathsf{PKE}}$ as defined in Fig. 1 (with $N = \mathsf{poly}\,(\lambda)$ being an integer). One can also define a corresponding security notion for chosen plaintext attack (IK-CPA) by removing the decryption oracle.

4 Formulation of AC and OF

4.1 Anonymous Communication (AC) Protocols

We present a simple yet expressive formulation of anonymous communication protocols. An anonymous communication protocol is run within a network of

IK-CCA2$_{\mathsf{PKE}}^{\mathcal{A}}(1^\lambda)$	Corr$\mathcal{O}(k)$
Corrupt $:= \phi$, Challenge $:= \phi$	Corrupt $:=$ Corrupt $\cup \{k\}$
pp $\leftarrow$ PKE.Setup(1^λ)	**return** sk$_k$
$(\mathsf{pk}_k, \mathsf{sk}_k) \leftarrow$ PKE.KGen(pp) $\forall k \in [N]$	
$(\mathsf{st}, j_0, j_1, m^*) \leftarrow \mathcal{A}^{\mathsf{Corr}\mathcal{O}, \mathsf{Dec}\mathcal{O}}(\mathsf{pp}, \{\mathsf{pk}_k\}_{k=1}^N)$	Dec$\mathcal{O}(k, c)$
$b \leftarrow \{0,1\}$	**if** $(k,c) \in$ Challenge **then**
$c^* \leftarrow$ PKE.Enc(pk_{j_b}, m^*)	**return** $\perp$
Challenge $:= \{(j_0, c^*), (j_1, c^*)\}$	**else**
$b' \leftarrow \mathcal{A}^{\mathsf{Corr}\mathcal{O}, \mathsf{Dec}\mathcal{O}}(\mathsf{st}, c^*)$	**return** PKE.Dec(sk_k, c)
return $(b = b' \wedge j_0 \notin$ Corrupt $\wedge j_1 \notin$ Corrupt$)$	**endif**

Fig. 1. Experiment for IK-CCA2 security of public-key encryption (modified from Bellare *et al.* [6])

an arbitrary number of nodes. We consider a dynamic environment where the network topology can change over time, *i.e.*, both nodes and edges may be added or removed. We assume that this network is equipped with a (most likely non-anonymous) routing protocol, so that our anonymous protocol does not need to deal with the changes to the network topology, yet will work regardless of the changes. We model this by letting each participating node k in the protocol possess some auxiliary information aux$_k$ (*e.g.*, routing tables) maintained by some external mechanisms such as the underlying routing protocol.

Overview. To participate in the anonymous communication protocol, a node runs the key generation algorithm, without any coordination with any other node, to set up its public and secret keys. It then publishes its public key. We assume that the nodes maintain their auxiliary information (*e.g.*, routing table) and learn the public keys of each other through external mechanisms. For example, they can obtain public keys while learning the network topology using the underlying routing protocol. Alternatively, they might use private information retrieval (PIR) along with a public-key infrastructure to retrieve public keys on-demand yet anonymously (similar to using PIR to retrieve a few IP-addresses of onion-routers in the Tor network on-demand [22]). The participating nodes form a graph G of $N = \mathsf{poly}(\lambda)$ nodes.

Each sender node in the network can encapsulate a message, using its auxiliary information and the public key of the receiver, into a packet ready for forwarding. The creator of the packet or any intermediate node receiving the packet forwards it by running a forwarding algorithm. It takes as input a secret key and some auxiliary information, attempts to decrypt the packet, and outputs an outgoing packet and the index of the next hop regardless of whether the decryption is successful. Hopefully, the intended receiver will be one of the intermediate nodes to receive the packet. For anonymity, the packets and the forwarding pattern must not leak any information about the sender and the receiver. It is important to forward the packet regardless of whether the intermediate node happens to be the actual receiver. Otherwise, an adversary observing all traffic can notice the disappearance of the packet and discover the real receiver.

Syntax. An anonymous communication protocol $\mathsf{AC} = (\mathsf{Setup}, \mathsf{KGen}, \mathsf{Enc}, \mathsf{Fwd})$ is a tuple of PPT algorithms:

$\mathsf{pp} \leftarrow \mathsf{Setup}(1^\lambda)$: The probabilistic setup algorithm is run by a trusted party which initiates the network environment. It takes as input the security parameter 1^λ, and outputs a public parameter pp. We note that this is the only algorithm run by a trusted party, and is run once only for setting up the system. Standard practices such as distributed parameter generation can be adopted to reduce trust.

$(\mathsf{pk}, \mathsf{sk}) \leftarrow \mathsf{KGen}(\mathsf{pp})$: The probabilistic key generation algorithm is run by each node joining the network individually. It takes as input the public parameter pp, and outputs a public key pk and a secret key sk. The participating nodes form a graph G of $N = \mathsf{poly}(\lambda)$ nodes.

$p \leftarrow \mathsf{Enc}(j, \mathsf{pk}_j, \mathsf{PK}, m, \mathsf{aux}_i)$: The probabilistic encapsulation algorithm is run by a sender node i. It takes as input a receiver j, its public key pk_j along with some other public keys PK, a message m, and some auxiliary information aux_i of node i, and outputs a packet p.

$(\{p'_{k'}, k'\}_{k'}, M) \leftarrow \mathsf{Fwd}(\mathsf{sk}_k, \mathsf{PK}, P, \mathsf{aux}_k)$: The probabilistic forwarding algorithm is run by a sender node or any intermediate node k. It takes as input a secret key sk_k of node k, a sequence of public keys PK, a sequence of input packets P, and some auxiliary information aux_k of node k. It outputs a sequence of packets $p'_{k'}$ with next hops k', and a sequence of messages M (or $\perp$). If $p'_{k'} \neq \perp$, it is forwarded to next hop k' regardless of whether a valid message $m \in M$ is obtained. If the node declines to forward the packet, it outputs $(\perp, \perp, M)$.

In general, the packet encapsulation algorithm takes multiple public keys as input, while the packet-forwarding algorithm takes multiple public keys and multiple incoming packets as input. The former captures onion routing protocol and its variants which encrypt messages to a pre-defined route of intermediate routers in layers, while the latter captures, for example, Mixnets and its variants which shuffle and forward packets in batches. Moreover, the algorithm generates different outgoing packets to multiple next hops. This captures, for example, some anonymous communication protocols based on broadcasting. Yet, for our purpose, the rest of this paper will stick to the setting where the packet encapsulation algorithm does not take any extra public keys PK as input, *i.e.*,

$$p \leftarrow \mathsf{Enc}(j, \mathsf{pk}_j, m, \mathsf{aux}_i),$$

while the packet-forwarding algorithm does not take any public keys as input, but only a single incoming packet, and outputs a single outgoing packet and a single next hop, *i.e.*,

$$(p', k', m) \leftarrow \mathsf{Fwd}(\mathsf{sk}_k, p, \mathsf{aux}_k).$$

We note that all the discussions and definitions in the rest of the paper can be naturally extended to the more general syntax.

Correctness. Informally, AC is said to be correct if, for any packet generated under an honest execution of the protocol, the packet reaches the intended destination after a reasonable delay with a reasonably high probability. Furthermore, the Fwd algorithm always recovers the message encapsulated in the packet when it reached the intended destination.

The idea is tricky to formalize. An anonymous communication protocol could have a low probability of successful delivery (p_{success}) but a short expected delivery time when successful, while another could have a high p_{success} but a long expected delivery time. For the first case, the sender can always intentionally repeat or re-transmit periodically to make up for the low success probability[2]. Another tricky part is that a protocol might be efficient over some types of graphs but inapplicable to some others. For instance, the buses protocol only works on graphs with a circular path connecting all nodes.

We model this formally by lower-bounding p_{success} after T forwardings by ρ. For any graph G with $N = \mathsf{poly}(\lambda)$ nodes, let $\{\mathsf{aux}_k\}_{k=1}^N$ be a set of auxiliary information of the nodes. AC is said to be (T, ρ)-correct on G if, for security parameter $\lambda \in \mathbb{N}$, all sender i, all receiver j, all message m, all public parameter generated by $\mathsf{pp} \leftarrow \mathsf{Setup}(1^\lambda)$, all key pairs generated by $(\mathsf{pk}_k, \mathsf{sk}_k) \leftarrow \mathsf{KGen}(\mathsf{pp})$,

$$p_{\mathsf{success}} := \Pr[\mathrm{Correct}_{\mathsf{AC}}^T(1^\lambda, i, j, m, \{\mathsf{aux}_k, \mathsf{pk}_k, \mathsf{sk}_k\}_{k=1}^N) = 1] \geq \rho > 0$$

where the probability is taken over the randomness of the experiment $\mathrm{Correct}_{\mathsf{AC}}^T$ defined in Fig. 2.

We can observe that if AC is (T, ρ)-correct on G, then we must have $T \geq l$, where l is the longest of all shortest hop-length between any sender-receiver pair, and that AC must be also (T', ρ')-correct on G for any $T' \geq T$ and $0 < \rho' \leq \rho$. As baselines for comparison, the buses protocol is $(N, 1)$-correct while the broadcast protocol is $(l, 1)$-correct.

Anonymity. We aim to capture both sender and receiver anonymity in the most hostile environment. For receiver anonymity, we require that a packet leaks nothing about the receiver, neither from the encapsulated message nor the traffic pattern. This implies that a packet encapsulating any message is indistinguishable from each other, so that a sender can safely re-transmit a message or switch to a different message for whatever reasons. Note that the indistinguishability should hold even if the correspondence between messages and senders are known.

For sender anonymity, notice that an adversary observing all traffic must be able to tell the original sender of any packet. Thus, we instead require that when

---

[2] Re-transmission triggered by events depending on the protocol, *e.g.*, transmission failure or time-out (albeit it is unclear how to define or discover them), might compromise anonymity.

$\text{Correct}_{\text{AC}}^{T}(1^{\lambda}, i, j, m, \{\text{aux}_k, \text{pk}_k, \text{sk}_k\}_{k=1}^{N})$
$t \leftarrow 0, b \leftarrow 0$
$p \leftarrow \text{AC.Enc}(j, \text{pk}_j, m, \text{aux}_i)$
while $t < T$ **then**
$\quad t \leftarrow t + 1$
$\quad (p, i, m') \leftarrow \text{AC.Fwd}(\text{sk}_i, p, \text{aux}_i)$
$\quad$ **if** $i = j \ \wedge \ m = m'$ **then**
$\quad\quad b \leftarrow 1$
$\quad$ **endif**
endwhile
return b

$\text{Correct}_{\text{OF}}^{T}(1^{\lambda}, i, j, \{\text{aux}_k\}_{k=1}^{N})$
$t \leftarrow 0, b \leftarrow 0$
$h \leftarrow \text{OF.Enc}(j, \text{aux}_i)$
while $t < T$ **then**
$\quad t \leftarrow t + 1$
$\quad (h, i) \leftarrow \text{OF.Fwd}(h, \text{aux}_i)$
$\quad$ **if** $i = j$ **then**
$\quad\quad b \leftarrow 1$
$\quad$ **endif**
endwhile
return b

Fig. 2. Correctness experiments of AC and OF

multiple senders send out a set of messages to multiple receivers, no one can tell which message originates from which sender, even if the correspondence between messages and receivers are known.

In technical terms, we consider a security game played between a challenger and a powerful adversary which is able to observe all traffic, corrupt at most all but two of the nodes, and obtain decrypted messages even from non-corrupt nodes. The security game consists of three phases.

In the first phase, the adversary corrupts as many nodes as it wishes, controls and learns from how packets are routed. These are modeled by the corruption and forwarding oracles respectively.

In the second phase, the adversary produces two distinct tuples (*i.e.*, at least one component is different) each consisting of a receiver, a message, and the auxiliary input of a sender. Eventually, the challenger is going to create packets according to some parts of the specifications (in the form of sender-message-receiver pairs) of the adversary. So, we also require the adversary to produce a bit to choose that either the sender-message correspondence or the message-receiver correspondence is fixed. This can be thought of as a slot machine with two slots such that the adversary can control the outcome of either one of the slots. If the adversary chose to fix the sender-message correspondence, an extra restriction is imposed that both of the challenged receivers are not corrupted.

The challenger then picks a random bit to determine the remaining slot of the slot machine: To decide whether the challenge packets should be created according to the specification by the adversary, or the remaining part of the sender-message-receiver pairs should be flipped. For example, if the adversary chose to fix the sender-message correspondence, then the random bit picked by the challenger decides whether the message-receiver correspondence should be flipped, as depicted in Fig. 3. The challenger then returns both challenge packets to the adversary.

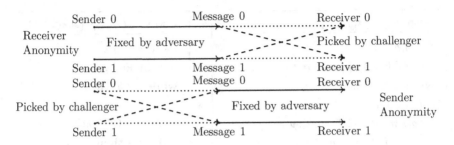

Fig. 3. An illustration of the IND-ANON$_{AC}^{\mathcal{A}}$ game

In the third phase, the adversary is again given access to the corruption and forwarding oracles. Naturally, if the adversary chose to fix the sender-message pairs, it is still not allowed to corrupt the challenge receivers. Also, when the forwarding oracle is queried with the challenge packets and receivers, no message will be decrypted (but the packet is still forwarded). These ensure that the adversary cannot win trivially. Finally, the adversary outputs a bit as a guess of whether the other part of the specification is flipped.

For any graph G with $N = \text{poly}(\lambda)$ nodes, let $\{\text{aux}_k\}_{k=1}^{N}$ be a set of auxiliary information of the nodes. AC is said to have indistinguishability of packets under anonymity attack (IND-ANON) if, for any security parameter $\lambda \in \mathbb{N}$, any PPT adversary $\mathcal{A}$ it holds that

$$|2\Pr[\text{IND-ANON}_{AC}^{\mathcal{A}}(1^{\lambda}, \{\text{aux}_k\}_{k=1}^{N}) = 1] - 1| \leq \text{negl}(\lambda)$$

where the probability is taken over the random coins of the experiment IND-ANON$_{AC}^{\mathcal{A}}$ defined in Fig. 4 and the adversary.

Remark. One can fit an onion routing protocol into our definition (by defining an Enc algorithm which encrypts a message to the routers along a random path in layers, and a Fwd algorithm which decrypts the outer-most layer and forwards the inner-layers to the next router). Yet, the routes always terminate at the receiver. An IND-ANON adversary can use the forwarding oracle to figure out the real receiver. So, an onion routing protocol would not satisfy our anonymity requirement.

Relaxations. The above definition takes away almost all quantitative information about the level of anonymity achieved. In practical scenarios, a more efficient protocol with weaker anonymity might be desirable. In this case, we need to know how weak the anonymity guaranteed actually is. To this end, we consider the following reasonable relaxations:

- q-bounded Collusion: The above definition essentially bounds the number of corrupted users by $N - 2$. In general, consider an adversary which only corrupts at most q users. The level of anonymity may depend on q. For example, one might consider q as a fraction N.

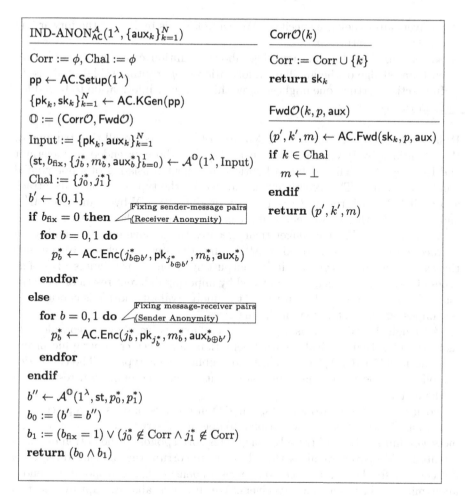

Fig. 4. Experiment for IND-ANON security of AC protocols

- 1-out-of-n Anonymity: The above definition models anonymity as a decision problem. Alternatively, we can model it as a search problem to capture anonymity of hiding within a group of n ($\leq N$) users: In the second phase, the adversary chooses to break either the receiver anonymity or the sender anonymity. For receiver anonymity, it outputs a set of n potential receivers $\{j_k^*\}_{k=1}^n$, a message m^*, and some auxiliary information of a sender aux^*. For sender anonymity, it outputs a target receiver j^*, a message m^*, and a set of auxiliary information for n potential senders $\{aux_k^*\}_{k=1}^n$. The challenger then picks randomly one of the n senders or receivers and outputs a challenge packet. The adversary wins if it guesses the choice of the challenger correctly.
- CPA-Anonymity: Our definition captures "CCA1-anonymity" since the forwarding oracle decrypts of the queried packet except for the challenge

receivers after they are specified. We can relax this by letting the forwarding oracle always return $m = \perp$.

- Secret Auxiliary Information: The above definition considers an adversary with knowledge of the auxiliary information (*e.g.*, routing table) of all users. In practical setting, one might assume this auxiliary information to be hidden from the adversary.

Relations to Other Notions of Anonymity. We first recall the structure of the definitions by Hevia and Micciancio [17], and those extended by Gelernter and Herzberg [16]. They define a hybrid experiment consisting of a polynomial number of rounds. The experiment is indexed by the type of anonymity attack and a bit b. In each round, the adversary produces two N-by-N matrices $M^{(0)}$ and $M^{(1)}$ where the (i, j)-th entry of the matrices specifies the message sent from node i to node j. The challenger then executes the anonymous communication protocol on messages contained in $M^{(b)}$. The adversary can choose to continue the experiment, or terminate it by outputting a bit b' as a guess of b. The types of anonymity attacks are captured by imposing different restrictions to the matrices $M^{(0)}$ and $M^{(1)}$. For instance, the unobservability notion is captured by *not* imposing any restriction on the matrices.

Although their definitions are rigorous and expressive, they are considerably more complex than typical security definitions for other cryptographic primitives, such as IND-CPA/CCA security for public-key encryption. The two major complicated aspects are the round-based nature and the unnatural restrictions to the matrices.

Focusing on these two aspects, our definition uses oracles to replace their round-based structure. Our definition also does not restrict the choice of senders, messages and receivers of the adversary. This corresponds to the notion of unobservability. Moreover, we allow the adversary to corrupt the challenged receivers if it chose to fix the message-receiver correspondence. This corresponds to sender anonymity against malicious receivers. We are thus able to capture the two strongest anonymity properties considered in the literature [16].

As our later constructions are inspired by the Drunk Motorcyclist protocol, it is also worth comparing our anonymity definition with that by Young and Yung [29]. They defined (receiver) anonymity and a rather unusual "blocking anonymity", for their DM protocol. While the adversary in the former is passive, the latter is able to block an arbitrary number of nodes in the network. Both types of adversaries can observe all traffic within the network. However, they are unable to maliciously inject, remove, or modify packets. The goal of the adversary against anonymity is to guess the identity of the real receiver out of all N possible choices, while the goal of the adversary against blocking anonymity is to block any subset of the N nodes so that the real receiver is in this subset.

We make the following observations. From the first glance, blocking anonymity appears to be a generalization of the receiver anonymity. Yet, we observe that their two anonymity notions are somewhat equivalent, up to the size of the blocking set. Suppose there exists an adversary against blocking anonymity, who outputs a set of nodes covering the target receiver with non-negligible probability, we can pick a random member of this set and break anonymity with non-negligible probability as well. Next, we observe that their anonymity definition actually corresponds to our 1-out-of-n anonymity definition. Finally, we remark that Young and Yung [29] did not consider sender anonymity.

4.2 Oblivious Forwarding (OF) Protocols

The forwarding pattern of a packet should not depend on the message content but rather the intended receiver. It is natural to separate the routing part of anonymous communication as an independent primitive. We formulate this idea as (receiver-)oblivious forwarding protocols[3].

Overview. An oblivious forwarding protocol is similar to an anonymous communication protocol, except that it only deals with the headers of the packets for routing. Given an intended receiver and some auxiliary information, the Enc algorithm creates a header containing the routing information. The Fwd algorithm creates headers for outgoing packets given an incoming header.

We emphasize again that, as in anonymous communication protocols, regardless of whether the actual receiver is an intermediate node or not, it always forwards the packet to the next hop.

Syntax. An oblivious forwarding protocol $\mathsf{OF} = (\mathsf{Enc}, \mathsf{Fwd})$ is a tuple of PPT algorithms defined as follows:

$h \leftarrow \mathsf{Enc}(j, \mathsf{aux}_i)$: The probabilistic encapsulation algorithm is run by a sender node i. It takes as input a receiver j and some auxiliary information aux_i of node i, and outputs a header h.

$(h', k') \leftarrow \mathsf{Fwd}(h, \mathsf{aux}_k)$: The probabilistic forwarding algorithm is run by a sender node or any intermediate node k. It takes as input an incoming header h and some auxiliary information aux_k of node k, and outputs an outgoing header h' and a next hop k'. If the node declines to forward the header, it outputs $(\bot, \bot)$.

Correctness. The correctness requirement of oblivious forwarding protocols is essentially the same as that of anonymous communication protocols, except that the former focuses only on the routing aspect.

For any graph G with $N = \mathsf{poly}(\lambda)$ nodes, let $\{\mathsf{aux}_k\}_{k=1}^N$ be a set of auxiliary information of the nodes. OF is said to be (T, ρ)-correct on G if, for security parameter $\lambda \in \mathbb{N}$, all sender i, all receiver j, it holds that

$$\Pr[\mathsf{Correct}_{\mathsf{OF}}^T(1^\lambda, i, j, \{\mathsf{aux}_k\}_{k=1}^N) = 1] \geq \rho > 0$$

where the probability is taken over the random coins of the experiment $\mathsf{Correct}_{\mathsf{OF}}^T$ defined in Fig. 2.

---

[3] Not to be confused with packet-oblivious forwarding in the network community.

Obliviousness. OF is said to be oblivious if, for any security parameter $\lambda \in \mathbb{N}$, any pair of receivers j_0 and j_1, and any auxiliary information aux_0 and aux_1, the distributions of the created headers from the Enc algorithm are identical, *i.e.*, $\mathsf{Enc}(j_0, \mathsf{aux}_0) \approx \mathsf{Enc}(j_1, \mathsf{aux}_1)$.

Although we consider perfect obliviousness in this work, one can relax it to statistical or computational obliviousness. We are however unaware of any possible construction, or the potential of efficiency benefits of such constructions. Alternative definitions, such as a game-based one similar to IND-ANON of anonymous communication protocols, are also possible.

4.3 Remarks on Dynamic Network Environments

In our formulation of anonymous communication and oblivious forwarding protocols, the network topology is assumed to be changing over time. In the syntax, it is captured by letting the encapsulation and forwarding algorithms take as input extra auxiliary information representing the view of the network topology. The security definitions also allow dynamic environments. In the anonymity definition, the adversary is allowed to submit any auxiliary information of its choice, both to the forwarding oracle and as challenges. In the obliviousness definition, this is captured by requiring the headers created with respect to any receivers and any auxiliary information have identical distributions.

Unfortunately, in the correctness definitions we fix the set of auxiliary information in advance, meaning that correctness is only guaranteed when the network is unchanged in T consecutive time steps. We could have extended the definition to allow a dynamic network, considering all possible combinations of auxiliary information per node per time step, as long as the corresponding graphs are still strongly connected in their respective time steps, but it would be difficult to prove a scheme to be correct under such a definition.

5   Generic Construction of AC

In this section, we show that anonymous communication protocols can be generically constructed from key-private public-key encryption and oblivious forwarding protocols. Recall that Young and Yung [29] pointed out the need of key-private public-key encryption in several existing anonymous communication protocols, our work here can be seen as formalizing and extending their idea. We also show that the Drunk Motorcyclist protocol is a special case of this generic construction. Since the obliviousness guaranteed by the oblivious forwarding protocols is information-theoretic, by plugging in an existing key-private public-key encryption secure under some intractability assumption into the generic construction, we obtain an anonymous communication protocol secure under the same assumption.

Intuitively, our construction works as follows. It encrypts the message to be encapsulated by key-private public-key encryption, and then precedes the ciphertext with the header produced by the oblivious forwarding protocol.

To forward a packet, a node attempts to decrypt the ciphertext, and forward the packet using the oblivious forwarding protocol regardless of the decryption result.

5.1 Formal Description

Let PKE = (Setup, KGen, Enc, Dec) be a public-key encryption scheme as defined in Sect. 3. Let OF = (Enc, Fwd) be an oblivious forwarding protocol as defined in Sect. 4. Figure 5 presents a generic construction of anonymous communication protocols.

The correctness of this generic construction follows directly from the correctness of the underlying building blocks. The key-privacy of PKE and obliviousness of OF provides anonymity.

Theorem 1. *Assume that* PKE *is correct and* OF *is* (T, ρ)*-correct. Then* AC *constructed in Fig. 5 is* (T, ρ)*-correct.*

Proving the above theorem is straightforward.

Theorem 2. *Assume that* PKE *is* (IK-CCA2)*-secure and* OF *is oblivious. Then* AC *constructed in Fig. 5 is* (IND-ANON)*-secure.*

To see the intuition behind Theorem 2, the key-private PKE hides the messages and their receivers encapsulated in the packets, so that no adversary can deduce any information compromising anonymity from ciphertexts. Thus, the only way to break anonymity is to observe or control the routing pattern. However, as the headers produced by OF is independent of its senders and receivers, the headers do not help the adversary in any way either. On the other hand, it is interesting that the seemingly complicated anonymity requirement of AC can be met by the one-line obliviousness requirement of OF.

Proof. (Theorem 2). Suppose $\mathcal{A}$ is a PPT adversary against the IND-ANON-security of the anonymous communication protocol. We wish to construct a PPT

AC.Setup(1^λ)	AC.Enc(j, pk$_j$, m, aux$_i$)	AC.Fwd(sk$_k$, p), aux$_k$)
pp ← PKE.Setup(1^λ)	h ← OF.Enc(j, aux$_i$)	**parse** p **as** (h, c)
return pp	c ← PKE.Enc(pk$_j$, m)	(h', k') ← OF.Fwd(h, aux$_k$)
	return $p := (h, c)$	m ← PKE.Dec(sk$_k$, c)
AC.KGen(pp)		**if** $(h', k') \neq (\bot, \bot)$ **then**
(pk, sk) ← PKE.KGen(pp)		**return** $((h', c), k', m)$
return (pk, sk)		**else**
		return $(\bot, \bot, m)$
		endif

Fig. 5. A generic construction of anonymous communication protocols

adversary $\mathcal{B}$ against the key-privacy of PKE. For this, we define the hybrids $\mathsf{Hyb}_{b'}$ for $b' = 0$ and 1 as follows:

- $\mathcal{B}$ setups the network environment as a graph G with N nodes. Let $\{\mathsf{aux}_k\}_{k=1}^N$ be a set of auxiliary information of the nodes. It receives pp and pk_k for $k \in [N]$ from the key-privacy challenger. It sends $\{\mathsf{pk}_k, \mathsf{aux}_k\}_{k=1}^N$ to $\mathcal{A}$.
- $\mathcal{B}$ answers queries to the CorrO oracle of IND-ANON by redirecting the request to the CorrO oracle of IK-CCA2. For queries (k, p, aux) to FwdO, $\mathcal{B}$ parses $p = (h, c)$, redirects (k, c) to the DecO oracle of IK-CCA2 to obtain m. $\mathcal{B}$ also runs $(h', k') \leftarrow \mathsf{OF.Fwd}(h, \mathsf{aux})$ and returns $((h', c), k', m)$ to $\mathcal{A}$.
- Eventually, $\mathcal{A}$ outputs b_{fix} and $\{(j_b^*, m_b^*, \mathsf{aux}_b^*)\}_{b=0}^1$. $\mathcal{B}$ picks a bit $\eta \leftarrow \{0, 1\}$. The bits b' (index of the hybrid) and η determine whether the specifications of $\mathcal{A}$ should be flipped.
- $\mathcal{B}$ reacts differently for the cases $b_{\mathrm{fix}} = 0$ and $b_{\mathrm{fix}} = 1$:
  - If $b_{\mathrm{fix}} = 0$, meaning that $\mathcal{A}$ chooses to attack the receiver anonymity, $\mathcal{B}$ sends (j_0, j_1, m_η) to the key-privacy challenger and receives from it a ciphertext c_η^*. Recall that b' is the index of the hybrid. To simulate the header, it runs $h_\eta^* \leftarrow \mathsf{OF.Enc}(j_{\eta \oplus b'}^*, \mathsf{aux}_\eta^*)$. To simulate the other packet, it runs $c_{1 \oplus \eta}^* \leftarrow \mathsf{PKE.Enc}(\mathsf{pk}_{j_{1 \oplus \eta \oplus b'}}, m_{1 \oplus \eta})$ and $h_{1 \oplus \eta}^* \leftarrow \mathsf{OF.Enc}(j_{1 \oplus \eta \oplus b'}^*, \mathsf{aux}_{1 \oplus \eta}^*)$.
  - If $b_{\mathrm{fix}} = 1$, meaning that $\mathcal{A}$ chooses to attack the sender anonymity, $\mathcal{B}$ computes $c_b^* \leftarrow \mathsf{PKE.Enc}(\mathsf{pk}_{j_b^*}, m_b^*)$ for $b = 0, 1$. It runs $h_b^* \leftarrow \mathsf{OF.Enc}(j_b^*, \mathsf{aux}_{b \oplus \eta \oplus b'}^*)$ for $b = 0, 1$.
- Finally, $\mathcal{B}$ sends (p_0^*, p_1^*) where $p_b^* = (h_b^*, c_b^*)$ to the adversary $\mathcal{A}$. $\mathcal{B}$ answers queries to CorrO and FwdO as before.
- The game terminates as the adversary $\mathcal{A}$ outputs a guess ξ. The adversary $\mathcal{A}$ wins if $\xi = \eta$.

We differentiate the cases between $b_{\mathrm{fix}} = 0$ and $b_{\mathrm{fix}} = 1$.

Case 1: $b_{\mathrm{fix}} = 0$. There are two differences between Hyb_0 and Hyb_1.

The first difference is that, $\mathcal{B}$ computes $h_{b \oplus \eta}^*$ from $j_{b \oplus \eta}$ in Hyb_0 and from $j_{1 \oplus b \oplus \eta}$ in Hyb_1. By the obliviousness of OF, the two methods of generating $h_{b \oplus \eta}^*$ are indistinguishable in the view of $\mathcal{A}$.

The second difference is that, $\mathcal{B}$ computes $c_{1 \oplus \eta}^*$ from $\mathsf{pk}_{j_{1 \oplus \eta}}$ in Hyb_0 and from pk_{j_η} in Hyb_1. Since η is chosen at random, the choice of m_η which is directed to the encryption oracle of the key-privacy challenger is random in the view of $\mathcal{A}$. Suppose $\mathcal{A}$ can distinguish between the two methods of generating $c_{1 \oplus \eta}^*$ with non-negligible advantage, then it has the same advantage in distinguishing the two methods of generating c_η^*, which breaks the IK-CCA2-security of PKE.

Case 2: $b_{\mathrm{fix}} = 1$. The only difference between Hyb_0 and Hyb_1 is that, $\mathcal{B}$ computes h_b^* from $\mathsf{aux}_{b \oplus \eta}^*$ in Hyb_0 while it computes h_b^* from $\mathsf{aux}_{1 \oplus b \oplus \eta}^*$ in Hyb_1. By the obliviousness of OF, the two methods of generating h_b^* are indistinguishable in the view of $\mathcal{A}$.

Therefore, in either case, Hyb_0 and Hyb_1 are indistinguishable in the view of $\mathcal{A}$. Moreover, in case 2, the bit b' is information theoretically hidden from $\mathcal{A}$ by the obliviousness of OF. Thus the advantage of $\mathcal{A}$ is zero. Suppose that $\mathcal{A}$ chooses $b_{\mathrm{fix}} = 0$. Denote the random bit chosen by the key-privacy challenger by b_{PKE}. When $b' = b_{\mathsf{PKE}} \oplus \eta$, which occurs with probability $\frac{1}{2}$, $\mathsf{Hyb}_{b'}$ is a perfect simulation of the IND-ANON security game. Conditioned on the above, suppose $\mathcal{A}$ breaks the IND-ANON security of AC with advantage ϵ, then $\mathcal{B}$ also has advantage ϵ in breaking the key-privacy of PKE. $\square$

5.2   Recasting the DM Protocol

Recall that in the DM protocol [29], a node encrypts its message to the intended receiver using a key-private public-key encryption scheme, and forwards the ciphertext to a random neighboring node. Upon receipt of a packet, a node copies the packet to a decryption queue and forwards the packet again to a random neighboring node. For each packet, this process is repeated until its time-to-live (TTL) value vanishes. Straightforwardly, the DM protocol can be seen as an anonymous communication protocol constructed from the above generic approach using an oblivious forwarding protocol $\mathsf{DM\text{-}OF} = (\mathsf{Enc}, \mathsf{Fwd})$ defined in Fig. 6a. The executions of the $\mathsf{DM\text{-}OF.Fwd}$ represent a simple random walk over a graph, at which a packet travels to each neighboring node with equal probability.

We consider two important parameters for random walk algorithms—the *hitting time*, which is the maximum of the expected time for traveling from any

---

$\mathsf{OF.Enc}(j, \mathsf{aux}_i = \mathcal{N}(i))$

return L

$\mathsf{OF.Fwd}(h = \ell, \mathsf{aux}_k = \mathcal{N}(k))$

if $\ell = 0$ **then**

    return $(\bot, \bot)$

endif

$k' \leftarrow \mathcal{N}(k)$

return $(\ell - 1, k')$

(a) DM-OF: A construction from DM [29], where $\mathcal{N}(k)$ denotes the set of neighboring nodes of node k

---

$\mathsf{OF.Enc}(j, \mathsf{aux}_i = (i, (p_{ij})_{j=1}^{N}, T^i))$

return $h := (\bot, L)$

$\mathsf{OF.Fwd}(h = (j', \ell), \mathsf{aux}_k = (k, (p_{kj})_{j=1}^{N}, T^k))$

if $\ell = 0$ **then**

    return $(\bot, \bot)$

elseif $j' = k \vee j' = \bot$ **then**

    $j' \leftarrow \chi_k$

endif

$k' := T^k[j']$

return $((j', \ell - 1), k')$

(b) $(P, \mathcal{T})$-OF: Our generic construction, where $P = (p_{kj})_{k,j=1}^{N}$ denotes a transition probability matrix, $\mathcal{T} = \{T^k\}_{k=1}^{N}$ denotes a set of routing tables, χ_k denotes a distribution over $[N]$ defined by $(p_{kj})_{j=1}^{N}$

Fig. 6. Constructions of OF protocols (L denotes a constant TTL value)

starting node to any destination), and *cover time*, which is the maximum of the expected time for traveling from any starting node to all other nodes at least once. It is well known that the hitting time and cover time of the simple random walk algorithm on general graphs, without using any topological information of the graph, are both $O(N^3)$ [8].

In the context of anonymous communication, there seems to be no reason to avoid using any topological information of the graph. Contrarily, the sender node should exploit this information as much as possible to improve the expected hitting time or probability of successful delivery given a fixed TTL value, as long as its routing strategy remains oblivious.

Intuitively, using a simple random walk algorithm and a fixed TTL value, it is less likely for a node in a more isolated area of a network to receive packets. This motivates us to design new routing strategies that make use of the topological information to improve efficiency.

6   Constructions of OF

In this section, we aim to construct oblivious forwarding protocols which exploit the topological information to improve efficiency. We first state a generic construction defined upon any transition probability matrix and routing tables of the network. From this generic construction, we can plug in the transition probability matrices of any random walk algorithms over graphs to obtain a class of oblivious forwarding protocols. For demonstration, we use the simple random walk algorithm and the β-random walk algorithm by Ikeda *et al.* [20] as examples.

Next, by introducing a convenient representation of the routing paths from a node as a "connectivity matrix" which is computed using partial topological information, we present our construction which maximizes the minimum probability of successful delivery (p_{success}) over all potential receivers for each fixed TTL value in one round, *i.e.*, during the transmission from one dummy receiver to another.

6.1   Generic Construction

Consider a network represented by a strongly connected directed graph. Typically, routing in such a network is performed in a distributed manner: Each node k maintains its routing table T^k mapping each destination to a next hop. Our strategy works as follows. Regardless of the intended destination, the sender node i chooses a dummy destination j' according to some distribution independent of the intended destination. The sender node and all intermediate nodes then just route the packet as a normal (non-anonymous) packet to the dummy destination j'. When the packet reaches the dummy destination j', node j' chooses another dummy destination as long as the TTL value is still positive.

Formally, let $P = (p_{kj})_{k,j=1}^N$ where $p_{kj} \geq 0$, $\sum_j p_{kj} = 1$, $k, j \in [N]$ be any transition probability matrix, and $\mathcal{T} = \{T^k\}_{k=1}^N$. Figure 6b defines the oblivious forwarding protocol $(P, \mathcal{T})$-OF = (Enc, Fwd), which is clearly oblivious as the header h output by Enc and the routing pattern (h', k') output by Fwd is independent of the receiver j. For correctness, we defer the analysis to specific constructions.

6.2   Construction from Any Random Walks

For a node k, $\deg(k)$ and $\mathcal{N}(k)$ are the out-degree and set of neighboring nodes of k respectively. It is obvious that the Drunk Motorcyclist protocol is a special case of the above generic construction, as stated in Lemma 1.

Lemma 1. *Let* $P_{\text{sim}} = (p_{kj})_{k,j=1}^{N}$ *and* $\mathcal{T}_{\text{sim}} = \{T^k\}_{k=1}^{N}$ *be defined as* $p_{kj} = \deg^{-1}(k)$ *if* $j \in \mathcal{N}(k)$, *and* $p_{kj} = 0$ *otherwise, and* $T^k[j] = j \; \forall j$ *respectively. Then* DM-OF $= (P_{\text{sim}}, \mathcal{T}_{\text{sim}})$-OF.

Alternatively, the transition probabilities may depend on the local topological information, *e.g.*, the degrees of the neighboring nodes. We consider the β-random walk algorithm designed by Ikeda *et al.* [20].

Definition 1. *The transition probability matrix* $P_\beta = (p_{kj})_{k,j=1}^{N}$ *of the* β-random walk algorithm is defined as $p_{kj} = \deg^{-\beta}(j)(\sum_{u \in \mathcal{N}(k)} \deg^{-\beta}(u))^{-1}$ *if* $j \in \mathcal{N}(k)$, *and* $p_{kj} = 0$ *otherwise.*

The β-random walk algorithm has hitting time and cover time equal to $O(N^2)$ and $O(N^2 \log N)$ respectively on general graphs when $\beta = \frac{1}{2}$ [20]. Therefore, in theory, β-OF $:= (P_\beta, \mathcal{T}_{\text{sim}})$-OF is asymptotically more efficient than DM-OF.

For random walks with transition probability matrix $P = (p_{kj})_{k,j=1}^{N}$, the (k,j)-th entry of P^ℓ gives the probability of reaching j from k in exactly ℓ steps. Thus, the (k,j)-th entry of $\sum_{\ell=1}^{T} P^\ell$ gives the probability of reaching j from k in no more than T steps. We therefore formulate the correctness of $(P, \mathcal{T})$-OF as follows.

Theorem 3. *Let* $T > 0$ *be a positive integer. Let* $\rho := \min_{k,j} \sum_{\ell=1}^{T} P^\ell$ *where* $P = (p_{kj})_{k,j=1}^{N}$ *is a transition probability matrix. Let* $\mathcal{T} = \{T^k\}_{k=1}^{N}$ *where* $T^k[j] = j \; \forall j$. *Then* $(P, \mathcal{T})$-OF *is* (T, ρ)-correct.

6.3   Optimizing p_{success}

Representation of the Routing Paths. To facilitate our discussion, we consider a typical network in which nodes route packets according to routing tables $\mathcal{T}_{\text{opt}}$ built using some distributed shortest path algorithm. Suppose the node k has a partial view of how packets will be routed to different destinations. More specifically, it has knowledge of some of the intermediate nodes along the path to each destination. These paths form a tree rooted at node k connecting all other nodes. Using this tree, we construct an N-by-N *connectivity matrix* A^k as follows: If node i is on the path from k to j, set $A^k(i,j) = 1$; Otherwise, set $A^k(i,j) = 0$.

The connectivity matrix A^k features an interesting structure. First, since node j must be on the path from k to j, $A^k(j,j) = 1$ for all j. Second, if node i is a leaf node of the tree, there is only one path from node k to node i. Thus, row i of A^k has only a single '1' which is $A^k(i,i)$. In other words, the 1-norm of row i is 1. Lastly, if node j is a neighbor of node k, there are no intermediate nodes along the path from node k to node j. Thus, column j of A^k has only a single '1' which is $A^k(j,j)$. In other words, the 1-norm of column j is 1.

For conciseness, we will drop the superscript k from A^k and simply write A when the context is clear.

The Construction. We aim to design a routing strategy which is independent of the intended receiver, and maximizes the probability that the most *unfortunate* node receiving its packets in one round. A node is considered the most *unfortunate* if, given a routing strategy, the node receives the packet with the lowest probability.

Formally, consider a sender node with transition probability $x = (x_i)_{i=1}^N$. Let A be the connectivity matrix of the node defined in Sect. 6.3. Then the i-th entry of Ax indicates the probability that node i belongs to the path from the sender to the dummy destination. Our task is to maximize the minimum of these probabilities, or

$$\max \left(\min_i (Ax)_i \right) \text{ s.t. } x \geq 0 \wedge \|x\|_1 = 1$$

where $x \geq 0$ means $x_i \geq 0 \ \forall i \in [N]$.

Intuitively, the optimal solution can be computed as follows. Consider the tree represented by A. Let $\hat{x}$ be the uniform distribution over the set of all leaf nodes. This can be computed by assigning equal weights to the i-th entry of $\hat{x}$ where the i-th row of A contains only a single 1, which is $A(i,i)$ (*i.e.*, node i is a leaf node). This ensures that the most unfortunate nodes (*i.e.*, the leaf nodes) receive their packets with a fair chance. We claim that $\hat{x}$ is an optimal solution to the problem.

Formally, the proposed solution $\hat{x}$ is given by $\hat{x}_i = \frac{1}{|I|}$ if $i \in I$, and $\hat{x}_i = 0$ otherwise, where $I = \{i : \|A_i\|_1 = 1\}$ and A_i is the i-th row of A.

Remark 1. Interestingly, simple random walk corresponds to assigning equal weights to $\hat{x}_j$ where the j-th *column* (instead of row) of A contains only a single 1, which is $A(j,j)$ (*i.e.*, node j is a neighboring node).

Proof of Optimality. The optimality of $\hat{x}$ is proven via standard arguments in linear optimization. Instead of proving the optimality of $\hat{x}$ directly, which is rather difficult, we construct a dual certificate for the primal solution $\hat{x}$. Then, by the LP Strong Duality Theorem [7, Theorem 4.4], $\hat{x}$ is an optimal solution to (2).

Lemma 2. *The optimal solution to*

$$\max \; (\min_{i} \; (Ax)_i) \; \text{s.t.} \; x \geq 0 \wedge \|x\|_1 = 1$$

where $x \geq 0$ *means* $x_i \geq 0 \; \forall i \in [N]$, *is given by* $\hat{x}_i = \frac{1}{|I|}$ *if* $i \in I$, *and* $\hat{x}_i = 0$ *otherwise, where* $I = \{i : \|A_i\|_1 = 1\}$ *and* A_i *is the i-th row of A.*

Proof. (Lemma 2) The equivalent model of (2) is

$$\min -p \; \text{s.t.} \; (Ax \geq pe) \wedge (e^T x = 1) \wedge (x \geq 0) \wedge (p \geq 0) \tag{1}$$

where $e = (1, 1, \ldots, 1)^T \in \mathbb{R}^N$.

To prove such $\hat{x}$ is an optimal solution to (2), we need to find a dual certificate for (1). That is, we need to find an optimal solution for the dual problem. The dual of the primal problem (1) is

$$\max d \; \text{s.t.} \; (A^T y \leq -de) \wedge (e^T y \geq 1) \wedge (y \geq 0) \tag{2}$$

Since $e^T x = 1 \geq 1$ and $x \geq 0$, let $y = \hat{x}$, we have

$$A^T \hat{x} = \sum_{i=1}^{N} A_i^T \hat{x}_i = \sum_{x_i \neq 0} \frac{1}{|I|} A_i^T + \sum_{x_i = 0} 0 \cdot A_i^T = \sum_{x_i \neq 0} \frac{1}{|I|} e_i \leq \frac{1}{|I|} e$$

where A_i is the i-th row of A, e_i is the i-th standard basis vector in $\mathbb{R}^N$.

From the calculation above, we can take $d = -\frac{1}{|I|}$. It is trivial that $Ax \geq \frac{1}{|I|} e$ and the objective value of (1) is $-p = -\frac{1}{|I|}$. Thus, we have found a feasible solution, $y = \hat{x}$, for the dual problem (2) such that the duality gap is zero, *i.e.*, $d = -p = -\frac{1}{|I|}$. By the LP Strong Duality Theorem [7, Theorem 4.4], $\hat{x}$ is an optimal solution to (1) and (2). □

Finally, we define a transition probability matrix $P_{opt} = (p_{kj})_{k,j=1}^N$ where $p_{kj} = \hat{x}_j^k \; \forall k, j$, and obtain our optimized scheme Opt-OF $= (P_{opt}, \mathcal{T}_{opt})$-OF.

It is not easy to formulate the correctness of this construction, at least not in a clean equation form as in the construction induced from random walks. The difficulties arise from that the dummy receiver which was chosen according to the transition probability matrix is not a neighbor of the sender. The packet may be forwarded multiple times by intermediate nodes until it reaches the dummy receiver. These intermediate nodes are not captured in the transition probability matrix. Moreover, the hop length to each dummy receiver may be different. Thus, a packet might reach several dummy receivers in an instance, while still not reaching the first dummy receiver in another. We would therefore only give a loose bound about the correctness of this construction.

Theorem 4. *Let l be the hop length of the longest shortest path in the graph G. Let I be the set of leaf nodes in the shortest path tree containing the longest shortest path. Then Opt-OF is $(l, \frac{1}{|I|})$-correct.*

6.4 Discussions

Optimality. We do not claim that the construction above is optimal over all possible oblivious forwarding protocols. Rather, it optimizes $p_{success}$ in one round, *i.e.*, during the transmission from one dummy receiver to another. Indeed, the optimal construction is arguably the one where a packet travels each of the nodes at least once with the minimal distance. Unfortunately, finding such a path is an NP-hard problem known as the traveling salesman problem.

On the other hand, optimizing the probability during the entire life span of a packet is undesirable for the following reasons: (1) The system to be optimized becomes too complicated to analyze, as it involves tensor product of L transition probability matrices, where L is the TTL value; (2) The sender might not trust the others to pick dummy receivers honestly. Thus it has the motivation to do the best it can.

Auxiliary Information. A sender might not have complete knowledge of the paths to all destinations, so some leaf nodes in the tree representing its routing paths may actually be intermediate nodes lying on other paths. Thus, assigning positive weight on these nodes would be a waste of effort. However, as the sender learns more about the paths, its strategy will only become better regarding $p_{success}$. We remark that the sender should only use side-channel information independent of the real receivers to update its auxiliary information. Otherwise, an adversary may be able to infer information about the real receivers from the auxiliary information. Note that the side-channel information can be tampered by the adversary without losing anonymity. This is captured in our anonymity game (Fig. 4) by allowing the adversary to query with and submit for challenge any auxiliary information of its choice.

7 How to Use Our **AC** Protocols?

As a quick summary, we have proposed a new formulation of anonymous communication protocols and provided several generic constructions. The functionality of an anonymous communication protocol itself is rather limited: It forwards packets randomly without guaranteeing delivery. We thus provide some guidelines about how to make the best use of these anonymous communication protocols below.

7.1 Network Environment

In the previous sections, we focus on strongly connected directed graphs, assuming each node in the graph to be both potential sender and receiver, and is willing to forward packets according to the anonymous communication protocol. This setting suffices for networks formed for specific uses (*e.g.*, P2P networks).

In the case of the internet, end users are connected to their local autonomous systems (AS), which are then interconnected to form the internet. While the end users are the potential senders and receivers, they are not supposed to forward packets to neighboring nodes. Instead, the majority of the routing tasks are performed between different ASes. Thus, it is reasonable to think of each AS as a node in an anonymous communication protocol. In this setting, an end user simply sends ordinary packets to its local AS, which then encapsulates the entire packet as a message using an anonymous communication protocol. Anonymity of the two end users at either side of a communication channel is guaranteed assuming only trusted local ASes of the senders and receivers. We think that this minimal level of trust is acceptable as the end users pay the ASes to subscribe for internet services.

7.2 How to Guarantee Successful Delivery?

Recall that a packet sent through an anonymous communication protocol is guaranteed to reach its intended destination only with a positive probability. Moreover, since the anonymous communication protocol is sender-anonymous even against malicious receivers, even the intended receiver itself does not know the sender. Thus, there is no indication of success delivery such as the acknowledgement (ACK) response. The sender can at best send out the message multiple times, using independent randomness for anonymity, so that hopefully the intended receiver can receive the message.

For reliable communication, the sender needs to concatenate its identity to the message, and send out the same message repeatedly, again using independent randomness for anonymity, until receiving an ACK from the receiver. This ACK response is again concatenated with the receiver identity, and is sent multiple times so that hopefully the sender can receive it. In short, we can imagine an anonymized TCP connection where each message is sent using an anonymous communication protocol.

7.3 Anonymity-Efficiency Trade-Off

It is of little doubt that anonymous communication protocols with the strongest possible anonymity are inefficient, which is shown under a more complex definition [16]. To benefit from the strong anonymity guarantee, one can consider a hybrid approach in practice: We still use low-latency anonymous communication protocols (e.g., Tor) to setup a fixed route between the sender and the receiver. However, at somewhere in the middle of the route, we use strong anonymous communication protocols within a small subgraph to "cut off" the link. We note that it requires a careful analysis of this approach, perhaps under relaxed definitions, to examine the actual gain of anonymity. We leave it as a future work.

8 Concluding Remark

We have presented simple yet expressive syntax and security definitions of anonymous communication protocols and oblivious forwarding protocols. For the former, we have proposed a generic construction from key-private public-key encryption and oblivious forwarding protocols. For the latter, we have proposed a generic construction from any random walk algorithm over graphs. Our work provides a modular way of constructing anonymous communication protocols and a simple way to analyze their anonymity.

Furthermore, we have specially designed a construction of oblivious forwarding protocols which optimizes the probability of successful delivery. Our experiment results suggest that our optimized construction performs significantly better in terms of the probability of successful delivery in graphs capturing the characteristics of real world networks. In contrast to some efficient yet not-so-anonymous communication protocols, our constructions provide anonymity even in the presence of a powerful adversary, which can observe and control all traffic in the network and corrupt all but two nodes. The strong anonymity comes at an efficiency cost due to the obliviousness of the routing strategy. We leave the study of the trade-off between anonymity and efficiency as a future work.

References

1. Backes, M., Clark, J., Kate, A., Simeonovski, M., Druschel, P.: BackRef: accountability in anonymous communication networks. In: Boureanu, I., Owesarski, P., Vaudenay, S. (eds.) ACNS 2014. LNCS, vol. 8479, pp. 380–400. Springer, Cham (2014). doi:10.1007/978-3-319-07536-5_23
2. Backes, M., Goldberg, I., Kate, A., Mohammadi, E.: Provably secure and practical onion routing. In: 25th IEEE Computer Security Foundations Symposium, Cambridge, MA, USA, 25–27 June 2012, pp. 369–385 (2012)
3. Backes, M., Kate, A., Manoharan, P., Meiser, S., Mohammadi, E.: AnoA: a framework for analyzing anonymous communication protocols. In: 2013 IEEE 26th Computer Security Foundations Symposium, New Orleans, LA, USA, 26–28 June 2013, pp. 163–178 (2013)
4. Backes, M., Kate, A., Meiser, S., Mohammadi, E.: (Nothing else) MATor(s): monitoring the anonymity of Tor's path selection. In: Proceedings of the 2014 ACM SIGSAC Conference on Computer and Communications Security, Scottsdale, AZ, USA, 3–7 November 2014, pp. 513–524 (2014)
5. Beimel, A., Dolev, S.: Buses for anonymous message delivery. J. Cryptol. 16(1), 25–39 (2003)
6. Bellare, M., Boldyreva, A., Desai, A., Pointcheval, D.: Key-privacy in public-key encryption. In: Boyd, C. (ed.) ASIACRYPT 2001. LNCS, vol. 2248, pp. 566–582. Springer, Heidelberg (2001). doi:10.1007/3-540-45682-1_33
7. Bertsimas, D., Tsitsiklis, J.: Introduction to Linear Optimization, 1st edn. Athena Scientific, Belmont (1997)
8. Brightwell, G., Winkler, P.: Maximum hitting time for random walks on graphs. Random Struct. Algorithms 1(3), 263–276 (1990)
9. Chaum, D.: Untraceable electronic mail, return addresses, and digital pseudonyms. Commun. ACM 24(2), 84–88 (1981)
10. Chaum, D.: The dining cryptographers problem: unconditional sender and recipient untraceability. J. Cryptol. 1(1), 65–75 (1988)
11. Corrigan-Gibbs, H., Boneh, D., Mazières, D.: Riposte: an anonymous messaging system handling millions of users. In: 2015 IEEE Symposium on Security and Privacy, SP 2015, San Jose, CA, USA, 17–21 May 2015, pp. 321–338 (2015)

12. Corrigan-Gibbs, H., Ford, B.: Dissent: accountable anonymous group messaging. In: Proceedings of the 17th ACM Conference on Computer and Communications Security, Chicago, Illinois, USA, 4–8 October 2010, pp. 340–350 (2010)

13. Corrigan-Gibbs, H., Wolinsky, D.I., Ford, B.: Proactively accountable anonymous messaging in verdict. In: Proceedings of the 22th USENIX Security, Washington, DC, USA, 14–16 August 2013, pp. 147–162 (2013)

14. Dingledine, R., Mathewson, N., Syverson, P.F.: Tor: the second-generation onion router. In: Proceedings of the 13th USENIX Security Symposium, 9–13 August 2004, San Diego, CA, USA, pp. 303–320 (2004)

15. Feigenbaum, J., Johnson, A., Syverson, P.F.: Probabilistic analysis of onion routing in a black-box model. ACM Trans. Inf. Syst. Secur. 15(3), 14 (2012)

16. Gelernter, N., Herzberg, A.: On the limits of provable anonymity. In: Proceedings of the 12th Annual ACM Workshop on Privacy in the Electronic Society, WPES 2013, Berlin, Germany, 4 November 2013, pp. 225–236 (2013)

17. Hevia, A., Micciancio, D.: An indistinguishability-based characterization of anonymous channels. In: Borisov, N., Goldberg, I. (eds.) PETS 2008. LNCS, vol. 5134, pp. 24–43. Springer, Heidelberg (2008). doi:10.1007/978-3-540-70630-4_3

18. Hirt, A., Jacobson Jr., M.J., Williamson, C.L.: A practical buses protocol for anonymous internet communication. In: Proceedings of the Third Annual Conference on Privacy, Security and Trust, 12–14 October 2005, The Fairmont Algonquin, St. Andrews, New Brunswick, Canada (2005)

19. Hirt, A., Jacobson Jr., M.J., Williamson, C.L.: Taxis: scalable strong anonymous communication. In: 16th International Symposium on Modeling, Analysis, and Simulation of Computer and Telecommunication Systems, Baltimore, Maryland, USA, 8–10 September 2008, pp. 269–278 (2008)

20. Ikeda, S., Kubo, I., Yamashita, M.: The hitting and cover times of random walks on finite graphs using local degree information. Theor. Comput. Sci. 410(1), 94–100 (2009)

21. Levine, B.N., Shields, C.: Hordes: a multicast-based protocol for anonymity. J. Comput. Secur. 10(3), 213–240 (2002)

22. Mittal, P., Olumofin, F.G., Troncoso, C., Borisov, N., Goldberg, I.: PIR-Tor: scalable anonymous communication using private information retrieval. In: Proceedings of the 20th USENIX Security Symposium, San Francisco, CA, USA, 8–12 August 2011

23. Panchenko, A., Niessen, L., Zinnen, A., Engel, T.: Website fingerprinting in onion routing based anonymization networks. In: Proceedings of the 10th Annual ACM Workshop on Privacy in the Electronic Society, WPES 2011, Chicago, IL, USA, 17 October 2011, pp. 103–114 (2011)

24. Pfitzmann, A., Hansen, M.: A terminology for talking about privacy by data minimization: anonymity, unlinkability, undetectability, unobservability, pseudonymity, and identity management, v0.34, August 2010. http://dud.inf.tu-dresden.de/litera tur/Anon_Terminology_v0.34.pdf

25. Reed, M.G., Syverson, P.F., Goldschlag, D.M.: Anonymous connections and onion routing. IEEE J. Sel. Areas Commun. 16(4), 482–494 (1998)

26. Reiter, M.K., Rubin, A.D.: Crowds: anonymity for web transactions. ACM Trans. Inf. Syst. Secur. 1(1), 66–92 (1998)

27. Syta, E., Corrigan-Gibbs, H., Weng, S.-C., Wolinsky, D., Ford, B., Johnson, A.: Security analysis of accountable anonymity in dissent. ACM Trans. Inf. Syst. Secur. 17(1), 4:1–4:35 (2014)

28. Wolinsky, D.I., Corrigan-Gibbs, H., Ford, B., Johnson, A.: Dissent in numbers: making strong anonymity scale. In: 10th USENIX Symposium on Operating Systems Design and Implementation, OSDI 2012, Hollywood, CA, USA, 8–10 October 2012, pp. 179–182 (2012)
29. Young, A.L., Yung, M.: The drunk motorcyclist protocol for anonymous communication. In: IEEE Conference on Communications and Network Security, CNS 2014, San Francisco, CA, USA, 29–31 October 2014, pp. 157–165 (2014)

Challenges with Assessing the Impact of NFS Advances on the Security of Pairing-Based Cryptography

Alfred Menezes[1(✉)], Palash Sarkar[2], and Shashank Singh[3]

[1] Department of Combinatorics and Optimization, University of Waterloo,
Waterloo, Canada
ajmeneze@uwaterloo.ca
[2] Applied Statistics Unit, Indian Statistical Institute, Kolkata, India
palash@isical.ac.in
[3] Inria, Nancy, France
sha2nk.singh@gmail.com

Abstract. In the past two years there have been several advances in Number Field Sieve (NFS) algorithms for computing discrete logarithms in finite fields $\mathbb{F}_{p^n}$ where p is prime and $n > 1$ is a small integer. This article presents a concise overview of these algorithms and discusses some of the challenges with assessing their impact on keylengths for pairing-based cryptosystems.

1 Introduction

A cryptographic pairing is a non-degenerate bilinear map $\hat{e} : \mathbb{G}_1 \times \mathbb{G}_2 \to \mathbb{G}_T$, where $\mathbb{G}_1$, $\mathbb{G}_2$, $\mathbb{G}_T$ are groups of the same prime order r. The pairing is symmetric if $\mathbb{G}_1 = \mathbb{G}_2$; otherwise it is asymmetric. Such pairings are generally constructed from elliptic curves E defined over a finite field $\mathbb{F}_q$ and having low embedding degree n. For symmetric pairings, $\mathbb{F}_q$ is either a characteristic-two or characteristic-three field (with $n = 4$ or $n = 6$) or a prime field (with $n = 2$). For asymmetric pairings, $\mathbb{F}_p$ is a prime field and n is small, e.g., $n \in \{2, 6, 12, 18, 24\}$. Here, $\mathbb{G}_1$ and $\mathbb{G}_2$ are order-r groups of $\mathbb{F}_{q^n}$-rational points on E, $\mathbb{G}_T$ is the order-r subgroup of $\mathbb{F}_{q^n}^*$, and the map $\hat{e}$ is derived from the Weil or Tate pairings.

Beginning in 2001 when Boneh and Franklin proposed their identity-based encryption scheme [11], pairings have become an indispensable instrument in the cryptographer's toolbox. Hundreds (if not thousands) of research papers have been written that use pairings to design protocols that achieve certain cryptographic or efficiency objectives that do not seem attainable with conventional cryptosystems such as RSA and elliptic curve cryptography (ECC). Among these applications are aggregate signature schemes, non-interactive zero-knowledge proof systems, certificateless encryption, attribute-based encryption, and searchable encryption.

A vast majority of research papers on pairing-based protocols treat the underlying pairing $\hat{e}$ as a black box, and emphasize reductionist security proofs for the

© Springer International Publishing AG 2017
R.C.-W. Phan and M. Yung (Eds.): LNCS 10311, Mycrypt 2016, pp. 83–108, 2017.
DOI: 10.1007/978-3-319-61273-7_5

protocols with respect to some hardness assumption on $\hat{e}$. An unfortunate consequence of this predominant point of view is that issues with functionality, efficiency and security of the pairing-based protocols have not been given the attention they deserve sometimes leading to misleading or incorrect claims. For example, beginning with the BLS signature scheme [12], many papers described protocols using so-called Type-2 asymmetric pairings whereby $\mathbb{G}_1 \neq \mathbb{G}_2$ and an efficiently-computable isomorphism ψ from $\mathbb{G}_2$ to $\mathbb{G}_1$ is known. However, a concrete analysis subsequently revealed that Type 2 pairings are inferior to their Type 3 counterparts with respect to functionality, efficiency and security, and therefore there is no reason to use them [14] (see also [15]). As a second example, consider the Boneh-Shacham group signature scheme with asymmetric pairings $\hat{e} : \mathbb{G}_1 \times \mathbb{G}_2 \to \mathbb{G}_T$ in which one needs to hash onto $\mathbb{G}_2$ and thereafter apply ψ to the resulting hash value [13]. This protocol, although 'provably secure', is not implementable since no construction of such a pairing is known. As a third example, we mention the bewildering array of contrived hardness assumptions that have been proposed in the literature in order to attain a reductionist security proof (see [32]). It is typically easy to prove that these assumptions are valid in the generic group model. However, their validity in practice is much more difficult to ascertain. Indeed, Cheon [16] showed that the so-called Strong Diffie-Hellman (SDH) problem that had been formulated by Boneh and Boyen [10] can be solved significantly faster than previous believed. Shortly after, Jao and Yoshida [24] showed that Cheon's SDH solver could be used to forge signatures for the Boneh-Boyen signature scheme.

More recently, confidence in the security of pairing-based protocols has been shaken because of spectacular advances in algorithms for solving the discrete logarithm problem (DLP) in $\mathbb{G}_T$, a problem whose intractability is necessary for the security of *all* pairing-based protocols. Most astonishingly, the DLP in small-characteristic finite fields can now be solved in quasi-polynomial time [3], thereby rendering insecure all protocols that use symmetric pairings derived from elliptic and hyperelliptic curves over small-characteristic fields. Moreover, numerous improvements to the Number Field Sieve for computing discrete logarithms in fields $\mathbb{F}_{p^n}$ where p is prime and $n > 1$ is small have been proposed [4,31], thereby appearing to decrease the security of popular asymmetric pairings including those derived from Barreto-Naehrig (BN) elliptic curves [8].

The purpose of this paper is to initiate an examination of the impact of the aforementioned NFS improvements on keylengths for protocols that employ asymmetric pairings. Of special interest are parameters for BN [8], BLS12 [7], KSS [30] and BLS24 [7] pairings that achieve the 128-bit and 192-bit security levels in light of the new DLP attacks. Table 1 lists the important parameters for these families of elliptic curves. All elliptic curves E are defined over a prime field $\mathbb{F}_p$. The group order $\#E(\mathbb{F}_p) = p+1-t$ is divisible by a prime r, and we set $\rho = \log p / \log r$. In order to achieve the ℓ-bit security level, one must select the parameter z so that the bitlength of r is at least 2ℓ (in order to resist Pollard's rho attack [36] on the DLP in $\mathbb{G}_1$), and so that the bitlength of p^n is sufficiently large to resist NFS attacks on the DLP in $\mathbb{F}_{p^n}^*$.

Table 1. Parameters for the BN, BLS12, KSS and BLS24 families of elliptic curves.

BN curves: $n = 12$, $\rho \approx 1$
$p(z) = 36z^4 + 36z^3 + 24z^2 + 6z + 1$
$r(z) = 36z^4 + 36z^3 + 18z^2 + 6z + 1$, $t(z) = 6z^2 + 1$
BLS12 curves: $n = 12$, $\rho \approx 1.5$
$p(z) = (z - 1)^2(z^4 - z^2 + 1)/3 + z$, $r(z) = z^4 - z^2 + 1$, $t(z) = z + 1$
KSS curves: $n = 18$, $\rho \approx 4/3$
$p(z) = (z^8 + 5z^7 + 7z^6 + 37z^5 + 188z^4 + 259z^3 + 343z^2 + 1763z + 2401)/21$
$r(z) = (z^6 + 37z^3 + 343)/343$, $t(z) = (z^4 + 16z + 7)/7$
BLS24 curves: $n = 24$, $\rho \approx 1.25$
$p(z) = (z - 1)^2(z^8 - z^4 + 1)/3 + z$, $r(z) = z^8 - z^4 + 1$, $t(z) = z + 1$

We find that the published analyses of the NFS algorithms are inherently asymptotic in nature, and that much more work remains to be done before the impact on keylengths can be determined with full confidence. In the meantime, implementers who wish to deploy pairing-based protocols are advised to make conservative parameter choices that ignore hidden constants in the running times of the NFS algorithms. Note that these hidden constants (most likely) have the effect of multiplying the running time by at least 1, so ignoring them results in an underestimation of the NFS running times.

The remainder of this paper is organized as follows. In Sect. 2 we give some examples of the difficulties and limitations of interpreting asymptotic results in practice. Concise overviews of the NFS and the Tower Number Field Sieve (TNFS) and their derivatives are presented in Sects. 3 and 4. In Sect. 5, we identify some hidden constants in the asymptotic analysis of the TNFS. The combined effect of these hidden constants is difficult to ascertain but can have a significant impact on the concrete running time of the algorithm. In Sect. 6, we consider the effect of one such constant, namely the constant that arises in the expression for the upper bound of the norm. Translating the effect of this constant into concrete running times yields several interesting observations on the practical efficiency of the algorithms. We make some concluding remarks in Sect. 7.

2 Pitfalls in Asymptotic Analysis

This section gives some examples of the difficulties and limitations of interpreting asymptotic results in practice.

2.1 Integer Factorization

The NFS for factoring integers N has running time $L_N(\frac{1}{3}, 1.923)$ [33]. Here, $L_N(a, c)$ with $0 < a < 1$ and $c > 0$ denotes the expression

$$O\left(\exp\left((c + o(1))(\log N)^a(\log\log N)^{1-a}\right)\right) \tag{1}$$

that is *subexponential* in $\log N$[1]. This running time expression hides a multiplicative constant. Moreover, an exact formula for the $o(1)$ term in the exponent is not known.

In the 1990's, there was considerable debate in standards forums about the RSA keylengths that were needed to ensure long-term security against NFS attacks. While experiments with factoring medium-sized N were being conducted, there was no consensus on how to scale the experimental results to large-sized N. In addition, the NFS has large storage needs and requires a large amount of RAM in order to perform sieving efficiently. Thus, since it is difficult to predict the cost and speed of hardware many years into the future, it was difficult to assess the true *cost* of running the NFS on large-sized N. Nonetheless, consensus was reached that the conservative approach to determining security levels for RSA would be to use the running time of the NFS as the sole measure. RSA keylength estimates that were made 15 years ago have survived with no changes. In particular, it has become widely accepted that RSA with moduli of bitlengths 1024, 2048, 3072, 7680, 15360, offers security levels of 80, 112, 128, 192, 256 bits, respectively [6].

2.2   Elliptic Curve Discrete Logarithm Problem

For any fixed $n \geq 4$, the Gaudry-Hess-Smart (GHS) Weil descent attack [21] for solving the elliptic curve discrete logarithm problem (ECDLP) in elliptic curves over characteristic-two finite fields $\mathbb{F}_{q^n}$ has running time

$$O(q^{2+\epsilon}) \text{ as } q \to \infty. \tag{2}$$

Consider the case of elliptic curves E over $\mathbb{F}_{2^{163}}$ where $\#E(\mathbb{F}_{2^{163}})$ is twice a prime. Pollard's rho attack takes time 2^{81} to compute logarithms in $E(\mathbb{F}_{2^{163}})$. One would expect that the GHS attack is not applicable since $q = 2$ is small. On the other hand, if one ignores the hidden constant and the ϵ term in (2) then one might conclude that by embedding $\mathbb{F}_{2^{163}}$ in $\mathbb{F}_{2^{25 \cdot 163}}$ (where we now have $q = 2^{25}$ and $n = 163$), the GHS attack would take time approximately 2^{50} and thus would be significantly faster than Pollard's rho method. However, the running time expression (2) hides a very bad dependency on n, namely a multiplicative constant $2^n!$. For $n = 163$, $2^n! \approx 2^{2^{170}}$ which makes it clear that the GHS attack is completely impractical for computing logarithms in elliptic curves over $\mathbb{F}_{2^{163}}$.

As another example, we mention Diem's striking result [18] (see also [19]). Let a and b be fixed positive real numbers with $a < b$. Then Diem proved that discrete logarithms in the group of rational points on any elliptic curve defined over $\mathbb{F}_{q^n}$ with $a \cdot \sqrt{\log q} \leq n \leq b \cdot \sqrt{\log q}$ can be solved in subexponential time

$$e^{O((\log q^n)^{2/3})}.$$

Now, a subexponential-time algorithm for solving the ECDLP could have devastating consequences for the security of conventional ECC whose *raison d'être*

---

[1] In this paper, $\log N$ and $\lg N$ are the logarithms of N to the base e and 2, respectively.

is the belief that the fastest algorithm for solving the ECDLP is Pollard's rho method which takes fully exponential time. However, Diem's algorithm is inherently asymptotic and it is generally accepted that it does not pose a threat to the security of ECC in practice where elliptic curves over prime fields or over prime-degree extensions of the field of two elements are employed.

2.3   Indistinguishability Obfuscation

In 2013, Garg et al. [20] gave the first provably-secure construction for a polytime indistinguishability obfuscator. The security proof was for all polytime adversaries under certain assumptions on the underlying cryptographic primitives. However, a concrete analysis undertaken by Mayo [34] highlighted the asymptotic nature of the scheme and its impracticality. Mayo considered the obfuscation of a circuit of depth ℓ^3 and size ℓ^5, where ℓ denotes the security parameter. When the Coron-Lepoint-Tibouchi multilinear map [17] is employed in the Garg et al. obfuscator, the size of the obfuscated circuit for $\ell = 128$ was estimated to be at least 2^{1357} bits. Thus, even though the Garg et al. obfuscator was provably secure and efficient within the "polynomial-time" paradigm, it is hopelessly impractical.

3   Overview of the Number Field Sieve

Let p be a prime, $n \geq 1$, and $Q = p^n$. Suppose that p is written as $p = L_Q(a, c_p)$ for real numbers $a, c_p > 0$. Depending on the value of a, finite fields $\mathbb{F}_Q$ are classified into the following types: small characteristic if $a \leq 1/3$; medium characteristic if $1/3 < a < 2/3$; boundary if $a = 2/3$; and large characteristic if $a > 2/3$.

For small-characteristic finite fields, there has been tremendous progress in the discrete logarithm computation. The approach has been based on the function field sieve algorithm and asymptotically the fastest known algorithm runs in quasi-polynomial time [3].

For the other classes of finite fields, i.e., those with medium to large characteristic, the Number Field Sieve is presently the state-of-the-art. The NFS was initially proposed for integer factorization [33]. Application of the NFS to DLP computation was first proposed by Gordon [22] who considered prime-order fields. Extensions to composite-order fields were made by Schirokauer [40]. For the case of prime-order fields, improvements were made by Joux and Lercier [27]. Joux, Lercier, Smart and Vercauteren [28] showed that the NFS is applicable to all finite fields. For fields where the prime p is of a special form, Joux and Pierrot [29] applied the special NFS to obtain improved complexity.

The NFS is an index-calculus algorithm having three main phases: (i) relation collection, (ii) linear algebra, and (iii) individual logarithm. Asymptotically, the times for the relation collection and the linear algebra phases dominate the time for the individual logarithm phase. The parameters are tuned so that the time for the relation collection phase is equal to the time for the linear algebra phase and this time is the asymptotic run time of the NFS.

Two number fields $\mathbb{K}_f = \mathbb{Q}[x]/(f)$ and $\mathbb{K}_g = \mathbb{Q}[x]/(g)$ are defined by choosing irreducible polynomials $f(x)$ and $g(x)$ over the integers. The required condition on $f(x)$ and $g(x)$ is that modulo p they have a common irreducible factor $\varphi(x)$ of degree n over $\mathbb{F}_p$. The field $\mathbb{F}_{p^n}$ is represented by $\varphi(x)$. Let γ be a generator of the multiplicative group of $\mathbb{F}_{p^n}$.

Let $\alpha, \beta \in \mathbb{C}$ and $m \in \mathbb{F}_{p^n}$ be roots of $f(x)$, $g(x)$ and $\varphi(x)$ respectively. The commutative diagram given in Fig. 1 shows two homomorphisms $\mathbb{K}_f \to \mathbb{F}_{p^n}$ and $\mathbb{K}_g \to \mathbb{F}_{p^n}$ given by $\alpha \mapsto m$ and $\beta \mapsto m$ respectively. This diagram explains the basic working of the NFS.

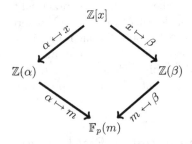

Fig. 1. The basic principle of NFS.

Instead of working over the whole number fields $\mathbb{K}_f$ and $\mathbb{K}_g$, one works over the corresponding rings of integers $\mathcal{O}_f$ and $\mathcal{O}_g$. The notion of norm of ideals allows one to define a suitable factor base, namely the prime ideals of $\mathcal{O}_f$ and $\mathcal{O}_g$ whose norms are at most some pre-specified smoothness bound B. The size of the factor base is $B^{1+o(1)}$, where B is chosen so as to balance the times for relation collection and linear algebra.

To generate relations, polynomials $\phi \in \mathbb{Z}[x]$ of degree at most $t - 1$ are considered. If the principal ideals $\phi(\alpha)\mathcal{O}_f$ and $\phi(\beta)\mathcal{O}_g$ are B-smooth, then such a ϕ yields a relation among the factor base elements. Formally, a relation is actually a linear relation between the discrete logarithms of certain elements of the field $\mathbb{F}_{p^n}$. Such discrete logarithms are called virtual logarithms[2]. The number of relations collected is a little more than B. This allows carrying out the linear algebra phase to compute the virtual logarithms of the factor base elements. The individual logarithm phase consists of finding the discrete logarithm of some element $h \in \mathbb{F}_{p^n}$. The task in this phase is to find an element of the form $h^i \gamma^j$ such that the principal ideal generated by the preimage of $h^i \gamma^j$ in $\mathcal{O}_f$ factors into prime ideals of degrees at most $t - 1$ and bounded norms. Then the

---

[2] The details are complicated and involve using the homomorphisms $\alpha \mapsto m$ and $\beta \mapsto m$ along with the class numbers and the torsion-free ranks of $\mathcal{O}_f$ and $\mathcal{O}_g$. We skip these details.

special-q technique is used to express the desired discrete logarithm in terms of the virtual discrete logarithms of factor base elements. Since these virtual discrete logarithms have already been computed, it is possible to finally obtain the desired discrete logarithm. We refer to [2] for more details on the relation collection phase and to [23] for the individual discrete logarithm phase.

3.1   Polynomial Selection and Sizes of Norms

The crucial step in relation collection is to obtain $\phi \in \mathbb{Z}[x]$ such that the ideals $\phi(\alpha)\mathcal{O}_f$ and $\phi(\beta)\mathcal{O}_g$ are both smooth. For ensuring this smoothness, it is sufficient to ensure that their norms, i.e., $\mathrm{Res}(f, \phi)$ and $\mathrm{Res}(g, \phi)$, are both B-smooth, where Res denotes the resultant. Let E be such that the coefficients of ϕ are in $\left[-\frac{1}{2}E^{2/t}, \frac{1}{2}E^{2/t}\right)$, whence $\|\phi\|_\infty \approx E^{2/t}$ and the number of polynomials ϕ that are considered for sieving is E^2. Here, the ℓ_∞ norm $\|\phi\|_\infty$ of the polynomial ϕ is the maximum of the absolute values of the coefficients of ϕ. For $p = L_Q(a, c_p)$ with $a > \frac{1}{3}$, the following can be shown (a more precise bound is provided later in the context of the TNFS):

$$|\mathrm{Res}(f, \phi)| = O\left(\left(\|f\|_\infty\right)^{t-1} E^{2(\deg f)/t}\right) \quad \text{and}$$

$$|\mathrm{Res}(g, \phi)| = O\left(\left(\|g\|_\infty\right)^{t-1} E^{2(\deg g)/t}\right), \tag{3}$$

yielding the norm bound

$$|\mathrm{Res}(f, \phi) \times \mathrm{Res}(g, \phi)| = O\left(\left(\|f\|_\infty \|g\|_\infty\right)^{t-1} E^{(\deg f + \deg g)2/t}\right). \tag{4}$$

The probability of B-smoothness of the product of norms (4) determines the cost of obtaining a single relation and hence the cost of relation collection. A suitable choice of B is made to balance this cost with the cost of the linear algebra step. Thus the value of the product of norms in (4) is crucial for determining the overall run time of the algorithm.

Note that the norm bound is determined by the degrees of f and g and their ℓ_∞ norms. So, to ensure that the norm bound is small, it is required that the degrees and ℓ_∞ norms of f and g are small. Ensuring both of these is a very difficult problem. In the literature several methods for polynomial selection have been proposed which provide polynomials with different trade-offs between degrees and ℓ_∞ norms. We briefly describe some of the important ones next.

JLSV1 [28]. Choose random polynomials $f_0(x)$ and $f_1(x)$ having small coefficients with $\deg(f_1) < \deg(f_0) = n$. Choose a random positive integer δ which is slightly greater than $\lceil\sqrt{p}\rceil$ and let (u, v) be a rational reconstruction of δ modulo p, i.e., $\delta \equiv u/v \pmod{p}$. More precisely, (u, v) is obtained as the first row on applying the LLL-reduction algorithm to the matrix

$$\begin{bmatrix} p & 0 \\ \delta & 1 \end{bmatrix}.$$

Let $f(x) = f_0(x) + \delta f_1(x)$ and $g(x) = v f_0(x) + u f_1(x)$ and $\varphi(x) = f(x)$ mod p. Repeat the above procedure until f and g are irreducible over $\mathbb{Z}$ and φ is irreducible over $\mathbb{F}_p$. For this method, the bound (4) is $O\left(E^{4n/t}Q^{(t-1)/n}\right)$.

GJL [2]. The basic Joux-Lercier method [27] works for prime fields. In [2], it was generalized to work over composite-order fields. Let $\varphi(x) = x^n + \varphi_{n-1}x^{n-1} + \cdots + \varphi_1 x + \varphi_0$ and $r \geq n$. Define an $(r+1) \times (r+1)$ matrix $M_{\varphi,r}$ whose rows are obtained from the coefficients of the polynomials $p, px, \ldots, px^{n-1}, \varphi(x), x\varphi(x), \ldots, x^{r-n}\varphi(x)$. The LLL algorithm is applied to $M_{\varphi,r}$. Let the first row of the resulting LLL-reduced matrix be $[g_0, g_1, \ldots, g_{r-1}, g_r]$ and let $g = \text{LLL}(M_{\varphi,r})$ denote the corresponding polynomial $g(x) = g_0 + g_1 x + \cdots + g_{r-1}x^{r-1} + g_r x^r$. By construction, $\varphi(x)$ is a factor of $g(x)$ modulo p.

The GJL procedure for polynomial selection is the following. Let $r \geq n$ and randomly choose a degree-$(r+1)$ polynomial $f(x)$ that is irreducible over $\mathbb{Z}$, has coefficients of size $O(\log p)$, and has a degree-n factor $\varphi(x)$ modulo p which is both monic and irreducible. The procedure is repeated until $g = \text{LLL}(M_{\varphi,r})$ is irreducible over $\mathbb{Z}$. In this case, the norm bound (4) is $O\left(E^{2(2r+1)/t}Q^{(t-1)/(r+1)}\right)$.

Conjugation [2]. Choose a random monic quadratic polynomial $\mu(x)$ having coefficients of size $O(\log p)$, and which is irreducible over $\mathbb{Z}$ but has a root t modulo p. Let (u, v) be a rational reconstruction of t modulo p. Choose random polynomials $g_0(x)$ and $g_1(x)$ with small coefficients with $\deg g_1 < \deg g_0 = n$. Let $g(x) = v g_0(x) + u g_1(x)$ and $f(x) = \text{Res}_y\left(\mu(y), g_0(x) + y \cdot g_1(x)\right)$. Repeat this until f and g are irreducible over $\mathbb{Z}$ and φ is irreducible over $\mathbb{F}_p$. In this case, the norm bound (4) is $O\left(E^{6n/t}Q^{(t-1)/(2n)}\right)$.

SS [37]. A general method (called Algorithm-$\mathcal{A}$) for polynomial selection was given in [37]. This method has two parameters d and r, where d is a divisor of n and $r \geq n/d$. The method uses the LLL algorithm in a more general manner than the GJL method. The norm bound (4) is $O\left(E^{2d(2r+1)/t}Q^{(t-1)/(d(r+1))}\right)$. Putting $d = n$ and $r = 1$ gives the Conjugation method, whereas putting $d = 1$ gives the GJL method. For $1 < d < n$ and also for $d = n$, $r > 1$, this method provides trade-offs which cannot be obtained using either the GJL or the Conjugation method.

3.2   Asymptotic Complexity

For each polynomial selection method, the norm bound (4) can be used to obtain a rough estimate of the efficiency of the resulting DLP computation. It is also possible to convert the norm bound into an asymptotic estimate of the run time. The details of how this can be done are a bit messy and so we skip the details. Instead, we just mention the final results.

Medium characteristic case: For $p = L_Q(a, c_p)$ with $a > 1/3$, the run time of the NFS with the Conjugation method is $L_Q(1/3, (96/9)^{1/3})$.

Boundary case: For $p = L_Q(2/3, c_p)$, the run time of the NFS with Algorithm-$\mathcal{A}$ is $L_Q(1/3, 2c_b)$, where

$$c_b = \frac{2r+1}{3c_p kt} + \sqrt{\left(\frac{2r+1}{3c_p kt}\right)^2 + \frac{kc_p(t-1)}{3(r+1)}} \tag{5}$$

with $k = n/d$. For $d = n$ and $r = k = 1$, this reduces to the complexity obtained by the Conjugation method. The best complexity that is obtained is $L_Q(1/3, (48/9)^{1/3})$. This complexity, however, is attained for only one particular value of c_p, namely $c_p = 12^{1/3} \approx 2.3$. As $c_p \to \infty$, the minimum complexity (taken over r, k and t) approaches $L_Q(1/3, (64/9)^{1/3})$ from below.

Large characteristic case: For $p = L_Q(a, c_p)$ with $a > 2/3$, the run time of the NFS with the GJL method is $L_Q(1/3, (64/9)^{1/3})$.[3]

Among the three cases, the best complexity is achieved in the boundary case for a specific c_p value.

Remark 1. The sharp distinction between the run times for the medium characteristic, boundary (with $c_p = 12^{1/3}$), and large characteristic cases highlights the inherent asymptotic nature of the analysis and the difficulty in deriving concrete run time estimates. In particular, without the benefit of extensive experimentation, it is not clear whether a concrete problem instance, e.g., with $p \approx 2^{256}$ and $n = 12$, falls within the medium characteristic, boundary, or large characteristic cases.

3.3 Multiple Number Field Sieve Algorithm

Using multiple number fields to obtain faster asymptotic complexity was suggested in [5,35]. Pierrot [35] provided a detailed analysis of the GJL and the Conjugation methods using multiple number fields. The MNFS variant of Algorithm-$\mathcal{A}$ was analyzed in [37]. The complexities of the MNFS algorithms for the different cases of $p = L_Q(a, c_p)$ are as follows: $L_Q(1/3, 2.156)$ for the medium characteristic case, $L_Q(1/3, 1.71)$ for the boundary case, and $L_Q(1/3, 1.90)$ for the large characteristic case. The complexity for the boundary case is obtained for only one value of c_p, namely $c_p \approx 2.12$.

3.4 Special Number Field Sieve Algorithm

Suppose that p can be written as $p = \Gamma(u)$ for some polynomial Γ of degree λ and having small coefficients so that $u = O(p^{1/\lambda})$. Note that the primes p in Table 1 are in this special form. Joux and Pierrot [29] showed how to modify the polynomial selection algorithm from [28] to obtain improved complexity. Choose an irreducible polynomial f of the form $f(x) = x^n + R(x) - u$ where $R(x)$ is a polynomial of small degree with coefficients from $\{0, \pm 1\}$. Let $g = \Gamma(x^n + R(x))$. Then $g(x) = \Gamma(f(x) + u) \equiv \Gamma(u) = p \pmod{f(x)}$ and so $g(x) - p$ is a multiple of $f(x)$ implying that $g(x)$ is a multiple of $f(x)$ modulo p. This choice of f and g ensures that $\deg f = n$, $\|f\|_\infty = O(p^{1/\lambda})$, $\deg g = \lambda n$, and $\|g\|_\infty = O((\log n)^\lambda)$.

---

[3] For comparisons with other run times, it is useful to note that $(96/9)^{1/3} \approx 2.201$, $(64/9)^{1/3} \approx 1.923$, $(48/9)^{1/3} \approx 1.747$, and $(32/9)^{1/3} \approx 1.526$.

The asymptotic complexities reported in [29] are the following. As before, let $p = L_Q(a, c_p)$.

Medium characteristic case: $L_Q\left(1/3, ((64/9) \cdot (\lambda+1)/\lambda)^{1/3}\right)$ for $1/3 \leq a < 2/3$.

Boundary case: $L_Q\left(1/3, ((32/9) \cdot (\lambda+1)/\lambda)^{1/3}\right)$ for $a = 2/3$;

Large characteristic case: $L_Q\left(1/3, (32/9)^{1/3}\right)$ for $2/3 < a < 1$.

Unlike the NFS, for SNFS the best complexity is achieved for the large characteristic case.

4   Overview of the Tower Number Field Sieve

The Tower Number Field Sieve (TNFS) algorithm was initially considered by Schirokauer [40] and was revisited by Barbulescu et al. [4]. The implications of this algorithm for improving the complexity of the medium prime case were pointed out by Kim and Barbulescu [31] which has led to several follow-up works [38,39]. Following [31] we denote these algorithms by 'extended TNFS' (exTNFS).

As we saw in Sect. 3.2, the Conjugation method from [2] resulted in the NFS complexity of the boundary case being smaller than the complexity of the medium prime case. Suppose that the extension degree n is composite and $n = \eta\kappa$ is a non-trivial factorization of n. Then $\mathbb{F}_{p^n}$ has a tower field representation $\mathbb{F}_{q^\kappa}$, where $q = p^\eta$. The main idea behind the complexity reduction for the medium prime case using a tower field representation is the following. If p and $Q = p^n$ are such that $p = L_Q(a, c_p)$ for $1/3 < a < 2/3$, then one can translate the problem to that of computing DLP in $\mathbb{F}_{q^\kappa}$ where $q = L_Q(2/3, c_q)$. This corresponds to the boundary case and so one benefits from the lower complexity of the boundary case for the medium prime case.

Here q is not prime and the characteristic of the field $\mathbb{F}_Q$ remains p irrespective of how the field is represented. Hence, strictly speaking, it is not correct to say that the medium characteristic case transforms to the boundary case. On the other hand, from the complexity point of view what matters are the norms of the polynomials and in that sense it is possible to obtain smaller norm bounds with the tower field representation than what could be done directly.

The basic idea of the exTNFS algorithm is the following. One starts with a monic polynomial $h(z)$ of degree η which is irreducible over $\mathbb{F}_p$ and hence is also irreducible over $\mathbb{Z}$. Let $\mathbb{F}_{p^\eta} = \mathbb{F}_p[z]/(h(z))$ and $R = \mathbb{Z}[z]/(h(z))$. Suppose f and g are polynomials in $R[x]$ whose leading coefficients are from $\mathbb{Z}$. It is required that both f and g are irreducible over R (this can be verified by testing irreducibility over $\mathbb{Q}[z]/(h(z))$ and over $\mathbb{F}_{p^\eta}$, and that f and g have a degree-κ common factor $\varphi(x)$ that is irreducible over $\mathbb{F}_{p^\eta}$. The field $\mathbb{F}_{p^n}$ is then realized as $\mathbb{F}_{p^\eta}[x]/(\varphi(x)) = (R/pR)[x]/(\varphi(x))$.

Let $\mathbb{K}_f$ and $\mathbb{K}_g$ be the number fields defined by f and g respectively. As in the case of the NFS (see Fig. 1), the above set-up provides two different decompositions of a homomorphism from $R[x]$ to $\mathbb{F}_{p^n}$. One of these goes through

$R[x]/(f(x))$ and the other goes through $R[x]/(g(x))$. Using this set-up it is possible to define a factor base and carry out the three main phases of NFS. Here also, the factor base consists of $B^{1+o(1)}$ elements for some smoothness bound B.

Sieving for relation collection is done using polynomials $\phi \in R[x]$ of degrees at most $t - 1$ with $\|\phi\|_\infty = E^{2/\eta t}$ so that the number of sieving polynomials is E^2. A sieving polynomial $\phi \in R[x]$ generates a relation if both the norms

$$N(\phi, f) = \mathrm{Res}_z(\mathrm{Res}_x(\phi(x), f(x)), h(z)) \quad \text{and}$$
$$N(\phi, g) = \mathrm{Res}_z(\mathrm{Res}_x(\phi(x), g(x)), h(z)) \tag{6}$$

are B-smooth. Note that in this case, f and g can be viewed as bivariate polynomials in x and z and hence the norm is obtained by taking resultants twice. Bounds on the norm are obtained from the bounds on resultants of bivariate polynomials [9].

The polynomial selection methods for NFS (see Sect. 3.1) translate to the exTNFS setting. Instead of providing the details of these methods, we provide a summary of recent work with a focus on the medium prime case.

Kim and Barbulescu [31]. This work chooses $f, g \in \mathbb{Z}[x]$, whence the degree-κ polynomial $\varphi(x) = \gcd(f(x), g(x))$ is over $\mathbb{F}_p$. The requirement that $\varphi(x)$ is irreducible over $\mathbb{F}_{p^\eta}$ imposes the condition that $\gcd(\eta, \kappa) = 1$. Hence, the Kim-Barbulescu method works only for composite non prime-power values of n. The actual polynomial selection is done using a translated version of the Conjugation method, resulting in the best achievable complexity for $p = L_Q(a, c_p)$, $1/3 < a < 2/3$ to be $L_Q(1/3, (48/9)^{1/3})$. This complexity, however, is not achieved for all values of p.

Sarkar and Singh [38]. This work described a polynomial selection method (called Algorithm-$\mathcal{C}$) in which the coefficients f and g are in R (and not necessarily in $\mathbb{Z}$), with the restriction that f is monic and the leading coefficient of g is in $\mathbb{Z}$. As a result, the restriction that $\varphi(x)$ has coefficients in $\mathbb{F}_p$ is also removed resulting in the removal of the $\gcd(\eta, \kappa) = 1$ constraint. This leads to a variant of the TNFS algorithm for the medium characteristic case with improved complexity for all composite n. For prime-power n, however, the minimum complexity obtained in [38] is larger than $L_Q(1/3, (48/9)^{1/3})$.

Jeong and Kim [25]. The Conjugation method was extended to the TNFS setting where the condition $\gcd(\eta, \kappa) = 1$ was not required. The best achievable complexity for $p = L_Q(a, c_p)$, $1/3 < a < 2/3$ was shown to be $L_Q(1/3, (48/9)^{1/3})$. For prime-power n, this improves upon the complexity achieved in [38].

Sarkar and Singh [39]. A polynomial selection algorithm, called Algorithm-$\mathcal{D}$, was described. This provides another translation of Algorithm-$\mathcal{A}$ to the TNFS setting without requiring the condition $\gcd(\eta, \kappa) = 1$. Special cases of Algorithm-$\mathcal{D}$ lead to the GJL and the Conjugation methods in both classical NFS and TNFS. As a result, Algorithm-$\mathcal{D}$ subsumes the Jeong-Kim polynomial selection method. The asymptotic complexity for the medium prime case can be described as follows. Let $p = L_Q(a, c_p)$ with $1/3 \le a < 2/3$ and suppose that $q = p^\eta$ can be written as $q = L_Q(2/3, c_\theta)$. Then the asymptotic complexity is $L_Q(1/3, 2c_b)$ where c_b is given by (5) with c_p replaced by c_θ. The minimum complexity is still $L_Q(1/3, (48/9)^{1/3})$ which is the same as that of the Jeong-Kim method and this complexity is attained for $c_\theta = 12^{1/3}$. However, improvements in asymptotic complexity are obtained for certain ranges of values of c_θ.

4.1   Multiple Number Field Sieve Algorithm

Multiple number fields can also be used with the TNFS [31]. As in the case of the NFS, this reduces the asymptotic complexity. The best achievable complexity for the medium prime case is $L_Q(1/3, 1.71)$.

4.2   Special Number Field Sieve Algorithm

As explained above, the main advantage of the TNFS method is to transform the problem for the medium characteristic case to that of the boundary characteristic for which the complexity is lower. In fact, it is also possible to transform to the large characteristic case. Carrying out this exercise for the SNFS (yielding the SexTNFS algorithm) leads to an asymptotic complexity of $L_Q(1/3, (32/9)^{1/3})$ for the medium prime case; this complexity is achieved for all medium primes unlike the case of TNFS. This works for composite n; in the case where n has a non-trivial factorization as $n = \eta\kappa$ with $\gcd(\eta, \kappa) = 1$, this complexity was reported in [31], whereas the same complexity was reported in [25] without the restriction $\gcd(\eta, \kappa) = 1$.

5   Asymptotic Analysis

In summary, the asymptotic run times of the NFS variants for computing discrete logarithms in $\mathbb{F}_Q$ in the medium characteristic case are $L_Q(1/3, c)$ where:

- $c = 2.201$ for the NFS (Sect. 3.2);
- $c = 2.156$ for the multiple NFS (Sect. 3.3);
- $c = 2.072$ for the special NFS with $\lambda = 4$ (Sect. 3.4);
- $c = 1.747$ for the exTNFS for some p (Sect. 4);
- $c = 1.71$ for the multiple exTNFS for some p (Sect. 4.1); and
- $c = 1.526$ for the SexTNFS (Sect. 4.2).

Asymptotic complexity analysis proceeds by ignoring various factors that do not have significant effect on the run time as $Q = p^n$ goes to infinity. In this section, we take a look at the different steps of this analysis with a view towards assessing whether the ignored factors can have a noticeable effect on the run time for concrete values of Q. At the same time, we consider issues of storage and observe that different operations which are assumed to asymptotically require $O(1)$ time, in practice have noticeable difference in their times.

We consider the TNFS setting where $Q = p^n$ and $n = \eta\kappa$ is a non-trivial factorization of n.

5.1   Bounds on Norms of Polynomials

The number fields $\mathbb{K}_f$ and $\mathbb{K}_g$ are defined using two polynomials $f(x)$ and $g(x)$ over $R = \mathbb{Z}[z]/(h(z))$ where $h(z)$ is a degree-η irreducible polynomial over $\mathbb{Z}$. The degrees and ℓ_∞ norms of f and g are the main factors that influence the running time. These quantities are determined based on the actual polynomial selection method that is employed. To make the ideas concrete, we work with a special case of Algorithm-$\mathcal{D}$ [37]. This special case is important since it is the TNFS variant of the Conjugation method proposed by Kim and Barbulescu [31].

Using random trials, one chooses a monic quadratic polynomial $A_1 \in \mathbb{Z}[y]$ having $O(\log p)$-size coefficients such that A_1 is irreducible over $\mathbb{Z}$ and has a factor $A_2(y) = y + \delta$ over $\mathbb{F}_p$. Further, using random trials, one chooses monic polynomials $C_0(x)$ and $C_1(x)$ over R with $\|C_i\|_\infty = O(1)$, $\deg C_0(x) = \kappa$, $\deg C_1(x) < \kappa$, and such that $f(x)$ and $g(x)$ are irreducible over R and $\varphi(x)$ is irreducible over $\mathbb{F}_{p^\eta} = \mathbb{F}_p[z]/(h(z))$ where $f(x) = \mathrm{Res}_x(A_1(y), C_0(x) + yC_1(x))$, $\varphi(x) = \mathrm{Res}_y(A_2(y), C_0(x) + yC_1(x)) \bmod p$, $\psi(x) = \psi_1 x + \psi_0$, and $g(x) = \mathrm{Res}_x(\psi(y), C_0(x) + yC_1(x))$. The integer coefficients ψ_1 and ψ_0 of $\psi(x)$ are obtained by a rational reconstruction of δ modulo p. From the bound on the first vector of an LLL-reduced basis and the bound on the shortest vector of a lattice, one obtains $\|\psi\|_\infty \leq 2p^{1/2}$.

Asymptotically, the above method for selecting f and g yields $\|f\|_\infty = O(\log p)$ and $\|g\|_\infty = O(Q^{1/2n})$. The contribution to $\|f\|_\infty$ and $\|g\|_\infty$ from the coefficients of $C_0(x)$ and $C_1(x)$ arising from the resultant computation are absorbed in the big-O notation.

5.2   Size of the Factor Base

The polynomials f and g define the two (tower) number fields $\mathbb{K}_f$ and $\mathbb{K}_g$ having ring of integers $\mathcal{O}_f$ and $\mathcal{O}_g$ respectively. The factor base consists of all prime ideals of $\mathcal{O}_f$ and $\mathcal{O}_g$ whose norms are at most B. From this, the factor base size is asymptotically $B^{1+o(1)}$. For concrete polynomials f and g, the actual number of prime ideals could have a small but noticeable difference from B.

5.3   Bounds on Norms of Ideals

Sieving is done using polynomials of degrees at most $t-1$. Consider the simplest and the most important case of $t = 2$. Then the sieving polynomials are linear polynomials $\phi \in R[x]$ with $\|\phi\|_\infty = E^{1/\eta}$. There are a total of E^2 sieving polynomials. A relation is obtained if the principal ideals generated by the images of ϕ in $\mathcal{O}_f$ and $\mathcal{O}_g$ are smooth over the factor base. This smoothness depends on the norms $N(f, \phi)$ and $N(g, \phi)$ whose bounds are given by (6).

Let $H = \|h\|_\infty$ and

$$\mathfrak{C}(\eta, s, H) = ((\eta - 1)(1 + s) + 1)^{\eta/2} \cdot (\eta+1)^{(\eta-1)(1+s)/2} \cdot H^{(\eta-1)(1+s)} \cdot ((s + 1)!\eta^s)^\eta .$$

The following bounds on the norms can be obtained:

$$N(f,\phi) \leq \mathfrak{C}(\eta, 2\kappa, H) \cdot E^{2\kappa} \cdot O\left((\log p)^{\eta}\right) \quad \text{and}$$

$$N(g,\phi) \leq \mathfrak{C}(\eta, \kappa, H) \cdot E^{\kappa} \cdot O\left(Q^{1/(2\kappa)}\right); \tag{7}$$

see Appendix A for details of the resultant calculations. In the asymptotic analysis these are written as $N(f,\phi) = E^{2\kappa} \cdot L_Q(2/3, o(1))$ and $N(g,\phi) = E^{\kappa}Q^{1/(2\kappa)} \cdot L_Q(2/3, o(1))$. In other words, one takes $\mathfrak{C}(\eta, 2\kappa, H) = L_Q(2/3, o(1))$ and $\mathfrak{C}(\eta, \kappa, H) = L_Q(2/3, o(1))$, and consequently their contribution to the overall running time $L_Q(1/3, c)$ is absorbed by the $o(1)$ term in the latter.

For concrete values, the factors that get absorbed in the $L_Q(2/3, o(1))$ expression can be very large. For $n = 12$, let $\eta = 4$ and $\kappa = 3$ and suppose that $H = 5$. Then $\mathfrak{C}(4,3,5) \approx 2^{91.5}$ and $\mathfrak{C}(4,6,5) \approx 2^{179.3}$. On the other hand, suppose we choose $\eta = 1$ and $\kappa = 12$ and $H = 5$ as before. These values of η and κ correspond to the usual NFS, i.e., we are not exploiting the tower structure. Then $\mathfrak{C}(1, 12, 5) \approx 2^{32.5}$ and $\mathfrak{C}(1, 24, 5) \approx 2^{83.7}$. While these are still large numbers, they are significantly smaller than the numbers obtained in the case of the tower representation.

5.4   Smoothness Probability from the Canfield-Erdős-Pomerance Theorem

The bounds obtained on the norms $N(f,\phi)$ and $N(g,\phi)$ are used to estimate the probability that a random sieving polynomial will provide a relation. The required condition is that the principal ideals generated by the images of ϕ in the two integer rings are smooth over the factor base. Two *assumptions* are used, which means that the entire analysis is heuristic.

1. It is *assumed* that the probability that the principal ideal generated by the image of ϕ in $\mathcal{O}_f$ factors over the factor base is the same as the probability that a random integer of size $N(f,\phi)$ is B-smooth. Similarly for $\mathcal{O}_g$.
2. The events that the two principal ideals generated by the images of ϕ are smooth over the factor base are independent.

The probability that a random integer of size $N(f,\phi)$ is B-smooth is obtained from the L-notation version of a theorem due to Canfield, Erdős and Pomerance. (See Theorem 15.2 of [26] for the statement of the theorem.) The smoothness probability can also be expressed in concrete terms. Let $\Psi(N, B)$ be the number of positive integers $\leq N$ which are B-smooth. Ignoring lower order terms, it can be shown that

$$\log\left(\frac{\Psi(N,B)}{N}\right) \approx -\frac{\log N}{\log B} \log\left(\frac{\log N}{\log B}\right). \tag{8}$$

Let $\Pi(N, B) = \Psi(N, B)/N$ be the probability that a random integer $\leq N$ is B-smooth. We are interested in

$$\Pi(N(f,\phi), B) \cdot \Pi(N(g,\phi), B). \tag{9}$$

The number of trials (i.e., the number of sieving polynomials to consider) to obtain a single relation is about

$$\mathfrak{n} = (\Pi(N(f,\phi), B) \cdot \Pi(N(g,\phi), B))^{-1}.$$

Since about B relations are required, the total number of trials to obtain all the relations is about $B\mathfrak{n}$. This is how the bounds on the norms determine the run time of the relation collection phase.

5.5 Balancing Costs

For the asymptotic analysis, the costs of the relation collection and the linear algebra phases are balanced. This balancing proceeds by imposing two conditions. Recall that the number of sieving polynomials is E^2 and the size of the factor base is $B^{1+o(1)}$. Sparse linear algebra requires time $B^{2+o(1)}$. Hence, the costs of the relation collection and linear algebra phases are balanced by setting $E = B$.

Note that this assumes that the total time for sieving with E^2 polynomials is equal to the time required for completing the linear algebra phase. While this is true in an asymptotic sense, in concrete terms the two costs can be significantly different. We note the following differences between these two tasks.

1. For both sieving and linear algebra, the basic operations are field operations in $\mathbb{F}_p$. However, the number of such field operations are different for the two tasks.
2. For linear algebra, it is required to perform read and write operations on a very large matrix. In practice, these read/write operations will incur a significant overhead.
3. The sieving step is parallelizable up to any extent. This is not true for the linear algebra step. The block Wiedemann algorithm can be parallelized, but this comes at the cost of additional memory requirements.

6 Concrete Analysis

We take some concrete values to get an idea of the effect of the constants in the norm bounds on the smoothness probability. Suppose $Q \approx 2^{3000}$, $n = 12$, $\eta = 4$, $\kappa = 3$, $H = 5$, whence $p \approx 2^{250}$. Assume that the factor base size is $B = 2^{64}$ so that the linear algebra phase requires approximately 2^{128} operations. As discussed in Sect. 5.5, E is taken to be equal to B so that E is also 2^{64}. Then taking the upper bounds in (7) to be the norm values and the hidden constants in the big-O notation to be 1, we obtain

$$N(f,\phi) \approx \mathfrak{C}(4,6,5)E^{2\kappa}(\log p)^\eta \approx 2^{593},$$
$$N(g,\phi) \approx \mathfrak{C}(4,3,5)E^\kappa Q^{1/(2\kappa)} \approx 2^{783},$$
$$\Pi(N(f,\phi), B) \approx \Pi(2^{593}, 2^{64}) \approx 2^{-29.8},$$
$$\Pi(N(g,\phi), B) \approx \Pi(2^{783}, 2^{64}) \approx 2^{-44.2},$$
$$\mathfrak{n} = (\Pi(N(f,\phi), B) \cdot \Pi(N(g,\phi), B))^{-1} \approx 2^{74}.$$

Table 2. Approximate run times of exTNFS and SexTNFS for $Q = p^n \approx 2^{3000}$ for several different values of n, with and without the constants. These constants are $\mathfrak{C}(\eta, \kappa, H)$ and $\mathfrak{C}(\eta, 2\kappa, H)$ for exTNFS, and $\mathfrak{C}(\eta, \kappa, H)$ and $\mathfrak{C}(\eta, \kappa\lambda, H)$ for SexTNFS.

n	Algorithm	(η, κ, λ)	With constants	Without constants
12	exTNFS	$(4, 3, -)$	2^{138}	2^{116}
	SexTNFS	$(6, 2, 4)$	2^{155}	2^{108}
18	exTNFS	$(6, 3, -)$	2^{154}	2^{118}
	SexTNFS	$(9, 2, 8)$	2^{279}	2^{160}
24	exTNFS	$(8, 3, -)$	2^{169}	2^{118}
	SexTNFS	$(12, 2, 10)$	2^{369}	2^{186}

Hence a single relation will be obtained after trying about 2^{74} sieving polynomials. On the other hand, if we ignore the factors $\mathfrak{C}(4, 6, 5)$ and $\mathfrak{C}(4, 3, 5)$, then $\mathfrak{n} \approx 2^{54}$. So, in this case the effect of the constants in the norm bounds is to increase the number of iterations for finding a single relation by a factor of about 2^{20}.

The number of iterations required to find a single relation affects the overall cost of the algorithm. The total number of iterations required to find the B required relations determines the cost of the relation collection phase. If we take the constants into account, the relation collection phase will have cost about $2^{74}B = 2^{138}$ for $B = 2^{64}$. On the other hand, if the constants are ignored, then the relation collection phase will have cost about $2^{54}B = 2^{118}$. In both cases, the linear algebra phase will have cost approximately $B^2 = 2^{128}$.

The above choice of B does not balance the costs of the relation collection and linear algebra phases. We have redone the calculations with $Q \approx 2^{3000}$ so as to balance these costs. The overall approximate costs of the algorithm are given by the values in the fourth column (if the constants are taken into consideration) and fifth column (if the constants are ignored) of the first row of Table 2. The size of the factor base is the square root of the overall cost.

A similar calculation can be done for SexTNFS; cf. Remark 2. In this case, however, the minimum complexity is not achieved for $\eta = 4$ and $\kappa = 3$. Instead the minimum is achieved for $\eta = 6$ and $\kappa = 2$. Since in this case η and κ are not coprime, this choice would not be allowed by the Kim-Barbulescu work [31], but would be permitted by [25]. The run times considering and ignoring constants are respectively given in the second row of Table 2.

We have performed similar calculations for $n = 18$ and $n = 24$. It turns out that the factorization of n that minimizes the exTNFS complexity is not the same as the factorization of n that minimizes the SexTNFS complexity. Table 2 provides the run times of exTNFS and SexTNFS both when the constants in the norm bound are taken into consideration and also when they are ignored. We observe the following from the values in Table 2.

1. If the constants are taken into consideration, then in each case the run time is significantly greater than if the constants are ignored. In particular, a 3000-bit Q does seem to provide at least 128-bit security for $n = 12$.
2. The concrete run time for exTNFS is smaller than that of SexTNFS. (The only exception to this is for $n = 12$ and when the constants in the norm bounds are ignored.) This is contrary to what one would expect from the asymptotic analysis in which the run time of SexTNFS is smaller than that of exTNFS; cf. Remark 2.
3. The asymptotic expression for the run time does not have any dependence on n and depends only on Q. This means that for a given Q, the run times will asymptotically be the same for all n. However, concrete values show a significant dependence on the value of n. For a fixed Q, as n increases there is a significant increase in the run time. This is because the constants in the norm bound depend on n and increase quite rapidly as n increases.
4. For exTNFS without constants, the run time does not vary much as n increases. This behaviour is not observed for SexTNFS.

Remark 2. For SexTNFS, the upper bounds on the norms are

$$N(f,\phi) \leq \mathfrak{C}(\eta, \kappa, H) \cdot E^{\kappa} \cdot p^{\eta/\lambda} \quad \text{and} \quad N(g,\phi) \leq \mathfrak{C}(\eta, \kappa\lambda, H) \cdot E^{\kappa\lambda} \cdot \|\Gamma\|_{\infty}^{n}. \quad (10)$$

These values should be compared with the norm bounds for exTNFS-Conj given by (7). The values of λ are 4, 6, 8 and 10 respectively for BN, BLS12, KSS and BLS24 curves. In the asymptotic complexity analysis, λ is treated as a constant and does not have a noticeable influence on the run time. On the other hand, in the concrete analysis, the actual value of λ has a noticeable effect on the upper bound on $N(g,\phi)$. This effect is present even if the constants $\mathfrak{C}(\eta, \kappa, H)$ and $\mathfrak{C}(\eta, \kappa\lambda, H)$ are ignored. Since a higher value of the norm bound implies a lower smoothness probability and hence a higher overall run time, the concrete run time for SexTNFS turns out to be greater than that of exTNFS.

Remark 3. Consider an asymmetric pairing derived from an ordinary elliptic curve over $\mathbb{F}_p$ with embedding degree n, whereby the target group is $\mathbb{G}_T = \mathbb{F}_Q^*$ with $Q = p^n$. For a fixed Q, as n increases, p decreases. Since the elliptic curve group is of size roughly p, the size of the elliptic curve group also decreases. Considering the 128-bit security level, the size of p going below 256 will violate Pollard-rho security. Hence, 128-bit security cannot be achieved by keeping Q at the 3000-bit level and simply increasing n beyond 12.

6.1   On the Tightness of the Norm Bounds

The norms $N(f,\phi)$ and $N(g,\phi)$ are expressed in terms of resultants. Upper bounds on these norms are known bounds on resultants [9] and are given by (7); let U_f and U_g be the upper bounds on $N(f,\phi)$ and $N(g,\phi)$. Let $V_f = E^{2\kappa} \cdot O\left((\log p)^\eta\right)$ and $V_g = E^{\kappa} \cdot O\left(Q^{1/(2\kappa)}\right)$. Note that V_f and V_g are not necessarily upper bounds on $N(f,\phi)$ and $N(g,\phi)$ since the constants $\mathfrak{C}(\eta, 2\kappa, H)$ and $\mathfrak{C}(\eta, \kappa, H)$ are absent.

Table 3. Upper bound and average values of norms for the 128-bit security level.

Method	n	η	κ	$\lg p$	$\lg B$	$\lg V_f$	$\lg \overline{N}_f$	$\lg U_f$	$\lg V_g$	$\lg \overline{N}_g$	$\lg U_g$	$-\lg \nu$	$-\lg \overline{\pi}$	$-\lg \mu$
exTNFS-gConj	12	4	3	384	70	452	458	603	978	992	1053	70	70	85
exTNFS-gConj	18	6	3	256	70	464	495	731	978	996	1113	71	72	98
exTNFS-gConj	24	8	3	256	81	545	617	937	1267	1302	1467	80	84	116
SexTNFS	12	6	2	384	64	543	717	902	711	655	806	64	70	99

Let $\mu = \varPi(U_f, B) \cdot \varPi(U_g, B)$ and $\nu = \varPi(V_f, B) \cdot \varPi(V_g, B)$ where B is the factor base size. Then μ is a lower bound on the probability of obtaining a single relation and μ^{-1} is an upper bound on the number of iterations required to obtain a single relation. The quantity ν is similar to that of μ except that V_f and V_g are used instead of U_f and U_g.

In [9], an example is provided to show that the resultant bounds are tight and in general cannot be improved. On the other hand, the question of whether the bounds are tight for the kinds of polynomials arising in the context of NFS algorithms deserves an answer. To determine this, we conducted some experiments. The generalized Conjugate method (gConj) [25] was implemented and the polynomials $h(z)$, $f(x)$ and $g(x)$ computed. This determines $H = \|h\|_\infty$, $\|f\|_\infty$ and $\|g\|_\infty$. Now choose a value for B and set $E = B$ so that the number of sieving polynomials $\phi(x)$ is B^2. We further set $t = 2$, i.e., only linear sieving polynomials were considered. Then the coefficients of a sieving polynomial $\phi(x)$ (considered as a bivariate polynomial in z and x) can take $B^{1/\eta}$ values. We chose 1000 random sieving polynomials and in each case computed the actual values of $N(f, \phi)$ and $N(g, \phi)$. From these two values, the smoothness probability $\pi = \varPi(N(f, \phi), B) \cdot \varPi(N(g, \phi), B)$ was computed. Let $\overline{N}_f$, $\overline{N}_g$ and $\overline{\pi}$ be the average of $N(f, \phi)$, $N(g, \phi)$ and π computed over the 1000 random ϕ's. A summary of these values is given in Table 3. The table also shows the results of a similar experiment conducted for the SexTNFS algorithm where BN curves were used.

In each case considered in Table 3, it turned out that taking $H = 2$ is sufficient. We have previously considered $H = 5$ and the reduced value of $H = 2$ results in slightly lower values for U_f and U_g. There are several points to note from the results in Table 3.

1. The average value $\lg \overline{N}_f$ is closer to $\lg V_f$ than to the known upper bound $\lg U_f$. Similarly for $\lg \overline{N}_g$ and $\overline{\pi}$.
2. Consider $\lg \overline{N}_f$. The average has been computed over 1000 iterations. The value of $\lg \overline{N}_f$ being substantially less than $\lg U_f$ indicates that polynomials ϕ such that $N(f, \phi)$ is close to U_f are not very common. On the other hand, this does not indicate that such polynomials do not occur at all. A total of B^2 sieving polynomials ϕ have to be considered. It is possible that a non-negligible fraction of these do have norms close to the upper bound. Our experiments only indicate that the fraction is less than $1/1000$. Thus, in the absence of further experimental data, one cannot completely disregard the role of the constants in the analysis.

3. In each case, B has been chosen so that $\lg B$ is roughly equal to μ. Further, in each case it turns out that $\bar{\pi}$ is at least μ. So, even if the actual norms behave like $\overline{N}_f$ and $\overline{N}_g$, choosing $\lg p$ as given provides at least 128 bits of security.

6.2 Deriving Group Sizes from the Asymptotic Run Time Expressions

In this subsection, we consider the question of deriving concrete group sizes from the asymptotic run time expressions. Following the ECRYPT recommendation [41, p. 26], consider a constant A and write the run time of an NFS algorithm as

$$A \cdot \exp\left((c + o(1))(\log Q)^{1/3}(\log\log Q)^{2/3}\right). \tag{11}$$

Again following [41], assume $o(1) = 0$. In [41], the constant A is determined in the following manner. It is mentioned that experience from available data points suggests that the resistance of RSA-512 is about 4 to 6 bits lower than that of DES. Plugging in $Q = 2^{512}$ and $c = (64/9)^{1/3}$ into (11) and setting the resulting expression equal to 2^{50}, one obtains $A \approx 2^{-14}$.

More generally, let $A = 2^{-d}$ and denote by $s(Q, c, d)$ the base-two logarithm of the expression in (11) with $o(1) = 0$. Then, we have

$$s(Q, c, d) = c(\lg e)(\log Q)^{1/3}(\log\log Q)^{2/3} - d. \tag{12}$$

Here Q is p^n and c is the second argument in the L-notation. As described above, the ECRYPT recommendation takes $d = 14$.

The task of deriving group sizes is the following. Given c, d, n and a target security level ℓ, find the minimum Q such that both $(\lg Q)/n \geq 2\ell$ and $s(Q, c, d) \geq \ell$ hold. The first condition ensures Pollard-rho security while the second condition ensures security against (exT)NFS attack. In the case where $\rho > 1$ (see Table 1), the Pollard-rho condition $(\lg Q)/n \geq 2\ell$ should be replaced with the condition $(\lg Q)/n \geq 2\ell\rho$.

Taking $c = (64/9)^{1/3}$ and $d = 14$, yields $\lg Q$ values of 3247, 7958, 15447 for $\ell = 128, 192, 256$, respectively. These values of $\lg Q$ are close to the ECRYPT recommendations of 3248, 7936, 15424. We note that taking $c = (64/9)^{1/3}$ and $d = 10$ yields $\lg Q$ values of 3034, 7587, 14889 for $\ell = 128, 192, 256$, respectively. Rounding up these $\lg Q$ values to the nearest integer multiple of 512 yields 3072, 7680, 15360, which are the NIST recommendations for prime-order fields at the 128, 192, 256-bit security levels [6].

Remark 4 (On the choice of d). As we have described earlier, the $L_Q(1/3, c)$ run time expression for exTNFS is obtained from the bounds $\mathfrak{C}(\eta, 2\kappa, H) \cdot E^{2\kappa} \cdot O((\log p)^\eta)$ and $\mathfrak{C}(\eta, \kappa, H) \cdot E^\kappa \cdot O(Q^{1/(2\kappa)})$ respectively on the norms $N(f, \phi)$ and $N(g, \phi)$. The asymptotic analysis considers $\mathfrak{C}(\eta, 2\kappa, H)$ and $\mathfrak{C}(\eta, \kappa, H)$ to be $L_Q(2/3, o(1))$ and ultimately the effect of these constants get absorbed in the $o(1)$ term in (11). At a later point, when we set $o(1)$ to be 0, we are in effect replacing the constants by 1. The values in Table 3 show that replacing the constants by 1 actually results in underestimates of the run time compared to what would be obtained from the actual values of the norms.

Choosing a positive value of d amounts to considering the actual run time to be lower than the run time predicted by values obtained from the asymptotic expression $L_Q(1/3, c)$ (with $o(1)$ assumed to be 0). Since the values obtained from the asymptotic run time expression are already lower than what would be obtained from the actual value of the norms, reducing these values further by choosing a value of d greater than 0 seems to be over-engineering. So, we suggest that the value of d be taken as 0 which would mean choosing $A = 1$ in (11).

6.3   The 128-Bit and 192-Bit Security Levels

In this section, we provide estimates of group sizes required to achieve a desired security level. These estimates depend on the values of the norms $N(f, \phi)$ and $N(g, \phi)$. One can work with the upper bounds on these norms. The upper bounds involve the constant terms which can be quite large. The experiments reported in Sect. 6.1 show that the actual values of the norms appear to be closer to the expressions for the upper bounds without the constants. So, we report group size estimates both with and without the constants in the norm bounds. The estimates obtained without considering the constants can be considered to be conservative estimates. The actual methodology for obtaining the estimates is described below.

For each choice of security level $\ell \in \{128, 192\}$, the value of n, the choice of curve, the choice of the algorithm (exTNFS or SexTNFS), and the choice of whether or not to use constants in the bounds, the following was done. For each possible non-trivial factor η of n, let $\lg p(\eta)$ denote the minimum value of $\lg p$ required to achieve security level ℓ. The maximum of $\lg p(\eta)$ over all possible non-trivial factors η of n is reported.

The values $\lg p(\eta)$ were determined as follows. The initial value of $\lg p$ was taken to be $2\rho\ell$ and the size B of the factor base was fixed to $2^{\ell/2}$. The joint smoothness probability (9) was computed and the value of $\lg p$ was incremented until for the first time the joint smoothness probability became lower than $2^{-\ell/2}$. This value was returned as $\lg p(\eta)$.

Once $\lg p$ was calculated, the complexity of each stage was determined as follows. With all other parameters fixed, the value of B was incremented until the smoothness probability became approximately equal to $1/B$. This balances the costs of the relation collection and the linear algebra stages.

The final results are given in Tables 4 and 5. Note that all estimates were generated using $H = 2$.

A reasonable conclusion is that BN curves with $\lg p = 383$ and $\lg Q = 4596$ offer (at least) 128 bits of security. With these parameters, there is a mismatch in security levels with BN curves ($\lambda = 4$) for $\mathbb{G}_1$ and $\mathbb{G}_T$—the former offers 191 bits of security, whereas the latter offers 128 bits. On the other hand, BLS12

Table 4. Approximate run times of exTNFS and SexTNFS for values of Q and n that achieve the 128-bit security level. The constants are $\mathfrak{C}(\eta, \kappa, H)$ and $\mathfrak{C}(\eta, 2\kappa, H)$ for exTNFS, and $\mathfrak{C}(\eta, \kappa, H)$ and $\mathfrak{C}(\eta, \kappa\lambda, H)$ for SexTNFS.

BN curves: $n = 12$, $\rho = 1$, $\lambda = 4$, $\|\Gamma\|_\infty = 36$						
Algorithm	Constants	η	κ	$\lg p$	$\lg Q$	$\lg(\text{run time})$
exTNFS	without	4	3	311	3732	128
exTNFS	with	4	3	256	3072	136
SexTNFS	without	6	2	383	4596	128
SexTNFS	with	6	2	256	3072	150

BLS12 curves: $n = 12$, $\rho \approx 1.5$, $\lambda = 6$, $\|\Gamma\|_\infty \approx 1$						
Algorithm	Constants	η	κ	$\lg p$	$\lg Q$	$\lg(\text{run time})$
exTNFS	without	4	3	384	4608	140
exTNFS	with	4	3	384	4608	156
SexTNFS	without	6	2	384	4608	132
SexTNFS	with	6	2	384	4608	189

KSS curves: $n = 18$, $\rho \approx 4/3$, $\lambda = 8$, $\|\Gamma\|_\infty \approx 2401/21$						
Algorithm	Constants	η	κ	$\lg p$	$\lg Q$	$\lg(\text{run time})$
exTNFS	without	6	3	342	6156	160
exTNFS	with	6	3	342	6156	184
SexTNFS	without	9	2	342	6156	170
SexTNFS	with	9	2	342	6156	274

BLS24 curves: $n = 24$, $\rho \approx 1.25$, $\lambda = 10$, $\|\Gamma\|_\infty \approx 1$						
Algorithm	Constants	η	κ	$\lg p$	$\lg Q$	$\lg(\text{run time})$
exTNFS	without	6	4	320	7680	172
exTNFS	with	6	4	320	7680	204
SexTNFS	without	12	2	320	7680	202
SexTNFS	with	12	2	320	7680	360

curves ($\lambda = 6$) with $\lg p = 384$ and $\lg Q = 4608$ do not have this mismatch—$\mathbb{G}_1$ and $\mathbb{G}_T$ both offer 128 bits of security (the former since the bitlength of r is approximately 256). Since KSS curves have $\rho \approx 4/3$ and BLS24 curves have $\rho \approx 1.25$, these curves with $\lg p = 342$ and $\lg p = 320$, respectively, offer (at least) 128 bits of security. In summary, if one is aiming for the 128-bit security level, then the bitlength of p should be at least 383, 384, 342 and 320 for BN, BLS12, KSS and BLS24 pairings.

For the 192-bit security level, the bitlength of p should be at least 1031, 1147, 597 and 480 for BLS, BLS12, KSS and BLS24 pairings. This should be contrasted with the pre-TNFS recommendations of 640, 640, 512 and 480 bits [1].

Table 5. Approximate run times of exTNFS and SexTNFS for values of Q and n that achieve the 192-bit security level. The constants are $\mathfrak{C}(\eta, \kappa, H)$ and $\mathfrak{C}(\eta, 2\kappa, H)$ for exTNFS, and $\mathfrak{C}(\eta, \kappa, H)$ and $\mathfrak{C}(\eta, \kappa\lambda, H)$ for SexTNFS.

BN curves: $n = 12$, $\rho = 1$, $\lambda = 4$, $\|\Gamma\|_\infty = 36$						
Algorithm	Constants	η	κ	$\lg p$	$\lg Q$	$\lg$(run time)
exTNFS	without	3	4	847	10164	192
exTNFS	with	3	4	728	8736	192
SexTNFS	without	6	2	1031	12372	192
SexTNFS	with	6	2	697	8364	192

BLS12 curves: $n = 12$, $\rho \approx 1.5$, $\lambda = 6$, $\|\Gamma\|_\infty \approx 1$						
Algorithm	Constants	η	κ	$\lg p$	$\lg Q$	$\lg$(run time)
exTNFS	without	3	4	847	10164	192
exTNFS	with	3	4	728	8736	192
SexTNFS	without	6	2	1147	13764	192
SexTNFS	with	6	2	576	6912	200

KSS curves: $n = 18$, $\rho \approx 4/3$, $\lambda = 8$, $\|\Gamma\|_\infty \approx 2401/21$						
Algorithm	Constants	η	κ	$\lg p$	$\lg Q$	$\lg$(run time)
exTNFS	without	3	6	512	9216	194
exTNFS	with	6	3	512	9216	214
SexTNFS	without	9	2	597	10746	192
SexTNFS	with	9	2	512	9216	281

BLS24 curves: $n = 24$, $\rho \approx 1.25$, $\lambda = 10$, $\|\Gamma\|_\infty \approx 1$						
Algorithm	Constants	η	κ	$\lg p$	$\lg Q$	$\lg$(run time)
exTNFS	without	6	4	480	11520	203
exTNFS	with	6	4	480	11520	231
SexTNFS	without	12	2	480	11520	214
SexTNFS	with	12	2	480	11520	366

Remark 5. We have reported group size estimates for several families of curves for the 128-bit and the 192-bit security levels. The methodology for obtaining these estimates is more general. It can be applied to other curve families and also the 256-bit security level.

Remark 6. Prior to the recent developments of the TNFS algorithm, BN curves with a 256-bit p (and consequently a 3072-bit Q) was considered to provide 128-bit security. Applying our methodology to such curves, we find the runtime estimates of exTNFS are 2^{136} and 2^{118} with and without constants respectively; and the runtime estimates of SexTNFS are 2^{150} and 2^{110} with and without constants respectively. So, a conservative estimate of the security of BN curves with a 256-bit prime is 110 bits.

7 Concluding Remarks

Our examination of the run times of recently-proposed improvements to the TNFS highlights their asymptotic nature. Much work remains to be done before the impact of these new algorithms on concrete keylengths for pairing-based cryptography can be determined with full confidence. Before this concrete analysis is completed, a conservative choice for BN pairings would be to increase the bitlength of p from 256 to 383 if one is aiming for the 128-bit security level. For BLS12, KSS and BLS24 pairings, there is no change in the pre-TNFS recommendations to use primes p of bitlength 384, 342 and 320, respectively, at the 128-bit security level. At the 192-bit security level, conservative choices for the bitlength p are 1031, 1147, 597 and 480 for BN, BLS12, KSS and BL24 pairings, respectively.

Acknowledgements. We thank the referees for their comments which helped improve the presentation of the paper.

A Calculations of Bounds on Resultants

Consider the setting of the TNFS with $Q = p^n$, $n = \eta\kappa$, h a degree-η irreducible polynomial in $\mathbb{Z}[z]$, $R = \mathbb{Z}[z]/(h(z))$, and $f, \phi \in R[x]$. Note that $\deg_z f = \deg_z \phi = \eta - 1$.

Let $\mathfrak{f}(z, x)$ be a bivariate polynomial with integer coefficients where $\mathfrak{f}_{i,j}$ is the coefficient of $x^i z^j$. Then $\|\mathfrak{f}\|_\infty = \max |\mathfrak{f}_{i,j}|$. Bounds on resultants of univariate and bivariate polynomials have been given in [9]. We summarize these below.

Let $a(u)$ and $b(u)$ be polynomials with integer coefficients. From [9], we have

$$|\mathrm{Res}_u(a(u), b(u))|$$
$$\leq (\deg(a) + 1)^{\deg(b)/2} \cdot (\deg(b) + 1)^{\deg(a)/2} \cdot \|a\|_\infty^{\deg(b)} \cdot \|b\|_\infty^{\deg(a)}. \tag{13}$$

Let $a(u, v)$ and $b(u, v)$ be polynomials with integer coefficients. Let $c(u) = \mathrm{Res}_v(a(u, v), b(u, v))$. Then

$$\|c\|_\infty \leq (\deg_v(a) + \deg_v(b))!$$
$$\cdot (\max(\deg_u(a), \deg_u(b)) + 1)^{\deg_v a + \deg_v b - 1} \cdot \|a\|_\infty^{\deg_v b} \cdot \|b\|_\infty^{\deg_v a}. \tag{14}$$

Bounds on $\mathrm{Res}_z(\mathrm{Res}_x(\phi(x), \mathfrak{f}(x)), h(z))$ can be derived by combining the bounds given by (13) and (14). Let $\mathfrak{c}(z) = \mathrm{Res}_x(\phi(x), \mathfrak{f}(x))$. The degree of $\mathfrak{c}(z)$ is given in [9] and from (14) we obtain $\|\mathfrak{c}\|_\infty$. These quantities are as follows:

$$\deg \mathfrak{c}(z) = (\deg_x \phi + \deg_x f) \cdot \max(\deg_z \phi + \deg_z f) = (\eta - 1)(\deg_x \phi + \deg_x f),$$
$$\|\mathfrak{c}\|_\infty \leq (\deg_x \phi + \deg_x \mathfrak{f})! \cdot (\max(\deg_z \phi, \deg_z \mathfrak{f}) + 1)^{\deg_x \phi + \deg_x \mathfrak{f} - 1}$$
$$\cdot \|\phi\|_\infty^{\deg_x \mathfrak{f}} \cdot \|\mathfrak{f}\|_\infty^{\deg_x \phi}$$
$$= (\deg_x \phi + \deg_x \mathfrak{f})! \cdot \eta^{\deg_x \phi + \deg_x \mathfrak{f} - 1} \cdot \|\phi\|_\infty^{\deg_x \mathfrak{f}} \cdot \|\mathfrak{f}\|_\infty^{\deg_x \phi}.$$

Using these values we obtain

$$\begin{aligned}
|\mathrm{Res}_z(\mathrm{Res}_x(\phi(x), \mathfrak{f}(x)), h(z))| &= |\mathrm{Res}_z(\mathfrak{c}(z), h(z))| \\
&\leq ((\eta - 1)(\deg_x \phi + \deg_x \mathfrak{f}) + 1)^{\eta/2} \cdot (\eta + 1)^{(\eta-1)(\deg_x \phi + \deg_x \mathfrak{f})/2} \quad (15) \\
&\quad \cdot \|\mathfrak{c}\|_\infty^\eta \cdot \|h\|_\infty^{(\eta-1)(\deg_x \phi + \deg_x \mathfrak{f})} \\
&\leq ((\eta - 1)(\deg_x \phi + \deg_x \mathfrak{f}) + 1)^{\eta/2} \cdot (\eta + 1)^{(\eta-1)(\deg_x \phi + \deg_x \mathfrak{f})/2} \quad (16) \\
&\quad \cdot \|h\|_\infty^{(\eta-1)(\deg_x \phi + \deg_x \mathfrak{f})} \cdot \left((\deg_x \phi + \deg_x \mathfrak{f})! \cdot \eta^{\deg_x \phi + \deg_x \mathfrak{f}-1} \right)^\eta \\
&\quad \cdot \|\phi\|_\infty^{\eta \deg_x \mathfrak{f}} \cdot \times \|\mathfrak{f}\|_\infty^{\eta \deg_x \phi}.
\end{aligned}$$

References

1. Aranha, D.F., Fuentes-Castañeda, L., Knapp, E., Menezes, A., Rodríguez-Henríquez, F.: Implementing pairings at the 192-bit security level. In: Abdalla, M., Lange, T. (eds.) Pairing 2012. LNCS, vol. 7708, pp. 177–195. Springer, Heidelberg (2013). doi:10.1007/978-3-642-36334-4_11
2. Barbulescu, R., Gaudry, P., Guillevic, A., Morain, F.: Improving NFS for the discrete logarithm problem in non-prime finite fields. In: Oswald, E., Fischlin, M. (eds.) EUROCRYPT 2015. LNCS, vol. 9056, pp. 129–155. Springer, Heidelberg (2015). doi:10.1007/978-3-662-46800-5_6
3. Barbulescu, R., Gaudry, P., Joux, A., Thomé, E.: A heuristic quasi-polynomial algorithm for discrete logarithm in finite fields of small characteristic. In: Nguyen, P.Q., Oswald, E. (eds.) EUROCRYPT 2014. LNCS, vol. 8441, pp. 1–16. Springer, Heidelberg (2014). doi:10.1007/978-3-642-55220-5_1
4. Barbulescu, R., Gaudry, P., Kleinjung, T.: The tower number field sieve. In: Iwata, T., Cheon, J.H. (eds.) ASIACRYPT 2015. LNCS, vol. 9453, pp. 31–55. Springer, Heidelberg (2015). doi:10.1007/978-3-662-48800-3_2
5. Barbulescu, R., Pierrot, C.: The multiple number field sieve for medium and high characteristic finite fields. LMS J. Comput. Math. 17, 230–246 (2014)
6. Barker, E.: Recommendation for key management, Part 1: General. NIST Special Publication 800-57, Part 1, Revision 4, January 2016
7. Barreto, P.S.L.M., Lynn, B., Scott, M.: Constructing elliptic curves with prescribed embedding degrees. In: Cimato, S., Persiano, G., Galdi, C. (eds.) SCN 2002. LNCS, vol. 2576, pp. 257–267. Springer, Heidelberg (2003). doi:10.1007/3-540-36413-7_19
8. Barreto, P.S.L.M., Naehrig, M.: Pairing-friendly elliptic curves of prime order. In: Preneel, B., Tavares, S. (eds.) SAC 2005. LNCS, vol. 3897, pp. 319–331. Springer, Heidelberg (2006). doi:10.1007/11693383_22
9. Bistritz, Y., Lifshitz, A.: Bounds for resultants of univariate and bivariate polynomials. Linear Algebra Appl. 432, 1995–2005 (2010)
10. Boneh, D., Boyen, X.: Strong signatures without random oracles and the SDH assumption in bilinear groups. J. Cryptol. 21, 149–177 (2008)
11. Boneh, D., Franklin, M.: Identity-based encryption from the weil pairing. In: Kilian, J. (ed.) CRYPTO 2001. LNCS, vol. 2139, pp. 213–229. Springer, Heidelberg (2001). doi:10.1007/3-540-44647-8_13
12. Boneh, D., Lynn, B., Shacham, H.: Short signatures from the Weil pairing. J. Cryptol. 17, 297–319 (2004)
13. Boneh, D., Shacham, H.: Group signatures with verifier-local revocation. In: 11th ACM Conference on Computer and Communications Security - CCS 2004, pp. 168–177 (2004)

14. Chatterjee, S., Menezes, A.: On cryptographic protocols employing asymmetric pairings - the role of ψ revisited. Discrete Appl. Math. **159**, 1311–1322 (2011)
15. Chatterjee, S., Menezes, A.: Type 2 structure-preserving signature schemes revisited. In: Iwata, T., Cheon, J.H. (eds.) ASIACRYPT 2015. LNCS, vol. 9452, pp. 286–310. Springer, Heidelberg (2015). doi:10.1007/978-3-662-48797-6_13
16. Cheon, J.H.: Security analysis of the strong Diffie-Hellman problem. In: Vaudenay, S. (ed.) EUROCRYPT 2006. LNCS, vol. 4004, pp. 1–11. Springer, Heidelberg (2006). doi:10.1007/11761679_1
17. Coron, J.-S., Lepoint, T., Tibouchi, M.: Practical multilinear maps over the integers. In: Canetti, R., Garay, J.A. (eds.) CRYPTO 2013. LNCS, vol. 8042, pp. 476–493. Springer, Heidelberg (2013). doi:10.1007/978-3-642-40041-4_26
18. Diem, C.: On the discrete logarithm problem in elliptic curves. Compositio Math. **147**, 75–104 (2011)
19. Diem, C.: On the discrete logarithm problem in elliptic curves II. Algebra Number Theory **7**, 1281–1323 (2013)
20. Garg, S., Gentry, C., Halevi, S., Raykova, M., Sahai, S., Waters, B.: Candidate indistinguishability obfuscation and functional encryption for all circuits. In: IEEE 54th Annual Symposium on Foundations of Computer Science (FOCS), pp. 40–49 (2013)
21. Gaudry, P., Hess, F., Smart, N.: Constructive and destructive facets of Weil descent on elliptic curves. J. Cryptol. **15**, 19–34 (2002)
22. Gordon, D.: Discrete logarithms in $GF(p)$ using the number field sieve. SIAM J. Discrete Math. **6**, 124–138 (1993)
23. Guillevic, A.: Computing individual discrete logarithms faster in $GF(p^n)$ with the NFS-DL algorithm. In: Iwata, T., Cheon, J.H. (eds.) ASIACRYPT 2015. LNCS, vol. 9452, pp. 149–173. Springer, Heidelberg (2015). doi:10.1007/978-3-662-48797-6_7
24. Jao, D., Yoshida, K.: Boneh-Boyen signatures and the strong Diffie-Hellman problem. In: Shacham, H., Waters, B. (eds.) Pairing 2009. LNCS, vol. 5671, pp. 1–16. Springer, Heidelberg (2009). doi:10.1007/978-3-642-03298-1_1
25. Jeong, J., Kim, T.: Extended tower number field sieve with application to finite fields of arbitrary composite extension degree. Cryptology ePrint Archive: Report 2016/526 (2016)
26. Joux, A.: Algorithmic Cryptanalysis. Chapman & Hall/CRC, Boca Raton (2009)
27. Joux, A., Lercier, R.: Improvements to the general number field sieve for discrete logarithms in prime fields. A comparison with the Gaussian integer method. Math. Comput. **72**, 953–967 (2003)
28. Joux, A., Lercier, R., Smart, N., Vercauteren, F.: The number field sieve in the medium prime case. In: Dwork, C. (ed.) CRYPTO 2006. LNCS, vol. 4117, pp. 326–344. Springer, Heidelberg (2006). doi:10.1007/11818175_19
29. Joux, A., Pierrot, C.: The special number field sieve in $\mathbb{F}_{p^n}$ – application to pairing-friendly construction. In: Cao, Z., Zhang, F. (eds.) Pairing 2013. LNCS, vol. 8365, pp. 45–61. Springer, Cham (2014). doi:10.1007/978-3-319-04873-4_3
30. Kachisa, E.J., Schaefer, E.F., Scott, M.: Constructing Brezing-Weng pairing-friendly elliptic curves using elements in the cyclotomic field. In: Galbraith, S.D., Paterson, K.G. (eds.) Pairing 2008. LNCS, vol. 5209, pp. 126–135. Springer, Heidelberg (2008). doi:10.1007/978-3-540-85538-5_9
31. Kim, T., Barbulescu, R.: Extended tower number field sieve: a new complexity for the medium prime case. In: Robshaw, M., Katz, J. (eds.) CRYPTO 2016. LNCS, vol. 9814, pp. 543–571. Springer, Heidelberg (2016). doi:10.1007/978-3-662-53018-4_20

32. Koblitz, N., Menezes, A.: The brave new world of bodacious assumptions in cryptography. Not. AMS **57**, 357–365 (2010)
33. Lenstra, A.K., Lenstra, H.W., Manasse, M.S., Pollard, J.M.: The number field sieve. In: Lenstra, A.K., Lenstra, H.W. (eds.) The Development of the Number Field Sieve. LNM, vol. 1554, pp. 11–42. Springer, Heidelberg (1993). doi:10.1007/BFb0091537
34. Mayo, K.: A primer on cryptographic multilinear maps and code obfuscation. M.Math. thesis, University of Waterloo (2015). http://hdl.handle.net/10012/9698
35. Pierrot, C.: The multiple number field sieve with conjugation and generalized Joux-Lercier methods. In: Oswald, E., Fischlin, M. (eds.) EUROCRYPT 2015. LNCS, vol. 9056, pp. 156–170. Springer, Heidelberg (2015). doi:10.1007/978-3-662-46800-5_7
36. Pollard, J.: Monte Carlo methods for index computation mod p. Math. Comput. **32**, 918–924 (1978)
37. Sarkar, P., Singh, S.: New complexity trade-offs for the (multiple) number field sieve algorithm in non-prime fields. In: Fischlin, M., Coron, J.-S. (eds.) EUROCRYPT 2016. LNCS, vol. 9665, pp. 429–458. Springer, Heidelberg (2016). doi:10.1007/978-3-662-49890-3_17
38. Sarkar, P., Singh, S.: A general polynomial selection method and new asymptotic complexities for the tower number field sieve algorithm. In: Cheon, J.H., Takagi, T. (eds.) ASIACRYPT 2016. LNCS, vol. 10031, pp. 37–62. Springer, Heidelberg (2016). doi:10.1007/978-3-662-53887-6_2
39. Sarkar, P., Singh, S.: A generalisation of the conjugation method for polynomial selection for the extended tower number field sieve algorithm. IACR Cryptology ePrint Archive: Report 2016/537 (2016)
40. Schirokauer, O.: Using number fields to compute logarithms in finite fields. Math. Comput. **69**, 1267–1283 (2000)
41. Smart, N. (ed.): ECRYPT II Yearly Report on Algorithms and Keysizes (2011–2012), 30 September 2012

Different Paradigms

Different Part figure

Key Recovery: Inert and Public

Colin Boyd[1], Xavier Boyen[2], Christopher Carr[1,2(✉)], and Thomas Haines[2]

[1] Norwegian University of Science and Technology, NTNU, Trondheim, Norway
ccarr@ntnu.no
[2] Queensland University of Technology, QUT, Brisbane, Australia

Abstract. We propose a public key infrastructure framework, inspired by modern distributed cryptocurrencies, that allows for tunable key escrow, where the availability of key escrow is only provided under strict conditions and enforced through cryptographic measures. We argue that any key escrow scheme designed for the global scale must be both *inert*—requiring considerable effort to recover a key—and *public*—everybody should be aware of all key recovery attempts. To this end, one of the contributions of this work is an abstract design of a proof-of-work scheme that demonstrates the ability to recover a private key for some generic public key scheme. Our framework represents a new direction for key escrow, seeking an acceptable compromise between the demands for control of cryptography on the Internet and the fundamental rights of privacy, which we seek to align by drawing parallels to the physical world.

1 Introduction

Key escrow was a popular research topic, and subject of contention, during the 1990s [2]—the era of the so-called crypto wars [12]. Recently, the crypto wars seem to have resumed, principally due to the "Snowden revelations" of global mass surveillance. Whilst security advocates and technical experts have spoken out against demands from government agencies to weaken, or even prevent, the use of cryptography, governments make an opposing case, demanding greater powers of surveillance in pursuit of terrorists and organised crime.

PROS AND CONS OF KEY ESCROW. Allowing a third party to hold unlimited escrow over cryptographic keys comes with serious security concerns. Indeed, even if the party holding the escrow is trustworthy, confidence that this party is completely secure against forms of malicious compromise, such as subversion or hacking may not be achievable, rendering the cryptographic keys vulnerable. Consequently, accepting key escrow amounts to placing broad and considerable trust in both the character and abilities of the party holding escrow.

Concerns surrounding the level of trust are not unfounded, as shown by recent examples of security breaches, directly compromising keys used by members of the public. For example, in 2015, the SIM card manufacturer Gemalto reportedly had the keys to millions of SIM cards compromised in a breach conducted by GCHQ and supported by the NSA [11].

© Springer International Publishing AG 2017
R.C.-W. Phan and M. Yung (Eds.): LNCS 10311, Mycrypt 2016, pp. 111–126, 2017.
DOI: 10.1007/978-3-319-61273-7_6

In a widely publicised study in 2015 [1], a group of 15 eminent cryptographers and technologists re-examined governmental access to data and communications, comparing the situation now with that in the 1990s. They concluded that law enforcement access "will open doors through which criminals and malicious nation-states can attack the very individuals law enforcement seeks to defend". They further argued that it would require unreasonable costs and lead to economic loss.

Arguably, governments have had far too much power over, and knowledge of, the communications of its average citizens. This has been recognised previously. In 2013, the United States President's Review Group on Intelligence and Communications Technologies [15] recommended that "... the US Government should:

1. Fully support and not undermine efforts to create encryption standards;
2. Not in any way subvert, undermine, weaken, or make vulnerable generally available commercial software; and
3. Increase the use of encryption and urge US companies to do so, in order to better protect data in transit, at rest, in the cloud, and in other storage."

On the other hand, there is an opinion that reasonable controls on cryptography are desirable. The oft presented motivation is the need to reveal the communications of those involved with, or suspected to be involved in, major criminal or terrorist activities. The same President's Review Group on Intelligence and Communications Technologies [15] also stated (Recommendation 20): "... the US Government should examine the feasibility of creating software that would allow the National Security Agency and other intelligence agencies more easily to conduct targeted information acquisition rather than bulk-data collection."

Bart Preneel in his 2016 IACR Distinguished Lecture, *The Future of Cryptography* [16], asked: What about the balance? He points out that privacy has both individual and collective dimensions. While intelligence agencies have arguably gone too far in their usage of technology, they may lose out in some scenarios. Preneel challenges the community to design better solutions.

COMPROMISE SOLUTIONS. During the first crypto-wars in the 1990s, cryptographers contributed a number of compromise key escrow solutions, aimed at balancing the power of law-enforcement agencies with various constraints. One such example was *partial key escrow* [4,5] which allows escrow agents to recover keys with moderate effort but makes the computational requirements of mass surveillance prohibitively costly. Another example was oblivious key escrow [6] in which users share their keys among a large number of unknown escrow agents who must cooperate in order to recover a key. Our proposal takes inspiration from both of these ideas.

During the current crypto-wars there has been little interest from the cryptographic community to explore compromise solutions. One exception is the cMix proposal of Chaum et al. [7], part of a practical project known as Privategrity.

Although the proposed system efficiently provides high level security and privacy properties, it relies on users sharing long-term keys with a fixed set of servers so that compromise of all servers reveals all user information. Such a level of trust, even if extensively distributed, has received widespread criticism. Note that compromise of all trusted nodes, and subsequent compromise of user secrets, if it should happen, can be undetectable. Importantly, we have already seen a willingness for international cooperation to record communications, and so it is not immediately obvious that this approach would achieve the desired results.

Our aim is to explore some ideas for better compromise solutions to key escrow. We are motivated by the observation that extraordinary access to personal property in the physical world does not receive the same debate and controversy as extraordinary access in the virtual world. By examining the differences between the virtual and physical world, we are led to propose some principles for key escrow which we believe can defuse much of the controversy behind earlier key recovery proposals. Furthermore, we observe that modern techniques, specifically based on ideas from decentralised cryptocurrencies, allow these principles to be realised in practice.

CONTRIBUTION. The main aim of this work is to focus the attention of the community in an area that is politically difficult, with an aim of offering responsible solutions to these problems. The technical contributions can be summarised as follows.

- New principles for the design of large scale escrow systems are proposed.
- Possible methods for implementing these principles are described.
- A generic method for creating a proof-of-work system based on a public key scheme is proposed.

2   Background

This section focuses on the building blocks of our proposal. The first two of these originate from the original crypto wars: oblivious key escrow and partial key escrow. The third element we introduce is a decentralised consensus mechanism conceived with modern cryptocurrencies.

2.1   Oblivious Key Escrow

The first of our three proposals described later has its roots in the notion of *oblivious key escrow* [6]. Blaze suggests a form of escrow in which keys are widely distributed amongst a large pool of escrow servers numbering thousands or millions. Key recovery can be performed by a threshold of escrow servers (perhaps numbering hundreds or thousands) following some pre-agreed policy. The key-sharing is oblivious in that the key owner does not know which subset of possible share holders were selected.

The system also allows for extraordinary access following a process which Blaze calls *angry mob cryptanalysis*: if enough servers of the system agree, they can ignore the policy and recover the secret key for other member of the scheme. The idea is that such a mob would only choose to do so if there was some mass consensus that key recovery is justified. This brings to mind the recent case where Apple and the FBI were in conflict over access to data in a convicted terrorist's iPhone [10]. With angry mob cryptanalysis, the decision in such cases could be made by wide consensus.

The principle behind Blaze's idea is that a party's secret key can be recovered when a large enough proportion of people demand it. The argument for this kind of key escrow is that in order to apply it, a vast number of people must be committed to finding a secret key that corresponds to a published public key. To get such a large body of people on board, it would be necessary to perform this process in the public eye. The argument is that even powerful agencies would have to announce their intentions in order to recover a key in reasonable time. The advantage here is removing the ability of a rogue state to simply recover the key through: ulterior means or coercion; via a legal process; or through simply amassing the power to do so. Crucial to this design is that the parameters are chosen so that coercion of such a large population of users is infeasible.

2.2   Partial Key Escrow

Shamir seems to have been the first to suggest the idea to escrow only part of the private key.[1] The principle is that law enforcement agencies with considerable computational power will be able to obtain some targeted keys, but will not have the resources available to perform mass surveillance, provided enough entropy remains in the unescrowed part of the key.

Using *weaker than usual* keys is another proposal that has been suggested for key escrow, with the argument that while there is less security on an individual level, by selecting appropriate parameters, it is infeasible to extract the keys for all.

However, partial and *weaker than usual* key escrow has seen a lot of criticism, with arguments against this approach reasonably stating arguments such as the following:

- One cannot make accurate assumptions on the computational resources of powerful agencies who may wish to undermine a key escrow system using weaker keys such as the NSA/FSB/MSS/GCHQ/ASD;
- Breakthroughs in cryptanalysis are unpredictable;
- Cryptanalysis techniques may not be made public, and there is little incentive to do so, as greater reward can be garnered through the private sale of cryptanalytic breakthroughs; and
- By the power of Moore's law, over time there will be a considerable reduction in the cost of recovering a key.

---

[1] Unpublished, but widely attributed [4,8].

2.3   Ensuring a Distributed User Base

One of the main achievements of Bitcoin, is enforcement of the principle first described in the original Bitcoin proposal [13], namely *one CPU, one vote*. Traditionally, peer-to-peer systems could be vulnerable to what are popularly referred to as Sybil vulnerabilities [9]. Sybil vulnerabilities often arise when a single party can easily and quickly create multiple pseudonymous identities. Malicious parties can perform a Sybil style attack in a threshold key escrow scheme by creating multiple pseudonymous identities—in order that they stand a greater chance of receiving more key shares.

Employing the one CPU, one vote principle attempts to solve the Sybil problem. In this scenario, it does not matter how many pseudonyms any participant in the scheme can create, what matters instead is how much work they can produce. Thus, any party wishing to pose as two entities, must provide the work of two entities, and so on linearly in the number of users one wishes to pose as. Inevitably, while one individual is able to pretend to be multiple entities, they cannot trivially increase their computational power beyond some reasonable margin.

In a distributed environment where there are multiple machines, all attempting to outpace the others, it soon becomes hard to implement any form of Sybil attack.

3   Design Principles

We are motivated by comparison between the process of cryptographic key recovery and the process of obtaining access to physical premises. This extends an analogy that is made between allowing extraordinary access to personal encrypted communications and giving access to personal property [1]. We observe that there are at least two important differences between the physical world and the cryptographic world which are not usually considered in such analogies.

1. Obtaining access to physical premises requires significant resources. This typically includes the presence of people and the use of physical equipment, both over a significant time.
2. Instances of access are difficult to hide from public view. Often they involve forced entry with multiple actors, and a form of recorded process.

Our thesis is that these two properties are inhibitors to abuse of extraordinary access, whether committed by law enforcement agencies or by legitimate property owners. Use of significant resources makes mass abuse without a valid target impossible. Public observation and records of access instances prevents covert abuse. Our aim, therefore, is to mimic these properties in the cryptographic world so that a more acceptable compromise can be reached. This leads us neatly to two design principles.

3.1 Inert and Public

Inert. Key recovery should cost something to those seeking to recover the key. Moreover, this cost should be measurable and increase with the number of keys to be recovered.

Public. The only viable way to recover keys should be with a publicly recorded process. Every instance of a key recovery attempt should be publicly known, and the record of key recovery instances should be infeasible to alter, hide or falsify.

As we have seen in recent years, both the properties of inertia and public accountability are almost nonexistent in online communication. With unencrypted communication, the effort required does not grow with the number of keys required to compromise. For example, consider the tapping of deep sea cables. The hard work required to intercept the communication happens once, and from then on the cost of interception is tiny for all new communications using that line. In contrast we advocate a fair cost for each communication recovery.

This example also shows how the public aspect is defeated. Tapping of communications can be achieved in a covert and undetectable fashion. At it stands, our knowledge of compromises comes mostly from whistle-blowers within the system who are often legally required to remain silent, and may have to break laws with heavy penalties in order to bring the activities to public attention.

Creating a system that is both public and inert is not easily achievable in the current Internet, but there are emerging technologies that are resolving some of the constraints. In order to describe the system we propose, we next define a space where certificates for escrowed keys can be placed.

3.2 Blockchains, Decentralised Ledgers and PKI

Traditional public key infrastructure (PKI) has been shown wanting for the end user, and we are still in a situation where, for the most part, secure online communication between two parties is "off" by default. This is especially true in email, where the difficulty of both obtaining and using public keys helps to prevent its widespread use [17,18]. Linking an identity with a public key is still a difficult problem, due to the methods for distribution and ways to assert key ownership. The web of trust model, while innovative, presents too great a challenge to the user to be effective as a world wide tool. On the other hand, over-reliance on centralised architecture is itself a major issue.

The emergence of decentralised ledger systems (or Bitcoin like systems), on the other hand, seems to provide a natural and, most importantly, a practical way to achieve the design goals since such systems are decentralised and inherently satisfy the properties of inertia and public verifiability. Indeed, using blockchains to construct public ledgers in which to store credentials or certificates for use in public key cryptography has been previously considered [19], along with alternative proposals for certified credentials within the Bitcoin system itself [3].

Our direction complements this work, considering the further requirement for key recovery under certain circumstances, whilst still maintaining strong levels of security for the majority of the system.

We have identified the following four properties as the main challenges in achieving a practical systems satisfying our goals.

Strong Keys. Generating keys that are currently strong does not prevent them from becoming weak in the (near) future. One problem is to create an escrow system that can hold up for an extended period.

Resistance to Sybil Vulnerability. Are the keys vulnerable to Sybil like attacks? In a key sharing mechanism, it is important that the vulnerability threshold to a Sybil attack is suitably measured.

Public. Attempted recovery of the key must be made public in order to be acceptable.

Inert. All mechanisms used to recover the keys require at least some amount of effort.

To satisfy the third property, we need to devise a way to allow for the users of the system to recover the secret keys for the public keys that are posted in the system. To do this, we devise a blockchain like public key infrastructure layer for users and devices, that acts in the middle ground, as well as allowing market forces to try their best. Blockchains have been proposed as a distributed public ledger before, and there are plenty of applications to choose from [13]. All we require is that the ledger is append only, and available to all members.

A central advantage of using a distributed decentralised ledger is that certificates can be uploaded to the system in a manner that resembles distributed public key infrastructure. A user asserting a key to the system will have it verified if it follows the rules of the system. The design is decentralised and traceable, so altering and faking certificates is difficult. Unlike the more centralised server storage approach, where compromising the key server gives control of the key server and allows revoking and creation of false keys, there is a much greater resistance in a proof–of–work based ledger model.

We propose two layers of infrastructure, a top layer, where authorities such as nation state, and perhaps well audited companies, remain, which we call *trusted roots*. The second layer is the key management layer, where the signed keys are appended to, and stored on, the global record.

For example, a user may want to claim their email address, sally@uni.gov along with registering their name Sally F and other supplementary information. This certificate is appended together and signed by the relevant authority that agrees that Sally is in fact the legitimate owner of the email address. By this method, there is a global consensus on trusted roots and their intermediate authorities across all platforms.

Assumption 1 (Trusted Roots). *All parties maintain an identical list of trusted root authorities, containing the root authorities, their corresponding public key, and some auxiliary information such as description and location.*

This design is similar to traditional public key infrastructure. However, this design incorporates both PKI and the assertion of certificates for individual users. These top layer authorities are axiom authorities of the system, designed to represent governmental bodies or *trusted* corporations, without which there would be no place to start. Notably, we make no judgement on whether this is ideal, however we seek to mimic the real world as closely as possible.

Assumption 2 (Certificate ledger). *The certificate ledger is append only and available to all parties. Specifically, all participants can write and read accurately to the certificate ledger. After some known period of time t, records within the ledger cannot be removed.*

4   Our Framework

We propose a framework that uses either one of, or a combination of, the oblivious key escrow method and the secret sharing of moderately weak keys. This leads to two alternative proposals and a third which combined the first two.

4.1   Proposal 1: Decentralised Oblivious Key Escrow

Our first proposal is to build a key escrow scheme using a distributed smart contract system. Specifically, the policy to enable release of keys, as described by Blaze [6], is embedded in a smart contract. Release of keys can then only occur when the policy is satisfied, as guaranteed by the integrity of the blockchain. This allows building oblivious key escrow into a transaction that can act as a credential. However, this requires an oblivious key escrow protocol to be secure in a form of white box execution. Users of the system would then be able to verify that the oblivious key escrow took place correctly and if so include the transaction in the system.

Using a Turing complete language, available in modern crypto-currencies like Ethereum [14], it is possible to programatically enforce a random choice of escrow servers. We note that, in the original oblivious key escrow paper [6], it is possible, without any risk of detection, for the sender to collaborate with the receivers to select only the receivers of their choice.

Proposal 1 is a method of oblivious escrow where key recovery is available according to some agreed and publicly checkable policy. The use of a distributed ledger enables the property of public accountability. However, this method does not require computational effort in order to effect key recovery.

4.2   Proposal 2: Partial Key Escrow

Our second proposal is to escrow parts of a key, and record them on a distributed ledger. The system therefore should act as a form of PKI, so that anyone can verify the correct association between a user and a public key.

We desire an efficient way to include a secret within a system such that the verifiers of the distributed ledger can quickly check that credentials are included.

In order to do that, we require that users select a public key of a specified length, such that it is *short enough* to recover the secret key if a considerable effort is applied for some length of time, yet *strong enough* to prevent recovery by reasonable computational resources.

Creating a good metric for the security of a public key cryptosystem of different lengths is challenging. Therefore we introduce a feedback loop mechanism between the security of the public key scheme used for key escrow and the proof–of–work system. This requires building an alternate proof-of-work system from a public key system in such a way that recovering the secret key for a given public key can be accurately quantified. The idea is to build a proof-of-work system that relies on the speed of finding secret keys to corresponding to registered public keys.

Quantifying Strength of Public Keys. This approach comes with another interesting feature. Since users are rewarded for their work on the proof-of-work system, there is an incentive for them to find the best algorithms and obtain the best hardware to recover the secret keys. This will be useful for determining the long term strength of any scheme deployed in this manner. If it is valuable to find weaknesses in a specific scheme, then weaknesses may be found more readily, and the absence of weaknesses being found indicates the security of that specific public key scheme. This creates a financial incentive to find weaknesses in keys, and increases the level of scrutiny of a public key scheme, from just a few interested parties, such as those interested in covert surveillance, and ones developing algorithms and software to sell to those agencies. It increases the scope to the general public, bringing the cryptanalysis of a public key scheme out into the open.

A solution to this problem is to create a public key based proof–of–work system. This means that a certain level of computational work has to be applied to a target credential in order to retrieve the secret key. This gives us a metric on the security of a public key scheme. Say, on average, every minute a secret key is found for a public key scheme with a certain level of security, then we can feed that back into the key generation process for the credential. If the key should be secure for a greater length of time, then this the key size should be scaled with respect to size of keys actually being recovered in the system.

4.3   Proposal 3: Decentralised Oblivious Partial Key Escrow

Proposals 1 and 2 each have certain benefits, but individually fall short of solving the full problem. However, by combining both proposals, we can eliminate the drawbacks of each approach.

Our third proposal is a system where to recover the secret for a user, both effort must be made and consensus must be reached. To recover a key, such that, a large population of users must agree to release a reduced key which in turn must be the target of significant computational resources. This guarantees both:

- *public accountability* since the oblivious key escrow is inherently public, and weak keys will not be publicly available; and

– *inertia* since all keys released are current in their security parameters and the oblivious key escrow can have security levels tuned for ongoing security.

This means that in the future the keys will not be trivially breakable unless one has previously mounted a Sybil attack on all keys. Therefore, it is necessary to mount the attack beforehand on all users in order to compromise the public property of the system. We summarise the proposals, and their advantages in Table 1.

Table 1. Comparison of main properties of the three proposals

	Partial escrow	Oblivious escrow	Prop. 1	Prop. 2	Prop. 3
Public	✗	✗	✓	✓	✓
Inert	✓	✓	✓	✓	✓
Future secure	✗	✓	✓	✗	✓
Sign up not required	✓	✗	✗	✓	✓[a]
Sybil resistant	✓	✗	✗	✓	✓
Traffic analysis resistant	✓	✗	✗	✓	✓

[a] There is a requirement for pre-registration for the oblivious part of the key escrow.

5 Methods for Implementation

Here we sketch ways to implement each of the three proposals. At present we are only in a position to outline a proof of concept. Detailed designs and experimental systems will require further work.

5.1 Implementing Proposal 1

Implementing oblivious key escrow as a smart contract requires working within the white box execution environment, where all execution is public. Doing any cryptography in this environment is hard since all keys used would be public. So, all values need to be encrypted by the relevant parties before submitting to the contract.

Implementing this first proposal can be done in solidity, the programming language for the distributed *smart contract* system Ethereum [14], using the code and protocol shown below. Note, that for simplicity, we have replaced the blind signatures and anonymous channel (as specified in the design of Blaze [6]) with the pseudo-anonymity of the block chain. It should however, be easy to reintroduce them.

```solidity
pragma solidity ^0.4.0;
contract ObliviousKeyEscrow {

    function randomGen(uint seed, uint max) constant returns (
        ↪ uint randomNumber) {
        return(uint(sha3(block.blockhash(block.number-1), seed
            ↪ ))%max);
    } //More secure randomness is preferable

    struct Receiver {
        uint publicKey; //Preferably new
        bool joined;
        bool chosen;
    }

    address sender; //The person wanting to escrow
    mapping(address => Receiver) receivers;
    mapping(uint => address) receiverNumbers;
    uint[] shares; //People who should hold shares
    uint[] encrypted_shares; //Encrypted shares of the key
    uint numReceivers;

    /// Create a new Oblivious Key Escrow
    function ObliviousKeyEscrow() {
        sender = msg.sender;
    }

    /// Join as a potential sender
    function Send(uint8 publicKey) {
        Receiver receiver = receivers[msg.sender];
        if (receiver.joined) return;
        receiverNumbers[numReceivers] = msg.sender;
        receiver.joined = true;
        receiver.publicKey = publicKey;
        numReceivers++;
    }

    /// Choose the receivers to receive key shares
    function ChooseReceivers(uint numShares) {
        if(msg.sender != sender) throw;
        if(numReceivers < numShares) return;
        uint seed = 0;
        for(uint i = 0; i < numReceivers; i++){
            seed += receivers[receiverNumbers[i]].publicKey;
        }//Calculate a seed
         for(i = 0; i < numShares; i++){
            uint randomNumber = randomGen(seed, numReceivers);
            seed += randomNumber;
            shares[i] = randomNumber;
```

```
        }
    }

    ///Send the key shares to the relevant parties
    function SendShares(uint [] tempEncryptedShares){
        encrypted_shares = tempEncryptedShares;
    }
}
```

A person choosing to escrow their key would put the contract on the block chain. They would then wait for senders to join by calling *Send*, after which they would call *ChooseReceivers* to securely, and publicly, choose random receivers. Once the receivers are chosen, the sender calls *SendShares* with the key shares encrypted to the relevant party's public key.

5.2 Implementing Proposal 2

First, we define general public key encryption and signature schemes.

Definition 1 (Public Key Encryption Scheme). *A public key encryption scheme is made up of a tuple of probabilistic polynomial–time algorithms (KeyGen, Enc, Dec) such that:*

$KeyGen(1^\lambda)$ *is the key generation algorithm, taking security parameter λ as input and producing a public key, secret key key tuple (pk, sk) respectively.*

$Enc(pk, m)$ *takes the public key pk and a message m, and outputs a ciphertext c.*

$Dec(sk, c)$ *takes the secret key sk and a ciphertext c, and outputs a plaintext m.*

Definition 2 (Signature Scheme). *A signature scheme is made up of a tuple of probabilistic polynomial–time algorithms (KeyGen, Sign, Ver) such that:*

$KeyGen(1^\lambda)$ *is the key generation algorithm, taking security parameter λ as input and producing a public key, secret key key tuple (pk, sk) respectively.*

$Sign(sk, m)$ *takes the secret key sk and some message m, and outputs a tag s.*

$Ver(pk, s, m)$ *takes the public key pk, the message m, and the tag s and returns either accept or reject.*

Providing correctness requirements, informally, for a public key scheme such that $(pk, sk) \leftarrow^r KeyGen(1^\lambda)$ for some security parameter λ, then the probability that

$\mathsf{Dec}(sk, \mathsf{Enc}(pk, m)) \neq m$ is a negligible of λ. Similarly for the signature aspect, for any m, the probability that $\mathsf{Ver}(pk, \mathsf{Sign}(sk, m), m)$ returns reject is also a negligible function of λ.

Remark 1 (Public Key Scheme). We call any overarching scheme that supports both signatures and encryption a public Key Scheme.

Definition 3 (Proof-of-Work Adaptable). *Let P be a public key scheme. For every $pk_i, pk_j \leftarrow^r \mathsf{KeyGen}(1^\lambda)$ such that $|pk_i| = |pk_j| = n$ and for all adversaries, the difference in expected computational steps between recovering a secret key for pk_i and recovering a secret key for pk_j is negligible in the security parameter λ.*

Furthermore, for any $pk_i \leftarrow^r \mathsf{KeyGen}(1^{\lambda'})$ and $pk_j \leftarrow^r \mathsf{KeyGen}(1^{\lambda''})$, where $d = |pk_j| - |pk_i|$, for any adversary A that can reliably recover a secret key for pk_i in t computational steps, then A can reliably recover a secret key for pk_j in $t + f(d) \geq t$ steps, for some monotonically increasing function f.

In other words, keys of equal size can be recovered in roughly the same number of steps, and that there is not some selection of weaker keys within the scheme that are easier to recover. It also ensures that the difficulty increases by a known factor on the length of the key, meaning we are able to extrapolate the security of larger keys, based on that of smaller keys.

Definition 4 (Full key space). *Let P be a public key scheme. We say that P has full key space in $[i, j]$ if for every binary string of length between and including i and j, then each public key is uniquely representable as such a string, and has a unique corresponding secret key, for some choice of security parameter λ.*

With the groundwork in place, it is possible to build a proof–of–work scheme based on a public keys scheme, providing it satisfies Definitions 1, 2, 3, and 4. The process for building this PoW system is described as follows:

1. Collect broadcast transactions (or credentials), and label them as x_i.
2. Take a unique reward value y, which is the information used to claim a reward.
3. Using a suitable hash function H, apply $c_i = H(x_i, c_{i-1}, y)$ and let pk be the first d bits of c_i, where c_{i-1} represents the previous state.
4. The challenge is to find a secret key sk corresponding to pk for the given public key scheme, such that $\mathsf{Ver}(pk, \mathsf{Sign}(sk, c_i), c_i)$ returns true.

Once such an sk is found, it can then be broadcast to claim the reward. This creates a chain-like consensus mechanism to be used as the backbone for the system. Now, when credentials (or transactions) are created, we insist that they are created using the same public key scheme for the consensus ledger mechanism, but with a higher level of security than creating the difficulty. With the properties defined for the public key scheme, we can pick a key that we know is quantifiably more challenging to recover than the amount of work on the system at a given time. Now, if a key ever needs to be recovered, the same amount of energy can be expended on the recovery of a key. However, with a large enough system it is necessary to engage with the community and have them recover the key in order to achieve recovery in a timely fashion.

There must be a way of adjusting the difficulty of the system depending on the rate at which keys are being recovered. This is possible in this framework as we can simply adjust the $pk = c[0, \ldots, d]$ for some maximum difficulty d, which can vary depending on the rate at which solutions are made available.

5.3   Implementing Proposal 3

Proposal three can be implemented by combining the implementations of proposal one and two. An important aspect here is the method used for the creation of credentials, as the corresponding secret key used in the credential creation must only be recoverable if both the instance of oblivious key escrow and the instance of scheme two is satisfied. Clearly these requirements are contradictory, as the secret keys chosen in scheme two are purposefully designed to be recoverable when a concerted computational effort is applied, whereas in scheme one, they are not. To solve this, we simply include two separate keys when creating a credential, where each key is implemented under either scheme. While this is not entirely elegant, we believe the extra burden on the key generator and the extra space requirement for storing the key are sufficiently small to allow for such a solution. Notably, this choice will double the number of authentication rounds, which may detract from applicability, but we consider to be a price worth paying for a PKI system which seeks to achieve simultaneous goals such as these.

6   Conclusion

The aim of this paper is to explore future directions in key escrow. We have presented three outline solutions, constructable in the no-man's-land between complete surveillance and complete security. There remain interesting open problems in this line of enquiry. From an applications perspective, a concrete proposal for a public key scheme that matches the criteria laid out in Definitions 1, 2, 3, and 4 is needed. Utilising quantum resistant public key schemes may be necessary for long term use. While we have no candidate construction, we note that for classical cryptography, there may be scope for schemes built on the discrete logarithm problem.

Of course, making the scheme truly public and inert requires a large user base, and wide adoption. For that reason, research into scale and usability of such a system would complement this work.

Leveraging Existing Technology. A slightly different, but otherwise parallel approach, would be to overlay the escrow credential management from proposal 2 directly on top of the pre-existing, and popular, Bitcoin system. The major advantage of doing this is that the inherent security of the system would come packaged with it. In principle, there is nothing stopping a system that utilises

Bitcoin's blockchain as the decentralised append-only ledger. This can allow anyone to announce their credential, along with a signature from a trusted axiom authority within a transaction. This is trivially possible on Bitcoin, and you could do this within a transaction to announce your public key. The open problem is how to store the secret key within the transaction, so that it could be recovered by finding a preimage of an output form the SHA-256 algorithm.

References

1. Abelson, H., Anderson, R.J., Bellovin, S.M., Benaloh, J., Blaze, M., Diffie, W., Gilmore, J., Green, M., Landau, S., Neumann, P.G., Rivest, R.L., Schiller, J.I., Schneier, B., Specter, M.A., Weitzner, D.J.: Keys under doormats. Commun. ACM **58**(10), 24–26 (2015)
2. Abelson, H., Anderson, R.J., Bellovin, S.M., Benaloh, J., Blaze, M., Diffie, W., Gilmore, J., Neumann, P.G., Rivest, R.L., Schiller, J.I., Schneier, B.: The risks of key recovery, key escrow, and trusted third-party encryption (1997)
3. Ateniese, G., Faonio, A., Magri, B., Medeiros, B.: Certified bitcoins. In: Boureanu, I., Owesarski, P., Vaudenay, S. (eds.) ACNS 2014. LNCS, vol. 8479, pp. 80–96. Springer, Cham (2014). doi:10.1007/978-3-319-07536-5_6
4. Bellare, M., Goldwasser, S.:. Verifiable partial key escrow. In: Richard Graveman et al. (ed) Proceedings of the 4th ACM Conference on Computer and Communications Security CCS 1997, pp. 78–91. ACM (1997)
5. Bellare, M., Rivest, R.L.: Translucent cryptography - an alternative to key escrow, and its implementation via fractional oblivious transfer. J. Cryptology **12**(2), 117–139 (1999)
6. Blaze, M.: Oblivious key escrow. In: Anderson, R. (ed.) IH 1996. LNCS, vol. 1174, pp. 335–343. Springer, Heidelberg (1996). doi:10.1007/3-540-61996-8_50
7. Chaum, D., Javani, F., Kate, A., Krasnova, A., de Ruiter, J., Sherman, A.T.: cMix: Anonymization by high-performance scalable mixing. IACR Cryptology ePrint Archive, 2016:8 (2016)
8. Denning, D.E., Branstad, D.K.: A taxonomy for key escrow encryption systems. Commun. ACM **39**(3), 34–40 (1996)
9. Douceur, J.R.: The sybil attack. In: Druschel, P., Kaashoek, F., Rowstron, A. (eds.) IPTPS 2002. LNCS, vol. 2429, pp. 251–260. Springer, Heidelberg (2002). doi:10.1007/3-540-45748-8_24
10. Hack, M.: The implications of Apple's battle with the FBI. Netw. Secur. **2016**(7), 8–10 (2016)
11. The Intercept: The great SIM heist (2015). https://theintercept.com/2015/02/19/great-sim-heist
12. WIRED: Todd Lappin. Winning the crypto wars (1997). www.wired.com/1997/05/cyber-rights-10/
13. Nakamoto, S.: Bitcoin: a peer-to-peer electronic cash system (2008). https://bitcoin.org/bitcoin.pdf
14. Ethereum Network: Ethereum: smart contract and decentralized application platform (2016). https://github.com/ethereum/wiki/wiki/White-Paper
15. President's Review Group on Intelligence, Communications Technologies, Clarke, R.A., Morell, M.J., Stone, G.R., Sunstein, C.R., Swire, P.P.: Liberty, security in a changing world: report and recommendations of the president's review group on intelligence and communications technologies (2013). http://www.whitehouse.gov/sites/default/files/docs/2013-12-12_rg_final_report.pdf

16. Bart Preneel: IACR distinguished lecture: the future of cryptography (2016). http://homes.esat.kuleuven.be/~preneel/preneel_iacr_dl_vienna2016.pdf
17. Ruoti, S., Andersen, J., Zappala, D., Seamons, K.E.: Why Johnny still, still can't encrypt: evaluating the usability of a modern PGP client. CoRR (2015)
18. Whitten, A., Tygar, J.D.: Why johnny can't encrypt: a usability evaluation of PGP 5.0. In: Treese, G.W. (ed) 8th USENIX. USENIX (1999)
19. Wilson, D., Ateniese, G.: From pretty good to great: enhancing PGP using bitcoin and the blockchain. In: Qiu, M., Xu, S., Yung, M., Zhang, H. (eds.) Network and System Security. LNCS, vol. 9408, pp. 368–375. Springer, Cham (2015). doi:10.1007/978-3-319-25645-0_25

Honey Encryption for Language
Robbing Shannon to Pay Turing?

Marc Beunardeau, Houda Ferradi, Rémi Géraud$^{(\boxtimes)}$, and David Naccache

École Normale Supérieure, Information Security Group, Paris, France
{marc.beunardeau,houda.ferradi,remi.geraud,david.naccache}@ens.fr

Abstract. Honey Encryption (HE), introduced by Juels and Ristenpart (Eurocrypt 2014, [12]), is an encryption paradigm designed to produce ciphertexts yielding plausible-looking but bogus plaintexts upon decryption with wrong keys. Thus brute-force attackers need to use additional information to determine whether they indeed found the correct key.

At the end of their paper, Juels and Ristenpart leave as an open question the adaptation of honey encryption to natural language messages. A recent paper by Chatterjee *et al.* [5] takes a mild attempt at the challenge and constructs a natural language honey encryption scheme relying on simple models for passwords.

In this position paper we explain why this approach cannot be extended to reasonable-size human-written documents *e.g.* e-mails. We propose an alternative solution and evaluate its security.

1 Introduction

Cryptography assumes that keys and passwords can be kept private. Should such secrets be revealed, any guarantee of confidentiality or authenticity would be lost. To that end, the set of possible secrets – the keyspace $\mathcal{K}$ – is designed to be very large, so that an adversary cannot possibly exhaust it during the system's lifetime.

In some applications however, the keyspace is purposely limited – for instance, passwords. In addition to the limited keyspace size, secret selection has a fundamental limitation: keys should be chosen uniformly at random – yet users routinely pick (the same) poor passwords. Consequently, key guessing is a guided process in which the adversary does not need to exhaust all possibilities. The deadly combination of low-entropy key generation and small keyspace make password-based encryption (PBE) particularly vulnerable [16].

The best security measure of a PBE is the min-entropy of the key distribution over $\mathcal{K}$:

$$\mu = -\log_2 \max_{k \in \mathcal{K}} p_k(k).$$

where p_k is the probability distribution of keys. The min-entropy captures how probable is the most probable guess. Conventional PBE schemes such as [24] can be broken with constant effort with probability $O(2^{-\mu})$, but μ is in practice

© Springer International Publishing AG 2017
R.C.-W. Phan and M. Yung (Eds.): LNCS 10311, Mycrypt 2016, pp. 127–144, 2017.
DOI: 10.1007/978-3-319-61273-7_7

very small: [2] reports $\mu < 7$ for passwords observed in a population of about 69 million users. If a message m were to be protected by such passwords, an adversary could easily recover m by trying the most probable passwords[1].

But how would the adversary *know* that the key she is trying is the correct one? A message has often some structure—documents, images, audio files for instance—and an attempt at decrypting with an incorrect key would produce something that, with high probability, does *not* feature or comply with this structure. The adversary can therefore tell apart a correct key from the incorrect ones, judging by how appropriate the decryption's output is. Mathematically, the adversary uses her ability to distinguish between the distribution of outputs for her candidate key k' and the distribution p_m of inputs she is expecting to recover.

Using such a distinguisher enables the attacker to try many keys, then select only the best key candidates. If there are not many possible candidates, the adversary can recover the plaintext (and possibly the key as well). In the typical case of password vaults, when one "master password" is used to encrypt a list of passwords, such an attack leads to a complete security collapse.

Example 1. Assume that we wish to AES-decrypt what we know is an English word protected with a small 4 digits key: $c \leftarrow \mathsf{Enc}_k(m)$. An efficient distinguisher is whether $m_{k'} \leftarrow \mathsf{Dec}_{k'}(c)$ is made of letters belonging to the English alphabet. For instance, if

$$c = \mathtt{0f\ 89\ 7d\ 66\ 8b\ 4c\ 27\ d7\ 50\ fa\ 99\ 0c\ 5a\ d6\ 11\ eb}$$

Then the adversary can distinguish between two candidate keys 5171 and 1431:

$$m_{5171} = \mathtt{48\ 6f\ 6e\ 65\ 79\ 00\ 00\ 00\ 00\ 00\ 00\ 00\ 00\ 00\ 00\ 00}$$
$$m_{1431} = \mathtt{bd\ 94\ 11\ 05\ a2\ e5\ a7\ c8\ 48\ 57\ 87\ 2a\ 88\ 52\ bc\ 7e}$$

Indeed, m_{5171} spells out 'Honey' in ASCII while m_{1431} has many characters that do not correspond to any letters. Exhausting all 4 digit keys yields only one message completely made of letters, hence $k = 5171$ and the adversary succeeded in recovering the plaintext m_{5171}.

To thwart such attacks, Juels and Ristenpart introduced Honey Encryption (HE) [12]. HE is an encryption paradigm designed to produce ciphertexts which, upon decryption with wrong keys, yields plausible-looking plaintexts. Thus brute-force attackers need to use additional information to decide whether they indeed found the correct key.

Mathematically, the decoding procedure in HE outputs candidate plaintexts distributed according to a distribution p_d close to the distribution p_m of real messages. This renders distinguishing attacks inoperant. The advantages of HE are discussed at length in [12] where the concept is applied to password-based encryption of RSA secret keys, credit card PINs and CVVs. In particular, HE does not reduce the security level of the underlying encryption scheme, but may act as an additional protection layer.

---

[1] Such passwords may be learnt from password leaks [11,21,22].

However, the applications of HE highlighted in [12] are very specific: Passwords protecting passwords (or passwords protecting keys). More precisely, low min-entropy keys protecting high min-entropy keys. The authors are wary not to extend HE to other settings and note that designing HE

> *"...for human-generated messages (password vaults, e-mail, etc.) (...) is interesting as a natural language processing problem."* [12]

To give a taste of the challenge, realizing HE as Juels and Ristenpart defined it is equivalent to modelling the probability distribution of human language *itself.* A more modest goal is to restrict to subsets of human activity where choices are more limited, such as passwords—this is indeed the target of a recent paper by Chatterjee, Bonneau, Juels and Ristenpart [5], which introduces encoders for human-generated passwords they call "natural language encoders" (NLE). Chatterjee *et al.*'s approach to language is to model the distribution of messages using either a 4-gram generative Markov model or a custom-trained probabilistic grammar model. This works reasonably well for passwords.

A natural question is therefore: Could the same techniques be extended or generalized to human-generated documents *in general?* Chatterjee *et al.* hint at it several times, but never actually take a leap: The core reason is that these approaches do not scale well and fail to model even simple sentences – let alone entire documents.

In this paper we give arguments why the approach of Chatterjee *et al.* does not extends, and give an alternative approach based on a corpus quotation distribution transforming encoding.

2   Preliminaries

Notations. We write $x \xleftarrow{D} X$ to denote the sampling of x from X according to a distribution D, and $x \xleftarrow{\$} X$ when D is the uniform distribution.

Message Recovery Attacks. Let $\mathcal{M}$ be a message space and let $\mathcal{K}$ be a key space. We denote by p_m the message distribution over $\mathcal{M}$, and by p_k the key distribution over $\mathcal{K}$. Let Enc be any encryption scheme. The *message-recovery advantage* of an adversary $\mathcal{A}$ against Enc is defined as

$$\mathbf{Adv}^{MR}_{\mathsf{Enc},p_m,p_k}(\mathcal{A}) = \Pr\left[\mathrm{MR}^{\mathcal{A}}_{\mathsf{Enc},p_m,p_k} = \mathsf{True}\right]$$

where the MR security game is described in Game 1. $\mathcal{A}$ may run for an unbounded amount of time, and make an unbounded number of queries to a random oracle.

This advantage captures the ability of an adversary knowing the distributions p_m, p_k to recover a message encrypted with Enc.

When key and message entropy are low, this advantage might not be negligible. However, using Honey Encryption, Juels and Ristempart show that $\mathcal{A}$'s advantage is bounded by $2^{-\mu}$, where $\mu = -\log\max_{k \in \mathcal{K}} p_k(k)$ is the min-entropy of the key distribution.

Game 1. Message recovery (MR) security game $\mathrm{MR}_{\mathsf{Enc},p_m,p_k}^{\mathcal{A}}$.

$K' \xleftarrow{p_k} \mathcal{K}$
$M' \xleftarrow{p_m} \mathcal{M}$
$C' \xleftarrow{\$} \mathsf{Enc}(K', M')$
$M \leftarrow \mathcal{A}(C')$
return $M == M'$

Distribution Transforming Encoding. HE relies on a primitive called the *distribution transforming encoding* (DTE). The DTE is really the central object of HE, which is then used to encrypt or decrypt messages. A DTE is composed of two algorithms, DTEncode and DTDecode which map messages into numbers in the interval $[0, 1]$ and back, *i.e.* such that

$$\forall M \in \mathcal{M}, \quad \mathsf{DTDecode}(\mathsf{DTEncode}(M)) = M.$$

More precisely, DTEncode: $\mathcal{M} \to [0, 1]$ is designed such that the output distribution of DTEncode is uniform over $[0, 1]$ when the input distribution over $\mathcal{M}$ is specified and known — in other terms, DTDecode samples messages in $\mathcal{M}$ according to a distribution p_d close to p_m, with

$$p_d(M) = \Pr\left[M' = M \mid x \xleftarrow{\$} [0, 1] \text{ and } M' \leftarrow \mathsf{DTDecode}(S)\right]$$

As such, DTEs cannot be arbitrary: They need to mimic the behaviour of the cumulative distribution function and its inverse. More precisely, the closeness of p_d and p_m is determined by the advantage of an adversary $\mathcal{A}$ in distinguishing the games of Figs. 1 and 2:

$$\mathsf{Adv}_{\mathsf{DTE},p_m}^{\mathcal{A}} = \left| \Pr\left[\mathsf{SAMP1}_{\mathsf{DTE},p_m}^{\mathcal{A}} = 1\right] - \Pr\left[\mathsf{SAMP0}_{\mathsf{DTE}}^{\mathcal{A}} = 0\right] \right|$$

$\mathcal{A}$ is provided with either a real message and its encoding, or a fake encoding and its decoding. $\mathcal{A}$ outputs 1 or 0 depending on whether it bets on the former or the latter. A perfectly secure DTE is a scheme for which the indistinguishability advantage is zero even for unbounded adversaries (this is equivalent to $p_d = p_m$).

$x' \xleftarrow{\$} [0, 1]$
$M' \leftarrow \mathsf{DTDecode}(x')$
$b \xleftarrow{\$} \mathcal{B}(M', x')$
return b

$M' \xleftarrow{p_m} \mathcal{M}$
$x' \xleftarrow{\$} \mathsf{DTEncode}(M')$
$b \xleftarrow{\$} \mathcal{B}(M', x')$
return b

Fig. 1. SAMP0$_{\mathsf{DTE}}^{\mathcal{B}}$ **Fig. 2.** SAMP1$_{\mathsf{DTE},p_m}^{\mathcal{B}}$

Having good DTEs is the central aspect of building a Honey Encryption scheme as well as the main technical challenge. Given a good DTE, the honey encryption and decryption of messages is provided by a variation of the "DTE-then-encrypt" construction described in Figs. 3 and 4 where some symmetric encryption scheme (ESEncode, ESDecode) is used. In the "DTE-then-encrypt" paradigm, a message is first transformed by the DTE into an integer x in some range, and x (or rather, some binary representation of x) is then encrypted with the key. Decryption proceeds by decrypting with the key, then reversing the DTE.

$\text{HEnc}^{ES}(K, M)$
$x \leftarrow \text{DTEncode}(M)$
$C \leftarrow \text{ESEncode}(x, K)$
return C

$\text{HDec}^{H}(K, C)$
$x \leftarrow \text{ESDecode}(K, C)$
$M \leftarrow \text{DTDecode}(x)$
return M

Fig. 3. Algorithm HEnc^{ES}

Fig. 4. Algorithm HDec^{ES}

2.1   Natural Language Encoding

Chaterjee *et al.* [5] developed an approach to generating DTEs based on two natural language models: an n-gram Markov model, and a custom probabilistic grammar tree.

Markov Model. The n-gram model is a local description of letters whereby the probability of the next letter is determined by the $n - 1$ last letters:

$$\Pr[w_1 \cdots w_k] = \prod_{i=1}^{k} \Pr\left[w_i \mid w_{i-(n-1)} \cdots w_{i-1}\right]$$

It is assumed that these probabilities have been learnt from a large, consistent corpus.

Such models are language-independent, yet produce strings that mimic the local correlations of a training corpus — but, as Chomsky pointed out [6–8], the output of such models lacks the long-range correlations typical of natural language. The latter is not an issue though, as Chatterjee *et al.* train this model on passwords.

The model can be understood as a directed graph where vertices are labelled with n-grams, and edges are labelled with the cumulative probability from some distinguished root node. To encode a string it suffices to encode the corresponding path through this graph from the root — and decoding uses the input as random choices in the walk. Encoding and decoding can be achieved in time linear in message size.

Grammar Model. Probabilistic context-free grammars (PCFG) are language-dependent models that learn from a tagged corpus a set of grammatical rules, and then use these rules to generate syntactically possible sentences. PCFGs are a compact way of representing a distribution of strings in a language.

132   M. Beunardeau et al.

Although it is known that context-free grammars do not capture the whole breadth of natural language, PCFGs are a good starting point, for such grammars are easy to understand, and from a given probabilistic context-free grammar, one can construct compact and efficient parsers [15]. The Stanford Statistical Parser, for instance, has been used by the authors to generate parse trees in this paper.

Mathematically, a probabilistic context-free grammar G is a tuple of the form $(N, T, R, P, \text{ROOT})$ where N are non-terminal symbols, T are terminal symbols (disjoint from N), R are production rules, P is the set of probabilities on production rules and ROOT is the start symbol. Every production rule is of the form $A \rightarrow b$, where $A \in N$ and $b \in (T \cup N)^*$.

Figure 5 shows a parse tree aligned with a sentence. Some grammatical rules can be read at every branching: S $\rightarrow$ NP VP, NP $\rightarrow$ DT VBN NN, NP $\rightarrow$ DT NN, *etc.*

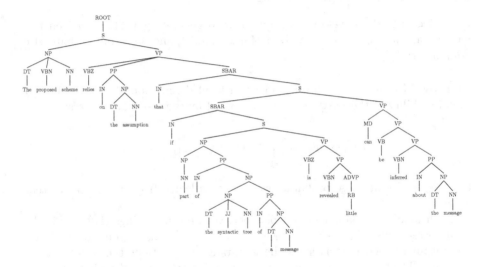

Fig. 5. Syntactic tree of an example sentence.

Chaterjee *et al.* [5] rely on a password-specific PCFGs [11, 14, 17, 21, 22] where grammatical roles are replaced by *ad hoc* roles.

The DTE encoding of a string is the sequence of probabilities defining a parse tree that is uniformly selected from all parse trees generating the same string (see *e.g.* Fig. 6, which provides an example of two parse trees for a same sentence, amongst more than 10 other possibilities). Decoding just emits the string indicated by the encoded parse tree.

In the probabilistic context, the probability of each parse tree can be estimated. A standard algorithm for doing so is due to Cocke, Younger, and Kasami (CYK) [9, 13, 23].

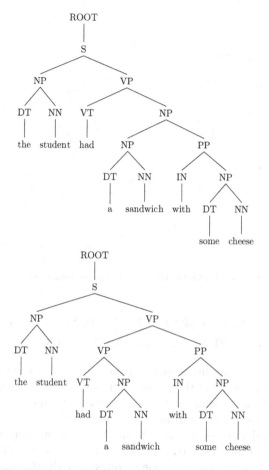

Fig. 6. Two possible derivations of the same sentence. Note that these derivations correspond to two possible meanings which are not identical.

Generalized Grammar Model. The generalized idea relies on the assumption that if part of the syntactic tree of a message is revealed, little can be inferred about the message. To understand the intuition, consider the syntactic tree of the previous sentence (clause) shown in Fig. 7.

As we can see, words are tagged using the *clause level*, *phrase level* and *word level* labels listed in Appendix A.

The idea underlying syntactic honey encryption consists in revealing a rewritten syntactic tree's word layer while encrypting words[2]. The process starts by a syntactic analysis of the message allowing to extract the plaintext's syntactic tree. This is followed by a projection at the word level. When applied to the previous example, we get the projection denoted by S (hereafter called *skeleton*):

---

[2] We stress that unlike e.g. Kamouflage [1] which deals with passwords, syntactic honey encyrption applies to natural language.

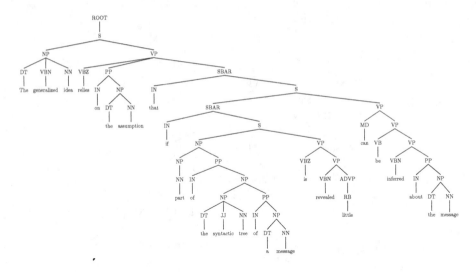

Fig. 7. Syntactic tree of an example sentence.

$$S = \text{DT VBN NN VBZ IN DT NN IN IN NN IN DT JJ}$$
$$\text{NN IN DT NN VBZ VBN RB MD VB VBN IN DT NN}$$

Given a clause, we can automatically associate each word s_i to a label L_i[3]. For instance, if the third word of the clause is "relies", then $L_3 \leftarrow$ VBZ. We denote by R_i the rank of the skeleton's i-th word in the dictionary of the category L_i. Finally we denote by $|X|$ the cardinality of the set X.

To map our ordered wordlist into a single integer, we note that because in the above example there are 5 DTs, 3 VBNs, 6 NNs, 2 VBZs, 6 INs, and 1 JJ, RB, MD and 1 VB, our specific clause is one amongst exactly B syntactically correct messages where:

$$B = |\text{DT}|^5|\text{VBN}|^3|\text{NN}|^6|\text{VBZ}|^2|\text{IN}|^6|\text{JJ}||\text{RB}||\text{MD}||\text{VB}|$$

We can thus map a clause skeleton into $\mathbb{N}$ by writing:

$$e \leftarrow \sum_{i=0}^{k-1} R_i \prod_{j=0}^{i-1} |L_j|$$

where, by typographic convention, $L_{-1} = 1$.

To get back the original clause, given e and the skeleton, we use the algorithm of Fig. 8.

---

[3] Note that such a skeleton might be ambiguous in certain constructions, for instance in sentences such as *"Time flies like an arrow; fruit flies like a banana"*.

Decoding
$\ell \leftarrow |L_0|$
for $i \leftarrow 0$ to $k-1$
$\quad R_i \leftarrow e \bmod \ell$
$\quad e \leftarrow (e - R_i)/\ell$
$\quad \ell \leftarrow \ell \times |L_{i+1}|$
$\quad \text{word}_i \leftarrow \text{Dictionary}_{L_i}(R_i)$

Fig. 8. Decoding algorithm.

The skeleton is transferred in clear:

$$s = \text{DT VBN NN VBZ IN DT NN IN IN NN IN DT JJ NN}$$
$$\text{IN DT NN VBZ VBN RB MD VB VBN IN DT NN}$$

Note that there is no need to tune precisely the plaintext size of the underlying block cipher because the decoding process for e stops automatically when i reaches $k-1$. In other words, we can randomize encryption at little cost by replacing e by $e + \mu B$ for some random integer μ.

The number e is then honey encrypted, thus attempting to protect the actual content of the plaintext sentence.

3   Limitations of Honey Encryption

As observed by [12], HE security is threatened when $\mathcal{A}$ has some side information about the target message. This puts strong constraints on HE's applicability to situations such as protecting RSA or HTTPS private keys. A second limitation is that the HE construction assumes that the key and message distributions are independent. When these distributions are correlated, $\mathcal{A}$ can identify a correct message by comparing that message with the decryption key that produced it. Similarly, encrypting two correlated messages under the same key enables $\mathcal{A}$ to identify correct messages.

Finally, constructing a DTE requires knowing the distribution p_m of messages in $\mathcal{M}$. As we will argue, this turns out to be extremely difficult to evaluate when $\mathcal{M}$ becomes a large enough space, such as human-generated messages (emails, etc.). In those cases, it might even turn out that adversaries know p_m *better than users*.

The methods described in Sect. 2.1 apply reasonably well to short passwords, but as we will now argue they cannot scale to deal with natural language as used in real-world scenarios such as: e-mails and written documents. The reason is threefold: First the methods require a huge amount of context-relevant information; Second, even when this information is available, the methods of [5] fail to produce convincing honey messages, *i.e.* messages that fool *automated tools* in telling them apart from real messages with high probability; Third, natural language HE may actually leak information about the underlying message.

Scaling NLE. The models developed for passwords in [5] can be extended: Markov models for instance can be configured to generate arbitrary-length messages. Instead of letters, such models can be trained to produce words, in accordance with some known distribution of n-grams. But while there are only a few English letters, a recent study of the English language [19] counts more than a million individual words in usage.

As a result assuming we use one hundredth of the English language, the memory required to store an n-gram database is of the order of $10^{4n} \approx 2^{13n}$. That becomes a problem not only in terms of storage, but also when access latency is taken into account. Applying directly the method of [5] to words (using $n = 5$) would require knowing, storing, and sharing 2^{65} bytes of data[4]. The real issue however is that measuring accurately 5-grams usage is extremely difficult in practice, so that most of this impossibly large database is essentially unknown[5].

Using grammars is one way to avoid this combinatorial explosion by keeping a simple and compact model of language. To that end, a sentence is parsed to reveal its grammatical structure as in Figs. 5 and 6. Each word is labelled with an indication of its grammatical role (see Appendix A).

A sentence is therefore uniquely represented by a list of grammatical tags, and a list of integers denoting which word is used. The idea behind syntactic honey encryption consists in revealing the tags but honey encrypting the words. By construction, generated honey messages have the same syntax as the original message, which makes decryption with a wrong key yield an often *plausible* plaintext. For instance, a sentence such as s_1 = "Secure honey encryption is hard" could be honey decrypted as Chomsky's famous sentence s_2 = "Colorless green ideas sleep furiously" [6], illustrating a sentence that is grammatically correct while being semantically void. Here s_1 and s_2 share the same syntax. To use this algorithm the communicating parties must agree on a dictionary that includes a set of labels and a parsing algorithm.

There are however two structural limitations to this grammatical approach. First, revealing the syntactic structure of a message leaks information. This is a very big deviation from classical cryptography, since it has always been taken for granted -for obvious reasons- that a ciphertext should not leak anything but the length of the underlying plaintext. On a more practical note unless the message is long enough, there might be only very few possible sentences with that given syntax. Second, a grammar is language-dependent — and furthermore, to some extent, there is variability within a given language[6]. The consequence of an inaccurate or incorrect tagging is that upon honey decoding, the sentence might be noticeably incorrect from the suitable linguistic standpoint.

---

[4] This is conceptually similar to Borges' famous library [3,4].

[5] See for instance http://www.ngrams.info/.

[6] An extreme example is William Shakespeare's use of inversion as a poetic device: *"If't be so, For Banquo's issue have I fil'd my mind,/ For them the gracious Duncan have I murther'd,/Put rancors in the vessel of my peace"* (*MacBeth*, III.1.8).

This opens yet another research avenue. Automatically translate the sentence into an artificially created language where syntactic honey encryption would be very efficient. For instance translate French to Hindi, then perform honey encryption on the Hindi sentence.

Quality of NLE. The question of whether a honey message is "correct" in a given linguistic context can be rephrased: Is it possible, to an adversary having access to a large corpus (written in the same language), to distinguish honey messages from the legitimate plaintext?

It turns out that the two approaches to modelling natural language provide two ways to construct a distinguisher: We can compare a candidate to a reference, either statistically or syntactically. But we can actually do both *simultaneously*: We can use Web search engines to assess how often a given sentence or word is used[7]. This empirical measure of probability is interesting in two respects: First, an adversary may query many candidates and prune those that score badly; Second, the sender cannot learn enough about the distribution of *all* messages using that "oracle" to perform honey encryption.

The situation is that there is a measurable distance between the model (used by the sender) of language, and language itself (as can be measured by *e.g.* a search engine). Mathematically, the sender assumes an approximate distribution p_m on messages which is different from the real-world distribution $\widehat{p}_m$. Because of that, a good DTE in the sense of Figs. 1 and 2 would, in essence, yield honey messages that follow p_m and not $\widehat{p}_m$. An adversary capable of distinguishing between these distributions can effectively tell honey messages apart.

What is the discrepancy between p_m and $\widehat{p}_m$? Since $\widehat{p}_m$ measures real-world usage, we can make the hypothesis that such messages correspond to human concerns, *i.e.* that they carry some meaning — in one word, what distinguishes p_m from $\widehat{p}_m$ is *semantics*.

Leaking Information. Another inherent limitation of HE is precisely that decryption of uniformly random ciphertexts produces in general the most probable messages. There are many situations in which linguistic constraints force a certain structure on messages, *e.g.* the position of a verb in a German sentence. Consequently, there might be enough landmarks for a meaningful reconstruction (see also [20]).

To thwart such reconstruction attacks, it is possible to consider phrase-level defences. Such defences imply modifying the syntactic tree in a way which is both reversible and indistinguishable from other sentences of the language. Phrase-level defences heavily depend on the language used. For instance the grammar of Latin, like that of other ancient Indo-European languages, is highly inflected; consequently, it allows for a large degree of flexibility in choosing word order. For example, *femina togam texuit*, is strictly equivalent to *texuit togam femina* or

---

[7] We may assume that communication with such services is secure, *i.e.* confidential and non-malleable, for the sake of argument.

togam texuit femina. In each word the desinence (also called ending or suffix): *-a*, *-am* and *-uit*, and not the position in the sentence, marks the word's grammatical function. This specific example shows that even if the target language allows flexibility in word order, this flexibility does not necessarily imply additional security. Semitic languages, such as Arabic or Hebrew, would on the contrary offer very interesting phrase-level defences. In semitic languages, words are usually formed by associating three, four or five-consonant verbs to structures. In Hebrew for example the structure *mi□□a□a* corresponds to the place where action takes place. Because the verb *drš* means *to teach* (or *preach*), and because the verb *zrk* means *to throw* (or *project*), the words *midraša*[8] and *mizraka* respectively mean "school" and "water fountain" (the place that projects (water)). This structure which allows, in theory, to build $O(ab)$ terms using $O(a)$ verbs and $O(b)$ and thus turns out to be HE-friendly.

4   Corpus Quotation DTE

We now describe an alternative approach which is interesting in its own right. Instead of targeting the whole breadth of human language, we restrict users to only quote from a known public document[9].

   The underlying intuition is that, since models fail to capture with enough detail the empirical properties of language, we should think the other way around and start from an empirical source directly. As such, the corpus quotation DTE addresses the three main limitations of HE highlighted in Sect. 3: It scales, it produces realistic sentences (because they are actual sentences), and it does not leak structural information.

   Consider a known public string $\mathfrak{M}$ (the "corpus"). We assume that $\mathcal{M}$ consists in contiguous sequence of words sampled from $\mathfrak{M}$, *i.e.* from the set of substrings of $\mathfrak{M}$. To build a DTE we consider the problem of mapping a substring $m \in \mathcal{M}$ to $[0, 1]$.

Interval Encoding of Substrings. Let M be the size of $\mathfrak{M}$, there are $|\mathcal{M}| = M(M-1)/2$ substrings denoted $m_{i,j}$, where i is the starting position and j is the ending position, with $i \leq j$. Substrings of the form $m_{i,i}$ are 1-letter long.

   The DTE encoding of $m \in \mathcal{M}$ is a point in a sub-interval of $[0, 1]$, whose length is proportional to the probability $p_m(m)$ of choosing m. If p_m is uniform over $\mathcal{M}$, then all intervals have the same length and are of the form

$$I_k = \left]\frac{2k}{M(M-1)}, \frac{2(k+1)}{M(M-1)}\right].$$

where k is the index of $m \in \mathcal{M}$ for some ordering on $\mathcal{M}$. Decoding determines which I_k contains the input and returns k, from which the original substring can be retrieved. For more general distributions p_m, each substring $m_{i,j}$ is mapped to an interval whose size depends on $p_m(m)$.

---

[8] The Arabic equivalent is *madrasa*.

[9] The way some characters do in Umberto Eco's novel, *Il pendolo di Foucault*[10].

Length-Dependent Distributions. Let's consider the special case where $p_m(m)$ depends only on the length of m. We will therefore consider the function $p : [1, M] \longrightarrow [0, 1]$ giving the probability of a substring of a given length. This captures some properties of natural languages such as Zipf's law [18]: Short expressions and words are used much more often than longer ones. Note that part of this is captured by the fact that there are fewer long substrings than short ones.

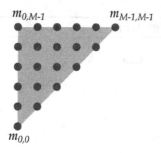

Fig. 9. Triangle representation T of the substrings $\mathcal{M} \subseteq \mathfrak{M}$. Substrings along right diagonals have equal length. The top-left point represents the entire corpus $\mathfrak{M}$.

Thus the encoding of a message $m_{i,j}$ is a random point in an interval of size $\ell(j - i)$ proportional to $p_m(m_{i,j}) = p(j - i)$:

$$\ell(k) = \frac{p(k)}{L}, \qquad L = \sum_{k=1}^{M}(M - k)p(k).$$

This ensures that

$$\sum_{k=1}^{M}(M - k)\ell(k) = 1.$$

The intervals associated to each substring are defined as follows. First, substrings $m_{i,j}$ are mapped via the map $\tau : m_{i,j} \mapsto (i, j)$ to a triangle (see Fig. 9):

$$T = \{(i, j) \mid j \geq i \in [0, M - 1]\} \subset \mathbb{N}^2.$$

Then points in T are mapped to $[0, 1]$ using the function:

$$\Phi : (i, j) \mapsto (i - 1)\ell(\text{diag}(i, j)) + \sum_{k=1}^{\text{diag}(i,j)-1} k\ell(k)$$

where $\text{diag}(i, j) = M - 1 - (j - i)$ indicates on which upright diagonal (i, j) is. All in all, a substring $m_{i,j}$ is encoded using the following algorithm:

$$\mathsf{DTEncode} : m_{i,j} \mapsto (\Phi + \epsilon\ell \circ \text{diag})\,(\tau(m_{i,j}))$$

where ϵ is sampled uniformly at random from $[0, 1]$.

Encoding can be understood as follows: Substrings of equal length k are mapped by τ to points along a diagonal of constant $k = j - i$. The first diagonal is the whole corpus $\mathfrak{M}$ and the only substring of length M. The $(M-1-k)$-th diagonal is the set of substrings $\{m_{i,i+k} \mid i \in [0, M-1-k]\}$ of length k. Decoding is achieved by Algorithm 1, which takes a number $x \in [0,1]$ and returns the position $(i,j) = \Phi^{-1}(x)$ of the corresponding substring by determining the position in T. The idea is to count the segment length before x. At each iteration we update the segment length and the current position in the diagonal.

Algorithm 1. Position of $\Phi^{-1}(x)$

Input: $x \in [0,1]$
Output: $(a,b) \in [|0, M|]^2$ such that $\Phi(a,b) = x$
  $i \leftarrow 0$
  $j \leftarrow 0$
  $k \leftarrow M$
  while $i < x$ **do**
    $i \leftarrow i + \ell(k)$
    $j \leftarrow j + 1$
    if $j \geq M - k + 1$ **then**
      $j \leftarrow 0$
      $k \leftarrow k - 1$
    end if
    return $(j-1, M+j-k-1)$
  end while

This decryption algorithm is linear in the number of substrings, *i.e.* it runs in time $O(M^2)$. We can speed things up using pre-computations, Algorithms 2 and 3 run in $O(M)$ time and memory.

Algorithm 2. Pre-computation

Output: vector V such that intervals in $[V[i], V[i+1]]$ are the intervals of length $\ell(i)$
  let $V[1..M]$ be a vector of length M
  for $i \leftarrow 1$ to M **do**
    $V[i] \leftarrow V[i-1] + (M-i+1)\ell(i)$
  end for
  return V

Algorithm 3. Fast Decryption

Input: $x \in [0,1], V$ the result of Algorithm 2.
Output: $(a,b) \in [|0, M|]^2$ such that $\Phi(a,b) = x$
  $i \leftarrow 1$
  while $V[i] < x$ **do**
    $i++$
  end while
  $j \leftarrow (x - V[i])/\ell(i)$
  return $(j-1, M-i-1)$

---

5   Further Research

This work opens a number of interesting research directions:

Machine to Human HE: Search engines, and more generally computational knowledge engines and answer engines such as Wolfram Alpha[10] provide users with structured answers that mimic human language. These algorithms generate messages using well-defined algorithmic process having a precise probability distribution which DTEs can be better modelled. Such sentences are hence likely to be safer to honey encrypt.

Automated Plaintext Pre-Processing: A more advanced, yet not that unrealistic option consists in having a machine understand a natural language sentence m and re-encode m as a humanly understandable yet grammatically and syntactically simplified sentence m' having the same meaning for a human. Such an ontology-preserving simplification process will not modify the message's meaning while allowing the construction better DTEs.

Adding Syntactic Defenses: This work was mostly concerned by protecting messages at the *word* level. It is however possible to imagine the adding of defenses at the clause and at the phrase levels. Two simple clause-level protections consist in adding decoy clauses to the message, and shuffling the order of clauses in the message. Both transforms can be easily encoded in the ciphertext by adding an integer field interpreted as the rank of a permutation and a binary strong whose bits indicate which clauses should be discarded. Decryption with a wrong key will yield a wrong permutation and will remove useful skeletons from the message. It should be noted that whilst the permutation has very little cost, the addition of decoy skeletons impacts message length. It is important to use decoy skeletons that are indistinguishable from plausible skeletons. To that end the program can either pick skeletons in a huge database (*e.g.* the web) or generate them artificially.

Adding Phrase-Level Defenses: Adding phrase-level defenses is also a very interesting research direction. A simple way to implement phrase-level defenses consists in adding outgrowths to the clause. An outgrowth is a collection of fake elements added using a specific rewriting rule. Note that information cannot be removed from the sentence. Here is an example of scrambling using outgrowths: the original clause m_0 is the sentence "During his youth Alex was tutored by a *skilled* architect until the age of 16". The syntactic tree of m_0 is:

---

[10] www.wolframalpha.com.

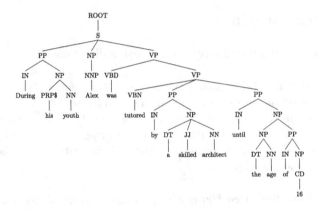

The skeleton of m_0 is IN PRP\$ NN NNP VBD VBN IN DT JJ NN IN DT NN IN CD.

Now consider the following rewriting rules:

PRP\$ NN $\longrightarrow$ PRP\$ JJ NN

DT NN $\longrightarrow$ DT JJ NN

IN DT JJ NN $\longrightarrow$ IN DT NN CC IN DT JJ NN

We can apply these rules to m_0 to obtain:

m_0 IN PRP\$ NN NNP VBD VBN IN DT JJ NN IN DT NN IN CD

$m_1 \leftarrow r_1(m_0)$ IN PRP\$ JJ NN NNP VBD VBN IN DT JJ NN IN DT NN IN CD

$m_2 \leftarrow r_2(m_1)$ IN PRP\$ JJ NN NNP VBD VBN IN DT JJ NN IN DT JJ NN IN CD

$m_3 \leftarrow r_3(m_2)$ IN PRP\$ JJ NN NNP VBD VBN IN DT NN CC IN DT JJ NN IN DT JJ NN IN CD

m_3 is a plausible skeleton that could have corresponded to the clause: "During his early youth Alex was tutored by a linguist and by a skilled architect until the approximate age of 16":

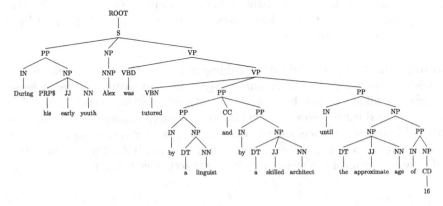

It remains to show how to reverse the process to recover the original skeleton m_0. To that end, we include in the ciphertext a binary string indicating which outgrowths should be removed. Removal consists in scanning m_0 and identifying what could have been the result of rewriting. Scanning reveals one potential application of rule 1 (namely "his early youth"), two potential applications of rule 2 ("a skilled architect" and "the approximate age") and one potential application of rule 2 ("by a linguist and by a skilled architect"). Hence 4 bits suffice to identify and remove the outgrowths.

A   Grammatical tags for English

See Table 1.

Table 1. Partial list of grammatical roles.

Clause Level	
S	Simple declarative clause
SBAR	Clause introduced by a (possibly empty) subordinating conjunction.
Phrase Level	
ADVP	Adverb phrase
NP	Noun phrase
PP	Prepositional phrase
VP	Verb phrase
Word Level	
CC	Conjunction, coordinating
DT	Determiner
IN	Preposition or subordinating conjunction
JJ	Adjective
MD	Modal
NN	Noun, singular or mass
PRP	Pronoun, personal
PRP$	Pronoun, possessive
RB	Adverb
VB	Verb, base form
VBN	Verb, past participle
VBZ	Verb, third person singular present

References

1. Bojinov, H., Bursztein, E., Boyen, X., Boneh, D.: Kamouflage: Loss-Resistant Password Management. In: Gritzalis, D., Preneel, B., Theoharidou, M. (eds.) ESORICS 2010. LNCS, vol. 6345, pp. 286–302. Springer, Heidelberg (2010). doi:10.1007/978-3-642-15497-3_18
2. Bonneau, J.: The science of guessing: analyzing an anonymized corpus of 70 million passwords, pp. 538–552 (2012)

3. Borges, J.L.: El Jardín de senderos que se bifurcan. Editorial Sur (1941)
4. Borges, J.L.: Ficcione. Editorial Sur (1944)
5. Chatterjee, R., Bonneau, J., Juels, A., Ristenpart, T.: Cracking-resistant password vaults using natural language encoders, pp. 481–498 (2015)
6. Chomsky, N.: Three models for the description of language. IRE Trans. Inf. Theory **2**(3), 113–124 (1956)
7. Chomsky, N.: On certain formal properties of grammars. Inf. Control **2**(2), 137–167 (1959)
8. Chomsky, N.: Syntactic structures. Walter de Gruyter, Berlin (2002)
9. Cocke, J.: Programming languages and their compilers: preliminary notes (1969)
10. Eco, U.: Il pendolo di Foucault. Bompiani (2011)
11. Jakobsson, M., Dhiman, M.: The benefits of understanding passwords. In: Traynor, P. (ed.) 7th USENIX Workshop on Hot Topics in Security, HotSec 2012, Bellevue, WA, USA, 7. USENIX Association (2012)., August 2012
12. Juels, A., Ristenpart, T.: Honey encryption: security beyond the brute-force bound, pp. 293–310 (2014)
13. Kasami, T.: An efficient recognition and syntax analysis algorithm for context-free languages. Technical report, DTIC Document (1965)
14. Kelley, P.G., Komanduri, S., Mazurek, M.L., Shay, R., Vidas, T., Bauer, L., Christin, N., Cranor, L.F., Lopez, J.: Guess again (and again and again): measuring password strength by simulating password-cracking algorithms, pp. 523–537
15. Klein, D., Manning, C.D.: Accurate unlexicalized parsing. In: Proceedings of the 41st Annual Meeting on Association for Computational Linguistics, vol. 1, pp. 423–430. Association for Computational Linguistics (2003)
16. Li, Z., He, W., Akhawe, D., Song, D.: The emperor's new password manager: security analysis of web-based password managers. In: Fu, K., Jung, J. (eds.) Proceedings of the 23rd USENIX Security Symposium, San Diego, CA, USA, 20–22 August, pp. 465–479. USENIX Association (2014)
17. Ma, J., Yang, W., Luo, M., Li, N.: A study of probabilistic password models. In: 2014 IEEE Symposium on Security and Privacy, SP 2014, Berkeley, CA, USA, 18–21 May, pp. 689–704. IEEE Computer Society (2014)
18. Manning, C.D., Schütze, H.: Foundations of statistical natural language processing. MIT Press, Cambridge (2001)
19. Michel, J.B., Shen, Y.K., Aiden, A.P., Veres, A., Gray, M.K., Pickett, J.P., Hoiberg, D., Clancy, D., Norvig, P., Orwant, J.: Quantitative analysis of culture using millions of digitized books. Science **331**(6014), 176–182 (2011)
20. Rayner, K., White, S.J., Johnson, R.L., Liversedge, S.P.: Reading wrods with jubmled lettres there is a cost. Psychol. Sci. **17**(3), 192–193 (2006)
21. Veras, R., Collins, C., Thorpe, J.: On semantic patterns of passwords and their security impact. In: The 21st Annual Network and Distributed System Security Symposium, NDSS 2014, San Diego, California, USA, 23–26 February 2014 (2014)
22. Weir, M., Aggarwal, S., de Medeiros, B., Glodek, B.: Password cracking using probabilistic context-free grammars, pp. 391–405 (2009)
23. Younger, D.H.: Recognition and parsing of context-free languages in time n^3. Inf. Control **10**(2), 189–208 (1967)
24. Kaliski, B.: PKCS #5: Password-based cryptography specification version 2.0. RFC 2898 (Informational). Internet Engineering Task Force, September 2000. http://www.ietf.org/rfc/rfc2898.txt

Randomized Stopping Times and Provably Secure Pseudorandom Permutation Generators

Michal Kulis[1], Pawel Lorek[2], and Filip Zagorski[1(✉)]

[1] Department of Computer Science,
Faculty of Fundamental Problems of Technology,
Wroclaw University of Science and Technology, Wroclaw, Poland
filip.zagorski@pwr.edu.pl
[2] Faculty of Mathematics and Computer Science, Mathematical Institute,
Wroclaw University, Wroclaw, Poland

Abstract. Conventionally, key-scheduling algorithm (KSA) of a cryptographic scheme runs for predefined number of steps. We suggest a different approach by utilization of randomized stopping rules to generate permutations which are indistinguishable from uniform ones. We explain that if the stopping time of such a shuffle is a *Strong Stationary Time* and bits of the secret key are not reused then these algorithms are immune against *timing attacks*.

We also revisit the well known paper of Mironov [15] which analyses a card shuffle which models KSA of RC4. Mironov states that expected time till reaching uniform distribution is $2nH_n - n$ while we prove that $nH_n + n$ steps are enough (by finding a new strong stationary time for the shuffle).

Nevertheless, both cases require $O(n \log^2 n)$ bits of randomness while one can replace the shuffle used in RC4 (and in Spritz) with a better shuffle which is optimal and needs only $O(n \log n)$ bits.

Keywords: Pseudo-random permutation generator · Markov chains · Mixing time · Stream cipher · Timing-attacks

1 Introduction

The applicability of card shuffles to cryptography was noticed many years ago by *e.g.*, Naor [17] for Thorp shuffle. The shuffles can be categorized into two groups. The first one are the *oblivious* shuffles, meaning that the trajectory of a card during the shuffle can be traced without tracing trajectories of other cards. Thus oblivious shuffles can be seen as block ciphers. The other group of card shuffles – *non-oblivious* shuffles require tracing all the cards in order to trace a single one. Since one needs to trace each of the n cards, straightforward application of non-oblivious shuffles as block ciphers would be inefficient. But non-oblivious

Authors were supported by Polish National Science Centre contract number DEC-2013/10/E/ST1/00359.

© Springer International Publishing AG 2017
R.C.-W. Phan and M. Yung (Eds.): LNCS 10311, Mycrypt 2016, pp. 145–167, 2017.
DOI: 10.1007/978-3-319-61273-7_8

shuffles are used in cryptographic schemes anyway, just in a slightly different role – often as a building block of a stream cipher.

Let us use the naming convention used by RC4 – a stream cipher designed in 1987 by Ronald Rivest. There is also a long line of stream ciphers: RC4A [23], Spritz [20], RC4+ [21], VMPC [25] – all of them are very similar – they are composed from two algorithms:

1. KSA (Key Scheduling Algorithm) uses a secret key to transform identity permutation of n cards into some other permutation (one can model KSA as a card shuffle).
2. PRGA (Pseudo Random Generation Algorithm) starts with a permutation generated by KSA and outputs random bits from it updating permutation at the same time.

Thus, KSAs of all aforementioned algorithms (RC4, RC4A, Spritz, RC4+, VMPC) can be seen as performing some card shuffling, where a secret key corresponds to/replaces randomness. If we consider a version of the algorithm with purely random secret key of infinite length then we indeed consider a card shuffling procedure. Following [15], we call such a version of the algorithm an *idealized version*. In the case of KSA used by RC4 the idealized version (mathematical model) of the card shuffle is called *Cyclic-to-Random Transpositions* shuffle which indeed is an example of *non-oblivious* shuffle. Recently, in 2013, Rivest and Schuldt presented a new version of an RC4-like cipher (Spritz [20]) which has a new sponge-like KSA which performs more complicated shuffle: $6N$ steps of *Cyclic-to-Random Transpositions* (as part of Whip procedure, see Fig. 7; compared to only N steps of in RC4) and in between, partial sorting (so called Crush) of elements in the internal state is performed twice (after 2-nd and 4-th shuffling).

The KSAs of mentioned ciphers perform shuffling for some **predefined number of steps**. The security of such a scheme is mainly based on analyzing idealized version of the algorithm and corresponds to the "quality of a shuffling". Roughly speaking, shuffling is considered as a Markov chain on permutations, all of them converge to uniform distribution (perfectly shuffled cards). Then we should perform as many steps as needed to be close to this uniform distribution, what is directly related to the so-called *mixing time*. This is one of the main drawbacks of RC4: it performs *Cyclic-to-Random Transpositions* for n steps, whereas the mixing time is of order $n \log n$.

There is a long list of papers which point out weaknesses of the RC4 algorithm. Attacks exploited both weaknesses of PRGA and KSA or the way RC4 was used in specific systems [4,9,10,12,13]. As a result, in 2015 RC4 was prohibited in TLS by IETF, Microsoft and Mozilla.

In the paper we use so-called **Strong Stationary Times** (SST) for Markov chains. The main area of application of SSTs is studying the rate of convergence of a chain to its stationary distribution. However, they may also be used for *perfect sampling* from stationary distribution of a Markov chain, consult [19] (on Coupling From The Past algorithm) and [8] (on algorithms involving Strong Stationary Times and Strong Stationary Duality).

1.1   Our Contribution

(1) Strong stationary time based KSA algorithm(s). Instead of running a KSA algorithm (*i.e.*, performing the shuffle) for some pre-defined number of steps, we make it randomized (Las Vegas algorithm). To be more specific we suggest utilization of so-called **Strong Stationary Times** (SST) for Markov chains. We use SST to obtain samples from uniform distribution on all permutations (we actually perform perfect sampling). We show benefits of such approach:

1. Use of SST may allow to close the gap between theoretical models and practice. As a result of Mironov's [15] work, one knows that idealized version of RC4's KSA would need keys of length $\approx$23 037 in order to meet the mathematical model. In fact one may use a better shuffling than the one that is used in RC4 *i.e.*, *time-reversed riffle shuffle* which requires (on average) 4096 bits – not much more than 2048 bits which are allowed for RC4 (see Sect. 4.1).
2. Coupling methods are most commonly used tool for studying the rate of convergence to stationarity for Markov chains. They allow to bound so-called *total variation distance* between the distribution of given chain at time instant k and its stationary distribution. However, the (traditional) security definitions require something "stronger". It turns out that bounding *separation distance* is what one actually needs. It fits perfectly in the notion of Strong Stationary Times we are using (see Sect. 4.2).
3. By construction, the running time of our model is key dependent. In extreme cases (very unlikely) it may leak some information about the key but it **does not leak any** information about the resulting permutation to an adversary. We also discuss how one can mask such a leakage (see Sect. 4.3).

(2) Better SST for RC4's KSA. Our complementary contribution (Sect. 5.2) is the analysis of RC4 showing a new upper bound on number of steps the algorithm should perform. Similarly as in [15], we propose SST which is valid for *Cyclic-to-Random Transpositions* and *Random-to-Random Transpositions*, for the latter one we calculate the mixing time, which is however "faster" than the one given in [15]. It is known that *Random-to-Random Transpositions* card shuffling needs $\frac{1}{2}n\log n$ steps to mix. It is worth mentioning that although *Random-to-Random Transpositions* and *Cyclic-to-Random Transpositions* are similar in the *spirit* it does not transfer automatically that the latter one also needs $\frac{1}{2}n\log n$ steps to mix. Mironov [15] states that expected time till reaching uniform distribution is upper bounded by $2nH_n - n$, we show in Lemma 3 that the expected running time for this SST in *Random-to-Random Transpositions* is equal to:

$$E[T] = nH_n + n + O(H_n)$$

and empirically check that the result is similar for *Cyclic-to-Random Transpositions*. This directly translates into the required steps that should be performed by RC4's KSA.

(3) Note on Spritz construction. We also have a look at Spritz (Sect. 6), a newer sponge-like construction. We explain why the *Sign distinguisher* attack cannot be successful and provide arguments why the KSA algorithm in Spritz may not perform enough steps.

2 Preliminary

Throughout the paper, let $\mathcal{S}_n$ denote a set of all permutations of a set $\{1, \ldots, n\} =: [n]$.

2.1 Markov Chains and Rate of Convergence

Consider ergodic Markov chain $\mathbf{X} = \{X_k, k \geq 0\}$ on finite state space $\mathbb{E} = \{0, \ldots, M-1\}$ with stationary distribution ψ. Let $\mathcal{L}(X_k)$ denote the distribution of a chain at time instant k. By the *rate of convergence* we understand the knowledge on how fast a distribution of a chain converges to its stationary distribution. We have to measure it according to some distance *dist*. Define

$$\tau_{mix}^{dist}(\varepsilon) = \inf\{k : dist(\mathcal{L}(X_k), \psi) \leq \varepsilon\},$$

which is called *mixing time* (w.r.t. given distance *dist*). In our case the state space is the set of all permutations of $[n]$, *i.e.*, $\mathbb{E} := \mathcal{S}_n$. The stationary distribution is the uniform distribution over $\mathbb{E}$, *i.e.*, $\psi(\sigma) = \frac{1}{n!}$ for all $\sigma \in \mathbb{E}$. In most applications the mixing time is defined w.r.t. total variation distance:

$$d_{TV}(\mathcal{L}(X_k), \psi) = \frac{1}{2} \sum_{\sigma \in \mathcal{S}_n} \left| Pr(X_k = \sigma) - \frac{1}{n!} \right|.$$

The **separation distance** is defined by

$$sep(\mathcal{L}(X_k), \psi) := \max_{\sigma \in \mathbb{E}} \left(1 - n! \cdot Pr(X_k = \sigma)\right).$$

It is relatively easy to check that $d_{TV}(\mathcal{L}(X_k), \psi) \leq sep(\mathcal{L}(X_k), \psi)$.

Strong Stationary Times. The definition of separation distance fits perfectly into notion of Strong Stationary Time (SST) for Markov chains. This is a probabilistic tool for studying the rate of convergence of Markov chains allowing also *perfect sampling*. We can think of stopping time as of a running time of algorithm which observes Markov chain $\mathbf{X}$ and which stops according to some stopping rule (depending only on the past).

Definition 1. *Random variable T is a randomized stopping time if it is a running time of the* RANDOMIZED STOPPING TIME *algorithm.*

Algorithm. RANDOMIZED STOPPING TIME
___
1: $k := 0$
2: $coin := $ `Tail`
3: **while** $coin == $ `Tail` **do**
4:    At time $k, (X_0, \ldots, X_k)$ was observed
5:    Calculate $f_k(X)$, where $f_k : (X_0, \ldots, X_k) \to [0, 1]$
6:    Let $p = f_k(X_0, \ldots, X_k)$. Flip the coin resulting in `Head` with probability p and in `Tail` with probability $1 - p$. Save result as $coin$.
7:    $k := k + 1$
8: **end while**
___

Definition 2. *Random variable T is **Strong Stationary Time (SST)** if it is a randomized stopping time for chain $\mathbf{X}$ such that:*

$$\forall (i \in \mathbb{E}) \ Pr(X_k = i | T = k) = \psi(i).$$

Having SST T for chain with uniform stationary distribution lets us bound the following (see [3])

$$sep(\mathcal{L}(X_k), \psi) \leq Pr(T > k). \tag{1}$$

We say that T is an optimal SST if $sep(\mathcal{L}(X_k), \psi) = Pr(T > k)$.

2.2 Distinguishers and Security Definition

We consider three distinguishers: Sign distinguisher and Position distinguisher which are exactly the same as defined in [15], we also consider Permutation distinguisher – we consider his advantage in a traditional cryptographic security definition. That is, Permutation distinguisher is given a permutation π and needs to decide whether π is a result of a shuffle or if it is a permutation selected uniformly at random from the set of all permutations of a given size. The important difference is that the upper bound on Permutation distinguisher's advantage is an upper bound on <u>any</u> possible distinguisher – even those which are not bounded computationally.

Position distinguisher

Input: S, t, table $(p_{i,j}^{(t)})$, threshold A
Output: b
$p := 0$
for i:=0 to $n - 1$ **do**
   $p := p + \log(np)_{i,S[i]}^{(t)}$
   if $p < A$
      then return *false*
      else return *true*

Sign distinguisher

Input: S, t
Output: b
if sign$(S) = (-1)^t$
   then return *false*
   else return *true*

Position Distinguisher. Because of the nature of the process (in fact both: *Random-to-Random Transpositions* and *Cyclic-to-Random Transpositions*), the probability that ith card is at jth position depends on t: $p_{i,j}^{(t)} = P(S[j] = i$ at time t) which can be pre-computed according to recursion: $p_{i,j}^{(0)} = \begin{cases} 1 & \text{if } i = j \\ 0 & \text{otherwise} \end{cases}$

and for $t > 0$:

$$p_{i,j}^{(t)} = \begin{cases} p_{i,j}^{(t-1)}\left(1 - \frac{1}{n}\right) + \frac{1}{n}p_{t_0,j}^{(t-1)} & \text{if } i \neq t_0, \\ \frac{1}{n} & \text{otherwise,} \end{cases}$$

where $t_0 = t \mod n$. The advantage of Position distinguisher dissolves in time – we will upper bound the time needed for this distinguisher to lose his advantage.

Sign Distinguisher. For a permutation π which has a representation of non-trivial transpositions $\pi = (a_1b_1)(a_2b_2)\ldots(a_mb_m)$ the sign is defined as: $sign(\pi) = (-1)^m$. So the value of the sign is $+1$ whenever m is even and is equal to -1 whenever m is odd.

Permutation Distinguisher. The distinguishability game for the adversary is as follows:

Definition 3. *The permutation indistinguishability* Shuffle$_{S,\mathcal{A}}(n, r)$ *experiment.*

Algorithm. Shuffle$_{S,\mathcal{A}}(n,r)$

Let S be a shuffling algorithm which in each round requires m bits.

1. S is initialized with:
   (a) a key generated uniformly at random $\mathcal{K} \sim \mathcal{U}(\{0,1\}^{rm})$,
   (b) $S_0 = \pi_0$ (identity permutation)
2. S is run for r rounds: $S_r := S(\mathcal{K})$ and produces a permutation π_r.
3. We set:
   - $c_0 := \pi_{rand}$ a random permutation from uniform distribution is chosen,
   - $c_1 := \pi_r$.
4. A challenge bit $b \in \{0,1\}$ is chosen at random, permutation c_b is sent to the Adversary.
5. Adversary replies with b'
6. The output of the experiment is defined to be 1 if $b' = b$, and 0 otherwise.

In the case when adversary wins the game (if $b = b'$) we say that $\mathcal{A}$ succeeded. Adversary wins the game if she can distinguish the random permutation from the permutation being a result of the PRPG algorithm.

Definition 4. *A shuffling algorithm S generates indistinguishable permutations if for all adversaries $\mathcal{A}$ there exists a negligible function* negl *such that*

$$Pr\left[\textit{Shuffle}_{S,\mathcal{A}}(n,r) = 1\right] \leq \frac{1}{2} + \textit{negl}(n).$$

The above translates into:

Definition 5. *A shuffling algorithm* S *generates indistinguishable permutations if for any adversary* $\mathcal{A}$ *there exists a negligible function* **negl** *such that:*

$$\left| \Pr_{K \leftarrow \{0,1\}^{KeyLen}} [\mathcal{A}(S(K)) = 1] - \Pr_{R \leftarrow \mathcal{U}(S_n)} [\mathcal{A}(R) = 1] \right| \leq negl(n)$$

3 Related Work

3.1 RC4 Algorithm

RC4 is a stream cipher, its so-called *internal state* is (S, i, j), where S is a permutation of $[n]$ and i, j are some two indices. As input it takes $L-$byte message $m_1, \ldots, m_L$ and a secret key K and returns ciphertext $c_1, \ldots, c_L$. The initial state is the output of KSA. Based on this state PRGA is used to output bits which are XORed with the message. The actual KSA algorithm used in RC4 is presented in Fig. 1 together with its *idealized version* KSA* (where a secret key-based randomness is replaced with pure randomness) and our version of the algorithm KSA** (where, in addition to KSA*, it does not run pre-defined number of steps, but the number depends on a key and is determined by some stopping procedure ST). The details on KSA** will be given in Sect. 4.1.

KSA(K)	KSA*	KSA**

for $i := 0$ to $n - 1$ **do**
    $S[i] := i$
end for
 	
$j := 0$
for $i := 0$ to $n - 1$ **do**
    $j := j + S[i] + K[i \bmod l]$
    swap($S[i], S[j]$)
end for
$i, j := 0$
	for $i := 0$ to $n - 1$ **do**
	    $S[i] := i$
	end for
	
	for $i := 0$ to $n - 1$ **do**
	    $j :=$**random**(n)
	    swap($S[i], S[j]$)
	end for
	$i, j := 0$
		for $i := 0$ to $n-1$ **do**
		    $S[i] := i$
		end for
		
		while $(\neg$ ST$)$ **do**
		    $j :=$**random**(n)
		    swap($S[i], S[j]$)
		    $i := i + 1 \bmod n$
		end while

Fig. 1. KSA of RC4 algorithm and its idealized version KSA*. The KSA** has some additional procedure ST (stopping time) which is computed during the execution of the algorithm (for original RC4 simply ST is: stop after n steps).

A closer look at KSA* reveals that it is actually so-called *Cyclic-to-Random Transpositions.* If we identify elements $[n]$ with cards then we do the following: at step t exchange card $t \bmod n$ with randomly chosen one. Throughout the paper, let $\mathbf{Z} = \{Z\}_{t \geq 0}$ denote the chain corresponding to this shuffling and let $\mathcal{L}(Z_t)$ denote the distribution of the chain at time t.

3.2 Sign Distinguisher for RC4's KSA

It was observed in [15] that the sign of the permutation at the end of KSA algorithm is not uniform. And as a conclusion it was noticed that the number of discarded shuffles (by PRGA) must grow at least linearly in n. Below we present this result obtained in a different way than in [15], giving the exact formula for advantage at any step t. This form will be used by us to draw conclusions about Spritz algorithm in Sect. 6.1.

One can look at the sign-change process for the *Cyclic-to-Random Transpositions* as follows: after the table is initialized, sign of the permutation is +1 since it is identity so the initial distribution is concentrated in $v_0 = (Pr(\text{sign}(Z_0) = +1)$, $Pr(\text{sign}(Z_0) = -1)) = (1,0)$.

Then in each step the sign is unchanged if and only if $i = j$ which happens with probability $1/n$. So the transition matrix M_n of a sign-change process induced by the shuffling process is equal to:

$$M_n := \begin{pmatrix} \frac{1}{n} & 1 - \frac{1}{n} \\ 1 - \frac{1}{n} & \frac{1}{n} \end{pmatrix}.$$

This conclusion corresponds to looking at the distribution of the sign-change process after t steps: $v_0 \cdot M_n^t$, where v_0 is the initial distribution. The eigenvalues and eigenvectors of M_n are $(1, \frac{2-n}{n})$ and $(1,1)^T, (-1,1)^T$ respectively. The spectral decomposition yields

$$v_0 \cdot M_n^t = (1,0) \begin{pmatrix} 1 & -1 \\ 1 & 1 \end{pmatrix} \begin{pmatrix} 1 & 0 \\ 0 & \frac{2-n}{n} \end{pmatrix}^t \begin{pmatrix} \frac{1}{2} & -\frac{1}{2} \\ -\frac{1}{2} & \frac{1}{2} \end{pmatrix} = \left(\frac{1}{2} + \frac{1}{2} \left(\frac{2}{n} - 1 \right)^t, \frac{1}{2} - \frac{1}{2} \left(\frac{2}{n} - 1 \right)^t \right).$$

For $n = 256$ (which corresponds to the value of n used in RC4) and initial distribution being identity permutation after $t = n = 256$ steps one gets: $v_0 \cdot M_{256}^{256} = (0.567138, 0.432862)$.

In [13] it was suggested that the first 512 bytes of output should be dropped. The Fig. 3 in Appendix A presents the advantage ϵ of a sign-adversary after dropping k bytes of the output (so after $n + k$ steps of the shuffle, for the mathematical model).

3.3 Position Distinguisher for RC4's KSA

Mironov suggested analysis of *idealized* version of KSA algorithm. Being in permutation $S \in \mathcal{S}_n$ at step i, the idealized version swaps element $S[i]$ with purely random $S[j]$. Treating the permutation as a permutation of a deck of cards, this is exactly a known *Cyclic-to-Random Transpositions* card shuffling. On the other hand if both, $S[i]$ and $S[j]$ are chosen uniformly at random, the procedure is called *Random-to-Random Transpositions* card shuffling. It is known that Random Transposition requires around $\frac{1}{2}n \log n$ to reach uniform distribution, see [7]. Moreover, authors showed that "*most of the action*" actually happens at this step – the process exhibit so called cut-off phenomena. The analysis of Position distinguisher uses Strong Stationary Time (called *Strong Uniform Times* in [15]), based on Broder's construction for *Random-to-Random Transpositions*. Unfortunately Mironov's "estimate of the rate of growth of the strong uniform time T is quite loose" and results "are a far cry both from the provable upper and lower bounds on the convergence rate". He:

- proved an upper bound $O(n \log n)$. More precisely Mironov showed that there exists some positive constant c such that $P[T > cn \log n] \to 0$ when $n \to \infty$. Author experimentally checked that $P[T > 2n \lg n] < 1/n$ for $n = 256$ which corresponds to $P[T > 4096] < 1/256$.
- experimentally showed that $E[T] \approx 11.16n \approx 1.4n \lg n \approx 2857$ (for $n = 256$) – which translates into: on average one needs to drop ≈ 2601 initial bytes.

Later Mosel et al. [16] proved a matching lower bound establishing mixing time to be of order $\Theta(n \log n)$. However, the constant was not determined.

4   Randomized Stopping Times and Cryptographic Schemes

4.1   Strong Stationary Time Based **KSA** Algrorithms

We propose to use the $\mathsf{KSA}^{**}_{\mathrm{Shuffle,ST}}(n)$ algorithm which works as follows. It starts with identity permutation. Then at each step it performs some card shuffling procedure Shuffle. Instead of running it for a pre-defined number of steps, it runs until an event defined by a procedure ST occurs. The procedure ST is designed in such a way that it guarantees that the event is a Strong Stationary Time. At each step the algorithm uses new randomness – one can think about that as of an idealized version but when the length of a key is greater than the number of random bits required by the algorithm then we end up with a permutation which cannot be distinguished from a random (uniform) one (even by a computationally unbounded adversary).

Algorithm. $\mathsf{KSA}^{**}_{\mathrm{Shuffle,ST}}(n)$

Require: Card shuffling Shuffle procedure, stopping rule ST which is a Strong Stationary Time for Shuffle.

   for $i := 0$ to $n - 1$ **do**
      $S[i] := i$
   end for

   while $(\neg \, \mathsf{ST})$ **do**
      $\mathsf{Shuffle}(S)$
   end while

Notational convention: in $\mathsf{KSA}^{**}_{\mathrm{Shuffle,ST}}(n)$ we omit parameter n. Moreover, if Shuffle and ST are omitted it means that we use *Cyclic-to-Random Transpositions* as shuffling procedure and stopping rule is clear from the context (as in KSA** given earlier in Fig. 1). Note that if we use for stopping rule ST "stop after n steps" (which of course is *not* SST), it is equivalent to RC4's KSA* (also in Fig. 1).

Given a shuffling procedure `Shuffle` one wants to have a "fast" stopping rule ST (perfectly one wants an optimal SST which is stochastically the smallest). The stopping rule ST is a parameter, since for a given shuffling scheme one can come up with a better stopping rule(s). This is exactly the case with *Cyclic-to-Random Transpositions* and *Random-to-Random Transpositions*, we recall Mironov's [15] stopping rule as well as new "faster" rule called `StoppingRuleKLZ` is given (in Sect. 5.2).

4.2 RST and Security Guarantees

Coupling method is a commonly used tool for bounding the rate of convergence of Markov chains. Roughly speaking, a coupling of a Markov chain $\mathbf{X}$ with transition matrix $\mathbf{P}$ is a bivariate chain $(\mathbf{X'}, \mathbf{X''})$ such that marginally $\mathbf{X'}$ and $\mathbf{X''}$ are Markov chains with transition matrix $\mathbf{P}$ and once the chains meet they stay together (in some definitions this condition can be relaxed). Let then $T_c = \inf_k\{X'_k = X''_k\}$ *i.e.*, the first time chains meet, called *coupling time*. The *coupling inequality* states that $d_{TV}(\mathcal{L}(X_k), \psi) \leq Pr(T_c > k)$.

On the other hand separation distance is an upper bound on total variation distance, *i.e.*, $d_{TV}(\mathcal{L}(X_k), \psi) \leq sep(\mathcal{L}(X_k), \psi)$. At first glance it seems that it is better to directly bound d_{TV}, since we can have d_{TV} very small, whereas *sep* is (still) large. However, knowing that *sep* is small gives us much more than just knowing that d_{TV} is small, what turns out to be **crucial** for proving security guarantees (*i.e.*, Definition 5). In our case ($\mathbb{E} = \mathcal{S}_n$ and ψ is a uniform distribution on $\mathbb{E}$) having d_{TV} small, *i.e.*, $d_{TV}(\mathcal{L}(X_k), \psi) = \frac{1}{2}\sum_{\sigma \in \mathcal{S}_n}|Pr(X_k = \sigma) - \frac{1}{n!}| \leq \varepsilon$ **does not** imply that $|Pr(X_k = \sigma) - \frac{1}{n!}|$ is uniformly small (*i.e.*, of order $\frac{1}{n!}$). Knowing however that $sep(\mathcal{L}(X_k), \psi) \leq \varepsilon$ implies

$$\forall(\sigma \in \mathbb{E}) \ \left|Pr(X_k = \sigma) - \frac{1}{n!}\right| \leq \frac{\varepsilon}{n!}. \tag{2}$$

Above inequality is what we need in our security definitions and shows that the notion of separation distance is an adequate measure of mixing time for our applications.

It is worth noting that $d_{TV}(\mathcal{L}(X_k), \mathcal{U}(\mathbb{E})) \leq \varepsilon$ implies (see Theorem 7 in [2]) that $sep(\mathcal{L}(X_{2k}), \mathcal{U}(\mathbb{E})) \leq \varepsilon$. This means that proof of security which bounds directly total variation distance by ε would require twice as many bits of randomness compared to the result which guarantees ε bound on separation distance.

4.3 RST and Timing-Attacks

One of the most serious threats to any cryptographic scheme are side-channel attacks. One type of such attacks are *timing-attacks* where an attacker by observing the running time of the execution of a cryptosystem derives information about the key used. Timing attacks are especially powerful [1,22] since an attacker may perform them remotely, over the network (while most of other types of side-channel attacks can be performed only when an attacker is nearby). In order to limit threat of timing-attacks, attempts to implement constant-time cryptographic schemes are made. The problem is that such attempts are usually unsuccessful [18] even if the underlying architecture ("claims") allows for that [11,24].

The running time of an SST-based algorithm strictly depends on the secret key. However, in this section we explain why algorithms using randomized stopping times are immune to timing-attacks, we discuss separately security of two assets: (1) resulting permutation, (2) secret key.

Timing-Attacks and the Security of the Resulting Permutation. We already defined SST (Definition 2) in Sect. 2.1 but one can define SST differently.

Definition 6. *Random variable T is* **Strong Stationary Time (SST)** *if it is a randomized stopping time for chain X such that:*

$$X_T \text{ has distribution } \psi \text{ and is independent of } T.$$

Corrolary 1. *The information about the number of rounds that an SST-algorithm performs does not reveal any information about the resulting permutation.*

Corollary 1 comes from the fact that the Definition 2 which defines SST as a certain randomized stopping time is equivalent to the Definition 6 which defines SST as a variable independent of the resulting distribution. For the proof of the equivalence see [3].

Timing Attacks and the Security of the Secret Key. Unfortunately, although no information about the resulting permutation leaks, some information about the secret key may leak. Shuffling may reveal randomness through the running time (see Example 2 in Appendix D). In practical implementations, one may use some function of a key instead of pure randomness in each step. Then (at least) two following cases may happen:

1. Bits of the keystream are re-used: the running time of the algorithm (SST) may leak both: information about key and the information about permutation (compare with Example 1 in Appendix D).
2. Bits of the keystream are "fresh" (never re-used): the running time of the algorithm (SST) may leak information about the key but it does not leak any information about the produced permutation! (compare with the Example 2 in Appendix D).

Masking SST. One can prevent obtaining information about the secret key by timing-attacks by performing a simple masking. For a stopping rule ST that results in expected running time ET one runs the algorithm for at least ET steps even if the ST occurred earlier.

This eliminates very short executions which could reveal information about the key.

On the other hand, for practical implementation one may want to eliminate the extremely long executions. This can be done by letting the algorithm run for *e.g.*, $ET + c \cdot \sqrt{VarT}$ (where c is a parameter and $VarT$ is the variance for the ST).

5   (Not So) Random Shuffles of RC4 – Revisited

5.1   Mironov's Stopping Rule – Details

The goal of KSA of original RC4 is to produce a pseudorandom permutation of $n = 256$ cards. The original algorithm performs 256 steps of *Cyclic-to-Random Transpositions*. However it is known that the mixing time of *Cyclic-to-Random Transpositions* is $\Theta(n \log n)$. Then performing only 256 (*i.e.*, n) steps seems much too less. In fact, it is recommended to perform at least 3072 steps, see Mironov [15]. Generally, the more steps are performed, the closer to uniformity the final permutation is. Mironov considered idealized version of the algorithm together with the following marking rule:

> "At the beginning all cards numbered $0, \ldots, n - 2$ are *unchecked*, the $(n-1)^{th}$ card is *checked*. Whenever the shuffling algorithm exchanges two cards, $S[i]$ and $S[j]$, one of the two rules may apply before the swap takes place:
>
> a. If $S[i]$ is unchecked and $i = j$, check $S[i]$.
> b. If $S[i]$ is unchecked and $S[j]$ is checked, check $S[i]$.
>
> The event T happens when all cards become checked."

Then the author proves that this is a SST for *Cyclic-to-Random Transpositions* and shows that there exists constant c (can be chosen less than 30) such that $Pr[T > cn \log n] \to 0$ when $n \to \infty$. Empirically, for $n = 256$ he shows that $Pr[T > 2n \log n] < 1/n$. Note that this marking scheme is also valid for *Random-to-Random Transpositions* shuffling.

Lemma 1. *The expected running time of Random-to-Random Transpositions shuffling with Mironov stopping rule is:*

$$ET = 2nH_n - n + O(H_n).$$

Proof. We start with one card checked. When k cards are checked, then probability of checking another one is equal to $p_k = \frac{(n-k)(k+1)}{n^2}$. Thus, the time to check all the cards is distributed as a sum of geometric random variables and its expectation is equal to:

$$\sum_{k=1}^{n-1} \frac{1}{p_k} = 2 \frac{n^2}{n+1} H_n - n = 2nH_n - n + O(H_n).$$

5.2 Better Stopping Rule

We suggest another "faster" SST which is valid for both *Cyclic-to-Random Transpositions* and *Random-to-Random Transpositions*. We will calculate its expectation and variance for *Random-to-Random Transpositions* and check experimentally (see Appendix C) that it is similar if the stopping rule is applied to *Cyclic-to-Random Transpositions*. As a result (proof given at the end of this Section) we have:

Theorem 1. *Let $\mathcal{A}$ be an adversary. Let $K \in \{0,1\}^{rn}$ be a secret key. Let $S(K)$ be KSA^*_{RTRT} (i.e., with Random-to-Random Transpositions shuffling) which runs for*

$$r = n(H_n + 1) + \frac{\pi n}{2} \frac{1}{\sqrt{n!\varepsilon}}$$

steps with $0 < \varepsilon < \frac{1}{n!}$. Then

$$\left| \Pr_{K \leftarrow \{0,1\}^{rm}} [\mathcal{A}(S(K)) = 1] - \Pr_{R \leftarrow \mathcal{U}(S_n)} [\mathcal{A}(R) = 1] \right| \le \varepsilon$$

The stopping rule is given in StoppingRuleKLZ algorithm.

Algorithm. StoppingRuleKLZ

Input set of already marked cards $\mathcal{M} \subseteq \{1, \ldots, n\}$, round r, *Bits*
Output {YES,NO}

$j = $ n-value(*Bits*)
if there are less than $\lceil (n-1)/2 \rceil$ marked cards **then**
  if both $\pi[r]$ and $\pi[j]$ are unmarked **then**
    mark card $\pi[r]$
  end if
else
  if ($\pi[r]$ is unmarked and $\pi[j]$ is marked) OR ($\pi[r]$ is unmarked and $r = j$) **then**
    mark card $\pi[r]$
  end if
end if

if all cards are marked **then**
  STOP
else
  CONTINUE
end if

Lemma 2. *The resulting permutation of KSA^{**} with $ST =StoppingRuleKLZ$ has a uniform distribution over S_n.*

Proof. We will show that the running time of the algorithm is a SST, *i.e.*, that the card marking procedure specified in StoppingRuleKLZ is a SST for *Cyclic-to-Random Transpositions*. First phase of the procedure (*i.e.*, the case when there are less than $\lceil (n-1)/2 \rceil$ cards marked) is constructing a random permutation of marked cards by placing unmarked cards on randomly chosen unoccupied positions, this is actually first part of Matthews's marking [14] scheme. Second phase is simply a Broder's construction. Theorem 9 of [15] shows that this is a valid SST for *Cyclic-to-Random Transpositions*. Both phases combined produce a random permutation of all cards.

Remark 1. One important remark should be pointed. Full Matthews's marking [14] scheme is "faster" than ours. However, although it is a SST for *Random-to-Random Transpositions*, this is not SST for *Cyclic-to-Random Transpositions*.

Calculating ET or $VarT$ seems to be a challenging task. But note that marking scheme StoppingRuleKLZ also yields a valid SST for *Random-to-Random Transpositions*. In next Lemma we calculate ET and $VarT$ for this shuffle, later we experimentally show that ET is very similar for both marking schemes.

Lemma 3. *Let T be the running time of KSA^{**} with Random-to-Random Transpositions shuffling and ST=StoppingRuleKLZ. Then we have*

$$E[T] = nH_n + n + O(H_n),$$
$$Var[T] \sim \frac{\pi^2}{4}n^2, \tag{3}$$

where H_n is the $n-$th harmonic number and $f(k) \sim g(k)$ means that $\lim_{k \to \infty} \frac{f(k)}{g(k)} = 1$.

The details of the proof of the Lemma 3 are in Appendix B.

Proof. Define T_k to be the first time when k cards are marked (thus $T \equiv T_n$). Let $d = \lceil (n-1)/2 \rceil$. Then T_d is the running time of the first phase and $(T_n - T_d)$ is the running time of the second phase. Denote $Y_k := T_{k+1} - T_k$.

Assume that there are $k < d$ marked cards at a certain step. Then the new card will be marked in next step if we choose two unmarked cards what happens with probability: $p_a(k) = \frac{(n-k)^2}{n^2}$. Thus Y_k is a geometric random variable with parameter $p_a(k)$ and

$$E[T_d] = n^2 \left(H_n^{(2)} - H_{n-d}^{(2)} \right).$$

Now assume that there are $k \geq d$ cards marked at a certain step. Then, the new card will be marked in next step with probability:

$$p_b(k) = \frac{(n-k)(k+1)}{n^2}$$

and Y_k is a geometric random variable with parameter $p_a(k)$. Thus:

$$E[T_n - T_d] = nH_n - \frac{n}{n+1}H_n + \frac{n^2}{n+1}\left(H_{n-d} - H_d\right).$$

For variance we have:

$$Var[T_n] = Var[T_d] + Var[T_n - T_d] \sim \frac{\pi^2}{4}n^2.$$

$\square$

From Lemma 3 and Chebyshev's inequality we immediately have the following:

Corrolary 2. *Consider the chain corresponding to* KSA** *with Random-to-Random Transpositions shuffling and* ST=StoppingRuleKLZ. *Then we have*

$$\tau_{mix}^{sep}(\varepsilon) \leq n(H_n + 1) + \frac{\pi n}{2}\frac{1}{\sqrt{\varepsilon}}.$$

Proof (of Theorem 1). In Theorem 1 we perform *Random-to-Random Transpositions* for $r = \tau_{mix}^{sep}(n!\varepsilon)$ steps, *i.e.*, $sep(\mathcal{L}(X_r), \psi) \leq n!\varepsilon$ Inequality (2) implies that $|Pr(X_r = \sigma) - \frac{1}{n!}| \leq \varepsilon$ for any permutation σ and thus completes the proof. $\square$

5.3   Predefined Number of Steps vs SST-based Algorithms

There is a subtle difference between the randomized stopping time (like the one suggested in the paper) and an algorithm that performs a predefined number of steps. If one wants to achieve security level of *e.g.*, $\varepsilon = O(1/n^k)$, $k > 2$ then the number of steps that would assure that the advantage is smaller than ε would need to be equal to: $\tau_{mix}(\varepsilon) \leq n(H_n + 1) + \frac{\pi n}{2}n^{k/2} = O\left(n^{1+k/2}\right)$.

As we can see the estimated running time of *Cyclic-to-Random Transpositions* with both stopping rules is similar to the theoretical results for *Random-to-Random Transpositions*. Recall that for *Cyclic-to-Random Transpositions* it is known that the mixing time is of order $\Theta(n \log n)$, see [16], however the constant was not determined. Based on our new SST and simulations (see Appendix C) one can conjecture the following

Conjecture 1. The mixing time τ_{mix}^{sep} for *Cyclic-to-Random Transpositions* converges to $n \log n$ as $n \to \infty$.

6   A Note on Spritz

Spritz [20] is a new stream cipher that was proposed in 2014 by Rivest and Schuldt as a possible replacement for RC4.

The first cryptanalytic results were already achieved: inefficient state recovery attack [5] (with 2^{1400} complexity) later improved by [6] (with 2^{1247} steps). The second paper presents also a devastating distinguishing attacks of complexity $2^{44.8}$ (multiple key-IV setting) and $2^{60.8}$ (single key-IV setting).

Here we analyze the distribution of the internal state of Spritz after the main part of the scheme (procedure SHUFFLE()) is run. Similarly to the previous approach we replace the deterministic part of UPDATE() function: $j := k + S[j + S[i]]$, with its idealized version: $j := \mathbf{random}(n)$.

The definitions of Spritz' procedures that are of our interest are presented in Appendix E.

6.1   Sign Distinguisher

Although we did not find strong stationary time for "KSA" part of Spritz algorithm, one can easily notice that the Sign Distinguisher has no advantage at all. This property is achieved thanks to Crush procedure. During this procedure, the table S is partially sorted *i.e.*, elements at positions v and $n-1-v$ (for $v = 0 \ldots \lfloor N/2 \rfloor - 1$) are swapped whenever $S[v] > S[n-1-v]$. So this corresponds to multiplying the sign process by:

$$M_{crush} := \begin{pmatrix} \frac{1}{2} & \frac{1}{2} \\ \frac{1}{2} & \frac{1}{2} \end{pmatrix}.$$

If Spritz is used as stream cipher, as part of Squeeze procedure, at least one call to Shuffle is made. So the distribution of sign can be described as

$$v_0 \cdot M_n^{2n} M_{crush}^{n/2} M_n^{2n} M_{crush}^{n/2} M_n^{2n} = \left(\frac{1}{2}, \frac{1}{2} \right).$$

This means that advantage of Sign Distinguisher for Spritz equals to 0.

6.2   Position Distinguisher

Let us recall what was one of the main drawbacks of the original RC4: it performed n steps instead of $cn \log n$. The underlying mathematical model is simply a *Cyclic-to-Random Transpositions* card shuffling. This is somehow similar to *Random-to-Random Transpositions* for which it takes of order of $n \log n$ steps. More exactly, there is so-called cutoff phenomena at $\frac{1}{2} n \log n$. Roughly speaking, lower and upper bounds are of this order. Analysis of *Cyclic-to-Random Transpositions* seemed to be harder, recently [16] the matching lower bound was established showing that mixing time is of order $\Theta(n \log n)$.

Recall that Spritz performs: in total $6n$ steps of *Cyclic-to-Random Transpositions* (as part of Whip procedure) and partial sorting of elements (Crush procedure) in the internal state is performed twice (after 2-nd and 4-th shuffling). This is of course more complicated shuffling than just repeating *Cyclic-to-Random Transpositions*.

Clever use of Crush lets Spritz to get rid of the Sign Distinguisher but at the same time it seems that it may badly influence the mixing time.

Imagine that there exists some SST which during the Spritz execution performs marking of the elements. Marked elements satisfy property that their mutual position is equally distributed. Now, take a look at the step when Crush is performed and there are two marked elements at positions v and $N-1-v$. Then after Crush their relative position will be uniquely determined! This observation suggests that mixing time for Spritz would be greater than $n \log n$.

7   A Note on Optimal Shuffling

Cyclic-to-Random Transpositions is the shuffle used in RC4 and in Spritz. To reach stationarity (*i.e.*, produce random permutation), as we shown, one needs to perform $O(n \log n)$ steps. In each step we use a random number from interval $[0, \ldots, n-1]$, thus this shuffling requires $O(n \log^2 n)$ random bits.

One can ask the following question: Is this shuffling optimal in terms of required bits? The answer is no. The entropy of the uniform distribution on $[n]$ is $O(n \log n)$ (since there are $n!$ permutations of $[n]$), thus one could expect that optimal shuffling would require this number of bits.

We will shortly describe (time reversal of) *Riffle Shuffle*, for details see [2]. For a given permutation $\sigma \in \mathcal{S}_n$ we assign each element a random bit. Then we put all the elements (cards) with assigned bit 0 to the top *keeping* their relative ordering. The following is a known SST for this shuffle: At the beginning all $\binom{n}{2}$ pairs of cards are unmarked. At each step one marks a pair (i, j) if elements i and j were assigned different bits. Let T be the first time all pairs are marked.

The above SST of *Riffle Shuffle* has $ET = 2 \lg n$, at each step n random bits are used, thus this shuffling requires $2n \lg n$ random bits, matching the requirement of optimal shuffle (up to a constant).

In this paper we mainly focused on RC4 and thus on *Cyclic-to-Random Transpositions* shuffle. However, we wanted to point out that using *Riffle Shuffle* (or other shuffling schemes) can result in better efficiency of the whole scheme.

8 Conclusions

We presented the benefits of using Strong Stationary Times in cryptographic schemes (pseudo random permutation generators). These algorithms have a "health-check" built-in and guarantee the best possible properties (when it comes to the quality of randomness of the resulting permutation). We showed that use of SST does not lead to timing attacks. We showed that algorithms using SST achieve better security guarantees than any algorithm which runs predefined number of steps.

	RC4	Mironov	KSA**	RS
#bits used	40 to 2 048			
#bits asymptotics		$(2nH_n - 1)\lg n$	$n(H_n + 1)\lg n$	$2n \lg n$
#bits required		23 037	14 590	4 096

Fig. 2. Comparison between number of bits used by RC4 (40 to 2048) and required by mathematical models (Mironov [15] and ours) versus length of the key for the time-reversed riffle shuffle. *Bits asymptotics* approximates the number of fresh bits required by the mathematical model (number of bits required by the underlying Markov chain to converge to stationary distribution). *Bits required* is (rounded) value of *# bits asymptotics* when $n = 256$.

Complementarily, we proved better bound for the mixing-time of the *Cyclic-to-Random Transpositions* shuffling process which is used in RC4 and showed that different, more efficient shuffling methods (*i.e.,* time reversal of *Riffle Shuffle*) may be used as KSA. This last observation shows that the gap between mathematical model (4096 bits required) and reality (2048 allowed as maximum length of RC4) is not that big as previously thought (bound of 23037 by Mironov [15]).

Appendix

A Sign Distinguisher Advantage

k	$+1$	-1	ϵ
0	.5671382998250798	.4328617001749202	$2^{-3.89672}$
256	.509015	.490985	$2^{-6.79344}$
512	.5012105173235390	.4987894826764610	$2^{-9.69016}$
768	.500163	.499837	$2^{-12.5869}$
1024	.5000218258757580	.4999781741242420	$2^{-15.4836}$
2048	.5000000070953368	.4999999929046632	$2^{-27.0705}$
4096	.5000000000000007	.4999999999999993	$2^{-50.2442}$
8192	.5	.5	$2^{-96.5918}$

Fig. 3. The advantage (ϵ) of Sign distinguisher of RC4 after discarding initial k bytes.

B Detailed Proof of Lemma 3

Proof. Define T_k to be the first time when k cards are marked (thus $T \equiv T_n$). Let $d = \lceil (n-1)/2 \rceil$. Then T_d is the running time of the first phase and $(T_n - T_d)$ is the running time of the second phase. Denote $Y_k := T_{k+1} - T_k$.

Assume that there are $k < d$ marked cards at a certain step. Then the new card will be marked in next step if we choose two unmarked cards what happens with probability:

$$p_a(k) = \frac{(n-k)^2}{n^2}.$$

Thus Y_k is a geometric random variable with parameter $p_a(k)$ and

$$E[T_d] = \sum_{k=0}^{d-1} E[Y_k] = \sum_{k=0}^{d-1} \frac{1}{p_a(k)} = \sum_{k=0}^{d-1} \frac{n^2}{(n-k)^2} = n^2 \sum_{k=n-d+1}^{n} \cdot \frac{1}{k^2}$$

$$= n^2 \left(H_n^{(2)} - H_{n-d}^{(2)} \right) = n^2 \left(\frac{1}{n} + O\left(\frac{1}{n^2}\right) \right) = n + O(1).$$

Now assume that there are $k \geq d$ cards marked at a certain step. Then, the new card will be marked in next step with probability:

$$p_b(k) = \frac{(n-k)(k+1)}{n^2}$$

and Y_k is a geometric random variable with parameter $p_a(k)$. Thus:

$$E[T_n - T_d] = \sum_{k=0}^{d-1} E[Y_k] = \sum_{k=d}^{n-1} \frac{1}{p_b(k)} = \sum_{k=d}^{n-1} \frac{n^2}{(n-k)(k+1)}$$

$$= \frac{n^2}{n+1} \sum_{k=d}^{n-1} \left(\frac{1}{n-k} + \frac{1}{k+1} \right) = \frac{n^2}{n+1} \left(\sum_{k=1}^{n-d} \frac{1}{k} + \sum_{k=d+1}^{n} \frac{1}{k} \right)$$

$$= \frac{n^2}{n+1} (H_{n-d} + H_n - H_d) = \frac{n^2}{n+1} H_n + \frac{n^2}{n+1} (H_{n-d} - H_d)$$

$$= nH_n - \frac{n}{n+1} H_n + \frac{n^2}{n+1} (H_{n-d} - H_d) = nH_n + O(H_n) + O(1) = nH_n + O(H_n).$$

For variance we have:

$$Var[T_d] = \sum_{k=0}^{d-1} Var[Y_k] = \sum_{k=0}^{d-1} \frac{1 - p_a(k)}{(p_a(k))^2} = \sum_{k=0}^{d-1} \frac{1 - \frac{(n-k)^2}{n^2}}{\left(\frac{(n-k)^2}{n^2} \right)^2} \approx \int_0^{\frac{n}{2}} \frac{1 - \frac{(n-x)^2}{n^2}}{\left(\frac{(n-x)^2}{n^2} \right)^2} dx = \frac{4}{3} n.$$

$$Var[T_n - T_d] = \sum_{k=d}^{n-1} Var[Y_k] = \sum_{k=d}^{n-1} \frac{1 - p_b(k)}{(p_b(k))^2} = \sum_{k=d}^{n-1} \frac{1 - \frac{(n-k)(k+1)}{n^2}}{\frac{(n-k)^2(k+1)^2}{n^4}}$$

$$= n^2 \sum_{k=d}^{n-1} \frac{n(n-1) + k(1-n) + k^2}{(n-k)^2(k+1)^2} \approx n^2 \cdot \frac{1}{2} \sum_{k=0}^{n-1} \frac{n(n-1) + k(1-n) + k^2}{(n-k)^2(k+1)^2}$$

$$\approx \frac{n^2}{2} \left[\left(\frac{2}{n} - \frac{4}{n^2} \right) H_n + 3H_n^{(2)} \right] \sim \frac{\pi^2}{4} n^2.$$

Finally

$$Var[T_n] = Var[T_d] + Var[T_n - T_d] \sim \frac{\pi^2}{4} n^2.$$

$\square$

C   Experimental Results

The expected running time of $\mathsf{KSA}^{**}_{\mathrm{RTRT}}$ (*i.e.*, with *Random-to-Random Transpositions* shuffling) is known:

- with ST=StoppingRuleKLZ it is $n(H_n + 1)$
- with stopping rule used in [15] it is $2nH_n - n$

For both stopping rules applied to *Cyclic-to-Random Transpositions* no precise results on expected running times are known. Instead we estimated them via simulations, simply running 10.000 of them. The results are given in Fig. 4.

		StoppingRuleKLZ	Mironov's ST
	$n = 256$	1811	2854
ET	$n = 512$	3994	6442
	$n = 1024$	8705	14324
	$n = 256$	111341	156814
$VarT$	$n = 512$	438576	597783
	$n = 1024$	1759162	2442503

	$n(H_n + 1)$	$2nH_n - n$
$n = 256$	1823.83	2879.66
$n = 512$	4002.05	6468.11
$n = 1024$	8713.39	14354.79

Fig. 4. Simulations' results for Mironov's and StoppingRuleKLZ stopping rules.

D   Timing-Attacks and KSA**

Example 1 (Top to random shuffle T2R – timing attack – re-used randomness).
Consider algorithm $KSA^{**}_{T2R,ST}$ with shuffling procedure corresponding to *Top-To-Random* card shuffling (put the card $S[1]$ which is currently on top to the position j defined by the randomness in the current round) and following stopping time ST:

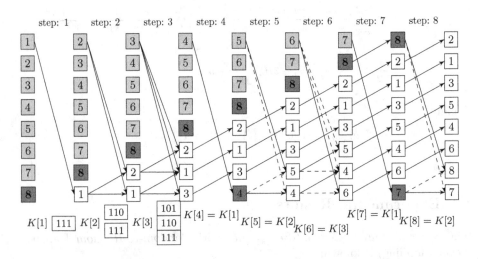

Fig. 5. Example run of *top-to-random* shuffle with "reused randomness" which is taken from the key K of the length equal to three 8-value bytes (9-bits). Let us assume that the running time of SST was exactly 8. Conditioning on the number of steps one can figure out that the first word of the key must be equal to 111 while the second part $K[2] \in \{110, 111\}$ with the same probability. Finally $K[3] \in \{101, 110, 111\}$ and for the step i: $K[i] = K[i \bmod 3]$. So now, instead of possible $2^9 = 512$ permutations which can be generated from 9 bits of key, only 6 are possible (based on the fact that SST has stopped exactly after 8 steps).

- before the start of the algorithm, mark the last card (*i.e.*, the card n is marked[1]),
- stop one step after the marked card reaches the top of the deck.

The sample execution of the algorithm is given in Fig. 5.

Example 2 (Top-to-random – timing attack – fresh randomness). Let us now consider a very similar situation with one important difference. Now no portion of the key is re-used. An example run of the algorithm is presented on the Fig. 6. Based on the knowledge on the number of performed steps, one can learn some information about the secret key (*i.e.*, $K[1] = 000$, $K[2]$ is either 110 or 111) but still no adversary can learn anything about the resulting permutation because any information is generated with exactly the same probability.

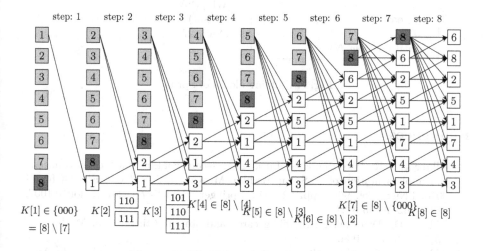

Fig. 6. Example run of *top-to-random* shuffle with "fresh randomness" taken from the key K of the length equal to 8 8-value bytes (24-bits). Conditioning on the number of steps of the SST (in this case 8) one can find out that: out of the possible 2^{24} keys only (exactly) 8! keys are possible (due to the fact that SST stopped after 8 steps) and every of 8! permutations are possible (based on the fact that SST has stopped exactly after 8 steps) – moreover each permutation with exactly the same probability. SST leaks bits of the key *i.e.*, $K[1] = 111$ but does not leak any information about the produced permutation.

---

[1] It is known [2] that optimal SST for *top to random* initially marks second from the bottom card.

E   Spritz Definition

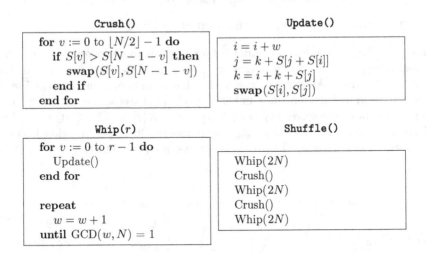

Fig. 7. Building blocks of Spritz

References

1. Albrecht, M.R., Paterson, K.G.: Lucky microseconds: a timing attack on Amazon's *s2n* implementation of TLS. In: Fischlin, M., Coron, J.-S. (eds.) EUROCRYPT 2016. LNCS, vol. 9665, pp. 622–643. Springer, Heidelberg (2016). doi:10.1007/978-3-662-49890-3_24

2. Aldous, D., Diaconis, P.: Shuffling cards and stopping times. Am. Math. Mon. **93**(5), 333–348 (1986)

3. Aldous, D., Diaconis, P.: Strong uniform times and finite random walks. Adv. Appl. Math. **8**, 69–97 (1987)

4. AlFardan, N., Bernstein, D.J., Paterson, K.G., Poettering, B., Schuldt, J.C.N.: On the security of RC4 in TLS. In: Presented as Part of the 22nd USENIX Security Symposium (USENIX Security 13), Washington, D.C., pp. 305–320. USENIX (2013)

5. Ankele, R., Kölbl, S., Rechberger, C.: State-recovery analysis of spritz. In: Lauter, K., Rodríguez-Henríquez, F. (eds.) LATINCRYPT 2015. LNCS, vol. 9230, pp. 204–221. Springer, Cham (2015). doi:10.1007/978-3-319-22174-8_12

6. Banik, S., Isobe, T.: Cryptanalysis of the full spritz stream cipher. In: Peyrin, T. (ed.) FSE 2016. LNCS, vol. 9783, pp. 63–77. Springer, Heidelberg (2016). doi:10.1007/978-3-662-52993-5_4

7. Diaconis, P., Shahshahani, M.: Generating a random permutation with random transpositions. Zeitschrift fur Wahrscheinlichkeitstheorie und Verwandte Gebiete **57**(2), 159–179 (1981)

8. Fill, J.A.: An interruptible algorithm for perfect sampling via Markov chains. Ann. Appl. Probab. **8**(1), 131–162 (1998)

9. Fluhrer, S., Mantin, I., Shamir, A.: Weaknesses in the key scheduling algorithm of RC4. In: Vaudenay, S., Youssef, A.M. (eds.) SAC 2001. LNCS, vol. 2259, pp. 1–24. Springer, Heidelberg (2001). doi:10.1007/3-540-45537-X_1
10. Fluhrer, S.R., McGrew, D.A.: Statistical analysis of the alleged RC4 keystream generator. In: Goos, G., Hartmanis, J., Leeuwen, J., Schneier, B. (eds.) FSE 2000. LNCS, vol. 1978, pp. 19–30. Springer, Heidelberg (2001). doi:10.1007/3-540-44706-7_2
11. Ge, Q., Yarom, Y., Cock, D., Heiser, G.: A Survey of Microarchitectural Timing Attacks and Countermeasures on Contemporary Hardware. IACR Eprint (2016)
12. Golić, J.D.: Linear statistical weakness of alleged RC4 keystream generator. In: Fumy, W. (ed.) EUROCRYPT 1997. LNCS, vol. 1233, pp. 226–238. Springer, Heidelberg (1997). doi:10.1007/3-540-69053-0_16
13. Mantin, I., Shamir, A.: A practical attack on broadcast RC4. In: Matsui, M. (ed.) FSE 2001. LNCS, vol. 2355, pp. 152–164. Springer, Heidelberg (2002). doi:10.1007/3-540-45473-X_13
14. Matthews, P.: A strong uniform time for random transpositions. J. Theoret. Probab. 1(4), 411–423 (1988)
15. Mironov, I.: (Not so) random shuffles of RC4. In: Yung, M. (ed.) CRYPTO 2002. LNCS, vol. 2442, pp. 304–319. Springer, Heidelberg (2002). doi:10.1007/3-540-45708-9_20
16. Mossel, E., Peres, Y., Sinclair, A.: Shuffling by semi-random transpositions. In: Foundations of Computer Science, pp. 572–581 (2004)
17. Naor, M., Reingold, O.: On the construction of pseudo-random permutations. In: Proceedings of the Twenty-Ninth Annual ACM Symposium on Theory of Computing - STOC 1997, pp. 189–199. ACM Press, New York (1997)
18. Pereida García, C., Brumley, B.B., Yarom, Y.: Make sure DSA signing exponentiations really are constant-time
19. Propp, J.G., Wilson, D.B.: Exact sampling with coupled Markov chains and applications to statistical mechanics. Random Struct. Algorithms 9, 223–252 (1996)
20. Schuldt, J.C.N., Rivest, R.L.: Spritz-a spongy RC4-like stream cipher and hash function (2014)
21. Paul, S., Preneel, B.: A new weakness in the RC4 keystream generator and an approach to improve the security of the cipher. In: Roy, B., Meier, W. (eds.) FSE 2004. LNCS, vol. 3017, pp. 245–259. Springer, Heidelberg (2004). doi:10.1007/978-3-540-25937-4_16
22. Standaert, F.-X., Pereira, O., Yu, Y., Quisquater, J.-J., Yung, M., Oswald, E.: Leakage resilient cryptography in practice. In: Sadeghi, A.-R., Naccache, D. (eds.) Towards Hardware-Intrinsic Security, pp. 99–134. Springer, Heidelberg (2010). doi:10.1007/978-3-642-14452-3_5
23. Maitra, S., Paul, G.: Analysis of RC4 and proposal of additional layers for better security margin. In: Chowdhury, D.R., Rijmen, V., Das, A. (eds.) INDOCRYPT 2008. LNCS, vol. 5365, pp. 27–39. Springer, Heidelberg (2008). doi:10.1007/978-3-540-89754-5_3
24. Yarom, Y., Genkin, D., Heninger, N.: CacheBleed: a timing attack on OpenSSL constant time RSA. CHES (2016)
25. Zoltak, B.: VMPC one-way function and stream cipher. In: Roy, B., Meier, W. (eds.) FSE 2004. LNCS, vol. 3017, pp. 210–225. Springer, Heidelberg (2004). doi:10.1007/978-3-540-25937-4_14

Cryptofication

A Virtual Wiretap Channel for Secure Message Transmission

Setareh Sharifian(✉), Reihaneh Safavi-Naini, and Fuchun Lin

Department of Computer Science, University of Calgary, Calgary, Canada
ssharifi@ucalgary.ca

Abstract. In the Wyner wiretap channel a sender is connected to a receiver and an eavesdropper through two noisy channels. It has been shown that if the noise in the eavesdropper channel is higher than the receiver's channel, information theoretically secure communication from Alice to Bob, without requiring a shared key, is possible. The approach is particularly attractive noting the rise of quantum computers and possibility of the complete collapse of todays' cryptographic infrastructure. If the eavesdropper's channel is noise free however, no secrecy can be obtained. The iJam protocol, proposed by Gollakota and Katabi, is an interactive protocol over noise free channels that uses friendly jamming by the receiver to establish an information theoretically secure shared key between the sender and the receiver. The protocol relies on the Basic iJam Transmission protocol (BiT protocol) that uses properties of OFDM (Orthogonal Frequency-Division Multiplexing) to create uncertainty for Eve (hence noisy view) in receiving the sent information, and use this uncertainty to construct a secure key agreement protocol. The protocol has been implemented and evaluated using extensive experiments that examines the best eavesdropper's reception strategy. In this paper we develop an abstract model for BiT protocol as a *wiretap channel* and refer to it as a *virtual wiretap channel*. We estimate parameters of this virtual wiretap channel, derive the secrecy capacity of this channel, and design a secure message transmission protocol with provable semantic security using the channel. Our analysis and protocol gives a physical layer security protocol, with provable security, that is implementable in practice (BiT protocol has already been implemented).

1 Introduction

Wireless communication provides flexible communication for mobile users, and with the increasing number of sensors and growth of the Internet of Things (IoT), will soon become the dominant form of communication. Wireless communication is vulnerable to passive eavesdropping. Wired Equivalent Privacy (WEP) is a security algorithm that was introduced in mid nineties to provide security for wireless access points, and was later replaced by Wi-Fi Protected Access (WPA) protocol [1]. Other communication security protocols such as Secure Socket Layer (SSL) [2] and Secure Shell (SSH) [3] are used for providing

© Springer International Publishing AG 2017
R.C.-W. Phan and M. Yung (Eds.): LNCS 10311, Mycrypt 2016, pp. 171–192, 2017.
DOI: 10.1007/978-3-319-61273-7_9

secure services over network. All these protocols rely on public key infrastructure to establish secure shared key between the sender and the receiver. Shor [23] proposed a quantum algorithm that efficiently solves the discrete logarithm and integer factorization problems, rendering today's public key infrastructure completely insecure if a quantum computer is invented. With advances in quantum technologies and projection of 10 years [9] to the development of such computers, the need and interest in the development of quantum-resistant cryptographic systems is rapidly growing.

In this paper we consider information theoretically secure communication systems that is secure against an adversary with unlimited computational power. Information theoretic security against a passive eavesdropper can be achieved using one-time-pad. This assumes sender and receiver share a secret key that is uniformly random and is of the same length as the message. The key must be chosen afresh for every message. These requirements severely limit the application of one-time-pad in practice. Wyner [33] proposed an ingenious model for information theoretically secure communication that is particularly suited for securing wireless communication. In Wyner wiretap model, a sender Alice is connected to a receiver Bob over a *main channel*. The eavesdropper, called Eve, receives the communication from the sender through a second channel referred to as the *wiretapper channel*. Wyner proved that as long as the wiretapper channel a degraded version of the main channel (or more generally noisier than the main channel), there exists an encoding method that provides information theoretic security for the receiver against Eve. A *wiretap code* is a randomized code that is used by the sender to encode the message. Wiretap channel allows quantum-resistant security using physical layer properties of the communication channels, complementing security that is provided at the higher layers of protocol stack layers using traditional cryptographic protocols. Security definition of wiretap channels has been strengthened over time with the latest security notion being semantic security: the strongest security notion for message confidentiality. Wiretap channels, however, rely on noise in the channel and need a correct estimate of noise in the wiretapper channel.

In [13], an innovative interactive physical layer protocol for key establishment over *noiseless channel* with security against a passive eavesdropper was introduced. The protocol was implemented and shown to provide security in practice, by measuring the received signal at Eve, and using the best decoding strategies to recover the sent information at Eve. The protocol uses cooperative jamming where the receiver sends a jamming signal that is combined with the sender's signal at Eve and creates an uncertain view of the communication for Eve, and uses that for providing security. One can see the approach as the sender and the receiver cooperatively creating a *virtual wiretap channel* and use that to establish a shared key.

In this paper we follow this intuition and model the main building block of iJam, referred to as *Basic iJam Transmission protocol (BiT protocol)*, as a virtual wiretap channel, and use it to provide efficient quantum-resistant secure message transmission with provable security.

1.1 Our Work

BiT protocol uses a coordinated jamming signal of the receiver to construct a noisy view of transmission for Eve. This is achieved by the sender repeating its transmitted information block in two consecutive subintervals, and the receiver randomly jamming one of the time samples of the two subintervals. Coordinated jamming ensures that the receiver is able to perfectly receive time samples that allow them to reconstruct a complete copy of the sent information block, while Eve will have a combination of jammed and unjammed samples which results in an uncertain view. This is shown to be achievable using appropriate choices of modulation and transmission technique (OFDM and 2^q-QAM modulation - See Sect. 2 for description).

We analyze BiT and show how it can be modelled as a virtual wiretap channel. Since the receiver is able to perfectly recover the transmitted information block, the virtual wiretap channel has a noiseless main channel. We estimate parameters of this channel and use them to compute the secrecy capacity of the virtual wiretap channel, that gives the best asymptotic efficiency for message transmission over this channel.

The modelling also allows us to adapt existing constructions of wiretap codes for providing message secrecy. We show how to use the wiretap encoding (seeded encryption) scheme of [6] to encode messages and then transmit the codeword using information block coding of the BiT protocol. The BiT protocol creation of a virtual wiretap channel ensures the seeded encryption will result in message transmission with information theoretic semantic security. The protocol achieves optimal efficiency asymptotically. The system thus provides provable quantum-resistant security, and is implementable in practice (thanks to starting from an already implemented protocol).

In Sect. 6 we show how this interpretation of BiT (a mechanism to add uncertainty in Eve's view) can be used to extend application of physical layer security protocols that use wiretap model. In particular we consider a setting where transmission in the physical channel from the sender to the receiver, is corrupted by Additive White Gaussian Noise (AWGN), but Eve has a noise free channel to Eve. Using known results for wiretap channels, secure communication using wiretap codes in this setting, is impossible. Using BiT protocol in this setting however, introduces uncertainty in Eve's view and so can enable secure communication. Figure 3 shows how to effectively use the BiT protocol to create a virtual wiretap channel when both the main channel and the wiretapper channel are noisy. The noise in the main channel is the physical noise, while the noise in the wiretapper channel is the result of the BiT protocol. Alice can send secret messages to Bob as long as the virtual wiretap channel is a stochastically degraded broadcast channel.

1.2 Related Work

Wiretap channel model was proposed by Wyner [33]. The model has attracted the attention of theoreticians and practitioners, resulting in a large body of work on the topic. A number of generalization of the mode has been proposed

[8,18,19], and the notion of security has been strengthened [6,21] over years, bringing it on par with the the strongest notion of security in cryptography. It has been proved that secure communication is possible if the eavesdropper's channel (signal reception ability) is worse than the receiver's [8]. There are efficient constructions of wiretap codes [6,20,31], with the more recent ones using a modular approach that can be used with any error correcting code.

Physical layer security protocols constructed by injecting jamming signal in the eavesdropper's view [11,17,27]. Showed that cooperative jamming can increase secrecy capacity [28–30]. In a general cooperative jamming setting, a trusted *helper* jams the transmitted signal. The legitimate receiver has some information about the jamming signal which is their advantage over the eavesdropper who is entirely oblivious to the jamming signal. This results in an inferior channel for the eavesdropper and so allows secure communication in presence of the eavesdropper. This type of jamming has also been referred to as, "helping" [26], or "friendly" [15] jamming. BiT protocol uses a variation of friendly jamming in which the receiver plays the role of the trusted helper.

BiT protocol [13] was used to construct a secret key agreement protocol (called iJam). The iJam key agreement uses multiple invocations of BiT protocol to establish a secret key that is generated as the XOR of multiple random strings, each transmitted in one invocation of BiT. The security of iJam has been experimentally evaluated.

Organization. Section 2 gives background and an outline of the BiT protocol. Section 3 is an example that motivates our approach, to modeling BiT as a virtual wiretap channel. In Sect. 4, we give our model of BiT as a virtual wiretap channel when the transmission from the sender to the receiver is noise free. Section 5 is a physical layer protocol for message transmission using a known seeded encryption algorithm and the BiT protocol. In Sect. 6, we study the case when the transmission from sender to receiver is corrupted by AWGN. Conclusion and future works are given in Sect. 7. In Appendix A, we provide approximation data and graphs of the information rate for the message transmission protocol in Sect. 5. In Appendix B, we provide an example of the noisy virtual main channel and virtual wiretapper channel of Sect. 6.

2 Preliminaries and Notations

We use uppercase letters X to denote random variables and bold lowercase letters to denote their corresponding realization. By $\Pr[X = \mathbf{x}]$ we mean the probability that X takes the value $\mathbf{x}$. This is also shown as $P_X(\mathbf{x})$. Calligraphic letters $\mathcal{X}$ denote sets, and $|\mathcal{X}|$ denotes the cardinality (number of elements) of a set. For two random variables X and Y, P_{XY} denotes their joint distribution, $P_{X|Y}$ denotes their conditional distribution, and P_X denotes X's marginal distribution. All *logs* are in base 2 and $\|$ is used to denote concatenation of two binary strings. For a random variable $X \in \mathcal{X}$, Shannon entropy is given by $H(X) = -\sum_{\mathbf{x} \in \mathcal{X}} P_X(\mathbf{x}) \log P_X(\mathbf{x})$. For two random variables $X \in \mathcal{X}$ and $Y \in \mathcal{Y}$ with joint probability distribution $P_{XY}(\mathbf{x}, \mathbf{y})$ and conditional probability distribution $P_{X|Y}(\mathbf{x}|\mathbf{y})$, the *conditional entropy* $H(X|Y)$ is defined as

$$H(X|Y) = -\sum_{\mathbf{x}\in\mathcal{X}} \sum_{\mathbf{y}\in\mathcal{Y}} P_{XY}(\mathbf{x}, \mathbf{y}) \log P_{X|Y}(\mathbf{x}|\mathbf{y}),$$

and the *mutual information* between the two is given by $I(X;Y) = H(X) - H(X|Y)$. The min-entropy of a random variable $X \in \mathcal{X}$, denoted by $H_\infty(X)$, is given by $H_\infty(X) = -\log(\max_{\mathbf{x}}(P_X(\mathbf{x})))$. The *statistical distance* between two random variables $X, Y \in \mathcal{X}$ is defined by,

$$SD(X,Y) \triangleq \frac{1}{2} \sum_{\mathbf{x}\in\mathcal{X}} |Pr(X = \mathbf{x}) - Pr(Y = \mathbf{x})|.$$

A communication channel is modelled as a probabilistic function that maps an input alphabet $\mathcal{X}$ to an output alphabet $\mathcal{Y}$. The channel $\mathsf{W}(X) = Y$ takes input $X \in \mathcal{X}$, and outputs $Y \in \mathcal{Y}$. The probability distribution of Y depends on the distributions of X and the probabilistic function $W(\cdot)$. In many communication systems input and/or output of the channel take values from real numbers. These are called *continuous channels*. An AWGN channel is a continuous channel in which the random variables X and Y corresponding to the input and output of the channel respectively, are related as $Y = X + N$, where N is the noise and is a random variable that is drawn from a zero-mean Gaussian distribution with variance $\frac{N_0}{2}$; that is, $\mathcal{N}(0, \frac{N_0}{2})$. If the noise variance is zero or the input is unconstrained, there exist an infinite subset of inputs that are distinguishable at the output with arbitrarily small error probability. However, in practice the variance is always non-zero and the input is always power limited. The input signal energy for each bit of the transmitted information block is denoted by E_b. This constrains the input signal energy and power. In a discrete channel W the input and output alphabets are discrete sets. The channel is specified by a *transition probability matrix* $\mathbf{P_W}$, where rows and columns are labelled by the input and output alphabets, respectively, and entries are conditional probabilities, $\mathbf{P_W}[\mathbf{x}, \mathbf{y}] = p_{\mathbf{xy}} = P_r(Y = \mathbf{y}|X = \mathbf{x})$. A channel is called *strongly symmetric* if the rows of the transition matrix are permutations of one another, and so is the case for the columns. The channel $\mathsf{W}(\cdot)$ is *symmetric* if there exists a partition of the output set $\mathcal{Y} = \mathcal{Y}_1 \cup \cdots \cup \mathcal{Y}_n$, such that for all i, the sub-matrix $\mathbf{P}_{W_i} = \mathbf{P_W}[\mathcal{X}, \mathcal{Y}_i]$ is strongly symmetric.

Wiretap Channel Model. In the general wiretap model, also called *broadcast model* [8], a sender is connected to the receiver through the *main* channel $\mathsf{W}_1 : \mathcal{X} \rightarrow \mathcal{Y}$, and to the eavesdropper through a second channel $\mathsf{W}_2 : \mathcal{X} \rightarrow \mathcal{Z}$, called the *wiretapper channel*. Thus, $\mathsf{WT} : \mathcal{X} \rightarrow \mathcal{Y} \times \mathcal{Z}$. In the Wyner's original model, the wiretapper channel is a *degraded* version of the main channel, and the Markov chain $X \rightarrow Y \rightarrow Z$ holds. We consider the original Wyner wiretap model. The goal of wiretap channel coding is to provide communication secrecy and reliability. Efficiency of wiretap codes is measured by the information rate, which is the number of information bits that can be transmitted reliably and secretly, per usage of the wiretap channel (One can also use a normalized form $R/\log|\Sigma|$ of the communication rate (cf [14]), where Σ is the code alphabet. For example, the information rate of linear codes is usually defined as the ratio of the code dimension to the block length.) The information rate of wiretap codes is upper bounded by the *secrecy capacity* C_s of the wiretap channel.

Theorem 1 [18]. *The secrecy capacity of Wyner wiretap channel when* W_1 *and* W_2 *are symmetric is given by,*

$$C_s = C_{W_1} - C_{W_2},$$

where C_{W_1} *and* C_{W_2} *are (reliability) channel capacities of* W_1 *and* W_2.

Since the capacity of a broadcast channel depends on the conditional marginal distributions only [7], the above capacity result also holds for a stochastically degraded broadcast channel, which is defined below.

Definition 1. *A broadcast channel* $\mathcal{X} \to \mathcal{Y} \times \mathcal{Z}$ *with conditional marginals* W_1 : $\mathcal{X} \to \mathcal{Y}$ *and* $W_2 : \mathcal{X} \to \mathcal{Z}$ *is said to be stochastically degraded if there exists a third channel* $W_3 : \mathcal{Y} \to \mathcal{Z}$ *such that,*

$$\mathbf{P}_{W_2}[\mathbf{x}, \mathbf{z}] = \sum_{\mathbf{y} \in \mathcal{Y}} \mathbf{P}_{W_3}[\mathbf{y}, \mathbf{z}] \mathbf{P}_{W_1}[\mathbf{x}, \mathbf{y}], \tag{1}$$

or equivalently

$$\mathbf{P}_{W_2} = \mathbf{P}_{W_3} \times \mathbf{P}_{W_1}.$$

2.1   QAM and OFDM

OFDM is a multicarrier modulation scheme which is widely used in modern wireless technologies and standards such as 4G mobile communications, WiMax, LTE and 802.11 a/g/n [22]. In OFDM many narrowband signals at different frequencies, each carry a small amount of information (number of bits). The narrowband signals may use modulations such as Quadrature Amplitude Modulation (QAM) which can be expressed as,

$$s(t) = A_I \cos 2\pi f_c t - A_J \sin 2\pi f_c t, \quad 0 < t < T,$$

where A_I and A_J are the amplitude for in-phase and quadrature phase components, f_c is the carrier frequency, and T is the symbol time duration. The *OFDM signal* is constructed at the transmitter by, (i) taking N (for example N = 64 in 802.11) QAM modulated signals, and (ii) applying Inverse Fast Fourier Transform (IFFT) to obtain OFDM time samples that will be sent over the channel. For N carrier frequencies, let $\mathbf{a}_k$ denote the OFDM time sample in the k-th time interval and obtained using IFFT:

$$\mathbf{a}_k = \sum_{n=0}^{N-1} \mathbf{A}_n e^{i2\pi kn/N} \qquad k = 0, 1, \dots, N-1, \tag{2}$$

where $\mathbf{A}_n$ is a complex number. Each *OFDM symbol* consists of N *time samples* $(\mathbf{a}_0, \mathbf{a}_1, \ldots, \mathbf{a}_{N-1})$. The transmitted signal is a sequence of OFDM time samples, each with Gaussian distribution. This is because each OFDM sample is a linear combination of N modulated *signals,* which because of central limit theorem results in a Gaussian distribution.

2.2   iJam and Basic iJam Transmission Protocol

iJam [13] is a protocol for key agreement between two parties, and uses Basic iJam Transmission (BiT) protocol as a subprotocol. Our focus is on BiT protocol. BiT protocol is a protocol between a sender and a receiver who also takes the role of a jammer, resulting in outputs for the receiver and the eavesdropper. The sender sends each OFDM symbol twice (the symbol and its identical copy) in two consecutive subintervals. Thus the time interval for sending an OFDM symbol twice of a subinterval (effectively doubling the sending time). An OFDM symbol is received as a sequence of time samples. The receiver randomly jams a time sample in the original symbol in the first subinterval, or its copy in the second subinterval. Jamming is by sending a Gaussian distributed jamming signal with the same distribution as the sent time samples, over the channel. The receiver will receive unjammed (clean) time samples of the two subintervals, and reconstructs the OFDM symbol with perfect fidelity.

2.3   Eavesdropper Strategies

In BiT, the sent time sample and the jamming signal will be combined at Eve's receiver. Thus for each OFDM symbol, Eve will receive two copies, each consisting of some jammed and some clean time samples. The eavesdropper can use different decoding strategies. They may treat the jamming signal as noise and try to decode in presence of jamming; or they can implement interference cancellation or joint decoding in an attempt to simultaneously decode the jamming signal and the original transmission. In [13] authors discuss strategies that can be used for the receiver's jamming signal to reduce detectability of the jammed samples. For example the jammer can transmit at an excessively high rate in an attempt to remove the possibility of joint decoding. This is because according to multiuser information theory, decoding multiple signals is impossible if the total information rate is outside the capacity region [32].

3   BiT as a Virtual Wiretap Channel – An Example

BiT is an interactive physical layer protocol between Alice and Bob, that takes input from Alice and Bob, and generates outputs for Bob and Eve. Alice's input is an information signal consisting of two copies of an input block of information bits; Bob's input is a coordinated jamming signal. The output of Bob is a block of information bits sent by Alice, and Eve's output is an element of Alice's space of block of information bits. We use a small example to provide intuition for our approach. In Example 1, we consider a scenario where Alice wants to send a 2-bit information block $\mathbf{x}$. Let $\mathbf{x}_s$ denote a 4-QAM modulated signal that carries the information block $\mathbf{x}$. For this small example, the OFDM symbol consists of only one signal ($N = 1$) and there is only a single time sample. Alice's input to the BiT protocol is two copies of the OFDM symbol (in this case $\mathbf{x}_s$), i.e. $(\mathbf{x}_s, \mathbf{x}_s)$, that are sent in two consecutive time subintervals. Bob's coordinated jamming signal is sent coordinated with Alice's transmission: Bob randomly chooses one of the two subintervals, corresponding to the two copies, and send their jamming signal in that time slot. For example, when Bob jams the second time slot, their jamming signal is $(\text{-}, \mathbf{J}'_s)$.

Bob will receive the signal corresponding to the unjammed time slot and will obtain the information block $\mathbf{x}$. Continuing with the above example, if Bob's input to the BiT protocol is $(\text{-}, \mathbf{J}'_s)$, he receives $(\mathbf{x}_s, \text{-})$.

Eve will receive a combination of the signals sent by Alice and Bob, $V_s = (\mathbf{x}_s, \mathbf{x}_s + J_s)$ where J_s is the jamming signal that is received by the Eve's antenna. If Eve cannot sufficiently distinguish the jammed signal from the unjammed one, the result will likely to cause an error in decoding. We denote Eve's decoder output by $\mathbf{z}$.

The above protocol can be seen as creating a wiretap (broadcast) channel from Alice to Bob and Eve, that can be described by the probability distribution $\Pr(\mathbf{y}, \mathbf{z}|\mathbf{x})$ where $\mathbf{x}, \mathbf{y}$, and $\mathbf{z}$ are the input of Alice, and outputs of Bob and Eve, respectively, as information blocks. Since $\mathbf{y} = \mathbf{x}$, the channel is characterized by $\Pr(\mathbf{z}|\mathbf{x})$ which represents the cumulative effect of detection of jamming, and decoding error caused by the received signal J_s.

Example 1. Let $\mathbf{x}$ be a 2-bit information block that is sent using BiT protocol and 4-QAM modulation with frequency f_1. Figure 1 shows the transmission of information block $\mathbf{x} = 00$ using BiT protocol. The process of Eve constructing their view of the channel is represented using a graph. In the graph, the physical output of the BiT protocol at Eve's side is a pair of signals denoted by V_s. One of the two signals is jammed and Eve tries to figure out which one. If Eve fails to distinguish the jammed signal from the clean one, V_s is decoded across one of the two edges labelled by $V = (\mathbf{x} \oplus J)\|\mathbf{x}$ and $V = \mathbf{x}\|(\mathbf{x} \oplus J)$, respectively. The list of 4-tuples following these edges, represent Eve's decoder's outputs after receiving the signal pairs and assuming the jammed subinterval is not detected. The next set of edges represent Eve's decision of information block based on the decoder's output. Note that when the decoder output is (0000), Eve decides correctly. In all other cases, Eve might make an error. For simplicity we assumed if the decoder's output of the two subintervals are different, Eve randomly chooses one of the two (they know one of the two are correct). The receiver, who is also the jammer, can always perfectly locate the unjammed subinterval and hence have perfect reception $\mathbf{y} = \mathbf{x} = 00$. In the following we provide more details on how the probability of Eve's outputting a particular information block can be obtained.

Eve receives two copies of the OFDM (here a 4-QAM) symbol denoted by V_s. Eve may use various decoding approaches to distinguish the jammed signal from the unjammed one. If Eve can detect the jammed subinterval (e.g. high

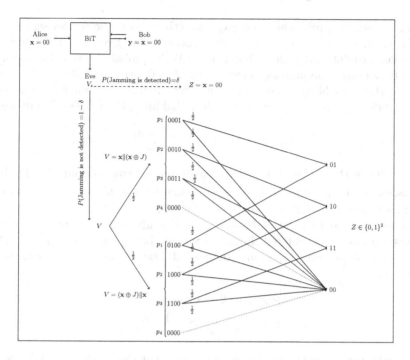

Fig. 1. BiT when a single 4-QAM (OFDM with $N = 1$) is used.

reception power), she can distinguish the jammed subinterval and can correctly receive the sent information block: they will simply discard the jammed subinterval and decode the unjammed one. Suppose Eve detects the correct jammed signal with probability $0 < \delta < 1$ (the dashed arrow in Fig. 1). This will create output $Z = \mathbf{x} = 00$ for Eve. If Eve's decoder cannot detect the jammed signal, the best thing she can do is to decode each OFDM symbol and then use the information about BiT protocol (repeated symbol) to find the sent information block. Eve's OFDM symbol decoder takes V_s and outputs either $V = \mathbf{x} \| (\mathbf{x} \oplus J)$ or $V = (\mathbf{x} \oplus J) \| \mathbf{x}$, depending on the receiver's choice of the jammed subinterval. Here J is a 2-bit random variable capturing the effect of the jamming on Eve's OFDM symbol decoding. The random variable J depends on the jamming signal power, the location of the adversary, and Eve's decoding capabilities, and does not depend on the sent OFDM symbol. Let $P[J = \alpha]$ denote the probability that jamming creates an offset α to the original information block. In our example, we set $P[J = 01] = p_1$, $P[J = 10] = p_2$, $P[J = 11] = p_3$ and $P[J = 00] = p_4$. To find the original transmitted information block, the adversary maps $V \in \{0, 1\}^4$ to $Z \in \{0, 1\}^2$. When $J = 00$, V consists of two identical information blocks and so is correctly mapped to the transmitted information block, $\mathbf{x}$ (dotted arrows in Fig. 1). When $J \neq 00$, Eve randomly chooses the decoded OFDM symbol of one of the two subintervals for $Z = \mathbf{z}$. One can use other distributions to choose the output OFDM symbol that better models the adversary's receiver.

To summarise, the probability that Eve correctly outputs the correct sent information block $\mathbf{x} = 00$ consists of, (i) the probability of Eve correctly detecting the jammed subinterval with probability δ, (ii) the probability that Eve cannot successfully detect the jammed interval, but $J = 00$ with probability $(1 - \delta)p_4$ and, (iii) the probability of jamming is not detected, $J \neq 00$ but Eve's guess of the sent information block is correct with probability $(1 - \delta)\frac{1-p_4}{2}$. Therefore:

$$P[Z = 00 | X = 00] = \delta + (1 - \delta)p_4 + (1 - \delta)\frac{1 - p_4}{2} = \delta + (1 - \delta)\frac{1 + p_4}{2}.$$

Next we study the probability of Eve having an incorrect output. To simplify the discussion, let $p_1 = p_2 = p_3 = \frac{(1-p_4)}{3}$. Then for any $\mathbf{x}' \in \{0,1\}^2$ such that $\mathbf{x}' \neq \mathbf{x}$, we have $P[Z = \mathbf{x}' | X = 00] = (1 - \delta)\frac{1-p_4}{6}$.

For any $\mathbf{x} \in \{0,1\}^2$, the probability that the adversary obtains the correct information block is calculated similar to $\mathbf{x} = 00$. Let $\eta = \delta + (1 - \delta)\frac{1+p_4}{2}$. The result of the above process specifies the probabilities of the wiretapper channel as follows:

$$P[Z = \mathbf{x} | X = \mathbf{x}] = \eta,$$
$$P[Z = \mathbf{x} | X \neq \mathbf{x}] = \frac{1 - \eta}{3}.$$

Thus the transition matrix of the virtual wiretapper channel W is as follows.

$$\mathbf{P_W} = \begin{bmatrix} \eta & \frac{1-\eta}{3} & \frac{1-\eta}{3} & \frac{1-\eta}{3} \\ \frac{1-\eta}{3} & \eta & \frac{1-\eta}{3} & \frac{1-\eta}{3} \\ \frac{1-\eta}{3} & \frac{1-\eta}{3} & \eta & \frac{1-\eta}{3} \\ \frac{1-\eta}{3} & \frac{1-\eta}{3} & \frac{1-\eta}{3} & \eta \end{bmatrix},$$

In summary, using BiT results in Eve receiving the information block $\mathbf{x}$ through a probabilistic channel with output $Z \in \{0,1\}^2$, resulting in a wiretapper channel that is noisier than the main channel (which is noiseless), hence enabling secure communication.

Remark 1. According to [13], when the three conditions described in Sects. 2.1, 2.2 and 2.3 are met, we can have $\eta < 1$ (the above example does not satisfy Sect. 2.1, so $\eta = 1$). We use the above example for the purpose of illustrating ideas.

4 Virtual Wiretap Channel Model

In the following we extend the above ideas to the general case where a complex OFDM signal is used.

Eavesdropper's View. Consider an OFDM signal with N frequencies where each signal uses 2^q-QAM modulation. Let $X \in \{0,1\}^{Nq}$ denote the information block that is transmitted using an OFDM symbol $(\mathbf{a}_0, \mathbf{a}_1, \cdots, \mathbf{a}_{N-1})$. By invoking BiT protocol, for each information block, $2N$ time samples are generated and sent over $2N$ consecutive time intervals. Eve receives $2N$ time samples. For two corresponding samples, one is a clean sample and the other is the jammed one. Let $V_s \in \mathbb{C}^{2N}$ be the random variable representing the $2N$ time samples. The received signal is mapped into an Nq-bit information block using the following eavesdropper decision unit (the includes their jamming detection, OFDM decoder and information block decision).

$$\mathsf{E} : \mathbb{C}^{2N} \to \{0,1\}^{Nq}.$$

There are two cases.

1. *Recovery of the information block is successful.* The adversary can correctly detect all N jammed samples, for example by examining the received signal power [24]. Using all the correct time samples, the adversary correctly recovers the OFDM symbol and the information block respectively. There are two other cases in which information block recovery is successful. One is when the jamming signal does not change any of the time samples and the other case is when the adversary's random guess for the clean sample is correct for all the clean samples. Let $\eta, 0 < \eta < 1$ denote the probability that the adversary recovers the information block correctly.

2. *Recovery of the information block fails.* If the adversary cannot correctly detect even one of the jammed time samples, because of the use of FFT on the time samples, all the recovered frequency samples will be affected and the recovered information block will be incorrect. We simplicity of calculations, we assume Eve outputs any of the incorrect information blocks from the set $\{0,1\}^{Nq}\backslash\{X\}$, with the same probability. That is each possible incorrect $2^{Nq} - 1$ string occurs with probability $\frac{1-\eta}{2^{Nq}-1}$. As noted earlier this can be replaced by other distributions that better estimates Eve's reception.

Let the random variable $Z \in \{0,1\}^{Nq}$ denote the information block that is output by Eve's decision unit E; that is, $Z = \mathsf{E}(V)$. We refer to Z as *Eve's view*. The conditional distribution of the Eve's view of the sent information block X is denoted by $Z|X$ and is given as follows.

$$P[Z = \mathbf{x}|X = \mathbf{x}] \simeq \eta,$$

$$P[Z = \mathbf{x}|X \neq \mathbf{x}] \simeq \frac{1-\eta}{2^{Nq}-1}.$$

Thus we have a virtual noisy channel $W : \{0,1\}^{Nq} \to \{0,1\}^{Nq}$ with transition matrix,

$$\mathbf{P_W} = \begin{bmatrix} \eta & \frac{1-\eta}{2^{Nq}-1} & \cdots & \frac{1-\eta}{2^{Nq}-1} \\ \frac{1-\eta}{2^{Nq}-1} & \eta & \cdots & \frac{1-\eta}{2^{Nq}-1} \\ \vdots & \vdots & \ddots & \vdots \\ \frac{1-\eta}{2^{Nq}-1} & \frac{1-\eta}{2^{Nq}-1} & \cdots & \eta \end{bmatrix}. \tag{3}$$

We call this channel a *virtual wiretapper channel* from the sender to Eve, represented by $Z = \mathsf{W}(X)$.

Receiver's View. The receiver always knows the unjammed time sample and so is effectively connected to the sender via a noiseless main channel.

Definition 2. *Let η denote the probability that Eve correctly recovers an information block that is sent using a Basic iJam Transmission (BiT) that uses OFDM with N-frequencies, each using 2^q-QAM. We define a virtual wiretap channel and denote it by $BiT_{\eta,q}^N$. This wiretap channel has noiseless main channel and the transition probability matrix of the wiretapper channel given in (3).*

Theorem 2. *The secrecy capacity of $BiT_{\eta,q}^N$ wiretap channel is given by:*

$$C_s(BiT_{\eta,q}^N) = -\{\eta \log \eta + (1 - \eta) \log \frac{1 - \eta}{(2^{Nq} - 1)}\}.$$

Proof. According to Definition 5, channel $W(\cdot)$ is symmetric and degraded with respect to the noiseless main channel.

The secrecy capacity of the wiretap channel is given by Theorem 1.

$$C_s = H(X|Z) - H(X|Y) = H(X|Z),$$

where X is uniform, and Y and Z are the output of the main channel and the wiretapper channel, respectively. Note that in the above equation $H(X|Y) = 0$ because the main channel is noiseless. Using the transition probability matrix in Definition 5, we have,

$$\begin{aligned} H(X|Z) &= \sum_{z \in \{0,1\}^{Nq}} P[Z = z] H(X|Z = z) \\ &= -\{\eta \log \eta + (1 - \eta) \log \frac{1 - \eta}{(2^{Nq} - 1)}\}. \end{aligned} \tag{4}$$

$\square$

5   Secure Message Transmission Using BiT

BiT had been introduced [13] to construct a key agreement protocol. Using the above model we construct a secure message transmission protocol with *provable security*. We will use capacity-achieving wiretap coding construction in [4] that provides semantic security, and has efficient encryption and decryption functions. The wiretap construction in [6] is for binary input symmetric channels. The q-ary channel alphabet is from [4, Sect. 5.5] and its extension [5].

5.1   A Semantically Secure Wiretap Code

The construction is a seeded encryption and uses an invertible extractor.

Definition 3 [10]. *A function* $\mathsf{EXT} : Sds \times \{0,1\}^n \to \{0,1\}^\ell$ *is a (d, ϵ)-strong, average-case extractor if, $SD((\mathsf{EXT}(S,X), Z, S); (U, Z, S)) \le \epsilon$ for all pairs of correlated random variables (X, Z) over $\{0,1\}^n \times \{0,1\}^*$, assuming $\tilde{H}_\infty(X|Z) \ge d$.*

Seeded Encryption. For a public uniformly distributed random variable $S \in Sds$ and an arbitrarily distributed message $M \in \{0,1\}^b$, the seeded encryption function $\mathsf{SE} : Sds \times \{0,1\}^b \to \{0,1\}^{nNq}$, outputs a ciphertext $\mathsf{SE}(S, M)$. The corresponding seeded decryption function is $\mathsf{SDE} : Sds \times \{0,1\}^{nNq} \to \{0,1\}^b$ such that for all $S \in Sds$ and $M \in \{0,1\}^b$ we have $\mathsf{SDE}(S, \mathsf{SE}(S, M)) = M$.

Inverting Extractors. The function $\mathsf{INV} : \{0,1\}^r \times Sds \times \{0,1\}^b \to \{0,1\}^{nNq}$ is an inverter for the extractor $\mathsf{EXT}(\cdot, \cdot)$ in Definition 3, if for a uniform $R \in \{0,1\}^r$ and for all $S \in Sds$ and $Y \in \{0,1\}^b$, the random variable $\mathsf{INV} : (S, R, Y)$ is uniformly distributed over all preimages of Y under $\mathsf{EXT}(S, \cdot)$.

Let $Sds = \{0,1\}^{nNq} \backslash 0^{nNq}$. For inputs $S \in Sds$ and $X \in \{0,1\}^{nNq}$ and $nNq > b$, the function $\mathsf{EXT} : Sds \times \{0,1\}^{nNq} \to \{0,1\}^b$ is defined as follows.

$$\mathsf{EXT}(S, X) = (S \odot X)|_b,$$

where $\odot$ denotes multiplication over $\mathbb{F}_2^{nNq} = \{0,1\}^{nNq}$, and $X|_b$ denotes the first n bits of X. An efficient inverter for $\mathsf{EXT}(S, X)$ is given by $\mathsf{INV}(S, R, M) = S^{-1} \odot (M\|R)$, where S^{-1} denotes the multiplicative inverse of S in $\mathbb{F}_2^{nNq}$ and R is a uniformly distributed variable over $\{0,1\}^{n-b}$. For the message block $M \in \{0,1\}^b, S \in Sds$, and $R \xleftarrow{\$} \{0,1\}^r$, the seeded encryption function $\mathsf{SE}(S, M)$ is defined as follows.

$$X = \mathsf{SE}(S, M) = \mathsf{INV}(S, R, M) = S^{-1} \odot (M\|R).$$

5.2   Using the Wiretap Construction with $\mathsf{BiT}_{\eta,q}^N$

Let ENC denote the construction that uses wiretap coding for $\mathsf{BiT}_{\eta,q}^N$.

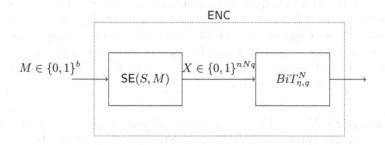

Fig. 2. Secure message transmission based on BiT protocol

As illustrated in Fig. 2, the encryption block ENC consists of two sub-blocks:

1. A seeded wiretap encryption code $\mathsf{SE} : Sds \times \{0,1\}^b \to \{0,1\}^{nNq}$ that encrypts each information block of size b bits into a codeword of size nNq bits.

2. The $\mathrm{BiT}_{\eta,q}^N$ block that breaks the codeword into Nq-bit units, and sends it using BiT protocol.

To capture efficiency of the proposed message transmission protocol, we define the communication rate $\mathcal{R}$ of the system as the number of transmitted bits that are sent with security and reliability, in each application of $\mathrm{BiT}_{\eta,q}^N$. This is similar to the definition of rate in wiretap channel literature (cf [8]).

Definition 4. *The rate of the message transmission protocol over $BiT_{\eta,q}^N$ in Fig. 2 is $\mathcal{R} = \frac{b}{n}$.*

The rate of the ENC block in Fig. 2 asymptotically approaches the secrecy capacity of the virtual wiretap channel $\mathrm{BiT}_{\eta,q}^N$. The construction provides semantic security and reliability. The codeword length from $\mathsf{SE}(S, M)$ is $nNq = b + r$, where b is the total length of the message and r is the length of the concatenated random string. For σ bit semantic security, the length of r is given in [25] as recalled below.

$$r = \left\lceil 2(\sigma + 1) + \sqrt{n}\log(2^{Nq} + 3)\sqrt{2(\sigma + 3)} + (n)\psi(\mathsf{W}) \right\rceil,$$

where $\psi(\mathsf{W}) = |\log \mathcal{Z}| - H(\mathsf{W}) = Nq - H(X|Z)$ in the above equation. Secrecy capacity of $\mathrm{BiT}_{\eta,q}^N$ for $N = 64$ and various values of η and q, are given in the Appendix A.

6   BiT Over Noisy Receiver Channel

In Wyner wiretap model the secrecy capacity is zero when the main channel is noisy while the eavesdropper's channel is noise free. That is one cannot expect any secure communication from Alice to Bob. BiT creates a virtual wiretap channel for Eve when the physical channel between Alice and Bob is noise free. In the following we will show that when receiver's physical channel is corrupted by Additive White Gaussian Noise (AWGN) (while the eavesdropper's physical channel remains noise free), BiT can be used to introduce noise in the Eve's channel and so make secure communication possible. Figure 3 shows application of BiT when the main channel is corrupted by AWGN.

Eavesdropper's View. The eavesdropper's channel is the same as in Sect. 4, created by the BiT protocol. This is because the noise only affects transmission in the main channel. Eve receives $V_s = (\mathbf{x_s} \oplus J_s)\|\mathbf{x_s}$ or $V_s = \mathbf{x_s}\|(\mathbf{x_s} \oplus J_s)$, and the eavesdropper channel transition probability is given by (3).

Receiver's View. The receiver channel, however, is corrupted by AWGN. We first consider the effect of AWGN on a *single 2^q-QAM signal* (i.e., OFDM with a single frequency) and then generalize it to an OFDM with N frequencies.

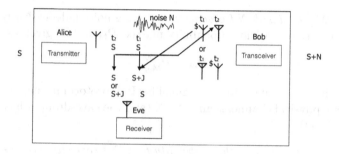

Fig. 3. BiT protocol when Bob's physical channel is noisy

Let AWGN($\cdot$) denote the AWGN channel where a noise is added to the input. Bob knows which subinterval is jammed. Therefore, his reception is one OFDM symbol corrupted by the AWGN noise, that is

$$AWGN(\mathbf{x}_s) = \mathbf{x}_s + N_s,$$

where N_s denotes the random signal corresponding to the white Gaussian noise. Let B($\cdot$) be the function that maps Bob's received signal to an Nq-bit string. The virtual main channel from Alice to Bob is defined as,

$$Y = M(X) = B(AWGN(\mathbf{x}_s)).$$

Let the transition probability matrix of a 2^q-QAM signal that is corrupted by AWGN be denoted by $\mathbf{P}_{M,q}$. Using the error probability calculation of BPSK in [12] Chap. 6.1.2, the 4-QAM transition probability matrix will be given as:

$$\mathbf{P}_{M,2} = \begin{bmatrix} (1-P_b)(1-P_b) & P_b(1-P_b) & P_b(1-P_b) & P_b^2 \\ P_b(1-P_b) & (1-P_b)(1-P_b) & P_b^2 & P_b(1-P_b) \\ P_b(1-P_b) & P_b^2 & (1-P_b)(1-P_b) & P_b(1-P_b) \\ P_b^2 & P_b(1-P_b) & P_b(1-P_b) & (1-P_b)(1-P_b) \end{bmatrix},$$

where the probability P_b is computed as follows.

$$P_b = Q(\sqrt{\frac{E_b}{N_0}}),$$

where E_b is the energy-per-bit of the input signal, $\frac{N_0}{2}$ is the variance of the AWGN, and $Q(z)$ is the probability that a Gaussian random variable x with mean 0 and variance 1 takes a value larger than z, namely,

$$Q(z) = P[x > z] = \int_z^\infty \frac{1}{2\pi} e^{-x^2/2} dx.$$

The function $Q(\cdot)$ can be efficiently computed using approximations such as the one in [16].

For *OFDM signal with N frequencies*, assuming noise independently corrupts each frequency the transition probability matrix, $\mathbf{P_M}$ will be given as,

$$\mathbf{P_M} = \mathbf{P}_{\mathsf{M},q}^{\otimes N}. \tag{5}$$

We thus have a virtual wiretap channel for BiT protocol in the setting where the receiver's physical channel is an AWGN (and the eavesdropper has noise free physical channel).

Definition 5. *Let η denote the probability that Eve correctly recovers an information block that is sent using a Basic iJam Transmission (BiT) that uses OFDM with N-frequencies, each using 2^q-QAM. We define a virtual wiretap channel for the setting where the receiver's physical channel is an AWGN and denote it by AWGN-BiT$_{\eta,q}^N$. This wiretap channel has a noisy main channel with transition probability matrix given by (5) and a wiretapper channel with transition probability matrix given by (3).*

Theorem 3. *The secrecy capacity of AWGN-BiT$_{\eta,q}^N$ is given by,*

$$C_s = C_\mathsf{M} - C_\mathsf{W},$$

if the matrix $\mathbf{R} = \mathbf{P_W} \times \mathbf{P_M}^{-1}$ is the transition probability matrix of a channel, namely, $\mathbf{R}$ satisfies the following two conditions,

1. *$\mathbf{R}$ does not have any negative component,*
2. *The sum of the components in each row of $\mathbf{R}$ is equal to 1.*

Remark 2. Condition 1 in Theorem 3 can be satisfied by imposing a relation between η (the parameter characterizing the virtual wiretapper channel W) and P_b (the parameter characterizing the virtual main channel M). Condition 2 can be verified directly by computation. We provide more details by giving an example for $N = 1$ case in Appendix B.

Proof. From $\mathbf{R} = \mathbf{P_W} \times \mathbf{P_M}^{-1}$, we have

$$\mathbf{R} \times \mathbf{P_M} = \mathbf{P_W}.$$

Conditions 1 and 2 are sufficient to ensure that $\mathbf{R}$ is a transition probability matrix for a channel and so using Definition 1, $\mathbf{P_W}$ is a stochastically degraded channel with respect to $\mathbf{P_M}$. The rest of the proof follows from Theorem 1. □

7   Conclusion and Future Works

BiT uses an innovative way of coordinated jamming to construct a virtual wiretap channel and enable information theoretically secure communication without a shared key. We showed how to model BiT as a virtual wiretap channel, estimate its parameters, and use the model to design a provably secure message transmission protocol.

BiT is a subprotocol of iJam protocol that had been implemented and experimentally analyzed. By formal modelling of BiT protocol and developing a provably secure message transmission scheme based on that, we have effectively constructed a keyless information theoretically secure message transmission system that can be used in practice.

Our scheme asymptotically achieves the secrecy capacity of the virtual wiretap channel. The primary assumption underlying our modelling is that the decoding error probability of Eve can be estimated. This probability depends on factors such as sender and receiver (jamming) signal power, the location and receiving equipments of the eavesdropper. An interesting direction for future work would be to design protocols that are more robust to correct estimation of the error probability. Extending our analysis and approach to other physical layer security protocols is also an interesting direction for future work.

Appendix A: Achievable Transmission Rate Using $\text{BiT}_{q,\eta}^N$

For a noise free main channel, the secrecy capacity of $\text{BiT}_{q,\eta}^N$ is given by:

$$C_s(\text{BiT}_{\eta,q}^N) = -\{\eta \log \eta + (1-\eta) \log \frac{1-\eta}{(2^{Nq}-1)}\}.$$

Figure 4 shows the rate of communication when, the information block length is Nq bits, $q = 2, 3$ and 4, and $N = 64$. The graphs show the achievable rates for $\sigma = 128$ semantic security, and $\eta = 0.2$ (upper graph) and $\eta = 0.4$ (lower graph). The figures show that the achievable secrecy rate and secrecy capacity decreases as η grows. This is expected because higher η means that the adversary has a better chance of correctly decoding the jammed signal.

Appendix B: BiT over Noisy Receiver Channel—An Example

In this section we derive a sufficient relation between P_b and η so that the virtual wiretap channel is a stochastically degraded broadcast channel. Following Sect. 3, the transition matrix of the virtual wiretapper channel W for $q = 2$ is given by:

$$\mathbf{P_W} = \begin{bmatrix} \eta & \frac{1-\eta}{3} & \frac{1-\eta}{3} & \frac{1-\eta}{3} \\ \frac{1-\eta}{3} & \eta & \frac{1-\eta}{3} & \frac{1-\eta}{3} \\ \frac{1-\eta}{3} & \frac{1-\eta}{3} & \eta & \frac{1-\eta}{3} \\ \frac{1-\eta}{3} & \frac{1-\eta}{3} & \frac{1-\eta}{3} & \eta \end{bmatrix},$$

where $u = \frac{1-\eta}{3}$, and $v = \eta - \frac{1-\eta}{3} = \frac{4\eta-1}{3}$. Note that the sum of each row is $4u + v = 1$. On the other hand, we can compute:

$$\mathbf{P_M^{-1}} = \frac{1}{(1-2P_b)^2} \cdot$$
$$\begin{pmatrix} (1-P_b)(1-P_b) & -P_b(1-P_b) & -P_b(1-P_b) & P_b^2 \\ -P_b(1-P_b) & (1-P_b)(1-P_b) & P_b^2 & -P_b(1-P_b) \\ -P_b(1-P_b) & P_b^2 & (1-P_b)(1-P_b) & -P_b(1-P_b) \\ P_b^2 & -P_b(1-P_b) & -P_b(1-P_b) & (1-P_b)(1-P_b) \end{pmatrix}.$$

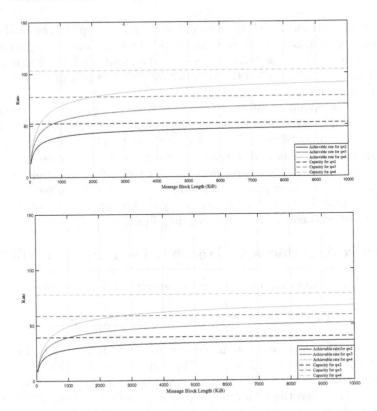

Fig. 4. The secrecy rate and capacity (bits per channel use) for $N = 64$ and different values of q for $\eta = 0.2$ (upper graph) and $\eta = 0.4$ (lower graph).

Let $a = 1 - P_b$ and $b = P_b$. The above matrix can be written as:

$$\mathbf{P}_M^{-1} = \frac{1}{(a-b)^2} \cdot \begin{pmatrix} a^2 & -ab & -ab & b^2 \\ -ab & a^2 & b^2 & -ab \\ -ab & b^2 & a^2 & -ab \\ b^2 & -ab & -ab & a^2 \end{pmatrix}.$$

The sum of entries of each row is given by, $\frac{1}{(a-b)^2}(a^2 - 2ab + b^2) = 1$. The following is used to prove the required relation.

Lemma 1. *Let there be two matrices*

$$A = \begin{bmatrix} a_{11} & a_{12} & \dots & a_{1n} \\ a_{21} & a_{22} & \dots & a_{2n} \\ \vdots & \vdots & & \vdots \\ a_{n1} & a_{n2} & \dots & a_{nn} \end{bmatrix}, B = \begin{bmatrix} b_{11} & b_{12} & \dots & b_{1n} \\ b_{21} & b_{22} & \dots & b_{2n} \\ \vdots & \vdots & & \vdots \\ b_{n1} & b_{n2} & \dots & b_{nn} \end{bmatrix}.$$

If $\sum_{j=1}^{n} a_{ij} = 1$ and $\sum_{j=1}^{n} b_{ij} = 1$ for any $i \in [n]$, then $\sum_{j=1}^{n}(AB)_{ij} = 1$, for any $i \in [n]$.

Proof. For any $i \in [n]$,

$$\sum_{j=1}^{n}(AB)_{ij} = \sum_{j=1}^{n}\left(\sum_{k=1}^{n} a_{ik}b_{kj}\right)$$
$$= \sum_{k=1}^{n} a_{ik} \cdot \left(\sum_{j=1}^{n} b_{kj}\right)$$
$$= \sum_{k=1}^{n} a_{ik}$$
$$= 1.$$

□

Lemma 2. *The virtual wiretap channel is a stochastically degraded broadcast channel if $P_b \leq \frac{1-\sqrt{\frac{4\eta-1}{3}}}{2}$ and $\eta > \frac{1}{4}$.*

Proof. The virtual wiretap channel is a stochastically degraded broadcast channel if there exists a matrix $\mathbf{R}$ such that $\mathbf{P_W} = \mathbf{P_M} \times \mathbf{R}$, and $\mathbf{R}$ is a channel transition matrix; that is, has non-negative entries and each row sums to 1. Using the matrices $\mathbf{P_M}$ and $\mathbf{P_W}$ above, we have:

$$\mathbf{R} = \mathbf{P_W} \times \mathbf{P_M^{-1}}$$
$$= \frac{1}{(a-b)^2}\begin{bmatrix} u(a-b)^2+va^2 & u(a-b)^2-vab & u(a-b)^2-vab & u(a-b)^2+vb^2 \\ u(a-b)^2-vab & u(a-b)^2+va^2 & u(a-b)^2+vb^2 & u(a-b)^2-vab \\ u(a-b)^2-vab & u(a-b)^2+vb^2 & u(a-b)^2+va^2 & u(a-b)^2-vab \\ u(a-b)^2+vb^2 & u(a-b)^2-vab & u(a-b)^2-vab & u(a-b)^2+va^2 \end{bmatrix}.$$

Using Lemma 1, entries in each row of $\mathbf{R}$ sum to 1.

To ensure entries of $\mathbf{R}$ are all non-negative, we first note that $u(a-b)^2+va^2 > 0$ and $u(a-b)^2+vb^2 > 0$. So the virtual wiretap channel is a stochastically degraded broadcast channel if $u(a-b)^2 - vab \geq 0$ and so:

$$u(a-b)^2 - vab \geq 0 \Leftrightarrow ua^2 + ub^2 - (2u+v)ab \geq 0$$
$$\Leftrightarrow ua^2 + ub^2 - (2u+1-4u)ab \geq 0$$
$$\Leftrightarrow ua^2 + ub^2 - (1-2u)ab \geq 0$$
$$\Leftrightarrow u(a+b)^2 - ab \geq 0$$
$$\Leftrightarrow u - ab \geq 0$$
$$\Leftrightarrow P_b^2 - P_b + u \geq 0,$$

where $4u+v = 1$ and $a+b = 1$ are repeatedly invoked to simplify the expressions. The solution to the above inequality depends on the determinant $1 - 4u$. When $1 - 4u > 0$, we have

$$P_b^2 - P_b + u \geq 0 \Leftrightarrow \left(P_b - \frac{1-\sqrt{1-4u}}{2}\right)\left(P_b - \frac{1+\sqrt{1-4u}}{2}\right) \geq 0$$
$$\Leftrightarrow \left(P_b - \frac{1-\sqrt{v}}{2}\right)\left(P_b - \frac{1+\sqrt{v}}{2}\right) \geq 0$$
$$\Leftrightarrow \left(P_b - \frac{1-\sqrt{\frac{4\eta-1}{3}}}{2}\right)\left(P_b - \frac{1+\sqrt{\frac{4\eta-1}{3}}}{2}\right) \geq 0$$
$$\Leftrightarrow P_b \leq \frac{1-\sqrt{\frac{4\eta-1}{3}}}{2} \text{ or } P_b \geq \frac{1+\sqrt{\frac{4\eta-1}{3}}}{2}.$$

By assumption, $P_b \in [0, \frac{1}{2}]$ and so $P_b \leq \frac{1-\sqrt{\frac{4\eta-1}{3}}}{2} = \frac{1}{2} - \sqrt{\frac{4\eta-1}{12}}$.   □

Example 2. Let $P_b = 0.1$ and Let $\eta = 0.55$. Therefore,

$$\mathbf{P}_\mathsf{M} = \begin{bmatrix} 0.81 & 0.09 & 0.09 & 0.01 \\ 0.09 & 0.81 & 0.01 & 0.09 \\ 0.09 & 0.01 & 0.81 & 0.09 \\ 0.01 & 0.09 & 0.09 & 0.81 \end{bmatrix}$$

and

$$\mathbf{P}_\mathsf{W} = \begin{bmatrix} 0.55 & 0.15 & 0.15 & 0.15 \\ 0.15 & 0.55 & 0.15 & 0.15 \\ 0.15 & 0.15 & 0.55 & 0.15 \\ 0.15 & 0.15 & 0.15 & 0.55 \end{bmatrix} .$$

Therefore

$$\mathbf{R} = \mathbf{P}_\mathsf{W} \times \mathbf{P}_\mathsf{M}^{-1} = \begin{bmatrix} 0.66 & 0.094 & 0.094 & 0.156 \\ 0.094 & 0.66 & 0.156 & 0.094 \\ 0.094 & 0.156 & 0.66 & 0.094 \\ 0.156 & 0.094 & 0.094 & 0.66 \end{bmatrix} .$$

$\mathbf{R}$ is the transition probability matrix of a virtual channel that confirms $\mathbf{P}_\mathsf{W}$ is degraded with respect to $\mathbf{P}_\mathsf{M}$. The secrecy capacity in this example is

$$C_s = C_\mathsf{M} - C_\mathsf{W} = (2 - 0.7624) - (2 - 1.1515) = 0.3891.$$

References

1. 802.1x & WPA settings. https://www.ietf.org/mail-archive/web/ietf/current/msg32026.html
2. The secure sockets layer (SSL) protocol version 3.0. https://tools.ietf.org/html/rfc6101
3. SSH protocol architecture. https://www.ietf.org/proceedings/52/I-D/draft-ietf-secsh-architecture-11.txt
4. Bellare, M., Tessaro, S.: Polynomial-time, semantically-secure encryption achieving the secrecy capacity. arXiv preprint arXiv:1201.3160 (2012)
5. Bellare, M., Tessaro, S., Vardy, A.: A cryptographic treatment of the wiretap channel. arXiv preprint arXiv: 1201.2205 (2012)
6. Bellare, M., Tessaro, S., Vardy, A.: Semantic security for the wiretap channel. In: Safavi-Naini, R., Canetti, R. (eds.) CRYPTO 2012. LNCS, vol. 7417, pp. 294–311. Springer, Heidelberg (2012). doi:10.1007/978-3-642-32009-5_18
7. Bergmans, P.: Random coding theorem for broadcast channels with degraded components. IEEE Trans. Inf. Theory **19**(2), 197–207 (1973)
8. Csiszár, I., Körner, J.: Broadcast channels with confidential messages. IEEE Trans. Inf. Theory **24**(3), 339–348 (1978)
9. Dickerson, K.: Microsoft lab predicts we'll have a working 'hybrid' quantum computer in 10 years, October 2015. http://www.techinsider.io/microsoft-hybrid-quantum-computer-2015-10

10. Dodis, Y., Reyzin, L., Smith, A.: Fuzzy extractors: how to generate strong keys from biometrics and other noisy data. In: Cachin, C., Camenisch, J.L. (eds.) EUROCRYPT 2004. LNCS, vol. 3027, pp. 523–540. Springer, Heidelberg (2004). doi:10.1007/978-3-540-24676-3_31

11. Dong, L., Han, Z., Petropulu, A.P., Poor, H.V.: Cooperative jamming for wireless physical layer security. In: 2009 IEEE/SP 15th Workshop on Statistical Signal Processing, pp. 417–420. IEEE (2009)

12. Goldsmith, A.: Wireless Communications. Cambridge University Press, Cambridge (2005)

13. Gollakota, S., Katabi, D.: Physical layer wireless security made fast and channel independent. In: 2011 Proceedings IEEE INFOCOM, pp. 1125–1133. IEEE (2011)

14. Guruswami, V.: Bridging Shannon and Hamming: list error-correction with optimal rate (2010)

15. Han, Z., Marina, N., Debbah, M., Hjørungnes, A.: Physical layer security game: interaction between source, eavesdropper, and friendly jammer. EURASIP J. Wirel. Commun. Netw. **2009**(1), 1 (2010)

16. Karagiannidis, G.K., Lioumpas, A.S.: An improved approximation for the Gaussian Q-function. IEEE Commun. Lett. **11**(8) (2007)

17. Lai, L., El Gamal, H.: The relay-eavesdropper channel: cooperation for secrecy. IEEE Trans. Inf. Theory **54**(9), 4005–4019 (2008)

18. Leung-Yan-Cheong, S.: On a special class of wiretap channels (Corresp.). IEEE Trans. Inf. Theory **23**(5), 625–627 (1977)

19. Leung-Yan-Cheong, S., Hellman, M.: The Gaussian wire-tap channel. IEEE Trans. Inf. Theory **24**(4), 451–456 (1978)

20. Mahdavifar, H., Vardy, A.: Achieving the secrecy capacity of wiretap channels using polar codes. IEEE Trans. Inf. Theory **57**(10), 6428–6443 (2011)

21. Muramatsu, J., Miyake, S.: Construction of wiretap channel codes by using sparse matrices. In: 2009 IEEE Information Theory Workshop (2009)

22. Schulze, H., Lüders, C.: Theory and Applications of OFDM and CDMA: Wideband Wireless Communications. Wiley, Hoboken (2005)

23. Shor, P.W.: Polynomial-time algorithms for prime factorization and discrete logarithms on a quantum computer. SIAM Rev. **41**(2), 303–332 (1999)

24. Strasser, M., Danev, B., Čapkun, S.: Detection of reactive jamming in sensor networks. ACM Trans. Sens. Netw. (TOSN) **7**(2), 16 (2010)

25. Tal, I., Vardy, A.: Channel upgrading for semantically-secure encryption on wiretap channels. In: 2013 IEEE International Symposium on Information Theory Proceedings (ISIT), pp. 1561–1565. IEEE (2013)

26. Tang, X., Liu, R., Spasojevic, P., Poor, H.V.: The Gaussian wiretap channel with a helping interferer. In: 2008 IEEE International Symposium on Information Theory, pp. 389–393. IEEE (2008)

27. Tang, X., Liu, R., Spasojevic, P., Poor, H.V.: Interference assisted secret communication. IEEE Trans. Inf. Theory **57**(5), 3153–3167 (2011)

28. Tekin, E., Yener, A.: Achievable rates for the general Gaussian multiple access wire-tap channel with collective secrecy. arXiv preprint cs/0612084 (2006)

29. Tekin, E., Yener, A.: The Gaussian multiple access wire-tap channel: wireless secrecy and cooperative jamming. In: 2007 Information Theory and Applications Workshop, pp. 404–413. IEEE (2007)

30. Tekin, E., Yener, A.: The general Gaussian multiple-access and two-way wire-tap channels: achievable rates and cooperative jamming. IEEE Trans. Inf. Theory **54**(6), 2735–2751 (2008)

31. Thangaraj, A., Dihidar, S., Calderbank, A.R., McLaughlin, S.W., Merolla, J.-M.: Applications of LDPC codes to the wiretap channel. IEEE Trans. Inf. Theory **53**(8), 2933–2945 (2007)
32. Tse, D., Viswanath, P.: Fundamentals of Wireless Communication. Cambridge University Press, Cambridge (2005)
33. Wyner, A.D.: The wire-tap channel. Bell Syst. Tech. J. **54**(8), 1355–1387 (1975)

Necessary and Sufficient Numbers of Cards for Securely Computing Two-Bit Output Functions

Danny Francis[1], Syarifah Ruqayyah Aljunid[1], Takuya Nishida[1],
Yu-ichi Hayashi[2], Takaaki Mizuki[3(✉)], and Hideaki Sone[3]

[1] Graduate School of Information Sciences, Tohoku University,
6-3-09 Aramaki-Aza-Aoba, Aoba-ku, Sendai, Miyagi 980-8578, Japan
[2] Faculty of Engineering, Tohoku Gakuin University,
1-13-1 Chuo, Tagajo, Miyagi 985-8537, Japan
[3] Cyberscience Center, Tohoku University,
6-3 Aramaki-Aza-Aoba, Aoba-ku, Sendai, Miyagi 980-8578, Japan
tm-paper+card2out@g-mail.tohoku-university.jp

Abstract. In 2015, Koch *et al.* proposed a five-card finite-runtime committed protocol to compute securely the AND function, showing that their protocol was optimal: there is no protocol computing the AND function with four cards in finite-runtime fashion and committed format. Thus, necessary and sufficient numbers of cards for computing single-bit output functions are known. However, as for two-bit output functions, such an exact characterization is unknown. This paper gives a six-card (or less) protocol for each of all two-bit output functions and proves that our finite-runtime committed protocols are optimal by providing a lower bound. In other words, we give the necessary and sufficient number of cards for any two-bit output function to be computed by a finite-runtime committed protocol. Our lower bound can also be applied to any function which outputs more than two bits.

1 Introduction

Card-based cryptographic protocols perform secure multi-party computations based on simple playing cards. They are interesting because they can be implemented with physical objects, and also have been practically used for pedagogical purpose [3]. Such a protocol was described for the first time in 1989 at EURO-CRYPT '89, that is, den Boer [1] described a protocol based on a deck of five cards computing securely the AND Boolean function $(a, b) \mapsto a \wedge b$. Let us introduce this protocol. Each of two players is given two cards: one ♡ and one ♣. These two cards allow each player to encode a bit: 0 is encoded by ♣♡ and 1 is encoded by ♡♣. Each player chooses an input. According to his input bit b, Player 2 arranges his two cards, and puts them on the table face down after he reverses the order of these two cards. Next, a ♣ card is put on the stack, and then Player 1 puts his own cards in the right order on the stack according to his input bit a. Now, we have a stack

© Springer International Publishing AG 2017
R.C.-W. Phan and M. Yung (Eds.): LNCS 10311, Mycrypt 2016, pp. 193–211, 2017.
DOI: 10.1007/978-3-319-61273-7_10

from top to bottom, and Table 1 sums up what the possible stacks are. One can see easily that stacks leading to a zero output ($a \wedge b = 0$) are cyclic permutations of one another. If each player cuts the stack randomly, the final result $a \wedge b$ can be deduced by revealing the cards, and no one can know which inputs the players chose.

Table 1. Possible input sequences for den Boer's protocol

Player 1	Player 2	Stack from top to bottom
0	0	♣♡♣♡♣
0	1	♣♡♣♣♡
1	0	♡♣♣♡♣
1	1	♡♣♣♣♡

This first protocol was improved by Mizuki *et al.* at ASIACRYPT 2012 [4]. They proposed a protocol using only four cards to compute securely the AND function. This protocol is optimal because we need four cards to describe two input bits as

$$\underbrace{?\ ?}_{a}\ \underbrace{?\ ?}_{b}$$

and hence it uses the minimal number of cards.

The problem with these two protocols is that the final result is revealed at the end of the computation. Therefore, researches have been made on *committed* protocols that would not need to reveal anything at the end of the computation: the output is a *commitment* encoding the desired result, say

$$\underbrace{?\ ?}_{a \wedge b}.$$

Committed protocols are quite useful because we can use output commitments as inputs for another computation without seeing intermediate results. In terms of the number of required cards, the best finite-runtime committed protocol was given by Koch *et al.* at ASIACRYPT 2015 [2]. They proposed a finite-runtime committed protocol for the AND function using five cards, and proved that no finite-runtime committed protocol could use fewer cards. Therefore, the necessary and sufficient number of cards for producing a commitment to $a \wedge b$ in finite-runtime fashion is five. We write this as

$$\mathrm{MinDeck}^{\mathrm{FC}}((a, b) \mapsto a \wedge b) = 5,$$

the formal definition of which will be given in Sect. 3.

How about other functions such as the OR and XOR functions? As we will show in Sect. 3, we can easily characterize classes of two-bit input one-bit output functions based on the minimal numbers of required cards to compute them securely in finite-runtime fashion. That is, it is not much difficult for us to specify the number $\mathrm{MinDeck}^{\mathrm{FC}}(f)$ for every two-bit input one-bit output function f : $\{0,1\}^2 \to \{0,1\}$ by applying known results. More specifically, there are two such classes:

$$\mathbb{B}_4 = \{f : \{0,1\}^2 \to \{0,1\} \mid \mathrm{MinDeck}^{\mathrm{FC}}(f) = 4\}$$

and

$$\mathbb{B}_5 = \{f : \{0,1\}^2 \to \{0,1\} \mid \mathrm{MinDeck}^{\mathrm{FC}}(f) = 5\},$$

as we will present in Sect. 3.

Then, how about two-bit output functions such as the half-adder $(a, b) \mapsto (a \wedge b, a \oplus b)$? Unfortunately, no exact characterization of necessary and sufficient numbers of cards is known for two-bit output functions. More precisely, as for sufficient numbers, Nishida *et al.* [7] gave six-card finite-runtime committed protocols for the functions $(a, b) \mapsto (a \wedge b, \bar{a} \wedge b)$, $(a, b) \mapsto (a \wedge b, b)$ and $(a, b) \mapsto (a \wedge b, a \oplus b)$, but it is open to determine whether the number six is necessary or not.

In this paper, we will give a complete and comprehensive answer to this question: we give a finite-runtime committed protocol for each of all functions $\{0,1\}^2 \to \{0,1\}^2$ with six cards or less and show that our protocols are optimal by providing a lower bound. For example, for the above three functions, six is a necessary number of cards[1]: we will prove that for every "non-trivial" and "non-degenerate" two-bit output function f, $\mathrm{MinDeck}^{\mathrm{FC}}(f) = 6$. For all trivial or degenerate functions f, we will also specify the numbers $\mathrm{MinDeck}^{\mathrm{FC}}(f)$, which are either 4 or 5.

Let us give a practical use of a two-bit output function. Four friends, Alice, Bob, Charlie and Dave, want to create a company, but both Charlie and Dave want to be the CEO. They decide to vote. Obviously, Charlie and Dave will vote for themselves. Therefore, one of them must get the votes of both Alice and Bob to become the CEO. However, neither Alice nor Bob want their votes to be revealed if they choose a different person. They decide to vote secretly with cards. Alice and Bob choose a secret input bit: 0 if they want to vote for Charlie and 1 if they want to vote for Dave. Then, they just need to perform a secure protocol to compute the result of the function $(a, b) \mapsto (a \wedge b, \bar{a} \wedge b)$. If the result is $(0, 1)$, then Charlie becomes the CEO. If the result is $(1, 0)$, then Dave becomes the CEO. If the result is $(0, 0)$, then nobody is elected. We will show in this paper that they only need a deck of six cards to vote in finite-runtime committed fashion.

The remainder of this paper is organized as follows. In Sect. 2, we will define the model we use to represent card-based cryptographic protocols. In Sect. 3, we will give some preliminary observations that will be useful for the understanding of our results: we will deal with one-bit output functions and notably

---

[1] This paper completes the work of [7] proving that what has been suggested is optimal.

define "trivial" functions[2] and "degenerate" functions. In Sect. 4, we will give six-card protocols computing any "non-trivial non-degenerate" two-bit input two-bit output function. Eventually in Sect. 5, we will prove that for "non-trivial non-degenerate" functions, our protocols are optimal. We conclude the paper in Sect. 6.

2   Definitions

In this section, we will define the mathematical objects that we will use to model card-based cryptographic protocols. All the definitions that we give in this section come from previous papers.

2.1   Abstract Machine Based Model

Mizuki et al. [5] proposed a formalization of card-based protocols based on abstract machine. We will first introduce this model, as it will be widely used in this paper.

Let $\mathcal{S} = \{?, \heartsuit, \clubsuit\}$ be a set of symbols. The $\heartsuit$ actually corresponds to a red card, the $\clubsuit$ corresponds to a black card, and the ? corresponds to a back-side of any card.

Definition 1. *Let u and v be two symbols of $\mathcal{S}$ such that exactly one of them is equal to ? The expression $\frac{u}{v}$ denotes a* lying card. *If $u = ?$ (resp. $v = ?$) then it is called a* face-down *(resp. face-up) card. For a lying card $\frac{u}{v}$, $\mathrm{swap}(\frac{u}{v}) \overset{\mathrm{def.}}{=} \frac{v}{u}$ and $\mathrm{top}(\frac{u}{v}) \overset{\mathrm{def.}}{=} u$. An element of $\mathcal{S} \setminus \{?\}$ is called an* atomic card. *If α is an atomic card, then $\mathrm{atom}(\frac{\alpha}{?}) \overset{\mathrm{def.}}{=} \alpha$ and $\mathrm{atom}(\frac{?}{\alpha}) \overset{\mathrm{def.}}{=} \alpha$.*

Definition 2. *A* deck *is a multiset of atomic cards. We call a d-tuple of lying cards $s = (s_1, \ldots, s_d)$ a* sequence *from a deck $\mathcal{D}$ if $\mathcal{D} = [\mathrm{atom}(s_1), \ldots, \mathrm{atom}(s_d)]$. An* atomic sequence *is a sequence of atomic cards. An* atomic sequence from a deck $\mathcal{D}$ *is an atomic sequence $\alpha = (\alpha_1, \ldots, \alpha_d)$ such that $\mathcal{D} = [\alpha_1, \ldots, \alpha_d]$. The expression $\mathrm{AtSeq}^{\mathcal{D}}$ (or just AtSeq if there is no ambiguity) denotes the set of all the atomic sequences from a deck $\mathcal{D}$. We call a sequence $v = (v_1, \ldots, v_d)$ a* possible visible sequence *of $\mathcal{D}$ if there is a sequence $(s_1, \ldots, s_d)$ from $\mathcal{D}$ such that $(v_1, \ldots, v_d) = (\mathrm{top}(s_1), \ldots, \mathrm{top}(s_d))$. The expression $\mathrm{Vis}^{\mathcal{D}}$ (or just Vis if there is no ambiguity) denotes the set of all the possible visible sequences from $\mathcal{D}$.*

Let us now define which actions can be made on a sequence of cards. Kinds of actions were introduced in [5] but we will use the definitions that are given in [2] and we will adapt them to our needs.

In the following, $\mathfrak{S}_d$ denotes the symmetric group on a set of d elements.

---

[2] Throughout the paper, we call a Boolean function such as $f : \{0,1\}^n \to \{0,1\}^k$ simply a function, except for an "action function" that will appear in Definition 5 in Sect. 2.1.

Definition 3 (Shuffling). *Let $\mathcal{D}$ be a deck with $|\mathcal{D}| = d$. Let $\Pi \subseteq \mathfrak{S}_d$, and $\mathcal{F}$ a probability distribution on Π. Let $\boldsymbol{s}$ be a sequence from $\mathcal{D}$. Let $\mathrm{shuffle}_{\Pi,\mathcal{F}}(\boldsymbol{s}) \overset{def.}{=} \pi(\boldsymbol{s})$ with π drawn from Π according to distribution $\mathcal{F}$. This action is called a shuffle. If there is π in Π such that $\Pr(\mathrm{shuffle}_{\Pi,\mathcal{F}} = \pi) = 1$ then the shuffle is called a* deterministic shuffle[3]. *Otherwise it is called a* random shuffle.

Definition 4 (Turn). *Let $\mathcal{D}$ be a deck with $|\mathcal{D}| = d$. Let $T \subseteq \{1, \ldots, d\}$ and $\boldsymbol{s} = (s_1, \ldots, s_d)$ be a sequence of lying cards. Let $\mathrm{turn}_T(\boldsymbol{s}) \overset{def.}{=} (s_1', \ldots, s_d')$ where $s_i' = \mathrm{swap}(s_i)$ if $i \in T$ and $s_i' = s_i$ otherwise. This action is called a* turn *action. If s_i' is a face-up card for all $i \in T$, then it is called a* turn-up *action. If s_i' is a face-down card for all $i \in T$, then it is called a* turn-down *action.*

These are the two actions that we will use. Let us now define what a protocol is.

Definition 5 (Protocol). *Let $\mathcal{D}$ be a deck, U a set of sequences (U is the set of input sequences), Q a set of states containing one initial state and one final state q_f. Let A be an action function*

$$A : (Q \setminus \{q_f\}) \times \mathrm{Vis} \to Q \times \mathrm{Action}$$

where Action *is a set of tuples containing:*

- *(shuffle, $\Pi, \mathcal{F}$) with $\Pi \in \mathfrak{S}_{|\mathcal{D}|}$ and a probability distribution $\mathcal{F}$ on Π;*
- *(turn, T) with $T \subseteq \{1, \ldots, |\mathcal{D}|\}$;*
- *(result, $p_1, \ldots, p_l$), which occurs if and only if the first component of A's output is the final state q_f, specifies that $(s_{p_1}, \ldots, s_{p_l})$ is the final result given by the protocol, where $(s_1, \ldots, s_{|\mathcal{D}|})$ is the current sequence.*

We say that $\mathcal{P} = (\mathcal{D}, U, Q, A)$ is a $|\mathcal{D}|$-card protocol.

This paper aims at giving finite-runtime committed protocols. Finite-runtime means that the length of the executions of the protocol is bounded, as in the following definition. Let us extend the definition of a protocol for a one-bit output function given in [2] to multi-output functions.

Definition 6 (Finite-runtime committed protocol for a function). *Let $f : \{0,1\}^n \to \{0,1\}^k$ be a function. Then we say that $\mathcal{P} = (\mathcal{D}, U, Q, A)$ is a finite-runtime committed protocol for f if the following holds:*

- *the deck $\mathcal{D}$ contains at least $\max(\{n, k\})$ cards of each symbol;*
- *there is a one-to-one correspondence $U \to \{0,1\}^n$ and there is $\sigma \in \mathfrak{S}_{|\mathcal{D}|}$ such that for every input sequence $(s_1, \ldots, s_{|\mathcal{D}|})$ whose corresponding input is $b \in \{0,1\}^n$,*

$$(s_{\sigma(2i-1)}, s_{\sigma(2i)}) = \begin{cases} \left(\dfrac{?}{\clubsuit}, \dfrac{?}{\heartsuit} \right) & \text{if } b[i] = 0 \\[2ex] \left(\dfrac{?}{\heartsuit}, \dfrac{?}{\clubsuit} \right) & \text{otherwise} \end{cases}$$

for all $i \in \{1, \ldots, n\}$;

---

[3] A deterministic shuffle is just a permutation of a sequence.

– *the last action is always* (result, $p_1, \ldots, p_{2k}$) *such that, for every* $b \in \{0,1\}^n$,

$$(r_{p_{2i-1}}, r_{p_{2i}}) = \begin{cases} \left(\frac{?}{\clubsuit}, \frac{?}{\heartsuit} \right) & \text{if } f(b)[i] = 0 \\ \left(\frac{?}{\heartsuit}, \frac{?}{\clubsuit} \right) & \text{otherwise} \end{cases}$$

for all $i \in \{1, \ldots, k\}$, *where* $(r_1, \ldots, r_{|\mathcal{D}|})$ *is the final sequence;*
– *it terminates after a bounded number of actions;*
– *it is secure according to the following Definition 7.*

Koch *et al.* [2] gave a definition of a secure protocol.

Definition 7 (Secure protocol [2]). *Let* $\mathcal{P} = (\mathcal{D}, U, Q, A)$ *be a protocol. Let* Γ_0 *be a random variable with values in the set of input sequences* U. *Let* V *be a random variable for the visible sequence trace of the protocol execution. The protocol* $\mathcal{P}$ *is said to be* secure *if* Γ_0 *and* V *are stochastically independent.*

2.2 Koch's Graphs

Koch *et al.* [2] also came up with an elegant mathematical object that we will use in this paper. We will call it a *status* and define it as below. The following definition is the same as the definition of a "state" in [2]. We changed the name to avoid any confusion with the states that are mentioned in Definition 5.

Definition 8 (Status [2]). *Let* $\mathcal{P}$ *be a finite-runtime committed protocol for* $f : \{0,1\}^n \to \{0,1\}^k$ *and* V *be a visible sequence trace of* $\mathcal{P}$. *The status* S *of* $\mathcal{P}$ *belonging to* V *is the map* $S : \text{AtSeq} \to \mathbb{X}_n, \alpha \mapsto \Pr[\alpha|V]$ *where:*

– $\mathbb{X}_n$ *denotes the set of polynomials over the variables* X_b *for* $b \in \{0,1\}^n$ *of the form* $\sum_{b \in \{0,1\}^n} \beta_b X_b$ *for* $\beta_b \in [0,1]$. *We interpret these polynomials as probabilities which depend on the probabilities of the inputs* b, *symbolized by the variables* X_b *for* $b \in \{0,1\}^n$.
– *for* $\alpha \in \text{AtSeq}$, $\Pr[\alpha|V]$ *denotes the probability that the current atomic sequence is* α *given that the current visible sequence trace is* V.

This definition will be useful because it allows us to draw graphs thanks to which protocols can be easily understood, as will be seen in Sect. 4.1. We will refer to these graphs as *Koch's graphs* in the rest of the paper.

3 Preliminary Observations

In this section, we will first give necessary and sufficient numbers of cards for one-bit output functions $\{0,1\}^2 \to \{0,1\}$. Then, we will define "trivial" functions and "degenerate" functions, and deal with necessary and sufficient numbers of cards for computing these functions.

3.1   Necessary and Sufficient Numbers of Cards for One-Bit Output Functions

First of all, let us introduce a notation about necessary and sufficient numbers of cards, as follows.

Definition 9. *Let* $f : \{0,1\}^n \to \{0,1\}^k$ *be a function. If* d *is the minimal number for which there exists a* d*-card finite-runtime committed protocol for* f*, then we write* $\mathrm{MinDeck}^{\mathrm{FC}}(f) \overset{\mathrm{def.}}{=} d$.

This subsection deals with two-bit input one-bit output functions $\{0,1\}^2 \to \{0,1\}$; the number of these functions is equal to $2^4 = 16$. We give necessary and sufficient numbers of cards for all these functions as follows.

As mentioned in Sect. 1, Koch *et al.* [2] provided a five-card finite-runtime committed protocol for the AND function and proved that this protocol was optimal, *i.e.* $\mathrm{MinDeck}^{\mathrm{FC}}((a,b) \mapsto a \wedge b) = 5$. Because negating a commitment can be easily done (just by swapping the two cards), all functions in the set $\mathbb{B}_5$ defined in Table 2, including the OR function, can be optimally computed with five cards.

For the constant functions and the functions outputting one input or its negation (such as $(a,b) \mapsto a$ or $(a,b) \mapsto \bar{b}$), the computation is trivial: these six functions (see the first six functions in $\mathbb{B}_4$ defined in Table 2) can be computed with four cards, which is the minimal number. The remaining two functions $(a,b) \mapsto a \oplus b$ and $(a,b) \mapsto \overline{a \oplus b}$ can also be computed with four cards, because Mizuki *et al.* [6] gave a four-card finite-runtime committed protocol for the XOR function.

Thus, we have immediately the following proposition on two-bit input one-bit output functions.

Proposition 10. *For any function* f *in* $\mathbb{B}_4$*, we have* $\mathrm{MinDeck}^{\mathrm{FC}}(f) = 4$. *For any function* f *in* $\mathbb{B}_5$*, we have* $\mathrm{MinDeck}^{\mathrm{FC}}(f) = 5$.

Table 2. Sets $\mathbb{B}_4$ and $\mathbb{B}_5$

Set $\mathbb{B}_4$	Set $\mathbb{B}_5$
$(a,b) \mapsto 0$	$(a,b) \mapsto a \wedge b$
$(a,b) \mapsto 1$	$(a,b) \mapsto a \vee b$
$(a,b) \mapsto a$	$(a,b) \mapsto a \wedge \bar{b}$
$(a,b) \mapsto \bar{a}$	$(a,b) \mapsto a \vee \bar{b}$
$(a,b) \mapsto b$	$(a,b) \mapsto \bar{a} \wedge b$
$(a,b) \mapsto \bar{b}$	$(a,b) \mapsto \bar{a} \vee b$
$(a,b) \mapsto a \oplus b$	$(a,b) \mapsto \bar{a} \wedge \bar{b}$
$(a,b) \mapsto \overline{a \oplus b}$	$(a,b) \mapsto \bar{a} \vee \bar{b}$

3.2 Trivial and Degenerate Cases for Two-Bit Output Functions

Let us bring our attention to two-bit output functions. We will show later that most of two-bit input two-bit output functions cannot be computed with less than six cards. However, some functions that we will call "trivial" functions and "degenerate" functions can be computed with four or five cards.

First, we name functions such as $(a, b) \mapsto (b, \bar{a})$ as follows.

Definition 11 (Trivial function). *Let* $f : \{0, 1\}^n \to \{0, 1\}^k$ *with* $n \geq k$ *be a function. We say that the function* f *is* trivial *if there are* $\sigma \in \mathfrak{S}_n$ *and* $\varepsilon_i \in \{id, b \mapsto \bar{b}\}, 1 \leq i \leq k$, *such that* $f(x_1, \ldots, x_n) = (\varepsilon_1(x_{\sigma^{-1}(1)}), \ldots, \varepsilon_k(x_{\sigma^{-1}(k)}))$.

Note that one can easily construct a four-card finite-runtime committed protocol for the trivial function $(a, b) \mapsto (b, \bar{a})$ above (because it suffices to exchange the input commitments and apply the negation to the second one). Thus, generally, we have the following lemma.

Lemma 12. *Let* $f : \{0, 1\}^n \to \{0, 1\}^k$ *with* $n \geq k$ *be trivial. Then,* $\mathrm{MinDeck}^{\mathrm{FC}}(f) = 2n$.

Next, we consider "degenerate" functions.

Definition 13 (Degenerate function). *Let* $f : \{0, 1\}^n \to \{0, 1\}^k$ *be a function. We say that the function* f *is* degenerate *if at least one of its output bits is constant.*

If a two-bit input two-bit output function is degenerate, then we can regard it as a one-bit output function (because any constant commitment can be easily created after main computations). Thus, we have the following lemma (according to Table 2).

Lemma 14. *Let* $f : \{0, 1\}^2 \to \{0, 1\}^2$ *be a degenerate function. If* f *has a non-constant output belonging to* $\mathbb{B}_5$ *then* $\mathrm{MinDeck}^{\mathrm{FC}}(f) = 5$. *Otherwise,* $\mathrm{MinDeck}^{\mathrm{FC}}(f) = 4$.

The goal of this paper will be to prove that $\mathrm{MinDeck}^{\mathrm{FC}}(f) = 6$ for any non-trivial non-degenerate function $f : \{0, 1\}^2 \to \{0, 1\}^2$. Section 4 will show the sufficiency and Sect. 5 will show the necessity.

4 Sufficient Numbers of Cards for Two-Bit Output Functions

In this section, we design a six-card finite-runtime committed protocol for any non-trivial non-degenerate two-bit output function. That is, six is an upper bound on the number of required cards for two-bit output functions. We will prove in Sect. 5 that the sufficient condition is actually a necessary one by providing a lower bound.

4.1 Existing Protocols Used to Obtain Upper Bounds

First of all, let us define some basic functions we will use later. Let $g_1 : \{0,1\}^2 \to \{0,1\}^2$ be the function such that $g_1(a,b) = (a, a \oplus b)$. Let $g_2 : \{0,1\}^2 \to \{0,1\}^2$ be the function such that $g_2(a,b) = (a \wedge b, \bar{a} \wedge b)$. We will also use $g_3 = g_1 \circ g_2$, $g_4 = g_3 \circ g_1$, and $g_5 : \{0,1\} \to \{0,1\}^2$ such that $g_5(a) = g_1(a,0)$. Note that $g_3(a,b) = (a \wedge b, b)$ and $g_4(a,b) = (a \wedge \bar{b}, a \oplus b)$. Also, $g_5(a) = (a,a)$ corresponds to making two identical copied commitments to a.

Each of these five functions can be computed securely thanks to the protocols introduced in some previous papers [6,7], as will be see later. For each of these existing protocols, we give its Koch's graph as follows.

The first protocol we will use is the COPY protocol coming from [6] and described in Fig. 1. It produces one copy of a single input commitment with six cards:

$$\underbrace{\boxed{?}\ \boxed{?}}\ \boxed{\clubsuit}\ \boxed{\heartsuit}\ \boxed{\clubsuit}\ \boxed{\heartsuit} \overset{\text{COPY}}{\longrightarrow} \boxed{\clubsuit}\ \boxed{\heartsuit}\ \underbrace{\boxed{?}\ \boxed{?}}\ \underbrace{\boxed{?}\ \boxed{?}}.$$
$$\quad\quad a \qquad\qquad\qquad\qquad\qquad\quad a \qquad a$$

Its possible inputs are listed in Table 3, and hence the input set is

$$U = \left\{ \left(\frac{?}{\clubsuit}, \frac{?}{\heartsuit}, \frac{?}{\clubsuit}, \frac{?}{\heartsuit}, \frac{?}{\clubsuit}, \frac{?}{\heartsuit} \right), \left(\frac{?}{\heartsuit}, \frac{?}{\clubsuit}, \frac{?}{\clubsuit}, \frac{?}{\heartsuit}, \frac{?}{\clubsuit}, \frac{?}{\heartsuit} \right) \right\}.$$

Table 3. Possible input sequences for Mizuki *et al.* COPY protocol [6]

Input a	Sequence from left to right
0	♣♥♣♥♣♥
1	♥♣♣♥♣♥

This is a finite-runtime committed protocol for the function g_5 that we defined above.

The second protocol we will use is the AND protocol from [6]. This protocol is supposed to compute the AND function, but as it will be easy to see on its Koch's graph, it also provides a commitment to $\bar{a} \wedge b$:

$$\underbrace{\boxed{?}\ \boxed{?}}\ \boxed{\clubsuit}\ \boxed{\heartsuit}\ \underbrace{\boxed{?}\ \boxed{?}} \overset{\text{AND}}{\longrightarrow} \boxed{\clubsuit}\ \boxed{\heartsuit}\ \underbrace{\boxed{?}\ \boxed{?}}\ \underbrace{\boxed{?}\ \boxed{?}}.$$
$$\quad\quad a \qquad\qquad\qquad b \qquad\qquad\qquad\qquad a \wedge b \quad\ \bar{a} \wedge b$$

Thus, its possible inputs are listed in Table 4. The Koch's graph of that protocol is represented on Fig. 2. This protocol computes the function g_2 that we defined above.

The third protocol we will use is actually the COPY protocol proposed in [7]. The difference is that we add a second input commitment, which makes the input sequences set differ from the real COPY protocol. Therefore, the possible

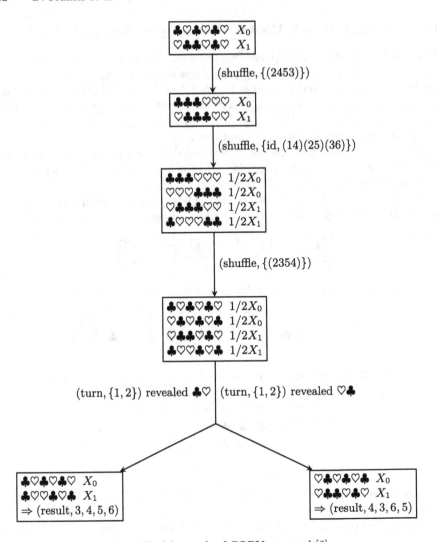

Fig. 1. Koch's graph of COPY protocol [6]

input sequences are the same as for the AND protocol we discussed above. The Koch's graph of that protocol is represented on Fig. 3.

This protocol provides commitments to a and $a \oplus b$:

Therefore it computes the function g_1.

As a result, we have six-card finite-runtime committed protocols for all functions from g_1 to g_5:

Table 4. Possible input sequences for Mizuki *et al.* [6] AND protocol

Input (a, b)	Sequence from left to right
$(0, 0)$	♣♡♣♡♣♡
$(0, 1)$	♣♡♣♡♡♣
$(1, 0)$	♡♣♣♡♣♡
$(1, 1)$	♡♣♣♡♡♣

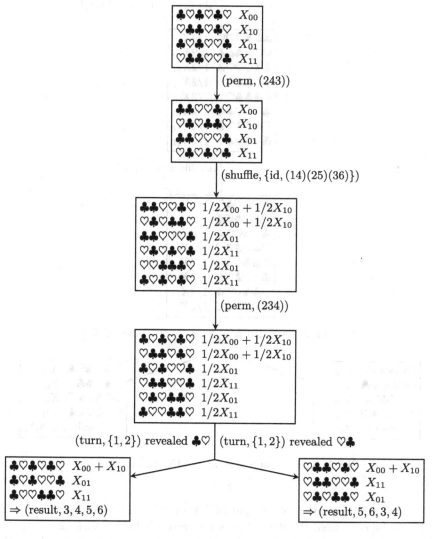

Fig. 2. Koch's graph of Mizuki *et al.* AND protocol [6]

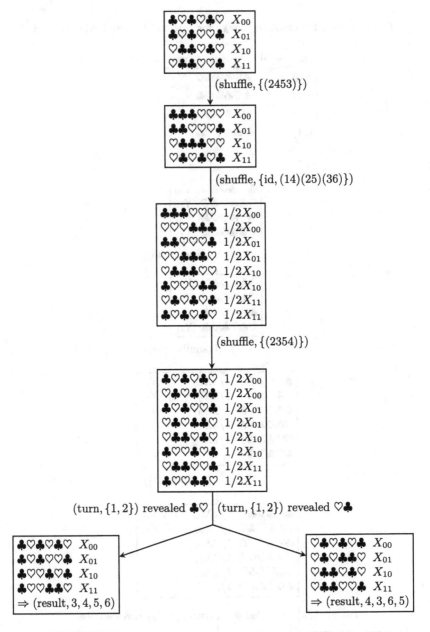

Fig. 3. Koch's graph of enhanced COPY protocol [7]

- g_1 can be computed thanks to the COPY protocol shown in Fig. 3;
- g_2 can be computed thanks to the AND protocol shown in Fig. 2;
- g_3 can be computed thanks to functions g_1 and g_2 (it corresponds to the improved AND from [7]);

- g_4 can be computed thanks to functions g_1 and g_3 (it corresponds to the improved half-adder from [7]);
- g_5 can be computed thanks to the COPY protocol shown in Fig. 3.

We will show in the following subsection that these protocols help us to make a six-card protocol for any function $\{0,1\}^2 \to \{0,1\}^2$.

4.2 Protocol

Any non-trivial non-degenerate function $f : \{0,1\}^2 \to \{0,1\}^2$ can be defined by choosing two functions among the following ones:

- Case 1: $(a,b) \mapsto a \wedge b$ and $(a,b) \mapsto \bar{a} \vee \bar{b}$;
- Case 2: $(a,b) \mapsto a \wedge \bar{b}$ and $(a,b) \mapsto \bar{a} \vee b$;
- Case 3: $(a,b) \mapsto \bar{a} \wedge b$ and $(a,b) \mapsto a \vee \bar{b}$;
- Case 4: $(a,b) \mapsto \bar{a} \wedge \bar{b}$ and $(a,b) \mapsto a \vee b$;
- Case 5a: $(a,b) \mapsto a$ and $(a,b) \mapsto \bar{a}$;
- Case 5b: $(a,b) \mapsto b$ and $(a,b) \mapsto \bar{b}$;
- Case 6: $(a,b) \mapsto a \oplus b$ and $(a,b) \mapsto \overline{a \oplus b}$.

Each function of a case is the negation of the other function of the same case: therefore one can be easily obtained from the other by reverting the order of the cards constituting commitments. Moreover, the computation of functions with two equal output bits is straightforward: we just compute one output bit and apply the COPY protocol to the commitment we obtained. Therefore, there are only fifteen non-trivial non-degenerate two-bit output functions to examine, as all other non-trivial non-degenerate two-bit output functions either have a straightforward six-card finite-runtime committed protocol or are deducible from these fifteen cases. Table 5 gives explicitly the computation of all these fifteen functions based on the above protocols for g_1, g_2, g_3 and g_4.

Table 5. Computations of two-bit output functions

Output 1	Output 2				
	Case 2 $a \wedge \bar{b}$	Case 3 $\bar{a} \wedge b$	Case 4 $\bar{a} \wedge \bar{b}$	Case 5 a	Case 6 $a \oplus b$
Case 1 $a \wedge b$	$g_2(b,a) =$ $(a{\wedge}b, a{\wedge}\bar{b})$	$g_2(a,b) =$ $(a \wedge b, \bar{a} \wedge b)$	$g_2 \circ g_1(a,\bar{b}) =$ $(a \wedge b, \bar{a} \wedge \bar{b})$	$g_3(b,a) =$ $(a \wedge b, a)$	$g_4(a,\bar{b}) =$ $(a{\wedge}b, a \oplus \bar{b})$
Case 2 $a \wedge \bar{b}$		$g_2 \circ g_1(a,b) =$ $(a \wedge \bar{b}, \bar{a} \wedge b)$	$g_2(a,\bar{b}) =$ $(a \wedge \bar{b}, \bar{a} \wedge \bar{b})$	$g_3(\bar{b},a) =$ $(a \wedge \bar{b}, a)$	$g_4(a,b) =$ $(a{\wedge}\bar{b}, a \oplus b)$
Case 3 $\bar{a} \wedge b$			$g_2(b,\bar{a}) =$ $(\bar{a} \wedge b, \bar{a} \wedge \bar{b})$	$g_3(b,\bar{a}) =$ $(\bar{a} \wedge b, \bar{a})$	$g_4(\bar{a},\bar{b}) =$ $(\bar{a} \wedge b, a \oplus b)$
Case 4 $\bar{a} \wedge \bar{b}$				$g_3(\bar{b},\bar{a}) =$ $(\bar{a} \wedge \bar{b}, \bar{a})$	$g_4(\bar{a},b) =$ $(\bar{a}{\wedge}\bar{b}, \overline{a \oplus b})$
Case 5 a					$g_1(a,b) =$ $(a, a \oplus b)$

Thus, we have constructed finite-runtime committed protocols for any two-bit output functions with six cards.

Lemma 15. *Let* $f : \{0,1\}^2 \to \{0,1\}^2$ *be a non-trivial non-degenerate function. Then* $\mathrm{MinDeck}^{\mathrm{FC}}(f) \leq 6$.

We will show in the next section that our protocols are optimal.

5   Optimality of Six-Card Protocols

In this section, we give a lower bound on the number of required cards: we prove that six is a necessary number of cards for non-trivial non-degenerate two-bit output functions.

5.1   Definitions

First of all, let us give some definitions that will be useful for our proof.

Definition 16 (Extension of [2]). *Let* $\mathcal{P}$ *be a finite-runtime committed protocol for a function* $f : \{0,1\}^n \to \{0,1\}^k$. *Let* α *be an atomic sequence, S a status of* $\mathcal{P}$ *belonging to a visible sequence trace and* $P = S(\alpha)$ *the polynomial representing the probability of* α *in S. Let* ω *be a possible output of f. If P contains only variables* X_b *with* $f(b) = \omega$, *then* α *is called an* ω*-sequence. If it contains at least two variables* X_{b_1} *and* X_{b_2} *with* $f(b_1) \neq f(b_2)$, *then* α *is called a* $\perp$*-sequence.*

Note that any finite-runtime committed protocol never produces a $\perp$-sequence [2].
    The difference between Definition 16 and the definition of [2] is that instead of considering only one-bit outputs, we consider k-bit outputs. If we replace k by 1 in our definition, we get back to the definition that can be found in [2].

Definition 17 (Execution of a protocol). *Let* $\mathcal{P} = (\mathcal{D}, U, Q, A)$ *be a finite-runtime protocol. An execution* $\mathcal{E}_{\mathcal{P}}$ *of* $\mathcal{P}$ *is a sequence* $(\xi_1, \ldots, \xi_p)$ *with* $\xi_i = (S_i, \mathrm{act}_i), 1 \leq i \leq p$, *such that:*

- S_1 *is the initial status of* $\mathcal{P}$;
- *for all* $i < p$, *status* S_{i+1} *comes from* S_i *based on action* act_i;
- S_p *is the final status.*

Definition 18 (Equivalent protocols). *Let* $\mathcal{P}_1 = (\mathcal{D}_1, U_1, Q_1, A_1)$ *and* $\mathcal{P}_2 = (\mathcal{D}_2, U_2, Q_2, A_2)$ *be two finite-runtime committed protocols for a function f. Then, we say that* $\mathcal{P}_1$ *and* $\mathcal{P}_2$ *are equivalent if* $\mathcal{D}_1 = \mathcal{D}_2$ *and* $U_1 = U_2$.

Definition 19 (Last meaningful action). *Let* $\mathcal{P}$ *be a finite-runtime protocol, and*

$$\mathcal{E}_{\mathcal{P}} = ((S_1, \mathrm{act}_1), \ldots, (S_p, \mathrm{act}_p))$$

an execution of $\mathcal{P}$ *with* $p \geq 2$. *The action* act_{p-1} *is called the last meaningful action.*

5.2 Proof

The most important idea behind our proof is, roughly speaking, that the last meaningful action of every execution of a finite-runtime committed protocol for a non-trivial function must be a turn action, and is stated in Lemma 27. The theorem we are going to prove states that there is no $(2k+1)$-card finite-runtime committed protocol for all non-trivial non-degenerate functions $f : \{0,1\}^n \rightarrow \{0,1\}^k$ with $n \leq k$. This theorem with the case of $n = k = 2$ implies $\mathrm{MinDeck}^{\mathrm{FC}}(f) \geq 6$.

To prove the theorem, we first show several lemmas. Lemma 20 derives a necessary condition for a status to be final. The Lemmas 21, 22, and 24 lead to Lemma 25, that derives equivalent protocols to obtain lower bounds. Note that these equivalent protocols do not violate the security notion defined in Defnition 7.

Lemma 20. *Let $\mathcal{P}$ be a $(2k+1)$-card finite-runtime committed protocol for a function $f : \{0,1\}^n \rightarrow \{0,1\}^k$ with $n \leq k$. Let $\mathcal{E}_\mathcal{P} = ((S_1, \mathrm{act}_1), \ldots, (S_p, \mathrm{act}_p))$ be an execution of $\mathcal{P}$. Then, every status S_i, $1 \leq i \leq p$, must contain at least one ω-sequence for each ω in $f(\{0,1\}^n)$; in particular, the final status S_p must contain exactly one ω-sequence for each ω in $f(\{0,1\}^n)$.*

Proof. If there is $b \in \{0,1\}^n$ such that

$$\frac{\partial}{\partial X_b} \sum_{\alpha \in \mathrm{AtSeq}} S(\alpha) \neq 1$$

for a status S, then the protocol is not secure: it contradicts the definition of a secure protocol, that states that the visible sequence traces and the input sequences must be independent. Therefore, every X_b for $b \in \{0,1\}^n$ must appear in all statuses, and hence, there is at least one ω-sequence for all $\omega \in f(\{0,1\}^n) \subseteq \{0,1\}^k$ in every possible status.

Assume now that $\alpha_1 = (\alpha_1^{(1)}, \alpha_1^{(2)}, \ldots, \alpha_1^{(2k+1)})$ and $\alpha_2 = (\alpha_2^{(1)}, \alpha_2^{(2)}, \ldots, \alpha_2^{(2k+1)})$ are two ω-sequences belonging to the final status S_p with $\omega \in \{0,1\}^k$. The action act_p produces k commitments, whose values are the same regardless of the atomic sequence. Without loss of generality, we can assume that $\mathrm{act}_p = (\mathrm{result}, 1, 2, \ldots, 2k)$. As the commitments are the same for α_1 and α_2, we have $\alpha_1^{(i)} = \alpha_2^{(i)}$ for all i in $\{1, \ldots, 2k\}$. Moreover, because α_1 and α_2 are atomic sequences from the same deck, we have $\alpha_1^{(2k+1)} = \alpha_2^{(2k+1)}$ and consequently $\alpha_1 = \alpha_2$. Thus, S_p must contain exactly one ω-sequence for each ω in $f(\{0,1\}^n)$. $\square$

Lemma 21. *Let $f : \{0,1\}^n \rightarrow \{0,1\}^k$ be a function and $\mathcal{P}$ be a finite-runtime committed protocol for f. Then, there is an equivalent protocol to $\mathcal{P}$ that never uses a turn-up action such that the number of atomic sequences remain unchanged.*

Proof. If a turn-up action leaves the numbers of atomic sequences unchanged, it means that the card that was revealed was the same for every possible sequence. Therefore, this action can be skipped. $\square$

Lemma 22. *Let $f : \{0,1\}^n \to \{0,1\}^k$ be a function and $\mathcal{P}$ be a finite-runtime committed protocol for f. Then, there is an equivalent protocol for f using no deterministic shuffle.*

Before proving the lemma, let us introduce some additional notations.

Definition 23. *For an action* act *on a sequence of size d and $\sigma \in \mathfrak{S}_d$, we define the action $\sigma(\mathrm{act})$ as follows:*

- *if* $\mathrm{act} = (\mathrm{shuffle}, \Pi, \mathcal{F})$*, then $\sigma(\mathrm{act}) \stackrel{\mathrm{def.}}{=} (\mathrm{shuffle}, \Pi\sigma, \mathcal{F}')$ with $\mathcal{F}'$ such that for all random variable $\pi \in \Pi'$ verifying $\pi \sim \mathcal{F}'$, $\pi\sigma^{-1} \sim \mathcal{F}$;*
- *if* $\mathrm{act} = (\mathrm{turn}, T)$*, then $\sigma(\mathrm{act}) \stackrel{\mathrm{def.}}{=} (\mathrm{turn}, \sigma^{-1}(T))$;*
- *if* $\mathrm{act} = (\mathrm{result}, p_1, \ldots, p_i)$*, then $\sigma(\mathrm{act}) \stackrel{\mathrm{def.}}{=} (\mathrm{result}, \sigma^{-1}(p_1), \ldots, \sigma^{-1}(p_i))$.*

Proof (Proof of Lemma 22). Let $\mathcal{P} = (\mathcal{D}, U, Q, A)$ be a finite-runtime committed protocol. Let us assume that there exist a visible sequence $\boldsymbol{v_1} \in \mathrm{Vis}$, a permutation $\pi \in \mathfrak{S}_{|\mathcal{D}|}$ and states $q_1, q_2 \in Q$ such that $A(q_1, \boldsymbol{v_1}) = (q_2, (\mathrm{shuffle}, \{\pi\}))$. Then, we can define an equivalent protocol $\mathcal{P}' = (\mathcal{D}, U, Q, A')$ such that:

- $A'(q_1, \boldsymbol{v_1}) = (q_3, \pi(\mathrm{act}))$ with $(q_3, \mathrm{act}) = A(q_2, \pi(\boldsymbol{v_1}))$;
- $A'(q, \boldsymbol{v}) = A(q, \boldsymbol{v})$ if $(q, \boldsymbol{v}) \neq (q_1, \boldsymbol{v_1})$.

Therefore, it is always possible to build an equivalent protocol containing no deterministic shuffle. $\qquad\square$

Lemma 24. *Let $f : \{0,1\}^n \to \{0,1\}^k$ be a function and $\mathcal{P}$ be a finite-runtime committed protocol for f. Then, there is an equivalent protocol for f using no random shuffle leaving the number of atomic sequences equal to $|f(\{0,1\}^n)|$.*

Proof. Assume that an execution of a protocol contains a random shuffle $(\mathrm{shuffle}, \Pi, \mathcal{F})$ leaving the numbers of atomic sequences equal to $|f(\{0,1\}^n)|$ for a given status. Then, by Lemma 20, the status must have exactly one ω-sequence for each $\omega \in f(\{0,1\}^n)$, and hence the shuffle can be replaced by the deterministic shuffle $(\mathrm{shuffle}, \{\pi\})$ with $\pi \in \Pi$. We conclude thanks to Lemma 22. $\qquad\square$

Lemmas 21, 22 and 24 immediately imply the following one.

Lemma 25. *Let $f : \{0,1\}^n \to \{0,1\}^k$ be a function. If $\mathcal{P}$ is a finite-runtime committed protocol for f, then there exists an equivalent protocol $\mathcal{P}'$ such that for any execution $\mathcal{E}_{\mathcal{P}'} = ((S_1, \mathrm{act}_1), \ldots, (S_p, \mathrm{act}_p))$ of $\mathcal{P}'$:*

- *if there is $i \in \{1, \ldots, p-1\}$ such that act_i is a turn up, then the number of atomic sequences in S_{i+1} is strictly lower than the number of atomic sequences in S_i;*
- *if there is $i \in \{1, \ldots, p-1\}$ such that act_i is a shuffle, then act_i is not deterministic and the number of atomic sequences in S_{i+1} is strictly bigger than $|f(\{0,1\}^n)|$.*

We will call such a protocol a *strictly random* protocol.

Lemma 26. *Let $f : \{0,1\}^n \to \{0,1\}^k$ be a function. If there is a $(2k+1)$-card strictly random finite-runtime committed protocol $\mathcal{P}$ for f such that at least one execution of $\mathcal{P}$ contains no random shuffle, then f is either trivial or degenerate.*

Proof. Note that if an execution of a protocol computing a function $\{0,1\}^n \to \{0,1\}^k$ contains no random shuffle, then it means that the protocol, from the initial state to the final state, is deterministic, as a turn cannot reveal an input commitment. Therefore, it means that no execution of $\mathcal{P}$ contains a random shuffle.

Let us assume now that there is such a protocol $\mathcal{P}$ for a function $f : \{0,1\}^n \to \{0,1\}^k$. Let $f_i : \{0,1\}^n \to \{0,1\}, 1 \le i \le k$ be functions such that for all $x \in \{0,1\}^n$, $(f_1(x),\dots,f_k(x)) = f(x)$. Let us assume that there is $i \in \{1,\dots,k\}$ such that f_i is non-constant. Let $(s_1,\dots,s_{2n+1})$ be the input sequence corresponding to an input $x = (x_1,\dots,x_n)$. Then, as the executions of $\mathcal{P}$ contain no shuffle, there are $p,q \in \{1,\dots,2k+1\}$ such that (s_p, s_q) is a commitment to an input x_m and either (s_p, s_q) or (s_q, s_p) is a commitment to $f_i(x)$. Moreover, all commitments to inputs $x_1,\dots,x_n$ must remain at the end of the protocol. Therefore, if $n = k$, then f is a trivial function. Otherwise, *i.e.* $n < k$, there is a permutation $\sigma \in \mathfrak{S}_k$ such that $x \mapsto (f_{\sigma(1)}(x),\dots,f_{\sigma(n)}(x))$ is a trivial function and $x \mapsto (f_{\sigma(n+1)}(x),\dots,f_{\sigma(k)}(x))$ is a constant function. $\square$

Lemma 27. *Let $\mathcal{P}$ be a $(2k+1)$-card strictly random finite-runtime committed protocol for a non-trivial non-degenerate function $f : \{0,1\}^n \to \{0,1\}^k$ with $n \le k$. For every execution of $\mathcal{P}$, the last meaningful action is a turn.*

Proof. Let $\mathcal{P}$ be a $(2k+1)$-card strictly random finite-runtime committed protocol for a non-trivial non-degenerate function $f : \{0,1\}^n \to \{0,1\}^k$ with $n \le k$. Then, at least one random shuffle exists and hence, Lemma 26 allows us to say that the last meaningful action exists for every execution. By Lemma 20, we have $|f(\{0,1\}^n)|$ atomic sequences in the last status. According to Lemma 25, and as $\mathcal{P}$ is strictly random, the last meaningful action cannot be a shuffle. It means that it must be a turn. $\square$

Theorem 28. *There is no $(2k+1)$-card finite-runtime committed protocol for any non-trivial non-degenerate function $f : \{0,1\}^n \to \{0,1\}^k$, with $n \le k$.*

Proof. Suppose for a contradiction that there exists a $(2k+1)$-card finite-runtime committed protocol $\mathcal{P} = (\mathcal{D}, U, Q, A)$ for a non-trivial non-degenerate function $f : \{0,1\}^n \to \{0,1\}^k$, with $n \le k$. Without loss of generality, we may assume that $\mathcal{D}$ contains $k\ \heartsuit$ and $k+1\ \clubsuit$, and $\mathcal{P}$ is strictly random. Then, by Lemma 27, the last meaningful action of any execution of our protocol is a turn. Consider last two statuses $(S_{p-1}, \mathrm{act}_{p-1})$ and (S_p, act_p) during an execution of $\mathcal{P}$. If the last turn reveals a $\heartsuit$, then S_p cannot be a final status one because the protocol could not produce k commitments. Therefore, that turn must reveal a $\clubsuit$ with probability 1, which is contrary to the assumption that $\mathcal{P}$ is strictly random. $\square$

Theorem 28 along with Lemma 15 implies the following corollary.

Corollary 29. *Let $f : \{0,1\}^2 \to \{0,1\}^2$ be a non-trivial non-degenerate function. Then,* $\mathrm{MinDeck}^{FC}(f) = 6$.

All protocols for non-degenerate non-trivial functions presented in Sect. 4 use only *uniform* shuffles: for a given shuffle, every possible permutation is drawn with the same probability. Moreover, if Π is the set of permutations of a shuffle of such a protocol and $\pi_1, \pi_2 \in \Pi$, then $\pi_1 \circ \pi_2 \in \Pi$ and $\pi_2 \circ \pi_1 \in \Pi$. It means that all shuffles are also *closed*. Koch *et al.* explained in [2] that uniform closed shuffles were easy to implement practically. It means that our result is still true if we restrict to simple uniform closed shuffles.

6 Conclusion

In this paper we gave a six-card finite-runtime committed protocol for every non-trivial non-degenerate two-bit output function, and we gave a proof that our protocols were optimal. We also gave optimal protocols for trivial or degenerate functions. Therefore, we gave the necessary and sufficient numbers of cards for any two-bit input two-bit output function. As the proof we presented in Sect. 5 covers more than only proving the optimality of our protocols, we hope that it will be useful to prove the optimality of other multi-output protocols. Moreover, this paper addressed only finite-runtime protocols. As there is a four-card Las Vegas committed AND protocol [2], it is an interesting open problem to characterize necessary and sufficient numbers of cards for Las Vegas committed protocols for multi-output functions. Another intriguing direction is to consider encoding schemes other than the two-card encoding as in [8]. It would also be interesting to use decks containing more than two colors.

Acknowledgment. We thank the anonymous referees, whose comments have helped us to improve the presentation of the paper. This work was supported by JSPS KAK-ENHI Grant Number 26330001.

References

1. den Boer, B.: More efficient match-making and satisfiability: the five card trick. In: Quisquater, J.-J., Vandewalle, J. (eds.) EUROCRYPT 1989. LNCS, vol. 434, pp. 208–217. Springer, Heidelberg (1990). doi:10.1007/3-540-46885-4_23
2. Koch, A., Walzer, S., Härtel, K.: Card-based cryptographic protocols using a minimal number of cards. In: Iwata, T., Cheon, J.H. (eds.) ASIACRYPT 2015. LNCS, vol. 9452, pp. 783–807. Springer, Heidelberg (2015). doi:10.1007/978-3-662-48797-6_32
3. Marcedone, A., Wen, Z., Shi, E.: Secure dating with four or fewer cards. Cryptology ePrint Archive, Report 2015/1031 (2015)
4. Mizuki, T., Kumamoto, M., Sone, H.: The five-card trick can be done with four cards. In: Wang, X., Sako, K. (eds.) ASIACRYPT 2012. LNCS, vol. 7658, pp. 598–606. Springer, Heidelberg (2012). doi:10.1007/978-3-642-34961-4_36
5. Mizuki, T., Shizuya, H.: A formalization of card-based cryptographic protocols via abstract machine. Int. J. Inf. Secur. 13(1), 15–23 (2014)

6. Mizuki, T., Sone, H.: Six-card secure AND and four-card secure XOR. In: Deng, X., Hopcroft, J.E., Xue, J. (eds.) FAW 2009. LNCS, vol. 5598, pp. 358–369. Springer, Heidelberg (2009). doi:10.1007/978-3-642-02270-8_36

7. Nishida, T., Hayashi, Y., Mizuki, T., Sone, H.: Card-based protocols for any boolean function. In: Jain, R., Jain, S., Stephan, F. (eds.) TAMC 2015. LNCS, vol. 9076, pp. 110–121. Springer, Cham (2015). doi:10.1007/978-3-319-17142-5_11

8. Shinagawa, K., Mizuki, T., Schuldt, J.C.N., Nuida, K., Kanayama, N., Nishide, T., Hanaoka, G., Okamoto, E.: Multi-party computation with small shuffle complexity using regular polygon cards. In: Au, M.-H., Miyaji, A. (eds.) ProvSec 2015. LNCS, vol. 9451, pp. 127–146. Springer, Cham (2015). doi:10.1007/978-3-319-26059-4_7

Malicious Cryptography

Controlled Randomness – A Defense Against Backdoors in Cryptographic Devices

Lucjan Hanzlik, Kamil Kluczniak, and Mirosław Kutyłowski$^{(\boxtimes)}$

Faculty of Fundamental Problems of Technology,
Wrocław University of Science and Technology, Wrocław, Poland
{Lucjan.Hanzlik,Kamil.Kluczniak,Miroslaw.Kutylowski}@pwr.edu.pl

Abstract. Security of many cryptographic protocols is conditioned by quality of the random elements generated in the course of the protocol execution. On the other hand, cryptographic devices implementing these protocols are designed given technical limitations, usability requirements and cost constraints. This frequently results in black box solutions. Unfortunately, the black box random number generators enable creating backdoors. So effectively the signing keys may be stolen, authentication protocol can be broken enabling impersonation, confidentiality of encrypted communication is not guaranteed anymore.

In this paper we deal with this problem. The solution proposed is a generation of random parameters such that: (a) the protocols are backwards compatible (a protocol participant gets additional data that can be ignored), (b) verification of randomness might be executed any time without any notice, so a device is forced to behave honestly, (c) the solution makes almost no change in the existing protocols and therefore is easy to implement, (d) the owner of a cryptographic device becomes secured against its designer and manufacturer that otherwise might be able to predict the output of the generator and break the protocol. We give a few application examples of this technique for standard schemes.

Keywords: Cryptographic device · Pseudorandom number generator · Backdoor · Discrete logarithm · Signature · Audit · Provable security

1 Introduction

Secure Devices and Provable Security Requirements. During the last decade a number of *provably secure* schemes have been introduced, following a growing demand for strong security guarantees. However, the real impact on security of cryptographic products is somewhat limited, as many issues are just disregarded or replaced by assumptions declaring unconditional security and trustworthiness of some key components.

This research has been supported by the Polish National Science Centre, project HARMONIA, DEC-2013/08/M/ST6/00928.

© Springer International Publishing AG 2017
R.C.-W. Phan and M. Yung (Eds.): LNCS 10311, Mycrypt 2016, pp. 215–232, 2017.
DOI: 10.1007/978-3-319-61273-7_11

One of the crucial components of cryptographic systems are so-called *secure devices*. Indeed, most of the operations of cryptographic protocols (in particular those involving secret keys) are performed not by the users themselves (as it would follow from a standard description of cryptographic protocols with Alice, Bob, etc.), but by black box devices entrusted by the users. Without trustworthy devices the whole security of cryptographic systems is a myth – at least for most schemes used in practice today.

Generation of (pseudo)random numbers (frequently never shown in clear) is a major Achilles heel of many cryptographic protocols. Devices executing protocols take these numbers from sources like physical randomness (RNG), Pseudorandom Number Generators (PRNG) or from hybrid systems. Their quality can be checked via statistical tests (such as those recommended by NIST [18]). A negative test result indicates that the statistical properties of random parameters are far from being uniformly random. Then, as the assumptions of a cryptographic scheme are not fulfilled, the device should not be used anymore. However, a positive result of a statistical test has not much to do with a "provable security", as it is merely an anomaly detection procedure. Moreover, it is easy to construct a weak cryptographic random generator that passes these tests [21].

Kleptography. A manipulated PRNG might be installed on purpose as a backdoor. The kleptographic techniques (see [23] for a survey) enable leaking secret keys to passive observers by malicious modification of the procedure of generating random parameters. Moreover, the device neither uses any extra communication channel, nor creates an output that would differ from the not-manipulated one in a way detectable for anyone but the adversary holding a special secret key. Moreover, this is not stored by the PRNG, as it would be the case for simple subliminal channels. Notably, the design of Dual_EC_DRBG enables direct installation of such a backdoor [7,14,20].

Typically, a kleptographic attack is undetectable in a cryptanalytic way unless some cryptographic assumptions get broken. However, the failure of cryptographic assumptions would presumably mean also the collapse of the underlying cryptographic protocol. So in the cryptanalytic sense we have to do with a perfect attack. There have been some efforts to create methods detecting kleptographic manipulations by side channel information from the devices [15], however with more sophisticated solutions one can make kleptography resilient to this kind of analysis [24].

Real Randomness. In this situation the sources of real randomness coming from reliable physical sources might be regarded as a proper solution. Unfortunately, there are problems with this approach as well. First, due to price constraints one might be forced to deploy low quality solutions. This concerns in particular smart cards, electronic ID documents [3] (like biometric passports) and other mobile solutions. Another threat are hardware Trojans – hardware random number generators manufactured in a way that enables the attacker to predict the generator's output [22]. The bad news is that the attacked devices do not contain malicious code or any changes in electronic design that could be revealed during an audit of the device including destructive inspection of the layout. So in fact, the situation is even worse than for PRNGs, where audit procedures may have a real impact (even if they are evasive in some cases).

Initializing PRNGs and Related Security Problems. Now let us consider in more detail the options for initializing a PRNG:

Option 1: the manufacturer installs the seed,

Option 2: the user creates the seed by starting an initialization procedure executed internally by a PRNG device,

Option 3: the user uploads the seed to the PRNG device,

Option 4: the user uploads a part of the seed while the second part of the seed is installed by the manufacturer,

Option 5: the user and/or the manufacturer uploads the seed, however, during its operation the PRNG modifies its state according to some number of entropy bits.

Unfortunately, the manufacturer can gain full control over the PRNG in a way unobservable for the device users, regardless which option has been used:

Option 1: the manufacturer may retain a copy of the seed of the PRNG. Note that it is easy to hide the seeds so that no inspection of digital data stored by the manufacturer would reveal their presence. One of the tricks is to derive the seeds from digital signatures – the inspectors cannot force the manufacturer to create such signatures.

Option 2: initializing the seed might be a fake operation— whatever the user does the result is anyway the seed predetermined by the manufacturer.

Option 3: this option is not acceptable since now the user himself can retrieve the private signing key in case of standard schemes like DSA. One of the very basic requirements for a secure device is that nobody, even the owner, can derive the secrets stored in the device. On the other hand, if the user behaves correctly and immediately destroys the copy of the seed existing outside the device, then the black box device can cheat and retreat to the state known by the manufacturer.

Option 4: the device can cheat and use the whole seed known to the manufacturer. The attack will be undetectable, if the PRNG is properly designed and therefore immune against partial seed leakage.

Option 5: the device can cheat and gradually change its internal state converting it to a state which can be guessed by the manufacturer.

1.1 Previous Attempts to Prevent PRNG Backdoors

The problems of PRNG backdoors have not been covered yet in the literature in the way corresponding to the critical importance of the problem. In fact, the research in this area has been started from a different angle. The fail-stop [19] signatures have been designed to provide an undeniable cryptographic proof in case of signature forgery resulting from cryptanalytic attacks. However, fail-stop schemes do not protect at all against the seizure of a particular private key from a signature creation device.

The concept of forward security (see e.g. [1]) is to protect the private key by limiting the period of its usage without changing the public key. So a cryptanalytic attack or leaking the signing key has effects for signatures from a limited signing period. Again, this does not protect against attacks, where all secret material from the signature creation device falls into the hands of an adversary, unless the device updates the private key in cooperation with an external trusted party.

Intrusion resilience of cryptographic devices may be achieved by introducing multiple (independent) devices executing the same protocol. The idea is that when the devices come from different vendors, it is less likely that an attack succeeds, as it is more difficult to take control over both devices at the same time. The solution examples are home base techniques [12] and mediated signatures [5,17].

The approach from [4] attempts to guard against a secret key usage by the adversary: any attempt to use it results in a secret key exposure with a substantial probability. Yet another approach is to enforce a fair generation of random keys by imposing a kind of randomness verification. For example, [13] proposes a protocol KEGVER for generation of RSA keys where the modulus belongs to a small interval determined jointly by the protocol participants. This thwarts many attack scenarios for weakening the key during a black box generation process. On the other hand, this approach is useless for controlling devices such as smart cards, since a complicated two-party protocol is executed, where both parties are generating random numbers. As it is hard to assume that the user could choose such numbers manually, we retreat to the scenarios with a protocol executed by two devices owned by the user. However, in this case the solutions like [5,17] seem to be easier from the practical point of view. An approach to secure a Diffie-Hellman based authentication protocol against devices colluding with passive adversaries has been proposed in [9]. This method can be used only, when the randomness used can be later exposed to other parties of the protocol – just like for Diffie-Hellman key exchange protocol. It cannot be used to control the process of creating standard digital signatures where random parameters are involved.

A partial solution for the problem has been proposed in [16]. However, the major difference is that the user has no chance to check that the system is implemented in this way. Moreover, separation of hardware components for digital signatures is problematic, if the random exponent occurs both as an exponent and in a linear expression (like for the DSA signatures).

Our Contribution. In this paper we present a security mechanism against PRNG backdoors called *controlled randomness*. The idea is to add certain procedures to standard cryptographic primitives that allow the device owner to control the randomness without revealing the secret key used by the device. What is more, our mechanism limits the capabilities of the malicious manufacturer to reveal the output of the PRNG. The only way the adversary can reveal any information about the output of the PRNG is by using generic kleptographic attacks, e.g. execute parts of the protocol repeatedly, until the message generated reveals in some way a certain secret key bit.

In more details, the proposed security mechanism fulfills the following properties:

Backwards Compatibility: The new protocol should make minimal changes in the sense that after ignoring some messages or their certain parts we get the original scheme.

Immunity Against the Manufacturer: no effective attack can be performed by a device manufacturer knowing the internal state of the PRNG.

Immunity Against the User: The new mechanisms prevents the owner of the device to attack his own device.

Verifiability: The device owner can check that the security mechanism has been really deployed. For this purpose, only the standard device output should be used.

There are several scenarios of using a PRNG in cryptographic protocols:

Case 1: *the output r of the PRNG is presented to other protocol participants.*
For example, r might be a challenge in a challenge-response protocol.

Case 2: *the output k of the PRNG is used to compute $r = g^k$ for a known g, and r is presented to other protocol participants.*
In this scenario g is a generator of a group where Discrete Logarithm Problem is hard. The most prominent examples are Diffie-Hellman Key Exchange and signature schemes based on ElGamal.

Case 3: *the output k of the PRNG is used to compute $r = h^k$, and r is presented to other protocol participants, but h is not known to the adversary.*
This case occurs for Generic Mapping PACE algorithm [2] – a standard scheme developed for password authentication.

Case 4: *the output r of PRNG is used to generate key pairs in a deterministic way.*
An example of this situation is the RSA key generation procedure, where we first determine at random a starting point for the search and then continue testing in a deterministic way.

Case 1 has been considered in the literature without directly referring to the general problem. A typical solution is to use Hash(U), where U is some data shared by the device and the user controlling the device. If Hash behaves like a random function, then the challenge has the desired properties.

Case 4 seems to be a hard challenge. If we proceed similarly as in Case 1, then we reveal the starting point for searching for p for a RSA number that is finally constructed as $n = pq$. Then, however, the verifier could redo the search and factorize n.

In this paper we focus on solutions for Case 2, i.e. we apply our mechanism to discrete logarithm based signatures, ElGamal encryption and Diffie-Hellman key agreement. Of course, real applicability of any solution depends on its simplicity – even a moderately complicated solution may be hard to deploy in practice. Fortunately, it turns out that it is possible to achieve the stated goals using simple and standard operations.

Case 3 is more involved technically and is postponed to the follow up work.

Paper Organization. The rest of the paper is organized as follows: in Sect. 2 we describe the idea of the solution, in Sect. 3 we show how to use it for securing signature creation devices, in Sect. 4 we present its application to Diffie-Hellman Key Exchange, in Sect. 5 we focus on public key encryption.

2 Idea of Controlled Randomness

In this section we present our controlled randomness mechanism. We focus on the case, where the device outputs, as part of the cryptographic primitive, a pseudorandom parameter $r = g^k$, where k is the output of a PRNG, r is one of the device outputs, and g is a generator of a group $\mathcal{G}$ of a prime order q. Such parameter is used by many discrete logarithm based schemes and for most of them k must remain secret and should never leave a secure device (there are a few exceptions, see e.g. [9]).

We commence by presenting a way to redesign generation of r, so that the discrete logarithm of r cannot be derived by the adversary even if the output of the PRNG is known to him. On the other hand, there must exist some procedure that assures the user that the outputted r was in fact computed in this way.

Then, we make the first informal attempt to model our controlled randomness (CR) idea. We describe actors in the system, potential adversaries and attack scenarios. As it turned out a more formal but universal definition is a challenge. Thus, in this paper we opted to an informal definition and leave the formalization to future work.

2.1 Controlled Randomness

We begin our description by assuming that a cryptographic device contains the following components:

- a PRNG P with a seed y installed by the manufacturer (or any generator with the output that may be guessed by an adversary with a non-negligible probability),
- a *blinding factor* $U = g^u$ installed in the device by its owner, where u is a secret of the owner never exposed to the device.

Generating r. Instead of taking k from the generator P, the device proceeds as follows:

- k_0 is taken as the output of P,
- $k_1 := \text{Hash}(U^{k_0})$, where Hash is a cryptographic hash function that returns the results in the range $[0, q - 1]$,
- $r' := g^{k_0}$, $r := (r')^{k_1}$.

(Alternatively, we may define $r := r' \cdot g^{k_1}$.) Then, instead of outputting r alone, the device presents both r and so called *control data* r'.

In cases when we would like to prevent generation of the same r by just repeating the same output from P, we modify slightly the definition of k_1:

$$k_1 := \text{Hash}(U^{k_0}, i)$$

where i is the counter value incremented by one at every protocol execution. In this case the value of the counter is one of the control parameters, which now take the form (r', i).

Verification of Pseudorandom Parameters. The verification procedure can be performed by the user holding the secret exponent u. On input r and control parameters (r', i), he performs the following steps:

- $\lambda := \text{Hash}((r')^u, i)$
- if $r \neq (r')^\lambda$, then consider the device as *faulty* or *malicious*.

Setup. We may assume that the PRNG is out of control of the user of the device - it comes to the user ready to use. On the other hand, the blinding factor U is installed by the user in the device after getting it from the vendor. The user first generates u at random, computes $U := g^u$ and installs U in the device. The key u must be retained for verification purposes and <u>must not</u> be uploaded to the device. Also, U must be chosen so that the device cannot derive u (e.g. the range of u must be big enough to prevent computing the discrete logarithm of U).

Presumably, U is generated with another device. The exponent u must be kept secret from the controlled device and the manufacturer, but leaking it to other parties does not create a security threat. In order to achieve dynamic protection, the owner of the device can always change the blinding factor. So, after loosing the secret u, the best choice is to install a new blinding factor.

2.2   Outline of the Security Model

While the formal security model depends very much on a concrete application case, we discuss here an informal definition.

Actors of the Scheme. We consider the following actors of the protocols concerned:

device: it is a secure device implementing a cryptographic protocol, and acting on behalf of its owner. The device implements some secrets (according to the protocol description) as well as the blinding factor U.

Alice, the device Owner: she holds the device and controls its output/input – possibly using other devices (e.g. a PC equipped with a smart card reader). She is responsible for:
  – setting the blinding factor U of the device,

- filtering the output of the device and removing the control parameters before forwarding them to other protocol participants,
- running the verification procedure of the device using secret key u and the control parameters.

Manufacturer Mallet: Mallet is a party that has access to all inputs and outputs of the device after filtering them by Alice as well as its initial state. Mallet knows the output of the PRNG of the device at each moment.

We skip considering the other protocol participants (e.g. verifiers of electronic signatures or partners in authentication protocols) as from their point of view the protocol execution is exactly the same as for the same scheme executed without CR. Moreover, we may assume that such participants are controlled by Mallet.

Protocol Lifecycle. There are the following events in the lifecycle of a device implementing CR:

device initialization: Mallet creates the device including all internal data except for the private data generated by the user – e.g. private signing keys. The blinding factor U is not initialized.

blinding factor update: Alice asks the device to perform an update and sends the update parameters of her choice. The device performs the update.

protocol execution: the original protocol is executed using the device. The only difference is that the implementation of the protocol by the device is modified and that the output of the device is filtered by Alice.

device verification: Alice, the device holder, takes the device output created during a protocol execution and executes the offline control procedure. Note that the controlled device is not aware about the verification.

Note that we assume that the only party interacting with the device is its owner. There is no direct interaction between Mallet and the device.

Adversaries. There are the following types of adversaries:

device: it may attempt to circumvent the protection offered by CR and break the protocol in cooperation with Mallet. However, no direct communication with Mallet is possible – only truncated output of the device is available for Mallet.

Alice: she may herself attempt to break security of the device (e.g. learn the secret signing key implemented in the device in order to create clone devices or to deny former signatures by leaking secret keys). Note that potentially the control parameters may ease the attack as they are not included in the original (allegedly secure) protocols.

Mallet: he may control and manipulate PRNG, however, he has no access to the blinding factor entered by the user. He may observe the protocol execution except for the control data filtered out by Alice.

Observe that an external observer has less knowledge than Mallet, so it suffices to show protocol security against Mallet.

3   Signatures with Controlled Randomness

In this section we present how to use our controlled randomness mechanism with standard signature schemes. We begin with the observation that many signature schemes based on difficulty of the discrete logarithm problem follow a common approach. Namely, the following steps are executed:

– a number k is chosen at random, $r := g^k$ for a fixed group generator g,
– the signature is computed as $\mathsf{F}(r, k, m, x)$, where m is a message to be signed, x is the signing key, and F is a deterministic function.

This concerns in particular ElGamal signatures, Schnorr signatures and DSA. We shall modify the first part of the scheme – generation of r – and add certain control parameters. As a case study we describe the use of CR with Schnorr signatures.

Schnorr Signature

Recall that for a private key x and public key $y = g^x$, the signature for a message m is created as follows:

$$k := \mathrm{PRNG}()$$
$$r := g^k$$
$$e := \mathrm{Hash}(m\|r)$$
$$s := (k - x \cdot e) \bmod q$$

(g is an element of a prime order q).
    The signature (s, e) is verified as follows:

1. r is reconstructed as $r_v = g^s y^e$,
2. the signature verifies positively, if $e = \mathrm{Hash}(m\|r_v)$.

The *Schnorr signature with CR* for a message m and for a blinding factor U is created as follows:

$$k_0 := \mathrm{PRNG}()$$
$$r' := g^{k_0}$$
$$k_1 := \mathrm{Hash}(U^{k_0}, i)$$
$$k := k_0 \cdot k_1$$
$$r := g^k$$
$$e := \mathrm{Hash}(m\|r)$$
$$s := (k - x \cdot e) \bmod q$$

For the signature (s, e), the control data are (r', i). The verification procedure follows the generic idea for randomness of the form $r = g^k$, described in Sect. 2 The only additional step is reconstruction of r as $g^s y^e$ (which is a part of the standard verification of Schnorr signatures). We now show that this construction is secure against the attacks described in Sect. 2.2.

3.1 Security Against Mallet

Here we show that our main goal is achieved - the adversary knowing the output of PRNG cannot forge a signature, even if before he can request a number of signatures for messages of his choice. Our security argument is based on a subcase of the Correlated-input Secure Hash Function assumption [10]. Now let us recall its full version:

Definition 1. *A hash function Hash is called* correlated-input secure, *if for arbitrary Boolean circuits* $C_1, \ldots, C_n$ *which satisfy:*

- *each* C_i *has high min-entropy output distribution for uniform random input distribution,*
- *for* $i \neq j$, *and* r *chosen uniformly at random,* $C_i(r) = C_j(r)$ *happens with a negligible probability.*

there is no efficient distinguisher that can distinguish with a non-negligible probability between $Hash(C_n(r))$ *and an* R *chosen uniformly at random from sequences of the same length as the hash values given the input consisting of* $Hash(C_1(r))$, *..., $Hash(C_{n-1}(r))$ computed for r chosen at random.*

Let us observe that if Hash is correlated-input secure, then in particular no efficient algorithm can distinguish between the inputs consisting of values $(Hash(C_1(r)), \ldots, Hash(C_n(r)))$ for a randomly chosen r from the inputs consisting of n random strings of the same length. This is the main property required for our security argument. We also confine ourselves to very specific functions C_i. Namely, we consider the following game:

Special Correlated Hash Values Game

> choose $s_1, \ldots, s_n \leq q$
> choose j
> choose U at random
> for $i = 1$ to n, put $h_i^{(0)} := Hash(U^{s_i}, j + i)$
> for $i = 1$ to n, choose $h_i^{(1)}$ at random
> choose $b \in \{0, 1\}$ at random
> $\hat{b} := \mathcal{A}(s_1, \ldots, s_n, j, h_1^{(b)}, \ldots, h_n^{(b)})$

The adversary $\mathcal{A}$ wins the game if $b = \hat{b}$. The advantage of $\mathcal{A}$ is measured as $|p - \frac{1}{2}|$, where p is the probability to win the game by $\mathcal{A}$.

Note that in the above game almost all values are known to $\mathcal{A}$, only U remains hidden. Moreover, we know the relationship between the first arguments of the Hash function in terms of the exponents $s_1, \ldots, s_n$.

Assumption 1. *The advantage of the adversary $\mathcal{A}$ in the Special Correlated Hash Values Game is negligible.*

Theorem 2. *For the Schnorr signature scheme, Mallet cannot distinguish between signatures created by a device implementing CR from the signatures created with the same signing key by a device with the standard implementation (without CR). In the first case Mallet is given the output of the PRNG, in the second case Mallet is given a random output.*

Proof. In order to prove Theorem 2 we have to observe that the adversary has a negligible advantage in the following game:

Game 0

$\mathcal{M}$ chooses $k_0^{(1)}, \ldots, k_0^{(n)} \le q$

$\mathcal{M}$ chooses chooses j

choose U at random

for $i = 1$ to n, put $k_1^{(i)} := \text{Hash}(U^{k_0^{(i)}}, j + i)$

choose $b \in \{0, 1\}$ at random

for $i = 1$ to n, put

$\quad k^{(i)} := k_0^{(i)} \cdot k_1^{(i)}$ if $b = 0$

$\quad$ choose $k^{(i)}$ at random, if $b = 1$

create signatures $\text{sign}_1, \ldots, \text{sign}_n$ using parameters $k^{(1)}, \ldots, k^{(n)}$

$\quad$ for messages chosen by $\mathcal{M}$

$\hat{b} := \mathcal{M}(\text{sign}_1, \ldots, \text{sign}_n, j, k_0^{(1)}, \ldots, k_0^{(n)})$

$\mathcal{M}$ wins the game, if $b = \hat{b}$.

We can rewrite this game as follows:

Game 1

$\mathcal{M}$ chooses $k_0^{(1)}, \ldots, k_0^{(n)} \le q$

$\mathcal{M}$ chooses j

~~choose U at random~~

for $i = 1$ to n, ~~$k_1^{(i)} := \text{Hash}(U^{k_0^{(i)}}, j + i)$~~   choose $k_1^{(1)}, \ldots, k_1^{(n)} \le q$ at random

choose $b \in \{0, 1\}$ at random

for $i = 1$ to n, put

$\quad k^{(i)} := k_0^{(i)} \cdot k_1^{(i)}$ if $b = 0$

$\quad$ choose $k^{(i)}$ at random, if $b = 1$

create signatures $\text{sign}_1, \ldots, \text{sign}_n$ using parameters $k^{(1)}, \ldots, k^{(n)}$

$\quad$ for messages chosen by $\mathcal{M}$

$\hat{b} := \mathcal{M}(\text{sign}_1, \ldots, \text{sign}_n, j, k_0^{(1)}, \ldots, k_0^{(n)})$

Of course, any non-negligible difference between advantage of the adversary in Game 0 and Game 1 could be used to construct a distinguisher breaking Assumption 1. On the other hand, in Game 1 the probability distribution of $k^{(1)}, \ldots, k^{(n)}$ is uniform regardless of the value of b. Therefore for Game 1 the advantage of the adversary is 0. □

In principle, the control parameters might ease forging a signature by an adversary holding a device. (In this case by *forging* we mean creating a signature outside the device.) Fortunately, it follows directly from Theorem 2 that this is not the case: if there is an effective forgery algorithm, then we can feed it with random control data and the forgery should work in exactly the same way.

Corollary 1. *If Mallet can forge a new signature based on signatures created by a device implementing CR, where he has access to the output of the PRNG, then he can forge a signature in case of the signatures created by a device implementing the standard scheme without CR and without access to the randomness created by PRNG in the device.*

More generally, by Theorem 2, there is no attack against Schnorr signature with CR that would not work for the regular Schnorr signature, as such an attack could be used as an distinguisher contradicting the statement of Theorem 2.

3.2 Security Against the User

In case of CR, the signer holding a signing device should be regarded as an adversary aiming, for instance, to extract the private signing key from the device. This has to be considered separately, since the signer gets control data that are not available for a recipient of his signatures. There may be different kinds of the attack, but what we really would like to protect is the ability to create valid signatures solely by the signing device. Therefore we define the following game:

User Forgery Game:

Phase 1. Alice interacts with the device asking for signatures of the message of her choice. If the signature scheme implements CR, then Alice can update the blinding factor arbitrarily and receives signatures together with the corresponding control data.

Phase 2. Alice has to present a signature s that has not been created by the device.

Alice wins the User Forgery Game, if the signature s yields a positive verification result. We show the following result:

Theorem 3. *If there is an adversary that wins the User Forgery Game for a signature scheme with CR with a non negligible probability in the Random Oracle Model, then there is an adversary that wins the User Forgery Game for the same signature scheme without CR also with a non negligible probability.*

Proof. Given an input for the User Forgery Game in the standard setting, we expand the input in order to provide the control data. We will show that for a signature that uses a random element r, the control data can be simulated by programming the random oracle. Namely, the simulation can be done as follows:

1. choose $U = g^u$ as a blinding factor for this step,
2. choose $k_1 < q$ at random,
3. put $r' := r^{k_1^{-1}} \bmod q$ as the control data,
4. in the hash oracle table put $k_1 = \text{Hash}(r'^u, i)$, where i is the signature sequential number.

Note that there is no chance for a conflict while inserting the hash values in the hash table, as each sequential number is used exactly once.   □

3.3 Security Against the device

The device may deviate from the protocol and in this way attempt to cheat the user in cooperation with Mallet. First, let us note that the value obtained from PRNG can be set freely by the device and that this value is shared with Mallet.

In the standard setting the `device` can deviate from the protocol and create a kleptographic channel. This channel enables to leak the signing key with just two signatures. Thereby, any security claims are illusory as long as we do not have real control over the manufacturer and over the delivery chain (the honest devices of the manufacturer can be exchanged by the malicious ones – as the devices are black boxes, it is hard to see the difference unless an additional protection level is implemented).

The standard kleptographic construction cannot be repeated for the signatures with CR, however – as presumably for all randomized signature schemes – there is a possibility to create a channel that leaks a few bits. The `device` shares a "public" key Z with Mallet, where $Z = g^z$. Then the value r leaking a string ω is determined as follows:

> **repeat until** ω is the suffix of Z^k represented in binary
> $\quad k_0$ is taken as the output of PRNG()
> $\quad k_1 := \text{Hash}(U^{k_0}, i)$
> $\quad k := k_0 \cdot k_1$
> $\quad r' := g^{k_0}$
> $\quad r := g^k$

Mallet can reconstruct ω from r^z. Of course, the string ω cannot be too long, as it would increase the signature creation time. Practically, it is possible for just a few bits (if any), as it is very hard to hide the increased computation time (and other stochastic properties of the computation time). In order to prevent any precomputations on the side of the `device` we may introduce an additional parameter c set by the user together with the signature request. Then, we would modify the computation of k_1 in the following way:

$$k_1 := \text{Hash}(U^{k_0}, c, i)$$

Even a short c would help very much as generally the memory on a signing device is very limited.

Below we show that the above method is essentially all the `device` can do in order to deviate from the protocol. The proof is based on KEA1 assumption [8], which informally states that: if there exists an algorithm that takes as input (g, g^a) and outputs $(g^r, (g^a)^r)$, then there exists an extractor that on the same input returns r with a non-negligible probability.

Proposition 1. *In the random oracle model, assuming KEA1, while computing r' and r the* `device` *must derive k_0 before k_1 with probability $1 - \varepsilon$, where ϵ is negligible.*

Proof. According to the definition of k_1 and the Random Oracle Model, the `device` must first compute all arguments before computing the value k_1 satisfying the equality $k_1 = \text{Hash}(J, i)$, where J should be equal to U^{k_0}. Otherwise the value k_1 may appear in the `device` with a negligible probability only. Note that the signer recomputes k_1 as $\text{Hash}((r')^u, i)$, it follows that the `device` has to use $J = (r')^u$. If we denote, $r' = g^{k_0}$, then it follows that $J = U^{k_0}$.

By KEA1 assumption, if the device can create the values $r' = g^{k_0}$ and $J = U^{k_0}$, then it must know k_0 (as there exists an extractor) at the moment of creation of these values. So, indeed k_0 is known before U^{k_0} and thereby before k_1.   □

Proposition 1 in fact says that whatever the device tries to do, essentially it must follow the protocol: first determine k_0 (maybe in some malicious way) and then derive the remaining values in a deterministic way.

Now assume that the device wishes to create r having some particular property P in order to leak some information. For this purpose the device may attempt to choose the value k_0 accordingly. However, in the Random Oracle Model the only way to learn whether $P(r)$ holds is via derivation of the pseudorandom value $k_1 = \text{Hash}(U^{k_0}, i)$. So, for each k_0 the probability that $P(r)$ holds can be treated as a random experiment with the success probability equal to the probability that a randomly chosen r satisfies the property P. Thereby, the device has to follow the procedure described above for creating a few leakage bits.

4   Diffie-Hellman Key Exchange

Just as for signature schemes, CR can be directly implemented in case of Diffie-Hellman key exchange protocol and ElGamal encryption scheme. This is quite important, since the Diffie-Hellman protocol is a key component of many other complex schemes.

This concerns some protocols for significant practical importance such as Extended Access Control (EAC) and Password Authenticated Connection Establishment (PACE) securing data exchange with biometric passports and electronic identity documents [6,11][1]. Both protocols utilize the Diffie-Hellman protocol as subroutines. For instance, PACE v2 is based on double execution of Diffie-Hellman key exchange.

4.1   Diffie-Hellman Protocol

Recall that during Diffie-Hellman protocol (DH) the device A of Alice sends to (the device of) Bob a value $Y_A = g^{y_A}$, where y_A is chosen at random. If the random exponent y_A can be guessed or derived by the adversary, then the whole security of the key exchange collapses, as the adversary can compute the shared key. In the DH protocol with CR the device A of Alice executes the following operations:

1. choose k at random (take the output from the PRNG),
2. $preY_A := g^k$,

---

[1] EAC and PACE are protocols of high importance for the security of biometric passports and electronic identity documents. EAC is a standardized authenticated key exchange protocol, which goal is to authenticate the identity document and the terminal against each other. PACE is a password authenticated key exchange protocol which secures the transmission between an identity document and the reader.

3. $k' := \text{Hash}(U^k, i)$,
4. $Y_A := (preY_A)^{k'}$,
5. $y_A := k \cdot k' \bmod q$, where q is the order of the group where the Diffie-Hellman protocol is executed.

Afterwards, the value Y_A is presented by the device A together with the control value $preY_A$ and the sequential number i.

Alice (or more precisely her another device) standing in between device A and Bob executes the control step by checking that

$$Y_A = (preY_A)^{\text{Hash}((preY_A)^u, i)},$$

where u is the secret key for the blinding factor. Neither $preY_A$ nor i is needed for the device of Bob, so only Y_A is forwarded to Bob.

Note that Bob need not to implement CR, in order for Alice to be able to run CR. Moreover, Bob even will not be able to recognize whether Alice is running CR or not.

4.2   Security Sketch for DH with CR

Session Key Secrecy Against Alice. Alice may attempt to compute the session key K and to continue the connection without the device A. There are attack scenarios where such hijacking attack would make sense. Note that apart from Y_A Alice knows $U, preY_A$ and can compute k'.

Deriving the session key K means deriving $(Y_B)^{k \cdot k'}$, so it is equivalent to getting $(Y_B)^k$ by Alice. We claim that an algorithm deriving $(Y_B)^k$ could be used to solve the Computational Diffie-Hellman (CDH) problem. Indeed, if we have an instance (g, h_1, h_2) of the CDH problem, then we may treat h_1 as $preY_A$, h_2 as Y_B, set the key u at random and create the case for the adversary by setting $k' := \text{Hash}((preY_A)^u)$ and $Y_A := preY_A^{k'}$.

Session Key Secrecy Against Malicious Manufacturer. In this case the adversary knows Y_A, Y_B, and k as the output of PRNG (hence he knows also $preY_A = g^k$). He has no access to k', u and U. The adversary's goal is to compute $K = (Y_B)^{k \cdot k'}$.

In order to argue that this is impossible we again have to use hash functions that are *correlated-input secure* - as discussed in Sect. 3.1. We may assume that an adversary can analyze the data from multiple sessions where the device uses the same blinding factor U of CR. Nevertheless, we claim that it is infeasible to distinguish between the correct session key K and a random key, even if the adversary knows the values of k' from all previous sessions.

Assume conversely that there is such a distinguisher $\mathcal{D}$. We build a case for the distinguisher where $C_i(U) = U^{k_i}$ – so U plays the role of r from Definition 1, where the values k_i are generated at random. For $i < n$ the elements k'_i are equal to $\text{Hash}(U^{k_i})$. The number k'_n is a candidate for $\text{Hash}(U^{k_n})$. Given k_i and k'_i for $i \leq n$, the elements $(preY_A)_i$, $(Y_A)_i$, $K_i = (Y_B)_i^{k_i \cdot k'_i}$ are created according to the definition.

Now, these data are given to the distinguisher $\mathcal{D}$ (of course except for U and the values k_i'). If k_n' was random, then so is the key K_n and the distinguisher would indicate that the key k_n' is incorrect.

5   ElGamal Public Key Encryption

Controlled randomness can be directly applied to enhance security level of ElGamal Encryption. As an example let us consider ElGamal Key Encapsulation mechanism. Let us assume that Alice has to encrypt a message for Bob holding a private key x and the public key $y = g^x$. As before, the device of Alice contains the blinding factor $U = g^u$ and Alice holds u. The modified encryption procedure executed by the `device` looks as follows:

Encrypting M for Bob

1. choose k at random using a PRNG
2. $r' := g^k$
3. compute $k' := \mathrm{Hash}(U^k)$,
4. $r := (r')^{k'}$
5. encrypt M with symmetric key $K := \mathrm{Hash}_0(y^{k \cdot k'})$, where H_0 maps the elements of the group to the key space of the symmetric encryption scheme,
6. output r, the ciphertext $\mathrm{Enc}_K(M)$ and the control element r'.

The decryption procedure is standard and uses the key $K = \mathrm{Hash}_0(r^x)$ obtained via deencapsulation. The control mechanism follows the same steps as described in Sect. 3 for the Schnorr signature scheme.

Security of the modified scheme has to be considered for the scenario in which the malicious manufacturer can reconstruct the output of the PRNG of Alice and wants to break the ciphertext $\mathrm{Enc}_K(M)$. We follow the same argument as in case of the Diffie-Hellman protocol.

6   Final Remarks and Future Work

The most important property of the proposed mechanism is its relative simplicity. With almost no alternations in the existing schemes we can provide a significant improvement of the security level. The major advantage is that a device attempting to cheat never knows whether its user is performing the checks or not. Therefore it is hard to imagine that any vendor would dare to enable the device to deviate from the protocol. An interesting feature of our method is that it is based on techniques borrowed from kleptography. Now, the trick used previously for malicious purposes is applied to improve security.

In this paper, we have presented applications of controlled randomness to very basic cryptographic protocols. In reality, these schemes are used as components in more complicated schemes. Our security arguments do not immediately apply to those schemes, as we need a property analogous to universal composability.

There are also other complications. For example, the password authentication protocol PACE with generic mapping (as adopted by ICAO organization for biometric passports) executes Diffie-Hellman key exchange twice. However, for the second key exchange the base element is unknown even for the owner of the device. Therefore a more sophisticated solution has to be applied. However, the most significant practical problem is the effort necessary to redo the formal security proof for such a modified scheme.

References

1. Bellare, M., Miner, S.K.: A forward-secure digital signature scheme. In: Wiener, M. (ed.) CRYPTO 1999. LNCS, vol. 1666, pp. 431–448. Springer, Heidelberg (1999). doi:10.1007/3-540-48405-1_28
2. Bender, J., Fischlin, M., Kügler, D.: The PACE—CA protocol for machine readable travel documents. In: Bloem, R., Lipp, P. (eds.) INTRUST 2013. LNCS, vol. 8292, pp. 17–35. Springer, Cham (2013). doi:10.1007/978-3-319-03491-1_2
3. Bernstein, D.J., Chang, Y.-A., Cheng, C.-M., Chou, L.-P., Heninger, N., Lange, T., Someren, N.: Factoring RSA keys from certified smart cards: coppersmith in the wild. In: Sako, K., Sarkar, P. (eds.) ASIACRYPT 2013. LNCS, vol. 8270, pp. 341–360. Springer, Heidelberg (2013). doi:10.1007/978-3-642-42045-0_18
4. Błaśkiewicz, P., Kubiak, P., Kutyłowski, M.: Two-head dragon protocol: preventing cloning of signature keys. In: Chen, L., Yung, M. (eds.) INTRUST 2010. LNCS, vol. 6802, pp. 173–188. Springer, Heidelberg (2011). doi:10.1007/978-3-642-25283-9_12
5. Boneh, D., Ding, X., Tsudik, G., Wong, C.M.: Instantenous revocation of security capabilities. In: USENIX Security Symposium (2001)
6. BSI. Advanced Security Mechanisms for Machine Readable Travel Documents 2.11. Technische Richtlinie TR-03110-3 (2013)
7. Checkoway, S., Fredrikson, M., Niederhagen, R., Green, M., Lange, T., Ristenpart, T., Bernstein, D.J., Maskiewicz, J., Shacham, H.: On the practical exploitability of Dual EC DRBG in TLS implementations (2014)
8. Damgård, I.: Towards practical public key systems secure against chosen ciphertext attacks. In: Feigenbaum, J. (ed.) CRYPTO 1991. LNCS, vol. 576, pp. 445–456. Springer, Heidelberg (1992). doi:10.1007/3-540-46766-1_36
9. Gołębiewski, Z., Kutyłowski, M., Zagórski, F.: Stealing secrets with SSL/TLS and SSH – kleptographic attacks. In: Pointcheval, D., Mu, Y., Chen, K. (eds.) CANS 2006. LNCS, vol. 4301, pp. 191–202. Springer, Heidelberg (2006). doi:10.1007/11935070_13
10. Goyal, V., O'Neill, A., Rao, V.: Correlated-input secure hash functions. In: Ishai, Y. (ed.) TCC 2011. LNCS, vol. 6597, pp. 182–200. Springer, Heidelberg (2011). doi:10.1007/978-3-642-19571-6_12
11. ISO/IEC JTC1 SC17 WG3/TF5 for the International Civil Aviation Organization. Supplemental access control for machine readable travel documents. Technical report, 2014. version 1.1, April 2014
12. Itkis, G., Reyzin, L.: SiBIR: signer-base intrusion-resilient signatures. In: Yung, M. (ed.) CRYPTO 2002. LNCS, vol. 2442, pp. 499–514. Springer, Heidelberg (2002). doi:10.1007/3-540-45708-9_32
13. Juels, A., Guajardo, J.: RSA key generation with verifiable randomness. In: Naccache, D., Paillier, P. (eds.) PKC 2002. LNCS, vol. 2274, pp. 357–374. Springer, Heidelberg (2002). doi:10.1007/3-540-45664-3_26

14. King, C.: Dual_EC_DRBG output using untrusted curve constants may be predictable (2013). http://www.kb.cert.org/vuls/id/274923
15. Kucner, D., Kutyłowski, M.: Stochastic kleptography detection. In: Alster, K., Urbanowicz, J., Williams, H.C. (eds.) Public-Key Cryptography and Computational Number Theory (Warsaw 2000), pp. 137–149. Walter de Gruyter Inc., Birmingham (2001)
16. Kutyłowski, M., Hanzlik, L., Kluczniak, K., Kubiak, P., Krzywiecki, L.: Forbidden city model – towards a practice relevant framework for designing cryptographic protocols. In: Huang, X., Zhou, J. (eds.) ISPEC 2014. LNCS, vol. 8434, pp. 42–59. Springer, Cham (2014). doi:10.1007/978-3-319-06320-1_5
17. Nicolosi, A., Krohn, M.N., Dodis, Y., Mazières, D.: Proactive two-party signatures for user authentication. In: Proceedings of the Network and Distributed System Security Symposium, NDSS 2003, San Diego, California, USA. The Internet Society (2003)
18. NIST. Random Number Generation (2010)
19. Pfitzmann, B.: Digital Signature Schemes, General Framework and Fail-Stop Signatures, vol. 1100. Springer, Heidelberg (1996)
20. Shumow, D., Ferguson, N.: On the possibility of a back door in the NIST SP800-90 Dual EC PRNG. In: CRYPTO Rump Session Presentation (2007)
21. Wang, Y., Nicol, T.: Statistical properties of pseudo random sequences and experiments with PHP and Debian OpenSSL. In: Kutyłowski, M., Vaidya, J. (eds.) ESORICS 2014. LNCS, vol. 8712, pp. 454–471. Springer, Cham (2014). doi:10.1007/978-3-319-11203-9_26
22. Yilek, S., Rescorla, E., Shacham, H., Enright, B., Savage, S.: When private keys are public: results from the 2008 Debian OpenSSL vulnerability. In: Proceedings of the 9th ACM SIGCOMM Conference on Internet Measurement Conference, IMC 2009, pp. 15–27. ACM, New York (2009)
23. Young, A.L., Yung, M.: Malicious Cryptography - Exposing Cryptovirology. Wiley, Hoboken (2004)
24. Young, A.L., Yung, M.: A timing-resistant elliptic curve backdoor in RSA. In: Pei, D., Yung, M., Lin, D., Wu, C. (eds.) Inscrypt 2007. LNCS, vol. 4990, pp. 427–441. Springer, Heidelberg (2008). doi:10.1007/978-3-540-79499-8_33

Malware, Encryption, and Rerandomization – Everything Is Under Attack

Herman Galteland$^{(\boxtimes)}$ and Kristian Gjøsteen

Department of Mathematical Sciences,
NTNU – Norwegian University of Science and Technology, Trondheim, Norway
{herman.galteland,kristian.gjosteen}@math.ntnu.no

Abstract. A malware author constructing malware wishes to infect a specific location in the network. The author will then infect n initial nodes with n different variations of his malicious code. The malware continues to infect subsequent nodes in the network by making similar copies of itself. An analyst defending M nodes in the network observes N infected nodes with some malware and wants to know if any sample is targeting any of his nodes. To reduce his work, the analyst need only look at unique malware samples. We show that by encrypting the malware payload and using rerandomization to replicate malware, we can make the N observed malware samples distinct and increase the analyst's work factor substantially.

Keywords: Malicious cryptography · Environmental keys · Rerandomization · Provable security

1 Introduction

Malware is software maliciously installed on a computer designed to give functionality and behavior desired by the *malware author*, but not by the legitimate computer owner.

Our goal is to study malware propagation and how to protect propagating malware from analysis. We will not study the construction of computer viruses or other types of malware, but rather how to construct a scheme designed to encrypt malware such that we can hide the intentions of the malware author.

1.1 Real World Examples

BurnEye [11] is a tool designed to protect binary files and is an example on how to protect malware. The tool adds three protective layers to a file: obfuscation, encryption, and a fingerprint layer. The latter layer ensures that the file can only be run on a specific computer that has the specifications stated by the fingerprint

---

K. Gjøsteen—This work is funded by Nasjonal sikkerhetsmyndighet (NSM), www.nsm.stat.no.

© Springer International Publishing AG 2017
R.C.-W. Phan and M. Yung (Eds.): LNCS 10311, Mycrypt 2016, pp. 233–251, 2017.
DOI: 10.1007/978-3-319-61273-7_12

layer. The encryption layer uses a user-chosen password as the encryption key such that the file can only be executed (or analyzed) by someone with the proper password.

Gauss [8] is an example of sophisticated malware that uses encryption to protect certain payloads. Gauss uses *environmental keys* to decrypt the payload, where an environmental key is a key that is generated from locally available data. The malware gathers local data on the infected computer and hashes it to create decryption keys, where the string of data that results in the correct key is selected by the malware author. The malicious code can only be executed when the correct key is produced, that is, when the malware infects the intended target. To our knowledge, the contents of the encrypted payloads of Gauss are still unknown.

1.2   Malware Propagation

Consider a malware author whose objective is to attack some specific location(s). The malware author's goals is to hide his intentions and identity. The malware author's adversary is an *analyst* observing and defending some network containing one or more of the malware author's target(s). The goal of the analyst is to detect malware targeting any part of the network he is protecting. He also wants to discover the intentions and the identity of the malware author. Hiding the mere existence of malware from the analyst is a distinct problem and not one we consider in the current work.

We use the following model to describe malware propagation (see also Fig. 1). The source S, the malware author, infects n initial nodes with (different variations of) his malware and they, in turn, will infect subsequent nodes in the network by making similar copies of themselves. Every direct link to the malware author increases the analyst's chances of discovering the malware author's identity, so to avoid identification, the malware author should perform as few initial infections as possible and use indirect paths to his intended target.

The analyst's job is to defend M nodes in the network from any possible malware threat and he has full knowledge of the environment he is protecting. By observing the wider network the analyst can find N malware samples.

The malware author will encrypt the malware payload to make the analyst's job harder. Encrypting the payload prevents reverse engineering of the malware code [4] and hides the intentions of the author. Since obfuscation is hard, we use encryption keys derived from environmental parameters, network triggers, or a combination of these [9]. Thus, the malware will have an encrypted payload, containing the malicious code, and a cleartext loader which gathers environmental parameters to generate decryption keys.

To generate the malware the author chooses environmental data corresponding to the intended target computer, hashes that data to create a secret key, and encrypts the payload using the key. The malware is then ready to be released. When the malware arrives on a new computer, the cleartext loader will determine the environmental data of the infected computer, hash the data to derive $L > 1$ keys, and try to decrypt the payload using the L derived keys. If the

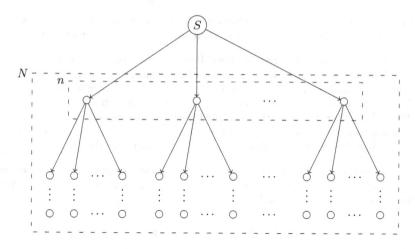

Fig. 1. Illustration of the malware infection paths

decryption is a success under one of the derived keys, the code will be executed. The cleartext loader makes copies of the malware and infect subsequent nodes.

Since only the malware author knows the secret key, only the malware author can create encryptions of the payload. This means that under our propagation model, there are at most n distinct encrypted payloads among the samples collected by the analyst. Each sample is encrypted and has a unknown target. If the analyst wants to be sure that none of these samples would attack the analyst's network, the analyst needs to do roughly L trial decryptions for each of his M nodes, which means that his work factor is nML.

Instead of making exact copies of the malware to replicate it, we want the loader to rerandomize [2,6] the encrypted payload using techniques from asymmetric cryptography. The rerandomization process takes as input an encrypted payload and some random numbers and produces a new encrypted payload that encrypts the same malicious code. Hence, the loader can produce several different-looking malware payloads to infect subsequent nodes, without knowledge of the secret key. This process is described in Fig. 2.

To fully utilize the rerandomization process, we want it to produce the payloads such that any two malware samples are indistinguishable. If the analyst is unable to distinguish between malware samples then, essentially, there are N unique variations of the malware in the network. This means that the analyst need to do L trial decryptions of N samples for M different nodes to ensure that none of the malware samples are targeting any of his nodes. This will increase the workload to NML. Since the malware creates new different variations of itself, the malware author can now choose n to be small and possibly significantly reducing the risk of detection.

Now, imagine a world filled with hundreds of different types of malware: all are encrypted, use rerandomization, are of the same size, use the same loader, and otherwise look the same. Every new malware sample an analyst would discover

When the malware, in the form of a cleartext loader and an encrypted payload, arrives on a new host, the cleartext loader is executed and performs the following steps:

1. The encrypted payload is rerandomized before it is stored on the host.
2. The loader scans the host environment and determines the environmental data.
3. The loader hashes the environmental data to produce one or more keys.
4. The loader tries to decrypt the encrypted payload with each key.
5. If the decryption succeeds, the decrypted payload is executed.
6. The malware may also attempt to infect some other host, in which case the encrypted payload is rerandomized before it is transmitted to the new host.

Note that the malware will certainly use some polymorphic engine and other standard malware techniques in order to provide a basic level of protection for the cleartext loader and the encrypted payload.

Fig. 2. The malware attack process.

needs to be analyzed. The analyst cannot be certain of whether a new sample corresponds to one he has previously determined is no threat, or a genuinely new piece of malware. Note that this requires the various malware authors to agree on standard payload sizes and a standard loader. If they do not, then the analyst can use these pieces of information to classify samples.

The limitations with our scheme is that the analyst can always guess, or predict, the target of the malware author. Also, if the malware reaches its target, the payload will be decrypted and executed. If the analyst notices the attack, he will be able to deduce the environmental key and thus be able to decrypt the payload. This seems impossible to avoid.

Another limitation is that once an analyst discovers the key used for one sample, he can easily discover all other samples corresponding to that key. However, the malware author will hope that different analysts are unwilling to reveal that they are under attack (they somehow consider this fact sensitive) and that they therefore do not share discovered keys. This means that one analyst's success may not make all the other analyst's work easier.

1.3 Related Work

Traditionally, cryptography has been developed and used as a defense against attackers. However, it is clear that cryptography can also be of use to the attackers.

Young and Yung where the first to raise the concern about malicious use of cryptography (cryptovirology) [13] and have several works related to malware construction and propagation: A virus capable of encrypting files on the victim's computer and hold them for ransom [12]. A mobile program that carries a rerandomizable ciphertext, which enables anonymous communication, where the program takes random walks through a network in a system called Feralcore, and, at each node, the ciphertext is rerandomized [14]. Utilizing a mix network to mix programs and propagate malware [13].

The mix network and the mobile program mentioned use the idea of universal re-encryption, by Golle et al. [6], to re-encrypt ciphertexts. The process transforms the ciphertexts into a new ciphertext that encrypts the same message and do not require knowledge about the public key. Similar to universal re-encryption is the notion of rerandomization by Canetti et al. [2].

Filiol showed that by encrypting malware payload [3,4] one can prevent anyone from analyzing the code and reverse engineer it, possibly using the environmental keys of Riordan and Schneier [9] as the encryption key. Similar to Riordan and Schneier, secure triggers [5,7] are used to keep certain content private until some particular event occurs.

1.4   Overview

In Sect. 2.1 we describe the cryptosystem designed to encrypt and rerandomize malware payload. In Sect. 2.2 we construct a basic scheme based on ElGamal. In Sect. 2.3 we show that the basic scheme is secure by using games, where the adversary is asked to distinguish between ciphertexts encrypting the same message and ciphertexts encrypting two different messages. That is, we will simulate whether the analyst is able to distinguish malware samples. In Sect. 2.4 we construct an extended scheme based on the basic scheme, which is capable of encrypting longer messages. In Sect. 2.5 we show that the extended scheme is secure using games. The procedure is similar to the security proof of the basic scheme.

2   Rerandomizable Encryption

We will in this section describe and construct the encryption scheme designed to encrypt and rerandomize malware payload. We will construct two (similar) example schemes, as proofs of concept, and show that it is hard to distinguish between encrypted payload samples.

As a simplification we will denote payload as messages, encrypted payload as ciphertexts, replication of malware as rerandomization of ciphertexts, and environmental derived keys as keys.

2.1   Preliminary

In our scheme we have an algorithm $\mathcal{E}$ encrypting messages, an algorithm $\mathcal{D}$ decrypting ciphertexts, and an algorithm $\mathcal{R}$ rerandomizing ciphertexts.

Encryption. For a message m and a key k the encryption algorithm $\mathcal{E}(k, m)$ outputs a ciphertext c.

Decryption. For a ciphertext c and a key k the decryption algorithm $\mathcal{D}(k, c)$ either outputs a message m or a special symbol indicating decryption failure.

Rerandomization. For a ciphertext c, encrypting a message m, the rerandomize algorithm $\mathcal{R}(c)$ outputs a ciphertext c', encrypting the same message m.

We want the output distribution of the rerandomize algorithm to be computationally indistinguishable from the output distribution of the encryption algorithm. That is, it should be hard to determine if two different ciphertexts encrypts the same message or not. We also want the system to be correct, that is, we should almost always be able to decrypt all ciphertexts output by the encryption algorithm. Since the output distribution of the encryption and rerandomize algorithms are computationally indistinguishable, ciphertexts output by the rerandomize algorithm will also almost always be correct.

Correctness. If c was output from $\mathcal{E}(k, m)$ then $\mathcal{D}(k, c)$ will always output m except with negligible probability.

Rerandomization. If c was output by $\mathcal{E}(k, m)$ then the output distribution of $\mathcal{R}(c)$ should be computationally indistinguishable from the output distribution of $\mathcal{E}(k, m)$.

We will not always be able to apply an arbitrary number of rerandomizations to a ciphertext without getting decryption errors, which we will see is the case in Sect. 2.4.

The security requirements of our cryptosystem reflects the intentions of the malware author. It should be difficult to guess the malware author's target, and it should be hard to determine if two ciphertexts are the encryption of the same message or not.

Key Indistinguishability. It should be hard to say something about which key a ciphertext has been encrypted under.

Indistinguishability. It should be hard to decide if two ciphertexts, encrypted under the same key, decrypts to the same message or not.

2.2   Basic Scheme

We will construct a basic scheme based on the ElGamal cryptosystem over a group G of prime order p generated by g. The basic scheme is essentially the same as the encryption scheme proposed by Golle et al. [6]. The algorithms of the scheme is the following.

Encryption. For a message $m \in G$ and a key $k \in \{1, \ldots, p-1\}$ pick $r, s \in \{1, \ldots, p-1\}$ uniformly and output

$$c = (x, y, z, w) = (g^r, g^{kr}, g^s, g^{ks}m).$$

Decryption. For a ciphertext $c = (x, y, z, w)$ and a key $k \in \{1, \ldots, p-1\}$ check if $x^k = y$. If not, output a symbol indicating decryption failure. If it is, output

$$m = z^{-k}w.$$

Rerandomize. For a ciphertext $c = (x, y, z, w)$ pick $r', s' \in \{1, \ldots, p-1\}$ uniformly and output

$$c' = (x', y', z', w') = (x^{r'}, y^{r'}, zx^{s'}, wy^{s'}).$$

Note that if $c = (x, y, z, w)$ was output by the encryption algorithm then there exists parameters r, s, and k, and a message m such that

$$c = (x, y, z, w) = (g^r, g^{kr}, g^s, g^{ks}m).$$

With input c the rerandomize algorithm will output a $c' = (x', y', z', w')$ where

$$x' = x^{r'} = g^{rr'},$$
$$y' = y^{r'} = g^{krr'},$$
$$z' = zx^{s'} = g^s g^{rs'} = g^{s+rs'},$$
$$w' = wy^{s'} = g^{ks} g^{krs'}m = g^{k(s+rs')}m.$$

That is, $c' = (g^{rr'}, g^{krr'}, g^{s+rs'}, g^{k(s+rs')}m)$. Since $r \neq 0$, we get that $s + rs'$ can take any value modulo p except s and all values are equally probable. Hence, we get that the output distribution of the encryption and rerandomize algorithms are computationally indistinguishable. Note that the ciphertext c' has the same form as a ciphertext output by the encryption algorithm, that is, $(g^{\hat{r}}, g^{\hat{k}\hat{r}}, g^{\hat{s}}, g^{\hat{k}\hat{s}}m)$, for some parameters $\hat{r}$, $\hat{s}$, and $\hat{k}$, and message m.

We can now show the correctness of the decryption algorithm. Note that for all ciphertexts $c = (x, y, z, w)$ we have that $x^k = (g^r)^k = g^{kr} = y$, which is true for ciphertexts output by both the encryption and rerandomize algorithms. We can therefore retrieve the message m by computing

$$z^{-k}w = (g^s)^{-k}g^{ks}m = g^{-ks+ks}m = m.$$

Thus the decryption algorithm is correct.

It is possible to extend the basic scheme by encrypting several messages under the same key. For a set of messages $m_1, m_2, \ldots, m_n$, we can encrypt them as

$$(g^r, g^{kr}, g^{s_1}, g^{ks_1}m_1, g^{s_2}, g^{ks_2}m_2, \ldots, g^{s_n}, g^{ks_n}m_n)$$

for a key k, and variables $s_1, s_2, \ldots, s_n$, and r. This is not a very efficient method, and we will in Sect. 2.4 construct a different extended scheme by using techniques from symmetric cryptography. In the next section we will show that the basic scheme is secure.

2.3   Security of the Basic Scheme

We will in this section use games to show that the basic scheme is secure given that it is hard to guess which environmental key the ciphertexts are encrypted under.

The key encrypting the malware payload is derived from environmental parameters sampled by the loader. From the adversary's perspective, the collection of sampled parameter types can be considered as a probability space of possible decryption keys. We will denote this space by D. If the size of D is large then the adversary is less likely to guess the correct decryption key, where the size of D is determined by, most notably, the number of different parameters the loader is gathering.

We want to show that the adversary is unable to distinguish between ciphertexts and that his advantage is determined by D, that is, the probability of the adversary guessing the correct key. To do so, we will use a sequence of games [10]. In our games we will start with simulating an experiment where we ask the adversary to differentiate between two cases; ciphertexts encrypting different messages, and ciphertexts encrypting the same message.

Experiment. Given two ciphertext c_1, and c_2, decide either

$$c_1 = \mathcal{E}(k_1, m_1) \qquad\qquad c_1 = \mathcal{E}(k_1, m_1)$$
$$c_2 = \mathcal{E}(k_2, m_2) \quad \text{or} \quad c_2 = \mathcal{R}(c_1)$$

for some messages m_1, m_2 and keys k_1, k_2.

We can show that the security of the scheme can be based on the hardness of the Decisional Diffie-Hellman (DDH) problem [1] in the random oracle model. The DDH problem is to distinguish between tuples of the form (g, g^a, g^b, g^{ab}) and tuples of the form (g, g^a, g^b, g^c), for some $a, b, c \in \{1, \ldots, p-1\}$. Where the DDH assumption states that the DDH problem is hard to solve.

To create the encryption keys, we will use a hashing oracle to hash elements drawn from the probability space D. We will denote the hashing oracle by H, where it should be impossible to get any information about the input by looking that the output of the oracle.

Game 0. In the first game we will follow the experiment. If $b = 0$, we will encrypt the two given messages under two different keys. If $b = 1$, we will encrypt only one of the messages and rerandomize the resulting ciphertext. In both cases, we send the ciphertexts to the adversary, who replies with a bit b' and the game ends. The full procedure of Game 0 can be seen in Fig. 3.

Let E_0 denote the event that $b = b'$ in Game 0.

Game 1. We will stop the game if the adversary guesses one of the keys correctly. If the adversary gives either u_1 or u_2 in one of its queries the oracle will: flip a coin, $b' \xleftarrow{r} \{0,1\}$, output b', and stop the game. We will denote this event by F_1.

Let E_1 denote the event that $b = b'$ in Game 1. Unless the event F_1 occurs Game 1 behaves just like Game 0. Thus we have that $E_0 \wedge \neg F_1 \iff E_1 \wedge \neg F_1$ and by the difference lemma we get that

$$|\Pr[E_0] - \Pr[E_1]| \leq \Pr[F_1].$$

Game 2. Since the adversary can no longer use the oracle to get any information about the keys without stopping the game, we are essentially drawing our keys randomly from a set. That is, we draw $k_1, k_2 \xleftarrow{r} \{1, \ldots, p-1\}$ uniformly and we will no longer query the hashing oracle. Note that since the adversary can still query the hashing oracle, we still need to draw samples from the space D to check if the adversary is guessing the keys correctly.

Let E_2 denote the event that $b = b'$ in Game 2. Since the adversary can no longer get any information about the environmental keys from the hashing

Game 0:
$$u_1, u_2 \leftarrow D, \; k_1 \leftarrow H(u_1), \; k_2 \leftarrow H(u_2), \; b \xleftarrow{r} \{0,1\}$$
Get m_1, m_2 from A

If $b = 0$ do:
$$r, r', s, s' \xleftarrow{r} \{1, \ldots, p-1\},$$
$$c_1 \leftarrow (x, y, z, w) = (g^r, g^{k_1 r}, g^s, g^{k_1 s} m_1)$$
$$c_2 \leftarrow (x', y', z', w') = (g^{r'}, g^{k_2 r'}, g^{s'}, g^{k_2 s'} m_2)$$
Send c_1, c_2 to A

If $b = 1$ do:
$$r, r', s, s' \xleftarrow{r} \{1, \ldots, p-1\},$$
$$c_1 \leftarrow (x, y, z, w) = (g^r, g^{k_1 r}, g^s, g^{k_1 s} m_1)$$
$$c_2 \leftarrow (x', y', z', w') = (x^{r'}, y^{r'}, zx^{s'}, wy^{s'})$$
Send c_1, c_2 to A

Get b' from A

Fig. 3. Game 0 of the basic scheme

oracle without stopping the game, the keys used are, essentially, some random group elements. Hence, $\Pr[E_2] = \Pr[E_1]$.

Game 3. We change how we compute the tuples such that we the encryption algorithm do not require the keys as input. To do so we will precompute the tuples before we receive the messages. That is, for some uniform $s, s' \in \{1, \ldots, p-1\}$ and keys k_1, k_2, we will compute

$$(x, y, z, w) = (g, g^{k_1}, g^s, g^{k_1 s})$$
$$(x', y', z', w') = (g, g^{k_2}, g^{s'}, g^{k_2 s'})$$

before we receive the messages m_1 and m_2.

In the case $b = 0$, we will encrypt the two messages using the precomputed tuples. That is, we will pick a random element per message, r and r', and compute

$$c_1 = (x^r, y^r, z, wm_1) = (g^r, g^{k_1 r}, g^s, g^{k_1 s} m_1),$$
$$c_2 = (x'^{r'}, y'^{r'}, z', w'm_2) = (g^{r'}, g^{k_2 r'}, g^{s'}, g^{k_2 s'} m_2).$$

In the case $b = 1$, we will encrypt one message and rerandomize the computed ciphertext. To encrypt m_1 we pick a random element r and compute

$$c_1 = (x^r, y^r, z, wm_1) = (g^r, g^{k_1 r}, g^s, g^{k_1 s} m_1).$$

To rerandomize $c_1 = (\hat{x}, \hat{y}, \hat{z}, \hat{w})$ we draw some uniform element r' and s', as usual, and compute

$$c_2 = (\hat{x}^{r'}, \hat{y}^{r'}, \hat{z}\hat{x}^{s'}, \hat{w}\hat{y}^{s'}) = (x^{rr'}, y^{rr'}, zx^{rs'}, wy^{rs'} m_1)$$
$$= (g^{rr'}, g^{k_1 rr'}, g^{s+rs'}, g^{k_1(s+rs')} m_1).$$

242   H. Galteland and K. Gjøsteen

Let E_3 denote the event that $b = b'$ in Game 3. The output distribution of the encryption algorithm in Game 2 and in Game 3 are exactly the same, similarly for the rerandomization algorithm. Therefore, we have that $\Pr[E_3] = \Pr[E_2]$.

Game 4. We will change the way we create the second tuple, which we use to encrypt the second message in the case $b = 0$. Now we will only make one tuple for the first key k_1 and use the first tuple to create the second. Let $(x, y, z, w) = (g, g^{k_1}, g^s, g^{k_1 s})$ be the first tuple, the second tuple will then be

$$
\begin{aligned}
(x', y', z', w') &= (x, x^a y^c, z x^b, w^c z^a y^{cb} x^{ab}) \\
&= (g, g^{a+ck_1}, g^{b+s}, g^{(a+ck_1)(b+s)})
\end{aligned}
$$

for some uniformly sampled $a, b, c \in \{1, \ldots, p-1\}$. Note that the second tuple will still be computed before we receive the messages.

Let E_4 be the event that $b = b'$ in Game 4. Since the new tuple results in the same output space when it is used for encrypting messages we get that $\Pr[E_4] = \Pr[E_3]$.

Game 5. We will in this game change the rerandomize algorithm. The output of the encryption and the rerandomize algorithm can be seen as two vectors, (x, y) and (z, w). For the encryption algorithm the first vector will always stay in the subgroup of $G \times G$ generated by (g, g^{k_1}), for a key k_1, and the second will always stay in the same coset of this subgroup. The first vector in the output of the rerandomize algorithm also stay the subgroup $G \times G$ generated by (g, g^{k_1}), however the second does not stay in the same coset. That is, the output of the rerandomize algorithm looks like

$$
(g^{rr'}, g^{k_1 rr'}, g^{s+rs'}, g^{k_1(s+rs')} m)
$$

for some r, r', s and s', where the sum of $s + rs'$ cannot be equal to s since none of the variables used in the algorithm can be zero. Therefore, there is a statistical difference of $1/p$ between the output distributions. We will instead compute the rerandomization of the first ciphertext (in the case $b = 1$) as

$$
(g^{rr'}, g^{k_1 rr'}, g^{s+rs'+\tilde{s}}, g^{k_1(s+rs'+\tilde{s})} m_1),
$$

where $\tilde{s}$ is a uniform element in $\{1, \ldots, p-1\}$. The new sum $s + rs' + \tilde{s}$ can now be any value in $\{1, \ldots, p-1\}$, and all values are equally probable.

Let F_5 be the event that $s + rs' + \tilde{s} = s$, and let E_5 be the event that $b = b'$ in Game 5. Unless F_5 occurs, Game 4 and Game 5 behaves the same, that is, $E_4 \wedge \neg F_5 \iff E_5 \wedge \neg F_5$ and by the difference lemma we get that

$$
|\Pr[E_4] - \Pr[E_5]| \le \Pr[F_5] = \frac{1}{p}.
$$

Game 6. In the last game, we will turn the first tuple into the form $(g, g^{a'}, g^{b'}, g^{c'})$, for some uniform elements $a', b', c' \in \{1, \ldots, p-1\}$. The second tuple will then look like

$$
(g, g^{a+a'c}, g^{b+b'}, g^{ab+ab'+a'bc+cc'}).
$$

Algorithm $B((x, y, z, w))$:
$$u_1, u_2 \xleftarrow{r} \mathcal{D}, \; b \xleftarrow{r} \{0, 1\}$$
$$a, b, c \xleftarrow{r} \{1, \dots, p-1\}$$
$$(x', y', z', w') = (x, x^a y^c, z x^b, w^c z^a y^{cb} x^{ab})$$
Get m_1, m_2 from A

If $b = 0$ do:
$$r, r' \xleftarrow{r} \{1, \dots, p-1\}$$
$$c_1 \leftarrow (x^r, y^r, z, w m_1)$$
$$c_2 \leftarrow (x'^{r'}, y'^{r'}, z', w' m_2)$$
Send c_1, c_2 to A

If $b = 1$ do:
$$r, r', s', \tilde{s} \xleftarrow{r} \{1, \dots, p-1\}$$
$$c_1 \leftarrow (x^r, y^r, z, w m_1)$$
$$c_2 \leftarrow (x^{rr'}, y^{rr'}, z x^{rs'+\tilde{s}}, w y^{rs'+\tilde{s}} m_1)$$
Send c_1, c_2 to A

Get b' from A

Fig. 4. Algorithm B

Both the encryption and rerandomization algorithms will output ciphertexts which can result in any group element when decrypted.

Let E_6 be the event that $b = b'$ in Game 6. Since we are using uniform variables in the tuples the encryption and rerandomization algorithms are, essentially, one-time pads. Hence, we get that $\Pr[E_6] = 1/2$.

To show the connection to the previous game we will use algorithm B, see Fig. 4. We claim that $|\Pr[E_5] - \Pr[E_6]| = \mathrm{Adv}_{\mathrm{ddh}}^{\mathrm{ind\text{-}cpa}}$, the DDH-advantage (with respect to indistinguishably under chosen plaintext attack, that is, semantic security). The input to the algorithm B is a tuple (x, y, z, w) which looks like (g, g^a, g^b, g^c), for some a, b, and c, where c can be equal to ab. Therefore, the algorithm will simulate Game 5 and Game 6 depending on its input. When the input is on the form (g, g^a, g^b, g^{ab}), the algorithm will proceed just as in Game 5, and therefore

$$\Pr[B(g, g^a, g^b, g^{ab}) = 1 \mid a, b \xleftarrow{r} \{1, \dots, p-1\}] = \Pr[E_5].$$

If the input is on the form (g, g^a, g^b, g^c) the algorithm proceed as in Game 6 and we get that

$$\Pr[B(g, g^a, g^b, g^c) = 1 \mid a, b, c \xleftarrow{r} \{1, \dots, p-1\}] = \Pr[E_6],$$

where the DDH-advantage of B is equal to $|\Pr[E_5] - \Pr[E_6]|$.

Recap. We can now use the results from the games to bound the advantage of the adversary.

$$\begin{aligned}
\mathrm{Adv}(A) &= |\Pr[E_0] - 1/2| \\
&= |\Pr[E_0] - \Pr[E_1] + \Pr[E_1] - \Pr[E_2] + \Pr[E_2] \\
&\quad - \Pr[E_3] + \Pr[E_3] - \Pr[E_4] + \Pr[E_4] \\
&\quad - \Pr[E_5] + \Pr[E_5] - \Pr[E_6] + \Pr[E_6] - 1/2| \\
&\le |\Pr[E_0] - \Pr[E_1]| + |\Pr[E_4] - \Pr[E_5]| + |\Pr[E_5] - \Pr[E_6]| \\
&\le \Pr[F_1] + \frac{1}{p} + \mathrm{Adv}_{\mathrm{ddh}}^{\mathrm{ind\text{-}cpa}}.
\end{aligned}$$

By the DDH assumption the DDH advantage is negligible. Therefore, for a large enough p, we get that the advantage of our adversary is determined by the probability that A guesses or predicts the correct key, that is, determined by the probability space D.

2.4   Extended Scheme

We will extend the basic scheme to longer messages by representing them as bit strings. This change will also reduce the number of rerandomizations we can perform on a ciphertext. Therefore, we need to relax the requirements of the cryptosystem slightly. The construction in this section is very similar to the hybrid scheme by Golle et al. [6].

Correctness. If c was produced by iteratively applying $\mathcal{R}$ to the output of $\mathcal{E}(k, m)$ at most n times, then $\mathcal{D}(k, c)$ will never output the failure symbol and output m except with negligible probability.

We will require a pseudorandom function $f : G \to \{0, 1\}^N$ mapping group elements to bit strings of length N, for some large $N \in \mathbb{N}$. We let f_L denote the truncation of the output to L bits, for $L < N$. We will assume that group elements can be encoded as bit strings of length $l/2$.

Encryption. For a message $m \in \{0, 1\}^L$ and a key $k \in \{1, \ldots, p - 1\}$ pick $r, s \in \{1, \ldots, p - 1\}$ and $\gamma \in G$ uniformly, output

$$c = g^r ||g^{kr}||g^s||g^{ks}\gamma|| \left(f_{L+l(n+1)+1}(\gamma) \oplus (m||1||0^{l(n+1)}) \right).$$

Decryption. For a ciphertext $c = x||y||b_0'$ and a key $k \in \{1, \ldots, p - 1\}$ check if $x^k = y$. If not, output a symbol representing decryption failure. If it is, let $b_0' = z_0||w_0||b_0$ and compute

$$b_1' = f_{|b_0|}(z_0^{-k} w_0) \oplus b_0.$$

If the result b_1' ends in $l' \ge l$ zeros, then the message is the result minus the tail of zeros and exactly one 1. If the result does not end with a tail of l' zeros, then interpret b_1' as $z_1||w_1||b_1$ and repeat the procedure. If the decryption algorithm is repeated $n + 1$ times, output a symbol representing decryption failure.

Rerandomization. For a ciphertext $c = x||y||b_\alpha||b_\beta$, where b_β is the last l bits, pick $r', s' \in \{1, \ldots, p-1\}$ and $\gamma' \in G$ uniformly, output

$$c' = x^{r'}||y^{r'}||x^{s'}||y^{s'}\gamma'|| \left(f_{|b_\alpha|}(\gamma') \oplus b_\alpha\right).$$

Note that, before applying the rerandomize algorithm, b_α looks like

$$g^s||g^{ks}\gamma|| \left(f_{L+ln+1}(\gamma) \oplus (m||1||0^{ln})\right)$$

for some $s \in \{1, \ldots, p-1\}$, key k, and $\gamma \in G$. The l last bits we discard, i.e., b_β, is an "encryption" of l zeros. We can therefore only perform n rerandomizations on a ciphertext before we get decryption failure, that is, there are no tail of zeros left for the decryption algorithm to detect. However, we get that the length of the ciphertext is preserved.

We will now show the correctness of the decryption algorithm. If $c = x||y||b_0'$ was output from the encryption algorithm, we have that $x^k = g^{kr} = y$. Hence, we can write b_0' as $z||w||b_0$, and compute

$$f_{|b_0|}(z^{-k}w) \oplus b_0 = f_{L+l(n+1)+1}(g^{-ks}g^{ks}\gamma) \oplus f_{L+l(n+1)+1}(\gamma)$$
$$\oplus (m||1||0^{l(n+1)})$$
$$= (m||1||0^{l(n+1)}).$$

Since the result ends with a tail of $l' \geq l$ zeros the output message is m.

Let c be a ciphertext that was produced by iteratively applying the rerandomize algorithm to the output of $\mathcal{E}(k, m)$ t times, where $1 \leq t \leq n$. Write c as $x||y||b_t'$, where $x = g^{r_1 \cdots r_{t+1}}$, $y = g^{k(r_1 \cdots r_{t+1})}$, and b_t' looks like

$$(g^{r_1 \cdots r_t})^{s'}||(g^{k(r_1 \cdots r_t)})^{s'}\gamma_t|| \left(f_{L+l(n+1-t)+1}(\gamma_t) \oplus b_{t-1}'\right)$$

for some $s', r_1, \ldots, r_{t+1} \in \{1, \ldots, p-1\}$, key k, and group element $\gamma_t \in G$. Observe that for all $1 \leq t \leq n$, we have that $x^k = y$. Hence, we can write $b_t' = z_t||w_t||b_t$ and compute

$$f_{|b_t|}(z_t^{-k}w_t) \oplus b_t = f_{L+l(n+1-t)+1}(g^{-ks'(r_1 \cdots r_t)}g^{ks'(r_1 \cdots r_t)}\gamma_t)$$
$$\oplus f_{L+l(n+1-t)+1}(\gamma_t) \oplus b_{t-1}'$$
$$= b_{t-1}'$$

where b_{t-1}' does not end with a tail of $l' \geq l$ zeros (except with negligible probability), since the ciphertext is also encrypted once using with the encryption algorithm (in addition to the t rerandomizations). Therefore, let $b_{t-1}' = z_{t-1}||w_{t-1}||b_{t-1}$ and repeat the process t more times. In the last iteration, we will perform the decryption on a bit string which looks like $z_0||w_0||b_0$, where we now have that b_0 looks like

$$f_{L+l(n+1-t)+1}(\gamma_0) \oplus (m||1||0^{l(n+1-t)})$$

which we know decrypts to the message m. That is, the decryption algorithm is correct.

2.5   Security of the Extended Scheme

We will in this section show that the adversary is unable to distinguish between encrypted ciphertexts and that his advantage is determined by D, that is, the probability of the adversary guessing the correct key. As in the proof of the basic scheme, we will use games to simulate the same experiment. Since the two example schemes are similar the games will be too.

Game 0. In the first game we will simulate the experiment. We ask the adversary to differentiate between ciphertexts encrypting two different messages and ciphertexts encrypting the same message. The full procedure of the game can be seen in Fig. 5.

Let E_0 be the event that $b = b'$ in Game 0.

Game 1. We will stop the game if the adversary guesses one of the keys correctly. If the adversary sends either u_0 or u_1 in one of its queries, the oracle will flip a coin, $b' \xleftarrow{r} \{0,1\}$, output b', and stop the game. We will denote this event by F_1.

Let E_1 denote the event that $b = b'$ in Game 1. If the event F_1 does not occur then Game 0 and Game 1 are equal. That is, $E_0 \wedge \neg F_1 \iff E_1 \wedge \neg F_1$ and by the difference lemma we get that

$$| \Pr[E_0] - \Pr[E_1]| \leq \Pr[F_1].$$

Game 2. Since the adversary can no longer use the oracle to get any information about the keys without stopping the game, we are essentially drawing our keys

Game 0:
$u_1, u_2 \xleftarrow{r} D,\ k_1 \leftarrow H(u_1),\ k_2 \leftarrow H(u_2),\ b \xleftarrow{r} \{0,1\}$
Get m_1, m_2 from A

If $b = 0$ do:
    $r, r', s, s', \gamma, \gamma' \xleftarrow{r} \{1, \ldots, p-1\}$
    $c_1 \leftarrow g^r ||g^{k_1 r}||g^s||g^{k_1 s}\gamma|| \left(f_{L+l(n+1)+1}(\gamma) \oplus (m_1 ||1||0^{l(n+1)}) \right)$
    $c_2 \leftarrow g^{r'} ||g^{k_2 r'}||g^{s'}||g^{k_2 s'}\gamma'|| \left(f_{L+l(n+1)+1}(\gamma') \oplus (m_2 ||1||0^{l(n+1)}) \right)$
    Send c_1, c_2 to A

If $b = 1$ do:
    $r, r', s, s', \gamma, \gamma' \xleftarrow{r} \{1, \ldots, p-1\}$
    $c_1 \leftarrow g^r ||g^{k_1 r}||g^s||g^{k_1 s}\gamma|| \left(f_{L+l(n+1)+1}(\gamma) \oplus (m_1 ||1||0^{l(n+1)}) \right)$
    Write c_1 as $x||y||b_\alpha||b_\beta$, where b_β is the last l bits
    $c_2 \leftarrow x^{r'} ||y^{r'}||x^{s'} ||y^{s'}\gamma'|| \left(f_{|b_\alpha|}(\gamma') \oplus b_\alpha \right)$
    Send c_1, c_2 to A

Get b' from A

Fig. 5. Game 0 of the extended scheme

at random from a set. That is, we draw $k_1, k_2 \xleftarrow{r} \{1, \ldots, p-1\}$ uniformly, and we will no longer query the hashing oracle. Note that since the adversary can still query the hashing oracle, we still need to draw samples from the space D to check if the adversary is guessing the correct keys.

Let E_2 denote the event that $b = b'$ in Game 2. Since the adversary can no longer get any information about the environmental keys from the hashing oracle without stopping the game, the keys used are, essentially, some random group elements. Hence, $\Pr[E_2] = \Pr[E_1]$.

Game 3. We change how we compute the tuples such that we the encryption algorithm do not require the keys as input. We will therefore precompute the tuples before we receive the messages. That is, for some uniform $s, s' \in \{1, \ldots, p-1\}$ and key k_1, k_2, we will compute

$$(x, y, z, w) = (g, g^{k_1}, g^s, g^{k_1 s})$$
$$(x', y', z', w') = (g, g^{k_2}, g^{s'}, g^{k_2 s'})$$

before we receive the messages m_1 and m_2.

In the case $b = 0$, we will encrypt the two messages using the precomputed tuples. That is, we will pick r, r', γ and γ' uniformly, and compute

$$
\begin{aligned}
c_1 &= x^r ||y^r||z||w\gamma|| \left(f_{L+l(n+1)+1}(\gamma) \oplus (m_1||1||0^{l(n+1)}) \right) \\
&= g^r ||g^{k_1 r}||g^s||g^{k_1 s}\gamma|| \left(f_{L+l(n+1)+1}(\gamma) \oplus (m_1||1||0^{l(n+1)}) \right), \\
c_2 &= x'^{r'} ||y'^{r'}||z'||w'\gamma'|| \left(f_{L+l(n+1)+1}(\gamma') \oplus (m_2||1||0^{l(n+1)}) \right) \\
&= g^{r'} ||g^{k_2 r'}||g^{s'}||g^{k_2 s'}\gamma'|| \left(f_{L+l(n+1)+1}(\gamma') \oplus (m_2||1||0^{l(n+1)}) \right).
\end{aligned}
$$

In the case $b = 1$, we will encrypt one message and rerandomize the computed ciphertext. That is, to encrypt, we pick r and γ uniformly, and compute

$$
\begin{aligned}
c_1 &= x^r ||y^r||z||w\gamma|| \left(f_{L+l(n+1)+1}(\gamma) \oplus (m_1||1||0^{l(n+1)}) \right) \\
&= g^r ||g^{k_1 r}||g^s||g^{k_1 s}\gamma|| \left(f_{L+l(n+1)+1}(\gamma) \oplus (m_1||1||0^{l(n+1)}) \right).
\end{aligned}
$$

To rerandomize let $c_1 = x^r ||y^r||b_\alpha||b_\beta$, where b_β is the last l bits, pick two element r', s' and a group element γ', as usual, and compute

$$
\begin{aligned}
c_2 &= x^{rr'} ||y^{rr'}||x^{rs'}||y^{rs'}\gamma'|| (f_{|b_\alpha|}(\gamma') \oplus b_\alpha) \\
&= g^{rr'} ||g^{k_1 rr'}||g^{rs'}||g^{k_1 rs'}\gamma'|| (f_{|b_\alpha|}(\gamma') \oplus b_\alpha).
\end{aligned}
$$

Let E_3 denote the event that $b = b'$ in Game 3. The output distribution of the encryption algorithm in Game 2 and in Game 3 are exactly the same, similarly for the rerandomization algorithm. Therefore, we have that $\Pr[E_3] = \Pr[E_2]$.

Game 4. We will only make one tuple and use it to create the second one. The first tuple will be $(x, y, z, w) = (g, g^{k_1}, g^s, g^{k_1 s})$ and the second tuple looks will then be

$$(x', y', z', w') = (x, x^a y^c, zx^b, w^c z^a y^{cb} x^{ab})$$
$$= (g, g^{a+ck_1}, g^{b+s}, g^{(a+ck_1)(b+s)})$$

for some uniformly sampled $a, b, c \in \{1, \ldots, p-1\}$.

Let E_4 be the event $b = b'$ in Game 4. Since the new tuple results in the same output space when it is used for encrypting messages we get that $\Pr[E_4] = \Pr[E_3]$.

Game 5. We will now make the first tuple to have the form $(g, g^{a'}, g^{b'}, g^{c'})$, for some uniform $a', b', c' \in \{1, \ldots, p-1\}$, where the second tuple will look like

$$(g, g^{a+a'c}, g^{b+b'}, g^{ab+ab'+a'bc+cc'}).$$

Let E_5 be the event $b = b'$ in Game 5. We will use algorithm B, see Fig. 6, to show that $|\Pr[E_4] - \Pr[E_5]|$ is equal to the DDH-advantage. If the input of the algorithm is a tuple on the form (g, g^a, g^b, g^{ab}), then the algorithm proceed as in Game 4. If the tuple is on the form (g, g^a, g^b, g^c), then the algorithm proceed as in Game 5. Therefore, the DDH-advantage is equal to $|\Pr[E_4] - \Pr[E_5]|$.

Game 6. In the last game, we will sample a function h from a family Γ of all functions from G to $\{0,1\}^N$ instead of using the function f. We want to show that the pseudorandom function (PRF) f can reliably hide the message. The PRF-advantage of an efficient adversary is defined by his ability to distinguishing the

---

Algorithm $B((x, y, z, w))$:
  $u_1, u_2 \xleftarrow{r} \mathcal{D}, \ b \xleftarrow{r} \{0,1\}$
  $a, b, c \xleftarrow{r} \{1, \ldots, p-1\}$
  $(x', y', z', w') = (x, x^a y^c, zx^b, w^c z^a y^{cb} x^{ab})$
  Get m_1, m_2 from A

  If $b = 0$ do:
    $r, r'\gamma, \gamma' \xleftarrow{r} \{1, \ldots, p-1\}$
    $c_1 \leftarrow x^r ||y^r||z||w\gamma|| \left(f_{L+l(n+1)+1}(\gamma) \oplus (m_1||1||0^{l(n+1)}) \right)$
    $c_2 \leftarrow x'^{r'} ||y'^{r'}||z'||w'\gamma'|| \left(f_{L+l(n+1)+1}(\gamma') \oplus (m_2||1||0^{l(n+1)}) \right)$
    Send c_1, c_2 to A

  If $b = 1$ do:
    $r, r's', \gamma, \gamma' \xleftarrow{r} \{1, \ldots, p-1\}$
    $c_1 \leftarrow x^r ||y^r||z||w\gamma|| \left(f_{L+l(n+1)+1}(\gamma) \oplus (m_1||1||0^{l(n+1)}) \right)$
    Write c_1 as $x^r ||y^r||b_\alpha||b_\beta$, where b_β is the last l bits
    $c_2 \leftarrow x^{rr'} ||y^{rr'}||x^{rs'}||y^{rs'}\gamma'|| \left(f_{|b_\alpha|}(\gamma') \oplus b_\alpha \right)$
    Send c_1, c_2 to A

  Get b' from A

Fig. 6. Algorithm B

Algorithm $B'((x, y, z, w))$:

$u_1, u_2 \xleftarrow{r} \mathcal{D}, b \xleftarrow{r} \{0, 1\}, h \leftarrow \Gamma$

$a, b, c \xleftarrow{r} \{1, \ldots, p-1\}$

$(x', y', z', w') = (x, x^a y^c, z x^b, w^c z^a y^{cb} x^{ab})$

Get m_1, m_2 from A

If $b = 0$ do:

$r, r', \gamma, \gamma' \xleftarrow{r} \{1, \ldots, p-1\}$

$c_1 \leftarrow x^r || y^r || z || w\gamma || \left(h_{L+l(n+1)+1}(\gamma) \oplus (m_1 || 1 || 0^{l(n+1)}) \right)$

$c_2 \leftarrow x'^{r'} || y'^{r'} || z' || w'\gamma' || \left(h_{L+l(n+1)+1}(\gamma') \oplus (m_2 || 1 || 0^{l(n+1)}) \right)$

Send c_1, c_2 to A

If $b = 1$ do:

$r, r', s', \gamma, \gamma' \xleftarrow{r} \{1, \ldots, p-1\}$

$c_1 \leftarrow x^r || y^r || z || w\gamma || \left(h_{L+l(n+1)+1}(\gamma) \oplus (m_1 || 1 || 0^{l(n+1)}) \right)$

Write c_1 as $x^r || y^r || b_\alpha || b_\beta$, where b_β is the last l bits

$c_2 \leftarrow x^{rr'} || y^{rr'} || x^{rs'} || y^{rs'} \gamma' || \left(h_{|b_\alpha|}(\gamma') \oplus b_\alpha \right)$

Send c_1, c_2 to A

Get b' from A

Fig. 7. Algorithm B'

function f from any function h sampled from Γ. The PRF-advantage of the adversary is negligible assuming the function f is pseudorandom. Just like for f, we will denote h_L as the truncation of the output to L bits.

Let E_6 be the event $b = b'$ in Game 6. From Game 5 we have that the new tuples looks like $(g, g^{a'}, g^{b'}, g^{c'})$, for some random variables $a', b',$ and c'. Hence, we will not be able to retrieve γ when we try to decrypt the ciphertext encrypting it. Since we are now using any function h, with the random group element γ, to encrypt the message m we are essentially XOR-ing a random bit string to the message. Therefore, the output ciphertexts of the encryption and rerandomization algorithms can be any random bit string and we get that $\Pr[E_6] = 1/2$.

By using the algorithm B', as seen in Fig. 7, we can show that the difference between Game 5 and Game 6 is equal to the PRF-advantage. The algorithm draws a function h from the family Γ, which may be equal to f. Hence, we get that the PRF-advantage is

$$|\Pr[B'((x, y, z, w)) = 1 \mid f \leftarrow \Gamma] - \Pr[B'((x, y, z, w)) = 1 \mid h \leftarrow \Gamma]|$$

which is equal to $|\Pr[E_5] - \Pr[E_6]|$.

Recap. We can now use the results from the games to bound the advantage of the adversary.

$$\begin{aligned}
\mathrm{Adv}(A) &= |\Pr[E_0] - 1/2| \\
&= |\Pr[E_0] - \Pr[E_1] + \Pr[E_1] - \Pr[E_2] + \Pr[E_2] \\
&\quad - \Pr[E_3] + \Pr[E_3] - \Pr[E_4] + \Pr[E_4] \\
&\quad - \Pr[E_5] + \Pr[E_5] - \Pr[E_6] + \Pr[E_6] - 1/2| \\
&\leq |\Pr[E_0] - \Pr[E_1]| + |\Pr[E_4] - \Pr[E_5]| + |\Pr[E_5] - \Pr[E_6]| \\
&\leq \Pr[F_1] + \mathrm{Adv}_{\mathrm{ddh}}^{\mathrm{ind\text{-}cpa}}(A) + \mathrm{Adv}_{\mathrm{prf}}(A).
\end{aligned}$$

Assuming that f is pseudorandom the PRF advantage is negligible, and the DDH assumption states that the DDH-advantage is negligible. Therefore, the advantage of the adversary is determined by the probability that the adversary guesses or predicts the correct key, that is, determined by the probability space D.

Acknowledgments. We would like to thank Adam Young for helpful discussions and comments. We would also like to thank the anonymous reviewers for helpful comments.

References

1. Boneh, D.: The decision Diffie-Hellman problem. In: Buhler, J.P. (ed.) ANTS 1998. LNCS, vol. 1423, pp. 48–63. Springer, Heidelberg (1998). doi:10.1007/BFb0054851
2. Canetti, R., Krawczyk, H., Nielsen, J.: Relaxing chosen-ciphertext security. Cryptology ePrint Archive, Report 2003/174 (2003). http://eprint.iacr.org/
3. Filiol, E.: Strong cryptography armoured computer viruses forbidding code analysis: the bradley virus. Research Report RR-5250, INRIA (2004)
4. Filiol, E.: Malicious cryptography techniques for unreversable (malicious or not) binaries. CoRR, abs/1009.4000 (2010)
5. Futoransky, A., Kargieman, E., Sarraute, C., Waissbein, A.: Foundations and applications for secure triggers. Cryptology ePrint Archive, Report 2005/284 (2005). http://eprint.iacr.org/
6. Golle, P., Jakobsson, M., Juels, A., Syverson, P.: Universal re-encryption for mixnets. In: Okamoto, T. (ed.) CT-RSA 2004. LNCS, vol. 2964, pp. 163–178. Springer, Heidelberg (2004). doi:10.1007/978-3-540-24660-2_14
7. Hohl, F.: Time limited blackbox security: protecting mobile agents from malicious hosts. In: Vigna, G. (ed.) Mobile Agents and Security. LNCS, vol. 1419, pp. 92–113. Springer, Heidelberg (1998). doi:10.1007/3-540-68671-1_6
8. Kaspersky Lab Global Research and Analysis Team. Gauss: Abnormal distribution. In-depth research analysis report, KasperSky Lab, 9 August 2012. http://www.securelist.com/en/analysis/204792238/gauss_abnormal_distribution
9. Riordan, J., Schneier, B.: Environmental key generation towards clueless agents. In: Vigna, G. (ed.) Mobile Agents and Security. LNCS, vol. 1419, pp. 15–24. Springer, Heidelberg (1998). doi:10.1007/3-540-68671-1_2
10. Shoup, V.: Sequences of games: a tool for taming complexity in security proofs. Cryptology ePrint Archive, Report 2004/332 (2004)
11. Skoudis, E., Zeltser, L.: Malware: Fighting Malicious Code. Prentice Hall PTR, Upper Saddle River (2003)

12. Young, A., Yung, M.: Cryptovirology: extortion-based security threats and countermeasures. In: Proceedings of the IEEE Symposium on Security and Privacy, pp. 129–140, May 1996
13. Young, A., Yung, M.: Malicious Cryptography: Exposing Cryptovirology. Wiley, Hoboken (2004)
14. Young, A., Yung, M.: The drunk motorcyclist protocol for anonymous communication. In: 2014 IEEE Conference on, Communications and Network Security (CNS), pp. 157–165, October 2014

Protecting Electronic Signatures in Case of Key Leakage

Mirosław Kutyłowski[1]([✉]), Jacek Cichoń[1], Lucjan Hanzlik[1],
Kamil Kluczniak[1], Xiaofeng Chen[2], and Jianfeng Wang[2]

[1] Faculty of Fundamental Problems of Technology,
Wrocław University of Science and Technology, Wrocław, Poland
{miroslaw.kutylowski,jacek.cichon,lucjan.hanzlik,
kamil.kluczniak}@pwr.edu.pl
[2] State Key Laboratory of Integrated Service Networks (ISN),
Xidian University, Xi'an, People's Republic of China
{xfchen,jfwang}@xidian.edu.cn

Abstract. We present a protection mechanism against forgery of electronic signatures with the original signing keys. It works for standard signatures based on discrete logarithm problem such as DSA. It requires only a slight modification of the signing device – an implementation of an additional hidden evidence functionality.

We assume that neither verification mechanism can be altered nor extra fields can be added to the signature (both as signed and unsigned fields). Therefore, the old software for signature verification can be used without any change. On the other hand, if a forged signature emerges, the signatory may prove its inconsistency with a probability close to 1.

Unlike fail-stop signatures, our method works not only against cryptanalytic attacks, but it is primarily designed for the case when the adversary gets the original signing key stored by the signing device of the user.

Unlike cliptographic constructions designed to defend against malicious implementations, we consider catastrophic situation when the key has been already compromised.

The technical idea we propose is an application of kleptography for good purposes. It is simple enough, efficient and almost self-evident to be ready for implementation of cryptographic smart cards of moderate storage and computational capabilities.

Unfortunately, we have also to bring into attention that our scheme has a dark side and it can be used for leaking the keys via the recent *subversion-resistant* signatures by A. Russell, Q. Tang, M. Yung and H.-Sh.Zhou.

This research has been supported by the Polish National Science Centre grant OPUS, no 2014/15/B/ST6/02837 and Polish-Chinese cooperation venture of Xidian University and Wrocław University of Science and Technology on Secure Data Outsourcing in Cloud Computing.

© Springer International Publishing AG 2017
R.C.-W. Phan and M. Yung (Eds.): LNCS 10311, Mycrypt 2016, pp. 252–274, 2017.
DOI: 10.1007/978-3-319-61273-7_13

1   Introduction

In practice, undeniability of electronic signatures is based on the following assumptions:

1. Creation of a valid signature is possible only with the secret key corresponding to the public key used during the verification procedure,
2. The private key is stored only in a so called *signature creation device* which is under a sole control of the signatory,
3. The link between the public key and the signatory is reliably confirmed by means of a public key infrastructure.

Violation of any of these assumptions means that a signature cannot be trusted.

Note that the sole control condition has to indicate that no one but the signatory can activate the device for signing. This does not mean that the signatory has open access to the device memory, its secret keys etc. Indeed, such an access would mean that the signatory can export the key from the device and therefore we could not assume that the private signing key is stored on the device only.

Unfortunately, testing whether the above mentioned conditions are fulfilled might be far beyond the possibilities of a normal user entrusting the signatures:

1. Validity of the first condition depends on the state of the art of cryptanalysis. However, one cannot expect that a user can evaluate state-of-the-art of the research in this area and exclude possibility of a forgery. Note that we are not talking only about the academic research with publicly available results, but also about cyber war research.
2. The second condition is very hard to check, even for the signatory himself. First, the hardware manufacturer may include trapdoors enabling to reveal the private key stored in the signature creation device. There is a wide range of methods that can be applied for this purpose: kleptographic code (see e.g. [16] and its bibliography list), weaknesses of (pseudo)random number generators, hardware Trojans (see e.g. [1]), side channel information leakage (see e.g. [7]). Such trapdoors can be also created by simple implementation errors. However, even the signatory has no access to the internal state of the device, so such trapdoors cannot be detected by a direct device inspection.

   Certification and audit procedures aim to prevent existence of such trapdoors in certified products, but it has been pointed out that certification authorities might be coerced by state authorities to install certain trapdoors for the sake of national security. However, these trapdoors can be used without any control and even endanger public security.

   Some defense against these threats is possible – e.g. cliptographic techniques [19] aim to protect against malicious software implementations, including the key generation process.
3. Creating a reliable, efficient and cheap PKI infrastructure is a problem itself. Last not least, one has to be sure that the most advanced players cannot create rogue certificates. From the past experience (see [12]), we know that this cannot be completely excluded.

In this paper we focus on the second condition. One line of R&D work is to provide secure hardware so that there are neither trapdoors nor malicious ways to extract the signing secret keys against the will of the signatory and the manufacturer. However, it is extremely hard to evaluate effectiveness of the current solutions for the sheer reason that the most dangerous and powerful adversaries may treat their capabilities as top secrets.

Key Generation Process. One of the Achilles heels of the signature schemes currently implemented on the smart cards is secret key generation. There are a few options, each of them involves critical threats:

1. **Key generated by a service provider and installed on the signature creation device:** one cannot prove that the service provider does not retain a copy of the key for malicious purposes. If they are later used only in a few cases, then this does not even endanger the reputation of the service as forgery might be hard to prove.

   Unfortunately, the current legal regulations do not ease the situation: e.g. the recent eIDAS framework of European Union [14] admits creation and keeping backup copies of secret signing keys as a service. It is required that *appropriate security level is assured*, however it does not mean that a rogue service provider cannot break these rules. Even worse, the current state-of-the-art of cryptographic research provides tools to erase all traces of misbehavior.

2. **Key generated by the user and installed on the signature creation device:** this option is frequently forbidden by legal rules: the device must not enable existence of the secret key outside of it. The only exception is generating key by a trusted service provider as mentioned by the first option. Note that admitting the user to install the key creates plenty of opportunities to steal this key by rogue software on the computer used to generate and upload the key. The user is likely to be unaware of this fact.

3. **Key generated itself by the signature creation device:** the procedure to create cryptographic keys on the signing device might be a fake one. A rogue device might "generate" the key that already has been stored on the device or is predictable by the manufacturer. As in response the device owner obtains only a public key, it is infeasible to detect such misbehavior.

We may conclude that the provider of the signature creation devices has quite realistic ways to gain access to the signing keys contained in these devices and thereby has a real opportunity to forge signatures.

Of course, there are some relatively simple solutions that would make such forgeries much harder or even impossible. For example, a multi-party key generation protocol executed by the signature creation device and another independent device of this user would help a lot (by "independent" we mean in particular "coming from a different manufacturer that cannot be coerced by the same authority"). Also using two different signature creation devices and a certificate stating that a signature is valid only when co-signed by the second device would provide more security in the practical sense.

Unfortunately, these simple solutions are not really pragmatic. Our experience is that in practice all solutions that require major changes in the already deployed systems have no or very low chances to be accepted by the industry and decision makers. As a reaction to such proposals we expect rather arguments to ignore "theoretical and purely academic threats not justified by a realistic risk analysis". If a new method is fully compatible with the already available and deployed products, then the chances to be accepted increase a lot.

Our Goal. We aim to create security mechanisms against forging signatures by an adversary having access to the private signing keys. Such an adversary can create signatures that cannot be distinguished by the standard means from the signatures created by the private key owner. As it seems to be hard to change the verification process already described in the standards, our goal is to

- detect signatures created with stolen signing keys,
- create an evidence that can be used to convince a judge about the forgery without invalidating the remaining signatures.

We aim to address the situation where:

- the manufacturer provides a device with a signing key already installed, or
- the key generation procedure is executed by the (black box) device, but we have no insight into this process, or
- some components are ill implemented (e.g. the (pseudo)random number generator has low entropy output), and therefore the signing key is weak by design.

For this purpose we enable the user to install an auxiliary parameter on the signature creation device after he gets the device to his hands. An important feature is that this parameter is not known to the manufacturer or the party delivering the device. As the signing device is usually supposed only to create signatures, it might be hard in practice to create a secret channel to the manufacturer that would enable to leak the signing key.

A naïve way to implement this idea would be to use an extra secret as a seed for a PRNG creating random exponent k i.e. in case of DSA signatures. Revealing the seed would immediately enable to distinguish the forged signatures from the signatures created by the card. Unfortunately, the user himself would be able to derive the secret key installed by the device, which contradicts the basic requirements e.g. from [14]. Indeed, in case of the signature creation devices the legitimate user of the device should be considered as an adversary, since in practice he might be tempted to find an evidence that his device has been tampered and therefore to claim that some of his past signatures have been forged.

One might hope that fail-stop signatures [11] solve the stated problem. Unfortunately, this is not true, as fail-stop signatures protect solely against cryptanalytic attacks. They are useless against an adversary holding the original signing keys.

Assumptions. We aim to provide a solution that would be relatively easy to deploy. Therefore we assume that:

1. The current signature standards should be used in an essentially unchanged form; only details of implementation should provide additional features for the device holder. Nothing should change for a verifier of electronic signatures.
2. The solution does not require any additional interaction during device delivery from the manufacturer to the end user. We are focused on the situation that the signature creation device is delivered ready to use with the signing key and a certificate for this key already preinstalled.
3. The solution should be implementable on devices such as smart cards with moderate resources (computing speed and memory size). These devices are assumed to be tamper proof (if they are not, then they cannot be used as signature creation devices for the general reasons).
4. For the signatory, creating a signature should be as convenient as in case of the standard signature creation devices.
5. In case of a forgery with original keys, detection probability should be high enough to discourage such behavior.

Example Application Scenario. A typical application case we have in mind is the following. A financial institution or an organization of financial institutions issues electronic ID cards for their customers. The ID cards enable creation of electronic signatures used to authenticate documents provided by the customers. In order to ease the card usage and simplify the logistics, the ID cards are delivered ready to use. As the card issuer holds the most relevant data about the customer enabling him to monitor card usage, it does not make sense to delegate the cards management chores (such as keeping revocation lists) to a third party Certificate Authority. However, if there is a dispute between the customer and the financial institution, the customer may challenge the signatures created by himself and claim that the institution has retained his signing key and used it to forge these documents. This is a crucial issue and the solution proposed in this paper aims to prevent such a situation.

Our Contribution and Paper Overview. We present a generic solution for randomized signature schemes based on the Discrete Logarithm Problem, including in particular ElGamal, DSA, ECDSA, and Schnorr signatures. The mechanism is based on a pair of hidden keys that enable to fish out the signatures created by the legitimate signature creation device (this device stores the public control key for signature creation). Our solution enables separation of trust: the hidden keys might be uploaded by the owner of the device.

In Sect. 2 we present our generic construction. In Sect. 3 we provide arguments showing that the modified schemes are secure just as the underlying signature schemes. Section 4 shows a dark side of the scheme – usage for leaking the key without using random numbers generated on the smart card. In Sect. 5.1 we discuss the implementation problems and in particular time complexity for signature creation on a smart card. In Sect. 5.2 we present and discuss a proof of concept implementation. Section 6 is devoted to the related previous work.

2   Scheme Description

2.1   Outline of the Solution

In our scheme we redesign the life cycle of secure signature creation devices. The following parties are involved:

Manufacturer: he creates a signature creation device - both the hardware and the software installed there. In particular, he either personalizes the device and installs there the private signing keys on behalf of the final user, or installs software responsible for key generation by the device.

Signatory: the person that holds the signature creation device and controls the physical access to it after getting it from the manufacturer. Apart from the signing activities, he is involved in interactions with his signature creation device aiming to protect against an adversary, who may potentially gain access to his signing key. In particular, the signatory holds an additional private *hidden control key* assigned to the signature creation device. He initializes the device by installing the corresponding hidden public key on his signing device. He keeps this key confidential and shows it to nobody but to the judge[1].

In case of a fraud with the original signing keys, the signatory interacts with the judge and proves forgery.

Verifier: a recipient of digitally signed documents. He runs the standard signature verification procedure to check validity of a signature.

Adversary: a party attempting to create a signature that would be attributed to the device of the attacked signatory. After delivering the device to the signatory, the adversary cannot directly interact with the device, but we assume that he holds the signing key stored in this device. He may also see the signatures created by the device and even request signatures under messages of his choice.

Judge: is a party that resolves disputes about claimed signature forgeries. For this purpose he interacts with the signatory.

In fact, unlike in the standard cryptographic literature, the signature creation device should be regarded as a scheme participant. In particular, it interacts with the signatory in a strictly defined way. So in particular, it should be tamper resistant. Its life cycle consists of the following phases:

Phase 1: device creation and delivery. The signature creation device is produced and delivered to the signatory. We assume that its whole memory contents at that time is known to the adversary and that the device might be personalized for the signatory.

Phase 2: hidden keys initialization. A fresh signature creation device gets the hidden public key installed by the signatory. (In Sect. 6, we indicate that it is possible to change the hidden keys, however this cannot be done completely freely.) The signatory does not certify the hidden public key, however

---

[1] Or even not to the judge, if he is only semi-trusted.

it might be helpful to register the signature creation device. According to our scheme, for this purpose he deposits a number of signatures created by the so initialized device.

Phase 3: regular usage. During this phase the signatory uses the device and creates digitally signed documents. At the same time the adversary may forge signatures using for this purpose the original signing keys of the signatory. Such forged signatures are used by the adversary for presumably malicious purposes.

Phase 4: forgery detection and proof. This phase may occur, when the signatory becomes aware of valid signatures that are attributed to him, but which have not been created by using his signature creation device. In this case the signatory interacts with a judge. (The adversary is not involved in this phase, since in general neither the signatory nor the judge can indicate who is the adversary.) Based on the undeniable cryptographic proof the judge can declare the challenged signature(s) as not created by the device of the signatory and therefore attributed to an unknown adversary.

Phase 5: termination of use. In this phase device usage is terminated; possibly all already existing signatures or some of them get revoked. This phase normally follows Phase 4, when the judge decides that the presented signature has been forged. In a standard situation, termination is due to aging of the signature creation device.

2.2   Preliminaries

First let us discuss a few components used by the proposed scheme.

Signature schemes involved: Our method will apply to signature schemes based on Discrete Logarithm Problem for a cyclic group $\mathcal{G}$ of a prime order q. For $\mathcal{G}$ we shall use the multiplicative notation, however the scheme can be applied for elliptic curve signatures in exactly the same way. The solution works for signature schemes such as ElGamal, DSA, Schnorr, where one of the signature parameters available for the verifier is an element $r = g^k$, where k is created at random.

Hidden control keys: The main feature of signature creation devices executing our scheme is that they contain an extra public key $V = g^v \in \mathcal{G}$. Each signature creation device implementing the scheme must have its own key V installed after handling the device to the signatory. The link between the device and V should be verifiable by a judge. Moreover, the hidden private key v should be known to the signatory, whose responsibility is to prove in the court that certain signatures have been forged, if this unfortunate case occurs. The hidden key v should not be installed on the signing device.

Data signed: the signature schemes concerned apply a hash function together with the core signing operations. According to the common practice, legal requirements and industrial standards, the hash function is applied not to a raw document M, but to M encoded together with some meta-data. The meta-data contain in particular the time of creating the signature (see for instance the popular XAdES standard and its QualifyingProperties: Signed-Properties: SigningTime). Following RFC 2985 [9] we assume that the signing time is given according to the ISO/IEC 9594-8 standard UTC Time, and according to RFC 2630 [6] MUST include seconds. The following issues are important for our scheme:

- For a signing time T, we utilize the field denoting the seconds, say t. For simplicity we assume that $0 \leq t \leq 59$ and ignore the rare cases when a minute has 61 s.
- The signature creation device can adjust a signing time by causing a delay of, say, two seconds.
- Before the secret key is used by the signing procedure, a small preprocessing takes place, allowing to adjust the parameter r to the signing time in a hidden way.

Truncated hash function TruncHash: Our scheme uses a hash function

$$\text{TruncHash} : \mathcal{G} \to \{0, \dots, 59\}$$

For the sake of security proof we have to assume that TruncHash is a pseudo-random function that can be modelled by a random oracle.

Note that for TruncHash we do not aim to achieve collision freeness or any property of this kind. What we really need is that a fair advantage in guessing the value of $\text{TruncHash}(x)$ indicates that its argument x has been presented first.

For the sake of a practical implementation, TruncHash could be taken as a truncated value of a hash function implemented on the smart card (e.g. SHA-256).[2]

2.3   Signing Scheme Procedures

In order to focus attention of the reader, we describe a scheme based on Schnorr signatures. We also skip most of the details that do not differ our solution from the standard one. Let $SignDev$ stand for a signing device.

I. GENERATING A KEY PAIR FOR A USER. We do not alter the way of generating the signing keys. We assume that after this phase:

- $SignDev$ contains a private signing key $x < q$ chosen at random,
- the public key $Y = g^x \in \mathcal{G}$ has been exported outside $SignDev$,
- $SignDev$ is in the state requiring installing the hidden control keys.

II. INSTALLING THE HIDDEN CONTROL KEYS. This procedure is executed by the signatory interacting with his device $SignDev$ with already instantiated private signing key x.

---

[2] There are some low level implementation issues, since hash function is typically hardware supported, but truncation presumably cannot be executed on the cryptographic co-processor and therefore might be relatively slow.

1. The signatory chooses the hidden secret key v, $v < q$, at random and computes $V := g^v \in \mathcal{G}$.
2. The signatory uploads V to $\mathcal{S}ign\mathcal{D}ev^3$. The device $\mathcal{S}ign\mathcal{D}ev$ stores V as its hidden key.
3. $\mathcal{S}ign\mathcal{D}ev$ enters the state in which it can be used only for creating signatures.
4. The signatory creates a few signatures (as described below) and deposits them by a third trusted party.

Remark 1. In practice, the signatory uses another (independent) device (presumably his PC) to create v and V. Insecurity of this auxiliary device may challenge the effectiveness of the forgery detection mechanism, however it does not endanger the underlying signature scheme. Involving a PC does not create an additional risk, as an adversary that controls it may replace the documents to be signed by the documents of his choice at the moment when the signatory authorizes $\mathcal{S}ign\mathcal{D}ev$ to create a signature.

III. SIGNING PROCEDURE. Creating a signature consists of two phases. The first phase can be executed in advance before the data to be signed (or its hash) are transmitted to the signature creation device. The second phase corresponds to the standard signing procedure.

First, for the reader's convenience we recall the Schnorr signing algorithm for a signing key x and a message M (note that M is the original document appended with the signature's meta-data, including in particular the signing time):

CREATION OF SCHNORR SIGNATURES
1. choose k at random
2. $r := g^k$
3. $e := \text{Hash}(M\|r)$
4. $s := (k + x \cdot e) \bmod q$
5. output (s, e) as a signature of M.

CREATION OF SCHNORR SIGNATURES - FORGERY EVIDENT VERSION

This procedure is executed by the device $\mathcal{S}ign\mathcal{D}ev$ in interaction with (the computer of) the signatory. (Of course, a signing request has to be authorized by the signatory – providing the correct PIN is a minimum requirement.)

Phase 1 (preprocessing):
1. create an empty array $A[0 \ldots 59]$
2. choose k at random
3. $U := V^k$
4. $i := 0$
5. repeat Δ times: // Δ *is a constant discussed later*
    5.1. $z := \text{TruncHash}(U, M)$

---

[3] In order to secure the logistic chain, the smart card may require presenting a one-time initialization PIN by the signatory.

5.2. $A[z] := i$

5.3. $i := i + 1$

5.4. $U := U \cdot V$

After Phase 1 the array A and the value k are retained for Phase 2.

Phase 2 (the main signing part):

1. let T be the signing time (in the UTC format), let t be the value of the seconds field of T
2. wait until $A[t]$ is nonempty
3. $r := g^{k+A[t]}$
4. having r already computed (see e.g. the first two steps of the Schnorr algorithm), proceed with the signature creation algorithm for the message to be signed appended with meta-data containing the signing time T.[4]

Remark 2. The choice of the parameter Δ will be discussed in Sect. 5.1. It must be large enough to fill majority of positions in the array A and thereby reduce the waiting time in the second step of Phase 2.

Remark 3. Note that if $A[t]$ is nonempty and contains $j < \Delta$, then the entry j has been inserted for $U = V^{k+j}$ and $t = \text{TruncHash}(U, M)$. Of course, during the first phase $A[t]$ might be overwritten many times, but it does not change the just mentioned property.

IV. VERIFICATION PROCEDURE. The verification procedure is unchanged and executed according to the standard version of the underlying signature.

Remark 4. Note that our signature creation process yields signatures in exactly the same format as in case of the underlying scheme. So, we do not need any modification of the verification software. In fact, we shall show that the verifier cannot recognize whether the signing procedure has been modified.

V. FORGERY DETECTION PROCEDURE. The test concerning a signature of M with the creation time T (as stated in the signed meta-data) is as follows:

1. reconstruct the value r, e.g. for a Schnorr signature (s, e), compute $r := g^s / Y^e$
2. check

$$\text{TruncHash}(r^v, M) \stackrel{?}{=} t \tag{1}$$

   where t denotes the value of the seconds field in T.
3. output `forgery`, if the equality (1) does not hold.

Remark 5. The detection procedure requires knowledge of v – the hidden control key. It must be kept secret by the signatory. Therefore there are two procedures: one concerning forgery detection and the second concerning proving forgery against a judge.

---

[4] In the cryptographic standards the signing time T is treated as meta-data and not a part of the message to be signed, in the cryptographic literature there is no such distinction and the message is understood as a message together with the meta-data.

Remark 6. There is a possibility of a false positive result – by pure luck the adversary creating a signature may use r such that equality (1) holds. This happens with probability $\approx \frac{1}{60}$ for each single signature. This probability is high from the point of view of cryptography, however in this case we are talking about disclosure of malicious behavior of the system provider, who gets into criminal charges with probability $\frac{59}{60}$, if he dares to use a stolen key even once! Note also that we are not talking about general probability to forge a signature, but about the probability to use a forged signature, when the private signing key has already been exposed to the adversary.

VI. FORGERY PROOF PROCEDURE. This procedure is executed by the signatory and the judge. First we describe the general framework and later describe the details of the core Proof Procedure:

1. The signatory presents the judge the hidden public key V.
2. The signatory presents the judge the registered signatures created right after installing the hidden control keys.
3. The judge and the signatory run interactively the Proof Procedure for these signatures and V. If the result is negative, then the signatory's claim gets rejected and the procedure terminates.
4. The signatory presents the judge the alleged forged signature S with the signature creation time T.
5. The judge and the signatory run the Proof Procedure for S. If the result is negative, then the signatory's claim gets rejected and the procedure terminates. Otherwise the judge declares S as forged.

PROOF PROCEDURE
Consider a signature S of a message M with a signing time T, with t denoting the seconds field, for the hidden public key V and the hidden private key v:

1. The judge and the signatory reconstruct the value r from the signature creation process of S – just as in case of a regular signature verification.
2. The signatory computes $u := r^v$ and presents u to the judge.
3. The judge rejects the forgery claim if $\text{TruncHash}(u, M) = t$.
4. The signatory and the judge perform an interactive zero knowledge proof of equality of discrete logarithms for the pairs (g, V) and (r, u). For instance, one may run a number of times the following standard procedure:
   (a) the signatory chooses σ at random and presents

$$v_1 = g^{v\sigma}, v_2 = r^{v\sigma}$$

   (b) the judge chooses a bit b at random,
   (c) if $b = 0$, then the signatory reveals σ and the judge checks that $v_1 = V^\sigma$, $v_2 = u^\sigma$,
   (d) if $b = 1$, then the signatory reveals $\delta = v\sigma$ and the judge checks that $v_1 = g^\delta, v_2 = r^\delta$.
5. If the equality of discrete logarithms test fails, then the judge rejects the forgery claim.

Remark 7. The above procedure can be modified so that V is not presented to the judge (this approach should be applied, if the judge is only semi-trusted.) In this case the signatory does not create a proof of equality of discrete logarithms for (g, V) and (r, u) as described above, but for $(r_0, u_0), \ldots, (r_m, u_m)$ and (r, u), where $(r_0, u_0), \ldots, (r_m, u_m)$ come from the signatures registered right after the personalization of the $SignDev$ device with the hidden keys or from the signatures created in front of the judge with $SignDev$.

Remark 8. One may speculate that the signatory might himself break into his signing device $SignDev$. Then having the secret signing key and the hidden key, the signatory would be able to create signatures that would be recognized as forged by the judge. However, according to the manufacturer's declaration $SignDev$ is tamper resistant.

3 Security of the Scheme

3.1 Resilience to Forgeries

Potentially, the special properties of the signatures created according to the scheme proposed in Sect. 2 might ease forging signatures by the third parties. Namely, we can consider the following game:

Game A

1. The adversary may request signatures of the messages of his choice at time of his choice. They have to be created by a device implementing the algorithm from Sect. 2.
2. The adversary presents a message and its signature S.

The adversary wins the game if S has not been created during step 1 and the standard verification of S yields a positive result.

Note that we do not demand that S would pass the forgery detection procedure. Moreover, the signature S may concern a message that has already been signed during Step 1 (re-signing an old message).

We can also define an analogous game for standard signature creation:

Game B

1. The adversary may request signatures of the messages of his choice at time of his choice.
2. The adversary presents a message and its signature S.

The adversary wins the game if S has not been created during step 1 and the standard verification of S yields a positive result.

Assume that an adversary $\mathcal{A}$ can win Game A. We use it to win Game B. Namely, we choose v at random and set $V = g^v$. Then we create the input for Game A. If the adversary of Game A asks for a signature over M at time T with t being the number of seconds, we use the oracle from Game B. When it returns a signature $(r, \ldots)$, then we compute $\mathrm{TruncHash}(r^v, M)$. If it equals t, then we pass the signature $(r, \ldots)$ to Game A. If not, then we repeat the request for the oracle of Game B, until this property is fulfilled. Since there are only 60 possible values of TruncHash, after a short time we get a signature that can be passed to Game A.

Obviously, the signature returned by the adversary $\mathcal{A}$ from Game A can be used as a response of the adversary in Game B. Thereby we get the following result:

Theorem 1. *If an adversary can win Game A (forge a signature based on a collection of signatures created according to the mechanism from Sect. 2), then an adversary can win Game B (forge a signature in the regular case).*

Remark 9. Note that the opposite direction is not immediate and may require extra assumptions: in the attack for Game A we create signatures that come from a subset of the set of all signatures. In theory, a successful attack from Game B could fail, if we are limited to such a subset.

3.2 Indistinguishability

Our second goal is to show that an adversary not knowing the control key $V = g^v$ cannot distinguish between the signatures passing the forgery detection test and signatures that would be found as forged. (This notion is basically the same as *indistinguishability against Subversion Attacks* from [18].) Namely, we consider the following game:

Game C

1. The adversary may request signatures of the messages of his choice at time of his choice. For each signature $(r_i, \dots)$ over M_i, he gets additionally z_i, where

$$z_i = \text{TruncHash}(r_i^v, M_i)$$

2. The challenger chooses a bit b at random.
3. The challenger presents a signature $(r, \dots)$ over M that has not been created during the first stage and a value z, which equals $\text{TruncHash}(r^v, M)$, if $b = 0$, and a random value $z \neq \text{TruncHash}(r^v, M)$, if $b = 1$.
4. The adversary returns a bit $\bar{b}$.

The adversary wins the game if $b = \bar{b}$.

In our proof we refer to Correlated-Input Secure Hash Function introduced in [5]. This model seems to better reflect the required properties than the Random Oracle Model and addresses directly the threats occurring in practice - including in particular our case. Moreover, for our purposes it suffices to use only a limited version of Correlated-Input Hash Function Assumption. Namely, we consider the following game:

Correlated TruncHash Values Game
    choose pairwise different elements $k_1, \dots, k_n \leq q$
      (an arbitrary strategy may be applied)
    choose $M_1, \dots, M_n \in \mathcal{G}$
      (an arbitrary strategy may be applied)

choose V at random

$h_i := \text{TruncHash}(V^{k_i}, M_i)$ for $i = 1$ to n,

choose M and $k \neq k_1, \ldots, k_n$

$h_{n+1}^{(0)} := \text{TruncHash}(V^k, M)$

choose $h_{n+1}^{(1)} \in \{0, \ldots, 59\} \setminus \{h_{n+1}^{(0)}\}$ at random

choose $b \in \{0, 1\}$ at random

$\hat{b} := \mathcal{A}(k_1, \ldots, k_n, k, M_1, \ldots, M_n, M, h_1, \ldots, h_n, h_{n+1}^{(b)})$.

$\mathcal{A}$ wins the game if $b = \hat{b}$. The advantage of $\mathcal{A}$ is defined as $p - \frac{1}{2}|$ where p is the probability to win the game by $\mathcal{A}$.

Note that in the above game the only value not available to $\mathcal{A}$ is V. We know exactly the relationship between the first arguments of TruncHash due to the knowledge of the exponents $k_1, \ldots, k_n, k$. However, the hash function TruncHash hides the arguments used, so we cannot derive V^k. The game describes the chances of the adversary to deduce the control value for r, given the correct values for some other $r_1, \ldots, r_n$.

Assumption 2. (REDUCED VERSION OF CORRELATED-INPUT HASH FUNCTION ASSUMPTION). The advantage of the adversary $\mathcal{A}$ in the Correlated Hash Values Game is negligible.

Note that the adversary can win the Correlated TruncHash Values Game, if he can win Game C. Indeed, the adversary may choose at random a signing key x. Then, knowing the exponents $k_1, \ldots, k_n, k$ he may create the corresponding signatures for $M_1, \ldots, M_n, M$. Therefore we may immediately conclude as follows:

Theorem 3. *Each adversary has a negligible advantage in Game C, if the (reduced) Correlated-Input Hash Function Assumption holds for the function TruncHash.*

4 The Dark Side of the Scheme

In this section we remark that the core mechanism of the scheme presented in Sect. 2 can be used for evil purposes as well. Jumping ahead to the related work (Sect. 6), let us recall that there have been attempts to eliminate software subversion attacks on signature schemes [18,19]. Their general recommendation is to use deterministic signature schemes and to involve in certain way a hashing function during key generation process. In this way we defend ourselves against choosing the keys that are advantageous for the adversary.

Unfortunately, the proposed approach does not fully implement the concept "nothing up my sleeve" and the security claims from [19] turn out to be incomplete. Let us sketch shortly how to leak the key with their scheme:

- There is no room for randomness during execution of the protocol, however the user cannot fully control the time of signature creation. A malicious software can postpone signature creation by, say, 1 or 2 s. This cannot be really observed by the user and can be attributed to many hardware issues. (For instance, a smart card may suffer from communication errors and attribute the delays to them).

- The time delay may be aimed to create a subliminal channel: the device tries to adjust some number of bits to leak a key. Say, it takes the signature s (created honestly according to the protocol) and computes $D := \text{Hash}(S, K)$, where K is a trapdoor key. Then it takes, say, the last 8 bits of D and treats them as an 8-bit address a. If the 9-th bit of S is the same as the ath bit of the 256-bit signing key, then the signature S is called *good*.

- The signing procedure is adjusted as follows:
  (1) the device creates a signature S taking into account the current time T. If S is *good*, then it releases S. Otherwise it waits one second and goes to step 2.
  (2) the device creates the next signature S' and releases it no matter whether it is *good* or not.

- The adversary collects a number of signatures created by the device. Each signature indicates the value of one bit of the signing key – but of course only the values coming from *good* signatures are true. Therefore for each key position the adversary gathers the values indicated by the signatures. As at average 75% of all signatures are *good*, with enough signatures it is possible to recover the signing key taking into account relevant statistics and enhancing them by brute force at positions where the statistics do not deliver a firm answer.

Note that if the signature creation device has to implement the scheme presented in Sect. 2, then installing the key leaking procedure described above becomes more problematic. Indeed, leaking the key bits requires adjusting the signing time and thereby creates some delays. However, this must be implemented on top of the procedure described in Sect. 2 which already causes some delays. So together the delay might be too high and easily observable.

5   Implementation Issues

When electronic signatures for real world applications are concerned, we have to take care about practical feasibility of the proposed solution. For instance, a protocol using non-standard cryptographic operations not implemented on the smart cards available on the market has almost no chance to be deployed in practice due to high costs of redesign of the card's cryptographic coprocessors.

For this reason we discuss the problems of a time delay introduced by the scheme from Sect. 2. This is a crucial issue, since there is a strict limit for a signature creation process – it should not exceed 2–3 s. Otherwise, the users get annoyed by the long processing time and generally do not accept the solution. (Interestingly, reducing the processing time does not make sense either, since the user should observe that the smart card is performing some computations.)

In order to check computational complexity in reality, we have also implemented our solution on MULTOS smart cards (the cards where the operating system enables access to low level operations for the programmer). The results show that no technical problems emerge for our scheme.

5.1   Empty Places in Array A

Creating a signature according to the procedure described in Sect. 2 at time t fails, if $A[t]$ is empty. In this case, the signature creation process gets postponed 1 s and the next attempt occurs at time $t + 1$.

From the mathematical point of view we have the following problem. There are n bins and k balls. Each ball is placed in one of the bins, the target bin is chosen uniformly at random, independently of other balls. We define the event $\mathcal{E}_i^{(n,k)}$ meaning that the bin $i \bmod n$ is empty after inserting all balls. Then let $p_{n,k,a}$ denote the probability

$$\Pr[\mathcal{E}_{i+1}^{(n,k)} \wedge \mathcal{E}_{i+2}^{(n,k)} \wedge \ldots \wedge \mathcal{E}_{i+a}^{(n,k)}] = \left(1 - \tfrac{a}{n}\right)^k$$

The probability of hitting a non-empty position in the first trial is therefore $\Pr[\neg \mathcal{E}_i^{(n,k)}] = 1 - (1 - \tfrac{1}{n})^k$. For $n = 60$ and $k = 120$ this equals approximately $1 - 1/e^2 \approx 0.864665$. The concrete values for $p_{60,120,a}$ are as follows:

a	1	2	3	4	5
$p_{60,120,a}$	0.133	$1.7 \cdot 10^{-2}$	$2.1 \cdot 10^{-3}$	$2.5 \cdot 10^{-4}$	$2.92 \cdot 10^{-5}$

Let us also consider

$$L_{i,n,k} = \min\left\{a : \neg\left(\mathcal{E}_{i+1}^{(n,k)} \wedge \mathcal{E}_{i+2}^{(n,k)} \wedge \ldots \wedge \mathcal{E}_{i+a}^{(n,k)}\right)\right\}$$

That is, $L_{i,n,k}$ corresponds to the number of steps required to find a nonempty position in the array A starting at position $i + 1$. The probability distribution of $L_{i,n,k}$ does not depend on i, so we will write $L_{n,k}$ instead of $L_{i,n,k}$. Then

$$\Pr[L_{n,k} > a] = \Pr[\mathcal{E}_1^{(n,k)} \wedge \mathcal{E}_2^{(n,k)} \wedge \ldots \wedge \mathcal{E}_a^{(n,k)}]$$

Then the expected value of $L_{n,k}$ can be computed as follows:

$$\begin{aligned} \mathrm{E}[L_{n,k}] &= 1 + \sum_{a=1}^{n} \Pr[L_{n,k} > a] = 1 + \sum_{a=1}^{n-1} \left(1 - \tfrac{a}{n}\right)^k \\ &= 1 + \tfrac{1}{n^k} \sum_{a=1}^{n-1}(n - a)^k = 1 + \tfrac{1}{n^k} \sum_{a=1}^{n-1} a^k \end{aligned}$$

Obviously, if $k_1 < k_2$, then $\mathrm{E}[L_{n,k_1}] > \mathrm{E}[L_{n,k_2}]$. Moreover, $\lim_{k\to\infty} \mathrm{E}[L_{n,k}] = 1$. The expression $1 + \tfrac{60}{k+1}$ is a quite good upper approximation of $\mathrm{E}[L_{60,k}]$ (see Fig. 1).

For $n = 60$ and $k = 1, \ldots, 240$ we may derive the following concrete values:

k	10	60	120	240	500
$\mathrm{E}[L_{60,k}]$	5.96..	1.56..	1.15..	1.01..	1.00022..

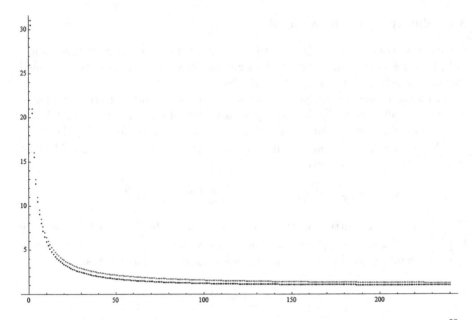

Fig. 1. The expected value $E[L_{60,k}]$ (black points) and its approximation by $1 + \frac{60}{k+1}$ (red points). The value of k is on the x-axis. (Color figure online)

In particular it means that for $k = 120$ the array A is almost filled and the expected position of the first nonempty entry is in most cases the starting position.

We have executed a number of trial executions of the preprocessing stage (implemented in Python) for $\Delta = 100$. We have ran three series of 50 executions of the preprocessing stage and we have counted the number of blocks of empty entries of a given length in the array A. (For this purpose we assume that the entries $A[59], A[0]$ are the neighboring ones.)

- In the first series of 50 executions we have observed:
  - 360 blocks of length 1 (i.e. empty entries preceded and followed by non-empty entries),
  - 64 blocks of 2 empty entries,
  - 10 blocks of 3 empty entries,
  - 1 block of 4 empty entries,
  - 2 blocks of 5 empty entries.

  The average number of empty blocks in a single execution therefore equals 7.2 blocks of length 1, 1.28 blocks of length 2, 0.2 blocks of length 3, 0.02 blocks of length 4, 0.04 blocks of length 5.
- In the second series of 50 executions we have observed: 375, 68, 17, 3 empty blocks of the length, respectively, 1, 2, 3, 4.
- In the third series of 50 executions we have observed: 376, 75, 11, none, 2 empty blocks of the length, respectively, 1, 2, 3, 4, 5.

5.2   MULTOS Trial Implementation

In order to estimate complexity of the proposed scheme in a real environment we have implemented the most critical part – the preprocessing phase – on the standard MULTOS smart cards. Note that this is the only potential bottleneck of the scheme. Indeed, the verification procedure is exactly the same as in the standard case, while forgery detection and proof are executed only occasionally (anyway, they are quite efficient). We have used the 2048-bit MODP group with a 224-bit prime order subgroup defined in RFC 5114. Below we include the source code of this implementation.

Note that this implementation uses only the high level API of the MULTOS standard. So in some sense we perform the worst case experiment, as we use modular arithmetic instead of more efficient elliptic curves. This influences both the time complexity of standard operations used for signature generation, as well as auxiliary operations such as an implementation of TruncHash (in the code presented below this is SHA-1 followed by reduction modulo 60). Of course, a low level implementation should be much more efficient.

```
#include"main.h"

#define FIELD_LEN 256
#define MESS_LEN 256
#define EXP_LEN 28
#define ARR_LEN 60

#pragma melpublic
BYTE tmp[FIELD_LEN+MESS_LEN];
BYTE k[EXP_LEN];
BYTE zb[20];
BYTE kk[8];
struct
{
   BYTE buffer[ARR_LEN];
} apdu;

#pragma melstatic
BYTE p[] = {0xAD,0x10,0x7E,0x1E,0x91,...};
BYTE g[] = {0xAC,0x40,0x32,0xEF,0x4F,...};
BYTE V[] = {0x88,0x31,0x1a,0xf3,0x91,..};
BYTE v[] = {0x80,0x1C,0x0D,0x34,0xC5,..};
BYTE mess[] = {0x01,0x02,0x03,0x04,0x05,..};

void main(void) {
int i=0,j=0;
int delta = 100;
DWORD z;
BYTE t;

   multosGetRandomNumber(kk);
```

```
memcpy(k,kk,8);
multosGetRandomNumber(kk);
memcpy(k+8,kk,8);
multosGetRandomNumber(kk);
memcpy(k+16,kk,8);
multosGetRandomNumber(kk);
memcpy(k+24,kk,4);
multosModularExponentiation(EXP_LEN,FIELD_LEN,
k,p,V,tmp);
memcpy(tmp+FIELD_LEN,mess,MESS_LEN);
for(i=0;i<delta;i++){
multosSHA1(FIELD_LEN+MESS_LEN,tmp,zb);
t = zb[0];
__push(t);
t = zb[1];
__push(t);
t = zb[2];
__push(t);
t = zb[3];
__push(t);
__code (__STORE, &z, 4);
z = z
apdu.buffer[z] = i;
multosModularMultiplication(FIELD_LEN,p,tmp,V);
}
multosExitLa(ARR_LEN);
}
```

The detailed results regarding the time required to execute Phase 1 for the above code on MULTOS are given in Table 1. Note that the time lower than 2 s is acceptable in a standard setting (the total signature creation time presumably below 3 s). We have to note that the preprocessing phase can be executed in parallel when the user is asked to confirm his will to create a signature – in this case the time overhead is zero, as the user needs a few econds to push the button. The most important message from the experiment is that there is no memory bottleneck on the card to execute the scheme.

Table 1. Timing details for particular steps of the preprocessing phase, the results are given as average over 50 trial executions. For the experiment, each trial involves $\Delta = 100$ (the smart card inserts data in the table A 100 times)

Preprocessing step	Execution time in ms (average over 50 trials)
2	5
3	401
5.1–5.3	1025
5.4	293
Total	1724

6   Related Work

Probably the first attempt to protect the signatories against powerful adversaries capable of deriving signing keys from the public keys are *fail-stop signatures* [11]. In case of a forgery, the legitimate owner of a signing device can break a hard cryptographic problem (that he normally is unable to do) thereby providing an evidence of a forgery. The fail-stop signature schemes are based on the idea that for one public key there is a large number of corresponding secret keys and a cryptanalyst cannot say which of them is kept by the signature creation device of the legitimate signatory. Failure to guess this key and attempt to sign with a different key leads to an evidence mentioned above.

Unfortunately, as far as we know this idea has never been deployed in business practice. One of the reasons might be its target: fail-stop signatures aim to reveal successful cryptanalytic attacks, but do not address the case when the adversary has access to the original signing key (when for instance he retains the keys after generating them for the user). Even if the description of a scheme states that the private key is generated by the user (see e.g. [13]), this is done by the signature creation card, and the card may choose in a way predetermined by the card manufacturer. On the other hand, some of the solutions proposed assume existence of a party that indirectly gets access to the private keys of the users (see e.g. [15]). In our opinion this defers on threat but creates a new one, even more serious. Many of the proposed fail-stop signature schemes are one-time signatures, which makes them useless for standard applications. Last not least, according to the third law of Adi Shamir: "Cryptography is typically bypassed, not penetrated."

A somewhat similar, but simpler idea has been presented in [3]. The idea is to provide exactly two options for creating the ith signature, namely with either $r_i = g^{k_i}$ or $r_i' = g^{k_i'}$. Which option is used depends on the user. In case of device cloning the same random parameter k for computing $r = g^k$ can be used in two different signatures: one created by the legitimate device and one by the cloned device. This leads immediately to revealing the signing key and thereby invalidates all signatures created with this key, including the ones created by the clone. Thereby the cloned devices are of little use. Invalidating all signatures seems to be a drastic measure, however the adversary holding a copy of the private key can always do it by leaking somewhere this key.

An alternative approach based on monitoring creation of electronic signatures has been proposed in [4] as *mediated signatures*. The idea is that in an online environment we need not to rely solely on the signing device: for signature creation it is necessary to involve another party, called *mediator*, that holds a supplementary key for each user. Therefore, the mediator may monitor the activities of the signing device and may inform the user about each signature created by the device. The paper [4] describes a simple implementation based on the RSA signatures, while [10] provides a solution based on Schnorr signatures. This approach has been further extended in [2] by the concept of key evolution. This enables the mediator to detect that it is interacting with two different signing devices attributed to the same user.

Technically, the most related design has been proposed in [8], where the authors aim to detect omissions on the list of electronic signatures corresponding to a single device. This scheme aims to cope with the problem of hiding documents created by a given device leading to cases such as Enron bookkeeping frauds.

Recently, there have been a growing interest on protecting signature schemes against attacks based on malicious software. In [18] a general model for subversion attacks is considered – the one that enables the adversary to replace the original algorithm creating the signature by a malicious one. A general method to leak the key bits is presented (following well-known kleptographic attacks [17] and the algorithm from [8]). An important point is that the change should not detectable by the user not knowing the trapdoor information. The general recommendation of this paper is to use deterministic signature algorithms in order to remove any room for malicious substitution. While this is an convincing argument, it also provides a perfect framework for the adversary holding the signing key – in such a setting no cryptographic procedure can help the owner of the signing key to prove the forgery with stolen original keys.

The paper [19] extends [18] by considering the model where also key generation software may be subverted. The key invention is to design a subversion resilient one-way permutation with a trapdoor. It is based on the idea that first an index r is chosen at random – but instead of using r as an index to one-way permutation, the scheme takes the hash value $r' := h_{\mathrm{SPEC}}(r)$ for computing the index of a permutation in the family of one-way trapdoor permutations. Having a trapdoor permutation it becomes relatively straightforward to construct a subversion resistant signature scheme. However, we again have to stress that while this construction protects against malicious implementations and leaking the signing key via the signatures, it provides no defense against an adversary that gets the private key in a different way – e.g. via retaining r by the service provider. Also, there is a long way from the idea to a real deployment, since widely used standard solutions tend to persist even when facing serious security problems.

Final Remarks

The method presented in this paper is simple enough to be easily incorporated in signing devices as an optional mechanism against frauds concerning electronic signatures. It requires changing neither signature standards nor the software used for signature verification. This is a major practical advantage since otherwise necessity of modifications could prohibit deployment of countermeasures for years.

Deployment of the proposed security measures would substantially improve real reliability and undeniability of digital signatures and should be considered for next generation of cryptographic smart cards. In our opinion, a potential threat of forging signatures by a dishonest technology provider is one of the barriers for widespread use of electronic signatures for legal purposes. In the current situation we feel that it is quite controversial to recommend usage of legally binding electronic signatures.

Even if the probability $\frac{59}{60}$ of detecting a forgery with the original signing keys in a single signature may be viewed as relatively low, this is significantly better than the current state-of-the art. Indeed, today the owner of the signing key is in a legally hopeless situation, if the original keys are used by the forger. Note that for transactions of a high value the security policy may require signing the document several times. If for instance 10 signatures are required, then the probability of a successful attack performed by an adversary holding the original signing key is reduced to $\approx 2^{-59}$.

A nice feature of the method proposed is that it partially defends against weaknesses of (pseudo)random number generators used for signature creation by standard schemes. It does not prohibit an adversary to learn the signing key, however according to Theorem 3 it does not enable the adversary to learn the key V and therefore to create signatures that would pass the forgery detection test.

Of course, the proposed method has also some limitations. If an adversary gets full access to the memory of a signing device $SignDev$, including the hidden control key V installed there, then he will be able to mimic the operation of $SignDev$ and create signatures that are indistinguishable from the signatures created by $SignDev$. A potential solution to this problem would be to use a chain of the hidden control keys similar to the Lamport's chain of hash values. Namely, instead of a single (v, V) one could use (v_i, V_i) for $i = k, k-1, \ldots, 1$, where $v_{i+1} = v_i \cdot \text{Hash}(V_i)$. The signatory would have to keep v_1 only and from time to time to update the hidden public key V by replacing $V = V_j$ by $V = V_{j-1}$.

Finally, we have to warn the user that the proposed solution does not protect against an adversary who has gained control over his PC (or the browser). The problem is that the smart card has no display and the user cannot be sure which document has been signed following his request.

References

1. Becker, G.T., Regazzoni, F., Paar, C., Burleson, W.P.: Stealthy dopant-level hardware Trojans: extended version. J. Cryptogr. Eng. **4**(1), 19–31 (2014)
2. Błaśkiewicz, P., Kubiak, P., Kutyłowski, M.: Digital signatures for e-government - a long-term security architecture. In: Lai, X., Gu, D., Jin, B., Wang, Y., Li, H. (eds.) e-Forensics 2010. LNICSSITE, vol. 56, pp. 256–270. Springer, Heidelberg (2011). doi:10.1007/978-3-642-23602-0_24
3. Błaśkiewicz, P., Kubiak, P., Kutyłowski, M.: Two-head dragon protocol: preventing cloning of signature keys. In: Chen, L., Yung, M. (eds.) INTRUST 2010. LNCS, vol. 6802, pp. 173–188. Springer, Heidelberg (2011). doi:10.1007/978-3-642-25283-9_12
4. Boneh, D., Ding, X., Tsudik, G., Wong, C.M.: Instantenous revocation of security capabilities. In: USENIX Security Symposium (2001)
5. Goyal, V., O'Neill, A., Rao, V.: Correlated-input secure hash functions. In: Ishai, Y. (ed.) TCC 2011. LNCS, vol. 6597, pp. 182–200. Springer, Heidelberg (2011). doi:10.1007/978-3-642-19571-6_12
6. Housley, R.: RFC 2630: cryptographic message syntax. ftp://ftp.rfc-editor.org/in-notes/rfc2630.txt (1999)

7. Kocher, P.C.: Timing attacks on Implementations of Diffie-Hellman, RSA, DSS, and other systems. In: Koblitz, N. (ed.) CRYPTO 1996. LNCS, vol. 1109, pp. 104–113. Springer, Heidelberg (1996). doi:10.1007/3-540-68697-5_9

8. Kubiak, P., Kutyłowski, M.: Supervised usage of signature creation devices. In: Lin, D., Xu, S., Yung, M. (eds.) Inscrypt 2013. LNCS, vol. 8567, pp. 132–149. Springer, Cham (2014). doi:10.1007/978-3-319-12087-4_9

9. Nystrom, M., Kaliski, B.: RFC 2985: PKCS #9: selected object classes and attribute types version 2.0. ftp://ftp.rfc-editor.org/in-notes/rfc2985.txt (2000)

10. Nicolosi, A., Krohn, M.N., Dodis, Y., Mazières, D.: Proactive two-party signatures for user authentication. In: NDSS 2003, The Internet Society (2003)

11. Pedersen, T.P., Pfitzmann, B.: Fail-stop signatures. SIAM J. Comput. **26**(2), 291–330 (1997)

12. Stevens, M., Sotirov, A., Appelbaum, J., Lenstra, A., Molnar, D., Osvik, D.A., Weger, B.: Short chosen-prefix collisions for MD5 and the creation of a rogue CA certificate. In: Halevi, S. (ed.) CRYPTO 2009. LNCS, vol. 5677, pp. 55–69. Springer, Heidelberg (2009). doi:10.1007/978-3-642-03356-8_4

13. Susilo, W., Safavi-Naini, R., Gysin, M., Seberry, J.: A new and efficient fail-stop signature scheme. Comput. J. **43**(5), 430–437 (2000)

14. The European Parliament and European Council. Regulation (EU) no 910/2014 of the European Parliament and of the Council on electronic identification and trust services for electronic transactions in the internal market and repealing directive 1999/93/EC. Official Journal of the European Union L 257/73 (2014)

15. Yamakawa, T., Kitajima, N., Nishide, T., Hanaoka, G., Okamoto, E.: A short fail-stop signature scheme from factoring. In: Chow, S.S.M., Liu, J.K., Hui, L.C.K., Yiu, S.M. (eds.) ProvSec 2014. LNCS, vol. 8782, pp. 309–316. Springer, Cham (2014). doi:10.1007/978-3-319-12475-9_22

16. Young, A., Yung, M.: Kleptography from standard assumptions and applications. In: Garay, J.A., Prisco, R. (eds.) SCN 2010. LNCS, vol. 6280, pp. 271–290. Springer, Heidelberg (2010). doi:10.1007/978-3-642-15317-4_18

17. Young, A., Yung, M.: Kleptography: using cryptography against cryptography. In: Fumy, W. (ed.) EUROCRYPT 1997. LNCS, vol. 1233, pp. 62–74. Springer, Heidelberg (1997). doi:10.1007/3-540-69053-0_6

18. Ateniese, G., Magri, B., Venturi, D.: Subversion-resilient signature schemes. In: Ray, I., Li, N., Kruegel, C. (eds.) CCS 2015, pp. 364–375. ACM, New York (2015)

19. Russell, A., Tang, Q., Yung, M., Zhou, H.-S.: Cliptography: clipping the power of kleptographic attacks. IACR Cryptology ePrint Archive vol. 2015, p. 695 (2015)

Advances in Cryptanalysis

A New Test Statistic for Key Recovery Attacks Using Multiple Linear Approximations

Subhabrata Samajder[✉] and Palash Sarkar

Applied Statistics Unit, Indian Statistical Institute,
203, B.T.Road, Kolkata 700108, India
{subhabrata_r,palash}@isical.ac.in

Abstract. The log-likelihood ratio (LLR) and the chi-squared distribution based test statistics have been proposed in the literature for performing statistical analysis of key recovery attacks on block ciphers. A limitation of the LLR test statistic is that its application requires the full knowledge of the corresponding distribution. Previous work using the chi-squared approach required *approximating* the distribution of the relevant test statistic by chi-squared and normal distributions. Problematic issues regarding such approximations have been reported in the literature. Perhaps more importantly, both the LLR and the chi-squared based methods are applicable only if the success probability P_S is greater than 0.5. On the other hand, an attack with success probability less than 0.5 is also of considerable interest. This work proposes a new test statistic for key recovery attacks which has the following features. Its application does not require the full knowledge of the underlying distribution; it is possible to carry out an analysis using this test statistic without using any approximations; the method applies for all values of the success probability. The statistical analysis of the new test statistic follows the hypothesis testing framework and uses Hoeffding's inequalities to bound the probabilities of Type-I and Type-II errors.

Keywords: Multiple linear cryptanalyis · LLR statistic · Chi-squared statistic · Hoeffding inequality

1 Introduction

Consider the setting of multiple linear cryptanalysis of block ciphers. Statistical analyses of such attacks proceed by identifying a suitable test statistic. In purely statistical terms, the setting is as follows. Let $X_1, \ldots, X_N$ be independent and identically distributed random variables taking values from the set $\{0, 1\}^\ell$. The distribution of the X_j's is either a distribution $\tilde{p} = (p_0, \ldots, p_{2^\ell - 1})$ or it is the uniform distribution on $\{0, 1\}^\ell$. For $\eta \in \{0, 1\}^\ell$, let Q_η be the random variable which counts the number of j's such that $X_j = \eta$. The following test statistics have been used in the literature on block cipher cryptanalysis. Assume $\ell > 1$.

$$\text{LLR} = \sum_{\eta=0}^{2^\ell - 1} Q_\eta \ln(2^\ell p_\eta); \quad \Lambda = 2^\ell N \sum_{\eta=0}^{2^\ell - 1} (Q_\eta/N - 2^{-\ell})^2.$$

© Springer International Publishing AG 2017
R.C.-W. Phan and M. Yung (Eds.): LNCS 10311, Mycrypt 2016, pp. 277–293, 2017.
DOI: 10.1007/978-3-319-61273-7_14

The LLR test statistic arises from the log-likelihood ratio while the distribution of Λ can be approximated by a chi-squared distribution. By the chi-squared test statistic, we will mean Λ. Approximate expressions for data complexities of key recovery attacks using the LLR and the chi-squared test statistics have been obtained in [14]. Both the LLR and the chi-squared test statistics have some limitations which are mentioned below.

Knowledge of the Distribution: To apply the LLR test statistic, it is required to have *full* knowledge of the probability distribution $\tilde{p}$. In many situations, this information may be difficult to obtain. The distribution $\tilde{p}$ is uncovered by a detailed analysis of the block cipher and for $\ell > 1$, obtaining the full distribution $\tilde{p}$ may not be possible. In such situations, it is not possible to apply the LLR test statistic.

To apply the chi-squared test statistic, the knowledge of $\tilde{p}$ is not required. The analysis needs to only unearth the expected value of the test statistic which is one of the factors that determine the number of plaintext-ciphertext pairs required to mount the attack. So, to apply an analysis based on the chi-squared test statistic, the requirement from the analysis of the block cipher is substantially lower than that required from the LLR test statistic.

Approximation Issues: For both the LLR and the chi-squared test statistics, the analysis in [14] approximates the corresponding distributions by normal. This involves an error in approximation which has not been studied in details. For the chi-squared based test statistic, this issue has been briefly noted in the literature [14,15]. For detailed analysis of problems arising from normal approximations we refer to [27].

Works Only for High Success Probability: The success probability of a key recovery attack is the probability that the target sub-key is indeed recovered by the algorithm. While it is good to have high success probabilities, from a cryptanalytic point of view, low success probabilities are also meaningful. For example, an attack with success probability 0.1 has a 10% chance of success. Such an attack should be considered to be a valid attack. It is helpful for a cryptanalyst to determine the amount of plaintext-ciphertext pairs required to achieve a certain success probability. The LLR and the chi-squared based approaches have serious limitations with respect to this requirement. Both the approaches are applicable only for success probabilities greater than 0.5. So, if one wishes to obtain an estimate of data complexity for an attack with 10% chance of success, then there is nothing in the literature which allows doing this.

Our Contributions. In this work, we propose to perform a statistical analysis which overcomes the previously mentioned limitations. This requires a suitable test statistic.

Our first choice is the chi-squared test statistic. For this, we considered the possibility of performing an analysis without making any approximations. We follow the hypothesis testing framework. An approach for avoiding approximations in this framework has been outlined in [26]. The idea is to apply the Hoeffding bounds to upper bound the probabilities of Type-I and Type-II errors. This requires expressing the test statistic as a sum of *independent* random variables. Unfortunately, for the chi-squared test statistic, this does not seem to be possible.

Since neither the LLR nor the chi-squared test statistics seem to apply, we propose a new test statistic. For $\eta \in \{0,1\}^{\ell}$, let $\underline{\eta}$ denote the integer whose binary representation is η. Let d be a positive real number. We propose the test statistic $T = \sum_{\eta \in \{0,1\}^{\ell}} \underline{\eta}^d Q_\eta$. The computation of this statistic does not require information about $\tilde{p}$. Let μ_0 (resp. μ_1) be the expectation of T when the X_j's follow $\tilde{p}$ (resp. the uniform distribution). If $\mu_0 \neq \mu_1$, then T can be used to carry out a key recovery attack. The requirement from the analysis of the internal structure of the block cipher is to obtain (an estimate of) μ_0. Given the value of μ_0, it is possible to obtain an expression for the data complexity (i.e., the number of plaintext-ciphertext pairs) required to attain the parameters of a successful attack.

The statistical analysis that we perform does not require us to make any approximations. It is possible to express T as a sum of independent random variables. So, the Hoeffding bounds can be used to bound the probabilities of the Type-I and Type-II errors.

The theoretical analysis holds for any positive d. The question that arises is what value of d should be used in practice. An important point to keep in mind is that for the chosen value of d, it should be possible to estimate the value of μ_0. Based on experiments, we suggest that the value of d should be taken to be 1.

We have evaluated the obtained bound using known linear approximations for the block cipher SERPENT. For success probabilities at most 0.5, there is no prior result in the literature to which we can compare. For success probabilities greater than 0.5, the values of the bounds turn out to be higher than the approximate values obtained using the chi-squared test statistic.

Note that the minimum required data complexity to achieve a certain value of the success probability is not known. Our method provides an upper bound on this minimum data complexity while the chi-squared method provides an approximate value where the error in approximation is not known. So, the data complexity of the chi-squared method cannot be taken to be the correct value and then the bound obtained by our method is criticised for being an over-estimate. It is possible that the chi-squared method grossly under-estimates the minimum required data complexity.

A work of independent interest would be to simulate an attack on some particular (toy) cipher to determine the required data complexity and then compare it to the bound that we obtain and the approximate value obtained from the chi-squared method. If done in a comprehensive manner this could be an interesting exercise, but, one that we feel is not directly related to the contribution of the paper. We have obtained a theoretical bound which holds for all ciphers. This in itself should be of some intrinsic interest. Note that the upper bound on the data complexity obtained in this paper depends only on the value of μ_0, provided such an estimate exists and that the estimate is an *accurate* one.

280   S. Samajder and P. Sarkar

While we do not make any claims that the bound is tight, we do note that carrying out an experiment for one particular cipher will not establish the bound to be loose. There may be ciphers for which the bound is loose while there could be other ciphers for which the bound is tight. Further, it is difficult to extrapolate results on data complexity obtained from simulation of an attack on some toy cipher to results about much higher data complexity on more complex and real-life ciphers. More work is required to establish the tightness (or not) of the bounds obtained here. We refer to [28] for a discussion on this issue. Another equally important issue is to be able to propose another test statistic which shares the advantages of the one that we use and for which it is possible to obtain lower values of the data complexity.

Previous and Related Work. Linear cryptanalysis was proposed by Matsui in [21] as an attack on DES and involved a single linear approximation of the cipher. Later, in [22], Matsui used two linear approximations (which were assumed to be independent) to improve the attack. Independently, Kaliski and Robshaw [20] extended Matsui's attack involving single linear approximations to multiple linear cryptanalysis using $\ell \geq 1$ independent linear approximations. The approximations that were considered had certain restrictions. It was assumed that the ℓ linear approximations have a common data mask (i.e., plaintext and ciphertext mask) but different key masks.

In [3], Biryukov et al. gave a more general method for multiple linear cryptanalysis without any assumption on the corresponding linear approximations. Their analysis, though, still assumed the linear approximations to be independent. Analysis under the independence assumption was also done independently by Junod and Vaudenay in [19] in the context of distinguishing attacks. Further work on distinguishing attacks without the independence assumption was carried out in [1,2,10,18]. Murphy [24] argued that the independence assumption need not be valid.

Junod [17] gave a detailed analysis of Matsui's ranking method [21,22]. This work introduced the notion of ordered statistics in linear cryptanalysis. This was further developed by Selçuk in [29], where he used a well known asymptotic result from the theory of ordered statistic to arrive at the expression for success probability for both single linear and differential cryptanalysis.

The test statistic used in [1,2,18] was the log-likelihood ratio (LLR). The chi-squared test statistic was initially used by Handschuh and Gilbert [11] for the cryptanalysis of the SEAL encryption algorithm. Later Johansson and Maximov [16] gave an explicit analysis of the success and the error probabilities in the context of their attack on the stream cipher Scream. The idea of Selçuk's order statistics based approach has been combined with the LLR and the chi-squared test statistics to obtain expressions for data complexities of multiple linear cryptanalysis [14].

A related line of work considered the situation where the correlation for a linear approximation depends on the key. This line of research originates from the work of Daemen and Rijmen [9] and was explicitly put in the context of linear cryptanalysis in [6] for single linear cryptanalysis and in [5] for multi-dimensional linear cryptanalysis.

In this paper, we have not considered the issue of key-dependent correlations. The problems with the use of normal approximations for linear cryptanalysis without key-dependent behaviour (reported in [27]) also extend to the case of key-dependent correlations. Further, there are several additional subtleties which need to be properly handled. Carefully analysing the setting of key-dependent correlations without approximations requires a separate comprehensive treatment. The goal of the present paper, on the other hand, is primarily to show how several limitations of previous statistical methods for analysing multiple linear cryptanalysis can be overcome. We believe that the usefulness of this contribution can be assessed independently of the issue of key-dependent correlations.

2   Multiple Linear Cryptanalysis

Let $E : \{0,1\}^k \times \{0,1\}^n \mapsto \{0,1\}^n$ be a block cipher, and so for each $K \in \{0,1\}^k$, $E_K(\cdot) \triangleq E(K, \cdot)$ is a bijection from $\{0,1\}^n$ to itself. Here, K is called the secret key, the n-bit input to E_K is called the plaintext and the n-bit output of E_K is called the ciphertext.

Usual constructions of block ciphers involve a simple round function parameterised by a round key which is iterated over several rounds. The round keys are produced by applying an expansion function, called the key scheduling algorithm, to the secret key K. Denote the round keys by $k^{(0)}, k^{(1)}, \dots$ and round functions by $R_{k^{(0)}}^{(0)}, R_{k^{(1)}}^{(1)}, \dots$. Also, let $K^{(i)}$ denote the concatenation of the first i round keys, i.e., $K^{(i)} = k^{(0)} \,||\, \cdots \,||\, k^{(i-1)}$ and $E_{K^{(i)}}^{(i)}$ denote the composition of the first i round functions, i.e.,

$$E_{K^{(0)}}^{(0)} = R_{k^{(0)}}^{(0)}; \quad E_{K^{(i)}}^{(i)} = R_{k^{(i-1)}}^{(i-1)} \circ \cdots \circ R_{k^{(0)}}^{(0)} = R_{k^{(i-1)}}^{(i-1)} \circ E_{k^{(i-1)}}^{(i-1)}; i \geq 1.$$

Suppose that an attack targets $r + 1$ rounds. For a plaintext P, we denote by B the output after r rounds, i.e., $B = E_{K^{(r)}}^{(r)}(P)$ and we denote by C the output after $r + 1$ rounds, i.e., $C = E_{K^{(r+1)}}^{(r+1)}(P) = R_{k^{(r)}}^{(r)}(B)$.

Block cipher cryptanalysis starts off with a detailed analysis of the block cipher. This results in one or possibly more relations between the plaintext P, the input to the last round B and possibly the expanded key $K^{(r)}$. In case of linear cryptanalysis these relations are linear in nature and are of the following form:

$$\langle \Gamma_P^{(i)}, P \rangle \oplus \langle \Gamma_B^{(i)}, B \rangle = \langle \Gamma_K^{(i)}, K^{(r)} \rangle; \quad i = 1, 2, \dots, \ell;$$

where $\Gamma_P^{(i)}, \Gamma_B^{(i)} \in \{0,1\}^n$ and $\Gamma_{K^{(r)}}^{(i)} \in \{0,1\}^{nr}$ denotes the plaintext mask, the mask to the input of the last round and the key mask respectively. A linear relation of the above form is called a linear approximation of the block cipher. Such linear approximations usually hold with some probability which is taken over the uniform random choices of the plaintext P. Obtaining such relations and their joint distribution is not a trivial task and requires a lot of ingenuity and experience. They form the basis on which the statistical analysis of block ciphers are built. If $\ell > 1$, the attack is called a multiple linear cryptanalysis and if $\ell = 1$, we call the attack single linear cryptanalysis, or simply, linear cryptanalysis. Define $L_i \triangleq \langle \Gamma_P^{(i)}, P \rangle \oplus \langle \Gamma_B^{(i)}, B \rangle$; for $i = 1, 2, \ldots, \ell$.

Inner Key Bits: Let $z_i = \langle \Gamma_K^{(i)}, K^{(r)} \rangle$; $i = 1, \ldots, \ell$. Note that for a fixed but unknown key $K^{(r)}$, z_i represents a single unknown bit. Denote by $z = (z_1, \ldots, z_\ell)$ the collection of the bits arising in this manner. Since, the ℓ key masks $\Gamma_K^{(1)}, \ldots, \Gamma_K^{(\ell)}$ are known, the tuple z is determined only by the unknown but fixed $K^{(r)}$. Hence, there is no randomness either of $K^{(r)}$ or z. The bits of z are called the inner key bits.

Target Sub-key Bits: Any linear relation of the form above, between P and B, usually involves only a subset of the bits of B. When $\ell > 1$, several (or multiple) relations between P and B are known. In such cases, it is required to consider the subset of the bits of B which covers all the relations. In order to obtain these bits from the ciphertext C it is required to partially decrypt C by one round. This involves a subset of the bits of the last round key $k^{(r)}$. We call this the target sub-key. The goal of linear cryptanalysis is then to find the correct value of the target sub-key using the ℓ linear approximations and their joint distributions. We denote the number of bits in the target sub-key by m. In other words, these m key bits are sufficient to partially decrypt C by one round and obtain the bits of B involved in any of the ℓ linear approximations. Notice that there are 2^m possible choices of the target sub-key out of which only one is correct. The purpose of the attack is to identify the correct key. For convenience of notation, we will denote the correct choice of the target sub-key as κ^*.

Setting of the Attack: The block cipher is instantiated with an unknown, but, fixed key. It is assumed that N independent and uniform random plaintexts are chosen and the corresponding ciphertexts under fixed key are obtained. Denote the plaintext-ciphertext pairs as (P_j, C_j); $j = 1, 2, \ldots, N$. For each choice κ of the target sub-key, it is possible for the attacker to partially decrypt each C_j by one round to obtain $B_{\kappa,j}$; $j = 1, 2, \ldots, N$. Note that $B_{\kappa,j}$ is dependent on κ even though C_j may not. Clearly, if the choice of κ is correct, then the C_j's depend on κ. On the other hand, for an incorrect choice of κ, C_j has no relation with κ.

Statistical analysis proceeds by defining a test statistic T_κ for each choice κ of the target sub-key. This provides 2^m random variables of the type T_κ. The distribution of T_κ depends on whether κ is the correct choice or, it is an incorrect choice. Under the usual wrong key hypothesis [12], it is assumed that the distributions of all the T_κ's for incorrect choices of κ's are the same.

Suppose that the plaintext P is uniformly distributed. Since, each round function is a bijection, the uniform distribution of P also induces a uniform distribution on B. By definition, L_i is a binary random variable taking values from the set $\{0, 1\}$. Also from the discussion above it is clear that the source of randomness of L_i comes from the randomness of P. Define the random variable X as $X = (L_1, \ldots, L_\ell)$. Then X is a random variable distributed over $\{0, 1\}^\ell$.

Joint Distribution Parameterised by Inner Key Bits: The distribution of the random variable $X = (L_1, \ldots, L_\ell)$ is the following. For $\eta \in \{0,1\}^\ell$ and $z \in \{0,1\}^\ell$,

$$p_z(\eta) = \Pr[L_1 = \eta_1 \oplus z_1, \ldots, L_\ell = \eta_\ell \oplus z_\ell] = \frac{1}{2^\ell} + \epsilon_\eta(z); \qquad (1)$$

where $-1/2^\ell \leq \epsilon_\eta(z) \leq 1 - 1/2^\ell$. Denote by $\tilde{p}_z = (p_z(0), p_z(1), \ldots, p_z(2^\ell - 1))$ the corresponding probability distribution, where the integers $\{0, 1, \ldots, 2^\ell - 1\}$ are identified with the set $\{0,1\}^\ell$. For each choice of z, we obtain a different but related distribution. Let, $z' = z \oplus \beta$ for some $\beta \in \{0,1\}^\ell$, then it is easy to verify that $\epsilon_\eta(z') = \epsilon_{\eta \oplus \beta}(z)$, which implies that $p_{z \oplus \beta}(\eta) = p_z(\eta \oplus \beta)$. Let, $\tilde{p}$ denote the probability distribution $\tilde{p}_{0^\ell}$, i.e., $\tilde{p} \stackrel{\Delta}{=} \tilde{p}_{0^\ell}$. Write $\tilde{p} = (p_0, \ldots, p_{2^\ell - 1})$, so that for all $\eta \in \{0,1\}^\ell$, $p_\eta \stackrel{\Delta}{=} p(\eta) = 1/2^\ell + \epsilon_\eta$.

For $\kappa \in \{0, 1, \ldots, 2^m - 1\}$, $j = 1, \ldots, N$ and $i = 1, \ldots, \ell$, define $L_{\kappa,j,i} = \langle \Gamma_P^{(i)}, P_j \rangle \oplus \langle \Gamma_B^{(i)}, B_{\kappa,j} \rangle$; $X_{\kappa,j} = (L_{\kappa,j,1}, \ldots, L_{\kappa,j,\ell})$; and

$$Q_{\kappa,\eta} = \#\{j \in \{1, 2, \ldots, N\} : X_{\kappa,j} = \eta\}. \qquad (2)$$

Note that $Q_{\kappa,\eta}$ is the number of times η appears among the random variables $X_{\kappa,1}, \ldots, X_{\kappa,N}$. Suppose z is the correct choice of the inner key bits. Then for the correct choice of the target sub-key (i.e., $\kappa = \kappa^*$) the random variable $Q_{\kappa,\eta}$ follows $\mathrm{Bin}(N, p_z(\eta))$, whereas for the incorrect choice of the target sub-key (i.e., $\kappa \neq \kappa^*$) the random variable $Q_{\kappa,\eta}$ follows $\mathrm{Bin}(N, 2^{-\ell})$. Denote the uniform distribution over the set $\{0,1\}^\ell$ by $p_\$ = (2^{-\ell}, \ldots, 2^{-\ell})$.

Success Probability and Advantage of An Attack: Two important parameters which are relevant to a key recovery attack are the success probability and the (expected) advantage. The success probability is the probability that the correct value of the target sub-key is recovered in the attack. The advantage of an attack is a, if a fraction 2^{-a} of all possible 2^m values of the target sub-key are reported as candidate values. So, for an attack with advantage a, the size of the list of candidate keys is 2^{m-a}.

3 Drawbacks of Previously Proposed Statistics

As mentioned in the introduction, two test statistics have been proposed earlier [14] for performing statistical analysis of key recovery attacks on block ciphers. In this section, we briefly review these statistics and point out certain drawbacks.

Log-Likelihood Ratio Test Statistic: The LLR test statistic has been used for key recovery attacks as well as distinguishing attacks in several works in the literature [1,4,14,26]. One drawback of this statistics is that to compute it, the full knowledge of $\tilde{p}$ is required. This is evident from the expression of the LLR test statistic. In many situations, such complete knowledge of the joint distribution of the multiple linear approximations may not be available. In such cases, it will not be possible to compute the value of LLR_κ.

The analysis in [14] provides an expression for the data complexity in terms of the success probability and the advantage. This expression is stated to be valid only for success probability greater than 0.5.

Chi-Squared Test Statistic: Recall from (2) that for a choice κ of the target sub-key and for $\eta \in \{0,1\}^\ell$, $Q_{\kappa,\eta}$ is the number of times η occurs among the random variables $X_{\kappa,1}, \ldots, X_{\kappa,N}$. Define a test statistic Λ_κ in the following manner:

$$\Lambda_\kappa = 2^\ell N \sum_{\eta=0}^{2^\ell-1} (Q_{\kappa,\eta}/N - 2^{-\ell})^2. \tag{3}$$

For the correct choice κ^* of the target sub-key bits, the right hand side of (3) involves $Q_{\kappa^*,\eta}$ whose distribution depends on the inner key bits z. Due to the relation $p_{z\oplus\beta}(\eta) = p_z(\eta \oplus \beta)$, the distribution of Λ_{κ^*}, however, does not depend on z.

To apply the chi-squared test statistic, it is not required to know the full distribution of the underlying probability distribution. Statistical analysis using this test statistic has been carried out in [14] in the following manner. The distribution of $Q_{\kappa,\eta}$ follows a binomial for both correct and incorrect choices of κ. The binomial can be approximated using a normal distribution and then the distribution of Λ_κ approximately follows a chi-squared distribution for both correct and incorrect choices of κ. There is, however, the issue of error in approximation which has not been properly analysed. This issue of error in approximation has been briefly mentioned in [14,15,27] and has been analysed in details in [27] where several shortcomings have been pointed out.

The data complexity for the chi-squared test statistic was given by Hermelin et al. in [14]. It was shown that for "large" values of a and $P_S > 0.5$, the data complexity, which we denote by N_Λ, is approximately

$$N_\Lambda = \frac{2\sqrt{2^\ell - 1}\Phi^{-1}(1 - 2^{-a}) + 4\left(\Phi^{-1}(2P_S - 1)\right)^2}{C(\tilde{p})}; \tag{4}$$

where $C(\tilde{p}) = \sum_{\eta=0}^{2^\ell-1}(p_\eta - 2^{-\ell})/2^{-\ell}$.

A reduced round linear cryptanalysis of SERPENT was earlier reported in [8] using a set of linear approximations [7]. Out of these, a subset of 64 linear approximations was later used in [13,14] to perform multidimensional linear cryptanalysis on SERPENT using the LLR and the chi-squared test statistics. It happens so that this subset can be generated by 10 linear approximations called the basis linear approximations and can be used to recover 10 bits of the last round key. Thus, for this particular experiment, $\ell = 10$ and $m = 10$.

It was pointed out in [27], that for a χ^2 approximation of the distribution of the test statistic Λ to be valid, the corresponding distributions under both the null and the alternate hypotheses need to satisfy the following two conditions for all $\eta \in \{0,1\}^{\ell}$: $| p_{\eta}(1-p_{\eta}) - q_{\eta}(1-q_{\eta}) | < p_{\eta}(1-p_{\eta})$; and $\frac{p_{\eta}(1-p_{\eta})-q_{\eta}(1-q_{\eta})}{p_{\eta}(1-p_{\eta})} \approx 0$. We checked whether these conditions hold for the linear approximations of the reduced round block cipher SERPENT reported in [7]. The total number of linear approximation required to generate the full probability distribution for the correct key is $2^{10} - 1 = 1023$. Out of these, only 64 are given in [7]. To find the full probability distribution for the correct key, two methods were suggested in [13]. We have used the second method, where the correlations of the remaining $1023 - 64 = 959$ approximations are assumed to be zero. The Walsh transform method of [25] was then used on these approximations to get the joint distribution.

For the joint distribution of the reduced round SERPENT it was found that for all η, $| p_{\eta}(1-p_{\eta}) - q_{\eta}(1-q_{\eta}) |$ is indeed less than $p_{\eta}(1-p_{\eta})$. The maximum value of the ratio $| (p_{\eta}(1 - p_{\eta}) - q_{\eta}(1 - q_{\eta})) | /(p_{\eta}(1 - p_{\eta}))$ is 0.0049. So, the χ^2 approximation is valid provided that the value 0.0049 is *assumed* to be sufficiently close to zero. The effect of this assumption on the final expression for the data complexity is not known. This is one of the several approximations that is required to obtain the chi-squared based data complexity expression. We refer to [27] for more details.

A question then arises as to whether it is possible to use the chi-squared test statistic to obtain an expression for the data complexity *without* using any approximation. Such an approach has been shown to be successful for the LLR test statistic [26] through the application of the Hoeffding bounds. This requires expressing the test statistic as a sum of independent random variables. However, Λ_{κ} is the sum of 2^{ℓ} random variables where these individual random variables are determined by $Q_{\kappa,\eta}$, $\eta \in \{0,1\}^{\ell}$. The $Q_{\kappa,\eta}$'s are dependent as $\sum_{\eta \in \{0,1\}^{\ell}} Q_{\kappa,\eta} = N$. So, the Hoeffding bound does not apply directly. Further, there does not seem any other way to write Λ_{κ} as the sum of independent random variables.

4   A New Test Statistic

Let d be a positive integer and consider the following test statistic.

$$T_{\kappa} = \sum_{\eta \in \{0,1\}^{\ell}} \underline{\eta}^d Q_{\kappa,\eta}. \tag{5}$$

Let μ_0 be the expectation of T_{κ} for the correct choice of κ and let μ_1 be the expectation of T_{κ} for an incorrect choice of κ. Then

$$\mu_1 = E[T_{\kappa}] = \sum_{\eta \in \{0,1\}^{\ell}} \underline{\eta}^d E[Q_{\kappa,\eta}] = N2^{-\ell} \sum_{\eta \in \{0,1\}^{\ell}} \underline{\eta}^d; \tag{6}$$

$$\mu_0 = E[T_{\kappa^*}] = \sum_{\eta \in \{0,1\}^{\ell}} \underline{\eta}^d E[Q_{\kappa^*,\eta}] = \mu_1 + N \sum_{\eta \in \{0,1\}^{\ell}} \underline{\eta}^d \epsilon_{\eta}. \tag{7}$$

So, $\mu_0 - \mu_1 = N \sum_{\eta \in 0,1^{\ell}} \underline{\eta}^d \epsilon_{\eta}$. One can now aim to design a statistical analysis which attempts to recover κ^* by exploiting the difference in the two expectations.

While doing this, we would like to avoid making any approximations. We next show how both of these aims can be achieved.

Recall that for a fixed κ, the random variables $X_{\kappa,1}, \ldots, X_{\kappa,N}$ are independent. The test statistic given by (5) can be rewritten in the following manner.

$$T_\kappa = \sum_{\eta \in \{0,1\}^\ell} \underline{\eta}^d Q_{\kappa,\eta} = \sum_{j=1}^N \underline{X}_{\kappa,j}^d. \tag{8}$$

This enables writing T_κ as the sum of independent random variables. The computation of T_κ can be done in $O(N)$ time using any one of the two expressions. This computation does not require the knowledge of the ϵ_η's.

Consider the following test of hypothesis:

Hypothesis Test-1:

H_0: "κ is correct" versus H_1: "κ is incorrect."

Decision rule:

Case $\mu_0 > \mu_1$: Reject H_0 if $T_\kappa \le t, \forall z \in \{0,1\}^\ell$; where $t \in (\mu_1, \mu_0)$;
Case $\mu_0 < \mu_1$: Reject H_0 if $T_\kappa \ge t, \forall z \in \{0,1\}^\ell$; where $t \in (\mu_0, \mu_1)$.

Proposition 1. *Let $0 < \alpha, \beta < 1$. In Hypothesis Test-1, it is possible to choose t such that for*

$$N \ge \frac{(2^\ell - 1)^{2d}(\sqrt{\ln(1/\alpha)} + \sqrt{\ln(1/\beta)})^2}{2 \left(\sum_{\underline{\eta}=0}^{2^\ell-1} \underline{\eta}^d \epsilon_\eta \right)^2} \tag{9}$$

the probabilities of the Type-I and Type-II errors are upper bounded by α and β respectively.

The proof follows by applying Hoeffding's bound (see Appendix A) to upper bound the probabilities of the type-I and type-II errors, and thereafter eliminating the threshold parameter t. The proof is given in Appendix B.

Let $\mu_1' = 2^{-\ell} \sum_{\eta \in \{0,1\}^n} \underline{\eta}^d$ and $\mu_0' = \sum_{\eta \in \{0,1\}^n} \underline{\eta}^d(2^{-\ell} + \epsilon_\eta)$. Then $\mu_0' - \mu_1' = \sum_{\eta \in \{0,1\}^n} \underline{\eta}^d \epsilon_\eta$ and so (9) can be written as

$$N \ge \frac{(2^\ell - 1)^{2d}(\sqrt{\ln(1/\alpha)} + \sqrt{\ln(1/\beta)})^2}{2 (\mu_0' - \mu_1')^2}.$$

Thus, although (9) suggests that it is necessary to know all the ϵ_η's to get a lower bound of N, it is actually not the case. It suffices to have a good estimate of μ_0' which is just the expected value of the random variable $\underline{X}_{\kappa^*,1}^d$. (Note that $\underline{X}_{\kappa^*,1}^d, \ldots, \underline{X}_{\kappa^*,N}^d$ are identically distributed.)

Relating to Success Probability and Expected Advantage: By definition, the success probability is $1 - \Pr[\text{Type-I error}]$. So, if α is an upper bound on the probability of the type-I error, then $P_S = 1 - \alpha$ is a lower bound on the success probability.

An incorrect value of κ is reported as a candidate key if a Type-II error occurs. Since there are a total of $2^m - 1$ incorrect values of the target sub-key, the expected number of wrong values reported as candidate keys is $\beta(2^m - 1)$. Equating to 2^{m-a} gives $\beta = 2^{-a} \times 2^m/(2^m - 1)$.

In the expression for the data complexity N, we may replace α by $1 - P_S$ and β by $2^{-a} \times 2^m/(2^m - 1)$. This provides an expression for the data complexity required to attain success probability at least P_S and advantage at least a.

Nature of the Bound: Proposition 1 shows a lower bound on the data complexity required for ensuring a certain minimum success probability and a certain minimum advantage. This lower bound is with respect to Hypothesis Test-1 which in particular means that the test statistic T_κ is used. We note, on the other hand, that there is a possibility of using some other test statistic for which the required data complexity is lower. This means that taken over all possible test statistics, the data complexity expression in Proposition 1 is actually an upper bound on the minimum data complexity that is required to achieve given values of success probability and advantage.

Attack Procedure: The actual application of the attack will be as follows. Given P_S and a, determine α $(= 1 - P_S)$ and β $(= 2^{-a} \times 2^m/(2^m - 1))$; then determine N as given by the right hand side of (9). From α and N determine t (given by (13) or (14) of Appendix B). Once t is determined, Hypothesis Test-1 can be performed. Suppose that $\mu_0 > \mu_1$, the other case being similar. Initialise a list $\mathcal{L}$ to be empty. For each choice κ of the target sub-key, compute T_κ; if $T_\kappa > t$, append κ to $\mathcal{L}$. At the end, $\mathcal{L}$ contains the set of candidate keys.

The above procedure does not require knowledge of $\tilde{p}$ to apply the test. Only the knowledge of μ_0 is required to obtain an estimate of the data complexity N.

Choice of d: The theory described above works for all positive d. We suggest the use of $d = 1$. The rationale behind such a choice is given in Appendix C.

5   Experimental Results for SERPENT

We compare the bound on the data complexity given by (9) to that of the approximate data complexity of the Λ-test statistic given by [14, Eq. (18)] and reproduced in (4) for the reduced round block cipher SERPENT. The distribution used for all the computations in this section is the one discussed in Sect. 3. The comparison presented in this section has been broadly classified into two groups, one where P_S has been fixed to 0.95 and the other where experiments have been conducted for different values of P_S.

Fixed P_S: For this experiment, the value of P_S was fixed to 0.95. The bound given by (9) with $d = 1$ and the approximate value given by (4) were then computed for $a = 1, 2, \ldots, 10$. Table 1 summarises the output of the experiment. The last column of the Table gives the ratio of the two data complexities. From the Table, it is clear that approximate estimate obtained from the Λ test statistic is lower than the upper bound obtained from the new method.

We note that the minimum data complexity required to achieve success probability 0.95 and advantage a is not known. While N_X is an upper bound, N_Λ is an approximation where the error in approximation is not known. At present, it is not possible to say anything more than this.

Table 1. Values of N_X and N_Λ for the joint distribution of SERPENT with a ranging from 1 to 10 and $P_S = 0.95$.

a	N_X (9)	N_Λ (4)	N_X/N_Λ
1	2.79×10^{10}	1.25×10^6	22246.87
2	3.59×10^{10}	9.48×10^6	3783.91
3	4.27×10^{10}	1.53×10^7	2793.44
4	4.89×10^{10}	2.0×10^7	2449.77
5	5.47×10^{10}	2.4×10^7	2283.17
6	6.03×10^{10}	2.75×10^7	2190.17
7	6.56×10^{10}	3.07×10^7	2134.69
8	7.08×10^{10}	3.37×10^7	2100.80
9	7.58×10^{10}	3.64×10^7	2080.41
10	8.07×10^{10}	3.90×10^7	2068.96

Table 2. Values of N_X and N_Λ for the joint distribution of SERPENT with $P_S = 0.1, 0.2, \ldots, 0.9$ and $a = 5$. In the table, n.a. denotes "not applicable."

P_S	N_X (9)	N_Λ (4)
0.10	2.03×10^{10}	n.a.
0.20	2.31×10^{10}	n.a.
0.30	2.56×10^{10}	n.a.
0.40	2.81×10^{10}	n.a.
0.50	3.08×10^{10}	n.a.
0.60	3.37×10^{10}	2.33×10^7
0.70	3.71×10^{10}	2.28×10^7
0.80	4.16×10^{10}	2.28×10^7
0.90	4.84×10^{10}	2.33×10^7

Varying P_S: We computed the value of N_X for different values of $P_S = 0.1, 0.2, \ldots, 0.9$ for the same joint distribution of SERPENT. For this experiment we fixed $a = 5$. Table 2 reports the results of the experiment. From the table, it can be seen that the data complexity N_X increases as P_S increases, which is what one would expect. But, the data complexity N_Λ first increases then decreases even for $P_S > 0.5$. This anomalous behaviour is due to the approximations used in deriving the expression for N_Λ.

6   Conclusion

The paper considered the problem of statistical analysis of attacks on block ciphers in the situation where the LLR test statistic cannot be applied. The other aspect considered was to follow the approach in [26] towards a rigorous analysis without using any approximations. We first considered the chi-squared based test statistic and argued that this test statistic is not amenable to analysis using our approach.

To resolve the problem, we introduced a new test statistic using which an attack can be applied without the full knowledge of the underlying probability distribution. Also, the resulting statistical framework can be analysed rigorously without making any approximations. The obtained expression for data complexity was compared to the *approximate* expression for data complexity for the chi-squared test statistic using known linear approximations for the block cipher SERPENT. As expected, the data complexity of the new test statistic turns out to be higher. This shows that if one wishes to follow a rigorous approach, then one would have to be satisfied with a conservative estimate of the data complexity.

An important aspect of our analysis is that it allows obtaining estimates of the data complexity for all possible values of the success probability. This is in contrast to previous work which required the success probability to be greater than half.

A   Hoeffding Inequality

We briefly recall Hoeffding's inequality for sum of independent random variables. The result can be found in standard texts such as [23].

Theorem 1 (Hoeffding Inequality). *Let,* $X_1, X_2, \ldots, X_\lambda$ *be a finite sequence of independent random variables, such that for all* $i = 1, \ldots, \lambda$, *there exists real numbers* $a_i, b_i \in \mathbb{R}$, *with* $a_i < b_i$ *and* $a_i \leq X_i \leq b_i$. *Let* $X = \sum_{i=1}^{\lambda} X_i$. *Then for any positive* $t > 0$,

$$\Pr[X - E[X] \geq t] \leq \exp\left(-\frac{2t^2}{D_\lambda}\right) \tag{10}$$

$$\Pr[X - E[X] \leq -t] \leq \exp\left(-\frac{2t^2}{D_\lambda}\right) \tag{11}$$

$$\Pr[|\,X - E[X]\,| \geq t] \leq 2\exp\left(-\frac{2t^2}{D_\lambda}\right); \tag{12}$$

where $D_\lambda = \sum_{i=1}^{\lambda}(b_i - a_i)^2$.

B   Proof of Propositon 1

We provide the proof for the case $\mu_0 > \mu_1$ with the other case being similar. Recall that $\underline{X}^d_{\kappa,1}, \ldots, \underline{X}^d_{\kappa,N}$ are N independently and identically distributed random variables such that for all $j = 1, \ldots, N$

$$v_{\min} = 0 \leq \underline{X}^d_{\kappa,j} \leq (2^\ell - 1)^d = v_{\max}.$$

Let, $v = v_{\max} - v_{\min} = (2^\ell - 1)^d$. Thus Hoeffding bounds (see Sect. A) can be used on the sum of independently and identically distributed random variables $T_\kappa = \sum_{j=1}^{N} \underline{X}^d_{\kappa,j}$; where $D_N = Nv^2$.

The probabilities of Type-I and Type-II errors are then given by

$$\Pr[\text{Type-I Error}] = \Pr[T_\kappa \leq t \mid H_0 \text{ holds}] = \Pr[T_\kappa - \mu_0 \leq -(\mu_0 - t)|H_0 \text{ holds}]$$
$$\leq \exp\left(-\frac{2(\mu_0 - t)^2}{Nv^2}\right); \quad [\text{By 11}].$$

$$\Pr[\text{Type-II Error}] = \Pr[T_\kappa > t \mid H_1 \text{ holds}] = \Pr[T_\kappa - \mu_1 > t - \mu_1] \mid H_1 \text{ holds}]$$
$$\leq \exp\left(-\frac{2(t - \mu_1)^2}{Nv^2}\right); \quad [\text{By 10}].$$

Let,

$$\alpha = \exp\left(-\frac{2(\mu_0 - t)^2}{Nv^2}\right); \quad \beta = \exp\left(-\frac{2(t - N\mu_1)^2}{Nv^2}\right).$$

Then, using the fact that $\mu_1 < t < \mu_0$, we get

$$\sqrt{2}t = \sqrt{2}\mu_0 - v\sqrt{2N\ln(1/\alpha)} \qquad (13)$$

$$\sqrt{2}t = \sqrt{2}\mu_1 + v\sqrt{N\ln(1/\beta)}. \qquad (14)$$

Eliminating t from the above two equations and using the expressions for μ_0, μ_1 and v, we get the expression given by the right hand side of (9). For any N greater than this value, the probabilities of Type-I and Type-II errors will be at most α and β respectively. □

C   Choice of d

There are two factors that need to be kept in mind while choosing a appropriate value of d.

1. The value of d has an effect on the data complexity. So, one should try to choose a value of d which minimises the data complexity.
2. For the chosen value of d, it should be possible to obtain an estimate of μ_0 through the analysis of the block cipher.

Regarding the first point, there does not seem to be a way to formally prove that one particular value of d will minimise the data complexity. Instead, we provide intuitive explanations and experimental evidence.

The statistic $T_\kappa = \sum_{j=1}^{N} \underline{X}_{\kappa,j}^d$. As d goes to zero, $X_{\kappa,j}^d$ goes to 1 and so the effect of $X_{\kappa,j}$ diminishes. Further, as $d \to 0$, $(2^\ell - 1)^d \to 1$ and $\underline{\eta}^d \to 1$ for all $\eta \in \{0,1\}^\ell$. So, the numerator of the data complexity expression given by (9) goes to a constant and the denominator goes to $\sum_{\eta \in \{0,1\}^\ell} \epsilon_\eta$. By definition, the later sum is 0. So, as $d \to 0$, the data complexity expression given by (9) goes to infinity. Experiments confirm this behaviour.

Based on the above, we do not consider values of $d < 1$. For values of $d = 1, \ldots, 100$, we have run experiments with the known linear approximations of SERPENT and have observed that the minimum data complexity is attained for $d = 1$ and $d = 2$. The values are shown in Table 3. To decide between these two values, we consider the second point mentioned above. Intuitively, it is easier to obtain the value of μ_0 for $d = 1$ than for $d = 2$. So, we suggest using $d = 1$ for defining the test statistic T_κ.

Negative Values of d: Most of the theory that has been developed also works for negative values of d. The only problem is that for $\underline{\eta} = 0$, the value of $\underline{\eta}^d$ is undefined. This defect can be rectified by defining T_κ to be $\sum_{j=1}^{N}(1 + \underline{X}_{\kappa,j})^d$. Working out the details of this test statistic leads to $v = |2^{\ell d} - 1|$ and $|\mu_0 - \mu_1| = \sum_{\eta \in \{0,1\}^\ell}(1 + \underline{\eta})^d \epsilon_\eta$. The value of v does not depend on the sign of d. Suppose

Table 3. Table showing the minimum data complexity over different values of d for the linear approximations of SERPENT with a ranging from 1 to 10.

a	Minimum data complexity	
	Value of d	Data complexity
1	1, 2	2.79×10^{10}
2	1, 2	3.59×10^{10}
3	1, 2	4.27×10^{10}
4	1, 2	4.89×10^{10}
5	1, 2	5.47×10^{10}
6	1, 2	6.03×10^{10}
7	1, 2	6.56×10^{10}
8	1, 2	7.08×10^{10}
9	1, 2	7.58×10^{10}
10	1, 2	8.07×10^{10}

$d > 0$, then the value of $|\mu_0 - \mu_1|$ with d is greater than the value of $|\mu_0 - \mu_1|$ with $-d$. As a result, the data complexity with d is lesser compared to the data complexity for $-d$. Due to this reason, we have not considered negative values of d.

References

1. Baignères, T., Junod, P., Vaudenay, S.: How far can we go beyond linear cryptanalysis? In: Lee, P.J. (ed.) ASIACRYPT 2004. LNCS, vol. 3329, pp. 432–450. Springer, Heidelberg (2004). doi:10.1007/978-3-540-30539-2_31
2. Baignères, T., Sepehrdad, P., Vaudenay, S.: Distinguishing distributions using chernoff information. In: Heng, S.-H., Kurosawa, K. (eds.) ProvSec 2010. LNCS, vol. 6402, pp. 144–165. Springer, Heidelberg (2010). doi:10.1007/978-3-642-16280-0_10
3. Biryukov, A., De Cannière, C., Quisquater, M.: On multiple linear approximations. In: Franklin, M. (ed.) CRYPTO 2004. LNCS, vol. 3152, pp. 1–22. Springer, Heidelberg (2004). doi:10.1007/978-3-540-28628-8_1
4. Blondeau, C., Gérard, B., Nyberg, K.: Multiple differential cryptanalysis using, and X^2 statistics. In: Visconti, I., Prisco, R. (eds.) SCN 2012. LNCS, vol. 7485, pp. 343–360. Springer, Heidelberg (2012). doi:10.1007/978-3-642-32928-9_19
5. Blondeau, C., Nyberg, K.: Joint data and key distribution of simple, multiple, and multidimensional linear cryptanalysis test statistic and its impact to data complexity. Des. Codes Crypt. 1–31 (2016). doi:10.1007/s10623-016-0268-6, ISSN: 1573-7586
6. Bogdanov, A., Tischhauser, E.: On the wrong key randomisation and key equivalence hypotheses in Matsui's algorithm 2. In: Moriai, S. (ed.) FSE 2013. LNCS, vol. 8424, pp. 19–38. Springer, Heidelberg (2014). doi:10.1007/978-3-662-43933-3_2
7. Collard, B., Standaert, F.-X., Quisquater, J.-J.: (2008). http://www.dice.ucl.ac.be/fstandae/PUBLIS/50b.zip. Accessed 30 July 2014

8. Collard, B., Standaert, F.-X., Quisquater, J.-J.: Experiments on the multiple linear cryptanalysis of reduced round serpent. In: Nyberg, K. (ed.) FSE 2008. LNCS, vol. 5086, pp. 382–397. Springer, Heidelberg (2008). doi:10.1007/978-3-540-71039-4_24

9. Daemen, J., Rijmen, V.: Probability distributions of correlation and differentials in block ciphers. J. Math. Crypt. JMC **1**(3), 221–242 (2007)

10. Gérard, B., Tillich, J.-P.: On linear cryptanalysis with many linear approximations. In: Parker, M.G. (ed.) IMACC 2009. LNCS, vol. 5921, pp. 112–132. Springer, Heidelberg (2009). doi:10.1007/978-3-642-10868-6_8

11. Handschuh, H., Gilbert, H.: χ^2 cryptanalysis of the SEAL encryption algorithm. In: Biham, E. (ed.) FSE 1997. LNCS, vol. 1267, pp. 1–12. Springer, Heidelberg (1997). doi:10.1007/BFb0052330

12. Harpes, C., Kramer, G.G., Massey, J.L.: A generalization of linear cryptanalysis and the applicability of Matsui's piling-up lemma. In: Guillou, L.C., Quisquater, J.-J. (eds.) EUROCRYPT 1995. LNCS, vol. 921, pp. 24–38. Springer, Heidelberg (1995). doi:10.1007/3-540-49264-X_3

13. Hermelin, M., Cho, J.Y., Nyberg, K.: Multidimensional linear cryptanalysis of reduced round serpent. In: Mu, Y., Susilo, W., Seberry, J. (eds.) ACISP 2008. LNCS, vol. 5107, pp. 203–215. Springer, Heidelberg (2008). doi:10.1007/978-3-540-70500-0_15

14. Hermelin, M., Cho, J.Y., Nyberg, K.: Multidimensional extension of Matsui's Algorithm 2. In: Dunkelman, O. (ed.) FSE 2009. LNCS, vol. 5665, pp. 209–227. Springer, Heidelberg (2009). doi:10.1007/978-3-642-03317-9_13

15. Hermelin, M., Cho, J.Y., Nyberg, K.: Statistical tests for key recovery using multidimensional extension of Matsui's Algorithm 1. In: Handschuh, H., Lucks, S., Preneel, B., Rogaway, P. (ed.) Symmetric Cryptography, number 09031 in Dagstuhl Seminar Proceedings, Dagstuhl, Germany. Schloss Dagstuhl - Leibniz-Zentrum fuer Informatik, Germany (2009). http://drops.dagstuhl.de/opus/volltexte/2009/1954, ISSN: 1862-4405

16. Johansson, T., Maximov, A.: A linear distinguishing attack on scream. In: Proceedings 2003 IEEE International Symposium on Information Theory, p. 164. IEEE (2003)

17. Junod, P.: On the complexity of Matsui's attack. In: Vaudenay, S., Youssef, A.M. (eds.) SAC 2001. LNCS, vol. 2259, pp. 199–211. Springer, Heidelberg (2001). doi:10.1007/3-540-45537-X_16

18. Junod, P.: On the Optimality of linear, differential, and sequential distinguishers. In: Biham, E. (ed.) EUROCRYPT 2003. LNCS, vol. 2656, pp. 17–32. Springer, Heidelberg (2003). doi:10.1007/3-540-39200-9_2

19. Junod, P., Vaudenay, S.: Optimal key ranking procedures in a statistical cryptanalysis. In: Johansson, T. (ed.) FSE 2003. LNCS, vol. 2887, pp. 235–246. Springer, Heidelberg (2003). doi:10.1007/978-3-540-39887-5_18

20. Kaliski, B.S., Robshaw, M.J.B.: Linear cryptanalysis using multiple approximations. In: Desmedt, Y.G. (ed.) CRYPTO 1994. LNCS, vol. 839, pp. 26–39. Springer, Heidelberg (1994). doi:10.1007/3-540-48658-5_4

21. Matsui, M.: Linear cryptanalysis method for DES cipher. In: Helleseth, T. (ed.) EUROCRYPT 1993. LNCS, vol. 765, pp. 386–397. Springer, Heidelberg (1994). doi:10.1007/3-540-48285-7_33

22. Matsui, M.: The first experimental cryptanalysis of the data encryption standard. In: Desmedt, Y.G. (ed.) CRYPTO 1994. LNCS, vol. 839, pp. 1–11. Springer, Heidelberg (1994). doi:10.1007/3-540-48658-5_1

23. Mitzenmacher, M., Upfal, E.: Probability and Computing: Randomized Algorithms and Probabilistic Analysis. Cambridge University Press, Cambridge (2005)

24. Murphy, S.: The independence of linear approximations in symmetric cryptanalysis. IEEE Trans. Inform. Theory **52**(12), 5510–5518 (2006)
25. Nyberg, K., Hermelin, M.: Multidimensional walsh transform and a characterization of bent functions. In: Proceedings of the 2007 IEEE Information Theory Workshop on Information Theory for Wireless Networks, pp. 83–86 (2007)
26. Samajder, S., Sarkar, P.: Rigorous upper bounds on data complexities of block cipher cryptanalysis. IACR Cryptology ePrint Archive, 2015:916 (2015). http://eprint.iacr.org/2015/916
27. Samajder, S., Sarkar, P.: Another Look at Normal Approximations in Cryptanalysis. J. Math. Crypt. (2016). doi:10.1515/jmc-2016-0006
28. Samajder, S., Sarkar, P.: Can large deviation theory be used for estimating data complexity? Cryptology ePrint Archive, Report 2016/465 (2016). http://eprint.iacr.org/
29. Selçuk, A.A.: On probability of success in linear and differential cryptanalysis. J. Cryptol. **21**(1), 131–147 (2008)

Tuple Cryptanalysis: Slicing and Fusing Multisets

Marine Minier[1,2,3] and Raphaël C.-W. Phan[4(✉)]

[1] Université de Lorraine LORIA, UMR 7503, 54506 Vandoeuvre-lès-Nancy, France
[2] Inria, 54600 Villers-lès-Nancy, France
[3] CNRS, LORIA, UMR 7503, 54506 Villers-lès-Nancy, France
[4] Faculty of Engineering, Multimedia University (MMU), Cyberjaya, Malaysia
raphael@mmu.edu.my

Abstract. In this paper, we revisit the notions of Square, saturation, integrals, multisets, bit patterns and tuples, and propose a new Slice & Fuse paradigm to better exploit multiset type properties of block ciphers, as well as relations between multisets and constituent bitslice tuples. With this refined analysis, we are able to improve the best bounds proposed in such contexts against the following block ciphers: Threefish, PRINCE, PRESENT and RECTANGLE.

Keywords: Block ciphers · Square · Saturation · Integrals · Multisets · Bit patterns · Tuples · Bitslice · Slice & Fuse paradigm · Division property

1 Introduction

Cryptanalysis based on a *multiset* of related texts aims for some property of the multiset to pass through cipher components with a probability that is furthest away as possible from the uniform distribution. Given a bijective function $S : \{0,1\}^w \rightarrow \{0,1\}^w$, then with a multiset $\mathcal{M}$ of 2^w elements such that $\forall x_j, x_{j'} \in \mathcal{M}$, $x_j \neq x_{j'}$ (so-called a multiset with the permutation property P, or called an active set), we have that $\forall x_j, x_{j'} \in \mathcal{M}$, $S(x_j) \neq S(x_{j'})$, i.e. such a P multiset passes through S without having its property changed. Given such a P, its presence can be detected by checking if the exclusive-OR ($\bigoplus$) sum of all elements of this multiset equals zero. This was first observed by Knudsen in 1997 [3] as part of the analysis of the SQUARE cipher, hence subsequently it was de facto known in the cryptographic community as the *Square attack*.

The term multiset was in fact first used for this cryptanalysis context by Lucks [9], within a so-called *saturation attack* framework. This naming convention makes sense, because essentially, only the active set (only later known as the permutation P set) is a proper set, other types e.g. the even (E) set, the passive (also called constant C) set, are actually multisets because any distinct value could appear more than once in such sets. The saturation term is named after the so-called saturated property, which is used to denote a dth-order P multiset.

© Springer International Publishing AG 2017
R.C.-W. Phan and M. Yung (Eds.): LNCS 10311, Mycrypt 2016, pp. 294–320, 2017.
DOI: 10.1007/978-3-319-61273-7_15

Furthermore, the semi-saturated property was used to refer to a multiset with 2^{w-1} unique values instead of 2^w. Lucks noted that when viewed in terms of separate bit positions, one such bit position is fixed to a constant value while the other $w-1$ are allowed to vary. To our knowledge, this was the first mention of viewing multisets in terms of their bitwise channels.

Knudsen and Wagner later proposed a formalisation of this type of cryptanalysis within the group theoretic setting as the *Integral cryptanalysis* [11], focussing on how the $\oplus$ sum of a P multiset could propagate through different cipher components; presenting ways to determine such an integral sum for multisets of different types i.e. C, P and those whose integral sum equals some fixed value.

Meanwhile, Biryukov and Shamir [10] initiated a formal study of *multiset calculus* by considering how multiple words of multisets propagate through the substitution and affine layers of a cipher, i.e. the notion of multiple multiset words $\mathcal{M}_1 \ldots \mathcal{M}_\ell$ *composing* to a multi-word multiset. This type of analysis is crucial as a multiset input in any word of a block eventually spreads to other words through the diffusing properties of the affine layers. Nakahara et al. [13] at Mycrypt 2005 built on this notion by focussing on how an n-bit P word multiset $\mathcal{M}$ can be *decomposed* into its constituent subword multisets $\mathcal{V}_i$ each of length $w < n$. Z'aba et al. [17] took this decomposition notion further by focussing on constituent *bitslice* multisets i.e. $w = 1$. Indeed, when the cardinality of a bitslice multiset $\mathcal{V}$ is more than two, i.e. $|\mathcal{V}| > 2$, the elements of the bitslice multiset are then seen to form a bit pattern. Using such bit patterns that are either constant c, regularly alternating a_i or non-alternating b_i, they were able to trace how such patterns behave through the exclusive-OR (XOR) operation and through the Sbox layer.

Aumasson et al. [22] emphasized on the internal ordering of the elements within a multiset, therefore such ordered multisets are better known as *tuples*. Keeping track of ordered tuple elements is useful in order to better trace the effects of operations among elements of different multisets, as well as be exploited to cancel out differences among elements of different tuples.

More recently, Todo [34,35] generalized the integral cryptanalysis approach by taking the integral sum on the low-order polynomial subsets of all the elements in the output multisets, such that due to the higher-order differential type property linked to the algebraic normal form (ANF) representation of the cipher component, a zero sum is obtained; so-called the *division property*. The approach was then formalised by Boura and Canteaut [46] in relation to Reed-Muller codes, based on parity computation across different multiple bits of each multiset element, which is related to the ANF representation of functions and the algebraic degree. In contrast, our proposed approach considers each bitslice channel independently and the focus is on the internal ordering of such bitslice elements rather than on their integral sum.

The Slice-&-Fuse Paradigm. In this paper, we put forth a refined multiset calculus, wherein we propose new types of bit*slice* tuples to better represent the rich internal structures of multiset constituents, analyse the behaviour of such tuples through different cipher components including exclusive-OR, addition, AND and

Sboxes, and discuss how these constituent bitslice tuples can be recomposed (*fused*) to form structured multisets.

We demonstrate such tuple formulations on the ciphers Threefish, PRINCE, PRESENT and RECTANGLE. For Threefish, more structured multiset properties are detected analytically in contrast to best-known results in [22], while for PRINCE we are able to extend the multiset tracking by one more round compared to previous work [44]. For PRESENT and RECTANGLE, we need much smaller multisets (thus less data complexity) and/or are able to detect a sum property in more number of output bits compared to the literature. Moreover, as side results, in Appendix A we also improve the previous integral attacks against CRYPTON and mCrypton.

2   Multiset Calculus

2.1   Multiset and Bitslice Channels

A w-bit *multiset*

$$\mathcal{M} = \{x_j\}_{j=0}^{2^w-1} = \beta_{w-1}\dots\beta_0$$

is a collection of 2^w w-bit word elements. The multiset can also collectively be viewed as a concatenation of w bitslice channels β_i, $i = 0, \dots, w-1$, one for each bit position within the word; and where each channel comprises 2^w bit elements. The bit elements within these bitslice channels will follow some internal ordering, and thus for the rest of this paper, we will use the term *bit tuples* (these have also been called bit patterns [17]) to refer to such ordered bitslice channels.

By definition, the elements $\{x_j\}$ within a multiset can take the same w-bit values, therefore it is at times useful to explicitly denote, using the multiset notation, the multiplicity/frequency $m(x_j)$ of element values. Thus a multiset can alternatively be expressed as a set of unordered pairs of the form $\{x_j, m(x_j)\}$, where each distinct value of x_j is taken from a so-called root set $\mathcal{R}$. Obviously, $|\mathcal{M}| \geq |\mathcal{R}|$.

Definition 1 (Multiset Properties): Depending on the values of its elements, a multiset is said to have (one or more of) any of the following properties:

- P: $\forall x_j, x_{j'} \in \mathcal{M}$, $x_j \neq x_{j'}$; thus $m(x_j) = m(x_{j'}) = 1$.
  This is known as a permutation multiset, where all elements are distinct.
- C: $\forall x_j \in \mathcal{M}$, $x_j = c$ for some constant value c; thus $m(c) = 2^w$.
  This is called a constant multiset, where all elements equal some constant c.
- E^k: $\forall x_j \in \mathcal{M}$, $m(x_j) \bmod k = 0$, where k is of the form 2^n.
  For simplicity reasons, E^2 will be denoted by E. This is an even multiset, where each element value from $\mathcal{R}$ appears an even number of times in $\mathcal{M}$.
- B: $\bigoplus_{j=0}^{2^w-1} x_j = 0$.
  Multisets with this property are said to be balanced. Essentially, this property is detected by the existence of a zero integral $\bigoplus$ sum.
- A: $\sum_{j=0}^{2^w-1} x_j = 0$.
  This is the additive variant of the $\bigoplus$ zero sum.

- F: $\sum_{j=0}^{2^w-1} x_j = 2^{w-1}$.
  This is similar to A except that the sum is non-zero.
- Q: $\forall x_j \in \mathcal{M}, \forall x'_j \in \mathcal{M}', \bigoplus_{j=0}^{2^w-1} x_j = \bigoplus_{j=0}^{2^w-1} x'_j$.
  This equality property exists for two multisets when their respective sums are equal.

It has been observed [22] that these multiset properties are essentially of two categories:

- those characterized by the multiplicity of the elements: P,C,E
- those characterized by relations among the elements: B,A,F,Q

Definition 2 (Bit Tuples): A bitslice channel β can be (one or more of) any of the following bit tuples:

- c: a contiguous sequence of all '0' or all '1' bits, i.e. $c \in \{00\ldots00, 11\ldots11\}$.
- a_i: alternating segments of length 2^i.
  e.g. $a_1 = \langle 00\ 11\ 00\ 11\ldots 00\ 11\rangle$.
- $\hat{a}_i$: the segment dual of a_i such that their segment boundaries are out of sync by half of the segment length.
  e.g. $\hat{a}_1 = \langle 01\ 10\ 01\ 10\ 01\ 10\ldots 10\rangle$ or $\langle 10\ 01\ 10\ 01\ 10\ 01\ldots 01\rangle$.
  Note. For any a_i where $i \neq 0$, its segment dual $\hat{a}_i$ is palindromic.
- $\tilde{a}_i$: the cyclic variant(s) of a_i such that their segment boundaries are out of sync. Note that $\hat{a}_i$ is a special case of $\tilde{a}_i$.
  e.g. Given $a_2 = \langle 0000\ 1111\ 0000\ 1111\ldots 0000\ 1111\rangle$, then we could have $\tilde{a}_2 = \langle 000\ 1111\ 0000\ 1111\ldots 0000\ 1111\ 0\rangle$ or $\tilde{a}_2 = \langle 0\ 1111\ 0000\ 1111\ldots 0000\ 1111\ 000\rangle$.
- p_{ij}: palindromic dual to a_i with alternating flipped pattern segments of length 2^j. See Example 1(a).
- $\hat{p}_{ij}$: the segment dual of p_{ij} such that their segment boundaries are out of sync by half of the segment length. See Example 1(a).
- c_{ij}: complement dual to $\hat{a}_i$ with alternating flipped pattern segments of length 2^j.
  Note. Observe that c_{ij} is to $\hat{a}_i$ what p_{ij} is to a_i. See *Example* 1(b).
- m_{ij} (resp. s_{ij}): masked dual of a_i where every other contiguous sequence of 2^j bits is masked to '0' (resp. set to '1').
  e.g. $m_{02} = \langle 0000\ 0101\ 0000\ 0101\ldots\rangle$ and $s_{02} = \langle 1111\ 0101\ 1111\ 0101\ldots\rangle$.
- z_j: a bit tuple of pattern cycle length 2^{j+1}, that begins with contiguous '0' bits followed by 2^{j-1} '1' bit(s).
  e.g. $z_2 = \langle 00000011\ 00000011\ \ldots 00000011\rangle$.
- f_j: a bit tuple with alternating flipped pattern segments each of length 2^j.
  e.g. $f_3 = \langle 00101011\ 11010100\ \ldots 00101011\ 11010100\rangle$.
- l_j: a bit tuple with alternating pattern segments each of length 2^j comprising contiguous runlengths of '0's and '1's.
  e.g. $l_2 = \langle 0111\ 0111\ \ldots 0111\rangle$ or $l_2 = \langle 0001\ 0001\ \ldots 0001\rangle$.
- e: unlike the above tuples where bit elements within the tuple conform to some defined ordering, the e is just used to denote the property of a bit tuple such that its bit elements appear an even number of times.

Note that a_i and c were defined in [17]. The other bit tuple definitions are new and will be useful later when we trace how different bit tuples are changed by the various cipher operations e.g. ARX (addition, rotation, exclusive-OR) and AND.

Example 1(a): For bit tuple $a_0 = \langle 01010101 \ldots 01 \rangle$, its palindromic duals p_{0j} include:

- $p_{01} = \langle 01\ 10\ 01\ 10\ \ldots 01\ 10 \rangle$
- $p_{02} = \langle 0101\ 1010\ 0101\ 1010 \ldots 0101\ 1010 \rangle$
- $p_{03} = \langle 01010101\ 10101010\ 01010101\ 10101010 \ldots 01010101\ 10101010 \rangle$
- $p_{04} = \langle 0101010101010101\ 1010101010101010 \ldots 0101010101010101\ 1010101010101010 \rangle$.
- $\ldots$

and the corresponding segment duals $\hat{p}_{0j}$ of these p_{0j}, $j \neq 1$ are:

- $\hat{p}_{02} = \langle 01\ 1010\ 0101\ 1010 \ldots 0101\ 1010\ 01 \rangle$.
- $\hat{p}_{03} = \langle 0101\ 10101010\ 01010101\ 10101010 \ldots 01010101\ 10101010\ 0101 \rangle$.
- $\hat{p}_{04} = \langle 01010101\ 1010101010101010 \ldots 0101010101010101\ 1010101010101010\ 01010101 \rangle$.
- $\ldots$

Example 1(b): For the bit tuple $\hat{a}_1 = \langle 0\ 11\ 00\ 11\ 00\ \ldots\ 00\ 11\ 00\ 11\ 0 \rangle$, then its complement duals can be:

- $c_{12} = \langle 0110\ 1001\ 0110\ 1001 \ldots\ 0110\ 1001\ 0110\ 1001 \rangle$
- $c_{13} = \langle 01100110\ 10011001\ 01100110\ 10011001 \ldots 01100110\ 10011001 \rangle$
- $c_{14} = \langle 0110011001100110\ 1001100110011001\ 0110011001100110\ 1001100110011001 \ldots 0110011001100110\ 1001100110011001 \rangle$
- $\ldots$

Note. An a_i and any of its palindromic duals p_{ij} differ in half of their bits. Similarly, an $\hat{a}_i$ and any of its complement duals c_{ij} also differ in half of their bits. Since a_i and $\hat{a}_i$ are e (even), therefore their respective duals p_{ij} and c_{ij} are also e.

Definition 3 (Segment Length): The length ℓ_s of a segment within a multiset or bit tuple is defined as the number of times that a unique element value is repeated contiguously within the defined ordering of the multiset/tuple. Note that for a multiset to have property P, its ℓ_s needs to be 1.

Definition 4 (Pattern Cycle): The cycle of a multiset or bit tuple is defined as the point when a pattern of elements repeats. The number of elements traversed before this occurs is known as the cycle length ℓ_c. Note that for a multiset $\mathcal{M}$ to have property P, its ℓ_c needs to equal $|\mathcal{M}|$.

Note. A w-bit multiset $\mathcal{M}$ is said to have property P if its segment length $\ell_s = 1$ and its cycle length $\ell_c = |\mathcal{M}|$.   □

Example 2: Consider a 3-bit multiset, thus comprising 8 elements:

$$\mathcal{M} = \langle 000, 001, 010, 011, 100, 101, 110, 111 \rangle$$

The above multiset is said to have the permutation property P as all elements are distinct, each of multiplicity 1. $\mathcal{M}$ has segment length $\ell_s = 1$ and cycle length $\ell_s = |\mathcal{M}| = 8$. Observe that the multiset also has property B and F.

Denote $\mathcal{M} = \beta_2 \beta_1 \beta_0$. The bitslice channel β_0 of $\mathcal{M}$ corresponding to the least significant bit (LSB) position of any element of $\mathcal{M}$ is an a_0 bit tuple, while β_1 and β_2 are respectively of the form a_1 and a_2. The segment and cycle lengths of these bitslice channels are:

- $\ell_s(a_0) = \ell_s(\langle 01010101 \rangle) = 1, \ \ell_c(a_0) = \ell_c(\langle 01\ 01\ 01\ 01 \rangle) = 2,$
- $\ell_s(a_1) = \ell_s(\langle 00\ 11\ 00\ 11 \rangle) = 2, \ \ell_c(a_1) = \ell_c(\langle 0011\ 0011 \rangle) = 4,$
- $\ell_s(a_2) = \ell_s(\langle 0000\ 1111 \rangle) = 4, \ \ell_c(a_2) = \ell_c(\langle 00001111 \rangle) = 8.$ □

Multiset Equivalence. Note that our definition of the bit tuple is cyclic. For instance, if the elements in the multiset of Example 2 were permuted to be:

$$\mathcal{M}' = \langle 001, 010, 011, 100, 101, 110, 111, 000 \rangle$$

the bit tuples are still considered to be of the form $a_2 a_1 a_0$. We say that two multisets are equivalent if they have the same property, up to cyclic shift as in the previous example. For two multisets that are not equivalent in this sense, there is a need to distinguish between them. For this purpose, we can define more precisely the P property as it could be mapped into two different bit tuples that are not equivalent.

Definition 5: (Refined P Property): We refine the P property as follows:

- P_{ord}: the property P_{ord} corresponds to a multiset that could be written as the composition of bit tuples of the form $\ldots a_2 a_1 a_0$.
- P_{nord}: the property P_{nord} corresponds to a multiset that could not be written as the composition of bit tuples of the form $\ldots a_2 a_1 a_0$.

When required, we precise if P is of the form P_{ord} or P_{nord}.

Slicing a Multiset. To our knowledge, Nakahara et al. [13] at Mycrypt 2005 were the first to consider viewing a w-bit multiset in terms of the properties of its constituent v-bit ($v < w$) channels; i.e. *slicing* a multiset. This may allow to better trace how multiset properties propagate through cipher components.

For instance, removing any bitslice channel from the 3-bit P multiset $\mathcal{M}$ of Example 2 results in a 2-bit multiset that is no longer P but rather is E. Thus, we could directly state that any v-bit ($v < w$) subset $\mathcal{V}$ of a w-bit P multiset $\mathcal{M}$ cannot be P but must be E.

2.2 Fusing the Slices of a Multiset

Decomposing a w-bit multiset $\mathcal{M}$ into its constituent bitslice channels or some v-bit ($v < w$) subset $\mathcal{V}$ could allow to better track the changes in its properties through cipher round operations. After going through round operations, it is helpful to evaluate if the multiset retains its original properties e.g. P is preserved. For this, we need to recompose the subset $\mathcal{V}$, i.e. *fuse* some concatenation of bitslice channels back into the w-bit multiset $\mathcal{M}$ and to infer if a multiset property at the w-bit level is preserved or if a new multiset property is obtained.

The most structured property that has the ability to go through round operations is the property P, thus it is interesting to study its constituent bitslice channel tuples.

From Example 2, we see that bit tuples of the form $a_{i_1} a_{i_2} \ldots a_{i_w}$, for any pairwise indices $i_j \neq i_k$ ($j, k \in \{1, \ldots, w\}$), compose to a word multiset of property P. We are interested in other constituent bit tuples that compose to a P.

Example 3: To motivate the idea, consider the multiset $\mathcal{M}$ as below:

$$\mathcal{M} = \langle 011, 111, 101, 010, 001, 100, 000, 110 \rangle$$

At first glance, such a multiset has no obvious bit tuple form. To facilitate extracting its bit tuples, some prior reordering is helpful before bitslicing. We introduce some notation to keep track of the elementwise permutations ρ, to enable subsequent unwinding to the element ordering of the original, i.e. $\rho(\mathcal{M})$. For instance, permute $\mathcal{M}$ according to (1 3 4) and we get a multiset $\rho(\mathcal{M})$ with property P, of the form $a_0 a_2 a_2$:

$$\rho(\mathcal{M}) = \langle 010, 111, 011, 101, 001, 100, 000, 110 \rangle$$

This example highlights that a P can be formed by a_i's that may have the same index, e.g. in this case, there are two a_2s. However, from closer inspection, these two a_2s are not identical in form, in fact their segment boundaries are out of sync.

We define a pair of a_is whose segment boundaries are out of sync by half of the segment length, as *segment duals* of each other, denoted as $\langle a_i, \hat{a}_i \rangle$. Note that such pairwise duals are equal in half of their elements. In fact, when concatenated, these two bitslice channels form 2-bit elements that occur an even number of times.

With this in place, the above permuted multiset $\rho(\mathcal{M})$ is actually of the form $a_0 a_2 \hat{a}_2$.

It is worth to investigate what property is obtained when fusing (recomposing) two dual bit tuples to form a 2-bit multiset $\mathcal{V}$. Continuing from the above example, $\mathcal{V} = a_2 \hat{a}_2$ gives:

$$\langle 10, 11, 11, 01, 01, 00, 00, 10 \rangle$$

which is a 2-bit multiset with elements of segment length $\ell_s = 2$, and cycle length $\ell_c = 8$.

Since its cycle length is already 8, what remains is to cause its segment length to reduce to 1, in order for this multiset $\mathcal{V}$ to be used to form a multiset $\mathcal{M}$ of property P. To do this, we append another bitslice channel such that the resultant segment length ℓ_s reduces to 1.

Consider appending a bit tuple a_0 (note that this has segment length $\ell_s = 1$) to $\mathcal{V}$, thus we can get either $a_0a_2\hat{a}_2$, $a_2a_0\hat{a}_2$ or $a_2\hat{a}_2a_0$, as follows:

$$
\begin{array}{ccc}
010 & 100 & 100 \\
111 & 111 & 111 \\
011 & 101 & 110 \\
101 & 011 & 011 \\
001 & 001 & 010 \\
100 & 010 & 001 \\
000 & 000 & 000 \\
110 & 110 & 101,
\end{array}
$$

all of which are multisets of property P. Notice that appending such a bit tuple a_0 has caused the resultant segment length ℓ_s to become 1.

Consider instead, to append the bit tuple $\hat{a}_1$ (note that this has segment length $\ell_s = 2$) to $\mathcal{V}$, but such that its segment boundary is out of sync with the segment boundary of $\mathcal{V}$. Thus we get either $\hat{a}_1a_2\hat{a}_2$, $a_2\hat{a}_1\hat{a}_2$ or $a_2\hat{a}_2\hat{a}_1$:

$$
\begin{array}{ccc}
110 & 110 & 101 \\
111 & 111 & 111 \\
011 & 101 & 110 \\
001 & 001 & 010 \\
101 & 011 & 011 \\
100 & 010 & 001 \\
000 & 000 & 000 \\
010 & 100 & 100,
\end{array}
$$

all of which are multisets of property P. Notice that appending such a bit tuple, though of segment length $\ell_s = 2$, yet due to the out of sync in segment boundaries, causes the resultant segment length ℓ_s to be halved, i.e. reduces to 1. $\qquad\qquad\Box$

Example 4: Consider another example 3-bit P multiset:

$$\mathcal{M}' = \langle 011, 111, 110, 010, 001, 101, 100, 000 \rangle$$

which has the form $\hat{a}_1a_2a_1$. Focussing on the concatenation of the two dual bit tuples $\hat{a}_1a_1$ gives:

$$\langle 01, 11, 10, 00, 01, 11, 10, 00 \rangle$$

which is a 2-bit multiset with elements of segment length $\ell_s = 1$, and cycle length $\ell_c = 4$. As the segment length ℓ_s is already 1, what remains is to cause the cycle length ℓ_c to increase to $|\mathcal{M}'| = 8$, in order for the resultant multiset to have the P property. Therefore, the only appropriate bit tuple to append to this would be of the form a_2 which has cycle length $\ell_c = 8$. This is why the example $\mathcal{M}'$, which is of the form $\hat{a}_1 a_2 a_1$, is a P.   □

We now have the notations, definitions and criteria for representing, slicing (decomposing) and fusing (recomposing) multisets. These are crucial in order to facilitate the tracking of multiset & bit tuple properties through cipher rounds.

2.3   Tuples Through Cipher Operations

In this section, we analyze the propagation of bitslice tuples through some main primitive operations commonly used in block ciphers and hash functions; notably the exclusive-OR (XOR), the AND operation, and substitution boxes (Sboxes).

Tuples Through XOR. The emphasis we place here is on constituent tuples obtained from slicing the multisets that have property P, since this has the richest structure that can survive through round operations better than multisets of other properties. Therefore it is vital that we understand what happens to P inputs after going through XOR; the most complex being P $\oplus$ P.

As a P multiset comprises bitslice tuples of the form a_j, then any P $\oplus$ P will cause the following types of XOR between its a_j (or cyclic variants $\hat{a}_j, \tilde{a}_j$) bitslice tuples:

- $a_j \oplus a_j = c$
- $a_j \oplus \hat{a}_{j;\, j \neq 0} = a_{j-1}$
- $a_j \oplus a_{j-1} = \hat{a}_j$
- $a_i \oplus a_{j;\, j > i+1} = p_{ij}$
- $a_i \oplus \hat{a}_{j;\, j > i+1} = \hat{p}_{ij}$
- $\hat{a}_i \oplus a_{j;\, j > i+1} = c_{ij}$
- $a_j \oplus \hat{a}_{j-1} = \hat{p}_{(j-2)\, j}$

- $a_j \oplus \tilde{a}_{j;\, j \neq 0,1} = \ell_j$
- $a_i \oplus \tilde{a}_{j;\, j > i} = f_j$
- $\tilde{a}_i \oplus a_{j;\, j > i} = f_j$
- $\tilde{a}_i \oplus \tilde{a}_{j;\, j > i} = f_j$
- $\hat{a}_i \oplus \tilde{a}_{j;\, j > i} = f_j$
- $\tilde{a}_i \oplus \hat{a}_{j;\, j > i} = f_j$

Note that these properties are more refined than the ones observed in [17], and better allow to retain the internal rich structures within a P.

Furthermore, the following XOR relations can also be observed between other types of bitslice tuples:

- $a_j \oplus p_{(j-2)(j-1)} = \hat{p}_{(j-2)\, j}$

- $a_j \oplus p_{ik;\ i,k<j}$: this leads to the complement dual of p_{ik}, i.e. p_{ik} but with alternating flipped pattern segments of length 2^j

Note. All the above bit tuples have the bit tuple property e.

Tuples Through AND. As the AND operation serves as one of the primitives in recent ciphers such as PRINCE, SIMON, and SIMECK, as well as when Sbox output bits are expressed in ANF representation, we consider bit tuple propagations through this AND operation.

In more detail, we analyze the bit tuple interactions involving one or two a_j:

- $c \wedge a_j = a_j$ or $0\ldots0$ (i.e. the all '0' sequence)
- $a_j \wedge a_j = a_j$
- $a_j \wedge a_{j-1} = z_j$
- $a_i \wedge a_{j;j>i+1} = m_{ij}$
  Intuitively, this leads to a bit tuple that equals a_i except that every other contiguous sequence of length 2^j corresponding to '0' bits in a_j contains bits masked to '0'.

Note. All the above output bit tuples have even parity. Furthermore, their segment and/or pattern cycles are based on a_i, therefore their pattern boundaries are aligned with the segment boundaries (if any) of the a_i's.

Tuples Through Sboxes. Any output bit of an Sbox can be viewed as the output of a coordinate Boolean function constituting the Sbox, and can be expressed in ANF representation, thereby relating each output bit as a function of input bits.

While the ANF representation of Sboxes has been exploited in the literature to track the development of the algebraic degrees of Sbox outputs, in this paper we propose the novel approach of using the ANF to enable the tracking of how bit tuples propagate through an Sbox. Essentially, from the ANF, it can be observed that any Sbox output bit is the result of ANDing and then XORing the input bits. Therefore, the problem of tracking how bit tuples behave as they pass through an Sbox can be reduced to the problem of tracking how bit tuples propagate through AND and XOR operations.

Details of this approach of tracking bit tuples through Sboxes appear in the later subsections Sects. 3.3 and 3.4 on the PRESENT and RECTANGLE ciphers.

3   Multiset Properties Through Ciphers

3.1   Threefish and ARX

To concisely exemplify our approaches, notably in terms of the slicing of multisets and analysing their constituent bitslice tuples, we will first consider here the MIX operation within the round function of the Threefish block cipher [21], before moving on the other ciphers. MIX is defined as follows:

$$\mathsf{MIX}(x,y) = \langle x+y, (x+y) \oplus (y \ggg r)\rangle.$$

Note that this MIX function is essentially of the ARX form, i.e. addition (ADD), rotation (ROT), and exclusive-OR (XOR).

The multisets' behaviours as they propagate through the Addition, XOR and Rotation operations are shown in Table 1 as reported in [22]. Our aim here is to derive more internal structures and corresponding properties within these output multisets, based on our formulations of multisets and their bitslice channels.

Table 1. Truth tables of Addition (ADD), XOR, and Rotation (ROT) on multisets as reported in [22].

+	A	B	C	E	F	P
A	A	X	A	X	F	F
B	X	X	X	X	X	X
C	A	X	C	E	F	P
E	X	X	E	X	X	X
F	F	X	F	X	A	A
P	F	X	P	X	A	A

⊕	A	B	C	E	F	P
A	X	X	X	X	X	X
B	X	B	B	B	X	B
C	X	B	C	E	X	P
E	X	B	B	B	X	B
F	X	X	X	X	X	X
P	X	B	P	B	X	B

≫	A	B	C	E	F	P
n	X	B	C	E	X	P

Multisets Through Rotation (ROT). We commence by looking at the simplest operation in terms of influence on multiset properties, i.e. the rotation (ROT), also denoted by $\ggg$.

To see why $(\ggg P)$ still gives P, consider its bitslice channels, e.g. for a $P = a_2 a_1 a_0$, then we have: $(\ggg a_2 a_1 a_0) \to a_0 a_2 a_1 = P$.

Multisets Through Addition (ADD). In [22] it was shown that $P + P \to A$ based on detecting a zero sum with respect to modulo addition. Alternatively, using bitslice tuples enables to show why we get the property: $P + P = a_2 a_1 a_0 + a_2 a_1 a_0 \to a_1 a_0 c = E$.

More importantly, crucial to the analysis of multisets through MIX is the behaviour of $C + P$. W.l.o.g. consider a multiset P_1 comprising the bit tuples $a_2 a_1 a_0$, e.g. whose elements are ordered ascendingly. It is known that when added to a C, this results in a P_2 where the same ordering is preserved. However, if we focus on the constituent bitslice tuples, we see that these tuples may no longer have segment boundaries aligned with those of a_2, a_1, a_0, i.e. its bitslice tuples would be from the set $\in \{a_j, \hat{a}_{j;j \neq 0}, \tilde{a}_{j;j \neq 0,1}\}$.

Multisets Through Exclusive-OR (XOR). We consider the XOR of different types of P, or with A, as typically encountered when propagating multisets through ARX constructions, e.g. the MIX function of Threefish; in terms of the constituent bitslice channels of such multiset types. Without loss of generality, we describe the analysis with respect to 3-bit multisets for better clarity.

- $P \oplus P$: $a_2 a_1 a_0 \oplus a_2 a_1 a_0 \to ccc = C$.

  This case considers the XOR of P multisets comprising the same bitslice tuples. The next case considers the XOR of P multisets comprising different orderings of bitslice tuples caused by rotation of P.

- $P \oplus (\ggg P)$: $a_2 a_1 a_0 \oplus a_0 a_2 a_1 = \langle a_2 \oplus a_0, a_1 \oplus a_2, a_0 \oplus a_1 \rangle \rightarrow p_{02} \hat{a}_2 \hat{a}_1 = E$. It could be generalized to E^k for larger P sets.

- $P_1 \oplus P_2$: here we utilize the tuples formulation of this paper to show why in some cases, where different types of P structures can be precise in their bitslice tuples, that the XOR of two P can still produce a P. It remains an open problem whether all types of P can be similarly defined, including P_{nord} which is not captured in our tuples formulation. In our earlier work i.e. [22] Sect. 2.2, it was stated with an example that this was possible for non-trivial cases of P, though no detailed analysis could be provided as to why this behaviour exists because we lacked the formulations to precisely detail different types of P. It was also mentioned in Sect. 3.2 that properties observed from empirical results were stronger than analytical predictions because tracking analytically was difficult. By using the tuples formulation, one can answer the question as to why a P could be empirically detected after $P_1 \oplus P_2$ although state-of-the-art integral and multiset analysis techniques were to date not able to explain why it is a P. To show that we can get a P, consider $P_{ord,1} = a_2 a_1 a_0$ and $P_2 = a_0 \hat{a}_2 a_1$, then $P_{ord,1} \oplus P_2 = a_2 a_1 a_0 \oplus a_0 \hat{a}_2 a_1 = \langle a_2 \oplus a_0, a_1 \oplus \hat{a}_2, a_0 \oplus a_1 \rangle = p_{02} a_2 \hat{a}_1 = P_3$. This results in a permutation P multiset because the segment length ℓ_s of the multiset is 1 due to the tuple p_{02} while the cycle pattern length $\ell_c = 8$ due to the tuples a_2 and p_{02}.

- $A \oplus P$: $a_1 a_0 c \oplus a_2 a_1 a_0 = \langle a_1 \oplus a_2, a_0 \oplus a_1, c \oplus a_0 \rangle \rightarrow \hat{a}_2 \hat{a}_1 a_0 = P$, for the specific type of A that can be formulated as $a_1 a_0 c$ e.g. when it is produced from $P + P$. Note that such an A is also an E. To see this why this XOR results in a P, we can recall our discussions on fusing in Subsect. 2.2, notably the segment length ℓ_s of this multiset is 1, while its cycle length ℓ_c is 8 equaling the size of this multiset.

Table 2 summarizes the revised truth table for XOR based on our analysis in this subsection.

Multisets Through MIX. With the enriched multiset properties based on bitslice tuples as discussed above, we can more precisely trace the following multiset

Table 2. Revised truth table of XOR on multisets based on tuple formulations.

$\oplus$	A	B	C	E	F	P
A	X	X	X	X	X	P/X
B	X	B	B	B	X	B
C	X	B	C	E	X	P
E	X	B	B	B	X	P/B·
F	X	X	X	X	X	X
P	P/X	B	P	P/B	X	P/E/C

propagations through the MIX function of Threefish, w.l.o.g. consider 3-bit multitisets for simplicity of description:

$$\mathsf{MIX}(\mathsf{C},\mathsf{P}) = \langle \mathsf{C}+\mathsf{P}, (\mathsf{C}+\mathsf{P}) \oplus (\ggg \mathsf{P}) \rangle = \langle \mathsf{P}, \beta_2\beta_1\beta_0 \rangle$$

Consider a P multiset of the form $a_2a_1a_0$, thus $(\ggg \mathsf{P}) = a_0a_2a_1$. From our above discussion in the subsection for the case of multisets through ADD, we see that $\mathsf{C}+\mathsf{P} \to \mathsf{P} = \alpha_2\alpha_1a_0$ where $\alpha_2 \in \{a_2, \hat{a}_2, \tilde{a}_2\}$, $\alpha_1 \in \{a_1, \hat{a}_1\}$. Therefore, we can see that the output bit tuples $\beta_2\beta_1\beta_0$ would form the multiset $(\mathsf{C}+\mathsf{P}) \oplus (\ggg \mathsf{P})$ as follows:

- $\beta_2 = \alpha_2 \oplus a_0 =$
  - $\cdot\ a_2 \oplus a_0 = p_{02}$, or
  - $\cdot\ \hat{a}_2 \oplus a_0 = \hat{p}_{02}$, or
  - $\cdot\ \tilde{a}_2 \oplus a_0 = f_2$.
- $\beta_1 = \alpha_1 \oplus a_2 =$
  - $\cdot\ a_1 \oplus a_2 = \hat{a}_2$, or
  - $\cdot\ \hat{a}_1 \oplus a_2 = \hat{p}_{02}$.
- $\beta_0 = a_0 \oplus a_1 = \hat{a}_1$.

These results have been corroborated by experiments on up to 8-bit multisets. Examples of an output multiset $(\mathsf{C}+\mathsf{P}) \oplus (\ggg \mathsf{P}) = \beta_2\beta_1\beta_0$ with such properties are as follows:

- $\beta_2 = p_{02} = 0101\ 1010$ or $\hat{p}_{02} = 01\ 1010\ 01$ or $f_2 = 1011\ 0100$
- $\beta_1 = \hat{a}_2 = 00\ 1111\ 00$ or $\hat{p}_{02} = 01\ 1010\ 01$
- $\beta_0 = \hat{a}_1 = 01100110$

Thus, we can deduce more structure through MIX at its output, answering the question left partially open in previous work i.e. [22] where it was stated that dependencies between tuples of words were difficult to track analytically, after having remarked that there was a difference in precision between what could be empirically observed and what could be analytically predicted.

In addition, for the cases where the input multiset going into MIX is $\langle \mathsf{P}, \mathsf{P} \rangle$, e.g. when via chosen-ciphertext attacks such structures are input to MIX, we can obtain:

$$\mathsf{MIX}(\mathsf{P},\mathsf{P}) = \langle \mathsf{P}+\mathsf{P}, (\mathsf{P}+\mathsf{P}) \oplus (\ggg \mathsf{P}) \rangle = \langle \mathsf{E}, \mathsf{P}/\mathsf{E} \rangle$$

rather than the $\langle \mathsf{A}, \mathsf{B} \rangle$ previously observed. To see this, first note that for the left half of the output multiset, as per the subsection discussion in the case of multisets through Addition (ADD), we have $(\mathsf{P}+\mathsf{P}) \to \mathsf{E} = a_1a_0c$. For the right half $(\mathsf{P}+\mathsf{P}) \oplus (\ggg \mathsf{P})$ of the multiset, recall that $(\ggg \mathsf{P}) = a_0a_2a_1$. Thus we have the right output multiset as:

$$\langle a_1 \oplus a_0, a_0 \oplus a_2, c \oplus a_1 \rangle = \langle \hat{a}_1, p_{02}, a_1 \rangle = \mathsf{P}.$$

An example of when the right half of the output multiset becomes E is where $\mathsf{P} = a_3a_2a_1a_0$, such that $(\mathsf{P}+\mathsf{P}) \oplus (\ggg \mathsf{P}) = (a_2a_1a_0c) \oplus (a_0a_3a_2a_1) = \langle a_2 \oplus$

$a_0, a_1 \oplus a_3, a_0 \oplus a_2, c \oplus a_1\rangle = \langle p_{02}p_{13}p_{02}\hat{a}_1\rangle = \mathsf{E}$, since the same bit tuple p_{02} is present twice in the multiset, with the same boundary alignment. Alternatively, one could obtain a P multiset for other parameters, e.g. for different rotation amounts, for instance $(\mathsf{P} + \mathsf{P}) \oplus (\ggg_2 \mathsf{P}) = (a_2 a_1 a_0 c) \oplus (a_1 a_0 a_3 a_2) = \langle a_2 \oplus a_1, a_1 \oplus a_0, a_0 \oplus a_3, c \oplus a_2\rangle = \langle \hat{a}_2 \hat{a}_1 p_{03} a_2\rangle = \mathsf{P}$, where $\ggg_r$ denotes rotation to the right by r amounts.

These examples show that one could analytically deduce better properties for $\mathsf{MIX}(\mathsf{P},\mathsf{P})$, filling the gap between what could be observed via analysis and what was actually observed via experiments.

3.2 PRINCE

PRINCE is an SPN block cipher with a particular involutive structure for low latency [25]. Its 64-bit block can be represented as a 4×4 state of nibbles, which goes through initial keying, 5 forward SPN rounds, an unkeyed middle layer, and then 5 more backward SPN rounds before final keying.

A forward SPN round mainly consists of an Sbox layer S acting on nibbles and a diffusion layer $M = SR \circ M'$, before a typical keyed XOR operation; where SR is the AES ShiftRows operation and M' is an involutive operation acting on independent columns at a time. [44] observed that M' can be expressed in terms of bitwise equations as follows. For the leftmost and rightmost columns, the output nibbles (one in each row below, and each row comprising its constituent four bits) are of the form:

$$y_0^0 = x_1^0 \oplus x_2^0 \oplus x_3^0 \quad y_0^1 = x_0^1 \oplus x_2^1 \oplus x_3^1 \quad y_0^2 = x_0^2 \oplus x_1^2 \oplus x_3^2 \quad y_0^3 = x_0^3 \oplus x_1^3 \oplus x_2^3$$
$$y_1^0 = x_0^0 \oplus x_1^0 \oplus x_2^0 \quad y_1^1 = x_1^1 \oplus x_2^1 \oplus x_3^1 \quad y_1^2 = x_0^2 \oplus x_2^2 \oplus x_3^2 \quad y_1^3 = x_0^3 \oplus x_1^3 \oplus x_3^3$$
$$y_2^0 = x_0^0 \oplus x_1^0 \oplus x_3^0 \quad y_2^1 = x_0^1 \oplus x_1^1 \oplus x_2^1 \quad y_2^2 = x_1^2 \oplus x_2^2 \oplus x_3^2 \quad y_2^3 = x_0^3 \oplus x_2^3 \oplus x_3^3$$
$$y_3^0 = x_0^0 \oplus x_2^0 \oplus x_3^0 \quad y_3^1 = x_0^1 \oplus x_1^1 \oplus x_3^1 \quad y_3^2 = x_0^2 \oplus x_1^2 \oplus x_2^2 \quad y_3^3 = x_1^3 \oplus x_2^3 \oplus x_3^3$$

On the other hand, the output nibbles of the inner two columns are:

$$y_0^0 = x_0^0 \oplus x_1^0 \oplus x_2^0 \quad y_0^1 = x_1^1 \oplus x_2^1 \oplus x_3^1 \quad y_0^2 = x_0^2 \oplus x_2^2 \oplus x_3^2 \quad y_0^3 = x_0^3 \oplus x_1^3 \oplus x_3^3$$
$$y_1^0 = x_0^0 \oplus x_1^0 \oplus x_3^0 \quad y_1^1 = x_0^1 \oplus x_1^1 \oplus x_2^1 \quad y_1^2 = x_1^2 \oplus x_2^2 \oplus x_3^2 \quad y_1^3 = x_0^3 \oplus x_2^3 \oplus x_3^3$$
$$y_2^0 = x_0^0 \oplus x_2^0 \oplus x_3^0 \quad y_2^1 = x_0^1 \oplus x_1^1 \oplus x_3^1 \quad y_2^3 = x_0^2 \oplus x_1^2 \oplus x_2^2 \quad y_2^3 = x_1^3 \oplus x_2^3 \oplus x_3^3$$
$$y_3^0 = x_1^0 \oplus x_2^0 \oplus x_3^0 \quad y_3^1 = x_0^1 \oplus x_2^1 \oplus x_3^1 \quad y_3^2 = x_0^2 \oplus x_1^2 \oplus x_3^2 \quad y_3^3 = x_0^3 \oplus x_1^3 \oplus x_2^3$$

The middle layer is of the form: $S^{-1} \circ M' \circ S$, while a backward SPN round comprises the functions M^{-1} and S^{-1}.

Since PRINCE rounds essentially comprise the Sbox layer and the diffusion layer M, the entire cipher consists of only AND and exclusive-OR (XOR) operations, thus it is an AND-XOR structured cipher.

Some previous results on integral and bitslice tuple tracking through PRINCE have been proposed in [28,44]. More precisely, in [28,44] a classical integral property was presented as follows: with one active nibble (i.e. 2^4 chosen plaintexts) is transformed into a balanced property on each nibble after rounds of the form $SM, SM, SR^{-1}M'SR, M^{-1}S^{-1} \simeq SM, SM, SR^{-1}, S^{-1}$, which is seen effectively as 2.5 rounds since SR^{-1} offers little diffusion. This first-order integral property could be extended by one additional round SM at the beginning using a 4th-order integral property requiring 2^{16} chosen plaintexts. Another 1st-order integral was presented in [44] needing 2^4 texts, covering the rounds $SM, SM', S^{-1}M^{-1}$, which can be considered effectively as 2.5 rounds since M' does not provide full diffusion compared to M. Moreover, a bitslice tuple property with three particular active bits (i.e. requiring 2^3 chosen plaintexts) was also proposed for the rounds of the form SM, SM' in [44], which is effectively at most seen to be 2 rounds of PRINCE.

In fact, we found the following bitslice tuple property on 3 rounds of the form SM, SM and SM:

$$
\begin{array}{cccc}
cccc & cccc & cccc & ccca_2 \\
cccc & cccc & cccc & ccca_1 \\
cccc & cccc & cccc & cccc \\
cccc & cccc & cccc & ccca_0
\end{array}
$$

gives after three rounds of SM, SM and SM:

$$
\begin{array}{cccc}
eeee & eeee & eeee & eeee \\
eeee & eeee & eeee & eeee \\
eeee & eeee & eeee & eeee \\
eeee & eeee & eeee & eeee
\end{array}
$$

This property stays true for triplets of bits placed on the 3 least significant bits of the Sbox output.

In more detail, this property starts with an input multiset as follows:

$$
\begin{array}{cccc}
c\,c\,c\,c & c\,c\,c\,c & c\,c\,c\,c & c\,c\,c\,a_2 \\
c\,c\,c\,c & c\,c\,c\,c & c\,c\,c\,c & c\,c\,c\,a_1 \\
c\,c\,c\,c & c\,c\,c\,c & c\,c\,c\,c & c\,c\,c\,c \\
c\,c\,c\,c & c\,c\,c\,c & c\,c\,c\,c & c\,c\,c\,a_0
\end{array}
$$

- S in R1: $\rightarrow$

$$
\begin{array}{cccc}
c\,c\,c\,c & c\,c\,c\,c & c\,c\,c\,c & \alpha_2\,\alpha_2\,\alpha_2\,\alpha_2 \\
c\,c\,c\,c & c\,c\,c\,c & c\,c\,c\,c & \alpha_1\,\alpha_1\,\alpha_1\,\alpha_1 \\
c\,c\,c\,c & c\,c\,c\,c & c\,c\,c\,c & c\,\,c\,\,c\,\,c \\
c\,c\,c\,c & c\,c\,c\,c & c\,c\,c\,c & \alpha_0\,\alpha_0\,\alpha_0\,\alpha_0
\end{array}
$$

After the S in round 1 (R1), all output bits of the nibble for which one of the inputs received an a_j tuple become the α_j tuple, where $\alpha_j \in \{a_j, c\}$.

- M' in R1: $\rightarrow$

$$
\begin{array}{cccccccccccc}
c\,c\,c\,c & c\,c\,c\,c & c\,c\,c\,c & \alpha_{01} & \alpha_{02} & \alpha_{012} & \alpha_{12} \\
c\,c\,c\,c & c\,c\,c\,c & c\,c\,c\,c & \alpha_{12} & \alpha_{01} & \alpha_{02} & \alpha_{012} \\
c\,c\,c\,c & c\,c\,c\,c & c\,c\,c\,c & \alpha_{012} & \alpha_{12} & \alpha_{01} & \alpha_{02} \\
c\,c\,c\,c & c\,c\,c\,c & c\,c\,c\,c & \alpha_{02} & \alpha_{012} & \alpha_{12} & \alpha_{01}
\end{array}
$$

where α_{ijk} is shorthand for $\alpha_i \oplus \alpha_j \oplus \alpha_k$.

- SR in R1: $\rightarrow$

$$
\begin{array}{cccccccccccccccc}
c & c & c & c & c & c & c & c & c & c & c & c & \alpha_{01} & \alpha_{02} & \alpha_{012} & \alpha_{12} \\
c & c & c & c & c & c & c & c & \alpha_{12} & \alpha_{01} & \alpha_{02} & \alpha_{012} & c & c & c & c \\
c & c & c & c & \alpha_{012} & \alpha_{12} & \alpha_{01} & \alpha_{02} & c & c & c & c & c & c & c & c \\
\alpha_{02} & \alpha_{012} & \alpha_{12} & \alpha_{01} & c & c & c & c & c & c & c & c & c & c & c & c
\end{array}
$$

- S in R2: $\rightarrow$

$$
\begin{array}{cccccccccccc}
c\,c & c & c & c\,c & c & c & c\,c & c & c & & v_{42} & \\
c\,c & c & c & c\,c & c & c & & v_{42} & & c\,c & c & c \\
c\,c & c & c & & v_{42} & & c\,c & c & c & c\,c & c & c \\
& v_{42} & & c\,c & c & c & c\,c & c & c & c\,c & c & c
\end{array}
$$

where v_{ij} denotes a nibble of i bits taking only j values.

- M' in R2: $\rightarrow$

$$
\begin{array}{cccccccccccccccc}
c & \alpha_0 & \alpha_0 & \alpha_0 & \alpha_1 & c & \alpha_1 & \alpha_1 & \alpha_2 & c & \alpha_2 & \alpha_2 & c & \alpha_3 & \alpha_3 & \alpha_3 \\
\alpha_0 & c & \alpha_0 & \alpha_0 & \alpha_1 & \alpha_1 & c & \alpha_1 & \alpha_2 & \alpha_2 & c & \alpha_2 & \alpha_3 & c & \alpha_3 & \alpha_3 \\
\alpha_0 & \alpha_0 & c & \alpha_0 & \alpha_1 & \alpha_1 & \alpha_1 & c & \alpha_2 & \alpha_2 & \alpha_2 & c & \alpha_3 & \alpha_3 & c & \alpha_3 \\
\alpha_0 & \alpha_0 & \alpha_0 & c & c & \alpha_1 & \alpha_1 & \alpha_1 & c & \alpha_2 & \alpha_2 & \alpha_2 & \alpha_3 & \alpha_3 & \alpha_3 & c
\end{array}
$$

- SR in R2: $\rightarrow$

$$
\begin{array}{cccccccccccccccc}
c & \alpha_0 & \alpha_0 & \alpha_0 & \alpha_1 & c & \alpha_1 & \alpha_1 & \alpha_2 & c & \alpha_2 & \alpha_2 & c & \alpha_3 & \alpha_3 & \alpha_3 \\
\alpha_1 & \alpha_1 & c & \alpha_1 & \alpha_2 & \alpha_2 & c & \alpha_2 & \alpha_3 & c & \alpha_3 & \alpha_3 & \alpha_0 & c & \alpha_0 & \alpha_0 \\
\alpha_2 & \alpha_2 & \alpha_2 & c & \alpha_3 & \alpha_3 & c & \alpha_3 & \alpha_0 & \alpha_0 & c & \alpha_0 & \alpha_1 & \alpha_1 & \alpha_1 & c \\
c & \alpha_1 & \alpha_1 & \alpha_1 & c & \alpha_2 & \alpha_2 & \alpha_2 & \alpha_3 & \alpha_3 & \alpha_3 & c & \alpha_0 & \alpha_0 & \alpha_0 & c
\end{array}
$$

- S in R3: $\rightarrow$

$$
\begin{array}{cccccccccccccccc}
e & e & e & e & e & e & e & e & e & e & e & e & e & e & e & e \\
e & e & e & e & e & e & e & e & e & e & e & e & e & e & e & e \\
e & e & e & e & e & e & e & e & e & e & e & e & e & e & e & e \\
e & e & e & e & e & e & e & e & e & e & e & e & e & e & e & e
\end{array}
$$

To see this, note from our previous discussion of tuple propagations through the XOR operation, that the above bit tuples input to the Sbox S are of the form:

$\circ$ $\alpha_{01} = \alpha_0 \oplus \alpha_1 \in \{c \oplus c, a_0 \oplus c, c \oplus a_1, a_0 \oplus a_1\} = \{c, a_0, a_1, \hat{a}_1\}$

$\circ \ \alpha_{12} = \alpha_1 \oplus \alpha_2 \in \{c \oplus c, a_1 \oplus c, c \oplus a_2, a_1 \oplus a_2\} = \{c, a_1, a_2, \hat{a}_2\}$
$\circ \ \alpha_{02} = \alpha_0 \oplus \alpha_2 \in \{c \oplus c, a_0 \oplus c, c \oplus a_2, a_0 \oplus a_2\} = \{c, a_0, a_2, p_{02}\}$
$\circ \ \alpha_{012} = \alpha_0 \oplus \alpha_1 \oplus \alpha_2 \in \{c, a_0, a_1, a_2, \hat{a}_1, \hat{a}_2, p_{02}, \hat{p}_{02}\}.$

All these possible bit tuples are even (e) tuples. Therefore, as S is bijective, the output bit tuples will also be e. These will remain as e after the subsequent M layer.

- M in R3: $\rightarrow$

$$
\begin{array}{cccc cccc cccc cccc}
e\ e\ e\ e & e\ e\ e\ e & e\ e\ e\ e & e\ e\ e\ e \\
e\ e\ e\ e & e\ e\ e\ e & e\ e\ e\ e & e\ e\ e\ e \\
e\ e\ e\ e & e\ e\ e\ e & e\ e\ e\ e & e\ e\ e\ e \\
e\ e\ e\ e & e\ e\ e\ e & e\ e\ e\ e & e\ e\ e\ e
\end{array}
$$

Therefore, after 3 rounds of PRINCE, we have that all the output bits give a zero integral $\oplus$ sum.

Note that there exist many such properties positioned at different places of the block with the three bits aligned in a column and for different combinations of three rounds. Note that this property leads to be able to build a distinguisher with only 2^3 plaintexts on 3 rounds SM, SM and SM.

Moreover, and due to the structure of the PRINCE matrix, we could transform the 4-th order integral property on 4 rounds described in [28] requiring 2^{16} chosen plaintexts into a 3-th order integral property on 4 rounds requiring only 2^{12} chosen plaintexts as noticed in [45]. Indeed:

$$
\begin{array}{cccc}
C & C & C & C \\
C & C & C & C \\
C & C & C & C \\
C & C & C & A^4
\end{array}
$$

is computed, inverting one round, from:

$$
\begin{array}{cccc}
C & C & C & A^{12} \\
C & C & C & A^{12} \\
C & C & C & C \\
C & C & C & A^{12}
\end{array}
$$

Thus, this Integral property on 4 rounds could be extended by one more round at the beginning leading to a 5 rounds distinguisher with 2^{48} chosen plaintexts remarking that

$$
\begin{array}{cccc}
C & C & C & A^{12} \\
C & C & C & A^{12} \\
C & C & C & C \\
C & C & C & A^{12}
\end{array}
\quad \text{is obtained from} \quad
\begin{array}{cccc}
A^{48} & C & A^{48} & A^{48} \\
A^{48} & C & A^{48} & A^{48} \\
A^{48} & C & A^{48} & A^{48} \\
A^{48} & C & A^{48} & A^{48}
\end{array}
$$

and do not required the whole codebook but only 2^{48} chosen plaintexts. Thus, we are able to construct an Integral attack on 7 rounds with a time complexity of about 2^{48} encryptions to recover half of the key improving the state-of-the-art concerning the best Integral attack on PRINCE.

3.3 PRESENT

PRESENT [15] is a popularly analysed lightweight cipher, published in 2007 and having been cited for over 1000 times by the end of 2015. PRESENT has a block size of 64 bits, and consists of 31 rounds, where each round is simply involving an XOR with a round key, a layer S comprising 16 parallel applications of a 4×4 Sbox and a bit permutation layer L such that the output bits of each Sbox spread uniformly to bit locations which are 16 bits apart. We can thus view PRESENT as being an NX structured cipher, similar to PRINCE.

For our purpose, we will exploit the ANF of the Sbox outputs of PRESENT, which is listed as follows [30]:

$$y_3 = 1 \oplus x_0 \oplus x_1 \oplus x_3 \oplus x_1 x_2 \oplus x_0 x_1 x_2 \oplus x_0 x_1 x_3 \oplus x_0 x_2 x_3 \tag{1}$$

$$y_2 = 1 \oplus x_2 \oplus x_3 \oplus x_0 x_1 \oplus x_0 x_3 \oplus x_1 x_3 \oplus x_0 x_1 x_3 \oplus x_0 x_2 x_3 \tag{2}$$

$$y_1 = x_1 \oplus x_3 \oplus x_1 x_3 \oplus x_2 x_3 \oplus x_0 x_1 x_2 \oplus x_0 x_1 x_3 \oplus x_0 x_2 x_3 \tag{3}$$

$$y_0 = x_0 \oplus x_2 \oplus x_3 \oplus x_1 x_2 \tag{4}$$

Note that from the above, the algebraic degree of the coordinate Boolean functions of the PRESENT Sbox is 2 for the LSB y_0 and 3 for the rest.

We begin with a multiset of 2^4 elements such that the 4 rightmost bitslices collectively form a 4-bit P word, which the other bitslices are c. For compactness of notation, we denote by c^s the contiguous sequence of s bitslices each of which is a bit tuple of the form c.

- Input: c^{16}, c^{16}, c^{16}, $c^{12} a_3 a_2 a_1 a_0$
- S in R1: $\rightarrow c^{16}$, c^{16}, c^{16}, $c^{12} a_3 a_2 a_1 a_0$
  The bit tuples propagate unchanged through the Sbox layer in Round 1.
- L in R1: $\rightarrow c^{15} a_3$, $c^{15} a_2$, $c^{15} a_1$, $c^{15} a_0$
  The linear layer L in Round 1 moves the a_i bit tuples ($i \in \{0, \ldots, 3\}$) to other bitslice positions, such that there is one a_i bit tuple in each 16-bit word state of PRESENT.
- S in R2: $\rightarrow c^{12} \alpha_3^3 a_3$, $c^{12} \alpha_2^3 a_2$, $c^{12} \alpha_1^3 a_1$, $c^{12} \alpha_0^3 a_0$
  where $\alpha_i \in \{c, a_i\}$
  Remark. On input $c^3 a_i$, the output from the Sbox is $\alpha_i^3 a_i$, where each α_i is either a c tuple (all constant '0's or constant '1's), or a_i including its bitwise complement. This can be seen by analyzing the ANF of the Sbox, noting that in this case only the input bit x_0 varies as per the bit tuple sequence, while the other bits $x_1 x_2 x_3$ are constants. Therefore y_0 will be x_0 or its complement $\overline{x_0}$, thus will be an a_i tuple.
- L in R2: $\rightarrow (c^3 \alpha_3 \ c^3 \alpha_2 \ c^3 \alpha_1 \ c^3 \alpha_0)^3 \ c^3 a_3 \ c^3 a_2 \ c^3 a_1 \ c^3 a_0$
  The linear layer causes each 4-bit input to the Sbox in the next layer to be of the form $c^i \alpha_i$ or $c^3 a_i$.

- S in R3: $\to (\alpha_3^4 \, \alpha_2^4 \, \alpha_1^4 \, \alpha_0^4 \,)^3 \, \alpha_3^3 a_3 \, \alpha_2^3 a_2 \, \alpha_1^3 a_1 \, \alpha_0^3 a_0$
  As per the above analysis for S in $R2$, we have discussed how an input $c^3 a_i$ propagates through the Sbox. What remains is to analyze the propagation of the input $c^3 \alpha_i$ through the Sbox. Recall that $\alpha_i \in \{c, a_i\}$, thus an input $c^3 \alpha_i$ is in fact the union of two types of inputs, i.e. $c^3 a_i \cup c^3 c$. The output will therefore be $\alpha_i^3 a_i$ or c^4.
- L in R3: $\to (\alpha_3 \alpha_2 \alpha_1 \alpha_0)^{15} \, a_3 a_2 a_1 a_0$
  The linear layer permutes the bit tuples such that every Sbox input in the next layer is of the form $\alpha_3 \alpha_2 \alpha_1 \alpha_0$ except for the rightmost Sbox whose input is $a_3 a_2 a_1 a_0$.
- S in R4: $\to (e^4)^{15} \, a_3 a_2 a_1 a_0$
  An $\alpha_3 \alpha_2 \alpha_1 \alpha_0$ input to an Sbox comprises four bitslices α_i, each of which could be a c or a_i tuple. This means that each bitslice is either (in the case that it is c) a bit multiset with one unique element of multiplicity 2^4, or (in the case that it is a_i) a bit multiset with two unique elements ('0' and '1') each of multiplicty 2^3. In either case, the bit multiset exhibits even parity, i.e. its elements have even multiplicity. Composing the four bitslices back into the 4-bit word, the 4-bit input $\alpha_3 \alpha_2 \alpha_1 \alpha_0$ to the Sbox is therefore also a multiset of property E. Since the Sbox is bijective, therefore this E property is preserved through to the output. Denote its output by e^4 with each e to represent a bit tuple of even parity.
- L in R4: $\to e^{15} a_3, \, e^{15} a_2, \, e^{15} a_1, \, e^{15} a_0$ The linear layer causes the 4-bit inputs to the Sbox in the next layer to be of the form e^4.
- S in R5: $\to (?^3 e)^{16}$
  All four input bit tuples going into any Sbox have even parity (note that the same is also true for a_i).
  We focus on the LSB output y_0 of any Sbox, whose expression is given in Eq. (4). Notably, the only nonlinear term is the AND $(x_1 \wedge x_2)$ term between x_1 and x_2, which is in this case $(e \wedge e)$, leading to an output bit tuple that is also e, preserving the segment/pattern boundaries.
  As each of the other terms in y_0 i.e. x_0, x_2, x_3 to be XORed to $x_1 x_2$ also has even parity, therefore the output bit y_0 of the Sbox would have even parity.
- L in R5: $\to ?^{48} (e)^{16}$
  The linear layer moves all the LSB output bits y_0 of all 16 Sboxes to the rightmost 16 bit positions, therefore at the end of Round 5, all the 16 rightmost bit tuples have even parity, and thus are balanced, i.e. $\bigoplus = 0$.

We have empirically tested this for several runs, each time with 2^{16} random structures of plaintext multisets and keys: it is verified that the rightmost 16 bits after 5 rounds of PRESENT have zero integral $\bigoplus$ sum with probability 1.

This result contrasts with the best-known integrals for PRESENT reported in [30], where a 5-round and a 7-round integral were demonstrated such that a zero integral sum is detected only in the single rightmost bit after five (resp. seven) rounds.

On the contrary, our tuple integral based on a multiset of size 2^4, has 16 times more checkable bits that provide more bit filtering conditions, thus better filtration power during the key recovery stage.

Furthermore, the single-bit integrals in [30] were tracked based on the maximal algebraic degree d of the coordinate Boolean function producing the rightmost LSB at the final output. This is done by collecting enough different texts within a multiset such that there are more than 2^d texts in order to exploit the higher-order derivative property [2], which basically tests for a zero sum.

In contrast, our integral is tracked by bitslice tuples through the round operations, notably through the Sbox's output bits, and within the Sboxes through the AND and XOR operations that make up the ANF expressions of an Sbox's coordinate Boolean functions. Such a bit tuple approach thus enables to observe richer internal structures within the multisets and bitslice tuples.

3.4 RECTANGLE

RECTANGLE [42] is a bit-sliced lightweight cipher, such that in each round it comprises an XOR based key addition, and then just an Sbox layer S followed by bitwise rotation layer R, thus RECTANGLE can be viewed as an NRX cipher, i.e. involving the primitive operations AND, rotation (ROT) and exclusive-OR (XOR). It has a 64-bit block size and 25 number of rounds.

Zhang et al. [38] reported a 7-round integral distinguisher needing a multiset of 2^{36} elements. Kosuge et al. [39] reported an 8-round integral distinguisher requiring a huge multiset of 2^{60} elements. Both these integrals track that the integral $\bigoplus$ sum equals zero in some output bits.

In contrast, we demonstrate how tuples can apply to RECTANGLE by using much smaller multisets, and track more structures in the output bits in contrast to zero sums, i.e. we track the bitslice tuples.

The Sbox S of RECTANGLE can be represented in terms of its ANF as:

$$y_3 = x_1 \oplus x_3 \oplus x_0 x_2 \oplus x_0 x_3 \oplus x_1 x_2 \oplus x_1 x_2 x_3 \tag{5}$$

$$y_2 = 1 \oplus x_2 \oplus x_3 \oplus x_0 x_1 \oplus x_0 x_2 \oplus x_1 x_2 \oplus x_2 x_3 \oplus x_0 x_1 x_2 \tag{6}$$

$$y_1 = 1 \oplus x_0 \oplus x_1 \oplus x_2 \oplus x_1 x_3 \tag{7}$$

$$y_0 = x_0 \oplus x_2 \oplus x_3 \oplus x_0 x_1 \tag{8}$$

As an aside, note that the algebraic degree of the coordinate Boolean functions producing two of its output bits i.e. y_0 and y_1 is 2, versus degree of 3 for the other two output bits.

RECTANGLE's state can be represented as a rectangle of 4×16 bits as follows:

$x_{15}x_{14}x_{13}x_{12}$	$x_{11}x_{10}x_9x_8$	$x_7x_6x_5x_4$	$x_3x_2x_1x_0$
$x_{31}x_{30}x_{29}x_{28}$	$x_{27}x_{26}x_{25}x_{24}$	$x_{23}x_{22}x_{21}x_{20}$	$x_{19}x_{18}x_{17}x_{16}$
$x_{47}x_{46}x_{45}x_{44}$	$x_{43}x_{42}x_{41}x_{40}$	$x_{39}x_{38}x_{37}x_{36}$	$x_{35}x_{34}x_{33}x_{32}$
$x_{63}x_{62}x_{61}x_{60}$	$x_{59}x_{58}x_{57}x_{56}$	$x_{55}x_{54}x_{53}x_{52}$	$x_{51}x_{50}x_{49}x_{48}$

Consider if we had an input multiset of 2^4 elements of the following form:

$$
\begin{array}{cccc}
a_0ccc & cccc & cccc & cccc \\
a_1ccc & cccc & cccc & cccc \\
a_2ccc & cccc & cccc & cccc \\
a_3ccc & cccc & cccc & cccc
\end{array}
$$

We then observe its progagation as detailed below:

- S in R1: This propagates essentially unchanged through the Sbox layer in round 1, because we have four bit tuples $a_3a_2a_1a_0$ forming the leftmost column, entering one Sbox, and constant bit tuples entering the other Sboxes. Alternatively, viewed as 4-bit column-words, we have: PCCC CCCC CCCC CCCC

- R in R1: After bitwise rotation in round 1, we have:

$$
\begin{array}{cccc}
a_0\,c\,c\;c & c\;c\,c\,c & c\,c\,c\,c & c\,c\,c\;c \\
c\;c\,c\;c & c\;c\,c\,c & c\,c\,c\,c & c\,c\,c\,a_1 \\
c\;c\,c\;c & a_2\,c\,c\,c & c\,c\,c\,c & c\,c\,c\;c \\
c\;c\,c\,a_3 & c\;c\,c\,c & c\,c\,c\,c & c\,c\,c\;c
\end{array}
$$

- S in R2: At the leftmost column, we have $ccca_0$ entering the Sbox. Analyzing the Sbox's ANF, we see that the following bit tuples will be obtained at the output:

$$y_0 : \alpha_0 \in \{a_0, c\}; y_1 : a_0; y_2 : \alpha_0 \in \{a_0, c\}; y_3 : \alpha_0 \in \{a_0, c\}$$

Similarly, for the other input tuples with one a_j entering the Sbox, we have:

$$
x_3x_2x_1x_0 = cca_1c \rightarrow
\begin{cases}
y_0 : \alpha_1 \in \{a_1, c\} \\
y_1 : \alpha_1 \in \{a_1, c\} \\
y_2 : \alpha_1 \in \{a_1, c\} \\
y_3 : \alpha_1 \in \{a_1, c\}
\end{cases}
$$

$$
x_3x_2x_1x_0 = ca_2cc \rightarrow
\begin{cases}
y_0 : a_2 \\
y_1 : a_2 \\
y_2 : \alpha_2 \in \{a_2, c\} \\
y_3 : \alpha_2 \in \{a_2, c\}
\end{cases}
$$

$$
x_3x_2x_1x_0 = a_3ccc \rightarrow
\begin{cases}
y_0 : a_3 \\
y_1 : \alpha_3 \in \{a_3, c\} \\
y_2 : \alpha_3 \in \{a_3, c\} \\
y_3 : \alpha_3 \in \{a_3, c\}
\end{cases}
$$

- R in R2: Therefore, after going through rotation, we have:

$$
\begin{array}{cccc}
\alpha_0\,c\;c\;a_3 & a_2\,c\;c\;c & c\;c\,c\,c & c\,c\;c\;\alpha_1 \\
c\;c\,\alpha_3\,a_2 & c\;c\;c\;c & c\;c\,c\,c & c\,c\,\alpha_1\,a_0 \\
c\;c\;c\;\alpha_1 & \alpha_0\,c\;c\;\alpha_3 & \alpha_2\,c\,c\,c & c\;c\,c\;c \\
c\;c\,\alpha_1\,a_0 & c\;c\,\alpha_3\,a_2 & c\;c\,c\,c & c\;c\,c\;c
\end{array}
$$

- S in R3: Passing through a subsequent Sbox layer gives:

$\alpha_0\, c\, \alpha_1 \oplus \alpha_3\, e$	$\alpha_0 \oplus \alpha_2\, c\, \alpha_3\, \alpha_2 \oplus \alpha_3$	$\alpha_2\, c\, c\, c$	$c\, c\, \alpha_1$	$\alpha_0 \alpha_1$
$\alpha_0\, c\ \ \alpha_1 \alpha_3\ \ e$	$\alpha_0 \oplus \alpha_2\, c\, \alpha_3\, \alpha_2 \oplus \alpha_3$	$\alpha_2\, c\, c\, c$	$c\, c\, \alpha_1\, \alpha_0 \oplus \alpha_1$	
$\alpha_0\, c\, \alpha_1 \oplus \alpha_3\, e$	$\alpha_0 \alpha_2\ \ c\, \alpha_3\ \ \alpha_2 \alpha_3$	$\alpha_2\, c\, c\, c$	$c\, c\, \alpha_1$	$\alpha_0 \alpha_1$
$\alpha_0\, c\ \ \alpha_1 \alpha_3\ \ e$	$\alpha_0 \alpha_2\ \ c\, \alpha_3\ \ \alpha_2 \alpha_3$	$\alpha_2\, c\, c\, c$	$c\, c\, \alpha_1\, \alpha_0 \oplus \alpha_1$	

- R in R3: After rotation, we have:

α_0	c	$\alpha_1 \oplus \alpha_3$	e	$\alpha_0 \oplus \alpha_2$	c	α_3	$\alpha_2 \oplus \alpha_3\, \alpha_2$	c	c	c c c	α_1	$\alpha_0 \alpha_1$
c	$\alpha_1 \alpha_3$	e	$\alpha_0 \oplus \alpha_2$	c	α_3	$\alpha_2 \oplus \alpha_3$	α_2	c	c c	c c $\alpha_1\, \alpha_0 \oplus \alpha_1$	α_0	
c	c	α_1	$\alpha_0 \alpha_1$	α_0	c	$\alpha_1 \oplus \alpha_3$	e	$\alpha_0 \alpha_2$ c	α_3	$\alpha_2 \alpha_3\, \alpha_2$ c	c	c
c	α_1	$\alpha_0 \oplus \alpha_1$	α_0	c	$\alpha_1 \alpha_3$	e	$\alpha_0 \alpha_2$	c	α_3	$\alpha_2 \alpha_3\, \alpha_2$ c c	c	c

- S in R4: Going through another Sbox layer results in the following:

$\alpha_0\, \alpha_1 \alpha_3 * *$	$\alpha_0 \oplus \alpha_2\, \alpha_1 \alpha_3 * *$	$\alpha_0 \alpha_2\, \alpha_3\, \alpha_2 \alpha_3\, \alpha_2 \alpha_3$	$\alpha_2\, \alpha_1\, \alpha_0 \alpha_1$	$\alpha_0 \alpha_1$
$\alpha_0\, \alpha_1 \alpha_3 * *$	$\alpha_0 \oplus \alpha_2\, \alpha_1 \alpha_3 * *$	$\alpha_0 \alpha_2\, \alpha_3\, \alpha_2 \alpha_3\, \alpha_2 \alpha_3$	$\alpha_2\, \alpha_1\, \alpha_0 \oplus \alpha_1\, \alpha_0 \oplus a_0\alpha_1$	
$\alpha_0\, \alpha_1 \alpha_3 * *$	$\alpha_0 \alpha_2\ \ \alpha_1 \alpha_3 * *$	$\alpha_0 \alpha_2\, \alpha_3\, \alpha_2 \alpha_3\, \alpha_2 \alpha_3$	$\alpha_2\, \alpha_1\, \alpha_0 \alpha_1$	$\alpha_0 \alpha_1$
$\alpha_0\, \alpha_1 \alpha_3 * *$	$\alpha_0 \alpha_2\ \ \alpha_1 \alpha_3 * *$	$\alpha_0 \alpha_2\, \alpha_3\, \alpha_2 \alpha_3\, \alpha_2 \alpha_3$	$\alpha_2\, \alpha_1\, \alpha_0 \oplus \alpha_1\, \alpha_0 \oplus a_0\alpha_1$	

- R in R4: After rotation, we have:

α_0	$\alpha_1 \alpha_3$	$*$	$*$	$\alpha_0 \oplus \alpha_2\, \alpha_1 \alpha_3 *$	$*$	$\alpha_0 \alpha_2\, \alpha_3\, \alpha_2 \alpha_3\, \alpha_2 \alpha_3$	α_2	α_1	$\alpha_0 \alpha_1$	$\alpha_0 \alpha_1$
$\alpha_1 \alpha_3$	$*$	$*$	$\alpha_0 \oplus \alpha_2$	$\alpha_1 \alpha_3$	$*$	$* \alpha_0 \alpha_2\, \alpha_3\, \alpha_2 \alpha_3\, \alpha_2 \alpha_3$	α_2	α_1	$\alpha_0 \oplus \alpha_1\, \alpha_0 \oplus a_0\alpha_1$	α_0
α_2	α_1	$\alpha_0 \alpha_1$	$\alpha_0 \alpha_1$	α_0	$\alpha_1 \alpha_3 *$	$* \alpha_0 \alpha_2\, \alpha_1 \alpha_3$	$*$	$* \alpha_0 \alpha_2$	α_3	$\alpha_2 \alpha_3\, \alpha_2 \alpha_3$
α_1	$\alpha_0 \oplus \alpha_1$	$*$	α_0	$\alpha_1 \alpha_3$	$* * \alpha_0 \alpha_2\, \alpha_1 \alpha_3$	$*$	$* \alpha_0 \alpha_2$	α_3	$\alpha_2 \alpha_3\, \alpha_2 \alpha_3$	α_2

- S in R5: In Round 5, passing through the Sbox layer obtains:

$e * * *$	$\alpha_0 \oplus \alpha_2 * * *$	$\alpha_0 \alpha_2 * * *$	e	α_1	$\alpha_0 \alpha_1$	$\alpha_0 \alpha_1$
$e * * *$	$\alpha_1 \alpha_3\ \ * * *$	$\alpha_3\ \ * * *$	$e\, \alpha_0 \oplus \alpha_1\, \alpha_0 \oplus a_0\alpha_1$		α_0	
$* * * *$	α_0	$* * *$	$\alpha_0 \alpha_2 * * *$	e	α_3	$\alpha_2 \alpha_3$ $\alpha_2 \alpha_3$
$e * * *$	$\alpha_1 \alpha_3\ \ * * *$	$\alpha_1 \alpha_3 * * *$	$*$	$\alpha_2 \alpha_3$	$\alpha_2 \alpha_3$	α_2

To see why e tuples are obtained at the output of the Sbox in the leftmost column, note that the input tuples are of the form $(\alpha_1, \alpha_2, \alpha_1 \alpha_3, \alpha_0)$, and recall that $\alpha_i \in \{a_i, c\}$. Therefore, the possible input tuples to the Sbox are:

- $(c^s a^{4-s})$, $s \in \{0, \dots, 3\}$ where the composition of c and a tuples are in any order. This forms an E word into the Sbox, therefore the output will also be an E.
- (a_1, a_2, a_3, a_0): This is a P word into the Sbox, therefore the output will also be a P.
- (a_1, a_2, a_1, a_0): This is actually an E word, thus the output will be an E.
- $(a_1, a_2, a_1 a_3, a_0) = (a_1, a_2, m_{13}, a_0)$: Note from our previous discussion that $a_1 a_3$ gives m_{13} i.e. a tuple as like a_1 but where every other contiguous sequence of 2^3 bits is 0. Given such an input, then the output tuple from the Sbox is of the form: $(e * ee)$.

Similar arguments apply to the fourth rightmost column whose input tuple is $(\alpha_2, \alpha_1, \alpha_0 \alpha_1, \alpha_3)$. The crucial analysis is for the case where the input is $(a_3, a_0 a_2, a_1, a_2) = (a_3, m_{02}, a_1, a_2)$, which gives an output of $(* eee)$.

- R in R5: After another rotation, we obtain:

$e * * *$	$* * * *$	$* * * *$	$e * * *$
$* * * *$	$* * * *$	$* * * e$	$* * * e$
$e * * *$	$* * * *$	$* * * *$	$* * * *$
$* * * e$	$* * * *$	$* * * *$	$* * * *$

Thus, we have a 5-round integral distinguisher for RECTANGLE that requires only 2^4 texts, and the integral $\bigoplus$ zero sum can be detected in at least 6 bits. This has been corroborated by experiments, repeated for several runs. In fact, there is more structure existing in other bits that require much in-depth analysis, indeed empirical results show that the zero sum is detected in 22 bits after 5 rounds.

Analysing the rounds via tuples enables us to precisely track the evolution of multisets through more rounds than previously possible, because tuples allow to represent much richer structure than conventional integral/multiset analysis. This is why using only much smaller amounts of text (2^4 in our case for RECTANGLE) we can track at the bit level granularity for up to five rounds.

4   Concluding Remarks

The Slice-&-Fuse paradigm considered in this paper enables to move between word and bitslice granularities when tracking the evolution of multisets and their constituent tuples through cipher rounds. Towards that aim, our proposed new types of tuples capture more structures than previously known, that are inherent in multisets and yet which have largely remained unexplored until now. Open questions include whether much richer types of tuples exist as constituents of multisets, and developing advanced approaches to track the propagation of tuples through other cipher component operations including modulo multiplications that make up exotic ciphers such as XMX [4] as well as the more celebrated IDEA cipher [1].

Acknowledgements. We thank the anonymous Mycrypt 2016 reviewers and Ana Sălăgean for constructive and critical comments that have improved this paper. RP's work is supported in part by the Malaysian Ministry of Education's Fundamental Research Grant Scheme (FRGS) under the project *ProvAdverse*.

A   Integrals of Crypton and mCrypton

CRYPTON v1.0 and mCrypton are two block ciphers proposed by Lim et al. [7,14]. CRYPTON was one of the candidates of the AES competition, acting on 128-bit blocks under keys of length 128, 192 or 256 bits whereas mCrypton is its equivalent lightweight version acting on 64-bit blocks under keys of length 64, 96 or 128 bits. Both have the same design principle based on an SPN structure with an Sbox layer and a linear layer composed of a bit permutation and a matrix transposition. The bit permutation could be represented by a matrix multiplication where the MDS bound is not reached (few column elements have an input/output weight of 4 instead of 5 in the case of an MDS transformation). For the rest of this subsection, we will denote the block size as $16n$ with $n = 8$ for CRYPTON and $n = 4$ for mCrypton.

We are not able to directly exhibit bitslice properties in the cases of CRYPTON and mCrypton. We conjecture that this fact is linked with the inherent design of the ciphers: the Sbox layer and the bit permutation act on different word sizes leading to impeding the possible properties at bit level. However, we are able to improve the classical integral property used in [6] on CRYPTON and that also works on mCrypton: we start with one active word at the beginning (whereas the other words are constant) and that gives a zero integral $\oplus$ sum in each word after three rounds. As the linear layer does not have the MDS property, we are able to construct a new four-round integral property that works for both CRYPTON and mCrypton. More precisely, the following 3rd-order integral property holds on 4 rounds requiring only 2^{3n} chosen plaintexts. Indeed:

$$
\begin{array}{cccc}
A^{3n} & C & C & C \\
A^{3n} & C & C & C \\
A^{3n} & C & C & C \\
C & C & C & C
\end{array}
$$

gives after 4 rounds

$$
\begin{array}{cccc}
B & B & B & B \\
B & B & B & B \\
B & B & B & B \\
B & B & B & B
\end{array}
$$

Thus, we could improve in the case of CRYPTON the data/time complexity of the 6-round attack presented in [6] by a factor 2^{16} for the time complexity and by a factor 2^8 for the data complexity. Indeed, we add two rounds at the end of the previous 4-round integral property leading to guess 5 key words of the subkeys K^5 and K^6 for a cost of 2^{40} tests whereas to discard false alarms we need to test $5 \cdot 2^{24}$ chosen plaintexts. This attack could be easily adapted to mCrypton leading to an attack on 6 rounds using $5 \cdot 2^{12}$ chosen plaintexts with a time complexity of about 2^{20} tests.

References

1. Lai, X., Massey, J.L.: A proposal for a new block encryption standard. In: Damgård, I.B. (ed.) EUROCRYPT 1990. LNCS, vol. 473, pp. 389–404. Springer, Heidelberg (1991). doi:10.1007/3-540-46877-3_35

2. Knudsen, L.R.: Truncated and higher order differentials. In: Preneel, B. (ed.) FSE 1994. LNCS, vol. 1008, pp. 196–211. Springer, Heidelberg (1995). doi:10.1007/3-540-60590-8_16

3. Daemen, J., Knudsen, L., Rijmen, V.: The block cipher square. In: Biham, E. (ed.) FSE 1997. LNCS, vol. 1267, pp. 149–165. Springer, Heidelberg (1997). doi:10.1007/BFb0052343

4. M'Raïhi, D., Naccache, D., Stern, J., Vaudenay, S.: XMX: a firmware-oriented block cipher based on modular multiplications. In: Biham, E. (ed.) FSE 1997. LNCS, vol. 1267, pp. 166–171. Springer, Heidelberg (1997). doi:10.1007/BFb0052344

5. Lidl, R., Niederreiter, H.: Finite Fields. Encyclopedia of Mathematics and its Applications, vol. 20. Cambridge University Press, Cambridge (1997)

6. D'Halluin, C., Bijnens, G., Rijmen, V., Preneel, B.: Attack on Six Rounds of CRYP-TON. In: Knudsen, L. (ed.) FSE 1999. LNCS, vol. 1636, pp. 46–59. Springer, Heidelberg (1999). doi:10.1007/3-540-48519-8_4
7. Lim, C.H.: A revised version of CRYPTON: CRYPTON V1.0. In: Knudsen, L. (ed.) FSE 1999. LNCS, vol. 1636, pp. 31–45. Springer, Heidelberg (1999). doi:10.1007/3-540-48519-8_3
8. Ferguson, N., Kelsey, J., Lucks, S., Schneier, B., Stay, M., Wagner, D., Whiting, D.: Improved cryptanalysis of rijndael. In: Goos, G., Hartmanis, J., Leeuwen, J., Schneier, B. (eds.) FSE 2000. LNCS, vol. 1978, pp. 213–230. Springer, Heidelberg (2001). doi:10.1007/3-540-44706-7_15
9. Lucks, S.: The saturation attack — a bait for twofish. In: Matsui, M. (ed.) FSE 2001. LNCS, vol. 2355, pp. 1–15. Springer, Heidelberg (2002). doi:10.1007/3-540-45473-X_1
10. Biryukov, A., Shamir, A.: Structural cryptanalysis of SASAS. In: Pfitzmann, B. (ed.) EUROCRYPT 2001. LNCS, vol. 2045, pp. 395–405. Springer, Heidelberg (2001). doi:10.1007/3-540-44987-6_24
11. Knudsen, L., Wagner, D.: Integral cryptanalysis. In: Daemen, J., Rijmen, V. (eds.) FSE 2002. LNCS, vol. 2365, pp. 112–127. Springer, Heidelberg (2002). doi:10.1007/3-540-45661-9_9
12. Standaert, F.-X., Piret, G., Rouvroy, G., Quisquater, J.-J., Legat, J.-D.: ICE-BERG: An involutional cipher efficient for block encryption in reconfigurable hardware. In: Roy, B., Meier, W. (eds.) FSE 2004. LNCS, vol. 3017, pp. 279–298. Springer, Heidelberg (2004). doi:10.1007/978-3-540-25937-4_18
13. Nakahara, J., de Freitas, D.S., Phan, R.C.-W.: New multiset attacks on rijndael with large blocks. In: Dawson, E., Vaudenay, S. (eds.) Mycrypt 2005. LNCS, vol. 3715, pp. 277–295. Springer, Heidelberg (2005). doi:10.1007/11554868_20
14. Lim, C.H., Korkishko, T.: mCrypton – a lightweight block cipher for security of low-cost RFID tags and sensors. In: Song, J.-S., Kwon, T., Yung, M. (eds.) WISA 2005. LNCS, vol. 3786, pp. 243–258. Springer, Heidelberg (2006). doi:10.1007/11604938_19
15. Bogdanov, A., Knudsen, L.R., Leander, G., Paar, C., Poschmann, A., Robshaw, M.J.B., Seurin, Y., Vikkelsoe, C.: PRESENT: an ultra-lightweight block cipher. In: Paillier, P., Verbauwhede, I. (eds.) CHES 2007. LNCS, vol. 4727, pp. 450–466. Springer, Heidelberg (2007). doi:10.1007/978-3-540-74735-2_31
16. Knudsen, L.R., Rijmen, V.: Known-key distinguishers for some block ciphers. In: Kurosawa, K. (ed.) ASIACRYPT 2007. LNCS, vol. 4833, pp. 315–324. Springer, Heidelberg (2007). doi:10.1007/978-3-540-76900-2_19
17. Z'aba, M.R., Raddum, H., Henricksen, M., Dawson, E.: Bit-pattern based integral attack. In: Nyberg, K. (ed.) FSE 2008. LNCS, vol. 5086, pp. 363–381. Springer, Heidelberg (2008). doi:10.1007/978-3-540-71039-4_23
18. Sun, B., Qu, L., Li, C.: New cryptanalysis of block ciphers with low algebraic degree. In: Dunkelman, O. (ed.) FSE 2009. LNCS, vol. 5665, pp. 180–192. Springer, Heidelberg (2009). doi:10.1007/978-3-642-03317-9_11
19. Wei, Y., Sun, B., Li, C.: New integral distinguisher for rijndael-256. In: IACR ePrint Archive. Report 559 (2009). http://eprint.iacr.org/2009/559
20. Biryukov, A., Shamir, A.: Structural cryptanalysis of SASAS. J. Cryptol. 23(4), 505–518 (2010)
21. Ferguson, N., Lucks, S., Schneier, B., Whiting, D., Bellare, M., Kohno, T., Callas, J., Walker, J.: The Skein Hash Function Family, version 1.3. Submitted to NIST SHA-3 Competition Round 3 (2010)

22. Aumasson, J.-P., Leurent, G., Meier, W., Mendel, F., Mouha, N., Phan, R.C.-W., Sasaki, Y., Susil, P.: Tuple cryptanalysis of ARX with application to BLAKE and Skein. In: ECRYPT II Hash Workshop (Hash 2011) (2011)
23. Zhang, W., Su, B., Wu, W., Feng, D., Wu, C.: Extending higher-order integral: an efficient unified algorithm of constructing integral distinguishers for block ciphers. In: Bao, F., Samarati, P., Zhou, J. (eds.) ACNS 2012. LNCS, vol. 7341, pp. 117–134. Springer, Heidelberg (2012). doi:10.1007/978-3-642-31284-7_8
24. Sasaki, Y., Wang, L.: Meet-in-the-middle technique for integral attacks against feistel ciphers. In: Knudsen, L.R., Wu, H. (eds.) SAC 2012. LNCS, vol. 7707, pp. 234–251. Springer, Heidelberg (2013). doi:10.1007/978-3-642-35999-6_16
25. Borghoff, J., Canteaut, A., Güneysu, T., Kavun, E.B., Knezevic, M., Knudsen, L.R., Leander, G., Nikov, V., Paar, C., Rechberger, C., Rombouts, P., Thomsen, S.S., Yalçın, T.: PRINCE – a low-latency block cipher for pervasive computing applications. In: Wang, X., Sako, K. (eds.) ASIACRYPT 2012. LNCS, vol. 7658, pp. 208–225. Springer, Heidelberg (2012). doi:10.1007/978-3-642-34961-4_14
26. Lu, J., Wei, Y., Kim, J., Pasalic, E.: The higher-order meet-in-the-middle attack and its application to the camellia block cipher. In: Galbraith, S., Nandi, M. (eds.) INDOCRYPT 2012. LNCS, vol. 7668, pp. 244–264. Springer, Heidelberg (2012). doi:10.1007/978-3-642-34931-7_15
27. Peng, C., Zhu, C., Zhu, Y., Kang, F.: Practical symbolic computation in block cipher with application to present. In: IACR ePrint Archive. Report 587 (2012). http://eprint.iacr.org/2012/587
28. Jean, J., Nikolić, I., Peyrin, T., Wang, L., Wu, S.: Security analysis of PRINCE. In: Moriai, S. (ed.) FSE 2013. LNCS, vol. 8424, pp. 92–111. Springer, Heidelberg (2014). doi:10.1007/978-3-662-43933-3_6
29. Sasaki, Y., Wang, L.: Bitwise partial-sum on HIGHT: a new tool for integral analysis against ARX designs. In: Lee, H.-S., Han, D.-G. (eds.) ICISC 2013. LNCS, vol. 8565, pp. 189–202. Springer, Cham (2014). doi:10.1007/978-3-319-12160-4_12
30. Wu, S., Wang, M.: Integral attacks on reduced-round PRESENT. In: Qing, S., Zhou, J., Liu, D. (eds.) ICICS 2013. LNCS, vol. 8233, pp. 331–345. Springer, Cham (2013). doi:10.1007/978-3-319-02726-5_24
31. Todo, Y., Aoki, K.: FFT key recovery for integral attack. In: Gritzalis, D., Kiayias, A., Askoxylakis, I. (eds.) CANS 2014. LNCS, vol. 8813, pp. 64–81. Springer, Cham (2014). doi:10.1007/978-3-319-12280-9_5
32. Wang, Q., Liu, Z., Varıcı, K., Sasaki, Y., Rijmen, V., Todo, Y.: Cryptanalysis of reduced-round SIMON32 and SIMON48. In: Meier, W., Mukhopadhyay, D. (eds.) INDOCRYPT 2014. LNCS, vol. 8885, pp. 143–160. Springer, Cham (2014). doi:10.1007/978-3-319-13039-2_9
33. Lu, J., Wei, Y., Kim, J., Pasalic, E.: The higher-order meet-in-the-middle attck and its application to the camellia block cipher. Inf. Process. Lett. 527, 102–122 (2014)
34. Todo, Y.: Structural evaluation by generalized integral property. In: Oswald, E., Fischlin, M. (eds.) EUROCRYPT 2015. LNCS, vol. 9056, pp. 287–314. Springer, Heidelberg (2015). doi:10.1007/978-3-662-46800-5_12
35. Todo, Y.: Integral cryptanalysis on full MISTY1. In: Gennaro, R., Robshaw, M. (eds.) CRYPTO 2015. LNCS, vol. 9215, pp. 413–432. Springer, Heidelberg (2015). doi:10.1007/978-3-662-47989-6_20
36. Blondeau, C., Peyrin, T., Wang, L.: Known-key distinguisher on full PRESENT. In: Gennaro, R., Robshaw, M. (eds.) CRYPTO 2015. LNCS, vol. 9215, pp. 455–474. Springer, Heidelberg (2015). doi:10.1007/978-3-662-47989-6_22

37. Akshima, C.D., Ghosh, M., Goel, A., Sanadhya, S.K.: Improved meet-in-the-middle attacks on 7 and 8-round ARIA-192 and ARIA-256. In: Biryukov, A., Goyal, V. (eds.) Progress in Cryptology – INDOCRYPT 2015. LNCS, vol. 9462, pp. 198–217. Springer, Cham (2015). doi:10.1007/978-3-319-26617-6_11
38. Zhang, H., Wu, W., Wang, Y.: Integral attack against bit-oriented block ciphers. In: Kwon, S., Yun, A. (eds.) ICISC 2015. LNCS, vol. 9558, pp. 102–118. Springer, Cham (2016). doi:10.1007/978-3-319-30840-1_7
39. Kosuge, H., Tanaka, H., Iwai, K., Kurokawa, T.: Integral attack on reduced-round Rectangle. In: IEEE CSCloud 2015, pp. 68–73 (2015)
40. Wei, Y.: Bit-pattern based integral attack on Iceberg. In: INCOS 2015, pp. 370–373 (2015)
41. Sasaki, Y., Wang, L.: Bitwise partial-sum on HIGHT: a new tool for integral analysis against ARX designs. IEICE Trans. Fundam. **E98A**(1), 49–60 (2015)
42. Zhang, W., Bao, Z., Lin, D., Rijmen, V., Yang, B., Verbauwhede, I.: RECTANGLE: a bit-slice lightweight block cipher suitable for multiple platforms. Sci. China Inf. Sci. **58**(12), 1–15 (2015)
43. Dinur, I., Dunkelman, O., Kranz, T., Leander, G.: Decomposing the ASASA block cipher construction. In: IACR ePrint Archive. Report 507 (2015). http://eprint.iacr.org/2015/507
44. Morawiecki, P.: Practical attacks on the round-reduced PRINCE. In: IACR ePrint Archive. Report 245. (2015). http://eprint.iacr.org/2015/245
45. Posteuca, R., Negara, G.: Integral cryptanalysis of round-reduced PRINCE cipher. Proc. Rom. Acad. Ser. A **16**, 265–269 (2015)
46. Boura, C., Canteaut, A.: Another view of the division property. In: Robshaw, M., Katz, J. (eds.) CRYPTO 2016. LNCS, vol. 9814, pp. 654–682. Springer, Heidelberg (2016). doi:10.1007/978-3-662-53018-4_24

Improvements of Attacks on Various Feistel Schemes

Emmanuel Volte, Valérie Nachef[(✉)], and Nicolas Marrière

Department of Mathematics, University of Cergy-Pontoise,
CNRS UMR 8088, 2 Avenue Adolphe Chauvin, 95011 Cergy-Pontoise Cedex, France
{emmanuel.volte,valerie.nachef,nicolas.marriere}@u-cergy.fr

Abstract. In this paper, we use a tool that computes exact values for expectations and standard deviations of random variables involved in generic attacks on various Feistel-type schemes in order to get a better study of these attacks. This leads to the improvement of previous attacks complexities: either we need less messages than expected or we can attack more rounds. These improvements are given for different sizes of the inputs. We also show that for rectangle attacks, there are more differential paths than presented in previous attacks and this strengthens the attacks.

Keywords: Generic attacks on Feistel type schemes · Pseudo-random permutations · Differential cryptanalysis

1 Introduction

The DES cipher [1,2] is based on Classical Feistel constructions studied by Luby and Rackoff [13]. Although AES [8] has replaced DES, Feistel type structures are still widely used to design many block ciphers like GOST [20] or SIMON [5]. These constructions are based on the repetition of two elementary operations: the Xor between one part of the plaintext and the result of another part through a nonlinear function F, and a permutation layer. Depending on the choice of the permutation layer and the internal functions, it is possible to obtain different kinds of block ciphers. The concrete ciphers based on Feistel network have a public function F and often use a Xor between the input of F and a round key which is derived from a master key by a key schedule. For example, MARS [7] is based on unbalanced Feistel schemes with expanding functions [11,19,22,25], and SMS4 [12] on unbalanced Feistel schemes with contracting functions [16,18]. Alternating Feistel schemes alternate contracting and expanding rounds and they are used in the BEAR/LION block cipher [4]. There are also type-1, type-2 and type-3 Feistel schemes [9,10,15,27]. Type-1 Feistel schemes are used in CAST-256 [3] and type-2 Feistel schemes in RC-6 [21] and CLEFIA [23] for example.

Different kinds of attacks can be mounted on block ciphers: linear, differential, impossible differential, impossible boomerang, Meet-in-the-middle attacks. The key schedule can also be a way to attack a cipher. The most common attacks

© Springer International Publishing AG 2017
R.C.-W. Phan and M. Yung (Eds.): LNCS 10311, Mycrypt 2016, pp. 321–344, 2017.
DOI: 10.1007/978-3-319-61273-7_16

on Feistel schemes are differential attacks [6] based on the Xor operation. The cryptanalysis of such attacks examines how a difference between inputs can be detected on the outputs. In a classical differential attack, one has to distinguish a permutation computed by the studied scheme from a random permutation. There are some conditions which are more often verified with the cipher than the random permutation due to the structure of the cipher. We will provide a further analysis of this point that leads to the improvement of some attacks. The overall results are given in Table 1.

Let us call a pair of (plaintext, ciphertext) a point. According to the structure of the scheme, it is possible to choose equalities and non equalities (based on the Xor operation) that must be satisfied by φ points[1]. All these conditions together form what we call a mirror system. Depending on the conditions, a mirror system can represent a differential attack, an impossible one or even a boomerang one for $\varphi = 4$ for example. In a known plaintext attack (KPA) model with m points, we look for the number of points satisfying the mirror system. If the current scheme is chosen randomly in an uniform distribution of schemes from the same family and if the m points are chosen randomly in an uniform distribution of all set which contains m points, we obtain a random variable $\tilde{\mathcal{N}}$. If the points are randomly chosen and the same system is studied under the assumption that we have a random permutation instead of a permutation produced by a scheme, the corresponding variable is denoted by $\mathcal{N}$. Then, we can try to **distinguish** $\tilde{\mathcal{N}}$ from $\mathcal{N}$.

We can mount a distinguishing attack if either $\tilde{\mathcal{N}}$ is significantly greater than $\mathcal{N}$, or $|\mathbb{E}(\mathcal{N}) - \mathbb{E}(\tilde{\mathcal{N}})| \geq \sigma(\mathcal{N})$, where $\mathbb{E}$ denotes the expectation function and σ stands for the standard deviation. The attacks work thanks to the Chebychev formula, which states that for any random variable X, and any $\alpha > 0$, we have $\mathbb{P}(|X - \mathbb{E}(X)| \geq \alpha\sigma(X)) \leq \frac{1}{\alpha^2}$. Using this formula, it is then possible to construct a prediction interval for $\tilde{\mathcal{N}}$ for example, in which future computations will fall, with a good probability.

Computing expectations and variances of these variables is most of the time very tedious and it is hard to obtain exact values. In most papers, only estimations are given with the use of a $O()$ function. Generally, it is possible to obtain the beginning of the Taylor expansion of the value of the expectations and standard deviations, and we assume that the remainder is negligible.

We have built a tool that allows to compute these exact values. It provides a rational fraction in N (the number of values taken by a branch) and m. We have verified our tool with an external computation of the exact value by hand for N equal to 2 or 4. This tool is available on the Internet[2] and can be used to verify the results and to have all the details.

In this paper, we will use this tool without giving all the details of the computations, in order to show that once we are able to obtain the exact values

---

[1] $\varphi = 2$ most of the time, but there are some rectangle attacks with $\varphi = 4$. There can be many values for φ even odd ones [14,17].

[2] The project is at this link: https://github.com/CryptoCergy/project.

for expectations and standard deviation, most of the time, we can improve the results obtained through theoretical studies. More specifically, for small values of the parameters, like the bit size of a branch, we can get a better complexity or attack more rounds. We focus on the improvements of previous attacks. All the details for the computations are provided in an extended version of this paper [24].

The paper is organized as follows. In Sect. 2, we introduce our notations and give definitions. Section 3 provides general formulas for expectations and standard deviations. In Sect. 4, we give results presented in previous parers and precise the limitations of the attacks performed in those papers. We also explain how we use the values obtained by the computer program to improve previous attacks. In Sect. 5, we use the tool to improve previous attacks on several ciphers: generalized Feistel schemes of type 1, 2 and 3, unbalanced Feistel schemes with contracting or expanding functions. We have also made simulations of some attacks. We obtained values that confirm the results given by the program. These simulations are presented in Appendix B.

Table 1. Overall improvements on some KPA attacks on Feistel type schemes with 4 branches.

Scheme	Round	Size of a branch	Messages needed	
			Previous attack	New attack
Type-1	22	8	2^{32} [15]	2^{28}
	23	6	$(\times)$	2^{23}
		4	$(\times)$	2^{14}
Type-2	10	8	2^{32} [15]	2^{28}
	12	5	$(\times)$	2^{20}
Type-3	7	8	2^{28} [15]	2^{25}
	8	8	2^{32} [15]	2^{29}
	9	5	$(\times)$	2^{20}
		4	$(\times)$	2^{16}
Contracting Feistel	6	8	2^{16} [18] (*)	2^{17}
	7	8	2^{20} [18] (*)	2^{21}
Expanding Feistel	11	16	$2^{62.4}$ [25]	2^{62}
		8	$2^{31.2}$ [25]	2^{30}
	12	7	$(\times)$	2^{28}
		6	$(\times)$	2^{24}
		4	$(\times)$	2^{15}

$(\times)$ means nothing was known from the theoretical study
$(*)$ Theoretical study is not accurate enough on this attack

2   Notation

In the improved attacks that we will present in this paper, we want to distinguish a random permutation from kn bits to kn bits from a permutation produced by a Feistel-type scheme using random internal functions. We will use the following notations:

- k is the number of branches. We have $k \geq 2$.
- m is the number of messages known by the attacker.
- n is the number of bits for each branch. $n \geq 1$.
- $N = 2^n$ is the number of values a branch can take.
- $J = \{0,1\}^{kn}$ is the set of all messages. J is ordered in a natural way. $\text{card}(J) = |J| = N^k$.
- $(a_1, a_2, \ldots, a_p) \not\in E^p$ means that the a_i values are pairwise distinct elements of E.
- $\mathcal{B}_{kn}$ denotes the set of bijections from J to J.
- Ψ^r is the set of the Feistel schemes (for a certain type: balanced Feistel scheme, unbalanced Feistel scheme with expanding or contracting functions, generalized Feistel scheme of type 1, 2 or 3) with r **rounds**.
- When a bijection f is given, a couple $(I, S) \in J \times J$ (input/output) where $S = f(I)$, is called **a point**. When we have φ points, they are denoted by $\big(I(1), S(1)\big), \ldots, \big(I(\varphi), S(\varphi)\big)$.
- I and S are divided in k branches of n bits. We set $I = [I_1, I_2, \ldots, I_k]$ and $S = [S_1, S_2, \ldots, S_k]$.
- The internal functions are denoted by F^1, F^2, $\ldots$, F^r.
- J_m is the set of all the subsets of J with a cardinal equal to m. If $M \in J_m$, we have $|M| = m$. We say that $M(i)$ is the i-the element of M (for the same order as J). With M and a given bijection f, we can define a set of points: $M_f = \{(I, f(I)) \mid I \in M\}$.
- A φ-**condition** is an equality or a non-equality between Xor of branches of two (or more) points $\big(I(1), S(1)\big), \big(I(2), S(2)\big), \ldots, \big(I(\varphi), S(\varphi)\big)$. The general form can be defined as follows:

$$\oplus_{i \in \Phi} \left(\oplus_{p \in C_1} I_p(i) \oplus_{q \in C_2} S_q(i) \right) \overset{=}{\underset{\neq}{}} 0$$

where Φ is a subset of $\{1, 2, \ldots, \varphi\}$, C_1 and C_2 are subsets of $\{1, \ldots, k\}$. Most often C_1 and C_2 have zero or one element.
- A φ-**attack** (A) is a finite set of φ-conditions.
- For an φ-attack (A), we define the random variable $\mathcal{N}$:

$$\mathcal{N} = \sum_{\substack{(i_1, \ldots, i_\varphi) \\ \not\in [1, m]^\varphi}} \mathcal{N}_{i_1, i_2, \ldots, i_\varphi}$$

where $\mathcal{N}_{i_1, i_2, \ldots, i_\varphi} = 1$ if the φ-attack works (all the conditions are realized) on points $\big(I(i_1), S(i_1)\big), \big(I(i_2), S(i_2)\big), \ldots, \big(I(i_\varphi), S(i_\varphi)\big)$ in M_f where $M \in_R J_m$ and $f \in_R \mathcal{B}_{kn}$. $\mathcal{N}_{i_1, i_2, \ldots, i_\varphi} = 0$ if the φ-attack does not work. In the same way we use the notations n, n' and n'' that are equal to 1 if the φ-attack works for the given points (each time we will precise the points) and 0 otherwise.

- For X, $Y \in J$, δ_{XY} is the Kronecker symbol equal to 1 if $X = Y$ and 0 otherwise.
- We write $\mathbf{I} = (I(1), \ldots, I(\varphi)) \in J^{\varphi}$. The same notation applies for the outputs.
- In the same way, we define $\tilde{\mathcal{N}}$ by choosing again $M \in_R J_m$ and $f \in_R \Psi^d$.
- We say that a φ-attack works if $|\mathbb{E}(\mathcal{N}) - \mathbb{E}(\tilde{\mathcal{N}})| \geq \sigma(\mathcal{N})$ (as explained in Sect. 1).

In order to make the notations more familiar, we provide a toy example.

Example:
$n = 1$, $k = 2$, $N = 2$, $N^k = 4$, $\varphi = 2$, $m = 2$, so $|J_m| = \binom{4}{2} = 6$.

$J = \{00, 01, 10, 11\}$. $J_m = \Big\{ \{00, 01\}; \{00, 10\}; \{00, 11\}; \{01, 10\}; \{01, 11\};$

$\{10, 11\} \Big\}$. We consider a classical Feistel scheme with $r = 1$:

$\psi_{F^1}([I_1, I_2]) = [S_1, S_2] = [I_2 \oplus F^1(I_1), I_1]$ where F^1 is one of the four possible functions from a set of 2 elements to a set of 2 elements.
We consider the following φ-attack $\mathcal{A} : I_1(i) \oplus I_1(j) \oplus S_1(i) \oplus S_1(j) = 0$. Then $\mathbb{E}(\mathcal{N}) = \frac{10}{9}$, $V(\mathcal{N}) = \frac{80}{81}$, $\mathbb{E}(\tilde{\mathcal{N}}) = \frac{4}{3}$, $V(\tilde{\mathcal{N}}) = \frac{8}{9}$.
In the general case, we found with our tool that $\mathbb{E}(\mathcal{N}) = \frac{(1 - 2N + N^3)m(m-1)}{(N^2 - 1)^2}$ and $\mathbb{E}(\tilde{\mathcal{N}}) = \frac{2m(m-1)}{N+1}$. For the variances, the general formula is very long, about thirty terms.

$$V(\mathcal{N}) = \frac{2(12m - 24N + 142N^4 - 102N^5 \ldots - 16N^5 m + N^{12})Nm(m-1)}{(N^2 - 3)^2(N^2 - 2)^2(N+1)^2(N^2 - 1)^2}$$

3 General Formulas for the Expectations and Standard Deviations

3.1 Computation of $\mathbb{E}(\mathcal{N})$

It is possible to obtain general formulas for expectations and standard deviations. Let (A) be a φ-attack and $\mathcal{N}$ the random variable: $\mathcal{N} = \sum_{\substack{(i_1, \ldots, i_{\varphi}) \\ \neq \in [1, m]^{\varphi}}} \mathcal{N}_{i_1, i_2, \ldots, i_{\varphi}}$.

Proposition 1. *Let* $P(N) = card\{\mathbf{I}, \mathbf{S} \neq \in J^{\varphi} \text{ satisfying } (A)\}$. *Then* $\mathbb{E}(\mathcal{N}) = \left(\frac{(N^k - \varphi)!}{N^k!} \right)^2 \frac{m!}{(m-\varphi)!} P(N)$

Remark 1. $P(N)$ will be the value returned by our computer program that gives the number of solutions of system (A).

Proof. We have: $\mathbb{E}(\mathcal{N}) = \dfrac{\displaystyle\sum_{f \in \mathcal{B}_{nk}} \sum_{M \in J_m} \sum_{\substack{(i_1, \ldots, i_{\varphi}) \\ \neq \in [1, m]^{\varphi}}} n_{i_1, \ldots, i_{\varphi}, f, M}}{|\mathcal{B}_{kn}| \times |J_m|}$ where $n_{i_1, \ldots, i_{\varphi}, f, M}$ is equal to 1 or 0 for the given f, M and $i_1, \ldots, i_{\varphi}$ whether or not the φ-attack is verified for the points $\big(M(i_1), f(M(i_1))\big), \ldots \big(M(i_\varphi), f(M(i_\varphi))\big)$. When $f \in \mathcal{B}_{kn}$ is given, we have:

$$\sum_{M \in J_m} \sum_{\substack{(i_1,\ldots,i_\varphi) \\ \notin [1,m]^\varphi}} n_{i_1,\ldots,i_\varphi,f,M} = \sum_{M \in J_m} \sum_{\substack{(i_1,\ldots,i_\varphi) \\ \notin [1,m]^\varphi}} \sum_{\mathbf{I} \neq J^\varphi} n' \prod_{u=1}^{\varphi} \delta_{I(u)M(i_u)}$$

$$= \sum_{\mathbf{I} \neq J^\varphi} n' \sum_{M \in J_m} \sum_{\substack{(i_1,\ldots,i_\varphi) \\ \notin [1,m]^\varphi}} \prod_{u=1}^{\varphi} \delta_{I(u)M(i_u)}$$

where $n' = n'_{I(1),\ldots,I(\varphi),f}$ is equal to 1 or 0 for the given $\mathbf{I}$ and f. And,

$$\sum_{M \in J_m} \sum_{\substack{(i_1,\ldots,i_\varphi) \\ \notin [1,m]^\varphi}} \prod_{u=1}^{\varphi} \delta_{I(u)M(i_u)} = \operatorname{card}\{M \in J_m \mid \{I(1),\ldots,I(\varphi)\} \subset M\}$$

$$= \frac{\binom{N^k - \varphi}{m - \varphi}}{|J_m|}|J_m| = \frac{\binom{N^k - \varphi}{m - \varphi}}{\binom{N^k}{m}}|J_m|$$

$$= \frac{(N^k - \varphi)!}{(m - \varphi)!(N^k - m)!} \times \frac{m!(N^k - m)!}{N^k!}|J_m|$$

$$= \frac{(N^k - \varphi)!}{N^k!}\frac{m!}{(m - \varphi)!}|J_m|$$

Thus,

$$\mathbb{E}(\mathcal{N}) = \frac{m!(N^k - \varphi)!}{|\mathcal{B}_{nk}|(m - \varphi)!N^k!} \sum_{f \in \mathcal{B}_{nk}} \sum_{\mathbf{I} \neq J^\varphi} n'_{I(1),\ldots,I(\varphi),f}$$

Furthermore:

$$n' = \sum_{\mathbf{S} \neq J^\varphi} n''_{I(1),\ldots,I(\varphi),S(1),\ldots,S(\varphi)} \prod_{u=1}^{\varphi} \delta_{S(u)f(I(u))}$$

where $n''_{I(1),\ldots,I(\varphi),S(1),\ldots,S(\varphi)}$ equal 1 or 0 for the given $\mathbf{I}$ and $\mathbf{S}$. So:

$$\sum_{f \in \mathcal{B}_{nk}} \sum_{\substack{\mathbf{I} \neq J^\varphi \\ \mathbf{S} \neq J^\varphi}} n'' \prod_{u=1}^{\varphi} \delta_{S(u)f(I(u))} = \sum_{\substack{\mathbf{I} \neq J^\varphi \\ \mathbf{S} \neq J^\varphi}} n'' \sum_{f \in \mathcal{B}_{nk}} \prod_{u=1}^{\varphi} \delta_{S(u)f(I(u))}$$

And,

$$\sum_{f \in \mathcal{B}_{kn}} \prod_{u=1}^{\varphi} \delta_{S(u)f(I(u))} = \operatorname{card}\{f \in \mathcal{B}_{nk} \mid \forall u \in \{1,\ldots,\varphi\},\ f(I(u)) = S(u)\}$$

$$= (N^k - \varphi)!$$

Finally,

$$
\begin{aligned}
\mathbb{E}(\mathcal{N}) &= \left(\frac{(N^k - \varphi)!}{N^k!}\right)^2 \frac{m!}{(m - \varphi)!} \sum_{\substack{\mathbf{I} \neq \in J^\varphi \\ \mathbf{S} \neq \in J^\varphi}} n''_{\mathbf{I},\mathbf{S}} \\
&= \left(\frac{(N^k - \varphi)!}{N^k!}\right)^2 \frac{m!}{(m - \varphi)!} \operatorname{card}\{\mathbf{I}, \mathbf{S} \neq \in J^\varphi \text{ satisfying } (A)\} \\
&= \left(\frac{(N^k - \varphi)!}{N^k!}\right)^2 \frac{m!}{(m - \varphi)!} P(N)
\end{aligned}
$$

as claimed. □

3.2 Computation of $\mathbb{E}(\tilde{\mathcal{N}})$

As previously, we have $1 \leq u \leq \varphi$.
The same computation for $\mathbb{E}(\tilde{\mathcal{N}})$ gives:

$$
\mathbb{E}(\tilde{\mathcal{N}}) = \frac{(N^k - \varphi)! \frac{m!}{(m-\varphi)!}}{N^k!} \frac{1}{|\Psi^d|} \sum_{\substack{\mathbf{I} \neq \in J^\varphi \\ \mathbf{S} \neq \in J^\varphi}} n''_{\mathbf{I},\mathbf{S}} \times \operatorname{card}\{f \in \Psi^d \mid \forall u, \ f(I(u)) = S(u)\}
$$

If $I(u)$ are distinct and f is a bijection with $f(I(u)) = S(u)$ for all u, then we have necessary $S(u)$ distinct, so we do not need to add this condition, and:

$$
\mathbb{E}(\tilde{\mathcal{N}}) = \frac{(N^k - \varphi)! \frac{m!}{(m-\varphi)!}}{N^k!} \frac{1}{|\Psi^d|} \underbrace{\sum_{\substack{\mathbf{I} \neq \in J^\varphi \\ \mathbf{S} \in J^\varphi}} n''_{\mathbf{I},\mathbf{S}} \times \operatorname{card}\{f \in \Psi^d \mid \forall u, \ f(I(u)) = S(u)\}}_{\Sigma_1}
$$

For the Feistel scheme used to go from $I(u)$ to $S(u)$, we only have to know some information from the internal functions $f^1, \ldots, f^d$. For this, we need to introduce new variables that are the outputs of the internal functions.

Definition 1. *For all* $u \in \{1, \ldots, \varphi\}$ *and all turn* $r \in \{1, \ldots, d\}$ $K_r(u) = f^r(I_1^{r-1}(u))$, *with the following rule:* $I_1^r(i) = I_1^r(j) \iff K_{r+1}(i) = K_{r+1}(j)$. *We define* $\mathbf{K} = K_r(u)_{1 \leq u \leq \varphi, \ 1 \leq r \leq d}$.

Taking into account this rule, how many different cases do we have to consider?

For each turn r, we consider the equalities between $I_1^{r-1}(1), \ldots, I_1^{r-1}(\varphi)$. This corresponds to the number of partition of a set of φ elements. The number of partition of a set of φ elements is the Bell number B_φ, and the number of partitions with p subsets is the second type number of Stirling $S(\varphi, p)$. Moreover, we have the formula:

$$
B_\varphi = \sum_{p=1}^{\varphi} S(\varphi, p)
$$

When we want to compute $S(\varphi, p)$, there exist several kinds of partitions. For example, if $\varphi = 5$ and $p = 3$, we have $\binom{5}{3} = 10$ ways to have a group of 3 and two groups of 1, and $5 \times 3 = 15$ ways to have two groups of 2 and one group of 1. So $S(5,3) = 25$. In that case, there are 2 kinds of partitions. Thus, in computing $S(\varphi, p)$, the number t will denote the chosen kind of partition. We obtain:

$$\Sigma_1 = \sum_{p_1=1}^{\varphi} \cdots \sum_{p_d=1}^{\varphi} \sum_{t_1=1}^{S(\varphi,p_1)} \cdots \sum_{t_d=1}^{S(\varphi,p_d)} \operatorname{card}\{I \not\in J^{\varphi}, f \in \Psi^d \text{ satisfying } (A)$$
$$\text{and induced equalities and non-equalities from } t_1, \ldots, t_d\}$$

Let (A') be the system derived from (A) where we have added the equalities and non-equalities induced by the choice of $p_1, \ldots, p_d$, and $t_1, \ldots, t_d$, and where we have replaced the values of $\mathbf{S}$ in function of $\mathbf{I}$ and $\mathbf{K}$. For this system, the variables are only $I_i(u)$ and $K_r(u)$ (in total $k\varphi + d\varphi$ variables).

$$\Sigma_1 = \sum_{(p_1,\ldots,p_d)} \sum_{(t_1,\ldots,t_d)} \frac{|\Psi^d|}{N^{p_1+p_2+\ldots p_d}} \operatorname{card}\{\mathbf{I} \not\in J^{\varphi}, \mathbf{K} \text{ satisfying } (A')\}$$
$$= \frac{|\Psi^d|}{N^{\varphi d}} \sum_{p_1,\ldots,p_d} \sum_{t_1,\ldots,t_d} \operatorname{card}\{\mathbf{I} \not\in J^{\varphi}, \mathbf{K}, K'_1, \ldots, K'_{\varphi d-p} \text{ satisfying } (A')\}$$

where K'_i are (artificial) other values taken by the round functions that enabled us to simplify the computation (we just increase by one the number of variables each time we have an equality between the variables $I_1^r(u)$). We can now state the result:

Proposition 2. *Let Q be the polynomial defined by:*

$$Q(N) = \sum_{(p_1,\ldots,p_d)} \sum_{(t_1,\ldots,t_d)} \operatorname{card}\{\mathbf{I} \not\in J^{\varphi}, \mathbf{K}, K'_1 \ldots K'_{\varphi d-p} \text{ satisfying } (A')\}.$$

Then: $\mathbb{E}(\tilde{N}) = \dfrac{(N^k - \varphi)!}{N^k!} \dfrac{\frac{m!}{(m-\varphi)!}}{N^{\varphi d}} Q(N).$

Remark 2. $Q(N)$ will be returned by our computer program that gives the number of solutions of the systems. Here d is the number of internal functions involved in the attacks. For the special case of the expanding schemes, we consider that each round function counts as $k-1$ internal functions, so $d = r \times (k-1)$, where r is the number of rounds.

3.3   Computing $V(\mathcal{N})$

Proposition 3. *Let $X = \sum_{i=1}^{n} X_i$ be a random variable, where each X_i follow a Bernoulli distribution. Then,* $V(X) = -\mathbb{E}(X)^2 + \sum_{i,j} \mathbb{E}(X_i X_j)$

Proof.

$$V(X) = \mathbb{E}(X^2) - \mathbb{E}(X)^2 = \mathbb{E}\left(\sum_{i=1}^{n} X_i^2 + \sum_{i\neq j} X_i X_j\right) - \mathbb{E}(X)^2$$

$$= \mathbb{E}(X) - \mathbb{E}(X)^2 + \sum_{i\neq j} \mathbb{E}(X_i X_j) = -\mathbb{E}(X)^2 + \sum_{i,j} \mathbb{E}(X_i X_j)$$

□

Since $\mathcal{N} = \sum_{i\neq\{1,\ldots,m\}^\varphi} \mathcal{N}_i$, we have the corollary:

Corollary 1.

$$V(\mathcal{N}) = -\mathbb{E}(\mathcal{N})^2 + \sum_{i,j} \mathbb{E}(\mathcal{N}_i\mathcal{N}_j) \text{ and } V(\tilde{\mathcal{N}}) = -\mathbb{E}(\tilde{\mathcal{N}})^2 + \sum_{i,j} \mathbb{E}(\tilde{\mathcal{N}}_i\tilde{\mathcal{N}}_j)$$

We now continue with the computation of $\sum_{i,j} \mathbb{E}(\mathcal{N}_i\mathcal{N}_j)$.

Proposition 4. *Let $P(N,m)$ be defined by:*

$$P(N,m) = \sum_{h=0}^{\varphi} \prod_{i=1}^{h}(N^k - 2\varphi + i)^2 \prod_{i=h+1}^{\varphi}(m - 2\varphi + i) \sum_{\substack{I,I'\neq J^\varphi \\ |I\cap I'|=h}} \sum_{\substack{S,S'\neq J^\varphi \\ S(i)=S'(i) \\ I(i)=I'(i)}} n''_{I,S} n''_{I',S'} \quad (1)$$

Then,

$$\sum_{i,j} \mathbb{E}(\mathcal{N}_i\mathcal{N}_j) = \frac{\prod_{i=1}^{\varphi}(m - i + 1)}{\prod_{i=1}^{2\varphi}(N^k - i + 1)^2} P(N,m) \quad (2)$$

Remark 3. $P(N,m)$ will be returned by the computer program.

Proof. See Appendix A. □

Corollary 2. *We have:* $V(\mathcal{N}) = -\mathbb{E}(\mathcal{N})^2 + \dfrac{\prod_{i=1}^{\varphi}(m - i + 1)}{\prod_{i=1}^{2\varphi}(N^k - i + 1)^2} P(N,m).$

Table 2 shows how many internal functions are used for r rounds, according to the type of the scheme, and the number of branches.

Table 2. Number d of internal functions involved in different schemes.

Scheme	Rounds	Branches	Internal functions
Classical	r	2	$d = r$
Type 1/contracting	r	k	$d = r$
Type 2	r	k (even)	$d = \frac{k}{2}r$
Type 3/expanding	r	k	$d = (k-1)r$

4   Previous Attacks and Their Limitations: Contribution of the Tool

4.1   Distinguishing Attacks

As mentioned in the introduction, the attacks considered here are successful when either $\tilde{\mathcal{N}}$ is significantly greater than $\mathcal{N}$ or when these quantities are of the same order but $|\mathbb{E}(\mathcal{N}) - \mathbb{E}(\tilde{\mathcal{N}})| \geq \sigma(\mathcal{N})$. We now precise this point.

More generally, we suppose X and X' are random independent variables that verify: $\mathbb{E}(X) - \mathbb{E}(X') \approx \sigma(X)$ and $\sigma(X) \approx \sigma(X')$.

We suppose also that X and X' follow a Gaussian distribution.

Let $Z = X - X'$. We have $Z \hookrightarrow \mathcal{N}(\sigma(X), \sqrt{2}\sigma(X))$ since $V(X - X') = V(X) + V(X') \approx 2V(X)$. The attacks works when $Z \geq 0$. We can compute that $P(Z \geq 0)$ is equal to 76%. If we generalize with $\mathbb{E}(X) - \mathbb{E}(X') \approx \alpha\sigma(X)$, then we have

α	$0,1$	$0,5$	1	$3,3$
$P(Z \geq 0) \approx$	$52,8\%$	$63,8\%$	76%	99%
$Advantage$	$5,6\%$	$27,6\%$	52%	98%

Thus asking for the condition $|\mathbb{E}(\mathcal{N}) - \mathbb{E}(\tilde{\mathcal{N}})| \geq \sigma(\mathcal{N})$ gives a successful attack. We point out that all our attacks, by hand, we obtain that $\sigma(\mathcal{N})$ and $\sigma(\tilde{\mathcal{N}})$ are of the same order. We have the algorithm to compute the standard deviation of $\sigma(\tilde{\mathcal{N}})$, but the program takes a lot of time, since we have to consider many more systems. However, if $\sigma(\mathcal{N})$ and $\sigma(\tilde{\mathcal{N}})$ are not of the same order, this will lead to an attack by the variance.

4.2   Previous Attacks and Their Limitations

All the attacks done previously [15,18,25] can be studied using the tool. This what is done in this paper for Feistel schemes of type 1, 2 and 3, unbalanced Feistel schemes with either expanding or contracting functions on several examples. By hand, it is tedious to compute $\mathbb{E}(\mathcal{N})$, $\mathbb{E}(\tilde{\mathcal{N}})$ and $\sigma(\mathcal{N})$. This comes from the fact that we have to compute the number of solutions of the so-called Mirror systems, i.e. systems of linear equalities and linear non-equalities. Most of the time, we look at the most significant equations, where we have the number of equalities as small a possible. This means that we can add one or two more equalities to the set on equalities imposed by the conditions of the attacks. This is why

Table 3. Complexities obtained from previous works.

Scheme	Complexity for r rounds	Max number of rounds - complexity	Reference
Type 1	$r \leq 21,\ 2^{(\frac{r-6}{4})n}$	$22 \rightarrow 2^{4n}$	[15]
Type 2	$7 \leq r \leq 9,\ 2^{(\frac{r-2}{2})n}$	$10 \rightarrow 2^{4n}$	[15]
Type 3	$r = 5, 6,\ 2^{(r-3)n}$	$7 \rightarrow 2^{4n}$	[15]
Contracting	$r = 5, 6,\ 2^{(\frac{r-2}{2})n}$	$7 \rightarrow 2^{5n}$	[18]
Expanding	$r \leq 10,\ 2^{(\frac{r+4}{2})n}$	$11 \rightarrow 2^{\frac{39}{10}n}$	[25]

the values of $\mathbb{E}(\mathcal{N})$, $\mathbb{E}(\tilde{\mathcal{N}})$ and $\sigma(\mathcal{N})$ are given with a $O()$ function. In Table 3, we give examples of complexities obtained in previous studies. We omit the $O()$ function. The values are given for schemes with $k = 4$ branches, since in Sect. 5, we will study this case. In Sect. 5, we give examples of attacks for which we get a better complexity or we can attack more rounds. It is not possible, due to the lack of space, to provide an exhaustive list of all the attacks, but it is most likely that thanks to the tool, we can obtain better complexities for some values of n.

4.3 Use of the Tool

As already explained, the exact computation of expectations and standard deviation, allows to know the $O()$ function and then to improve the attack, since we have a precise value of all quantities. Indeed, the conditions to have a successful attack are more accurate and according to the number of rounds, we obtain the number of messages needed to distinguish a random permutation from a permutation produced by the scheme. For attacks that use couples of input/output pairs (2-point attacks as defined in the next section), generally, the situation is as follows, we have $|\mathbb{E}(\mathcal{N}) - \mathbb{E}(\tilde{\mathcal{N}})| = O(\frac{m^2}{4^{\alpha n}})$ where α decreases when r increases since the differential path is longer. But $\sigma(\mathcal{N}) = O(\frac{m}{2^{\beta n}})$ and β does not change. This shows that for a given number of rounds, the attack is better for small values of n. The knowledge of the $O()$ functions is important in order to tighten the bounds.

5   Improvements of Previous Attacks on Different Types of Feistel Schemes

Most of concrete block ciphers have four branches (CLEFIA [23], SMS4 [12], MARS [7], RC6 [21]). This is why, in this section, we choose $k = 4$ and we will give examples of attacks that can be improved on several kind of Feistel schemes by using the exact values for expectations and standard deviations.

5.1   Notation for Feistel-Type Schemes

With Feistel type schemes, according to the structure of the scheme, internal variables are defined at each round. Specific equalities on these internal variables will allow the differential path to propagate. Thus, once equlities on the input variables are settled, the wanted conditions on the output variables will appear either at random or due to equalities satisfied by some internal variables. We consider two types of attacks: 2-point attacks, where we use plaintext/cipher-text pairs and φ-point attacks, where we use φ-tuples of plaintext/ciphertexts. These last attacks allow more possibilities of conditions on the inputs, the outputs and the internal variables. They were first introduced in [11] and then generalized in [19,25]. Since $k = 4$, the input and output are denoted by $[I_1, I_2, I_3, I_4]$ and $[S_1, S_2, S_3, S_4]$. For the message i, the input and the output are denoted by $[I_1(i), I_2(i), I_3(i), I_4(i)]$ and $[S_1(i), S_2(i), S_3(i), S_4(i)]$.

5.2   Examples of 2-Point Attacks

Differential Notation for 2-Point Attacks. We use plaintext/ciphertext pairs. In KPA, on the input variables, the notation $[\mathbf{0}, \mathbf{0}, \Delta_3^0, \Delta_4^0]$ means that the pair of messages (i, j) satisfies $I_1(i) = I_1(j)$, $I_2(i) = I_2(j)$, and $I_s(i) \oplus I_s(j) = \Delta_s^0$, $3 \leq s \leq 4$. The differential of the outputs i and j after round t is denoted by $[\Delta_1^t, \Delta_2^t, \Delta_3^t, \Delta_4^t]$. At each round, internal variables are defined by the structure of the scheme. In our attacks, we determine equalities that have to be satisfied by the inputs and the outputs. With a scheme, some equalites on the internal variables on some rounds will allow the differential path to propagates. On an intermediate round, when equalities on the internal variables are needed in order to get a differential characteristic, we use the notation $\boxed{0}$ to mean that the corresponding internal variables are equal in messages i and j. When we write 0, this means that the differential path propagates without any constraint on the internal functions, i.e. with probability 1.

Type-1 Feistel Schemes with $k = 4$ and $r = 22, 23$: "2-Point Attack". In Fig. 1, one round of a type-1 Feistel scheme is represented with $k = 4$. The theoretical study of [15] shows that, for $k = 4$, KPA can be mounted up to 22 rounds, where the maximal number of messages is needed, i.e. 2^{4n} messages. The attack for 22 rounds is given by

$$S_2(i) = S_2(j) \text{ and } I_1(i) \oplus I_1(j) = S_3(i) \oplus S_3(j)$$

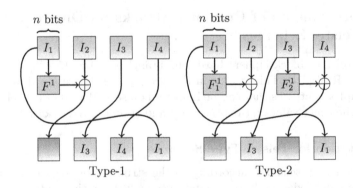

Fig. 1. One round of type-1 and type-2 Feistel schemes with $k = 4$

Table 4. Improvements on the attacks for type-1 Feistel schemes.

Round	n	m	$\mathbb{E}(\mathcal{N})$	$\mathbb{E}(\tilde{\mathcal{N}})$	$\sigma(\mathcal{N})$	$D(*)$
14	10	2^{19} (2^{20} [15])	262 143	263 426	724	552
14	8	2^{15} (2^{16} [15])	16 383	16 700	181	135
22	8	2^{28} (2^{32} [15])	10 995 115 · 10^5	10 995 137 · 10^5	1 482 503	704 854
23	6	2^{23} (×)	10 995 114 · 10^5	10 995 135 · 10^5	1 471 279	597 578
23	4	2^{14} (×)	16 776 192	16 782 642	5 608	842

$(*)\ D = |\mathbb{E}(\mathcal{N}) - \mathbb{E}(\tilde{\mathcal{N}})| - \sigma(\mathcal{N})$

We obtain $P(N) = N^8 - N^{10} - N^{11} + N^{14}$,
$Q(N) = -N^{37} + 5N^{38} - 5N^{39} - 29N^{40} + 134N^{41} - 301N^{42} + 449N^{43} - 476N^{44} + 351N^{45} - 161N^{46} + 33N^{47} + N^{50}$, and
$P(N, m) = -36N^8 m + 36N^8 m^2 + \cdots - 2N^{27} - 2N^{28} - N^{28}m + \dot{N}^{28}m^2 + 2N^{30}$.
If we use the exact values for $P(N)$, $Q(N)$ and $P(N, m)$, we obtain that this attack on 22 rounds may have a complexity better than 2^{4n} for some values of n. Moreover, depending on n, it is also possible to obtain a KPA on 23 rounds with the following attack: $I_1(i) \oplus I_1(j) = S_2(i) \oplus S_2(j)$ (the values of $P(N)$, $Q(N)$ and $P(N, m)$ are again computed with the tool). The results are summarized in Table 4. We write in parenthesis the value obtained from the theoretical study. The notation (×) means that nothing was known from the theoretical study.

Type-2 Feistel Schemes With $k = 4$ and $r = 10, 12$: "2-Point Attack". In Fig. 1, one round of a type-2 Feistel scheme is represented with $k = 4$.

The attack on 10 rounds is given by

$$I_1(i) = I_1(j) \text{ and } S_4(i) \oplus S_4(j) = I_2(i) \oplus I_2(j)$$

We have $P(N) = N^8 - N^{10} - N^{11} + N^{14}$ and
$$P(N, m) = -36N^8 m + 36N^8 m^2 - 180N^9 + 150N^9 m - 30N^9 m^2 + 246N^{10} \ldots$$
$$- 22N^{26} - 2N^{27} - 2N^{28} - N^{28}m + N^{28}m^2 + 2N^{30}$$

In order to have a better understanding of the attack, we provide in Table 5 an example of differential path, found by hand. However, the computer takes into account all the possibilities.

We obtain $Q(N) = -N^{33} + 5N^{34} - 4N^{35} - 35N^{36} + 149N^{37} - 321N^{38} + 464N^{39} - 482N^{40} + 352N^{41} - 161N^{42} + 33N^{43} + N^{46}$

The theoretical study of [15] showed that the attack on 10 rounds requires the maximal number of messages, i.e. 2^{4n} and that is was not possible to go further. However, for small values of n, by using the exact values, we obtained that it is possible to have a better complexity for 10 rounds and to attack 12 rounds. The attack for 12 rounds is given by $I_1(i) = I_1(j)$ and $I_2(i) \oplus I_2(j) = S_2(i) \oplus S_2(j)$. The results are shown in Table 6.

Type-3 Feistel schemes with $k = 4$ and r=7, 8, 9: "2-point attack". In Fig. 2, one round of a type-3 Feistel scheme is represented with $k = 4$.

Table 5. Attack on a type-2 Feistel scheme with $k = 4$ and 10 rounds.

Round	0	Δ_2^0	Δ_3^0	Δ_4^0
1	Δ_2^0	Δ_3^0	Δ_3^1	0
2	Δ_1^2	Δ_3^1	$\boxed{0}$	Δ_2^0
3	Δ_1^3	0	Δ_2^0	Δ_1^2
4	$\boxed{0}$	Δ_2^0	Δ_3^4	Δ_1^3
5	Δ_2^0	Δ_3^4	Δ_3^5	0
6	Δ_1^6	Δ_3^5	$\boxed{0}$	Δ_2^0
7	Δ_1^7	0	Δ_2^0	Δ_1^6
8	$\boxed{0}$	Δ_2^0	Δ_3^8	Δ_1^7
9	Δ_2^0	Δ_3^8	Δ_3^9	0
10	Δ_1^{10}	Δ_3^9	Δ_3^{10}	Δ_2^0

Table 6. Improvements on the attacks for type-2 Feistel schemes.

Round	n	m	$\mathbb{E}(\mathcal{N})$	$\mathbb{E}(\tilde{\mathcal{N}})$	$\sigma(\mathcal{N})$	$D(*)$
8	10	2^{27} (2^{30} [15])	$17\ 179\ 869 \cdot 10^3$	$181\ 800\ 081 \cdot 10^3$	185 363	26 861
8	8	2^{21} (2^{24} [15])	$67\ 108\ 828$	$67\ 121\ 949$	11 585	1 536
10	8	2^{28} (2^{32} [15])	$10\ 995\ 115 \cdot 10^5$	$10\ 995\ 137 \cdot 10^5$	1 482 559	704 799
12	5	2^{20} ($\times$)	$107\ 370\ 905$	$1\ 073\ 785\ 154$	45 610	30 488

(*) $D = |\mathbb{E}(\mathcal{N}) - \mathbb{E}(\tilde{\mathcal{N}})| - \sigma(\mathcal{N})$

For $k = 4$, theoretical results in [15] (see also Table 3) show that for 6 rounds, the complexity of the attacks is in $O(2^{3n})$, for 7 rounds, it is in $O(2^{\frac{7n}{2}})$, and for 8 rounds, the complexity is 2^{4n}. We reach the maximal number of messages and it is not possible to attack more rounds. We will show that once we are able to compute the exact values for the expectations and standard deviation, we can get improvements. As in previous attacks, $P(N)$, $Q(N)$ and $P(N, m)$ are returned by the computer program. We only state the values obtained for the expectations and standard deviations corresponding to specific values of n and m.

Attack on 6 Rounds. The attack is described by the following conditions:

$$\begin{cases} I_1(i) = I_1(j) \text{ and } I_2(i) = I_2(j) \text{ and } I_3(i) = I_3(j) \\ I_4(i) \oplus I_4(j) = S_1(i) \oplus S_1(j) \end{cases}$$

For $n = 8, 10$, the results are given in Table 7. The complexities are better than the theoretical ones obtained from Table 3.

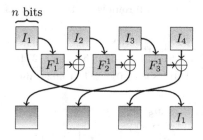

Fig. 2. First round for a type-3 Feistel scheme

Table 7. Attacks on 6 rounds for type-3 Feistel schemes.

Round	n	m	$\mathbb{E}(\mathcal{N})$	$\mathbb{E}(\tilde{\mathcal{N}})$	$\sigma(\mathcal{N})$	$|\mathbb{E}(\mathcal{N}) - \mathbb{E}(\tilde{\mathcal{N}})| - \sigma(\mathcal{N})$
6	10	2^{28} (2^{30} [15])	65 471	65 982	361	148
6	8	2^{25} (2^{28} [15])	16 318	16 824	186	323

Attack on 7 rounds. The attack is described by the following conditions:

$$\begin{cases} I_1(i) = I_1(j) \text{ and } I_2(i) = I_2(j) \text{ and } I_3(i) = I_3(j) \\ S_4(i) = S_4(j) \\ I_4(i) \oplus I_4(j) = S_1(i) \oplus S_1(j) \end{cases}$$

For $n = 8$, the results are given in Table 8. The complexity is better than expected: 2^{25} instead of 2^{28} (that was the theoretical value).

Table 8. Attacks on 7 rounds for type-3 Feistel schemes.

Round	n	m	$\mathbb{E}(\mathcal{N})$	$\mathbb{E}(\tilde{\mathcal{N}})$	$\sigma(\mathcal{N})$	$|\mathbb{E}(\mathcal{N}) - \mathbb{E}(\tilde{\mathcal{N}})| - \sigma(\mathcal{N})$
7	8	2^{25} (2^{28} [15])	1 019	1 102	45	37

Attack on 8 rounds. The attack is described by the following conditions:

$$\begin{cases} I_1(i) = I_1(j) \text{ and } I_2(i) = I_2(j) \text{ and } I_3(i) = I_3(j) \\ I_4(i) \oplus I_4(j) = S_4(i) \oplus S_4(j) \end{cases}$$

For $n = 8$, the results are given in Table 9. The complexity is better than expected: 2^{29} instead of 2^{32}, which is the maximal number of messages.

Table 9. Attacks on 8 rounds for type-3 Feistel schemes.

Round	n	m	$\mathbb{E}(\mathcal{N})$	$\mathbb{E}(\tilde{\mathcal{N}})$	$\sigma(\mathcal{N})$	$|\mathbb{E}(\mathcal{N}) - \mathbb{E}(\tilde{\mathcal{N}})| - \sigma(\mathcal{N})$
8	8	2^{29} (2^{32} [15])	66 846 719	66 867 827	11557	9549

Table 10. Attacks on 9 rounds for type-3 Feistel schemes.

Round	n	m	$\mathbb{E}(\mathcal{N})$	$\mathbb{E}(\tilde{\mathcal{N}})$	$\sigma(\mathcal{N})$	$|\mathbb{E}(\mathcal{N}) - \mathbb{E}(\tilde{\mathcal{N}})| - \sigma(\mathcal{N})$
9	4	2^{16} ($\times$)	61 440	62 055	339	274
9	5	2^{20} ($\times$)	1 015 808	1 017 281	1 402	69

n bits

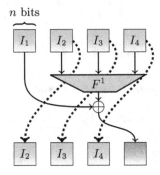

Fig. 3. One round of an unbalanced Feistel scheme with contracting functions

Attack on 9 rounds. The attack is described by the following conditions:

$$\begin{cases} I_1(i) = I_1(j) \text{ and } I_2(i) = I_2(j) \text{ and } I_3(i) = I_3(j) \\ I_4(i) \oplus I_4(j) = S_3(i) \oplus S_3(j) \end{cases}$$

For $n = 4, 5$, the results are given in Table 10.

Here for small values of n, we can attack one more round than with the theoretical study of [15].

Unbalanced Feistel Schemes with Contracting Functions. In Fig. 3 one round of an unbalanced Feistel scheme with contracting functions is represented with $k = 4$. Here we show on an example how the knowledge of the O functions determines the complexity of the attack. We take $n = 8$. As in previous attacks, $P(N)$, $Q(N)$ and $P(N, m)$ are returned by the computer program. We only state the value obtained for the expectations and standard deviations corresponding to specific values of n and m.

Attack on 6 Rounds. The attack is given by $I_4(i) = I_4(j)$ and $I_3(i) \oplus I_3(j) = S_1(i) \oplus S_1(j)$. The theoretical study of [18] shows that the complexity of the attack is $O(2^{2n})$. In fact, this O function is about $\sqrt{2}$. This factor is important since it shows that the complexity is not 2^{2n} as expected but a bit worse, as we can see in Table 11, where the first line shows that the theoretical study is not accurate enough:

Attack on 7 rounds. The attack is given by $I_4(i) \oplus I_4(j) = S_1(i) \oplus S_1(j)$. Again, the theoretical study of [18] shows that the complexity of the attack is $O(2^{\frac{5n}{2}})$. But this O function is about $\sqrt{2}$. Again this factor shows that the attack as

Table 11. Attacks on 6 rounds for unbalanced Feistel schemes with contracting functions.

Round	n	m	$\mathbb{E}(\mathcal{N})$	$\mathbb{E}(\tilde{\mathcal{N}})$	$\sigma(\mathcal{N})$	$\|\mathbb{E}(\mathcal{N}) - \mathbb{E}(\tilde{\mathcal{N}})\| - \sigma(\mathcal{N})$	Comment
6	8	2^{16}	65 534	65 788	362	-108	[18]
6	8	2^{17}	262 141	263 157	724	291	This paper

a complexity greater than expected by the theoretical study. The values are provided in Table 12, where the first line shows that the theoretical study is not accurate enough:

Table 12. Attacks on 7 rounds for unbalanced Feistel schemes with contracting functions.

Round	n	m	$\mathbb{E}(\mathcal{N})$	$\mathbb{E}(\tilde{\mathcal{N}})$	$\sigma(\mathcal{N})$	D (*)	Comment
7	8	2^{20}	4 294 963 200	4 295 028 224	92500	-27475	[18]
7	8	2^{21}	17 179 860 992	17 180 121 091	185001	75098	This paper

(*) $D = \|\mathbb{E}(\mathcal{N}) - \mathbb{E}(\tilde{\mathcal{N}})\| - \sigma(\mathcal{N})$

Remark 4. In [18], the condition is $O(2^{2n})$ for 6 rounds as shown in Table 3, but the $O()$ function is not evaluated. If we test $2^{2n} = 2^{16}$ messages when $n = 8$, we can see that the attack is not valid. The computation of the exact value of expectations and standard deviations shows that there is a factor of $\sqrt{2}$. Thus if we choose 2^{17} messages, then the attack is successful. The same phenomena appears for seven rounds. In [18], it is shown that it is possible to attack up to $2k - 1$ rounds and with a complexity in $O(2^{5n/2})$, but again 2^{20} messages is not successful and we need to take 2^{21} messages.

5.3   Example of a 4-Point Attack (Rectangle Attack)

Introduction to 4-Point Attacks. In [11,19,25], rectangle attacks (i.e. attacks that use φ-tuples of points, φ even) were investigated in order to get more efficient attacks on unbalanced Feistel scheme with expanding functions. These attacks allow to attack more rounds and there are more possibilities for differential paths than 2-point attacks. We now explain the notation for 4-point attacks, since these are the only attacks involved here. In 4-point attacks, there is a set of equalities (a mirror system) that have to be satisfied by 4-tuples of points (inputs/outputs). As explained before, the number of this 4-tuples will be different for a random permutation and for permutation produced by a scheme. This is due to the fact, that with a scheme, equalities on the internal variables produced during intermediate rounds will help the equalities to propagate. Thus, the equalities on the outputs will appear more frequently. We now explain what kind of equalities may be satisfy either by the inputs, or by the internal variables, or by the outputs in the case of 4-point attacks. We give them on the inputs. The same definition applies for the internal variables and the outputs.

1. "Horizontal equalities" on I_i: $I_i(1) = I_i(3)$, $I_i(2) = I_i(4)$.
2. "Vertical equalities" on I_i: $I_i(1) = I_i(2)$, $I_i(3) = I_i(4)$.
3. "Differential equalities" on I_i: $I_i(1) \oplus I_i(2) = I_i(3) \oplus I_i(4)$.

We notice that if we have $I_i(1) \oplus I_i(2) = I_i(3) \oplus I_i(4)$, then $I_i(1) = I_i(3)$ will imply $I_i(2) = I_i(4)$ and $I_i(1) = I_i(2)$ will imply $I_i(3) = I_i(4)$. Figure 4 shows why we choose the terms of "vertical" and "horizontal" equalities.

Horizontal conditions

| $I(1)$ | $S(1)$ | $I_1(1) = I_1(3)$ | $I(3)$ | $S(3)$ |

Vertical conditions $\quad$ $S_3(1){=}S_3(2)$ $I_2(1){=}I_2(2)$ $\qquad$ $S_3(3){=}S_3(4)$ $I_2(3){=}I_2(4)$

| $I(2)$ | $S(2)$ | - - - - - - - - - | $I(4)$ | $S(4)$ |

Fig. 4. Example of differential equalities

Unbalanced Feistel schemes with expanding functions with $k = 4$ and $r = 10, 11, 12$: "4-point attack". We provide in Fig. 5, the first round of an unbalanced Feistel scheme with expanding functions when $k = 4$.

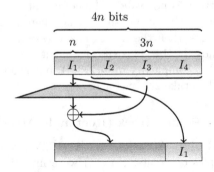

Fig. 5. One round of an unbalanced Feistel scheme with expanding functions

Better results on the differential paths when $r = 10$. We have m messages and we want to compute the expectation of the number $\mathcal{N}$ of 4-tuples (i, j, ℓ, p) of points satisfying the following relations:

$$\begin{cases} I_1(i) = I_1(j) \\ I_1(\ell) = I_1(p) \neq I_1(i) \\ I_2(i) = I_2(j) \\ I_2(\ell) = I_2(p) \neq I_2(i) \\ I_3(i) = I_3(j) \\ I_3(\ell) = I_3(p) \neq I_3(i) \\ I_4(i) \oplus I_4(j) = I_4(\ell) \oplus I_4(p) \neq 0 \end{cases} \qquad \begin{cases} S_1(i) \oplus S_1(j) = S_1(\ell) \oplus S_1(p) \neq 0 \\ S_2(i) = S_2(\ell) \\ S_2(j) = S_2(p) \neq S_2(i) \\ S_3(i) = S_3(\ell) \\ S_3(j) = S_3(p) \neq S_3(i) \\ S_4(i) = S_4(\ell) \\ S_4(j) = S_4(p) \neq S_4(i) \end{cases}$$

Here we have vertical equalities on the inputs, horizontal equalities on the outputs and differential equalities on both outputs and inputs.

We obtain: $P(N) = 4N^8 - 8N^9 + 4N^{10} - 4N^{12} + 8N^{13} - 4N^{14} + N^{16} - 2N^{17} + N^{18}$. With a scheme, the conditions on the outputs may appear at random or due to equalities satisfied by the internal variables (this is related to the behavior of the round functions). If we want to keep the differential equalities inside the path, we need to have either vertical or horizontal equalities satisfied by the internal variables. In Table 13, we show two differential paths that use horizontal and vertical equalities. To be more specific, when we write $[0, .\Delta_2^0, \Delta_3^0, \Delta_4^0]$, this means that, on the input variables, vertical equalities on the first coordinate and horizontal equalities on the second coordinate. The same notation applies to the internal variables and the output variables. The differential path is constructed such that we always keep the differential equalities. In the differential path, when a vertical equality is needed for the propagation of the differential path we indicate this by $\boxed{0}$. For a horizontal equality, we use •. We will write 0 and ., when these equalities propagate without any constraint on the internal functions.

Table 13. Example of two differential paths.

Path 1

round	0	0	0	Δ_4^0
1	0	0	Δ_4^0	0
2	0	Δ_4^0	0	0
3	•Δ_4^0	0	0	0
4	•Δ_1^4	Δ_2^4	Δ_3^4	.Δ_4^0
5	•Δ_1^5	Δ_2^5	.Δ_3^5	.Δ_1^4
6	$\boxed{0}$	.Δ_2^6	.Δ_5^6	.Δ_1^5
7	•Δ_2^6	Δ_3^6	Δ_1^5	0
8	•Δ_1^8	Δ_2^8	Δ_3^8	.Δ_2^6
9	•Δ_1^9	Δ_2^9	.Δ_3^9	.Δ_1^8
10	Δ_1^{10}	.Δ_2^{10}	.Δ_3^{10}	.Δ_1^9

Path 2

round	0	0	0	Δ_4^0
1	0	0	Δ_4^0	0
2	0	Δ_4^0	0	0
3	•Δ_4^0	0	0	0
4	•Δ_1^4	Δ_2^4	Δ_3^4	.Δ_4^0
5	$\boxed{0}$	Δ_2^5	.Δ_3^5	.Δ_1^4
6	$\boxed{0}$	Δ_3^5	Δ_1^4	0
7	•Δ_3^5	Δ_1^4	0	0
8	•Δ_1^8	Δ_2^8	Δ_3^8	.Δ_3^5
9	•Δ_1^9	Δ_2^9	.Δ_3^9	.Δ_1^8
10	Δ_1^{10}	.Δ_2^{10}	.Δ_3^{10}	.Δ_1^9

However, the study of all different paths obtained by our computer program shows that there are also another kind of equalities that we can call "diagonal equalities". This definition is taken according to Fig. 4. For example, on the inputs diagonal equalities are defined by: $I_i(1) = I_i(4)$ and $I_i(2) = I_i(3)$ on points numbered 1, 2 3 and 4. These kind of equalities were not used in the attacks of [25]. Diagonal equalities are denoted by $\star$ in the tables.

If we use these diagonal equalitites, we can modify the previous paths in the following way and obtain new paths. In Table 14, we give examples of differential paths obtained from the previous path 1 by introducing diagonal equalities.

Thanks to the computer program, it is possible to obtain all the possible differential paths, and this shows that all these attacks will succeed with high probability since for the same input and output equalitites, we get much more paths than expected.

Table 14. Path 1 with one, two or three $\star$ equalities.

round	0	0	0	Δ_4^0	round	0	0	0	Δ_4^0	round	0	0	0	Δ_4^0
1	0	0	Δ_4^0	0	1	0	0	Δ_4^0	0	1	0	0	Δ_4^0	0
2	0	Δ_4^0	0	0	2	0	Δ_4^0	0	0	2	0	Δ_4^0	0	0
3	$\star\Delta_4^0$	0	0	0	3	$\bullet\Delta_4^0$	0	0	0	3	$\star\Delta_4^0$	0	0	0
4	$\star\Delta_1^4$	Δ_2^4	Δ_3^4	$\star\Delta_4^0$	4	$\star\Delta_1^4$	Δ_2^4	Δ_3^4	$.\Delta_4^0$	4	$\star\Delta_1^4$	Δ_2^4	Δ_3^4	$\star\Delta_4^0$
5	$\bullet\Delta_1^5$	Δ_2^5	$\star\Delta_3^5$	$\star\Delta_1^4$	5	$\bullet\Delta_1^5$	Δ_2^5	Δ_3^5	$\star\Delta_1^4$	5	$\star\Delta_1^5$	Δ_2^5	$\star\Delta_3^5$	$\star\Delta_1^4$
6	$\boxed{0}$	Δ_2^6	Δ_5^6	$.\Delta_1^5$	6	$\boxed{0}$	Δ_2^6	Δ_5^6	$.\Delta_1^5$	6	$\boxed{0}$	$\star\Delta_2^6$	$\star\Delta_5^6$	$\star\Delta_1^5$
7	$\bullet\Delta_2^6$	Δ_3^6	Δ_1^5	0	7	$\bullet\Delta_2^6$	Δ_3^6	Δ_1^5	0	7	$\bullet\Delta_2^6$	Δ_3^6	Δ_1^5	0
8	$\bullet\Delta_1^8$	Δ_2^8	Δ_3^8	$.\Delta_2^6$	8	$\bullet\Delta_1^8$	Δ_2^8	Δ_3^8	$.\Delta_2^6$	8	$\bullet\Delta_1^8$	Δ_2^8	Δ_3^8	$.\Delta_2^6$
9	$\bullet\Delta_1^9$	Δ_2^9	$.\Delta_3^9$	$.\Delta_1^8$	9	$\bullet\Delta_1^9$	Δ_2^9	$.\Delta_3^9$	$.\Delta_1^8$	9	$\bullet\Delta_1^9$	Δ_2^9	$.\Delta_3^9$	$.\Delta_1^8$
10	Δ_1^{10}	$.\Delta_2^{10}$	$.\Delta_3^{10}$	$.\Delta_1^9$	10	Δ_1^{10}	$.\Delta_2^{10}$	$.\Delta_3^{10}$	$.\Delta_1^9$	10	Δ_1^{10}	$.\Delta_2^{10}$	$.\Delta_3^{10}$	$.\Delta_1^9$

Moreover, these diagonal equalities can also be set on the input and output variables. Thus for the same scheme, there are many attacks with the same complexity and for each attack, there are more paths than shown in previous studies. For example, for this attack on 10 rounds, we obtain 35 differential paths besides the random one. We have: $Q(N) = 48N^{114} - 320N^{115} + 876N^{116} - 1252N^{117} + 992N^{118} - 468N^{119} + 208N^{120} - 120N^{121} + 36N^{122}$. Here we will obtain that the expectation for a scheme is 36 times the expectation for a random permutation. Thus if $m \sim N^{7/2}$, i.e. $m \sim 2^{7n/2}$, then $\mathbb{E}(\mathcal{N})$ is close to 1 and $\mathbb{E}(\tilde{\mathcal{N}})$ is close to 36. Thus we can distinguish a permutation generated by F_4^{10} from a random permutation and attack is successful. Here there is no need to compute the standard deviation.

Better results for small values of n when $r = 11, 12$. The theoretical study of [25] shows that, for any k, it is possible to get KPA up to $3k - 1$ rounds, with a $2k + 2$-point attacks. This attack needs $2^{(k - \frac{1}{2k+2})n}$ messages. When $k = 4$, the exact computation allows to show that for small values of n, it is possible to obtain, for 11 rounds, a 4-point attack that has a better complexity than the 10-point attack of [25], and also that it is possible to attack 12 rounds instead of 11 rounds. For the 4-point attacks on 11 and 12 rounds, we choose the same relations as those for the attack on 10 rounds, but here we need to compute the standard deviation. Again, $P(N)$, $Q(N)$ and $P(N, m)$ are returned by the computer program. Results are summarized in Table 15.

Table 15. Improvements on the attacks for unbalanced Feistel schemes with expanding functions.

Round	n	m	$\mathbb{E}(\mathcal{N})$	$\mathbb{E}(\tilde{\mathcal{N}})$	$\sigma(\mathcal{N})$	$\|\mathbb{E}(\mathcal{N}) - \mathbb{E}(\tilde{\mathcal{N}})\| - \sigma(\mathcal{N})$
11	16	2^{62} ($2^{62.4}$ [25])	16 776 704	16 802 558	8191	17662
11	8	2^{30} ($2^{31.2}$ [25])	254	353	31	67
12	7	2^{28} ($\times$)	16 129	16 414	254	31
12	6	2^{24} ($\times$)	3 969	4 243	126	148
12	4	2^{15} ($\times$)	14	27	7	6

6   Conclusion

In this paper, we showed that it is possible to improve some attacks on Feistel type schemes. We used a tool that allows to compute the exact value of the expectations and standard deviations. This could open a new way to an accurate investigation of attacks on concrete ciphers. Also, the "diagonal equalities" obtained for unbalanced Feistel schemes with expanding functions can be used for the more general rectangle attacks as studied in [25]. Moreover this tool can also be used to count the number of solutions of Mirror systems which can be also useful to get security bounds. However, there are still gaps between bounds obtained from security results (for example from [9,26]) and bounds given from the attacks.

A   Proof of Proposition 4

Proof. We have:

$$\sum_{i,j} \mathbb{E}(\mathcal{N}_i \mathcal{N}_j) = \mathbb{E}(\sum_{i,j} \mathcal{N}_i \mathcal{N}_j)$$

$$= \frac{\sum_{f \in \mathcal{B}_{kn}} \sum_{M \in J_m} \sum_{i,j} n_{i,f,M} n_{j,f,M}}{|\mathcal{B}_{kn}| \times |J_m|}$$

$$\sum_{M \in J_m} \sum_{i,j} n_{i,f,M} n_{j,f,M} = \sum_{M \in J_m} \sum_{i,j} \sum_{I \neq J^\varphi} \sum_{I' \neq J^\varphi} n_{i,f,M} n_{j,f,M} \prod_{u=1}^{\varphi} \delta_{I(u)M(i_u)} \delta_{I'(u)M(j_u)}$$

$$= \sum_{M \in J_m} \sum_{i,j} \sum_{I \neq J^\varphi} \sum_{I' \neq J^\varphi} n'_{I,f} n'_{I',f} \prod_{u=1}^{\varphi} \delta_{I(u)M(i_u)} \delta_{I'(u)M(j_u)}$$

$$= \sum_{I \neq J^\varphi} \sum_{I' \neq J^\varphi} n'_{I,f} n'_{I',f} \underbrace{\sum_{M \in J_m} \sum_{i,j} \prod_{u=1}^{\varphi} \delta_{I(u)M(i_u)} \delta_{I'(u)M(j_u)}}_{1 \text{ iff } I \subset M \text{ and } I' \subset M, 0 \text{ otherwise}}$$

$$= \sum_{I \neq J^\varphi} \sum_{I' \neq J^\varphi} n'_{I,f} n'_{I',f} \text{card}\{M \in J_m \mid I \subset M \text{ and } I' \subset M\}$$

$$= \sum_{h=0}^{\varphi} \sum_{\substack{I,I' \neq J^\varphi \\ |I \cap I'|=h}} n'_{I,f} n'_{I',f} \underbrace{\binom{N^k - 2\varphi + h}{m - 2\varphi + h}}_{\Lambda}$$

$$= \sum_{h=0}^{\varphi} \Lambda \sum_{\substack{I,I' \neq J^\varphi \\ |I \cap I'|=h}} \sum_{\substack{S,S' \neq J^\varphi \\ S(i)=S'(i) \\ I(i)=I'(i)}} n''_{I,S} n''_{I',S'} \prod_{u=1}^{\varphi} \delta_{S(u)f(I(u))} \delta_{S'(u)f(I'(u))}$$

So, $\displaystyle\sum_{f\in\mathcal{B}_{kn}}\sum_{M\in J_m}\sum_{i,j} n_{i,f,M}n_{j,f,M} =$

$$= \sum_{\substack{h=0}}^{\varphi} \Lambda \sum_{\substack{I,I'\notin J^\varphi \\ |I\cap I'|=h}} \sum_{\substack{S,S'\notin J^\varphi \\ S(i)=S'(i) \\ I(i)=I'(i)}} n''_{I,S}n''_{I',S'}\,\mathrm{card}\{f\in\mathcal{B}_{kn}\mid f(I)=S,f(I')=S'\}$$

$$= \sum_{\substack{h=0}}^{\varphi} \Lambda \sum_{\substack{I,I'\notin J^\varphi \\ |I\cap I'|=h}} \sum_{\substack{S,S'\notin J^\varphi \\ S(i)=S'(i) \\ I(i)=I'(i)}} n''_{I,S}n''_{I',S'}(N^k - 2\varphi + h)!$$

$$= \sum_{\substack{h=0}}^{\varphi} \frac{(N^k - 2\varphi + h)!^2}{(N^k - m)!(m - 2\varphi + h)!} \sum_{\substack{I,I'\notin J^\varphi \\ |I\cap I'|=h}} \sum_{\substack{S,S'\notin J^\varphi \\ S(i)=S'(i) \\ I(i)=I'(i)}} n''_{I,S}n''_{I',S'}$$

So,

$$\sum_{i,j}\mathbb{E}(\mathcal{N}_i\mathcal{N}_j) = \frac{m!}{(N^k!)^2} \sum_{\substack{h=0}}^{\varphi} \frac{(N^k - 2\varphi + h)!^2}{(m - 2\varphi + h)!} \sum_{\substack{I,I'\notin J^\varphi \\ |I\cap I'|=h}} \sum_{\substack{S,S'\notin J^\varphi \\ S(i)=S'(i) \\ I(i)=I'(i)}} n''_{I,S}n''_{I',S'}$$

$$= \frac{\displaystyle\prod_{i=1}^{\varphi}(m-i+1)}{\displaystyle\prod_{i=1}^{2\varphi}(N^k - i + 1)^2} \sum_{h=0}^{\varphi}\prod_{i=1}^{h}(N^k - 2\varphi + i)^2 \prod_{i=h+1}^{\varphi}(m - 2\varphi + i) \sum_{\substack{I,I'\notin J^\varphi \\ |I\cap I'|=h}} \sum_{\substack{S,S'\notin J^\varphi \\ S(i)=S'(i) \\ I(i)=I'(i)}} n''_{I,S}n''_{I',S'}$$

Thus

$$\sum_{i,j}\mathbb{E}(\mathcal{N}_i\mathcal{N}_j) = \frac{\displaystyle\prod_{i=1}^{\varphi}(m-i+1)}{\displaystyle\prod_{i=1}^{2\varphi}(N^k - i + 1)^2} P(N,m)$$

as claimed   □

B   Simulation of Some KPA Attacks

We have made simulations of several attacks for small values of n in order to confirm our main results (Table 16). The results of these simulations are consistent with the theoretical study. The process of the simulations is as follow: we choose a random instance of the studied scheme (a Feistel type scheme) and a random permutation (generated by a classical Feisel scheme with 20 rounds). Then we start the attack for m messages and we count the number of plaintext/ciphertext pairs that verify the relations involved for the studied scheme and for the permutation. Finally, we repeat these steps 50 times in order to compute the mean value for the studied Feistel scheme and for the permutation and the standard deviation for the permutation.

Table 16. Simulation 50 times.

Feistel type	r	n	m	$\mathbb{E}(\mathcal{N})$	$\mathbb{E}(\tilde{\mathcal{N}})$	$\sigma(\mathcal{N})$	$D(*)$	Detection(%)
1 in Table 4	23	4	2^{14}	16776899	16783946	4583	2461	68
2 in Table 6	12	5	2^{20}	1073699658	1073787202	46537	41007	84
3 in Table 10	9	4	2^{16}	61532	61968	367	69	82

(*) $D = |\mathbb{E}(\mathcal{N}) - \mathbb{E}(\tilde{\mathcal{N}})| - \sigma(\mathcal{N})$

References

1. Encryption algorithm for computer data protection. Technical report Federal Register **40**(52) 12134. National Bureau of Standards, March 1975
2. Notice of a proposed federal information processing data encryption. Technical report Federal Register, vol. 40(149), p. 12607. National Bureau of Standards, August 1975
3. Adams, C., Heys, H., Tavares, S., Wiener, M.: The CAST-256 encryption algorithm. Technical report. AES Submission (1998)
4. Anderson, R., Biham, E.: Two practical and provably secure block ciphers: BEAR and LION. In: Gollmann, D. (ed.) FSE 1996. LNCS, vol. 1039, pp. 113–120. Springer, Heidelberg (1996). doi:10.1007/3-540-60865-6_48
5. Beaulieu, R., Shors, D., Smith, J., Treatman-Clark, S., Weeks, B., Wingers, L.: The SIMON and SPECK families of lightweight block ciphers. Cryptology ePrint Archive: 2013/404: Listing for 2013
6. Biham, E., Shamir, A.: Differential cryptanalysis of DES-like cryptosystems. J. Cryptol. **4**(1), 3–72 (1991)
7. Burwick, C., Coppersmith, D., D' Avignon, E., Gennaro, R., Halevi, S., Jutla, C., Matyas Jr., S.M., O' Connor, L., Peyravian, M., Safford, D., Zunic, N.: MARS - a candidate cipher for AES. Technical report. AES Submission (1998)
8. Daemen, J., Rijmen, V.: The Design of Rijndael. Springer-Verlag, New York (2002)
9. Hoang, V.T., Rogaway, P.: On generalized feistel networks. In: Rabin, T. (ed.) CRYPTO 2010. LNCS, vol. 6223, pp. 613–630. Springer, Heidelberg (2010). doi:10.1007/978-3-642-14623-7_33
10. Ibrahim, S., Mararof, M.A.: Diffusion analysis of scalable Feistel networks. World Acad. Sci. Eng. Technol. **5**, 98–101 (2005)
11. Jutla, C.S.: Generalized birthday attacks on unbalanced Feistel networks. In: Krawczyk, H. (ed.) CRYPTO 1998. LNCS, vol. 1462, pp. 186–199. Springer, Heidelberg (1998). doi:10.1007/BFb0055728
12. Lu, S.-W.: SMS4 encryption algorithm for wireless networks. Cryptology ePrint Archive: 2008/329: Listing for 2008, Translated from Chinese by Whitfield Diffie and George Ledin
13. Luby, M., Rackoff, C.: How to construct Pseudorandom permutations from pseudorandom functions. SIAM J. Comput. **17**(2), 373–386 (1988)
14. Nachef, V., Patarin, J., Treger, J.: Generic attacks on Misty schemes. In: Abdalla, M., Barreto, P.S.L.M. (eds.) LATINCRYPT 2010. LNCS, vol. 6212, pp. 222–240. Springer, Heidelberg (2010). doi:10.1007/978-3-642-14712-8_14
15. Nachef, V., Volte, E., Patarin, J.: Differential attacks on generalized Feistel schemes. In: Abdalla, M., Nita-Rotaru, C., Dahab, R. (eds.) CANS 2013. LNCS, vol. 8257, pp. 1–19. Springer, Cham (2013). doi:10.1007/978-3-319-02937-5_1

16. Naor, M., Reingold, O.: On the construction of pseudorandom permutations: Luby-Rackoff revisited. J. Cryptol. **12**(1), 29–66 (1999)

17. Patarin, J.: Generic attacks on Feistel schemes. In: Boyd, C. (ed.) ASIACRYPT 2001. LNCS, vol. 2248, pp. 222–238. Springer, Heidelberg (2001). doi:10.1007/3-540-45682-1_14

18. Patarin, J., Nachef, V., Berbain, C.: Generic attacks on unbalanced feistel schemes with contracting functions. In: Lai, X., Chen, K. (eds.) ASIACRYPT 2006. LNCS, vol. 4284, pp. 396–411. Springer, Heidelberg (2006). doi:10.1007/11935230_26

19. Patarin, J., Nachef, V., Berbain, C.: Generic attacks on unbalanced feistel schemes with expanding functions. In: Kurosawa, K. (ed.) ASIACRYPT 2007. LNCS, vol. 4833, pp. 325–341. Springer, Heidelberg (2007). doi:10.1007/978-3-540-76900-2_20

20. Poschmann, A., Ling, S., Wang, H.: 256 Bit standardized crypto for 650 GE – GOST revisited. In: Mangard, S., Standaert, F.-X. (eds.) CHES 2010. LNCS, vol. 6225, pp. 219–233. Springer, Heidelberg (2010). doi:10.1007/978-3-642-15031-9_15

21. Rivest, R.L., Robshaw, M., Sidney, R., Yin, Y.L.: The RC6 Block Cipher. Technical report. AES Submission (1998)

22. Schneier, B., Kelsey, J.: Unbalanced Feistel networks and block cipher design. In: Gollmann, D. (ed.) FSE 1996. LNCS, vol. 1039, pp. 121–144. Springer, Heidelberg (1996). doi:10.1007/3-540-60865-6_49

23. Shirai, T., Shibutani, K., Akishita, T., Moriai, S., Iwata, T.: The 128-bit blockcipher CLEFIA (Extended Abstract). In: Biryukov, A. (ed.) FSE 2007. LNCS, vol. 4593, pp. 181–195. Springer, Heidelberg (2007). doi:10.1007/978-3-540-74619-5_12

24. Volte, E., Nachef, V., Marriere, N.: Automatic expectation and variance computing for attacks on Feistel schemes. Cryptology ePrint Archive: 2016/136: Listing for 2016

25. Volte, E., Nachef, V., Patarin, J.: Improved generic attacks on unbalanced feistel schemes with expanding functions. In: Abe, M. (ed.) ASIACRYPT 2010. LNCS, vol. 6477, pp. 94–111. Springer, Heidelberg (2010). doi:10.1007/978-3-642-17373-8_6

26. Yun, A., Park, J.H., Lee, J.: Lai-Massey scheme and Quasi-Feistel networks. Cryptology ePrint Archive: 2007/347: Listing for 2007

27. Zheng, Y., Matsumoto, T., Imai, H.: On the construction of block ciphers provably secure and not relying on any unproved hypotheses. In: Brassard, G. (ed.) CRYPTO 1989. LNCS, vol. 435, pp. 461–480. Springer, New York (1990). doi:10.1007/0-387-34805-0_42

Primitives and Features

Updatable Functional Encryption

Afonso Arriaga[✉], Vincenzo Iovino, and Qiang Tang

SnT, University of Luxembourg, Luxembourg, Luxembourg
{afonso.delerue,vincenzo.iovino}@uni.lu, tonyrhul@gmail.com

Abstract. Functional encryption (FE) allows an authority to issue tokens associated with various functions, allowing the holder of some token for function f to learn only $f(\mathsf{D})$ from a ciphertext that encrypts D. The standard approach is to model f as a circuit, which yields inefficient evaluations over large inputs. Here, we propose a new primitive that we call *updatable functional encryption* (UFE), where instead of circuits we deal with RAM programs, which are closer to how programs are expressed in von Neumann architecture. We impose strict efficiency constrains in that the run-time of a token $\overline{\mathsf{P}}$ on ciphertext CT is proportional to the run-time of its clear-form counterpart (program P on memory D) up to a *polylogarithmic* factor in the size of D, and we envision tokens that are capable to *update* the ciphertext, over which other tokens can be subsequently executed. We define a security notion for our primitive and propose a candidate construction from obfuscation, which serves as a starting point towards the realization of other schemes and contributes to the study on how to compute RAM programs over public-key encrypted data.

Keywords: Updatable functional encryption · RAM model · Persistent memory

1 Introduction

The concept of functional encryption (FE), a generalization of identity-based encryption, attribute-based encryption, inner-product encryption and other forms of public-key encryption, was independently formalized by Boneh, Sahai and Waters [8] and O'Neil [21]. In an FE scheme, the holder of a master secret key can issue tokens associated with functions of its choice. Possessing a token for f allows one to recover $f(\mathsf{D})$, given an encryption of D. Informally, security dictates that only $f(\mathsf{D})$ is revealed about D and nothing else.

Garg et al. [14] put forth the first candidate construction of an FE scheme supporting all polynomial-size circuits based on indistinguishability obfuscation (iO), which is now known as a central hub for the realization of many cryptographic primitives [22].

The most common approach is to model functions as circuits. In some works, however, functions are modeled as Turing machines (TM) or random-access machines (RAM). Recently, Ananth and Sahai [3] constructed an adaptively

© Springer International Publishing AG 2017
R.C.-W. Phan and M. Yung (Eds.): LNCS 10311, Mycrypt 2016, pp. 347–363, 2017.
DOI: 10.1007/978-3-319-61273-7_17

secure functional encryption scheme for TM, based on indistinguishability obfuscation. Nonetheless, their work does not tackle the problem of having the token update the encrypted message, over which other tokens can be subsequently executed.

In the symmetric setting, the notion of garbled RAM, introduced by Lu and Ostrovsky [19] and revisited by Gentry et al. [15], addresses this important use-case where garbled memory data can be reused across multiple program executions. Garbled RAM can be seen as an analogue of Yao's garbled circuits [23] (see also [6] for an abstract generalization) that allows a user to garble a RAM program without having to compile it into a circuit first. As a result, the time it takes to evaluate a garbled program is only proportional to the running time of the program on a random-access machine. Several other candidate constructions were also proposed in [10–12,16].

Desmedt et al. [13] proposed an FE with controlled homomorphic properties. However, their scheme updates and re-encrypts the entire data, which carries a highly inefficient evaluation-time.

OUR CONTRIBUTION. We propose a new primitive that we call *updatable functional encryption* (UFE). It bears resemblance to functional encryption in that encryption is carried out in the public-key setting and the owner of the master secret key can issue tokens for functions—here, modeled as RAM programs—of its choice that allow learning the outcome of the function on the message underneath a ciphertext. We envision tokens that are also capable to *update* the ciphertext, over which other tokens can be subsequently executed. We impose strict efficiency constrains in that the run-time of a token $\overline{P}$ on ciphertext CT is proportional to the run-time of its clear-form counterpart (program P on memory D) up to a *polylogarithmic* factor in the size of D. We define a security notion for our primitive and propose a candidate construction based on an instance of distributional indistinguishability (DI) obfuscation, a notion introduced by [5] in the context of point function obfuscation and later generalized by [2]. Recent results put differing-inputs obfuscation (diO) [1] with auxiliary information in contention with other assumptions [7]; one might question if similar attacks apply to the obfuscation notion we require in our reduction. As far as we can tell, the answer is negative. However, we view our construction as a starting point towards the realization of other updatable functional encryption schemes from milder forms of obfuscation.

2 Preliminaries

NOTATION. We denote the security parameter by $\lambda \in \mathbb{N}$ and assume it is implicitly given to all algorithms in unary representation 1^λ. We denote the set of all bit strings of length ℓ by $\{0,1\}^\ell$ and the length of a string a by $|a|$. We write $a \leftarrow b$ to denote the algorithmic action of assigning the value of b to the variable a. We use $\perp \notin \{0,1\}^*$ to denote a special failure symbol and ϵ for the empty string. A vector of strings $\mathbf{x}$ is written in boldface, and $\mathbf{x}[i]$ denotes its ith entry. The number of entries of $\mathbf{x}$ is denoted by $|\mathbf{x}|$. For a finite set X, we denote its cardinality by $|X|$ and the action of sampling a uniformly random element a from X by $a \leftarrow_\$ X$. If $\mathcal{A}$ is a probabilistic algorithm we write $a \leftarrow_\$ \mathcal{A}(i_1, i_2, \ldots, i_n; r)$ for the action of running $\mathcal{A}$ on inputs $i_1, i_2, \ldots, i_n$ with random coins r, and assigning the result to a. For a circuit C we denote its size by $|C|$. We call a real-valued function $\mu(\lambda)$ negligible if $\mu(\lambda) \in \mathcal{O}(\lambda^{-\omega(1)})$ and denote the set of all negligible functions by NEGL. We adopt the code-based game-playing framework [4]. As usual "ppt" stands for probabilistic polynomial time.

CIRCUIT FAMILIES. Let $\mathsf{MSp} := \{\mathsf{MSp}_\lambda\}_{\lambda \in \mathbb{N}}$ and $\mathsf{OSp} := \{\mathsf{OSp}_\lambda\}_{\lambda \in \mathbb{N}}$ be two families of finite sets parametrized by a security parameter $\lambda \in \mathbb{N}$. A circuit family $\mathsf{CSp} := \{\mathsf{CSp}_\lambda\}_{\lambda \in \mathbb{N}}$ is a sequence of circuit sets indexed by the security parameter. We assume that for all $\lambda \in \mathbb{N}$, all circuits in CSp_λ share a common input domain MSp_λ and output space OSp_λ. We also assume that membership in sets can be efficiently decided. For a vector of circuits $\mathbf{C} = [C_1, \ldots, C_n]$ and a message m we define $\mathbf{C}(m)$ to be the vector whose ith entry is $C_i(m)$.

TREES. We associate a tree T with the set of its nodes $\{\mathsf{node}_{i,j}\}$. Each node is indexed by a pair of non-negative integers representing the position (level and branch) of the node on the tree. The root of the tree is indexed by $(0, 0)$, its children have indices $(1, 0)$, $(1, 1)$, etc. A binary tree is *perfectly balanced* if every leaf is at the same level.

2.1 Public-Key Encryption

SYNTAX. A public-key encryption scheme $\mathsf{PKE} := (\mathsf{PKE.Setup}, \mathsf{PKE.Enc}, \mathsf{PKE.Dec})$ with message space $\mathsf{MSp} := \{\mathsf{MSp}_\lambda\}_{\lambda \in \mathbb{N}}$ and randomness space $\mathsf{RSp} := \{\mathsf{RSp}_\lambda\}_{\lambda \in \mathbb{N}}$ is specified by three ppt algorithms as follows. (1) $\mathsf{PKE.Setup}(1^\lambda)$ is the probabilistic key-generation algorithm, taking as input the security parameter and returning a secret key sk and a public key pk. (2) $\mathsf{PKE.Enc}(\mathsf{pk}, m; r)$ is the probabilistic encryption algorithm. On input a public key pk, a message $m \in \mathsf{MSp}_\lambda$ and possibly some random coins $r \in \mathsf{RSp}_\lambda$, this algorithm outputs a ciphertext c. (3) $\mathsf{PKE.Dec}(\mathsf{sk}, c)$ is the deterministic decryption algorithm. On input of a secret key sk and a ciphertext c, this algorithm outputs a message $m \in \mathsf{MSp}_\lambda$ or failure symbol $\perp$.

CORRECTNESS. The correctness of a public-key encryption scheme requires that for any $\lambda \in \mathbb{N}$, any $(\mathsf{sk}, \mathsf{pk}) \in [\mathsf{PKE.Setup}(1^\lambda)]$, any $m \in \mathsf{MSp}_\lambda$ and any random coins $r \in \mathsf{RSp}_\lambda$, we have that $\mathsf{PKE.Dec}(\mathsf{sk}, \mathsf{PKE.Enc}(\mathsf{pk}, m; r)) = m$.

SECURITY. We recall the standard security notions of *indistinguishability under chosen ciphertext attacks* (IND-CCA) and its weaker variant known as *indistinguishability under chosen plaintext attacks* (IND-CPA). A public-key encryption scheme PKE is IND-CCA secure if for every *legitimate* ppt adversary $\mathcal{A}$,

$$\mathbf{Adv}_{\mathsf{PKE}, \mathcal{A}}^{\mathrm{ind\text{-}cca}}(\lambda) := 2 \cdot \Pr[\text{IND-CCA}_{\mathsf{PKE}}^{\mathcal{A}}(\lambda)] - 1 \in \text{NEGL},$$

where game IND-CCA$_{\mathsf{PKE}}^{\mathcal{A}}$ described in Fig. 1, in which the adversary has access to a left-or-right challenge oracle (LR) and a decryption oracle (Dec). We say that

IND-CCA$_{\mathsf{PKE}}^{\mathcal{A}}(\lambda)$:	$\underline{LR(m_0, m_1)}$:	$\underline{Dec(c)}$:
$(\mathsf{sk}, \mathsf{pk}) \leftarrow_\$ \mathsf{PKE.Setup}(1^\lambda)$	$c \leftarrow_\$ \mathsf{PKE.Enc}(\mathsf{pk}, m_b)$	$m \leftarrow \mathsf{PKE.Dec}(\mathsf{sk}, c)$
$b \leftarrow_\$ \{0, 1\}$	$\mathsf{List} \leftarrow c : \mathsf{List}$	return m
$b' \leftarrow_\$ \mathcal{A}^{\mathsf{LR},\mathsf{Dec}}(1^\lambda, \mathsf{pk})$	return c	
return $(b = b')$		

Fig. 1. Game defining IND-CCA security of a public-key encryption scheme PKE.

$\mathcal{A}$ is legitimate if: (1) $|m_0| = |m_1|$ whenever the left-or-right oracle is queried; and (2) the adversary does not call the decryption oracle with $c \in \mathsf{List}$. We obtain the weaker IND-CPA notion if the adversary is not allowed to place any decryption query.

2.2 NIZK Proof Systems

SYNTAX. A non-interactive zero-knowledge proof system for an **NP** language $\mathcal{L}$ with an efficiently computable binary relation $\mathcal{R}$ consists of three ppt algorithms as follows. (1) NIZK.Setup(1^λ) is the setup algorithm and on input a security parameter 1^λ it outputs a common reference string crs; (2) NIZK.Prove(crs, x, w) is the proving algorithm and on input a common reference string crs, a statement x and a witness w it outputs a proof π or a failure symbol $\bot$; (3) NIZK.Verify(crs, x, π) is the verification algorithm and on input a common reference string crs, a statement x and a proof π it outputs either true or false.

PERFECT COMPLETENESS. Completeness imposes that an honest prover can always convince an honest verifier that a statement belongs to $\mathcal{L}$, provided that it holds a witness testifying to this fact. We say a NIZK proof is *perfectly complete* if for every (possibly unbounded) adversary $\mathcal{A}$

$$\mathbf{Adv}_{\mathsf{NIZK},\mathcal{A}}^{\mathrm{complete}}(\lambda) := \Pr\left[\mathrm{Complete}_{\mathsf{NIZK}}^{\mathcal{A}}(\lambda)\right] = 0,$$

where game $\mathrm{Complete}_{\mathsf{NIZK}}^{\mathcal{A}}(\lambda)$ is shown in Fig. 2 on the left.

COMPUTATIONAL ZERO KNOWLEDGE. The zero-knowledge property guarantees that proofs do not leak information about the witnesses that originated them. Technically, this is formalized by requiring the existence of a ppt simulator Sim := $(\mathsf{Sim}_0, \mathsf{Sim}_1)$ where Sim_0 takes the security parameter 1^λ as input and outputs a simulated common reference string crs together with a trapdoor tp, and Sim_1 takes the trapdoor as input tp together with a statement x for which it must forge a proof π. We say a proof system is *computationally zero knowledge* if, for every ppt adversary $\mathcal{A}$, there exists a ppt simulator Sim such that

$$\mathbf{Adv}_{\mathsf{NIZK},\mathcal{A},\mathsf{Sim}}^{\mathrm{zk}}(\lambda) := \left| \Pr\left[\mathrm{ZK\!-\!Real}_{\mathsf{NIZK}}^{\mathcal{A}}(\lambda)\right] - \left[\mathrm{ZK\!-\!Ideal}_{\mathsf{NIZK}}^{\mathcal{A},\mathsf{Sim}}(\lambda)\right] \right| \in \mathrm{NEGL} ,$$

where games $\mathrm{ZK\!-\!Real}_{\mathsf{NIZK}}^{\mathcal{A}}(\lambda)$ and $\mathrm{ZK\!-\!Ideal}_{\mathsf{NIZK}}^{\mathcal{A},\mathsf{Sim}}(\lambda)$ are shown in Fig. 3.

STATISTICAL SIMULATION SOUNDNESS. Soundness imposes that a malicious prover cannot convince an honest verifier of a false statement. This should be true even when the adversary itself is provided with simulated proofs. We say NIZK is *statistically simulation sound* with respect to simulator Sim if, for every (possibly unbounded) adversary $\mathcal{A}$

$$\mathbf{Adv}_{\mathsf{NIZK},\mathcal{A}}^{\mathsf{sound}}(\lambda) := \Pr\left[\mathsf{Sound}_{\mathsf{NIZK}}^{\mathcal{A},\mathsf{Sim}}(\lambda)\right] \in \mathrm{NEGL},$$

where game $\mathsf{Sound}_{\mathsf{NIZK}}^{\mathcal{A}}(\lambda)$ is shown in Fig. 2 on the right.

$\mathsf{Complete}_{\mathsf{NIZK}}^{\mathcal{A}}(\lambda)$:	$\mathsf{Sound}_{\mathsf{NIZK}}^{\mathcal{A}}(\lambda)$:
$\mathsf{crs} \leftarrow_{\$} \mathsf{NIZK}.\mathsf{Setup}(1^{\lambda})$	$\mathsf{crs} \leftarrow_{\$} \mathsf{NIZK}.\mathsf{Setup}(1^{\lambda})$
$(\mathsf{x}, w) \leftarrow_{\$} \mathcal{A}(1^{\lambda}, \mathsf{crs})$	$(\mathsf{x}, \pi) \leftarrow_{\$} \mathcal{A}(1^{\lambda}, \mathsf{crs})$
if $(\mathsf{x}, w) \notin \mathcal{R}$ return 0	return $(\mathsf{x} \notin \mathcal{L} \wedge$
$\pi \leftarrow_{\$} \mathsf{NIZK}.\mathsf{Prove}(\mathsf{crs}, \mathsf{x}, w)$	$\mathsf{NIZK}.\mathsf{Verify}(\mathsf{crs}, \mathsf{x}, \pi))$
return $\neg(\mathsf{NIZK}.\mathsf{Verify}(\mathsf{crs}, \mathsf{x}, \pi))$	

Fig. 2. Games defining the completeness and soundness properties of a non-interactive zero-knowledge proof system NIZK.

$\mathsf{ZK}\text{-}\mathsf{Real}_{\mathsf{NIZK}}^{\mathcal{A}}(\lambda)$:	$\mathsf{ZK}\text{-}\mathsf{Ideal}_{\mathsf{NIZK}}^{\mathcal{A},\mathsf{Sim}}(\lambda)$:
$\mathsf{crs} \leftarrow_{\$} \mathsf{NIZK}.\mathsf{Setup}(1^{\lambda})$	$(\mathsf{crs}, \mathsf{tp}) \leftarrow_{\$} \mathsf{Sim}_1(1^{\lambda})$
$b \leftarrow_{\$} \mathcal{A}^{\mathrm{Prove}}(1^{\lambda}, \mathsf{crs})$	$b \leftarrow_{\$} \mathcal{A}^{\mathrm{Prove}}(1^{\lambda}, \mathsf{crs})$
$\mathrm{Prove}(\mathsf{x}, w)$:	$\mathrm{Prove}(\mathsf{x}, w)$:
if $(\mathsf{x}, w) \notin \mathcal{R}$ return $\bot$	if $(\mathsf{x}, w) \notin \mathcal{R}$ return $\bot$
$\pi \leftarrow_{\$} \mathsf{NIZK}.\mathsf{Prove}(\mathsf{crs}, \mathsf{x}, w)$	$\pi \leftarrow_{\$} \mathsf{Sim}_2(\mathsf{crs}, \mathsf{tp}, \mathsf{x})$
return π	return π

Fig. 3. Games defining the zero-knowledge property of a non-interactive zero-knowledge proof system NIZK.

2.3 Collision-Resistant Hash Functions

A hash function family $\mathsf{H} := \{\mathsf{H}_{\lambda}\}_{\lambda \in \mathbb{N}}$ is a set parametrized by a security parameter $\lambda \in \mathbb{N}$, where each H_{λ} is a collection of functions mapping $\{0,1\}^m$ to $\{0,1\}^n$ such that $m > n$. The hash function family H is said to be collision-resistant if no ppt adversary $\mathcal{A}$ can find a pair of colliding inputs, with noticeable probability, given a function picked uniformly from H_{λ}. More precisely, we require that

$$\mathbf{Adv}_{\mathsf{H},\mathcal{A}}^{\mathsf{cr}}(\lambda) := \Pr[\mathsf{CR}_{\mathsf{H}}^{\mathcal{A}}(\lambda)] \in \mathrm{NEGL},$$

where game $\mathsf{CR}_{\mathsf{H}}^{\mathcal{A}}(\lambda)$ is defined in Fig. 4.

$$
\begin{array}{|l|}
\hline
\mathrm{CR}_{\mathsf{H}}^{\mathcal{A}}(\lambda): \\
\hline
h \leftarrow_{\$} \mathsf{H}_\lambda \\
(\mathsf{x}_0, \mathsf{x}_1) \leftarrow_{\$} \mathcal{A}(1^\lambda, h) \\
\text{return } (\mathsf{x}_0 \neq \mathsf{x}_1 \wedge h(\mathsf{x}_0) = h(\mathsf{x}_1)) \\
\hline
\end{array}
$$

Fig. 4. Game defining collision-resistance of a hash function family H.

2.4 Puncturable Pseudorandom Functions

A puncturable pseudorandom function family PPRF := (PPRF.Gen, PPRF.Eval, PPRF.Punc) is a triple of ppt algorithms as follows. (1) PPRF.Gen on input the security parameter 1^λ outputs a uniform element in K_λ; (2) PPRF.Eval is deterministic and on input a key $\mathsf{k} \in \mathsf{K}_\lambda$ and a point $\mathsf{x} \in \mathsf{X}_\lambda$ outputs a point $\mathsf{y} \in \mathsf{Y}_\lambda$; (3) PPRF.Punc is probabilistic and on input a $\mathsf{k} \in \mathsf{K}_\lambda$ and a polynomial-size set of points $\mathsf{S} \subseteq \mathsf{X}_\lambda$ outputs a punctured key k_S. As per [22], we require the PPRF to satisfy the following two properties:

FUNCTIONALITY PRESERVATION UNDER PUNCTURING: For every $\lambda \in \mathbb{N}$, every polynomial-size set $\mathsf{S} \subseteq \mathsf{X}_\lambda$ and every $\mathsf{x} \in \mathsf{X}_\lambda \setminus \mathsf{S}$, it holds that

$$
\Pr\left[\mathsf{PPRF.Eval}(\mathsf{k}, \mathsf{x}) = \mathsf{PPRF.Eval}(\mathsf{k}_\mathsf{S}, \mathsf{x}) \;\middle|\; \begin{array}{l} \mathsf{k} \leftarrow_{\$} \mathsf{PPRF.Gen}(1^\lambda) \\ \mathsf{k}_\mathsf{S} \leftarrow_{\$} \mathsf{PPRF.Punc}(\mathsf{k}, \mathsf{S}) \end{array} \right] = 1.
$$

PSEUDORANDOMNESS AT PUNCTURED POINTS: For every ppt adversary $\mathcal{A}$,

$$
\mathbf{Adv}_{\mathsf{PPRF}, \mathcal{A}}^{\mathsf{pprf}}(\lambda) := 2 \cdot \Pr[\mathsf{PPRF}_{\mathsf{PPRF}}^{\mathcal{A}}(\lambda)] - 1 \in \mathrm{NEGL},
$$

where game $\mathsf{PPRF}_{\mathsf{PPRF}}^{\mathcal{A}}(\lambda)$ is defined in Fig. 5.

$$
\begin{array}{|ll|}
\hline
\mathsf{PPRF}_{\mathsf{PPRF}}^{\mathcal{A}}(\lambda): & \mathsf{Fn}(\mathsf{x}): \\
\hline
(\mathsf{S}, \mathsf{st}) \leftarrow_{\$} \mathcal{A}_0(1^\lambda) & \text{if } \mathsf{x} \notin \mathsf{S} \text{ return } \mathsf{PPRF.Eval}(\mathsf{k}_\mathsf{S}, \mathsf{x}) \\
\mathsf{k} \leftarrow_{\$} \mathsf{PPRF.Gen}(1^\lambda) & \text{if } T[\mathsf{x}] = \bot \text{ then} \\
\mathsf{k}_\mathsf{S} \leftarrow_{\$} \mathsf{PPRF.Punc}(\mathsf{k}, \mathsf{S}) & \quad T[\mathsf{x}] \leftarrow_{\$} \mathsf{Y}_\lambda \\
b \leftarrow_{\$} \{0, 1\} & \text{if } b = 1 \text{ return } T[\mathsf{x}] \\
b' \leftarrow_{\$} \mathcal{A}_1^{\mathsf{Fn}}(1^\lambda, \mathsf{k}_\mathsf{S}, \mathsf{st}) & \text{else return } \mathsf{PPRF.Eval}(\mathsf{k}, \mathsf{x}) \\
\text{return } (b = b') & \\
\hline
\end{array}
$$

Fig. 5. Game defining pseudorandomness at punctured points of PPRF.

2.5 Obfuscators

SYNTAX. An obfuscator for a circuit family CSp is a uniform ppt algorithm Obf that on input the security parameter 1^λ and the description of a circuit $\mathsf{C} \in \mathsf{CSp}_\lambda$ outputs the description of another circuit $\overline{\mathsf{C}}$. We require any obfuscator to satisfy the following two requirements.

FUNCTIONALITY PRESERVATION: For any $\lambda \in \mathbb{N}$, any $C \in CSp_\lambda$ and any $m \in MSp_\lambda$, with overwhelming probability over the choice of $\overline{C} \leftarrow_s Obf(1^\lambda, C)$ we have that $C(m) = \overline{C}(m)$.

POLYNOMIAL SLOWDOWN: There is a polynomial poly such that for any $\lambda \in \mathbb{N}$, any $C \in CSp_\lambda$ and any $\overline{C} \leftarrow_s Obf(1^\lambda, C)$ we have that $|\overline{C}| \leq poly(|C|)$.

In this paper we rely on the security definitions of *indistinguishability obfuscation* (iO) [14] and *distributional indistinguishability* (DI). The latter definition was first introduced by [5] in the context of point function obfuscation and later generalized by [2] to cover samplers that output not only point circuits. We note that the work of [2] considers only *statistically* unpredictable samplers, which is a more restricted class of samplers, and therefore is a more amenable form of obfuscation. Unfortunately, for the purpose of proving the construction we present in Sect. 3.2 secure, we rely on a DI obfuscator against a computationally unpredictable sampler.

INDISTINGUISHABILITY OBFUSCATION (iO). This property requires that given any two functionally equivalent circuits C_0 and C_1 of equal size, the obfuscations of C_0 and C_1 should be computationally indistinguishable. More precisely, for any ppt adversary $\mathcal{A}$ and for any sampler $\mathcal{S}$ that outputs two circuits $C_0, C_1 \in CSp_\lambda$ such that $C_0(m) = C_1(m)$ for all inputs m and $|C_0| = |C_1|$, we have that

$$\mathbf{Adv}^{io}_{Obf,\mathcal{S},\mathcal{A}}(\lambda) := 2 \cdot \Pr[iO^{\mathcal{S},\mathcal{A}}_{Obf}(\lambda)] - 1 \in \text{NEGL},$$

where game $iO^{\mathcal{S},\mathcal{A}}_{Obf}(\lambda)$ is defined in Fig. 6 on the left.

DISTRIBUTIONAL INDISTINGUISHABILITY (DI). We define this property with respect to some class of unpredictable samplers $\mathbb{S}$. A sampler is an algorithm $\mathcal{S}$ that on input the security parameter 1^λ and possibly some state information st outputs a pair of vectors of CSp_λ circuits $(\mathbf{C}_0, \mathbf{C}_1)$ of equal dimension and possibly some auxiliary information z. We require the components of the two circuit vectors to be encoded as bit strings of equal length. $\mathcal{S}$ is said to be *unpredictable* if no ppt predictor with oracle access to the circuits can find a differing input m such that $\mathbf{C}_0(m) \neq \mathbf{C}_1(m)$. An obfuscator Obf is DI secure with respect to a class of unpredictable samplers $\mathcal{S}$ if for all $\mathcal{S} \in \mathbb{S}$ the obfuscations of $\mathbf{C}_0$ and $\mathbf{C}_1$ output by $\mathcal{S}$ are computationally indistinguishable. More precisely, for every $\mathcal{S} \in \mathbb{S}$ and every ppt adversary $\mathcal{A}$ we have that

$$\mathbf{Adv}^{di}_{Obf,\mathcal{S},\mathcal{A}}(\lambda) := 2 \cdot \Pr[DI^{\mathcal{S},\mathcal{A}}_{Obf}(\lambda)] - 1 \in \text{NEGL},$$

where game $DI^{\mathcal{S},\mathcal{A}}_{Obf}(1^\lambda)$ is defined in Fig. 6 (middle). Furthermore, we say sampler $\mathcal{S}$ is *computationally unpredictable* if for any ppt predictor $\mathcal{P}$

$$\mathbf{Adv}^{pred}_{\mathcal{S},\mathcal{P}}(\lambda) := \Pr\left[\text{Pred}^{\mathcal{P}}_{\mathcal{S}}(\lambda)\right] \in \text{NEGL},$$

where game $\text{Pred}^{\mathcal{P}}_{\mathcal{S}}(\lambda)$ is shown in Fig. 6 on the right.

$\mathrm{iO}_{\mathsf{Obf}}^{\mathcal{S},\mathcal{A}}(\lambda)$:	$\mathrm{DI}_{\mathsf{Obf}}^{\mathcal{S},\mathcal{A}}(\lambda)$:	$\mathrm{Pred}_{\mathcal{S}}^{\mathcal{P}}(\lambda)$:
$(\mathbf{C}_0, \mathbf{C}_1, z) \leftarrow_\$ \mathcal{S}(1^\lambda)$	$(\mathsf{st}, \mathsf{st}') \leftarrow_\$ \mathcal{A}_0(1^\lambda)$	$(\mathsf{st}, \mathsf{st}') \leftarrow_\$ \mathcal{P}_0(1^\lambda)$
$b \leftarrow_\$ \{0,1\}$	$(\mathbf{C}_0, \mathbf{C}_1, z) \leftarrow_\$ \mathcal{S}(1^\lambda, \mathsf{st})$	$(\mathbf{C}_0, \mathbf{C}_1, z) \leftarrow_\$ \mathcal{S}(\mathsf{st})$
$\overline{\mathbf{C}} \leftarrow_\$ \mathsf{Obf}(1^\lambda, \mathbf{C}_b)$	$b \leftarrow_\$ \{0,1\}$	$m \leftarrow_\$ \mathcal{P}_1^{\mathrm{Fn}}(1^\lambda, z, \mathsf{st}')$
$b' \leftarrow_\$ \mathcal{A}_1(1^\lambda, z, \overline{\mathbf{C}})$	$\overline{\mathbf{C}} \leftarrow_\$ \mathsf{Obf}(1^\lambda, \mathbf{C}_b)$	return $(\mathbf{C}_0(m) \neq \mathbf{C}_1(m))$
return $(b = b')$	$b' \leftarrow_\$ \mathcal{A}_1(1^\lambda, z, \mathsf{st}', \overline{\mathbf{C}})$	
	return $(b = b')$	$\underline{\mathrm{Fn}(m)}:$
		return $(\mathbf{C}_0(m))$

Fig. 6. Games defining iO and DI security of an obfuscator Obf, and unpredictability of a sampler $\mathcal{S}$.

2.6 RAM Programs

In the RAM model of computation, a program P has random-access to some initial *memory data* D, comprised of $|\mathsf{D}|$ *memory cells*. At each *CPU step* of its execution, P reads from and writes to a single memory cell *address*, which is determined by the previous step, and updates its internal state. By convention, the address in the first step is set to the first memory cell of D, and the initial internal state is empty. Only when P reaches the final step of its execution, it outputs a result y and terminates. We use the notation $\mathsf{y} \leftarrow \mathsf{P}^{\mathsf{D} \to \mathsf{D}^*}$ to indicate this process, where D^* is the resulting memory data when P terminates, or simply $\mathsf{y} \leftarrow \mathsf{P}^\mathsf{D}$ if we don't care about the resulting memory data. We also consider the case where the memory data *persists* between a sequential execution of n programs, and use the notation $(\mathsf{y}_1, \ldots, \mathsf{y}_n) \leftarrow (\mathsf{P}_1, \ldots, \mathsf{P}_n)^{\mathsf{D} \to \mathsf{D}^*}$ as short for $(\mathsf{y}_1 \leftarrow \mathsf{P}_1^{\mathsf{D} \to \mathsf{D}_1} ; \ldots ; \mathsf{y}_n \leftarrow \mathsf{P}_n^{\mathsf{D}_{n-1} \to \mathsf{D}^*})$. In more detail, a RAM program description is a 4-tuple $\mathsf{P} := (\mathcal{Q}, \mathcal{T}, \mathcal{Y}, \delta)$, where:

- $\mathcal{Q}$ is the set of all possible states, which always includes the empty state ϵ.
- $\mathcal{T}$ is the set of all possible contents of a memory cell. If each cell contains a single bit, $\mathcal{T} = \{0, 1\}$.
- $\mathcal{Y}$ is the output space of P, which always includes the empty output ϵ.
- δ is the transition function, modeled as a circuit, which maps $(\mathcal{Q} \times \mathcal{T})$ to $(\mathcal{T} \times \mathcal{Q} \times \mathbb{N} \times \mathcal{Y})$. On input an internal state $\mathsf{st}_i \in \mathcal{Q}$ and a content of a memory cell $\mathsf{read}_i \in \mathcal{T}$, it outputs a (possibly different) content of a memory cell $\mathsf{write}_i \in \mathcal{T}$, an internal state $\mathsf{st}_{i+1} \in \mathcal{Q}$, an address of a memory cell $\mathsf{addr}_{i+1} \in \mathbb{N}$ and an output $\mathsf{y} \in \mathcal{Y}$.

In Fig. 7 we show how program P is executed on a random-access machine with initial memory data D.

To conveniently specify the *efficiency* and *security* properties of the primitive we propose in the following section, we define functions runTime and accessPattern that on input a program P and some initial memory data D return the number of steps required for P to complete its execution on D and the list of addresses accessed during the execution, respectively. In other words, as per description in Fig. 7, runTime returns the value i when P terminates, whereas accessPattern returns List. More generally, we also allow these functions to receive as input a *set* of programs $(P_1, \ldots, P_n)$ to be executed sequentially on persistent memory, initially set to D.

$$
\boxed{
\begin{array}{l}
\text{EXECUTE } P^D: \\
\quad i \leftarrow 0; \quad \text{addr}_i \leftarrow 0; \quad \text{st}_i \leftarrow \epsilon; \quad y \leftarrow \epsilon; \quad \text{List} \leftarrow [] \\
\quad \text{while } (y = \epsilon) \\
\qquad /\!/ \text{ step } i \\
\qquad \text{List} \leftarrow \text{addr}_i : \text{List } /\!/ \text{ record the access pattern} \\
\qquad \text{read}_i \leftarrow D[\text{addr}_i] /\!/ \text{ read from memory} \\
\qquad (\text{write}_i, \text{st}_{i+1}, \text{addr}_{i+1}, y) \leftarrow \delta(\text{st}_i, \text{read}_i) \\
\qquad D[\text{addr}_i] \leftarrow \text{write}_i /\!/ \text{ write to memory} \\
\qquad i \leftarrow i + 1 \\
\quad \text{return } (y)
\end{array}
}
$$

Fig. 7. Execution of program P on a RAM machine with memory D.

3   Updatable Functional Encryption

We propose a new primitive that we call *updatable functional encryption*. It bears resemblance to functional encryption in that encryption is carried out in the public-key setting and the owner of the master secret key can issue tokens for functions of its choice that allows the holder of the token to learn the outcome of the function on the message underneath a ciphertext. Here, we model functions as RAM programs instead of circuits, which is closer to how programs are expressed in von Neumann architecture and avoids the RAM-to-circuit compilation. Not only that, we envision tokens that are capable to *update* the ciphertext, over which other tokens can be subsequently executed. Because the ciphertext evolves every time a token is executed and for better control over what information is revealed, each token is numbered sequentially so that it can only be executed *once* and *after all previous extracted tokens* have been executed on that ciphertext. Informally, the security requires that the ciphertext should not reveal more than what can be learned by applying the extracted tokens in order. As for efficiency, we want the run-time of a token to be proportional to the run-time of the program up to a *polylogarithmic* factor in the length of the encrypted message.

3.1   Definitions

SYNTAX. An updatable functional encryption scheme UFE for program family $\mathcal{P} := \{\mathcal{P}_\lambda\}_{\lambda \in \mathbb{N}}$ with message space $\mathsf{MSp} := \{\mathsf{MSp}_\lambda\}_{\lambda \in \mathbb{N}}$ is specified by three ppt algorithms as follows.

- UFE.Setup(1^λ) is the setup algorithm and on input a security parameter 1^λ it outputs a master secret key msk and a master public key mpk;

- UFE.TokenGen(msk, P, tid) is the token-generation algorithm and on input a master secret key msk, a program description $P \in \mathcal{P}_\lambda$ and a token-id tid $\in \mathbb{N}$, outputs a token (i.e. another program description) $\overline{P}_{tid}$;
- UFE.Enc(mpk, D) is the encryption algorithm and on input a master public key mpk and memory data $D \in MSp_\lambda$ outputs a ciphertext CT.

We do not explicitly consider an evaluation algorithm. Instead, the RAM program $\overline{P}$ output by UFE.TokenGen executes directly on memory data CT, a ciphertext resulting from the UFE.Enc algorithm. Note that this brings us close to the syntax of Garbled RAM, but in contrast encryption is carried out in the public-key setting.

CORRECTNESS. We say that UFE is correct if for every security parameter $\lambda \in \mathbb{N}$, for every memory data $D \in MSp_\lambda$ and for every sequence of polynomial length in λ of programs $(P_1, \ldots, P_n)$, it holds that

$$
\Pr\left[y_1 = y_1' \wedge \ldots \wedge y_n = y_n' \left|
\begin{array}{l}
(\text{msk}, \text{mpk}) \leftarrow_\$ \text{UFE.Setup}(1^\lambda) \\
\text{CT} \leftarrow_\$ \text{UFE.Enc}(\text{mpk}, D) \\
\text{for } i \in [n] \\
\quad \overline{P}_i \leftarrow_\$ \text{UFE.TokenGen}(\text{msk}, P_i, i) \\
(y_1, \ldots, y_n) \leftarrow (P_1, \ldots, P_n)^D \\
(y_1', \ldots, y_n') \leftarrow (\overline{P}_1, \ldots, \overline{P}_n)^{\text{CT}}
\end{array}
\right.\right] = 1.
$$

EFFICIENCY. Besides the obvious requirement that all algorithms run in polynomial-time in the length of their inputs, we also require that the run-time of token $\overline{P}$ on ciphertext CT is proportional to the run-time of its clear-form counterpart (program P on memory D) up to a polynomial factor in λ and up to a polylogarithmic factor in the length of D. More precisely, we require that for every $\lambda \in \mathbb{N}$, for every sequence of polynomial length in λ of programs $(P_1, \ldots, P_n)$ and every memory data $D \in MSp_\lambda$, there exists a fixed polynomial function poly and a fixed polylogarithmic function polylog such that

$$
\Pr\left[
\begin{array}{l}
\text{runTime}((\overline{P}_1, \ldots, \overline{P}_n), \text{CT}) \leq \\
\quad \text{runTime}((P_1, \ldots, P_n), D) \cdot \\
\quad \text{poly}(\lambda) \cdot \text{polylog}(|D|)
\end{array}
\left|
\begin{array}{l}
(\text{msk}, \text{mpk}) \leftarrow_\$ \text{UFE.Setup}(1^\lambda) \\
\text{CT} \leftarrow_\$ \text{UFE.Enc}(\text{mpk}, D) \\
\text{for } i \in [n] \\
\quad \overline{P}_i \leftarrow_\$ \text{UFE.TokenGen}(\text{msk}, P_i)
\end{array}
\right.\right] = 1.
$$

In particular, this means that for a program P running in sublinear-time in $|D|$, the run-time of $\overline{P}$ over the encrypted data remains sublinear.

SECURITY. Let UFE be an updatable functional encryption scheme. We say UFE is *selectively secure* if for any legitimate ppt adversary $\mathcal{A}$

$$
\mathbf{Adv}^{\text{sel}}_{\text{UFE}, \mathcal{A}}(\lambda) := 2 \cdot \Pr\left[\text{SEL}^{\mathcal{A}}_{\text{UFE}}(\lambda)\right] - 1 \in \text{NEGL},
$$

where game $\text{SEL}^{\mathcal{A}}_{\text{UFE}}(\lambda)$ is defined in Fig. 8. We say $\mathcal{A}$ is legitimate if the following two conditions are satisfied:

1. $(P_1, \ldots, P_n)^{D_0} = (P_1, \ldots, P_n)^{D_1}$
2. accessPattern$((P_1, \ldots, P_n), D_0) = $ accessPattern$((P_1, \ldots, P_n), D_1)$

These conditions avoid that the adversary trivially wins the game by requesting tokens whose output differ on left and right challenge messages or have different access patterns.

$$
\begin{array}{|l|}
\hline
\mathrm{SEL}^{\mathcal{A}}_{\mathsf{UFE}}(\lambda): \\
(\mathsf{D}_0, \mathsf{D}_1, (\mathsf{P}_1, ..., \mathsf{P}_n), \mathsf{st}) \leftarrow_{\$} \mathcal{A}_0(1^{\lambda}) \\
(\mathsf{msk}, \mathsf{mpk}) \leftarrow_{\$} \mathsf{UFE.Setup}(1^{\lambda}) \\
b \leftarrow_{\$} \{0, 1\} \\
\mathsf{CT} \leftarrow_{\$} \mathsf{UFE.Enc}(\mathsf{mpk}, \mathsf{D}_b) \\
\text{for } i \in [n] \\
\quad \overline{\mathsf{P}}_i \leftarrow_{\$} \mathsf{UFE.TokenGen}(\mathsf{msk}, \mathsf{P}_i) \\
b' \leftarrow_{\$} \mathcal{A}_1(\mathsf{CT}, (\overline{\mathsf{P}}_1, ..., \overline{\mathsf{P}}_n), \mathsf{st}) \\
\text{return } (b = b') \\
\hline
\end{array}
$$

Fig. 8. Selective security of an updatable FE scheme UFE.

3.2 Our Construction

The idea of our construction is the following. Before encryption we append to the cleartext the token-id of the first token to be issued, the address of the first position to be read and the initial state of the program. These values are all pre-defined at the beginning. We then split the data into bits and label each of them with a common random tag, their position on the array and a counter that keeps track of how many times that bit was updated (initially 0). Then, we build a Merkle tree over the labeled bits. Later, this will allow us to check the consistency of the data without having to read through all of it. It also binds a token-id, a read-position and a state to the data at a particular stage. Finally, we encrypt each node of the tree, twice, and attach a NIZK proof attesting that they encrypt the same content. Tokens include the decryption key inside their transition circuit in order to perform the computation over the clear data and re-encrypt the nodes at the end of each CPU step. These circuits are obfuscated to protect the decryption key, and the random coins necessary to re-encrypt come from a puncturable PRF. The proof then follows a mix of different strategies seen in [2,14,17,18,20].

- UFE.Setup(1^{λ}) samples two public-key encryption key pairs $(\mathsf{sk}_0, \mathsf{pk}_0) \leftarrow_{\$} \mathsf{PKE.Setup}(1^{\lambda})$ and $(\mathsf{sk}_1, \mathsf{pk}_1) \leftarrow_{\$} \mathsf{PKE.Setup}(1^{\lambda})$, a common reference string $\mathsf{crs} \leftarrow_{\$} \mathsf{NIZK.Setup}(1^{\lambda})$ and a collision-resistant hash function $\mathsf{H} \leftarrow_{\$} \mathsf{H}_{\lambda}$. It then sets constants (l_1, l_2, l_3) as the maximum length of token-ids, addresses and possible states induced by the supported program set $\mathcal{P}_{\lambda}$, respectively, encoded as bit-strings. Finally, it sets $\mathsf{msk} \leftarrow \mathsf{sk}_0$ and $\mathsf{mpk} \leftarrow (\mathsf{pk}_0, \mathsf{pk}_1, \mathsf{crs}, \mathsf{H}, (l_1, l_2, l_3))$ and outputs the key pair $(\mathsf{msk}, \mathsf{mpk})$.
- UFE.Enc(mpk, D) parses mpk as $(\mathsf{pk}_0, \mathsf{pk}_1, \mathsf{crs}, \mathsf{H}, (l_1, l_2, l_3))$ and appends to the memory data D the token-id 1, address 0 and the empty state ϵ, encoded as bit-stings of length l_1, l_2 and l_3, respectively: $\mathsf{D} \leftarrow (\mathsf{D}, 1, 0, \epsilon)$. (We assume from

now on that $|D|$ is a power of 2. This is without loss of generality since D can be padded.) UFE.Enc sets $z \leftarrow \log(|D|)$, samples a random string $\mathsf{tag} \leftarrow_\$ \{0,1\}^\lambda$ and constructs a perfectly balanced binary tree $T := \{\mathsf{node}^{(i,j)}\}$, where leafs are set as

$$\forall j \in \{0, \ldots, (|D| - 1)\}, \ \mathsf{node}^{(z,j)} \leftarrow (D[j], \mathsf{tag}, (z,j), 0)$$

and intermediate nodes are computed as

$$\forall i \in \{(z-1), \ldots, 0\}, \ \forall j \in \{0, \ldots, (2^i - 1)\},$$
$$\mathsf{node}^{(i,j)} \leftarrow (\mathsf{H}(\mathsf{node}^{(i+1,2j)}, \mathsf{node}^{(i+1,2j+1)})).$$

UFE.Enc then encrypts each node independently under pk_0 and pk_1, i.e.

$$\forall i \in \{0, \ldots, z\}, \ \forall j \in \{0, \ldots, (2^i - 1)\},$$
$$r_0^{(i,j)} \leftarrow_\$ \mathsf{RSp}_\lambda \ ; \ r_1^{(i,j)} \leftarrow_\$ \mathsf{RSp}_\lambda$$
$$\mathsf{CT}_0^{(i,j)} \leftarrow \mathsf{PKE.Enc}(\mathsf{pk}_0, \mathsf{node}^{(i,j)}; r_0^{(i,j)})$$
$$\mathsf{CT}_1^{(i,j)} \leftarrow \mathsf{PKE.Enc}(\mathsf{pk}_1, \mathsf{node}^{(i,j)}; r_1^{(i,j)})$$

and computes NIZK proofs that $\mathsf{CT}_0^{(i,j)}$ and $\mathsf{CT}_1^{(i,j)}$ encrypt the same content. More precisely,

$$\forall i \in \{0, \ldots, z\}, \ \forall j \in \{0, \ldots, (2^i - 1)\},$$
$$\pi^{(i,j)} \leftarrow_\$ \mathsf{NIZK.Prove}(\mathsf{crs}, x^{(i,j)}, (\mathsf{node}^{(i,j)}, r_0^{(i,j)}, r_1^{(i,j)})),$$

where $x^{(i,j)}$ is the NP statement

$$\exists (m, r_0, r_1) : \mathsf{CT}_0^{(i,j)} = \mathsf{PKE.Enc}(\mathsf{pk}_0, m; r_0) \wedge \mathsf{CT}_1^{(i,j)} = \mathsf{PKE.Enc}(\mathsf{pk}_1, m; r_1).$$

Finally, UFE.Enc lets

$$\mathsf{CT} := \{(\mathsf{CT}_0^{(i,j)}, \mathsf{CT}_1^{(i,j)}, \pi^{(i,j)})\},$$

which encodes a perfectly balanced tree, and outputs it as a ciphertext of memory data D under mpk.

- UFE.TokenGen$(\mathsf{msk}, \mathsf{mpk}, P, \mathsf{tid})$ parses $(\mathsf{pk}_0, \mathsf{pk}_1, \mathsf{crs}, \mathsf{H}, (l_1, l_2, l_3)) \leftarrow \mathsf{mpk}$, $(\mathcal{Q}, \mathcal{T}, \mathcal{Y}, \delta) \leftarrow P$ and $\mathsf{sk}_0 \leftarrow \mathsf{msk}$. It then samples a new puncturable PRF key $k \leftarrow_\$ \mathsf{PPRF.Gen}(1^\lambda)$. Next, it sets a circuit $\hat{\delta}$ as described in Fig. 9, using the parsed values as the appropriate hardcoded constants with the same naming. UFE.TokenGen then obfuscates this circuit by computing $\overline{\delta} \leftarrow_\$ \mathsf{Obf}(\hat{\delta})$. Finally, for simplicity in order to avoid having to explicitly deal with the data structure in the ciphertext, and following a similar approach as in [9], we define token $\overline{P}$ not by its transition function, but by pseudocode, as the RAM program that executes on CT the following:
  1. Set initial state $\mathsf{st} \leftarrow \epsilon$, initial address $\mathsf{addr} \leftarrow 0$ and empty output $y \leftarrow \epsilon$.

2. While ($y = \epsilon$)
    (a) Construct a tree $\overline{T}$ by selecting from CT the leaf at address addr and the last $(l_1 + l_2 + l_3)$ leafs (that should encode tid, addr and st if CT is valid), as well as all the necessary nodes to compute the hash values of their path up to the root.
    (b) Evaluate $(\overline{T}, \text{addr}, y) \leftarrow \overline{\delta}(\overline{T})$.
    (c) Update CT by writing the resulting $\overline{T}$ to it.
3. Output y.

Theorem 1. *Let* PKE *be an* IND-CCA *secure public-key encryption scheme, let* NIZK *be a non-interactive zero knowledge proof system with perfect completeness, computational zero knowledge and statistical simulation soundness, let* H *be a collision-resistant hash function family, let* PPRF *be a puncturable pseudorandom function and let* Obf *be an* iO-*secure obfuscator that is also* DI-*secure w.r.t. the class of samplers described in* Game$_4$. *Then, the updatable functional encryption scheme* UFE[PKE, NIZK, H, PPRF, Obf] *detailed in Sect. 3.2 is selectively secure (as per definition in Fig. 8).*

Proof (Outline). The proof proceeds via a sequence of games as follows.

Game$_0$: This game is identical to the real SEL game when the challenge bit $b = 0$, i.e. the challenger encrypts D_0 in the challenge ciphertext.

Game$_1$: In this game, the common reference string and NIZK proofs are simulated. More precisely, at the beginning of the game, the challenger executes $(\text{crs}, \text{tp}) \leftarrow_\$ \text{Sim}_0(1^\lambda)$ to produce the crs that is included in the mpk, and proofs in the challenge ciphertext are computed with Sim_1 and tp. The distance to the previous game can be bounded by the zero-knowledge property of NIZK.

Game$_2$: Let $T_0 := \{\text{node}_0^{(i,j)}\}$ be the perfectly balanced tree resulting from the encoding of D_0 with tag$_0$, and $T_1 := \{\text{node}_1^{(i,j)}\}$ the one resulting from the encoding of D_1 with tag$_1$, where (D_0, D_1) are the challenge messages queried by the adversary and $(\text{tag}_0, \text{tag}_1)$ are independently sampled random tags. In this game, $\text{CT}_1^{(i,j)}$ in the challenge ciphertext encrypts $\text{node}_1^{(i,j)}$; the NIZK proofs are still simulated. This transition is negligible down to the IND-CPA security of PKE.

Game$_3$: In this game we hardwire a pre-computed lookup table to each circuit $\widehat{\delta}_l$, containing fixed inputs/outputs that allow to bypass the steps described in Fig. 9. If the input to the circuit is on the lookup table, it will immediately return the corresponding output. The lookup tables are computed such that executing the tokens in sequence starting on the challenge ciphertext will propagate the execution over D_0 in the left branch and D_1 in the right branch. Because the challenge ciphertext evolves over time as tokens are executed, to argue this game hop we must proceed by hardwiring one input/output at the time, as follows: (1) We hardwire the input/output of the regular execution [iO property of Obf]; (2) we puncture the PPRF key of $\widehat{\delta}_l$ on the new hardwired

Hardcoded: Transition circuit δ, token-id tid^*, secret key sk_0, puncturable PRF key k, public keys pk_0 and pk_1, common reference string crs, hash function H and bit-length constants (l_1, l_2, l_3). **Input:** Tree $\overline{\mathsf{T}}$.

1. Verify the NIZK proof in each node of tree $\overline{\mathsf{T}}$, and decrypt the first ciphertext of each node with sk_0. Let T be the resulting decrypted tree.

   $\forall (i,j) \in \mathbb{N}^2 : \overline{\mathsf{node}}^{(i,j)} \in \overline{\mathsf{T}}$,
      parse $\overline{\mathsf{node}}^{(i,j)}$ as $(\mathsf{CT}_0^{(i,j)}, \mathsf{CT}_1^{(i,j)}, \pi^{(i,j)})$ or return $\perp$
      if $\mathsf{NIZK.Verify}(\mathsf{crs}, x^{(i,j)}, \pi^{(i,j)}) = \mathsf{false}$ return $\perp$
      $\mathsf{node}^{(i,j)} \leftarrow \mathsf{PKE.Dec}(\mathsf{sk}_0, \mathsf{CT}_0^{(i,j)})$
   let $\mathsf{T} := \{\mathsf{node}^{(i,j)}\}$

2. On the decrypted tree T, verify the path of each leaf up to the root (i.e. intermediate nodes must be equal to the hash of their children) and check that all leafs are marked with the same random tag.

   $z \leftarrow \max\{i \in \mathbb{N} : \mathsf{node}^{(i,j)} \in \mathsf{T}, \exists j \in \mathbb{N}\}$
   $\forall j \in \mathbb{N} : \mathsf{node}^{(z,j)} \in \mathsf{T}$,
      $\forall i \in \{(z-1), ..., 0\}$
         if $\mathsf{node}^{(i, \lfloor \frac{j}{2^{(z-i)}} \rfloor)} \neq \mathsf{H}(\mathsf{node}^{((i+1), 2\lfloor \frac{j}{2^{(z-i)}} \rfloor)}, \mathsf{node}^{((i+1), (2\lfloor \frac{j}{2^{(z-i)}} \rfloor +1))})$ return $\perp$
      parse $\mathsf{node}^{(z,j)}$ as $(\mathsf{value}^{(z,j)}, \mathsf{tag}^{(z,j)}, \mathsf{position}^{(z,j)}, \mathsf{counter}^{(z,j)})$ or return $\perp$
   if $\exists (j,j') \in \mathbb{N}^2 : \mathsf{node}^{(z,j)} \in \mathsf{T} \wedge \mathsf{node}^{(z,j')} \in \mathsf{T} \wedge \mathsf{tag}^{(z,j)} \neq \mathsf{tag}^{(z,j')}$ return $\perp$

3. Read the token-id, address and state of the current step encoded in tree T. Check that the token-id matches the one hardcoded in this token. Then, evaluate the transition circuit δ.

   read $(\mathsf{tid}, \mathsf{addr}, \mathsf{st})$ with fixed bit-length (l_1, l_2, l_3) from T or return $\perp$
   if $\mathsf{tid} \neq \mathsf{tid}^*$ return $\perp$
   $(\mathsf{value}^{(z, \mathsf{addr})}, \mathsf{st}, \mathsf{addr}, \mathsf{y}) \leftarrow \delta(\mathsf{st}, \mathsf{value}^{(z, \mathsf{addr})})$

4. If the transition circuit δ outputs some result y then increase the token-id and reset the internal state and address.

   if $\mathsf{y} \neq \epsilon$ then $\mathsf{tid} \leftarrow \mathsf{tid} + 1$; $\mathsf{st} \leftarrow 0$; $\mathsf{addr} \leftarrow 0$

5. Write the (possibly new) token-id, address and state to tree T, update the counters of leaf nodes and recompute the path of each leaf up to the root.

   write $(\mathsf{tid}, \mathsf{addr}, \mathsf{st})$ with fixed bit-length (l_1, l_2, l_3) to T
   $\forall j \in \mathbb{N} : \mathsf{node}^{(z,j)} \in \mathsf{T}, \mathsf{counter}^{(z,j)} \leftarrow \mathsf{counter}^{(z,j)} + 1$
   $\forall j \in \mathbb{N} : \mathsf{node}^{(z,j)} \in \mathsf{T}, \forall i \in \{(z-1), ..., 0\}$,
      $\mathsf{node}^{(i, \lfloor \frac{j}{2^{(z-i)}} \rfloor)} \leftarrow \mathsf{H}(\mathsf{node}^{((i+1), 2\lfloor \frac{j}{2^{(z-i)}} \rfloor)}, \mathsf{node}^{((i+1), (2\lfloor \frac{j}{2^{(z-i)}} \rfloor +1))})$

6. Re-encrypt all nodes of T (as before, encrypt under pk_0 and pk_1 and add NIZK proofs under crs). To extract the necessary random coins, we use the puncturable PRF under key k, providing as input the input of this circuit, i.e. $\overline{\mathsf{T}}$.

   $\forall (i,j) \in \mathbb{N}^2 : \mathsf{node}^{(i,j)} \in \mathsf{T}, (r_0^{(i,j)}, r_1^{(i,j)}, r_\pi^{(i,j)}) \leftarrow \mathsf{PPRF.Eval}(\mathsf{k}, (\overline{\mathsf{T}}, (i,j)))$
   $\forall (i,j) \in \mathbb{N}^2 : \mathsf{node}^{(i,j)} \in \mathsf{T}$,
      $\mathsf{CT}_0^{(i,j)} \leftarrow \mathsf{PKE.Enc}(\mathsf{pk}_0, \mathsf{node}^{(i,j)}; r_0^{(i,j)}); \mathsf{CT}_1^{(i,j)} \leftarrow \mathsf{PKE.Enc}(\mathsf{pk}_1, \mathsf{node}^{(i,j)}; r_1^{(i,j)})$
      $\pi^{(i,j)} \leftarrow \mathsf{NIZK.Prove}(\mathsf{crs}, x^{(i,j)}, (\mathsf{node}^{(i,j)}, r_0^{(i,j)}, r_1^{(i,j)}); r_\pi^{(i,j)})$

7. Finally, output the updated (encrypted) tree $\overline{\mathsf{T}}$, the address for next iteration and possibly the outcome of the token.

   return $(\overline{\mathsf{T}}, \mathsf{addr}, \mathsf{y})$

Fig. 9. Specification of circuit $\widehat{\delta}$, as part of our updatable functional encryption scheme.

input [functionality preservation under puncturing of PPRF + iO property of Obf]; (3) we replace the pseudorandom coins used to produce the hardwired output with real random coins [pseudorandomness at punctured points of PPRF]; (4) we use simulated NIZK proofs in the new hardwired output [zero-knowledge property of NIZK]; (5) we compute circuit δ_l independently on the right branch before encrypting the hardwired output [IND-CPA security of PKE].

Game$_4$: In all circuits $\widehat{\delta_l}$, we switch the decryption key sk_0 with sk_1 and perform the operations based on the right branch, i.e. we modify the circuits such that $\mathsf{node}^{(i,j)} \leftarrow \mathsf{PKE.Dec}(\mathsf{sk}_1, \mathsf{CT}_1^{(i,j)})$. This hop can be upper-bounded by the distributional indistinguishability of Obf. To show this, we construct an adversary $(\mathcal{S}, \mathcal{B})$ against the DI game that runs adversary $\mathcal{A}$ as follows.

Sampler $\mathcal{S}$ runs $\mathcal{A}_0$ to get the challenge messages $(\mathsf{D}_0, \mathsf{D}_1)$ and circuits δ_l. Then, it produces the challenge ciphertext (same rules apply on Game$_3$ and Game$_4$), and compute circuits $\widehat{\delta_l}$ according to rules of Game$_3$ (with decryption key sk_0) on one hand and according to rules of Game$_4$ (with decryption key sk_1) on the other. Finally, it outputs the two vectors of circuits and the challenge ciphertext as auxiliary information.

Adversary $\mathcal{B}$ receives the obfuscated circuits $\overline{\delta_l}$ either containing sk_0 or sk_1 and the challenge ciphertext. With those, it runs adversary $\mathcal{A}_1$ perfectly simulating Game$_3$ or Game$_4$. $\mathcal{B}$ outputs whatever $\mathcal{A}_1$ outputs.

It remains to show that sampler $\mathcal{S}$ is *computationally* unpredictable. Suppose there is a predictor Pred that finds a differing input for the circuits output by sampler $\mathcal{S}$. It must be because either the output contains a NIZK proof for a false statement (which contradicts the soundness property of NIZK), or there is a collision in the Merkle tree (which contradicts the collision-resistance of H), or the predictor was able to guess the random tag in one of the ciphertexts (which contradicts the IND-CCA security of PKE). Note that (1) the random tag is high-entropy, so lucky guesses can be discarded; (2) we cannot rely only on IND-CPA security of PKE because we need the decryption oracle to check which random tag the predictor was able to guess to win the indistinguishability game against PKE. We also rely on the fact that adversary $\mathcal{A}_0$ is legitimate in its own game, so the outputs in clear of the tokens are the same in Game$_3$ and Game$_4$.

Game$_5$: In this game, we remove the lookup tables introduced in Game$_3$. We remove one input/output at the time, from the last input/output pair added to the first, following the reverse strategy of that introduced in Game$_3$.

Game$_6$: Here, the challenge ciphertext is computed exclusively from D_1 (with the same random tag on both branches). This transition is negligible down to the IND-CPA security of PKE.

Game$_7$: In this game, we move back to regular (non-simulated) NIZK proofs in the challenge ciphertext. The distance to the previous game can be bounded by the zero-knowledge property of NIZK.

Game$_8$: We now switch back the decryption key to sk_0 and perform the decryption operation on the left branch. Since NIZK is statistically sound, the circuits are functionally equivalent. We move from sk_1 to sk_0 one token at the time.

This transition is down to the iO property of Obf. This game is identical to the real SEL game when the challenge bit $b = 1$, which concludes our proof.   □

It is easy to check that the proposed scheme meets the correctness and efficiency properties as we defined in Sect. 3.1 for our primitive. The size of the ciphertext is proportional to the size of the cleartext. The size expansion of the token is however proportional to the number of steps of its execution, as the circuit $\overline{\delta}$ must be appropriately padded for the security proof.

4   Future Work

The problem at hand is quite challenging to realize even when taking strong cryptographic primitives as building blocks. Still, one might wish to strengthen the security model by allowing the adversary to obtain tokens adaptively, or by relaxing the legitimacy condition that imposes equal access patterns of extracted programs on left and right challenge messages using known results on Oblivious RAM. We view our construction as a starting point towards the realization of other updatable functional encryption schemes from milder forms of obfuscation.

Acknowledgements. The authors would like to thank Karol Zebrowski for his contribution to an earlier version of this work. Afonso Arriaga is supported by the National Research Fund, Luxembourg, under AFR Grant No. 5107187, and by the Doctoral School of Computer Science & Computer Engineering of the University of Luxembourg. Vincenzo Iovino is supported by the National Research Fund, Luxembourg. Qiang Tang is supported by a CORE (junior track) grant from the National Research Fund, Luxembourg.

References

1. Ananth, P., Boneh, D., Garg, S., Sahai, A., Zhandry, M.: Differing-inputs obfuscation and applications. IACR Cryptology ePrint Archive, Report 2013/689 (2013)
2. Arriaga, A., Barbosa, M., Farshim, P.: Private functional encryption: indistinguishability-based definitions and constructions from obfuscation. In: Dunkelman, O., Sanadhya, S.K. (eds.) INDOCRYPT 2016. LNCS, vol. 10095, pp. 227–247. Springer, Cham (2016). doi:10.1007/978-3-319-49890-4_13
3. Ananth, P., Sahai, A.: Functional encryption for turing machines. In: Kushilevitz, E., Malkin, T. (eds.) TCC 2016. LNCS, vol. 9562, pp. 125–153. Springer, Heidelberg (2016). doi:10.1007/978-3-662-49096-9_6
4. Bellare, M., Rogaway, P.: The security of triple encryption and a framework for code-based game-playing proofs. In: Vaudenay, S. (ed.) EUROCRYPT 2006. LNCS, vol. 4004, pp. 409–426. Springer, Heidelberg (2006). doi:10.1007/11761679_25
5. Bitansky, N., Canetti, R.: On strong simulation and composable point obfuscation. J. Cryptol. **27**(2), 317–357 (2014)
6. Bellare, M., Hoang, V., Rogaway, P.: Foundations of garbled circuits. In: CCS 2012, pp. 784–796. ACM (2012)

7. Bellare, M., Stepanovs, I., Tessaro, S.: Contention in cryptoland: obfuscation, leakage and UCE. In: Kushilevitz, E., Malkin, T. (eds.) TCC 2016. LNCS, vol. 9563, pp. 542–564. Springer, Heidelberg (2016). doi:10.1007/978-3-662-49099-0_20

8. Boneh, D., Sahai, A., Waters, B.: Functional encryption: definitions and challenges. In: Ishai, Y. (ed.) TCC 2011. LNCS, vol. 6597, pp. 253–273. Springer, Heidelberg (2011). doi:10.1007/978-3-642-19571-6_16

9. Canetti, R., Chen, Y., Holmgren, J., Raykova, M.: Succinct adaptive garbled RAM. IACR Cryptology ePrint Archive, Report 2015/1074 (2015)

10. Canetti, R., Holmgren, J.: Fully succinct garbled RAM. In: ITCS 2016, pp. 169–178. ACM (2016)

11. Canetti, R., Holmgren, J., Jain, A., Vaikuntanathan, V.: Indistinguishability obfuscation of iterated circuits and RAM programs. IACR Cryptology ePrint Archive, Report 2014/769 (2014)

12. Canetti, R., Holmgren, J., Jain, A., Vaikuntanathan, V.: Succinct garbling and indistinguishability obfuscation for RAM programs. In: STOC 2015, pp. 429–437. ACM (2015)

13. Desmedt, Y., Iovino, V., Persiano, G., Visconti, I.: Controlled homomorphic encryption: definition and construction. IACR Cryptology ePrint Archive, Report 2014/989 (2014)

14. Garg, S., Gentry, C., Halevi, S., Raykova, M., Sahai, A., Waters, B.: Candidate indistinguishability obfuscation and functional encryption for all circuits. In: FOCS 2013, pp. 40–49. IEEE Computer Society (2013)

15. Gentry, C., Halevi, S., Lu, S., Ostrovsky, R., Raykova, M., Wichs, D.: Garbled RAM revisited. In: Nguyen, P.Q., Oswald, E. (eds.) EUROCRYPT 2014. LNCS, vol. 8441, pp. 405–422. Springer, Heidelberg (2014). doi:10.1007/978-3-642-55220-5_23

16. Gentry, C., Halevi, S., Raykova, M., Wichs, D.: Outsourcing private RAM computation. In: FOCS 2014, pp. 404–4013. IEEE Computer Society (2014)

17. Goyal, V., Jain, A., Koppula, V., Sahai, A.: Functional encryption for randomized functionalities. In: Dodis, Y., Nielsen, J.B. (eds.) TCC 2015. LNCS, vol. 9015, pp. 325–351. Springer, Heidelberg (2015). doi:10.1007/978-3-662-46497-7_13

18. Ishai, Y., Pandey, O., Sahai, A.: Public-coin differing-inputs obfuscation and its applications. In: Dodis, Y., Nielsen, J.B. (eds.) TCC 2015. LNCS, vol. 9015, pp. 668–697. Springer, Heidelberg (2015). doi:10.1007/978-3-662-46497-7_26

19. Lu, S., Ostrovsky, R.: How to garble RAM programs? In: Johansson, T., Nguyen, P.Q. (eds.) EUROCRYPT 2013. LNCS, vol. 7881, pp. 719–734. Springer, Heidelberg (2013). doi:10.1007/978-3-642-38348-9_42

20. Naor, M., Yung, M.: Public-key cryptosystems provably secure against chosen ciphertext attacks. In: STOC 1990, pp. 427–437. ACM (1990)

21. O'Neill, A.: Definitional issues in functional encryption. IACR Cryptology ePrint Archive, Report 2010/556 (2010)

22. Sahai, A., Waters, B.: How to use indistinguishability obfuscation: deniable encryption, and more. In: STOC 2014, pp. 475–484. ACM (2014)

23. Yao, A.: How to generate and exchange secrets. In: FOCS 1986, pp. 162–167. IEEE Computer Society (1986)

Linking-Based Revocation for Group Signatures: A Pragmatic Approach for Efficient Revocation Checks

Daniel Slamanig, Raphael Spreitzer[✉], and Thomas Unterluggauer

IAIK, Graz University of Technology, Graz, Austria
{daniel.slamanig,raphael.spreitzer,thomas.unterluggauerg}@iaik.tugraz.at

Abstract. Group signature schemes (GSS) represent an important privacy-enhancing technology. However, their practical applicability is restricted due to inefficiencies of existing membership revocation mechanisms that often place a too large computational burden and communication overhead on the involved parties. Moreover, it seems that the general belief (or unwritten law) of avoiding online authorities by all means artificially and unnecessarily restricts the efficiency and practicality of revocation mechanisms in GSSs. While a mindset of preventing online authorities might have been appropriate more than 10 years ago, today the availability of highly reliable cloud computing infrastructures could be used to solve open challenges. More specifically, in order to overcome the inefficiencies of existing revocation mechanisms, we propose an alternative approach denoted as *linking-based revocation* (LBR) which is based on the concept of controllable linkability. The novelty of LBR is its transparency for signers and verifiers that spares additional computations as well as updates. We therefore introduce dedicated revocation authorities (RAs) that can be contacted for efficient (constant time) revocation checks. In order to protect these RAs and to reduce the trust in involved online authorities, we additionally introduce *distributed controllable linkability*. Using latter, RAs cooperate with multiple authorities to compute the required linking information, thus reducing the required trust. Besides efficiency, an appealing benefit of LBR is its generic applicability to pairing-based GSSs secure in the BSZ model as well as GSSs with controllable linkability. This includes the XSGS scheme, and the GSSs proposed by Hwang et al., one of which has been standardized in the recent ISO 20008-2 standard.

1 Introduction

Group signature schemes (GSSs) [17] represent an important privacy-enhancing technology. Such schemes allow users to anonymously prove affiliation to a managed group by issuing so called group signatures. Although such signatures do not reveal the identity of the users, in case of dispute a dedicated opening authority is able to identify a signer. GSSs are especially attractive in scenarios like

© Springer International Publishing AG 2017
R.C.-W. Phan and M. Yung (Eds.): LNCS 10311, Mycrypt 2016, pp. 364–388, 2017.
DOI: 10.1007/978-3-319-61273-7_18

public transport or subscription-based services, where there is no need for service providers to uniquely identify single users, but only to ensure that they are allowed to use the respective services. Nevertheless, service providers typically want to ensure that group membership can be efficiently revoked. However, membership revocation is a non-trivial task as (1) users are anonymous and their privacy should be protected even in case of revocation, and (2) the revocation of one user should not affect the signing capabilities of other users. Even though membership revocation has gained increasing attention in the last decade [2,8,9,23,24,38,44], existing mechanisms place a computational burden and communication overhead on signers and verifiers.

While it seems that existing concepts aim to prevent online authorities by all means, we show that a *paradigm shift* towards online authorities—which we will protect by means of threshold cryptography—allows for the most efficient (constant time), and most generic revocation mechanism for existing GSSs. Given a signature in question and a revocation list, a revocation authority determines the revocation status of a signer by using a dedicated trapdoor to (anonymously) link the signature against a list of revoked members. Hence, this authority still preserves the signer's anonymity as it is strictly less powerful than an opening authority who can determine the actual identity of the signer. Our somewhat *unconventional* proposal of using an *online* revocation authority overcomes many issues of existing revocation mechanisms and, thus, we believe that our revocation mechanism represents a valuable addendum to the portfolio of revocation mechanisms.

Online Requirement. Although our approach relies on an online authority for revocation checks, we argue that today many devices are already connected to the Internet permanently and rely on the availability of cloud computing infrastructures. The establishment of the Internet of Things (IoT) requires devices being connected to the Internet, either via WiFi or even embedded SIM cards, for various (sometimes dubious) reasons. Nevertheless, irrespective of whether existing IoT devices provide any useful features, the point is that many devices are already interconnected among each other and also extensively use Internet services based on cloud computing infrastructures. Thus, we consider an online RA as absolutely reasonable. In cases where network connectivity and availability are not an issue, our proposed revocation mechanism provides significant advantages compared to other revocation mechanisms. Further, since signature verification is decoupled from the online revocation checks, these revocation checks can also be postponed in case the revocation authority might not be available. Besides, as we will discuss later, we protect the required trapdoor information by means of threshold cryptography in order to reduce the risk of its exposure.

Contributions. The contributions of this work can be summarized as follows.

- We suggest a paradigm shift towards online revocation checks for group signatures by relying on online revocation authorities (RAs). Although currently this seems to be prevented by all means, it, however, allows us to come up with a privacy-respecting constant-time revocation mechanism that can be generically applied to a large class of pairing-based GSSs. In doing so, we use the concept of controllable linkability for group signatures.

- We introduce the concept of distributed controllable linkability. This is a threshold variant of controllable linkability, which requires a predefined number t of linking authorities to cooperate in order to link signatures. Using this concept, we can reduce the trust in and also improve the robustness of RAs. Besides, it may be of independent interest as a feature on its own.
- We demonstrate the ease of applicability of the linking-based revocation mechanism using the well-known XSGS group signature scheme [21].

2   State-of-the-Art in Revocation and Motivation

Below we discuss efficiency considerations of existing revocation mechanisms using the metric of additional computations and updates required for signers as well as verifiers. Subsequently, R denotes the number of revoked members and N the number of group members.

Basic Approaches. The most basic approach is *reissuance-based revocation* [1] which requires all non-revoked members to receive new group signing keys. Similarly, *credential-update revocation* (CUR) [7] requires non-revoked members to update their group signing key on every revocation. Both mechanisms suffer from additional communication and computation overhead in case of frequent revocations. In particular, $\mathcal{O}(R)$ (multi-)exponentiations for signers.

Blacklist Revocation. *Certificate-based blacklist revocation* (BR-C) [10] requires signers to provide a zero-knowledge proof that they are not listed on a revocation list (RL), which means that signers/verifiers need to perform $\mathcal{O}(R)$ computations for every sign/verify operation and the signature size also increases linearly in R. Similar revocation mechanisms relying on revocation lists of private keys or signatures have been proposed for direct anonymous attestation (DAA) settings [11,12]. Again, the computational effort as well as the signature size increases linearly with the revocation list. *Accumulator-based blacklist revocation* (BR-A) [3,13,24] applies (universal) cryptographic accumulators to allow for a compact and constant-size representation of RL as well as constant-size proofs to prove (non-)membership. *Blacklists of ordered credential-identifier pairs* [42] also lead to constant costs for signers and verifiers. But signers need to fetch an updated RL in the size of $\mathcal{O}(R)$ (and $\mathcal{O}(N)$ in predecessor schemes [46,47]) on each revocation, and the public key size is $\mathcal{O}(N)$ (or $\mathcal{O}(\sqrt{N})$ with significantly higher signing costs).

Verifier-Local Revocation. In *verifier-local revocation* (VLR) [8,44,45,62], verifiers when given a signature in question test if a certain relation holds for the signature and each entry on RL. The validity of the relation indicates that the signer has been revoked. Consequently, verifiers need to update RL on every revocation and a check during verification costs $\mathcal{O}(R)$ (typically group operations or even pairing evaluations). Although [8] proposes a constant-time revocation check, it only works if the same message and the same randomness is used for all signatures. This, however, is only reasonable for specific applications like DAA. Other disadvantages of VLR are that signatures of revoked members become linkable for all verifiers, $i.e.$, it lacks backward unlinkability, and that anyone in possession of RL can link signatures of revoked users. Although this can be fixed and backward unlinkability can be added, e.g., by introducing time intervals [44], this still adds additional non-trivial overhead. Besides, [22] proposed time-token dependent linking which can also be applied for VLR. However, signatures become publicly linkable (without any trapdoor information) if users sign more than once per time period, a separate RL must be maintained for each time period, and RLs need to be recomputed entirely for each time period since the revocation tokens of users change in each time period. While one could simply encrypt these revocation tokens to be decryptable by revocation authorities only, $i.e.$, to prevent public linkability, this would increase the signature size in part due to additional zero-knowledge proofs. In contrast, our approach preserves the privacy of signers and does not increase the signature size as it relies on the information already available in standard group signature schemes. Chow et al. [18] proposed a similar concept for membership revocation in ID-based ring signatures, a related but different concept. *Group signatures with probabilistic revocation* (GSPR) [38] allow for constant-time revocation checks at the expense of probabilistic revocation guarantees. However, in contrast to VLR the signer has to perform $\mathcal{O}(m)$ expensive operations, where m is a fixed value representing the number of signatures that can be issued by a signer before signatures become publicly linkable. Consequently, there is a trade-off between the storage/computational requirements for signers and the requirement for performing the group setup phase again. Moreover, the size of the group public key is $\mathcal{O}(m)$ and the GM needs to process $\mathcal{O}(m)$ user-specific tokens in order to update RL.

Revocation Mechanisms for Standard Model GS. For the sake of completeness we want to mention that there are also various revocation mechanisms designed to be compatible with the Groth-Sahai proof system [28] (instead of relying on Σ-protocols and the random oracle model). State-of-the-art mechanisms are due to Libert, Peters, and Yung (LPY) [40], which rely on the ciphertext of a broadcast encryption scheme as a RL. Later, LPY [39] has been improved to achieve constant size group signing keys. Attrapadung et al. (AEHS) [2] further reduced the revocation list to a constant size. However, signature sizes for LPY are about 100 and 144 group elements respectively, and AEHS produces even larger signatures. Thus, we exclude these revocation mechanisms from our comparison below, since we put a focus on group signature schemes that allow signers to be executed in resource-constrained environments as will be discussed in our motivating example later in this section.

Our Proposal (LBR). Existing revocation mechanisms are either inefficient or, in the worst case, even impossible to be implemented in resource-constrained environments, which is why revocation has been identified as the major bottleneck of state-of-the-art GSSs [41]. We address this problem by introducing *linking-based revocation* (LBR). LBR allows for a constant-time revocation mechanism that can be generically applied to existing GSSs, and in particular PB-GSSs [21, 29–32, 49] following the sign-and-encrypt-and-prove (SEP) paradigm [14, 36]. Essentially, we rely on the feature of controllable linkability [6, 29–31, 56] that allows a dedicated entity to determine whether two signatures have been produced by the same (anonymous) signer. By replacing the used public key encryption scheme of a GSS with its AoN-PKEET* variant (cf. Sect. 3.1), this feature can be added to GSSs following the SEP paradigm generically. The idea of LBR is that, in analogy to the online certificate status protocol (OCSP) [54] which is widely used for certificate revocation checks in the PKIX [19] setting[1], an online party can be contacted for revocation checks. To obtain robustness against compromise of online authorities, we introduce the feature of *distributed controllable linkability*, which may be of independent interest. When applying distributed controllable linkability to revocation, it allows RAs to anonymously link a given signature—with the cooperation of at least two linking authorities—against anonymous revocation tokens on RL. An additional optimization even allows for constant-time revocation checks.

In contrast to existing revocation mechanisms, our mechanism is transparent for signers and verifiers. Most importantly, LBR is efficient in the sense that (1) no key updates or additional computations are required for signers, (2) no expensive local revocation checks are required for verifiers, and (3) neither the signature size nor the key size increases. While all existing revocation approaches require signers and/or verifiers to fetch (possibly large) RLs from time to time, our mechanism relies on an always-online authority that is available for revocation checks. Although an *online* authority for such tasks might be considered as being *unconventional* or impractical at first, we believe that such an always-online requirement for specific authorities is absolutely reasonable.[2] In order to protect these RAs against attacks, we distribute the linking trapdoor required for the revocation checks to multiple entities. Consequently, an attacker would have to corrupt multiple entities to recover the linking trapdoor.

Comparison. Table 1 compares existing revocation mechanisms for practical group signature schemes in the random oracle model regarding their efficiency and practicality. For each mechanism, we compare the memory overhead for the group public key (GPK) and the signature as well as the computational overhead for updating keys/credentials, signature generation, and signature verification. Furthermore, we indicate the amount of information that needs to be fetched by signers and verifiers in case of revocation. As some schemes require both signers and verifiers to fetch updates, we also indicate whether these updates must be

---

[1] A recent study [51] even states that OCSP is the most popular approach for revocation checks in the PKIX setting.

[2] An approach similar in spirit to our approach has recently also been discussed in the context of anonymous credential systems (cf. [60]).

synchronized, *i.e.*, whether both parties need to have the same update-version as otherwise valid signatures cannot be computed and verified. Last but not least, we indicate whether signers and verifiers must be online for the revocation mechanism to work, where $\Diamond$ means semi-online, *i.e.*, signers and verifiers can decide when to go online to fetch the necessary updates.

Table 1. Comparison of revocation mechanisms.

Type	Overhead memory		Overhead time			Updates		Synchronized	Online	
	GPK	Signature	Update (signer)	Sign	Verify	Signers	Verifiers		Signers	Verifiers
CUR [7,30]	–	–	$\mathcal{O}(R)$	–	–	$\mathcal{O}(R)$	$\mathcal{O}(1)$	✓	$\Diamond$	$\Diamond$
BR-C [10]	–	$\mathcal{O}(R)$	–	$\mathcal{O}(R)$	$\mathcal{O}(R)$	$\mathcal{O}(R)$	$\mathcal{O}(R)$	✓	$\Diamond$	$\Diamond$
BR-ID [42]	$\mathcal{O}(\sqrt{N})$	$\mathcal{O}(1)$	–	$\mathcal{O}(1)$	$\mathcal{O}(1)$	$\mathcal{O}(R)$	$\mathcal{O}(1)$	✓	$\Diamond$	$\Diamond$
BR-A [3,13,24]	–	$\mathcal{O}(1)$	–	$\mathcal{O}(1)$	$\mathcal{O}(1)$	$\mathcal{O}(1)$	$\mathcal{O}(1)$	✓	$\Diamond$	$\Diamond$
VLR [8]	–	–	–	–	$\mathcal{O}(R)$	–	$\mathcal{O}(R)$	–	✗	$\Diamond$
GSPR [38]	$\mathcal{O}(m)$	$\mathcal{O}(1)$	–	$\mathcal{O}(m)$	$\mathcal{O}(1)$	–	$\mathcal{O}(1)$	–	✗	$\Diamond$
LBR	–	–	–	–	$\mathcal{O}(1)$	–	–	–	✗	✓

Although VLR seems to provide similar advantages and features as LBR, our approach of LBR only allows RAs to link signatures, which is not the case within VLR as users can link signatures themselves. Thus, LBR overcomes the delicate issue of revoked members losing their anonymity. In addition, our proposed revocation mechanism can be generically applied to many PB-GSSs following the SEP paradigm, which covers a large class of state-of-the-art and practically efficient GSSs. As we will see below, our approach of LBR provides dedicated advantages and superior features for specific scenarios.

Motivating Example. As pairing-based cryptography has been optimized for resource-constrained devices [15,16,27,33,52,59], PB-GSSs have become entirely practical, at least when considering the performance of signature generation only. For instance, GSSs have been proposed as a privacy-preserving mechanism for public transport systems [34]. Their application allows passengers to anonymously prove possession of a valid ticket, but the service provider cannot identify passengers. Still, revocation of misbehaving passengers by invalidating tickets must be possible and these revocations should not affect other tickets in any way. Clearly, frequent updates through authenticated channels between the tickets and the service provider are impractical, as they would affect the valid tickets. Besides, performance is a crucial issue and, hence, the invalidation of one ticket should not lead to additional computations for the remaining (valid) tickets. While VLR might be a possible solution to overcome these problems, public transport systems usually support tickets with a limited validity, *i.e.*, 1-h tickets, daily tickets, monthly tickets, and yearly tickets. Such tickets must be immediately revoked as soon as their validity ends and, thus, immediate revocation of tickets must be efficiently possible. Hence, VLR still faces the following problems. (1) RLs lead to $\mathcal{O}(R)$ computational effort for verifiers which is especially daunting in case of large RLs, and (2) RLs change frequently and must be distributed in a timely manner, *i.e.*, immediately after the revocation of one ticket, to many verifiers. Clearly, LBR overcomes these issues as it gets rid of the computational overhead for signers as well as verifiers and the need to communicate any revocation updates to signers and verifiers. Applying LBR allows the service provider to implement an existing PB-GSS (e.g. [21,29–31,49]) as a means to prove possession of a valid ticket. Turnstiles and gates that check the validity of a ticket are connected to the revocation authority. For each ticket to be verified, turnstiles request the revocation check via specifically deployed RAs and depending on the returned decision, access is either granted or denied. Considering some of the biggest metro systems around the world with several hundred millions of served passengers per year, e.g., Beijing, Moscow, and NY City, the efficiency of the used revocation mechanism is of utmost importance. Besides, also the European Union demands for privacy protection of individuals and the principle of data minimization in transportation systems within the EU Directive 2010/40. Hence, GSSs will likely play an important role in the future and efficient revocation mechanisms will be required.

3   Preliminaries

Let $\mathbb{G}_1 = \langle g_1 \rangle$, $\mathbb{G}_2 = \langle \hat{g}_2 \rangle$, and $\mathbb{G}_T$ be cyclic groups of prime order p. We write elements in $\mathbb{G}_2$ as $\hat{g}, \hat{v}$, etc. A bilinear map $e : \mathbb{G}_1 \times \mathbb{G}_2 \to \mathbb{G}_T$ is a map, such that $e(u^a, \hat{v}^b) = e(u, \hat{v})^{ab}$ for all $u \in \mathbb{G}_1$, $\hat{v} \in \mathbb{G}_2$, and $a, b \in \mathbb{Z}_p$. Additionally, we require $e(g_1, \hat{g}_2) \neq 1$ and e to be efficiently computable. If $\mathbb{G}_1 = \mathbb{G}_2$, then e is called *symmetric* (Type 1) and *asymmetric* (Type 2 or Type 3) otherwise. For Type 2 pairings there is an efficiently computable isomorphism $\psi : \mathbb{G}_2 \to \mathbb{G}_1$, whereas for Type 3 pairings no such efficient isomorphism is known. We (informally) state the used assumptions below.

DDH. Let $\mathbb{G}$ be a cyclic group of prime order p and g a generator. The DDH assumption states that given (g, g^a, g^b, g^c) it is hard to decide whether $ab = c$.

(S)XDH. Let $\mathbb{G}_1$, $\mathbb{G}_2$, and $\mathbb{G}_T$ be three cyclic groups of prime order p and $e : \mathbb{G}_1 \times \mathbb{G}_2 \to \mathbb{G}_T$ a pairing. The (S)XDH assumption states that the DDH assumption holds in $\mathbb{G}_1$ (and $\mathbb{G}_2$).

A function $\epsilon : \mathbb{N} \to \mathbb{R}^+$ is called negligible if for all $c > 0$ there is a k_0 such that $\epsilon(k) < 1/k^c$ for all $k > k_0$. We use ϵ to denote such a negligible function.

3.1   All-or-Nothing Public Key Encryption with Equality Tests

All-or-nothing public key encryption with equality tests (AoN-PKEET) [57] allows entities in possession of a trapdoor to perform equality tests on ciphertexts without learning the underlying plaintexts. A modification denoted as AoN-PKEET* [56] allows IND-CPA security against outsiders and requires compatibility with efficient zero-knowledge proofs of knowledge (ZKPoK) of plaintexts. Such an AoN-PKEET* scheme (KeyGen, Enc, Dec, Aut, Com) is a conventional (at least IND-CPA secure) public key encryption scheme (KeyGen, Enc, Dec) (compatible with efficient ZKPoK) augmented by two algorithms Aut and Com.

KeyGen(1^λ): The key generation algorithm takes a security parameter λ, and generates a public-private key pair (pk, sk) used for encryption and decryption operations.

Enc(pk, m): The encryption algorithm takes the public key pk, and a message m, and returns the encryption c of m under pk.

Dec(sk, c): The decryption algorithm takes the private key sk, and a ciphertext c, and returns the message m.

Aut(sk): Takes the private decryption key sk of the public key encryption scheme and returns the trapdoor tk used for equality tests.

Com(T, T', tk): Takes two ciphertexts (T, T') and a trapdoor tk and returns true if T and T' encrypt the same (unknown) message and false otherwise.

Definition 1 [56]. *An AoN-PKEET* scheme is secure if it is sound, provides OW-CPA security against Type-I adversaries (trapdoor holders) and if the underlying encryption scheme provides IND-CPA/IND-CCA2 security against Type-II adversaries (outsiders).*

Soundness requires correctness of the public key encryption scheme and for all (pk, sk) $\leftarrow$ KeyGen(1^λ) one requires that Com(Enc(pk, m), Enc(pk, m'), Aut(sk)) = true if and only if $m = m'$. Subsequently, we provide definitions for OW-CPA as well as IND-CPA/IND-CCA2 security, where $\mathcal{M}$ denotes the message space of a scheme.

Definition 2 (OW-CPA **security**). *An AoN-PKEET* scheme is OW-CPA secure against Type-I adversaries, if for all PPT adversaries $\mathcal{A}$ and security parameters λ there is a negligible function ϵ such that:*

$$\Pr\left[\begin{array}{l} \text{(pk, sk)} \leftarrow \text{KeyGen}(1^\lambda), \text{tk} \leftarrow \text{Aut(sk)}, m \xleftarrow{R} \mathcal{M}, \\ c \leftarrow \text{Enc(pk, } m), m^* \leftarrow \mathcal{A}(\text{pk}, \text{tk}, c) \end{array} : m^* = m \right] \leq \epsilon(\lambda).$$

Definition 3 (IND-CPA/IND-CCA2 **security**). *An AoN-PKEET* scheme is IND-CPA/IND-CCA2 secure against Type-II adversaries, if for all PPT adversaries $\mathcal{A}$ and security parameters λ there is a negligible function ϵ such that:*

$$\Pr\left[\begin{array}{l} \text{(pk, sk)} \leftarrow \text{KeyGen}(1^\lambda), \text{tk} \leftarrow \text{Aut(sk)}, \\ (m_0, m_1, s) \leftarrow \mathcal{A}^{\mathcal{O}_1}(\text{pk}), b \xleftarrow{R} \{0,1\}, c \leftarrow \text{Enc(pk, } m_b) : b^* = b \\ b^* \leftarrow \mathcal{A}^{\mathcal{O}_2}(\text{pk}, c, s) \end{array} \right] \leq {}^1\!/_2 + \epsilon(\lambda)$$

where $m_0, m_1 \in \mathcal{M}$, and

$$\begin{array}{llll} \mathcal{O}_1(\cdot) = \perp & \text{and } \mathcal{O}_2(\cdot) = \perp & \text{for IND} - \text{CPA} \\ \mathcal{O}_1(\cdot) = \mathcal{O}_{\text{Dec}} & \text{and } \mathcal{O}_2(\cdot) = \mathcal{O}_{\text{Dec}} & \text{for IND} - \text{CCA2} \end{array}$$

and $\mathcal{O}_{\text{Dec}}$ represents the decryption oracle, and $\mathcal{A}$ is not allowed to query the decryption oracle for the challenge ciphertext c.

Instantiation Based on ElGamal Encryption. For reasons of simplicity we will demonstrate our approach based on ElGamal encryption (instead of its twin variants [25, 48, 53] as demonstrated in Sect. 6), but we stress that it also works for other encryption schemes used in the pairing setting like linear ElGamal [7] (and its corresponding twin variant) or Cramer-Shoup encryption [20]. Nevertheless, if AoN-PKEET* is instantiated with ElGamal encryption and assuming the private decryption key to be $\xi \in \mathbb{Z}_p$ and the corresponding public key $h = g^\xi \in \mathbb{G}_1$, the resulting ciphertext is $(T_1 = g^\alpha, T_2 = m \cdot h^\alpha) \in \mathbb{G}_1^2$ for a random $\alpha \in \mathbb{Z}_p$. Given two ciphertexts T and T', and the trapdoor key $\mathsf{tk} = (\hat{r}, \hat{s} = \hat{r}^\xi) \leftarrow \mathsf{Aut}(\xi)$ for a random $\hat{r} \in \mathbb{G}_2$, the equality test on the encrypted messages can be performed via the $\mathsf{Com}(T, T', \mathsf{tk})$ algorithm of the AoN-PKEET* scheme by evaluating whether the following holds: $e(T_2, \hat{r}) \cdot e(T_1, \hat{s})^{-1} = e(T_2', \hat{r}) \cdot e(T_1', \hat{s})^{-1}$.

3.2   Sign-and-Encrypt-and-Prove Paradigm

GSSs following the SEP paradigm consist of the following three building blocks: (1) a secure signature scheme $\mathcal{DS} = (\mathsf{KeyGen}_s, \mathsf{Sign}, \mathsf{Verify})$; (2) an at least IND-CPA secure public key encryption scheme $\mathcal{AE} = (\mathsf{KeyGen}_e, \mathsf{Enc}, \mathsf{Dec})$; and (3) non-interactive zero-knowledge proofs of knowledge (NIZKPKs). For schemes in the ROM latter are honest-verifier zero-knowledge proofs of knowledge made non-interactive using the Fiat-Shamir transform (denoted as signatures of knowledge (SoK) subsequently).

The group public key gpk consists of the public encryption key pk_e, and the signature verification key pk_s. The master opening key mok is the decryption key sk_e and the master issuing key mik is the signing key sk_s. During the joining procedure a user i sends $f(x_i)$ to the issuer, where $f(\cdot)$ is a one-way function applied to a secret x_i. The issuer returns a signature $\mathsf{cert} \leftarrow \mathsf{Sign}(\mathsf{sk}_s, f(x_i))$ as the user's certificate. A group signature $\sigma = (T, \pi)$ for a message M consists of a ciphertext $T \leftarrow \mathsf{Enc}(\mathsf{pk}_e, \mathsf{cert})$ and the following SoK π:

$$\pi \leftarrow \mathsf{SoK}\{(x_i, \mathsf{cert}) : \mathsf{cert} = \mathsf{Sign}(\mathsf{sk}_s, f(x_i)) \ \wedge \ T = \mathsf{Enc}(\mathsf{pk}_e, \mathsf{cert})\}(M).$$

The above description provides an intuition for the SEP approach and there exist different variations, e.g., sometimes cert is computed for x_i instead of $f(x_i)$ (which, however, does not yield constructions providing non-frameability), or T may represent an encryption of $f(x_i)$ or $g(x_i)$ for some one-way function $g(\cdot)$. This, however, is not important in the context of controllable linkability, where it is only required that T contains the encryption of a constant, per-user unique value cert.

3.3   Threshold Secret Sharing

A (t, n)-threshold secret sharing scheme allows to distribute a secret s to n parties in a way such that it requires the cooperation of at least t parties to recover s, while any set of up to $t-1$ parties learns nothing about s. An elegant and famous (t, n)-threshold scheme based on polynomial interpolation has been proposed by Shamir [55]. Here, to share a secret s, one chooses a random polynomial $F(x) = s + \sum_{\ell=1}^{t-1} a_\ell \cdot x^\ell$ of degree $t-1$ such that $F(0) = s$ over some finite field.

Shares are of the form $(i, F(i))$ and any subset of size at most $t - 1$ will learn no information about s. Given shares $(i, F(i)) \in \mathcal{I}$ such that $|\mathcal{I}| \geq t$, s can however be efficiently recovered via $s = F(0) = \sum_{i \in \mathcal{I}} c_i \cdot F(i)$, where $c_i = \prod_{j \in \mathcal{I}, j \neq i} \frac{j}{j-i}$ are the corresponding Lagrange coefficients. We will use this scheme to share a group element from a prime order group (cf. [4]).

3.4   Group Signatures with Controllable Linkability

Hwang et al. [29–31] introduced a model for GSSs with controllable linkability that builds upon the well-established BSZ model [5] (although, Hwang et al. use a weaker notion of anonymity). The group manager is logically split into (1) an opening authority capable of opening signatures, (2) an issuing authority capable of issuing signing keys to group members, and (3) a linking authority capable of linking signatures, *i.e.*, an authority that can determine whether or not two signatures have been issued by the *same anonymous* signer. However, the linking key does not allow to actually identify the signer, which means that the linking authority is strictly less powerful than the opening authority. We denote the keys of these authorities as master opening key (mok), master issuing key (mik), and master linking key (mlk), respectively. A GSS with controllable linkability is a tuple $\mathcal{GS\text{-}CL} = ($GkGen, UkGen, Join, Issue, GSig, GVf, Open, Judge, Link$)$ of PPT algorithms as defined in [5, 29–31] and recalled subsequently.

GkGen(1^λ): On input a security parameter λ, this algorithm generates the public parameters and outputs a tuple (gpk, mok, mik, mlk), representing the group public key, the master opening key, the master issuing key, and the master linking key.

UkGen(1^λ): On input a security parameter λ, this algorithm generates a user key pair (usk$_i$, upk$_i$).

Join(usk$_i$, upk$_i$): On input the user's key pair (usk$_i$, upk$_i$), this algorithm interacts with Issue and outputs the group signing key gsk$_i$ of user i.

Issue(gpk, mik, **reg**): On input the group public key gpk, the master issuing key mik, and the registration table **reg**, this algorithm interacts with Join to add user i to the group.

GSig(gpk, M, gsk$_i$): On input the group public key gpk, a message M, and a user's secret key gsk$_i$, this algorithm outputs a group signature σ.

GVf(gpk, M, σ): On input the group public key gpk, a message M, and a signature σ, this algorithm verifies whether the signature σ is valid with respect to the message M and the group public key gpk and outputs `true` if the verification succeeds and `false` otherwise.

Open(gpk, **reg**, M, σ, mok): On input the group public key gpk, the registration table **reg**, a message M and a valid signature σ corresponding to this message, and the master opening key mok, this algorithm returns the signer i together with a publicly verifiable proof τ attesting the validity of the claim and $\perp$ otherwise.

Judge(gpk, M, σ, i, upk$_i$, τ): On input the group public key gpk, a message M, a valid signature σ, the claimed signer i, the user's public key upk$_i$ as well as

374   D. Slamanig et al.

a proof τ, this algorithm returns `true` if τ is a valid proof that i produced σ and `false` otherwise.

Link(gpk, M, σ, M', σ', mlk): On input the group public key gpk, a message M and valid signature σ as well as a message M' and valid signature σ', and the master linking key mlk, this algorithm returns `true` if both signatures stem from the same signer and `false` otherwise.

Security Properties for GSs with Controllable Linkability. For a GSS with controllable linkability to be secure it needs to satisfy the following properties (cf. [5,29,30] for a formal description).

Anonymity: The identity of the signer can only be determined by the authority in possession of the master opening key.

Traceability: The opening authority must be able to open a valid signature and to prove the corresponding claim.

Non-frameability: An adversary should not be able to prove that an honest user generated a signature unless this user indeed produced this signature.

Linkability: The master linking key should neither be useful to gain any information for opening a signature nor for generating opening proofs. Furthermore, colluding parties—including users, the linker, and/or the opener—should not be able to generate pairs of messages and signatures with contradicting open and link decisions.

3.5   Concepts Related to Controllable Linkability

The following two concepts are related to controllable linkability and thus we discuss their suitability to implement linking-based revocation (LBR) subsequently. The first concept does not qualify for LBR due to privacy concerns (public linkability of signatures). The second concept qualifies for LBR, but the underlying linking mechanisms always requires computations linear in the number of revoked members. We briefly discuss both approaches below. For a more detailed comparison of these concepts we refer the interested reader to [56].

Linkable Group Signatures. Constructions of GSSs relying on tracing-by-linking [58,61] do not employ the SEP paradigm and, hence, no authority can open a given signature. Only if a member signs more than k times, signatures of this member become publicly traceable. Similarly, link-but-not-trace GSSs [43] allow to publicly link signatures, while opening requires all users to prove that a signature has not been produced by them (disavowing), which is clearly not possible for the member who actually produced this message. Both of these approaches allow to publicly link signatures and are thus not appropriate candidates for a revocation mechanism we are envisioning.

Traceable Signatures. Traceable signatures [35] are a variant of group signatures that allow the group manager to publish a tracing trapdoor for any group member that can be used to trace signatures of the respective member. Consequently, they could be used in a similar manner for revocation as controllable linkability. Namely, on revocation one could call the Reveal algorithm for the revoked user and provide the respective tracing trapdoor to the RA. Given a signature, the RA uses all the tracing trapdoors (representing the revocation list) and runs the Trace algorithm on the given signature and every tracing trapdoor from the list. However, while we achieve a constant-time revocation check, traceable signatures would always require a linear check.

4   Building Blocks for GSs with Linking-Based Revocation

Subsequently, we briefly outline the high-level idea of our proposed revocation mechanism. Afterwards, we introduce the necessary building blocks and modifications, *i.e.*, we show how to achieve constant-time revocation checks and we also introduce the feature of distributed controllable linkability.

4.1   High-Level Idea of GSs with Linking-Based Revocation

We recall that the generic compiler in [56] allows to add controllable linkability to PB-GSSs following the SEP paradigm that are secure in the BSZ [5] model. Essentially, this generic compiler replaces the used encryption scheme (used to encrypt membership certificates) with its AoN-PKEET* variant, which allows to determine whether two ciphertexts encrypt the same plaintext, without learning the plaintexts. Thereby, controllable linkability allows a dedicated authority to determine whether two signatures have been issued by the same *unknown* signer by performing an equality test on the encrypted membership certificates. We stress that the dedicated approaches to construct group signatures with controllable linkability in [29–31] implicitly use the same idea and thus can also be used in combination with our revocation approach. Technically, they do not directly apply AoN-PKEET* to the membership certificate, but another user-related value. Nevertheless, one can apply our subsequently discussed ideas analogously.

Based on the concept of controllable linkability, the idea of linking-based revocation is as follows. A verifier first verifies a given signature before contacting a dedicated RA for the revocation check. Note that this also means that the signature verification is decoupled from the actual revocation check and, thus, revocation checks can also be postponed as in case of OCSP requests in the PKIX setting. The RA is given the master linking key, a revocation list, e.g., a list of signatures of revoked members, and a signature in question. For the revocation check, the RA links the given signature against all entries on RL. If any of these signatures links, the corresponding signer has been revoked. Figure 1 illustrates this basic approach, which is, however, rather naive for the following reasons.

1. The revocation check is linear in the size of RL, *i.e.*, has cost $\mathcal{O}(R)$.
2. If an attacker compromises the RA and steals the linking key, she would be able to link any two signatures, which is clearly not desired.

Subsequently, we deal with these issues and gradually introduce the necessary modifications to achieve (1) constant-time revocation checks, and (2) to remove the single point of attack by distributing the linking key among multiple entities.

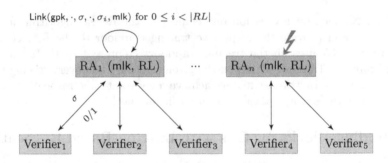

Link(gpk, ·, σ, ·, σ_i, mlk) for $0 \leq i < |RL|$

Fig. 1. Naive (insecure) instantiation of linking-based revocation: an attacker can steal the linking key mlk by compromising a single revocation authority.

4.2 Constant-Time Revocation Checks

To obtain constant-time revocation checks, we modify the $\mathsf{Com}(T, T', \mathsf{tk})$ algorithm of the AoN-PKEET* scheme in a way that it does no longer decide whether two ciphertexts encrypt the same message, but instead returns a value—the *revocation token*—that is computed from a given ciphertext T and the trapdoor tk. We stress that we cannot generically decide for any AoN-PKEET* whether this is possible, but it is in fact possible for all natural ElGamal-style AoN-PKEET* schemes in the pairing setting. For instance, for conventional ElGamal encryption this yields revocation tokens of the form $e(T_2, \hat{r}) \cdot e(T_1, \hat{s})^{-1} = e(m, \hat{r})$. Subsequently we denote such an invocation as $\mathsf{t} \leftarrow \mathsf{Com}(T, \bot, \mathsf{tk})$.

Security Definition. In order to reason about the security of such a mechanism when applied to revocation, we introduce the notion of token indistinguishability. Token indistinguishability considers an adversary that does not know the trapdoor tk but for a ciphertext $T = (T_1, T_2)$ on any message m observes tokens t of the form $e(T_2, \hat{r}) \cdot e(T_1, \hat{t})^{-1} = e(m, \hat{r})$. This information, however, does not allow the adversary to reason about any tokens seen in the future.

Definition 4 (Token Indistinguishability). *An* AoN-PKEET* *scheme is* T-IND, *if for all PPT adversaries* $\mathcal{A}$ *and security parameters* λ *there is a negligible function* ϵ *such that:*

$$\Pr \left[\begin{array}{l} (\mathsf{pk}, \mathsf{sk}) \leftarrow \mathsf{KeyGen}(1^\lambda), \mathsf{tk} \leftarrow \mathsf{Aut}(\mathsf{sk}), \\ s \leftarrow \mathcal{A}^{\mathcal{O}_{\mathrm{RMsg}}}(\mathsf{pk}), m \xleftarrow{R} \mathcal{M}, c \leftarrow \mathsf{Enc}(\mathsf{pk}, m), \\ b \xleftarrow{R} \{0,1\}, \mathsf{t}_0 \leftarrow \mathsf{Com}(c, \bot, \mathsf{tk}), \mathsf{t}_1 \xleftarrow{R} \mathcal{T} \\ b^* \leftarrow \mathcal{A}^{\mathcal{O}_{\mathrm{RMsg}}}(\mathsf{pk}, m, \mathsf{t}_b, s) \end{array} : b^* = b \right] \leq {}^1\!/_2 + \epsilon(\lambda)$$

where $\mathcal{M}$ *represents the message space,* $\mathcal{T}$ *represents the token space, and* $\mathcal{O}_{\mathrm{RMsg}}$ *represents the oracle to generate random messages and corresponding tokens* (m_i, t_i), *such that* $m_i \xleftarrow{R} \mathcal{M}$ *and* $\mathsf{t}_i \leftarrow \mathsf{Com}(\mathsf{Enc}(\mathsf{pk}, m_i), \bot, \mathsf{tk})$.

Lemma 1. *Under the DDH assumption,* AoN-PKEET* *based on ElGamal in* $\mathbb{G}_1$ *in an XDH setting is* T-IND.

Proof (Lemma 1). Given an adversary $\mathcal{A}$ that breaks the T-IND of AoN-PKEET*, we show how to construct an adversary $\mathcal{B}$ against DDH. Let (g, g^a, g^b, g^c) be a DDH instance given to $\mathcal{B}$. $\mathcal{B}$ randomly generates a private key sk and a corresponding public key pk, and sets tk $\leftarrow$ Aut(sk), *i.e.*, $\mathcal{B}$ implicitly sets $\hat{r} = \hat{g}^b$. $\mathcal{A}$ is now allowed to query $\mathcal{O}_{\mathsf{RMsg}}$, which $\mathcal{B}$ answers as $(g^{m_i}, e((g^b)^{m_i}, \hat{g}))$ for a random $m_i \in \mathbb{Z}_p$. Eventually, $\mathcal{A}$ receives the challenge $(g^a, e(g^c, \hat{g}))$ and outputs its guess. It is clear that if the DDH instance is valid, then the challenge represents a valid message-token tuple and is an independent and random element otherwise. Thus, we perfectly simulate the T-IND game for $\mathcal{A}$ and it is clear that whenever $\mathcal{A}$ breaks T-IND we can break DDH with the same probability. $\square$

4.3   Distributed Controllable Linkability

Due to the fact that our proposed revocation mechanism relies on an always-online revocation authority, the attack surface is significantly larger than in case of an offline authority. Essentially, we want to ensure that an attacker cannot steal the master linking key mlk by compromising such an authority. In order to prevent such a single point of failure, we thus introduce threshold AoN-PKEET* which then enables us to realize distributed controllable linkability. Thereby, we reduce the trust in the linking authority and also obtain more robustness. In a similar manner Ghadafi [26] introduced distributed tracing, such that multiple opening authorities must cooperate in order to open a signature.

The basic idea is to distribute the trapdoor tk $\leftarrow$ Aut(sk) of an AoN-PKEET* primitive among n entities using a (t, n)-secret sharing scheme. Then, the cooperation of at least t authorities is required to recover the trapdoor tk or to employ the trapdoor to perform equality tests on encrypted data.

Formal Model. We define threshold AoN-PKEET* as a tuple of algorithms $\mathcal{T}\text{-}\mathcal{PKEQ}^* = (\mathsf{KeyGen}, \mathsf{Enc}, \mathsf{Dec}, \mathsf{Aut}, \mathsf{DKAut}, \mathsf{TShare}, \mathsf{TSCom})$, where DKAut is an algorithm that computes the shares for the trapdoor key, TShare is an algorithm to compute the corresponding trapdoor shares for given ciphertexts, and TSCom is an algorithm to perform the plaintext equality test based on a given set of shares.

DKAut(tk, t, n): Takes a trapdoor key tk, a threshold t, and a number of total shares n, and returns trapdoor shares $(\mathsf{tk}_i)_{i=1}^n$, such that a subset of at least t entities is required to perform equality tests.

TShare(T, T', tk_i): Takes two ciphertexts (T, T') and a trapdoor share tk_i, and returns corresponding shares C_i and C_i' for the equality test.

TSCom($\{C_i, C_i'\}_{i \in \mathcal{I}}$): Given a set of shares $\{C_i, C_i'\}_{i \in \mathcal{I}}$ with $|\mathcal{I}| \geq t$, the algorithm combines the shares to perform the plaintext equality test and returns **true** if T and T' encrypt the same (unknown) message and **false** otherwise.

Similarly to how the Com($T, \perp$, tk) algorithm has been adapted to return an anonymous revocation token t for a given ciphertext T, the TShare and TSCom algorithms can be adapted to return appropriate shares and the corresponding

378   D. Slamanig et al.

revocation token, respectively. We denote an invocation that returns the corresponding shares as $\mathsf{TShare}(T, \perp, \mathsf{tk}_i)$ and an invocation that combines these shares to compute the revocation token as $\mathsf{TSCom}(\{C_i, \perp\})$.

Instantiation Based on ElGamal and Shamir's Secret Sharing. Again, we assume a conventional ElGamal-based AoN-PKEET* in a bilinear map setting, where the private key is $\mathsf{sk} = \xi \in \mathbb{Z}_p$ and the trapdoor for plaintext equality tests is $\mathsf{tk} = (\hat{r}, \hat{s} = \hat{r}^\xi) \in \mathbb{G}_2^2$. We omit the KeyGen, Enc, Dec algorithms for the sake of brevity and only present the relevant algorithms below.

$\mathsf{DKAut}(\mathsf{tk}, t, n)$: Given a trapdoor key $\mathsf{tk} = (\hat{r}, \hat{s} = \hat{r}^\xi)$, a threshold t, and a total number of shares n, it computes the shares $(\mathsf{tk}_i)_{i=1}^n$. Therefore, it computes a polynomial $F(x) = \hat{r} \cdot \prod_{\ell=1}^{t-1} \hat{f}_\ell^{x^\ell}$ for random $\hat{f}_\ell \in \mathbb{G}_2$, e.g., $\hat{f}_\ell = \hat{g}^r$ for random $r \in \mathbb{Z}_p$. Similarly, it computes a polynomial $G(x) = \hat{s} \cdot \prod_{\ell=1}^{t-1} \hat{g}_\ell^{x^\ell}$ for random $\hat{g}_\ell \in \mathbb{G}_2$. Finally, it returns the shares $(\mathsf{tk}_i = (i, \hat{r}_i \leftarrow F(i), \hat{s}_i \leftarrow G(i)))_{i=1}^n$.

$\mathsf{TShare}(T, T', \mathsf{tk}_i)$: Given two ciphertexts $(T, T') = ((T_1, T_2), (T_1', T_2'))$ as well as a share of the trapdoor tk_i, it computes and returns the comparison shares $C_i = e(T_2, \hat{r}_i)$ and $D_i = e(T_1, \hat{s}_i)$ and $C_i' = e(T_2', \hat{r}_i)$ and $D_i' = e(T_1', \hat{s}_i)$.

$\mathsf{TSCom}(\{C_i, D_i, C_i', D_i'\}_{i \in \mathcal{I}})$: Given a set of comparison shares $\{C_i, D_i, C_i', D_i'\}_{i \in \mathcal{I}}$ with $|\mathcal{I}| \geq t$, the algorithm combines the shares to perform the plaintext equality test. Therefore, it computes S and S' as follows:

$$S = \prod_{i \in \mathcal{I}} C_i^{L_i} \cdot \left(\prod_{i \in \mathcal{I}} D_i^{L_i} \right)^{-1} \qquad S' = \prod_{i \in \mathcal{I}} C_i'^{L_i} \cdot \left(\prod_{i \in \mathcal{I}} D_i'^{L_i} \right)^{-1}$$

where $L_i = \prod_{j \in \mathcal{I}} \frac{j}{j-i}$ for $j \neq i$ are the Lagrange coefficients. Finally, it returns true if $S = S'$ and false otherwise.

The correctness of the above construction can be seen by inspection. Furthermore, the notion of token indistinguishability (T-IND) as defined in Sect. 4.2 also holds for the threshold variant of AoN-PKEET*.

5 GSs with Linking-Based Revocation

We now specify a GSS with linking-based revocation as a tuple $\mathcal{GS\text{-}LBR} = $ (GkGen, UkGen, Join, Issue, GSig, GVf, Open, Judge, CheckStatus, Revoke). Next, we outline the algorithms that change due to our modifications as well as the additional algorithms CheckStatus and Revoke.

$\mathsf{GkGen}(1^\lambda, t, n)$: On input a security parameter λ, a threshold t, and a total number of shares n, the algorithm outputs a tuple $(\mathsf{gpk}, \mathsf{mok}, \mathsf{mik}, \mathsf{mlk}, (\mathsf{mlk}_i)_{i=1}^n)$. First, it runs $(\mathsf{pk}_e, \mathsf{sk}_e) \leftarrow \mathsf{KeyGen}(1^\lambda)$ of the (t, n)-threshold AoN-PKEET* scheme, sets $\mathsf{mok} = \mathsf{sk}_e$, and integrates pk_e into gpk. Then it runs $(\mathsf{tk}_{\mathsf{pub}}, \mathsf{tk}_{\mathsf{priv}}) \leftarrow \mathsf{Aut}(\mathsf{mok})$, sets the master linking key $\mathsf{mlk} = (\mathsf{tk}_{\mathsf{pub}}, \mathsf{tk}_{\mathsf{priv}})$, and integrates $\mathsf{tk}_{\mathsf{pub}}$ into mik. Furthermore, it generates the shares $\mathsf{mlk}_i \leftarrow \mathsf{DKAut}(\mathsf{mlk}, t, n)$ for the n distributed linking authorities. The rest remains unchanged.

GVf(gpk, M, σ): On input the group public key gpk, a message M, and a signature σ, the algorithm determines whether the signature σ is valid with respect to the message M and the group public key gpk. It returns **true** if the signature is valid and **false** otherwise.

CheckStatus(RL, $\mathcal{L}, \sigma$): On input a revocation list RL containing anonymous revocation tokens, a set of existing linking authorities $\mathcal{L}$, and a signature σ, this algorithm determines the revocation status of the signer corresponding to signature σ. In order to determine the revocation status, it interacts with t linking authorities $\mathcal{L}_i \in \mathcal{L}$ via the TShare algorithm and retrieves the corresponding shares for the computation of the revocation token. Afterwards, it uses the TSCom algorithm to combine these shares and to retrieve the final revocation token. If the revocation token exists on the RL, the signer has been revoked and it returns **true**. Otherwise, it has not been revoked and it returns **false**.

Revoke(gpk, mik, **reg**, RL, i): On input of the group public key gpk, the master issuing key mik, the registration table **reg**, the current revocation list RL, and a user i to be revoked[3], the algorithm computes the anonymous revocation token t_i corresponding to user i and adds it to the revocation list. It returns the updated revocation list RL = RL $\cup \{t_i\}$.

5.1 Discussion and Security

Computation of Revocation Tokens. Staying with our conventional ElGamal example and considering revocation tokens which are of the form $e(T_2, \hat{r}) \cdot e(T_1, \hat{s})^{-1} = e(m, \hat{r})$, we observe that these tokens can be computed in two different ways.

Given m and $\hat{r}$: Given a message m, e.g., a user's certificate, and $\hat{r}$ allows to compute revocation tokens $t = e(m, \hat{r})$. Thus, if the issuer is given access to $\hat{r}$, the revocation token t can be computed with the information available during the Issue algorithm and added to the registration table **reg**.

Given $\sigma = (T, \pi)$ and mlk = $(\hat{r}, \hat{r}^\xi)$: Given a signature $\sigma = (T, \pi)$ and the master linking key mlk, such a revocation token t can also be computed on the fly, i.e., $t \leftarrow \text{Com}(T, \perp, \text{mlk})$. Thus, such a token can be computed directly from a given signature which allows for anonymous revocation of users as signatures need not be opened before revocation.

Note that if the revocation tokens are precomputed and stored in the registration table **reg**, then an attacker who manages to get in possession of the registration table **reg** and RL (but not necessarily the mlk) can conceptually identify (open) all signers on RL as the same tokens can be found in **reg**. This is not possible in case the revocation tokens are computed on the fly as in this case the attacker cannot link entries on RL to entries in the registration table **reg** since the revocation tokens do not yield any useful information (cf. T-IND).

---

[3] Note that revocation can also be done based on a user's signature by means of mlk in which case the user's identity will not be required.

Revocation and Revocation Check. Eventually, in case of a revocation, the token—that can either be computed (1) by opening a signature first or (2) from the signature directly—is added to RL. The actual revocation check for a signature $\sigma = (T, \pi)$ then requires the computation of the token $t \leftarrow \mathsf{Com}(T, \bot, \mathsf{mlk})$ and a simple look-up operation, *i.e.*, checking whether or not $t \in \mathsf{RL}$, and consequently can be performed in time $\mathcal{O}(1)$.[4] Also note that, except for the party in possession of mlk, the anonymous revocation token does not yield any useful information and also does not endanger the privacy of signers.

Revocation Check with Threshold AoN-PKEET.* We point out that a RA in our setting does not hold the linking key mlk but always needs to contact a set of at least t linking authorities (over authenticated and confidential channels) to compute the required tokens. Thereby, we assume that no t linking authorities can be compromised. Consequently, in contrast to the naive approach, breaking into the (always-online) RA does not reveal the linking key. The only information that an attacker gains by compromising RAs is a list of revocation tokens t_i (and possibly the corresponding messages m_i and ciphertexts c_i). However, as already argued in Sect. 4.2, this does not allow the attacker to compromise the overall anonymity of the scheme as this information does not allow her to distinguish other tokens t from random. Although we only cover passive attacks, this is a reasonable model because in case a RA gets compromised, the corresponding authentication key will be revoked and replaced and the adversary can only behave passive.

Figure 2 illustrates the basic idea of our secure instantiation of linking-based revocation. A verifier first verifiers a given signature $\sigma = (T, \pi)$ via the GVf algorithm and afterwards wants to learn about the revocation status of this signature. Therefore, it interacts with a RA by means of CheckStatus. The RA interacts with t LAs in order to retrieve the corresponding shares by means of $\{C_i, \bot\} \leftarrow \mathsf{TShare}(T, \bot, \mathsf{mlk}_i)$ and, afterwards, RA employs the

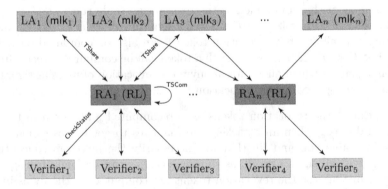

Fig. 2. Schematic of our secure instantiation of linking-based revocation.

---

[4] Hash tables allow to check whether or not $t \in \mathsf{RL}$ in constant time. For instance, employing cuckoo hashing [50] allows for a worst-case complexity of $\mathcal{O}(1)$.

$t \leftarrow \mathsf{TSCom}(\{C_i, \perp\}_{i \in \mathcal{I}})$ algorithm to combine these shares, which yields the revocation token t. After checking whether or not t exists on RL, the RA returns the corresponding decision via CheckStatus.

As the RL as well as the learned tokens t do not endanger the privacy of group members (cf. T-IND), the role of RAs can be distributed over multiple cloud services. Besides, we assume that no t LAs can be compromised at once and since the trapdoor shares mlk_i do not endanger the privacy of group members, these trapdoor shares mlk_i can also be safely distributed over multiple cloud services. Similar to traditional OCSP responses, the response from RAs must be signed. Although one could also employ the feature of verifiable controllable linkability [6] in order to prove that a given signature has or has not been revoked, this would add an additional overhead for the verifier, especially in case the signer has not been revoked. More specifically, if the signer is already revoked, the RA needs to perform a proof that the given signature can be linked to one specific entry on RL. In contrast, if the signer has not been revoked, the RA needs to prove that the given signature has not been produced by any entity on RL and, hence, a naive instantiation of verifiable controllable linkability means that the proof increases linearly with the size of RL. In addition, such a naive instantiation of verifiable controllable linkability would break backward unlinkability as verifiers would receive a proof that a specific signature links to a specific entry on RL (that must be publicly available in this case), which means that verifiers could link all signatures of revoked members. Although backward unlinkability can be achieved by using disjunctive zero-knowledge proofs (*i.e.*, OR proofs), such proofs also introduce non-trivial overhead for the involved entities. Thus, we suggest that RAs sign the returned response, similar to traditional OCSP responses. In order to reduce the communication and computational costs one could additionally employ distributed OCSP (D-OCSP) as proposed by Koga and Sakurai [37], such that all RAs have a different signing key although they share the same public key.

Security Model. In contrast to other revocation mechanisms, the RA in our setting only returns a boolean decision, *i.e.*, whether or not the signer has been revoked. Thus, unlike other revocation mechanisms, the RA does not reveal additional revocation-related information which would require to integrate the revocation feature into the formal security model of the GSS. Basically, LBR is an application of the feature of controllable linkability and, thus, we do not need to formally model our revocation mechanism in the security model. In fact, the role of the RA is already covered by the security model of group signatures with controllable linkability and our revocation mechanism is already considered in the original model (although this functionality is now implemented by the RA). Consequently, we also do not modify the security properties listed in Sect. 3.4. In addition, correctness of revocation follows from correctness of the GSS with controllable linkability.

6 Applying Linking-Based Revocation

For illustration purposes, we show how linking-based revocation can be applied to the *eXtremely Short Group Signature* scheme (XSGS) [21]. By means of the generic compiler [56], XSGS can be turned into a GSS with controllable linkability for free (without any overhead). Since LBR is transparent for signers, we only need to modify the GkGen algorithm. GkGen generates the additional linking key, which allows the revocation authority (in cooperation with multiple linking authorities) to perform the revocation check. Furthermore, we add the algorithms CheckStatus and Revoke according to the model for GSs with controllable linkability. The scheme is as follows (cf. [21] for more details):

GkGen($1^\lambda, t, n$): The master opening key is $\mathsf{mok} = (\xi_1, \xi_2)$ for two randomly chosen elements $\xi_1, \xi_2 \in \mathbb{Z}_p$. The master linking key is $\mathsf{mlk} = (\hat{r}, \hat{s} = \hat{r}^{\xi_1})$ for a randomly chosen element $\hat{r} \in \mathbb{G}_2$. The master issuing key consists of $\mathsf{mik} = (\gamma, \hat{r})$ for a randomly chosen element $\gamma \in \mathbb{Z}_p$. The master linking key mlk is distributed among n linking authorities via $\mathsf{mlk}_i = (\hat{r}_i, \hat{s}_i) \leftarrow \mathsf{DKAut}(\mathsf{mlk}, t, n)$. The group public key is $\mathsf{gpk} = (g_1, k, h = k^{\xi_1}, g = k^{\xi_2}, \hat{g}_2, \hat{w} = \hat{g}_2^\gamma) \in \mathbb{G}_1^4 \times \mathbb{G}_2^2$.

Issue($\mathsf{gpk}, \mathsf{mik}, \mathbf{reg}$): The Issue algorithm interacts with the Join algorithm to add a user to the group. Thereby, a user receives a membership certificate (A_i, x, y), such that $A_i^{x+\gamma} = g_1 h^y$, where $y \in \mathbb{Z}_p$ is known to the user only. The issuing authority adds all the relevant information to the registration table $\mathbf{reg}$.

GSig($\mathsf{gpk}, M, \mathsf{gsk}_i$): Given the group public key gpk, a message M, and a user's signing key $\mathsf{gsk}_i = (A, x, y) \in \mathbb{G}_1 \times \mathbb{Z}_p^2$, a signature is computed as follows. It randomly selects $\alpha, \beta \in \mathbb{Z}_p$ and encrypts the membership certificate:

$$T_1 = k^\alpha \qquad T_2 = Ah^\alpha \qquad\qquad T_3 = k^\beta \qquad T_4 = Ag^\beta$$

Then it sets $z = x\alpha + y$ and computes the non-interactive zero-knowledge proof of knowledge (α, β, x, z) as follows. It picks blinding values $r_\alpha, r_\beta, r_x, r_z \in \mathbb{Z}_p$ and computes the following values:

$$R_1 = k^{r_\alpha} \qquad R_3 = k^{r_\beta} \qquad R_4 = h^{r_\alpha}/g^{r_\beta}$$
$$R_2 = e(T_2, \hat{g}_2)^{r_x} \cdot e(h, \hat{w})^{-r_\alpha} \cdot e(h, \hat{g}_2)^{-r_z}$$
$$c = H(M, T_1, T_2, T_3, T_4, R_1, R_2, R_3, R_4)$$
$$s_\alpha = r_\alpha + c\alpha \qquad s_\beta = r_\beta + c\beta$$
$$s_x = r_x + cx \qquad s_z = r_z + cz$$

Finally, output the signature $\sigma = (T_1, T_2, T_3, T_4, c, s_\alpha, s_\beta, s_x, s_z)$.

GVf(gpk, M, σ): Given the group public key gpk, a message M and a corresponding signature σ, verification is performed by checking the following relations:

$$k^{s_\alpha} = R_1 \cdot T_1^c \qquad k^{s_\beta} = R_3 \cdot T_3^c \qquad h^{s_\alpha}/g^{s_\beta} = R_4 \cdot \left(\frac{T_2}{T_4}\right)^c$$
$$e(T_2, \hat{g}_2)^{s_x} \cdot e(h, \hat{w})^{-s_\alpha} \cdot e(h, \hat{g}_2)^{-s_z} = R_2 \cdot \left(\frac{e(g_1, \hat{g}_2)}{e(T_2, \hat{w})}\right)^c$$

If all of the above relations hold, the algorithm returns true and false otherwise.

CheckStatus(RL, $\mathcal{L}, \sigma$): Given a revocation list RL consisting of revocation tokens of revoked members, a list of available linking authorities $\mathcal{L}$, and a signature σ, it determines the revocation status of the signer corresponding to the signature σ in question. Therefore, it interacts with a subset of at least t linking authorities $\mathcal{I} \subseteq \mathcal{L}$, i.e., $|\mathcal{I}| \geq t$, to retrieve t comparison shares $\{C_i, D_i, \bot, \bot\} \leftarrow \mathsf{TShare}((T_1, T_2), \bot, \mathsf{mlk}_i)$ as follows:

$$C_i = e(T_2, \hat{r}_i) \qquad D_i = e(T_1, \hat{s}_i).$$

Based on these comparison shares, it computes the revocation token t via $t = \mathsf{TSCom}(\{C_i, D_i, \bot, \bot\}_{i \in \mathcal{I}})$ as follows:

$$L_i = \prod_{j \in \mathcal{I}, j \neq i} \frac{j}{j - i} \qquad t = \prod_{i \in \mathcal{I}} C_i^{L_i} \cdot \left(\prod_{i \in \mathcal{I}} D_i^{L_i}\right)^{-1}$$

Return true (revoked) if $t \in$ RL and false (not revoked) otherwise.

Revoke(gpk, mik, reg, RL, i): Given the group public key gpk, the master issuing key mik, the registration table reg, the current revocation list RL, and a user i to be revoked[5], the algorithm computes the revocation token $t_i = e(A_i, \hat{r})$ for user i and adds it to the revocation list, i.e., RL $=$ RL $\cup \{t_i\}$.

Security Analysis. We know that XSGS scheme is secure in the BSZ model (as shown in [21]) and in particular uses twin ElGamal encryption. We simply use its AoN-PKEET* version (according to the generic compiler [56]), which yields a secure group signature with controllable linkability. Token indistinguishability, as defined and shown for standard ElGamal in Sect. 4.2, also holds for twin ElGamal and consequently an adversary cannot learn anything from the anonymous revocation tokens on RL.

7 Conclusion

It is well known that revocation mechanisms represent the major bottleneck in group signature schemes (see e.g., [41]). However, the general belief that any online authority must be prevented unnecessarily restricts the efficiency and practicality of revocation in group signature schemes. In this paper, we showed that the major drawbacks of existing revocation mechanisms, e.g., additional computations/updates for signers and verifiers, can be overcome by a paradigm-shift towards the incorporation of an online revocation authority. Since many applications and services already rely on always-connected devices that permanently interact with cloud computing infrastructures, the introduction of such an

---

[5] Again, revocation can also be done based on a user's signature $\sigma = (T, \pi)$ by means of mlk in which case the user's identity will not be required.

online revocation authority is absolutely reasonable. Considering (1) the significant performance gain (constant-time revocation checks), (2) the transparency for signers as well as verifiers, and (3) the general applicability to well-established PB-GSSs, we claim that our approach advances the open issue of efficient membership revocation in the context of GSSs. Hence, linking-based revocation represents a valuable contribution to the existing portfolio of revocation mechanisms.

Acknowledgments. The authors would like to thank the anonymous reviewers for their valuable comments and suggestions. Daniel Slamanig has been supported by the H2020 project Prismacloud, grant agreement number 644962. Raphael Spreitzer and Thomas Unterluggauer have been supported by the European Commission through the FP7 program under project number 610436 (project MATTHEW).

References

1. Ateniese, G., Song, D.X., Tsudik, G.: Quasi-efficient revocation of group signatures. In: Blaze, M. (ed.) FC 2002. LNCS, vol. 2357, pp. 183–197. Springer, Heidelberg (2003). doi:10.1007/3-540-36504-4_14
2. Attrapadung, N., Emura, K., Hanaoka, G., Sakai, Y.: A revocable group signature scheme from identity-based revocation techniques: achieving constant-size revocation list. In: Boureanu, I., Owesarski, P., Vaudenay, S. (eds.) ACNS 2014. LNCS, vol. 8479, pp. 419–437. Springer, Cham (2014). doi:10.1007/978-3-319-07536-5_25
3. Au, M.H., Tsang, P.P., Susilo, W., Mu, Y.: Dynamic universal accumulators for DDH groups and their application to attribute-based anonymous credential systems. In: Fischlin, M. (ed.) CT-RSA 2009. LNCS, vol. 5473, pp. 295–308. Springer, Heidelberg (2009). doi:10.1007/978-3-642-00862-7_20
4. Baek, J., Zheng, Y.: Identity-based threshold decryption. In: Bao, F., Deng, R., Zhou, J. (eds.) PKC 2004. LNCS, vol. 2947, pp. 262–276. Springer, Heidelberg (2004). doi:10.1007/978-3-540-24632-9_19
5. Bellare, M., Shi, H., Zhang, C.: Foundations of group signatures: the case of dynamic groups. In: Menezes, A. (ed.) CT-RSA 2005. LNCS, vol. 3376, pp. 136–153. Springer, Heidelberg (2005). doi:10.1007/978-3-540-30574-3_11
6. Blazy, O., Derler, D., Slamanig, D., Spreitzer, R.: Non-interactive plaintext (in-)equality proofs and group signatures with verifiable controllable linkability. In: Sako, K. (ed.) CT-RSA 2016. LNCS, vol. 9610, pp. 127–143. Springer, Cham (2016). doi:10.1007/978-3-319-29485-8_8
7. Boneh, D., Boyen, X., Shacham, H.: Short group signatures. In: Franklin, M. (ed.) CRYPTO 2004. LNCS, vol. 3152, pp. 41–55. Springer, Heidelberg (2004). doi:10.1007/978-3-540-28628-8_3
8. Boneh, D., Shacham, H.: Group signatures with verifier-local revocation. In: Computer and Communications Security - CCS, pp. 168–177 (2004)
9. Bootle, J., Cerulli, A., Chaidos, P., Ghadafi, E., Groth, J.: Foundations of fully dynamic group signatures. In: Manulis, M., Sadeghi, A.-R., Schneider, S. (eds.) ACNS 2016. LNCS, vol. 9696, pp. 117–136. Springer, Cham (2016). doi:10.1007/978-3-319-39555-5_7
10. Bresson, E., Stern, J.: Efficient revocation in group signatures. In: Kim, K. (ed.) PKC 2001. LNCS, vol. 1992, pp. 190–206. Springer, Heidelberg (2001). doi:10.1007/3-540-44586-2_15

11. Brickell, E., Li, J.: Enhanced privacy ID from bilinear pairing for hardware authentication and attestation. In: Social Computing - SocialCom/Privacy, Security, Risk and Trust - PASSAT 2010, pp. 768–775 (2010)

12. Brickell, E., Li, J.: Enhanced privacy ID: a direct anonymous attestation scheme with enhanced revocation capabilities. IEEE Trans. Dependable Secure Comput. **9**, 345–360 (2012)

13. Camenisch, J., Lysyanskaya, A.: Dynamic accumulators and application to efficient revocation of anonymous credentials. In: Yung, M. (ed.) CRYPTO 2002. LNCS, vol. 2442, pp. 61–76. Springer, Heidelberg (2002). doi:10.1007/3-540-45708-9_5

14. Camenisch, J., Stadler, M.: Efficient group signature schemes for large groups. In: Kaliski, B.S. (ed.) CRYPTO 1997. LNCS, vol. 1294, pp. 410–424. Springer, Heidelberg (1997). doi:10.1007/BFb0052252

15. Canard, S., Coisel, I., de Meulenaer, G., Pereira, O.: Group signatures are suitable for constrained devices. In: Rhee, K.-H., Nyang, D.H. (eds.) ICISC 2010. LNCS, vol. 6829, pp. 133–150. Springer, Heidelberg (2011). doi:10.1007/978-3-642-24209-0_9

16. Canard, S., Desmoulins, N., Devigne, J., Traoré, J.: On the implementation of a pairing-based cryptographic protocol in a constrained device. In: Abdalla, M., Lange, T. (eds.) Pairing 2012. LNCS, vol. 7708, pp. 210–217. Springer, Heidelberg (2013). doi:10.1007/978-3-642-36334-4_14

17. Chaum, D., van Heyst, E.: Group signatures. In: Davies, D.W. (ed.) EUROCRYPT 1991. LNCS, vol. 547, pp. 257–265. Springer, Heidelberg (1991). doi:10.1007/3-540-46416-6_22

18. Chow, S.S.M., Susilo, W., Yuen, T.H.: Escrowed linkability of ring signatures and its applications. In: Nguyen, P.Q. (ed.) VIETCRYPT 2006. LNCS, vol. 4341, pp. 175–192. Springer, Heidelberg (2006). doi:10.1007/11958239_12

19. Cooper, D., Santesson, S., Farrell, S., Boeyen, S., Housley, R., Polk, W.: Internet X.509 Public Key Infrastructure Certificate and Certificate Revocation List (CRL) Profile. RFC 5280, RFC Editor, May 2008. http://www.rfc-editor.org/rfc/rfc5280.txt

20. Cramer, R., Shoup, V.: A practical public key cryptosystem provably secure against adaptive chosen ciphertext attack. In: Krawczyk, H. (ed.) CRYPTO 1998. LNCS, vol. 1462, pp. 13–25. Springer, Heidelberg (1998). doi:10.1007/BFb0055717

21. Delerablée, C., Pointcheval, D.: Dynamic fully anonymous short group signatures. In: Nguyen, P.Q. (ed.) VIETCRYPT 2006. LNCS, vol. 4341, pp. 193–210. Springer, Heidelberg (2006). doi:10.1007/11958239_13

22. Emura, K., Hayashi, T.: A light-weight group signature scheme with time-token dependent linking. In: Güneysu, T., Leander, G., Moradi, A. (eds.) LightSec 2015. LNCS, vol. 9542, pp. 37–57. Springer, Cham (2016). doi:10.1007/978-3-319-29078-2_3

23. Emura, K., Miyaji, A., Omote, K.: An r-hiding revocable group signature scheme: group signatures with the property of hiding the number of revoked users. J. Appl. Math. **2014**, 983040:1–983040:14 (2014)

24. Fan, C.-I., Hsu, R.-H., Manulis, M.: Group signature with constant revocation costs for signers and verifiers. In: Lin, D., Tsudik, G., Wang, X. (eds.) CANS 2011. LNCS, vol. 7092, pp. 214–233. Springer, Heidelberg (2011). doi:10.1007/978-3-642-25513-7_16

25. Fouque, P.-A., Pointcheval, D.: Threshold cryptosystems secure against chosen-ciphertext attacks. In: Boyd, C. (ed.) ASIACRYPT 2001. LNCS, vol. 2248, pp. 351–368. Springer, Heidelberg (2001). doi:10.1007/3-540-45682-1_21

26. Ghadafi, E.: Efficient distributed tag-based encryption and its application to group signatures with efficient distributed traceability. In: Aranha, D.F., Menezes, A. (eds.) LATINCRYPT 2014. LNCS, vol. 8895, pp. 327–347. Springer, Cham (2015). doi:10.1007/978-3-319-16295-9_18

27. Grewal, G., Azarderakhsh, R., Longa, P., Hu, S., Jao, D.: Efficient implementation of bilinear pairings on ARM processors. In: Knudsen, L.R., Wu, H. (eds.) SAC 2012. LNCS, vol. 7707, pp. 149–165. Springer, Heidelberg (2013). doi:10.1007/978-3-642-35999-6_11

28. Groth, J., Sahai, A.: Efficient non-interactive proof systems for bilinear groups. In: Smart, N. (ed.) EUROCRYPT 2008. LNCS, vol. 4965, pp. 415–432. Springer, Heidelberg (2008). doi:10.1007/978-3-540-78967-3_24

29. Hwang, J.Y., Chen, L., Cho, H.S., Nyang, D.: Short dynamic group signature scheme supporting controllable linkability. IEEE Trans. Inf. Forensics Secur. 10, 1109–1124 (2015)

30. Hwang, J.Y., Lee, S., Chung, B.H., Cho, H.S., Nyang, D.: Short group signatures with controllable linkability. In: LightSec, pp. 44–52. IEEE (2011)

31. Hwang, J.Y., Lee, S., Chung, B., Cho, H.S., Nyang, D.: Group signatures with controllable linkability for dynamic membership. Inf. Sci. 222, 761–778 (2013)

32. International Organization for Standardization (ISO): ISO/IEC 20008-2: Information technology - Security techniques - Anonymous digital signatures - Part 2: Mechanisms using a group public key, November 2013

33. Isern-Deyà, A.P., Huguet-Rotger, L., Payeras-Capellà, M., Mut-Puigserver, M.: On the practicability of using group signatures on mobile devices: implementation and performance analysis on the android platform. Int. J. Inf. Secur. 14, 335–345 (2015)

34. Isern-Deyà, A.P., Vives-Guasch, A., Puigserver, M.M., Payeras-Capellà, M., Castellà-Roca, J.: A secure automatic fare collection system for time-based or distance-based services with revocable anonymity for users. Comput. J. 56, 1198–1215 (2013)

35. Kiayias, A., Tsiounis, Y., Yung, M.: Traceable signatures. In: Cachin, C., Camenisch, J.L. (eds.) EUROCRYPT 2004. LNCS, vol. 3027, pp. 571–589. Springer, Heidelberg (2004). doi:10.1007/978-3-540-24676-3_34

36. Kiayias, A., Yung, M.: Group signatures with efficient concurrent join. In: Cramer, R. (ed.) EUROCRYPT 2005. LNCS, vol. 3494, pp. 198–214. Springer, Heidelberg (2005). doi:10.1007/11426639_12

37. Koga, S., Sakurai, K.: A distributed online certificate status protocol with a single public key. In: Bao, F., Deng, R., Zhou, J. (eds.) PKC 2004. LNCS, vol. 2947, pp. 389–401. Springer, Heidelberg (2004). doi:10.1007/978-3-540-24632-9_28

38. Kumar, V., Li, H., Park, J.J., Bian, K., Yang, Y.: Group signatures with probabilistic revocation: a computationally-scalable approach for providing privacy-preserving authentication. In: Computer and Communications Security - CCS 2015, pp. 1334–1345 (2015)

39. Libert, B., Peters, T., Yung, M.: Group signatures with almost-for-free revocation. In: Safavi-Naini, R., Canetti, R. (eds.) CRYPTO 2012. LNCS, vol. 7417, pp. 571–589. Springer, Heidelberg (2012). doi:10.1007/978-3-642-32009-5_34

40. Libert, B., Peters, T., Yung, M.: Scalable group signatures with revocation. In: Pointcheval, D., Johansson, T. (eds.) EUROCRYPT 2012. LNCS, vol. 7237, pp. 609–627. Springer, Heidelberg (2012). doi:10.1007/978-3-642-29011-4_36

41. Manulis, M., Fleischhacker, N., Felix Günther, F.K., Poettering, B.: Group signatures: authentication with privacy. Technical report, BSI - Federal Office for Information Security (2012)

42. Nakanishi, T., Fujii, H., Hira, Y., Funabiki, N.: Revocable group signature schemes with constant costs for signing and verifying. In: Jarecki, S., Tsudik, G. (eds.) PKC 2009. LNCS, vol. 5443, pp. 463–480. Springer, Heidelberg (2009). doi:10.1007/978-3-642-00468-1_26

43. Nakanishi, T., Fujiwara, T., Watanabe, H.: A linkable group signature and its application to secret voting. Trans. Inf. Process. Soc. Jpn. 40(7), 3085–3096 (1999)

44. Nakanishi, T., Funabiki, N.: Verifier-local revocation group signature schemes with backward unlinkability from bilinear maps. In: Roy, B. (ed.) ASIACRYPT 2005. LNCS, vol. 3788, pp. 533–548. Springer, Heidelberg (2005). doi:10.1007/11593447_29

45. Nakanishi, T., Funabiki, N.: A short verifier-local revocation group signature scheme with backward unlinkability. In: Yoshiura, H., Sakurai, K., Rannenberg, K., Murayama, Y., Kawamura, S. (eds.) IWSEC 2006. LNCS, vol. 4266, pp. 17–32. Springer, Heidelberg (2006). doi:10.1007/11908739_2

46. Nakanishi, T., Kubooka, F., Hamada, N., Funabiki, N.: Group signature schemes with membership revocation for large groups. In: Boyd, C., González Nieto, J.M. (eds.) ACISP 2005. LNCS, vol. 3574, pp. 443–454. Springer, Heidelberg (2005). doi:10.1007/11506157_37

47. Nakanishi, T., Sugiyama, Y.: A group signature scheme with efficient membership revocation for reasonable groups. In: Wang, H., Pieprzyk, J., Varadharajan, V. (eds.) ACISP 2004. LNCS, vol. 3108, pp. 336–347. Springer, Heidelberg (2004). doi:10.1007/978-3-540-27800-9_29

48. Naor, M., Yung, M.: Public-key cryptosystems provably secure against chosen ciphertext attacks. In: Symposium on the Theory of Computing - STOC, pp. 427–437 (1990)

49. Nguyen, L., Safavi-Naini, R.: Efficient and provably secure trapdoor-free group signature schemes from bilinear pairings. In: Lee, P.J. (ed.) ASIACRYPT 2004. LNCS, vol. 3329, pp. 372–386. Springer, Heidelberg (2004). doi:10.1007/978-3-540-30539-2_26

50. Pagh, R., Rodler, F.F.: Cuckoo hashing. J. Algorithms 51, 122–144 (2004)

51. Ponemon Institute LLC: 2015 PKI Global Trends Study (2015)

52. Potzmader, K., Winter, J., Hein, D.M., Hanser, C., Teufl, P., Chen, L.: Group signatures on mobile devices: practical experiences. In: Huth, M., Asokan, N., Čapkun, S., Flechais, I., Coles-Kemp, L. (eds.) Trust 2013. LNCS, vol. 7904, pp. 47–64. Springer, Heidelberg (2013). doi:10.1007/978-3-642-38908-5_4

53. Rackoff, C., Simon, D.R.: Non-interactive zero-knowledge proof of knowledge and chosen ciphertext attack. In: Feigenbaum, J. (ed.) CRYPTO 1991. LNCS, vol. 576, pp. 433–444. Springer, Heidelberg (1992). doi:10.1007/3-540-46766-1_35

54. Santesson, S., Myers, M., Ankney, R., Malpani, A., Galperin, S., Adams, C.: X.509 internet public key infrastructure online certificate status protocol - OCSP. RFC 6960, Internet Engineering Task Force (IETF), June 2013. https://www.ietf.org/rfc/rfc6960.txt

55. Shamir, A.: How to share a secret. Commun. ACM 22, 612–613 (1979)

56. Slamanig, D., Spreitzer, R., Unterluggauer, T.: Adding controllable linkability to pairing-based group signatures for free. In: Chow, S.S.M., Camenisch, J., Hui, L.C.K., Yiu, S.M. (eds.) ISC 2014. LNCS, vol. 8783, pp. 388–400. Springer, Cham (2014). doi:10.1007/978-3-319-13257-0_23

57. Tang, Q.: Public key encryption supporting plaintext equality test and user-specified authorization. Secur. Commun. Netw. 5, 1351–1362 (2012)

58. Teranishi, I., Furukawa, J., Sako, K.: k-times anonymous authentication (extended abstract). In: Lee, P.J. (ed.) ASIACRYPT 2004. LNCS, vol. 3329, pp. 308–322. Springer, Heidelberg (2004). doi:10.1007/978-3-540-30539-2_22
59. Unterluggauer, T., Wenger, E.: Efficient pairings and ECC for embedded systems. In: Batina, L., Robshaw, M. (eds.) CHES 2014. LNCS, vol. 8731, pp. 298–315. Springer, Heidelberg (2014). doi:10.1007/978-3-662-44709-3_17
60. Verheul, E.R.: Practical backward unlinkable revocation in FIDO, German e-ID, Idemix and U-Prove. IACR Cryptology ePrint Archive 2016/217 (2016)
61. Wei, V.K.: Tracing-by-linking group signatures. In: Zhou, J., Lopez, J., Deng, R.H., Bao, F. (eds.) ISC 2005. LNCS, vol. 3650, pp. 149–163. Springer, Heidelberg (2005). doi:10.1007/11556992_11
62. Zhou, S., Lin, D.: Shorter verifier-local revocation group signatures from bilinear maps. In: Pointcheval, D., Mu, Y., Chen, K. (eds.) CANS 2006. LNCS, vol. 4301, pp. 126–143. Springer, Heidelberg (2006). doi:10.1007/11935070_8

CARIBE: Cascaded IBE for Maximum Flexibility and User-Side Control

Britta Hale[✉], Christopher Carr, and Danilo Gligoroski

Department of Information Security and Communication Technology,
Norwegian University of Science and Technology, Trondheim, Norway
{britta.hale,christopher.carr,danilo.gligoroski}@ntnu.no

Abstract. Mass surveillance and a lack of end-user encryption, coupled with a growing demand for key escrow under legal oversight and certificate authority security concerns, raise the question of the appropriateness of continued general dependency on PKI. Under this context, we examine Identity-Based Encryption (IBE) as an alternative to public-key encryption. Cascade encryption, or sequential multiple encryption, is the concept of layering encryption such that the ciphertext from one encryption step is the plaintext of the next. We describe CARIBE, a cascaded IBE scheme, for which we also provide a cascaded CCA security experiment, IND-ID-C.CCA, and prove its security in the computational model. CARIBE combines the ease-of-use of IBE with key escrow, limited to the case when the entire set of participating PKGs collaborate. Furthermore, we describe a particular CARIBE scheme, CARIBE-S, where the receiver is a self-PKG – one of the several PKGs included in the cascade. CARIBE-S inherits IND-ID-C.CCA from CARIBE, and avoids key escrow entirely. In essence, CARIBE-S offers the maximum flexibility of the IBE paradigm and gives the users complete control without the key escrow problem.

1 Introduction

Mass surveillance has undoubtedly formed one of the most contentious turning points in modern Internet history. Fueled by the revelation of the **PRISM** surveillance program in 2013 [19], which collected Internet data from some of the biggest operators, including Microsoft™, Google™, and Yahoo™, and subsequent knowledge of programs such as **xkeyscore** [18] for collecting data en-mass as it is transferred, the essential need for encryption has never been more salient. Simultaneously, the case for *backdoored* encryption is also being argued, as governments fight for control of, potentially vital, information for stanching terrorist threats. Under this context we raise the question of optimal key management infrastructure, comparing identity-based encryption (IBE) as an alternative to the current public key infrastructure (PKI) to increase the ease of encryption use, while also presenting a new prospective on IBE – CARIBE – which provides limited key escrow and allows for end-user control of encryption.

© Springer International Publishing AG 2017
R.C.-W. Phan and M. Yung (Eds.): LNCS 10311, Mycrypt 2016, pp. 389–408, 2017.
DOI: 10.1007/978-3-319-61273-7_19

Fundamentally, a PKI system relies upon the receiver's precaution; whether or not they have set up a public/private key pair with the public key listed securely online for access by others who wish to communicate with them. Since powerful organizations can subvert these *secure* key directories [27] and most end users demonstrate a general apathy for establishing such keys to begin with [41], there is a clear need for a change to the conventional method. What the exigency of the situation demands is a system that provides secure ease-of-use encryption, does not rely upon a trusted third party, and yet allows for limited key escrow subject to the constraints of law. Notably, the need for limited key escrow is an axiomatically oblique path from the hither-to performed mass surveillance and pressure for backdoored products. As was recently stated in an open letter by U.S. congressmen Hurd and Lieu to the James Comey, Director of the FBI, "There is a difference between private companies assisting law enforcement and the government compelling companies to weaken their products..." [21].

Historically, IBE has been proposed and discussed as a means of encryption that is particularly user-friendly, as it is not necessary to have public or private keys established [37], yet it traditionally demands a trusted third party, namely a private-key generator (PKG). As with a CA in PKI, it is necessary to establish trusted parameters for the PKG, although it has been argued that this should not be as formidable a task as in PKI [17]. Developments in IBE have allowed for some preventative measures against a malicious PKG by distributing the key generation duties among multiple PKGs [25], but these designs demand that all involved PKGs must use the same IBE scheme which in turn could lead to a monopoly of the system. Realistically, it is even possible for groups of PKGs with the same IBE schemes in common to form coalitions, making collusion between PKGs easier and consequently increasing the risk of exposure for encrypted data. To our knowledge, no existing adaptation realizes the IBE ease-of-use, eliminates the demand for trust in a single third party, allows free choice in the combination of PKGs and thereby encryption schemes, and only provides key escrow under collaboration of the entire set of PKGs involved – all of which we address with CARIBE. Furthermore, if any receiver entity has offered itself as a PKG, a message sender may select that entity as one of the chosen PKGs, thereby bypassing escrow entirely on the message (CARIBE-S). Thus, for example, if various governments with sufficient resources offer themselves as PKGs under CARIBE-S, they also allow the option for communication partners to send escrow-prohibited messages to the selected government PKG entities; meanwhile messages to third parties can only be accessed under the combined agreement of all PKGs (for example, when legally required).

In order to address cogent issues in key management for data encryption, we propose applying IBE through the interface of cascade encryption, galvanizing it into a realistic scheme for response to modern issues. Even while allowing key escrow in the most extreme circumstances, and demanding no less than the participation of all key generators to achieve it, the freshly-interfaced IBE provides far greater power to the end user for selecting a trust model than has been previously proposed. Merging such varied cryptographic areas precisely reflects the spirit of the IACR *Copenhagen Resolution* [22]:

> Population-wide surveillance threatens democracy and human dignity. We call for expediting research and deployment of effective techniques to protect personal privacy against governmental and corporate overreach.

Markedly other, similar, infrastructures also exist, such as attribute-based encryption (anyone in possession of the correct set of attributes can decrypt) and, more generally, predicate-based encryption (anyone in possession of the correct set of attributes and a decryption key corresponding to a certain predicate can decrypt). While these are certainly interesting, we focus on IBE due to its resemblance to PKI, with encryption based upon fixed identities. However, it is straightforward to envisage that our results can be adapted for both attribute-based encryption and predicate-based encryption, with the same freedom-of-choice benefit of the trust model maintained for the end user.

1.1 Related Work

Identity-Based Cryptography. Identity-based cryptography has seen a considerable amount of development since it was first envisaged by Shamir in the mid 1980's [37]. Initial examples of schemes for identity-based cryptography are due to Cocks [11], Boneh and Franklin [5], and Sakai, Ohgishi and Kasahara [35]. Today, there are entire books devoted to the subject [8,24], as well as an extensive amount of research in various aspects of the field (for example, [5–7,11,12,17,20,25,35,37,42]).

Shamir argued for a public key encryption system conceptually similar to the postal system, albeit idealized, where a sender needs only the name and address of the recipient [37]. "IBE is a kind of public key encryption scheme where the public key of a user can be any arbitrary string – typically the e-mail address" [8], i.e. the recipient's identity, such as name, location, etc., becomes their public key. Enticingly, this approach offers some substantial advantages over the traditional PKI. One of the attributes that make IBE advantageous is its suitability for situations where network access is not continuous. Furthermore, and perhaps one of the more notable advantages of identity-based cryptography, is the nullification of the need for certificates and thereby the instantaneousness with which encryption can be performed without the requirement to obtain such a certificate. Joux [23] provides a broad, non-specific introduction to identity-based cryptography, relating it with other common public key practices. Examples of current IBE frameworks for scheme proposals include Full-Domain-Hash IBE, Exponent-Inversion IBE, and Commutative-Blinding IBE [6].

In 2008, some research was performed into the security of IBE when decryption keys are generated under the same ID from multiple PKGs [33]. Those security results are highly relevant to this work, and support the security analysis of our schemes. However, we do not limit our cascades to the use of one encryption scheme.

While the wider field of identity-based cryptography is of great interest, throughout this paper our focus will be on key management in the context and, in particular, identity-based encryption. Thus, we will generically refer to IBE and key management for IBE schemes. Identity-based signatures and the problem with the key escrow have been addressed in [43].

Cascade Encryption. Cascade encryption, or sequential multiple encryption, is the concept of layering encryption such that the ciphertext from one encryption step is the plaintext of the next. Essentially, an n-fold cipher cascade of Encrypt algorithms takes in a message m and outputs

$$(\text{Encrypt}_{k_n} \circ \cdots \circ \text{Encrypt}_{k_1})(m).$$

Ways of realizing a cascade cipher include increasing the number of rounds of a cipher, cascading encryption under different keys, and cascading actual ciphers. In the context of IBE in this paper, the latter is of particular interest due to the potential benefits of employing various PKGs. Essentially, while multiple cipher rounds may be generally beneficial for security, it is the benefit of key escrow without mandatory trust in one PKG that makes cascade encryption attractive in the context of IBE.

Previous work focusing on cascade encryption includes that of Even and Goldreich [15] which proved that in a 2-fold block-cipher cascade, the security of the cascade was reducible to the security of either of the component ciphers. Later work showed that the security of an n-cipher cascade was reducible to that of the first cipher in the cascade [30] (using weaker security assumptions about the individual ciphers than [15]), that triple-encryption (3-fold cipher cascade) for block ciphers provides a security improvement over single- or double-encryption [4,16], and describe generic CCA security for multiple encryption [13]. Furthermore, it has been information-theoretically demonstrated that an n-fold cascade of pseudo-random permutations (PRPs), for which the computational distinguishing advantage is bounded by $\epsilon < 1$ (ϵ-PRP), yields a $((n-(n-1)\epsilon)\epsilon^n + \nu)$-PRP for negligible function ν [39].

Despite the extensive research on multiple and cascade encryption, the application of an n-fold IBE-cipher cascade has not been addressed, nor have security considerations (CCA, etc.) been considered in this context. Since IBE already presents a possible alternative to PKI with some alluring benefits, the security of cascade encryption composed of IBE schemes, and the exact manner in which such a cascade can be realized, is particularly interesting. Moreover, cascade encryption with IBE goes beyond the encryption itself – in such a context, it is essential to consider collusion among the private key generators (PKGs) involved.

1.2   Our Contributions

We propose CARIBE, an IBE scheme that addresses the ease-of-use, eliminates the demand for trust in a single third party, allows free choice in the combination of PKGs and thereby encryption schemes, and only provides key escrow under collaboration of the entire set of PKGs involved. CARIBE addresses the cogent issues in key management for data encryption by applying IBE through the interface of cascade encryption. Even while allowing key escrow in the most extreme circumstances, and demanding no less than the participation of all key generators to achieve it, CARIBE provides far greater power to the end user for selecting a trust model than has been previously proposed.

Furthermore, we define IND-ID-CCA security, IND-ID-C.CCA, for the CARIBE environment. The experiment game for IND-ID-C.CCA expands on IND-ID-CCA by allowing for scheme cascades. Handling of the ID-based indistinguishability experiment presents special challenges in a cascade, and is therefore presented in formalized pseudo-code, as opposed to the ad-hoc discussion definitions historically presented for IND-ID-CCA.

We highlight a special instance of CARIBE, CARIBE-S, which inherits security from CARIBE but completely avoids key escrow. This is accomplished with a simple addition that was present in some earlier schemes [12, Sect. 1.1.2]: namely, that recipients of encrypted messages are themselves one of several cascaded PKGs. However, in contrast to previous proposals where recipients are also PKGs, our proposal argues that there is no need for involvement of any part of a public key infrastructure.

All CARIBE schemes apply a cascade of encryptions via multiple PKGs to eliminate the single point of failure that is inherent in traditional IBE. However, while generally with CARIBE there is a possibility for the principle of the key escrow to be realized under the assumption that all PKGs collude, in CARIBE-S that possibility is void due to the fact that one of the PKGs is the recipient itself. CARIBE schemes are not limited by the selection choice of concrete schemes involved. Unlike distributed IBE where all PKGs must operate under the same scheme, CARIBEs allow for the interaction of multiple IBE schemes. Notably, the benefit of this should not be under-estimated in the modern real-world context. Even as each PKG has freedom to use whatever IBE scheme it desires, a sender may either select PKGs based upon the combination of options given or upon a trust (or mutual distrust) foundation without regard to the corresponding IBE schemes being used. Thus, a sender may feasibly select rival PKGs to reduce the chances of collusion; for instance, PKGs implemented by, and operating under, the standards of competitive world powers [32, 34].

2   Background and Preliminaries

Identity-based cryptography falls within the scope of public key cryptography. Currently, public key systems rely almost completely on *certificate authorities* (CAs), employing certificate chaining to distribute, assert, and prove ownership of public keys. Popularly, this system is referred to as PKI [26]. Mao [29], as well as Katz and Lindell [26], offers a good foundational overview of PKI.

Structurally, identity-based cryptography diverges considerably from PKI and, as such, comes with certain advantages and disadvantages, particularly relating to key management. For a good introduction, Joux, via Joye and Neven

Table 1. Comparison of properties between PKI and IBE.

Architecture	PKI	IBE
Key management authority	CA	PKG
Key escrow	No	Yes
Certificate management	Yes	No
Non-interactive authentication	No	Yes
Always encrypt	No	Yes
Compromise of management authority is fatal	No	Yes

[24], gives a clean and concise introduction to identity-based cryptography, relating it with the certificated system encompassed within public key infrastructure. Table 1 compares the properties of PKI and IBE.

2.1 IBE Schemes with More Than One PKG

This section examines current proposals for IBE schemes employing more than one PKG, providing an overview of these architectures and highlighting the properties they possess.

Hierarchical IBE. Originally envisaged by Horwitz and Lynn, HIBE has parallels with the hierarchical nature of current PKI [20]. Like PKI it is comprised of root nodes, intermediate nodes, and users. Informally, the root PKG holds the master secret key masterkey, while an intermediate PKG holds its own identity ($ID_{PKG.i}$) and must request their own secret key from the root PKG. Similarly, the user has an identity (ID_{user}) and requests its secret key from the intermediate node. Keys at each stage are derived from functions on the keys at the higher level, as demonstrated in the following generalization for a hierarchy of n PKGs [20].

$$Root: \qquad f_1(\text{masterkey}, ID_{PKG.2}) = dkey_2 \rightarrow PKG.2$$
$$PKG.2: \qquad f_2(dkey_2, ID_{PKG.3}) = dkey_3 \rightarrow PKG.3$$
$$\vdots$$
$$PKG.n: \qquad f_n(dkey_n, ID_{User}) = dkey_{User} \rightarrow User$$

It is worth noting that the functions f_i, for $i \in \{1, n\}$, are known and that every intermediate PKG_i in the hierarchy, excepting the user, may have multiple descendants.

Gentry and Silverberg offered an improved HIBE scheme, presenting an instantiation of HIBE that is CCA-secure under the Bilinear Diffie-Hellman Problem and collusion resistant [17].

Other extensions of hierarchical IBE schemes exist, such as Multi-HIBE and Anonymous-HIBE. Multi-hierarchical offers forward security [42] and anonymous-hierarchical offers anonymous communication between sender and receiver [7].

Certificateless Public Key Cryptography. Certificateless public key cryptography (CL-PKC) (see [1,10]), was developed with the aim of finding public key schemes that are not dependent on certificates and do not have the key escrow property. As CL-PKC is claimed as an intermediate between standard PKI and the identity-based variant, we describe how it works on a high level.

CL-PKC requires two parties to generate public and private keys, where one is the end user. The PKG in this instance has a known public key ID_{PKG} and a master secret key. The user U has an identity ID_{User} and some secret information $SecInfo-U$ known only to themselves, with the public key and the secret key generated from these parameters.

$$PKG:\qquad f_1(\text{masterkey}, \mathsf{ID}_{User}) = dkey_{\tilde{U}} \rightarrow \text{User}$$
$$\text{User}:\qquad f_2(dkey_{\tilde{U}}, SecInfo\text{-}U) = dkey_U$$
$$\text{User}:\qquad f_3(SecInfo\text{-}U, \mathsf{ID}_{PKG}) = pkey_U$$

Again, the functions f_i are assumed to be known for all i, and $dkey_{\tilde{U}}$ represents the partial decryption key for user U, which must be combined with the users secret information to form the decryption key.

Generation of the public key can still be done prior to generation of the private key. Note that the public key is not computable from the identity of the user and therefore must also be pre-computed and made available publicly (though verification of the public key is no longer required). Consequently, due to the nature of this public key derivation, CL-PKC is no longer identity-based [1] and lacks the major advantage of the identity-based paradigm which allows any-time encryption without the receiver having to preform any set up.

Distributed IBE. Joux [23] advocates for a system with many, independent PKGs for nullifying the issue of compromise of a single PKG: "Such a scheme could mitigate the trust issues, at the cost of making the private key generation step heavier ..." Distributed IBE, as formally defined by Kate and Goldberg [25] does precisely that, with a form of threshold trust. Informally, an IBE scheme of n PKGs is (n,t)-distributed if no collusion of $x \leq t$ of the PKGs can compute the master key, for some threshold value $t \leq n$, where all n PKGs contain a share of a user's private key. In the distribution model, the n PKGs share parts of one master secret key.

An approach that is complementary to the distributed IBE approach was proposed by Chow in [9].

2.2   Identity-Based Encryption

Formally, we present the definition of an IBE scheme which serves as a grounding point for the work in this section. From IBE schemes, we build CARIBE – the scheme contribution of this paper – using generic, yet unspecified, number of n PKG.

Definition 1 (Identity-Based Encryption [12]). *Under a Private Key Generator (PKG), an identity-based encryption scheme IBE.$\mathcal{E}$ is a tuple of algorithms:*

- Setup(λ) $\overset{\$}{\to}$ (params, masterkey): *A probabilistic setup generation algorithm that takes as input a security parameter λ and outputs parameters* params *and a PKG master key* masterkey.

- Extract(params, masterkey, ID) $\overset{\$}{\to}$ *dkey: A probabilistic extraction algorithm that takes as input system parameters* params, *a PKG master key* masterkey, *and a public identity string* ID, *and outputs a decryption key dkey.*

- Encrypt(params, ID, m) $\overset{\$}{\to}$ *c: A probabilistic encryption algorithm that takes as input system parameters* params, *a public identity string* ID, *and a message $m \in \mathcal{M}$, and outputs a ciphertext $c \in \mathcal{C}$.*

- Decrypt(params, c, dkey) $\overset{\$}{\to}$ *m: A possibly probabilistic decryption algorithm that takes as input system parameters* params, *a ciphertext $c \in C$, and a private decryption key dkey, and outputs either a message $m \in \mathcal{M}$ or an error symbol $\bot$.*

In addition, it is required that if dkey $\leftarrow$ Extract(params, masterkey, ID), then

$$\forall m \in \mathcal{M} : \text{Decrypt}(\text{params}, \text{Encrypt}(\text{params}, \text{ID}, m), dkey) = m.$$

As an accepted assessment of security for public key encryption schemes, IND-CCA is also the criterion for security in the layered PKG setting. While IND-CCA security has been extensively handled before [3], and even the particular case of IND-ID-CCA for IBE described [12], a clear, formalized, pseudo-code definition for IND-ID-CCA has been lacking. Consequently, we unambiguously delineate the IND-ID-CCA experiment and adversary win conditions, corresponding to Definition 2, in Fig. 1. Notationally, we let Π denote the protocol employed by the PKG.

Definition 2. *Let Π be an identity-based encryption scheme according to Definition 1 and let $\mathcal{A}$ be an adversary algorithm. Then, for the IND-ID-CCA experiment given in Fig. 1,*

$$\text{Adv}_{\Pi,\mathcal{A}}^{IND\text{-}ID\text{-}CCA} = |\Pr[b = b'] - {}^{1}/{}_{2}|.$$

3   Cascade-Realized IBE – CARIBE

Employing a generic, finite number of n IBE schemes, as defined above, we describe Cascade-realized Identity-Based Encryption (CARIBE), an n-fold IBE-cipher cascade, in Definition 3. Saliently, the CARIBE definition does not restrict the type of encryption schemes being used. As a result, a CARIBE scheme $CARIBE.\mathcal{E}$ could literally consist of PKGs using n distinct IBE schemes, if such variation is desired or deemed necessary. Essentially, a CARIBE scheme works

$\mathsf{Exp}_{\Pi,\mathcal{A}}^{IND-ID-CCA}()$:

1: $(\mathsf{params}, \mathsf{masterkey}) \xleftarrow{\$} \mathsf{Setup}(\lambda)$
2: $\mathsf{ID.list_{ext}} \leftarrow \perp$
3: $\mathsf{ID.list_{enc}} \leftarrow \perp$
4: $S \leftarrow \emptyset$
5: $b \xleftarrow{\$} \{0, 1\}$
6: $\mathcal{A}^{\mathrm{Extract}(\cdot),\mathrm{Encrypt}(\cdot),\mathrm{Decrypt}(\cdot),\mathsf{params}}()$
7: $b' \xleftarrow{\$} \mathcal{A}^{\mathrm{Extract}(\cdot),\mathrm{Encrypt}(\cdot),\mathrm{Decrypt}(\cdot),\mathsf{params}}$

Oracle Extract(ID_i):

1: **if** $\mathsf{ID}_i \in \mathsf{ID.list_{enc}}$ **then**
2:     **return** $\perp$
3: $dkey_i \leftarrow \mathrm{Extract}(\mathsf{params}, \mathsf{masterkey}, \mathsf{ID}_i)$
4: $\mathsf{ID.list_{ext}} \leftarrow \mathsf{ID.list_{ext}} | \mathsf{ID}_i$
5: **return** $dkey_i$

Oracle Decrypt(ID_i, c_i):

1: **if** $(\mathsf{ID}_i, c_i) \in S$ **then**
2:     **return** $\perp$
3: $dkey_i \leftarrow \mathrm{Extract}(\mathsf{params}, \mathsf{masterkey}, \mathsf{ID}_i)$
4: $m_i \leftarrow \mathrm{Decrypt}(\mathsf{params}, c_i, dkey_i)$
5: **return** m_i

Oracle Encrypt(ID, m_0, m_1):

1: **if** $\mathsf{ID} \in \mathsf{ID.list_{ext}}$ **then**
2:     **return** $\perp$
3: $c^{(0)} \leftarrow \mathrm{Encrypt}(\mathsf{params}, \mathsf{ID}, m_0)$
4: $c^{(1)} \leftarrow \mathrm{Encrypt}(\mathsf{params}, \mathsf{ID}, m_1)$
5: **if** $c^{(0)} = \perp$ or $c^{(1)} = \perp$ **then**
6:     **return** $\perp$
7: $c_u := c^{(b)}$
8: $S \leftarrow S \cup \{(\mathsf{ID}, c_u)\}$
9: $\mathsf{ID.list_{enc}} \leftarrow \mathsf{ID.list_{enc}} | \mathsf{ID}$
10: **return** c_u

Fig. 1. IND-ID-CCA experiment for $IBE.\mathcal{E}$.

with cascaded encryption; however, unlike general cascade encryption schemes, multiple PKGs are involved and encryption is sequentially performed using the parameters params generated by each of them.

In the following definition it is required that each generated ciphertext is in the plaintext space of the next cipher in the encryption cascade. We make the practical assumption that the selected IBE schemes are compatible in this manner, either directly using the plaintext space as ciphertext space, or indirectly, by applying some transformation before encryption, which is known to all parties. Note that, as the outputs of all schemes can be interpreted as a sequence of bits, the composition of IBE schemes defined on different algebraic domains is possible.

Definition 3 (Cascade-Realized Identity-Based Encryption). *Under* n
*Private Key Generators (PKG), a cascade-realized identity-based encryption
scheme* $CARIBE.\mathcal{E}$ *is a tuple of algorithms that takes* n *IBE encryption schemes*
Π_i, *for* $i \in \{1, \ldots n\}$ *where* $n \geq 2$, *with the restriction that* $\mathcal{C}_i \subseteq \mathcal{M}_{i+1}$ *for*
$i \in \{1, \ldots n-1\}$, *i.e. that the ciphertext of each IBE scheme is in the plaintext
space of the next IBE scheme in the cascade. Algorithms for* $CARIBE.\mathcal{E} :=$
$C(\Pi_1, \ldots, \Pi_n)$ *are as follows:*

– $\mathrm{Setup}_S(\lambda_1, \ldots, \lambda_n)$:
  1: **for** $i \in \{1, \ldots, n\}$ **do**
  2:    $(\mathsf{params}_i, \mathsf{masterkey}_i) \leftarrow \mathrm{Setup}_i(\lambda_i)$
  3: **return** $(\{\mathsf{params}_i\}, \{\mathsf{masterkey}_i\})$

  *A probabilistic setup generation algorithm that takes as input an ordered
  sequence of* n *security parameters* λ_i *and outputs an ordered set of* n *parameters* $\{\mathsf{params}_i\}$ *and an ordered set of* n *PKG master keys* $\{\mathsf{masterkey}_i\}$, *for*
  $i \in \{1, \ldots n\}$.

– $\mathrm{Extract}_S((\{\mathsf{params}_i\}, \{\mathsf{masterkey}_i\}), \mathsf{ID})$:
  1: **for** $i \in \{1, \ldots, n\}$ **do**
  2:    $dkey_i \leftarrow \mathrm{Extract}_i(\mathsf{params}_i, \mathsf{masterkey}_i, \mathsf{ID})$
  3: **return** $\{dkey_i\}$

  A probabilistic extraction algorithm that takes as input a set of n *system
  parameters* $\{\mathsf{params}_i\}$, *a set of* n *PKG master keys* $\{\mathsf{masterkey}_i\}$, *and a public
  identity string* ID, *and outputs a set of* n *decryption keys* $\{dkey_i\}$, *for* $i \in$
  $\{1, \ldots n\}$.

– $\mathrm{Encrypt}_S(\{\mathsf{params}_i\}, \mathsf{ID}, m)$:
  1: $c_2 \leftarrow \mathrm{Encrypt}_1(\mathsf{params}_1, \mathsf{ID}, m)$
  2: **for** $i \in \{2, \ldots, n\}$ **do**
  3:    $c_{i+1} \leftarrow \mathrm{Encrypt}_i(\mathsf{params}_i, \mathsf{ID}, c_i)$
  4: $c \leftarrow c_{n+1}$
  5: **return** c

  A probabilistic encryption algorithm that takes as input an ordered set of n
  system parameters $\{\mathsf{params}_i\}$, *for* $i \in \{1, \ldots n\}$, *a public identity string* ID,
  and a message $m \in \mathcal{M}$, *and outputs a ciphertext* $c \in \mathcal{C}$. *The plaintext space of*
  $\mathrm{Encrypt}_S$ *is* $\mathcal{M} := \mathcal{M}_1$, *the plaintext space of* $\mathrm{Encrypt}_1$, *while the ciphertext
  space of* $\mathrm{Encrypt}_S$ *is* $\mathcal{C} := \mathcal{C}_n$, *the ciphertext space of* $\mathrm{Encrypt}_n$.

– $\mathrm{Decrypt}_S(\{\mathsf{params}_i\}, c, \{dkey_i\})$:
  1: $c_n \leftarrow c$
  2: $i \leftarrow n$
  3: **while** $i > 0$ **do**
  4:    $c_{i-1} \leftarrow \mathrm{Decrypt}_i(\mathsf{params}_i, c_i, dkey_i)$
  5:    $i \leftarrow i - 1$
  6: $m \leftarrow c_0$
  7: **return** m

  *A possibly probabilistic decryption algorithm that takes as input an ordered
  set of* n *system parameters* $\{\mathsf{params}_i\}$, *a ciphertext* $c \in C$, *and an ordered set*

of n *private decryption keys* $\{dkey_i\}$, *and outputs either a message* $m \in \mathcal{M}$ *or an error symbol* $\bot$.

In addition, it is required that if $\{dkey_i\} \leftarrow \text{Extract}_S(\{\text{params}_i\}, \{\text{masterkey}_i\}, \text{ID})$, *then*

$$\forall m \in \mathcal{M} : \text{Decrypt}_S(\{\text{params}_i\}, \text{Encrypt}(\{\text{params}_i\}, \text{ID}, m), \{dkey_i\}) = m.$$

Remark 1. Note that while we require at least two PKGs ($n \geq 2$), we do not *require* unique encryption schemes. Indeed, it is possible for any two PKGs, PKG$_i$ and PKG$_j$ with encryption schemes Π_i and Π_j, respectively, that $\Pi_i = \Pi_j$.

3.1   Security of CARIBE

CARIBE is essentially a type of cascade encryption scheme and as such presents a challenge in the context of CCA security. As noted in [13], any encryption cascade fails to provide CCA security, since an adversary that is in possession of the key of the outermost encryption layer could simply decrypt that layer and re-encrypt, yielding a new ciphertext to call the decryption oracle on, trivially breaking the CCA security. However, this is based upon the assumption that an adversary already has the key for the outermost encryption, and assuming such key possession is generally non-standard when considering CCA security. To avoid this fairly trivial break, we propose the following assumption when analyzing the CCA security of cascade schemes; a CCA security game based upon this assumption will be termed Cascade CCA (C.CCA):

C.CCA Assumption for Cascaded Encryption. *For a cascaded encryption scheme, with ciphertexts generated as* $c := E_{K_n}(\ldots (E_{K_2}(E_{K_1}(m))))$, *and an adversary* $\mathcal{A}$, *CCA security is analyzed under the assumption that* $\mathcal{A}$ *does not possess* K_n.

Note that if $\mathcal{A}$ possesses the encryption keys for the outermost r layers, and can win a CCA security game on an encryption scheme cascade of the remaining $n - r$ layers, then $\mathcal{A}$ can certainly win the CCA game on the entire scheme. Thus, the layers can be "peeled back" until $\mathcal{A}$ does not possess the outermost decryption key. While making this assumption is not ideal for security analysis, it is necessary to avoid the trivial break mentioned above and allows for realistic analyses to still be performed for cascaded schemes.

In addition to the CARIBE scheme description, we present the tailored IND-ID-C.CCA experiment in Fig. 2. While most schemes are analyzed under a general security experiment, an experiment for CARIBE itself is required as any CARIBE scheme is in fact a particular cascade selection of other IBE schemes. Adversarial advantage for the experiment is described in Definition 4.

Definition 4. *Let* C *be an algorithm taking as input* n *identity-based encryption schemes* Π_j, *for* $j \in \{1, \ldots, n\}$ *where* $n \geq 2$, *according to Definition 1 and yielding a cascade-realized identity-based encryption scheme* $C(\Pi_1, \ldots, \Pi_n)$, *per Definition 3, comprised of a new tuple of algorithms* (Setup$_S$, Extract$_S$, Encrypt$_S$,

Decrypt$_S$). *Let $\mathcal{A}$ be an adversary algorithm. Then, for the IND-ID-C.CCA experiment given in Fig. 2,*

$$\mathsf{Adv}_{C(\Pi_1,\dots,\Pi_n),\mathcal{A}}^{IND\text{-}ID\text{-}C.CCA} = |\Pr[b = b'] - {}^1\!/_2|.$$

As noted in Remark 1, we do not *require* unique encryption schemes Π, although at least two PKGs ($n \geq 2$) are required for the cascaded IND-ID-C.CCA definition.

Security for CARIBE depends upon the constituent IBE schemes involved, as each contributes to the security of the final ciphertext. As previously mentioned, it has been shown that the encryption security of a 2-fold cascade is reducible to the security of either of the composite ciphers [14,15]. Consequently, not only do we focus on the expanded n-fold case, but take into consideration the possibility of collusion. On a logical level, the ciphertext cannot be decrypted even if $n - 1$ of the n PKGs collude, so long as one PKG is honest. Essentially, this worst-case scenario would then be precisely equivalent to a basic, secure IBE scheme under an honest PKG. Thus, the distinction between a CARIBE and IBE scheme becomes one of existence; for IBE we demand that the IBE scheme in use is secure with an honest PKG, while for CARIBE it is only required that one such IBE scheme exists in the cascade. Succinctly, Theorem 1 provides a proof of this analysis, similar to that of [15] yet given in detail for IBE.

To enable the analysis, we operate in an execution environment with the following standard list of allowed adversarial queries:

- Extract$_S$($ID_i, \{1, \dots, n\}$): Operates as in Fig. 2.
- Encrypt$_S$($ID, m_0, m_1, \{1, \dots, n\}$): Operates as in Fig. 2.
- Decrypt$_S$($ID_i, c_i, \{1, \dots, n\}$): Operates as in Fig. 2.
- Corrupt$_S$(Π_j): This query returns masterkey$_j$. As an adversary $\mathcal{A}$ already has access to params$_j$, corrupting Π_j allows $\mathcal{A}$ to extract decryption keys $dkey_j$ for any ID_i. This query models $\mathcal{A}$'s ability to request the collusion of the PKG operating Π_j.

Note that we prove Theorem 1 for a static adversary which, although it may adaptively corrupt nodes at will, must commit to at least one honest PKG before the protocol run. Notably, this follows similarly to the chain of past IBE work, where a static adversarial model is common [25]. Our theorem, and its accompanying proof, are IBE-specific adaptions of a more general approach by Dodis and Katz for cascade encryption [13]. The proof is given in detail here to elucidate the adversarial power in the inter-workings of IBE, and how a challenger answers adversarial queries.

Theorem 1. *If IBE.$\mathcal{E}$ protocols Π_j, for $j \in \{0, \dots, n\}$, are combined to form CARIBE.$\mathcal{E} = C(\Pi_1, \dots, \Pi_n)$, the resulting CARIBE.$\mathcal{E}$ protocol will be IND-ID-C.CCA provided that there exists $\Pi_t \in \{\Pi_j\}$ that is IND-ID-CCA secure.*

Specifically, for any efficient adversary $\mathcal{A}$ that runs in time t and asks $q = q_{ext} + q_{enc} + q_{dec}$ queries, where q_{ext} are extraction queries, q_{enc} are encryption

$\mathsf{Exp}^{IND-ID-C.CCA}_{C(\Pi_1,\ldots,\Pi_n),\mathcal{A}}()$:

1: $(\{\mathsf{params}_j\}, \{\mathsf{masterkey}_j\}) \xleftarrow{\$} \mathsf{Setup}(\lambda_1,\ldots,\lambda_n)$
2: $\mathsf{ID.list}_{ext} \leftarrow \perp$
3: $\mathsf{ID.list}_{enc} \leftarrow \perp$
4: $S \leftarrow \emptyset$
5: $b \xleftarrow{\$} \{0,1\}$
6: $\mathcal{A}^{\mathsf{Extract}_S(\cdot),\mathsf{Encrypt}_S(\cdot),\mathsf{Decrypt}_S(\cdot),\{\mathsf{params}_j\}}()$
7: $b' \xleftarrow{\$} \mathcal{A}^{\mathsf{Extract}_S(\cdot),\mathsf{Encrypt}_S(\cdot),\mathsf{Decrypt}_S(\cdot),\{\mathsf{params}_j\}}$

Oracle $\mathsf{Extract}_S(\mathsf{ID}_i,\{1,\ldots,n\})$:

1: **if** $\mathsf{ID}_i \in \mathsf{ID.list}_{enc}$ **then**
2:     **return** $\perp$
3: $\{dkey_j\} \leftarrow \mathsf{Extract}_S(\{\mathsf{params}_j\}, \{\mathsf{masterkey}_j\}, \mathsf{ID}_i)$
4: $\mathsf{ID.list}_{ext} \leftarrow \mathsf{ID.list}_{ext}|\mathsf{ID}_i$
5: **return** $\{dkey_j\}_i$

Oracle $\mathsf{Decrypt}_S(\mathsf{ID}_i, c_i, \{1,\ldots,n\})$:

1: **if** $(\mathsf{ID}_i, c_i) \in S$ **then**
2:     **return** $\perp$
3: $\{dkey_j\}$
    $\leftarrow \mathsf{Extract}_S(\{\mathsf{params}_j\}, \{\mathsf{masterkey}_j\}, \mathsf{ID}_i)$
4: $m_i \leftarrow \mathsf{Decrypt}_S(\{\mathsf{params}_j\}, c_i, \{dkey_j\})$
5: **return** m_i

Oracle $\mathsf{Encrypt}_S(\mathsf{ID}, m_0, m_1, \{1,\ldots,n\})$:

1: **if** $\mathsf{ID} \in \mathsf{ID.list}_{ext}$ **then**
2:     **return** $\perp$
3: $c^{(0)} \leftarrow \mathsf{Encrypt}_S(\{\mathsf{params}_j\}, \mathsf{ID}, m_0)$
4: $c^{(1)} \leftarrow \mathsf{Encrypt}_S(\{\mathsf{params}_j\}, \mathsf{ID}, m_1)$
5: **if** $c^{(0)} = \perp$ or $c^{(1)} = \perp$ **then**
6:     **return** $\perp$
7: $c := c^{(b)}$
8: $S \leftarrow S \cup \{(\mathsf{ID}, c)\}$
9: $\mathsf{ID.list}_{enc} \leftarrow \mathsf{ID.list}_{enc}|\mathsf{ID}$
10: **return** c

Fig. 2. IND-ID-C.CCA experiment for $CARIBE.\mathcal{E}$.

queries, and q_{dec} are decryption queries, there exists adversaries $\mathcal{B}_j$ that run in time $t_\mathcal{B} \approx t$ and asks $q_\mathcal{B}$ queries, such that

$$\mathsf{Adv}^{IND-ID-C.CCA}_{C(\Pi_1,\ldots,\Pi_n)}(\mathcal{A}) \leq \mathsf{Adv}^{IND-ID-CCA}_{\Pi_t}(\mathcal{B}_t).$$

Proof. Let $\mathcal{A}$ be an challenger against the IND-ID-C.CCA security of $CARIBE.\mathcal{E} = C(\Pi_1, \ldots, \Pi_n)$ and let $\mathcal{B}$ be an adversary against the IND-ID-C.CCA of $IBE.\mathcal{E}$ for Π_t, with $\mathsf{Extract}_t$, $\mathsf{Encrypt}_t$, and $\mathsf{Decrypt}_t$ oracles running on params_t and $\mathsf{masterkey}_t$, corresponding to Π_t, as well as $\mathrm{Extract}_j$, $\mathrm{Encrypt}_j$, and $\mathrm{Decrypt}_j$ algorithms running on params_j and $\mathsf{masterkey}_j$ for $j \in \{1, \ldots, n\}/\{t\}$ which he uses to answer $\mathcal{A}$'s queries. Let $S_\mathcal{B}$ be a list of pairs (ID, c) which $\mathcal{B}$ sends back to $\mathcal{A}$ in response to $\mathsf{Encrypt}_S$ queries, maintained by $\mathcal{B}$ and initialized to $\perp$. If at any time $\mathcal{A}$ makes a $\mathsf{Corrupt}(\Pi_t)$ query, $\mathcal{B}$ gives up.

When $\mathcal{A}$ asks an $\mathsf{Extract}_S$ query on $(\mathsf{ID}_i, \{1, \ldots, n\})$, $\mathcal{B}$ calls his $\mathsf{Extract}_t$ oracle and $\mathrm{Extract}_j$ algorithms, and sends the agglomerated responses $\{dkey_j\}$ to $\mathcal{A}$.

When $\mathcal{A}$ asks a $\mathsf{Decrypt}_S$ query on a ciphertext c, $\mathcal{B}$ sequentially uses his $\mathrm{Decrypt}_j$ algorithms – starting with $\mathrm{Decrypt}_n$ on c, followed with $\mathrm{Decrypt}_{n-1}$ on the output of $\mathrm{Decrypt}_n$, and so forth for $t < j$. $\mathcal{B}$ then call $\mathsf{Decrypt}_t$ on the output of $\mathrm{Decrypt}_{t+1}$. Thereafter, $\mathcal{B}$ continues with his $\mathrm{Decrypt}_j$ calculations, starting with $\mathrm{Decrypt}_{t-1}$ on the output of $\mathsf{Decrypt}_t$, for $j < t$. Finally, the result of $\mathrm{Decrypt}_1$ is returned to $\mathcal{A}$.

Should $\mathcal{A}$ query $\mathsf{Encrypt}_S$ on $(\mathsf{ID}, m_0, m_1, \{1, \ldots, n\})$, $\mathcal{B}$ behaves as follows: $\mathcal{B}$ calculates $\mathrm{Encrypt}_1(\mathsf{params}_1, \mathsf{ID}, m_0) = c_1^{(0)}$ and $\mathrm{Encrypt}_1(\mathsf{params}_1, \mathsf{ID}, m_1) = c_1^{(1)}$, then $\mathrm{Encrypt}_j(\mathsf{params}_j, \mathsf{ID}, c_{j-1}^{(0)}) = c_j^{(0)}$ and $\mathrm{Encrypt}_j(\mathsf{params}_j, \mathsf{ID}, c_{j-1}^{(1)}) = c_j^{(1)}$ for $1 < j < t$. Then, $\mathcal{B}$ calls $\mathsf{Encrypt}_t(\mathsf{ID}, c_{t-1}^{(0)}, c_{t-1}^{(1)})$ to get c_t. Thereafter $\mathcal{B}$ continues calculating $\mathrm{Encrypt}_j(\mathsf{params}_j, \mathsf{ID}, c_{j-1}) = c_i$ for $t < j \leq n$. Finally, $\mathcal{B}$ sets $c := c_n$, $S \leftarrow S \cup (\mathsf{ID}, c)$, and passes c back to $\mathcal{A}$.

By the IND-ID-C.CCA success of $\mathcal{A}$ against $CARIBE.\mathcal{E}$, at some point $\mathcal{A}$ returns a correct bit guess b'. Hence, $\mathcal{B}$ also wins the IND-ID-CCA game against the $IBE.\mathcal{E}$ protocol Π_t with b'.   □

From Theorem 1 it can be observed that the CARIBE scheme is at least as secure as the strongest IBE protocol in the cascade. As mentioned before, should multiple PKG collude to determine the decryption key or plaintext, this implies that a CARIBE scheme is secure as long as at most $n-1$ of the n PKGs collude.

4   Cascade-Realized IBE with Self-PKG – CARIBE-S

While the PKGs used in CARIBE would naturally be expected to be separate from the users, special security considerations arise when we consider the possibility of a *self-PKG*, where a receiver also acts in the role of a PKG. This could be desirable in real-world implementation, for example, if the receiver is a government organisation with both the capabilities and legal authority to be a self-PKG. Although the receiver may trust its PKG, and even demand its use for all in-coming messages, the sender may not. In response to this situation, the sender may include the receiver's PKG along with other (trusted) PKGs under CARIBE-S, creating a mutually satisfactory trust relationship. Thus, while avoiding key escrow, CARIBE-S allows for a trust balance between sender and receiver.

Definition 5 (CARIBE-S). *We say that an entity is a member of a CARIBE-S scheme if it is one of n PKGs in the Definition 3.*

Naturally, a receiver cannot force a self-selection as a PKG under CARIBE. This is to be expected. As in CARIBE, the decision to use a CARIBE-S and avoid key-escrow is completely the sender's choice. However, it is dependent on the *a priori* action on a receiver's part to arrange PKG functionality. Consequently, the option of a CARIBE-S is also restricted to those receivers with the resources and forethought to form self-PKGs.

4.1 Security of CARIBE-S

From the security of CARIBE we have the following theorem.

Theorem 2 (Security of CARIBE-S). *If IBE.$\mathcal{E}$ protocols Π_j, for $j \in \{0, \dots, n\}$, are combined to form CARIBE.$\mathcal{E} = C(\Pi_1, \dots, \Pi_n)$, the resulting CARIBE.$\mathcal{E}$ protocol will be IND-ID-C.CCA provided that there exists $\Pi_t \in \{\Pi_j\}$ that is IND-ID-CCA secure.*

Specifically, for any efficient adversary $\mathcal{A}$ that runs in time t and asks $q = q_{ext} + q_{enc} + q_{dec}$ queries, where q_{ext} are extraction queries, q_{enc} are encryption queries, and q_{dec} are decryption queries, there exists adversaries $\mathcal{B}_j$ that run in time $\mathsf{t}_{\mathcal{B}} \approx \mathsf{t}$ and asks $q_{\mathcal{B}}$ queries, such that

$$\mathsf{Adv}_{C(\Pi_1, \dots, \Pi_n)}^{IND\text{-}ID\text{-}C.CCA}(\mathcal{A}) \leq \mathsf{Adv}_{\Pi_t}^{IND\text{-}ID\text{-}CCA}(\mathcal{B}_t).$$

Whilst a CARIBE scheme is secure as long as at most $n - 1$ of the n PKGs collude, in CARIBE-S, one of the PKGs which has joined the infrastructure, having provided its PKG parameters, is also the intended recipient of the encrypted messages. Thus, Corollary 1 follows directly from Theorem 2.

Corollary 1 (Security Against Collusion of CARIBE-S). *A CARIBE-S scheme is secure against collusion and thus is not a key escrow scheme.*

Table 3 in Appendix A presents a comparison of CARIBE and CARIBE-S with other IBE-composite schemes discussed in Sect. 2.1, under standard structure. Both CARIBE and CARIBE-S allow for composition across multiple IBE platforms – namely, each PKG can use a different IBE scheme for extraction, encryption, etc. In the interest of security, this yields logical real-world benefits in the case where preferred PKGs, known to be adverse to mutual collusion, insist on utilizing different IBE schemes.

4.2 Ciphertext Expansion in CARIBE/CARIBE-S

One natural consequence of cascaded encryption is the amplification of ciphertext expansion. Historically, ciphertext expansion would be a concern under slow transmission rates, and since it is already common in IBE schemes, further expansion from cascades would hardly have been welcomed. However, with increasing improvements in IBE scheme developments involving less expansion, as well as faster transmission rates than were previously available, this issue is not as imposing as it once was. Moreover, in the context of modern internet privacy concerns, it is more likely that users will be willing to trade factors like the convenience of fast transmission times, for increased privacy and security.

Since CARIBE is a composition of ciphers of the sender's choice, it is not possible to predict the relative ciphertext expansion in advance without knowing which ciphers, and in what order, the sender will select. Still, some basic observations can be noted. Unquestionably, the expansion involved in a CARIBE of several compositions of Cocks' IBE schemes [11] would be enormous, as ciphertext length in under single encryption is already several times larger than the plaintext length. Yet the ciphertext expansion for a regular, single Cocks scheme is daunting enough at the outset to be naturally prohibitive in practice. Meanwhile, in a CARIBE of n Boneh–Franklin [12] ciphers, a plaintext length $|m|$ would expand to $n \cdot |P| + |m|$, where $|P|$ is the length of a pre-selected element of a group $\mathbb{G}$, of large prime order q, which is part of a bilinear map. Markedly, such a linear expansion is hardly imposing. Aside from classic IBE ciphers, ciphertext expansion in modern IBE proposals lie spread on the scale between these examples, yet far closer to Boneh–Franklin efficiency than Cocks. With simultaneously increasing efficiency and security awareness, it is reasonable that expansion for CARIBE will be of limited concern.

5   Software Libraries for Implementation of CARIBE and CARIBE-S

Our proposals for CARIBE and CARIBE-S collect widely known and tested concepts and combine them in a novel way, expanding the horizons of IBE. This is not a new approach. We can refer to the well known concept of Bitcoin that had a similar approach [31] in the beginning.

In the modern Internet era, the crucial moment for a collection of concepts to become a widely used paradigm is linked to the existence of publicly available (preferably free) software tools and libraries. We argue that these types of conditions are maturing for CARIBE and CARIBE-S. In Table 2 we give a list of tools, packages or libraries that can perform the operations for Pairing-Based Cryptography – these are described in more detail below.

- The Pairing-Based Cryptography Library [28] is a free portable C library for rapid prototyping of pairing-based cryptosystems. It provides an interface to a cyclic group with a bilinear pairing. It has also Perl, Python, Java and C++ wrappers and bindings.
- MIRACL [36] is a C/C++ library that provides pairing-based cryptography primitives. It runs on different OS and CPU platforms but is especially efficient on x86 platforms due to hand-optimized assembly code for low-level arithmetic. It is free for noncommercial use.

Table 2. Software libraries and packages for Pairing-Based Cryptography

Name	License	Language
PBCl	GNU GPL	C
MIRACL	Free non noncommercial	C/C++
SAGE	GNU GPL	Python
PARI/GP	GNU GPL	C/GP
RELIC	GNU Lesser GPL	C
HP IBE	Yes	C/C++

- SAGE [38] is a powerful open-source computer algebra system. Its interface is Python based. It has huge number of cryptographic functions including some of the worlds fastest implementations of operations with elliptical curves.
- PARI/GP [40] is a computer algebra system designed for fast computations in number theory. It is implemented in C but it has also a script language called GP.
- RELIC [2] is a modern cryptographic meta-toolkit with emphasis on efficiency and flexibility. RELIC implements Elliptic curves over prime and binary fields (NIST curves and pairing-friendly curves), Bilinear maps and related extension fields. Its portable C produces very efficient codes in a range from tiny micro-controllers to powerful modern 64-bit CPUs.
- In February 2015, HP announced that they will acquire Voltage Security Inc. that previously offered the IBE Toolkit. Now that toolkit is commercially available as HP Identity-Based Encryption.

6 Conclusion

CARIBE-S is a completely key escrow-free variant of a CARIBE, combining IBE and cascade encryption. It synthesizes diverse ideas to provide more benefits and a higher level of security than have been achieved in other schemes.

Even though there is an inherent time-cost in multiple encryptions for CARIBE and CARIBE-S, the added security coupled with ever-increasing computing power, and publicly available open source or commercial software libraries, dilutes this drawback. Additionally, while the onus is on the receiver to obtain decryption keys from multiple authorities, the sender's ability to virtually select a security level for encryption through the PKGs of their choice – possibly based on the encryption type provided by the PKG – may commonly be seen as a sufficient time/security trade-off.

In comparison with other multi-PKG IBE variations, this end-user security selection power provides the catalyst for re-visiting cascade encryption in a previous un-investigated context. In terms of the modern world, where desire to legally access encrypted messages is paired with hostility among internet powers, the consolidation or avoidance of key escrow and leveraged distrust makes CARIBE – and therefore CARIBE-S – a viable option.

Acknowledgements. We would like to thank the anonymous reviewers of Mycrypt 2016 for their valuable comments and suggestions.

A   Scheme Comparison

Table 3. Comparison of properties among composition IBE schemes including CARIBE and CARIBE-S.

Topic	H-IBE	D-IBE	CARIBE	CARIBE-S
Key management authority	n PKGs, hierarchy	n PKGs, t-threshold	n PKGs	n PKGs
Key escrow	Yes	Limited to $t+1$ collusions	Limited n collusions	No
Certificates	No	No	No	No
Ciphertext expansion	Yes	Yes	Yes	Yes
Management authority has access to private keys	Yes for all PKGs	Under $t+1 \leq n$ collusions	Under n collusions	No
Compromise of management authority is fatal	Yes for any PKG in hierarchy	Under $t+1 \leq n$ collusions	Under n collusions	No
Incorporation of various encryption methods across PKGs possible	No	No	Yes	Yes

References

1. Al-Riyami, S.S., Paterson, K.G.: Certificateless public key cryptography. In: Laih, C.-S. (ed.) ASIACRYPT 2003. LNCS, vol. 2894, pp. 452–473. Springer, Heidelberg (2003). doi:10.1007/978-3-540-40061-5_29
2. Aranha, D.F., Gouvêa, C.P.L.: RELIC is an Efficient LIbrary for Cryptography. http://code.google.com/p/relic-toolkit/
3. Bellare, M., Desai, A., Pointcheval, D., Rogaway, P.: Relations among notions of security for public-key encryption schemes. In: Krawczyk, H. (ed.) CRYPTO 1998. LNCS, vol. 1462, pp. 26–45. Springer, Heidelberg (1998). doi:10.1007/BFb0055718
4. Bellare, M., Rogaway, P.: The security of triple encryption and a framework for code-based game-playing proofs. In: Vaudenay, S. (ed.) EUROCRYPT 2006. LNCS, vol. 4004, pp. 409–426. Springer, Heidelberg (2006). doi:10.1007/11761679_25
5. Boneh, D., Franklin, M.: Identity-based encryption from the weil pairing. SIAM J. Comput. **32**(3), 586–615 (2003)
6. Boyen, X.: A tapestry of identity-based encryption: practical frameworks compared. Int. J. Appl. Cryptogr. **1**(1), 3–21 (2008)

CARIBE: Cascaded IBE for Maximum Flexibility and User-Side Control   407

7. Boyen, X., Waters, B.: Anonymous hierarchical identity-based encryption (without random oracles). In: Dwork, C. (ed.) CRYPTO 2006. LNCS, vol. 4117, pp. 290–307. Springer, Heidelberg (2006). doi:10.1007/11818175_17
8. Chatterjee, S., Sarkar, P.: Identity-Based Encryption. Springer Science & Business Media, Berlin (2011)
9. Chow, S.S.M.: Removing escrow from identity-based encryption. In: Jarecki, S., Tsudik, G. (eds.) PKC 2009. LNCS, vol. 5443, pp. 256–276. Springer, Heidelberg (2009). doi:10.1007/978-3-642-00468-1_15
10. Chow, S.S.M., Boyd, C., Nieto, J.M.G.: Security-mediated certificateless cryptography. In: Yung, M., Dodis, Y., Kiayias, A., Malkin, T. (eds.) PKC 2006. LNCS, vol. 3958, pp. 508–524. Springer, Heidelberg (2006). doi:10.1007/11745853_33
11. Cocks, C.: An identity based encryption scheme based on quadratic residues. In: Honary, B. (ed.) Cryptography and Coding 2001. LNCS, vol. 2260, pp. 360–363. Springer, Heidelberg (2001). doi:10.1007/3-540-45325-3_32
12. Boneh, D., Franklin, M.: Identity-based encryption from the weil pairing. SIAM J. Comput. **32**(3), 586–615 (2003). Society for Industrial and Applied Mathematics
13. Dodis, Y., Katz, J.: Chosen-ciphertext security of multiple encryption. In: Kilian, J. (ed.) TCC 2005. LNCS, vol. 3378, pp. 188–209. Springer, Heidelberg (2005). doi:10.1007/978-3-540-30576-7_11
14. Even, S., Goldreich, O.: On the power of cascade ciphers. Technical report no. 275, Computer Science Department, Technion, Haifa, Israel, May 1983
15. Even, S., Goldreich, O.: On the power of cascade ciphers. In: Chaum, D. (ed.) Advances in Cryptology: Proceedings of CRYPTO 1983, pp. 43–50. Springer US, New York (1984)
16. Gaži, P., Maurer, U.: Cascade encryption revisited. In: Matsui, M. (ed.) ASIACRYPT 2009. LNCS, vol. 5912, pp. 37–51. Springer, Heidelberg (2009). doi:10.1007/978-3-642-10366-7_3
17. Gentry, C., Silverberg, A.: Hierarchical ID-based cryptography. In: Zheng, Y. (ed.) ASIACRYPT 2002. LNCS, vol. 2501, pp. 548–566. Springer, Heidelberg (2002). doi:10.1007/3-540-36178-2_34
18. Greenwald, G.: XKeyscore: NSA tool collects nearly everything a user does on the internet, 31 July 2013. http://www.theguardian.com/world/2013/jul/31/nsa-top-secret-program-online-data. Accessed 2 June 2015
19. Greenwald, G., MacAskill, E.: NSA Prism program taps in to user data of Apple, Google and others, 7 June 2013. http://www.theguardian.com/world/2013/jun/06/us-tech-giants-nsa-data. Accessed 2 June 2015
20. Horwitz, J., Lynn, B.: Toward hierarchical identity-based encryption. In: Knudsen, L.R. (ed.) EUROCRYPT 2002. LNCS, vol. 2332, pp. 466–481. Springer, Heidelberg (2002). doi:10.1007/3-540-46035-7_31
21. Hurd, W., Lieu, T.W.: Congressman Lieu Letter to FBI Director Comey on Encryption "Backdoor" Proposal, 1 June 2015. https://lieu.house.gov/media-center/. Accessed 2 June 2015
22. IACR: IACR Statement on Mass Surveillance: Copenhagen Resolution, 14 May 2014. http://www.iacr.org/misc/statement-May2014.html. Accessed 2 June 2015
23. Joux, A.: Introduction to Identity-Based Cryptography. Identity- Based Cryptography (2009)
24. Joye, M., Neven, G.: Identity-Based Cryptography, vol. 2. IOS Press, Amsterdam (2009)
25. Kate, A., Goldberg, I.: Distributed private-key generators for identity-based cryptography. In: Garay, J.A., Prisco, R. (eds.) SCN 2010. LNCS, vol. 6280, pp. 436–453. Springer, Heidelberg (2010). doi:10.1007/978-3-642-15317-4_27

26. Katz, J., Lindell, Y.: Introduction to Modern Cryptography. CRC Press, Boca Raton (2014)
27. Leavitt, N.: Internet security under attack: the undermining of digital certificates. Computer **44**(12), 17–20 (2011)
28. Lynn, B.: PBC library manual 0.5.11 (2006)
29. Mao, W.: Modern Cryptography: Theory and Practice. Prentice Hall PTR, Upper Saddle River (2004)
30. Maurer, M., Massey, J.: Cascade ciphers: the importance of being first. J. Cryptol. **6**(1), 55–61 (1993). Springer
31. Nakamoto, S.: Bitcoin: a peer-to-peer electronic cash system. Consulted **1**(2012), 28 (2008)
32. National Institute of Standards and Technology. http://www.nist.gov/. Accessed 2 June 2015
33. Paterson, K.G., Srinivasan, S.: Security and anonymity of identity-based encryption with multiple trusted authorities. In: Galbraith, S.D., Paterson, K.G. (eds.) Pairing 2008. LNCS, vol. 5209, pp. 354–375. Springer, Heidelberg (2008). doi:10.1007/978-3-540-85538-5_23
34. Popov, V., Kurepkin, I., Leontiev, S.: RFC 4357: Additional Cryptographic Algorithms for Use with GOST 28147-89, GOST R 34.10-94, GOST R 34.10-2001, and GOST R 34.11-94 Algorithms, January 2006
35. Sakai, R., Ohgishi, K., Kasahara, M.: Cryptosystems based on pairing. In: The 2000 Symposium on Cryptography and Information Security, Okinawa, Japan, pp. 135–148 (2000)
36. Scott, M.; MIRACL - Multiprecision Integer and Rational Arithmetic C/C++ Library (2007)
37. Shamir, A.: Identity-based cryptosystems and signature schemes. In: Blakley, G.R., Chaum, D. (eds.) CRYPTO 1984. LNCS, vol. 196, pp. 47–53. Springer, Heidelberg (1985). doi:10.1007/3-540-39568-7_5
38. Stein, W., Joyner, D.; Sage: system for algebra and geometry experimentation. Commun. Comput. Algebra (SIGSAM Bull.) (2005). http://sage.sourceforge.net
39. Tessaro, S.: Security amplification for the cascade of arbitrarily weak PRPs: tight bounds via the interactive hardcore lemma. In: Ishai, Y. (ed.) TCC 2011. LNCS, vol. 6597, pp. 37–54. Springer, Heidelberg (2011). doi:10.1007/978-3-642-19571-6_3
40. The PARI Group: Bordeaux. PARI/GP version 2.7.0 (2014). http://pari.math.u-bordeaux.fr/
41. Whitten, A., Tygar, J.D.: Why Johnny can't encrypt: a usability evaluation of PGP 5.0. In: Usenix Security, vol. 1999 (1999)
42. Yao, D., Fazio, N., Dodis, Y., Lysyanskaya, A.: ID-based encryption for complex hierarchies with applications to forward security and broadcast encryption. In: Proceedings of the 11th ACM conference on Computer and communications security, pp. 354–363. ACM (2004)
43. Yuen, T.H., Susilo, W., Mu, Y.: How to construct identity-based signatures without the key escrow problem. Int. J. Inf. Secur. **9**(4), 297–311 (2010)

Multi-authority Distributed Attribute-Based Encryption with Application to Searchable Encryption on Lattices

Veronika Kuchta[✉] and Olivier Markowitch

Department of Computer Science, Université Libre de Bruxelles, Brussels, Belgium
{veronika.kuchta,olivier.markowitch}@ulb.ac.be

Abstract. Many Internet users deploy several cloud services for storing sensitive data. Cloud services provide the opportunity to perform cheap and efficient storage techniques. In order to guarantee secrecy of uploaded data, users need first to encrypt it before uploading it to the cloud servers. There are also certain services which allow user to perform search operations according to certain attributes without revealing any information about the encrypted content. In the cryptographic community this service is known as the public key encryption with keyword search. In order to enable user control during performed search operations there exists an attribute-based encryption scheme that provides the required functionality. We introduce the first Key-Policy Multi-Authority Attribute-Based Encryption (KP-MABE) on lattices assuming existence of multiple servers, where each of these servers contributes to the decryption process by computing decryption shares using its own secret share. Furthermore we construct a Key-Policy Distributed Attribute-Based Searchable Encryption (DABSE) which is based on lattices and use the introduced KP-MABE as a building block for the transformation to DABSE. We prove our scheme secure against chosen ciphertext attacks under the assumption that the underlying KP-MABE is secure under the hardness of learning with errors (LWE) problem.

1 Introduction

Searchable Encryption. Cloud computing allows users to use big data storage and computation capabilities at a very low price. Cloud security became an appealing research topic for recent cryptographic community. Storing data on a cloud enables users to reduce purchase and maintaining cost of computing and storage tools which attracted a lot of attention from computer users. When personal and confidential data is outsourced to the public cloud the customers require a guarantee that their data is securely stored and is not vulnerable against adversaries who are interested in getting any information of customer's private data. Therefore several cryptographic encryption schemes are gaining interest for distinct applications in cloud security. One of such encryption schemes is searchable public key encryption that has been introduced by Boneh et al. [9]. Many further systems supporting keyword search have been

© Springer International Publishing AG 2017
R.C.-W. Phan and M. Yung (Eds.): LNCS 10311, Mycrypt 2016, pp. 409–435, 2017.
DOI: 10.1007/978-3-319-61273-7_20

developed to enable users to perform search operations over encrypted data without leaking any information about encrypted content. We distinguish between symmetric encryption [14,17,19,39]) and public-key encryption (e.g., [9,10,20,25]) techniques. Searchable encryption applied to the cloud setting allows an user to upload encrypted data together with encrypted keywords of this data such that the user can later perform any search operations using special trapdoors for the required keywords. Apart from offering search techniques over encrypted data there is a motivation to guarantee access control for cloud users. This functionality is given by the attribute-based encryption schemes which is the subject of our next paragraph.

Attribute-Based Encryption. Sahai and Waters [36] introduced the first construction of an attribute-based encryption (ABE), which allows fine-grained access control of encrypted data. Their idea was to associate ciphertexts and private keys with sets of descriptive attributes such that the decryption is possible only if the overlap of these two sets is sufficient. There are two flavors of attribute-based encryption, key-policy ABE (KP-ABE) and ciphertext-policy ABE (CP-ABE). The former handles ciphertexts which are annotated with attributes and private keys which are associated with access structures that specify which ciphertexts can be chosen to be decrypted by an user. A ciphertext-policy ABE was first introduced by Bethencourt et al. [8] and by Cheung et al. [16], which handled AND gates only. Publication that analyzes the first expressive construction was presented by Goyal et al. [22] in the standard model. Further standard model CP-ABE constructions were provided by Waters [41] and Lewko et al. [26]. In CP-ABE attribute sets are assigned to private keys, where the sender specifies an access policy such that receiver's attribute set can comply with it. Attrapadung et al. [5] constructed the first ABE scheme with constant-size ciphertexts. They also designed a CP-ABE system for threshold policies, where ciphertext and receivers private key have at least t attributes in common and t is specified by the sender. The ciphertext in their scheme is constant and the private key size is linear in the number of attributes. Goyal et al. [23] generalized those techniques from [36] and introduced a new technique where user's key is associated with a tree-access structure and the leaves are associated with attributes. A user is only able to decrypt a ciphertext if attributes associated with ciphertext satisfy key's access structure. This technique differs from secret-sharing schemes by the fact that any communication between different parties is forbidden. An ABE scheme which allows a group of authorities to distribute attributes was developed by Chase [15]. This multi-authority construction allows to corrupt any number of attribute authorities but guarantees security of encryption as long as not all required attributes can be obtained from those corrupted authorities. Boyen [12] introduced the first ABE construction based on lattices. Lattice is a mathematical construct that represents an important part of post-quantum cryptography because of its assumed resistance against quantum attacks as described in the following paragraph.

Lattice-Based Cryptosystems. Cryptographic schemes based on lattices have especially attractive features as stated in [30]. The best attacks of lattice-based schemes require exponential time in security parameter, even for a quantum adversary, where the classic factoring-based cryptographic schemes can be broken in subexponential time or even in polynomial time using quantum algorithms. In contrast to the latter, lattice-based schemes are especially efficient and simple in their implementation. Lattices were introduced to cryptography by the work of Ajtai [3] and became a valuable tool for quantum-resistant cryptography. A deep research has been made in lattice-based cryptosystem [18,31,34]. Recently, many lattice-based applications have been provided, such as identity-based encryption schemes [2,13,18], public-key encryption [4,32,33], signatures [11,29], attribute-based encryption [12], public-key encryption with keyword search [24]. Miccianchio and Peikert [30] provided new methods for trapdoor generation which make the constructions simple, efficient and easy in the implementation. We are going to use this very efficient technique in the construction of our schemes.

Our Contribution and Application. Searchable attribute-based encryption has significant relevance for cloud systems and their security. Attribute-based encryption allows an additional option for many applications of searchable encryption in cloud computing. It enables a data owner to control the search function of outsourced cloud data. Our main contribution is the construction of a key-policy attribute based encryption for multiple authorities (KP-MABE). It is the first lattice-based construction of this type involving both a distributed server setting and multiple authorities. In our construction we refer to the technique of Boyen [12] which we amend to the multi-authority and multi-server setting. Furthermore we achieve higher efficiency of our scheme than that one in [12] using the more efficient lattice trapdoors introduced by Micciancio and Peikert [30]. In contrast to [12], our construction works in multi-authority setting where the role of one attribute authority is distributed among N parties. Additionally, we distribute the role of one decryption server among n servers in order to distribute the only point of failure during the decryption process. Finally, we provide security definitions of our KP-MABE construction against chosen ciphertext attack. Besides indistinguishability property, our scheme also achieves anonymity of users. Zheng [42] introduced the first attribute-based keyword search scheme which is based on bilinear maps and is defined in a single server setting that can also be applied to cloud systems. Their construction guarantees the user that the search operation has been executed correctly by the cloud server. This functionality is enabled by designing a search result verification algorithm which makes the whole search process verifiable. Sun et al. [38] presented another attribute-based searchable encryption which enables authenticity check over the search results considering multiple users and a multiple-data-contributor search scenario involving only one cloud server which executes the search operation. Further solutions for searchable attribute-based encryption schemes have been provided in [27,40] which are all defined in a single server setting and are secure under number-theoretic assumptions. Whereas there exists other lattice-based scheme, known as "Attribute-Based Encryption for Circuits" introduced by [21] which supports larger class of policies containing the ones captured by Boolean formula - and it also uses most efficient lattice-trapdoor generation algorithm introduced in [30]. In contrast to their construction our scheme enjoys further useful improvements such as employing multiple attribute authorities and using distributed setting in order to improve the security of the construction.

A useful application to personal health records in cloud computing based on multi-authorities and multi-users attribute-based encryption presented by Li et al. [28] can profit by providing the users with an additional function enabling authorized search of uploaded data which uses the search operation of searchable public key encryption. The shortcoming of Li et al. [28] construction is the single point of failure which is represented by the one cloud server that stores the encrypted data containing personal health records and performs the required decryption process. Applying our KP-MABE construction to the health record system we are able to solve two shortcomings of the recent scheme in [28]. Regarding this application, our construction involves a search function, such that each authorized user, who wants to download or read outsourced health records, can first conduct a search operation on stored data according to some special keywords. The other solution of aforementioned shortcomings is the distribution of the single server role among a certain number of servers. Encrypted health records are uploaded together with encrypted keywords to each of the cloud servers. Once the user wants to read or download his or her encrypted health records, he prepares trapdoor shares of keywords and sends them to the servers. Using access control function of attribute-based encryption the servers first check whether the user is allowed to read and/or download the data and upon successfully passed control each server executes the search function using the trapdoor shares and encrypted keywords. The user is able to download required data upon receiving a certain threshold of valid search results from cloud servers. As another possible application of our scheme is an investigation cloud system that would allow a closer and more effective collaboration between distinct intelligence agencies. For example, Interpol is highly interested in developing more efficient investigation services which can serve all member countries using "Interpol cloud system".

2   Preliminaries

Lattices. Let $B = \{b_1, \ldots, b_n\} \subset \mathbb{R}^n$ be a basis of a lattice L which consists of n linearly independent vectors. The n−dimensional lattice L is then defined as $L = \sum_{i=1}^{n} \mathbb{Z}b_i$. The i-th minimum of lattice Λ, denoted by $\lambda_i(L)$ is the smallest radius r such that L contains i linearly independent vectors of norms $\leq r$. (The norm of vector b_i is defined as $\|b_i\| = \sqrt{\sum_{j=1}^{n} c_{i,j}^2}$, where $c_{i,j}, j \in \{1, \ldots, n\}$

are coefficients of vector b_i. We denote by $\lambda_1^\infty(L)$ the minimum distance measured in the infinity norm, which is defined as $\|b_i\|_\infty := \max(|c_{i,1}|, \dots, |c_{i,n}|)$. Additionally we recall $\|B\| = \max\|b_i\|$ and its fundamental parallelepiped is given by $P(B) = \left\{ \sum_{i=1}^n a_i b_i \mid \mathbf{a} \in [0,1)^n \right\}$. Given a basis B for a lattice L and a vector $\mathbf{a} \in \mathbb{R}^n$ we define $\mathbf{a} \bmod L$ as the unique vector in $P(B)$ such that $\mathbf{a} - (\mathbf{a} \bmod L) \in L$. If L is a lattice, its dual L^* is the lattice $\left\{ \hat{\mathbf{b}} \in \mathbb{R}^n \mid \forall \mathbf{b} \in L, \langle \hat{\mathbf{b}}, \mathbf{b} \rangle \in \mathbb{Z} \right\}$.

Integer Lattices. The following specific lattices contain $q\mathbb{Z}^m$ as a sub-lattice for a prime q. For $A \in \mathbb{Z}_q^{n \times m}$ and $s \in \mathbb{Z}_q^n$, define:

$$\Lambda_q(A) := \{e \in \mathbb{Z}^m \mid \exists s \in \mathbb{Z}_q^n, \text{ where } A^T s = e \mod q\},$$
$$\Lambda_q^\perp(A) := \{e \in \mathbb{Z}^m \mid Ae = 0 \mod q\},$$

Many lattice-based works rely on Gaussian-like distributions called Discrete Gaussians. In the following paragraph we recall the main notations of this distribution.

Discrete Gaussians. Let L be a subset of $\mathbb{Z}^m$. For a vector $c \in \mathbb{R}^m$ and a positive $\sigma \in \mathbb{R}$, define

$$\rho_{\sigma,c}(x) = \exp\left(-\pi \frac{\|x - c\|^2}{\sigma^2}\right) \quad \text{and} \quad \rho_{\sigma,c}(L) = \sum_{x \in L} \rho_{\sigma,c}(x).$$

The discrete Gaussian distribution over L with center c and parameter σ is given by $\mathcal{D}_{L,\sigma,c}(y) = \frac{\rho_{\sigma,c}(y)}{\rho_{\sigma,c}(L)}$, $\forall y \in L$. The distribution $\mathcal{D}_{L,\sigma,c}$ is usually defined over the lattice $L = \Lambda_q^\perp(A)$ for $A \in \mathbb{Z}_q^{n \times m}$. Gentry et al. [18] defined an algorithm for sampling from the introduced discrete Gaussian $\mathcal{D}_{\Lambda,\sigma,c}$ for a given basis B of the n dimensional lattice Λ with a Gaussian parameter $\sigma \geq \|B\| \cdot \omega(\sqrt{\log m})$, where $\tilde{B}$ is the orthonormal basis of B, defined as follows:

The Gram-Schmidt Norm of a Basis [2]. Let B be a set of n vectors $B = \{b_1, \dots, b_n\} \in \mathbb{R}^n$ with the following functionalities:

- $\|B\|$ denotes the norm of the longest vector in B, i.e. $\|B\| := \max_i \|b_i\|$ for $1 \leq i \leq m$.
- $\tilde{B} := \{\tilde{b}_1, \dots, \tilde{b}_m\} \subset \mathbb{R}^m$ is the Gram-Schmidt orthogonalization of the vectors $b_1, \dots, b_m$.

The Gram-Schmidt norm is denoted by $\left\|\tilde{B}\right\|$.

Ajtai [3] showed how to sample an uniform matrix $A \in \mathbb{Z}_q^{n \times m}$ with a basis B_A of $\Lambda_q^\perp(A)$ with low Gram-Schmidt norm.

Learning With Errors (LWE). The LWE problem, first introduced by Regev [33], relies on the Gaussian error distribution χ, which is given as $\chi = \mathcal{D}_{\mathbb{Z},s}$ over

414   V. Kuchta and O. Markowitch

integers. The LWE problem instance consists of access to a challenge oracle $\mathcal{O}$, which is either a purely random sampler $\mathcal{O}_r$ or a noisy pseudo-random sampler $\mathcal{O}_s$, with some random secret key $s \in \mathbb{Z}_q^s$. For positive integers n and $q \geq 2$, a vector $s \in \mathbb{Z}_q^n$ and error term $e \leftarrow \chi$, the LWE distribution $A_{s,\chi}$ is sampled over $\mathbb{Z}_q^n \times \mathbb{Z}_q$. Chosen a vector $a \in \mathbb{Z}_q^n$ uniformly at random it outputs the pair $(a, t = \langle a, s \rangle + e \mod q) \in \mathbb{Z}_q^n \times \mathbb{Z}_q$. A more detailed description of χ can be found in [33]. The sampling oracles work in the following way:

$\mathcal{O}_s$: outputs samples of the form $(a, t) = (a, as + e) \in \mathbb{Z}_q^n \times \mathbb{Z}_q$, where $s \in \mathbb{Z}_q^n$ is uniformly distributed value across all invocations and $e \in \mathbb{Z}_q$ is a fresh sample from χ.

$\mathcal{O}_r$: outputs truly random samples from $\mathbb{Z}_q^n \times \mathbb{Z}_q$.

Definition 1 (LWE-Problem). *For an integer q and error distribution χ, the goal of $LWE_{q,\chi}$ in n dimensions is to find $s \in \mathbb{Z}_q^n$ with overwhelming probability, given access to any arbitrary $poly(n)$ number of samples from $A_{s,\chi}$ for random s.*

Trapdoor Generation. Micciancio and Peikert [30] defined the new notion or trapdoor and the new method of trapdoor generation, which is simple, efficient and easy to implement. The standard notion of a strong trapdoor for strong lattices $\Lambda^\perp(A)$ is a short basis $S \in \mathbb{Z}^{m \times m}$, where the lattice is defined by a uniformly random matrix $A \in \mathbb{Z}_q^{n \times m}$. The main results of [30] are formulated in the following theorem:

Definition 2 (TrapGen Algorithm:). *Efficient trapdoor generation algorithm TrapGen proposed in [30] uses a matrix $G \in \mathbb{Z}_q^{n \times w}$ that admits efficient inversion and preimage sampling algorithms. We recall this algorithm in the following algorithm:*

Input: *A matrix $\bar{A} \in \mathbb{Z}_q^{n \times \bar{m}}$ for some $\bar{m} \geq 1$, invertible matrix $H \in \mathbb{Z}_q^{n \times n}$, and a distribution $\mathcal{D}$ over $\mathbb{Z}_q^{\bar{m} \times w}$. (If no particular $\bar{A}, H$ are given as input, then the algorithm may choose them itself, i.e. $\bar{A} \xleftarrow{r} \mathbb{Z}_q^{n \times \bar{m}}$ and $H = I$.*
Output: *A parity-check matrix $A = [\bar{A}|A_1] \in \mathbb{Z}_q^{n \times m}$, $m = \bar{m} + w$, trapdoor R with tag H.*
1. *Choose a matrix $R \in \mathbb{Z}_q^{\bar{m} \times w}$ from distribution $\mathcal{D}$.*
2. *Output $A = [\bar{A}|HG - \bar{A}R] \in \mathbb{Z}_q^{n \times m}$ and trapdoor $R \in \mathbb{Z}_q^{\bar{m} \times w}$.*

Definition 3 (Invert Algorithm:). *For our construction we need to learn how to use the trapdoor from above to solve LWE problem relative to A. As before, we recall the inversion algorithm [30] Invert, which on a given trapdoor R for $A \in \mathbb{Z}_q^{n \times m}$ and an LWE instance $b^t = s^t A + e^t \mod q$ for a short error vector $e \in \mathbb{Z}^m$, recovers s and e.*

Input: *An oracle $\mathcal{O}$ for inverting the function $g_G(\hat{s}, \hat{s})$ when $\hat{e} \in \mathbb{Z}^w$ is suitably small.*
1. *parity-check matrix $A \in \mathbb{Z}_q^{n \times m}$,*
2. *G-trapdoor $R \in \mathbb{Z}_q^{\bar{m} \times kn}$ for A with invertible tag $H \in \mathbb{Z}_q^{n \times n}$,*
3. *$b^t = g_A(s, e) = s^t A + e^t$ for any $s \in \mathbb{Z}_q^n$ and suitably small $e \in \mathbb{Z}^m$.*

Output: *The vectors s and e.*

1. *Compute $\hat{\boldsymbol{b}}^t = \boldsymbol{b}^t \begin{bmatrix} R \\ I \end{bmatrix}$.*

2. *Get $(\hat{s}, \hat{s}) \leftarrow \mathcal{O}(\hat{\boldsymbol{b}})$.*
3. *Return $s = H^{-t}\hat{s}$ and $e = b - A^t s$.*

In the following Definition 4 we provide a description of access policy.

Definition 4 (Access Structure [6]). *Let $\mathbb{P} = \{P_1, \ldots, P_n\}$ be a set of parties. A collection $\mathbb{A} \subseteq 2^{\mathbb{P}}$ is monotone if for all ω, ω' holds: if $\omega \in \mathbb{A}$ and $\omega' \subseteq \omega$, then $\omega \in \mathbb{A}$. An access structure (resp., monotone access structure) is a collection (resp. monotone collection) $\mathbb{A} \subseteq 2^{\mathbb{P}} - \{\emptyset\}$. The sets in $\mathbb{A}$ are called the authorized sets and the sets not in $\mathbb{A}$ are called the unauthorized sets.*

We assume that the mentioned parties $\{P_1, \ldots, P_n\}$ play the role of attributes and the access structure $\mathbb{A}$ contains authorized sets of attributes. In the following Definition 5 we recall the threshold version of access structure and linear secret sharing scheme (LSSS). In LSSS there is a trusted party, who knows a secret x and distributes it among n parties such that the secret can be reconstructed using linear combination of the shares and Lagrange coefficients.

Definition 5 (Linear Secret Sharing Scheme). *A linear secret sharing scheme Π over a set of parties $\mathcal{P}$ consists of an index map ρ and a share-generating matrix $L \in \mathbb{Z}_q^{l \times \theta}$ where l is the number of shares specified by the scheme, and θ depends on the structure of the scheme. For all $i \in 1, \ldots, n$, the function ρ maps the $i-th$ row of L to the corresponding party. For the matrix L maps an input θ-vector $\boldsymbol{v} = (s, r_1, \ldots, r_\theta)$, where $s \in \mathbb{Z}_q$ is the secret to be shared and $r_2, \ldots, r_\theta \in \mathbb{Z}_q$ are random, into an output l-vector $L\boldsymbol{v} = (s_1, \ldots, s_l)$ containing the shares of the secret s according to Π. The share $s_i = (L\boldsymbol{v})_i$ is assigned to party $\rho(i)$.*

3   Attribute-Based Encryption for Multiple Authorities Based on Lattices

In this section we present the first construction of a key-policy attribute-based encryption which employs multiple authorities (KP-MABE) and is defined on the aforementioned mathematical construct - lattices. The lattice-based construction of our scheme enjoys especial efficiency and is even assumed to be secure against quantum attacks. While the secret keys of our scheme are associated with access policies, the ciphertext is associated with attributes. As aforementioned in the introduction of this paper, we assume for our scheme existence of N attribute authorities and n decryption servers. We amend the technique by Boyen [12] to let the muliple attribute authorities collectively generate secret key for an user. Furthermore we modify Boyen's construction using the new trapdoors from [30] making our scheme more effective. The technique in [12] relies on ephemeral lattices for all secret key extractions, which means that the lattice is high-dimensional and allows encoding of key-policy attributes. In the following paragraph we the main idea and syntax of our construction.

The Scheme (Intuition). A key-policy MABE scheme consists of the following algorithms ABE.Setup, ABE.Extract, ABE.KeyDistr, ABE.Encrypt, ABE.ShareDec, ABE.Decrypt. We begin with the ABE.Setup algorithm which generates public key and master secret key for each attribute authority $i \in [N]$, such that the public key consists of a random Ajtai matrix $\bar{A}_i$ and a random vector u_i, where the secret key is associated with the corresponding trapdoor R_i (we define the dimensions of these keys in the detailed construction of the scheme, Sect. 3.1). The ABE.Extract algorithm generates for each of N attribute authorities a private key $\{sk_{A_i}\}_{i \in [N]}$ and a public key $\{pk_{A_i}\}_{i \in [N]}$ amending the technique from [12] and using trapdoor generation technique introduced by Micciancio and Peikert [30]. Upon applying secret sharing to these authority's secret keys, an user runs ABE.KeyDistr algorithm in order to generate secret shares $\{sk_{A_i,S_\iota}\}_{i \in [N], \iota \in [n]}$ corresponding to each attribute authority and each server. It means that this algorithm computes in total $N \times n$ secret shares, where each secret share key tuple is ordered by $\iota \in [n]$, i.e. $(sk_{A_1,S_1}, sk_{A_1,S_2}, \ldots, sk_{A_1,S_n})$ and corresponding to authority one, and so on until N authorities. The ABE.Encrypt algorithm computes a ciphertext of a message for a given set of attributes A_i. The user sends generated secret shares to the decryption servers, where each of these servers runs the ABE.ShareDec algorithm taking as input previously generated ciphertext in order to generate N decryption shares in total, using those secret shares received from user. Afterwards, upon receiving these decryption shares from the decryption server, an user runs the ABE.Decrypt algorithm taking as input at least t valid and previously computed decryption shares, a public key and finally decrypts the ciphertext. The above mentioned secret shares of each attribute authority are computed considering decryption policy which is represented by an LSSS and referring to technique in [12] it considers the given attributes of each user. We assert that there is a problem if different users collude and use their secret shares from different authorities in order to reconstruct the decryption keys. To prevent this collusion we use the idea from [15] introducing a global identifier GID_κ for each user κ, such that a user needs to present the same identifier to each attribute authority in order to receive the required number of authorities' secret shares. The shortcoming of this technique is that it hinders the users to keep their anonymity, which wouldn't make the scheme applicable to searchable encryption. Therefore, in order to enable the application to searchable encryption, we make an especially strong assumption that there is no collusion of users possible. We emphasize that it is still an open question to achieve all of the highly required properties such like avoiding collusion of users and achieving anonymity of the construction.

Definition 6 (Multi-authority Distributed Attribute-Based Encryption). *The multi-authority distributed attribute-based encryption consists of the following algorithms:*

ABE.Setup$(1^\lambda, N, n, t)$: *On input security parameter* 1^λ, *number of authorities,* N, *number of servers* n *and threshold parameters* t *it computes master public key* mpk_i *and master secret key* msk_i *for each authority* $i \in [N]$.

ABE.Extract$(mpk_i, msk_i, policy)$: *On input authority* i's *master keys* mpk_i, msk_i, *user's policy, each authority* $\mathcal{A}_i$ *computes a policy embed secret key* $sk_{\mathcal{A}_i}$. **ABE.KeyDistr**$(mpk_i, sk_{\mathcal{A}_i}, t, n)$: *On input authorities master public key* mpk, *the policy embed secret key* $sk_{\mathcal{A}_i}$, *the number of servers* n *and threshold parameter* t *it outputs* n *secret shares* $sk_{\mathcal{A}_i, S_\iota}$, *where* t-*out-of-*n *secret shares are required to reconstruct* $sk_{\mathcal{A}_i}$, *for* $\iota \in [n]$.

ABE.Encrypt$(\{mpk_i\}_{i \in [N]}, \{\mathbb{A}_i\}_{i \in [N]}, m)$: *On input each authority's master public key* $\{mpk_i\}_{i \in [N]}$, *attribute sets* $\{\mathbb{A}_{i \in [N]}\}$ *and a message* m *it computes a ciphertext* C_i.

ABE.ShareDec$(\{mpk_i\}_{i \in [N]}, sk_{\mathcal{A}_i, S_\iota}, policy, C_i)$: *On input each authority's master public key* $\{mpk_i\}_{i \in [N]}$, *secret share according to each attribute authority* $sk_{\mathcal{A}_i, S_\iota}$, *known policy, and ciphertext* C_i *it computes decryption shares* $\delta_{i,\iota}$.

ABE.Decrypt$(\{mpk_i\}_{i \in [N]}, \{\delta_{i,\iota}\}_{\iota \in \Omega}, \{C_i\}_{i \in [N]})$: *On input* $\{mpk_i\}_i$, *at least* t *valid decryption shares* $\{\delta_i\}_{i \in \Omega}$, *and ciphertext* $\{C_i\}_{i \in [N]}$, *where* $|\Omega| \geq t$ *is a set of indices, it decrypts and outputs* m *or* 0.

We prove our scheme secure against chosen-ciphertext attacks in chosen attribute model. It means that an adversary has to provide the index of an authority it wishes to attack and a target attribute set it wants to be challenged on. Unlike the already existing ABE schemes, we provide an additional property for our construction, which addresses the user's identity privacy. The latter becomes one of the main concerns in the cloud security nowadays. The users want the authorities not to learn anything about those identities, in other words they want to stay anonymous. In our scheme we approach anonymity during the secret key generation protecting the attribute set in encryption. This is a novelty which together with the multy-authority and distributed server setting makes our scheme advantageous in contrast to the already existing attribute-based encryptions.

Rouselakis [35] handled in his thesis the security definition of a ciphertext-policy multi-authority attribute-based encryption. We provide a definition of a key-policy multi-authority ABE with an extension to a multi-server setting. We assume an adversary who is able to corrupt up to $t - 1$ attribute authorities and additionally up to $t - 1$ decryption servers.

Remark 1. We note that our scheme requires decryption server which is not required by previous multi-authority ABE schemes [35]. The purpose of decryption server is to enable a transformation from multi-authority ABE scheme to the searchable encryption where the decryption server takes the role of a test server, as will be showed later in Definition 10.

Definition 7 (Indistinguishability and Anonymity Against CCA).
The KP-MABE scheme is indistinguishable and anonymous against chosen-ciphertext attacks (IND/ANO − MABE − CCA) *if all probabilistic polynomial time*

adversaries have at most a negligible advantage probability in winning the following game:

Init: *The adversary chooses a set J of $t-1$ decryption servers and a set I of $t-1$ attribute authorities that it wants to corrupt. Let $J = \{\iota_1, \ldots, \iota_{t-1}\} \subset \{1, \ldots, n\}$, $I = \{i_1, \ldots, i_{t-1}\} \subset \{1, \ldots, N\}$.*

Setup: *An adversary $\mathcal{A}_{ind/ano}$ generates a list of attribute sets for each authority AA_i, where $i \in [N]$ the index of N authorities. Additionally, he provides a list of corrupted authorities and servers. The challenger computes public key of each authority and secret keys of corrupted authorities and sends them to $\mathcal{A}_{ind/ano}$.*

Secret Key Queries: *On input $(policy_\kappa, \cdot)$ the adversary $\mathcal{A}_{ind/ano}$ can issue as many secret key queries as he wants to the attribute authorities. We assume that there is at least one honest attribute authority, such that the adversary cannot recover all secret keys and combine them into one, which he can use during the decryption process. Furthermore we assume, that the secret key queries corresponding to certain $policy_j$ do not satisfy the challenge $policy_\kappa^*$ of user κ. We also restrict the adversary to query the same authority twice.*

Secret Key Share Queries: *$\mathcal{A}_{ind/ano}$ issues secret key share queries on input $(policy_\kappa, \iota)$ according to chosen policies for $\iota \in [n], i \in [N]$ and κ denotes the index of an user. We note that $policy_\kappa$ has not been input of secret key query, such that the adversary cannot recover the missing secret share using the previously received secret key. That means that an adversary issues queries for an user κ and chosen $policy_\kappa$. We assume existence of a list K which consists of the entries $(policy_\kappa, \{sk_{A_i, S_\iota}\}_{i \in [N], \iota \in [n]})$. The oracle checks if $(policy_\kappa, \cdot) \in K$, if so it returns the corresponding secret share tuple $\{sk_{A_i, S_\iota}\}_{i \in [N], \iota \in [n]}$ to the adversary. Otherwise it invokes the ABE.Extract algorithm to compute a secret key sk_{A_i} and on this input it invokes the ABE.KeyDistr algorithm to generate secret shares of the given secret key. Finally it outputs $\{sk_{A_i, S_\iota}\}_{i \in [N], \iota \in [n]}$ to adversary. We allow up to $t-1$ secret key queries on the same global identity, i.e. the oracle can return secret share tuple $\{sk_{A_i, S_\iota}\}_{\iota \in [N], \iota \in [n]}$ for at most $t-1$ attribute authorities $i \in [N]$. If the number of queries exceeds $t-1$, the oracle returns $\perp$.*

Decryption Queries: *The adversary is allowed to issue queries for decryption for certain policies which do not satisfy the target list of attributes. On input $(C, \mathbb{A}_k)$ the decryption oracle checks if $(policy_\kappa, \cdot) \in K$. If so, the oracle takes t-out-of-N secret share tuples $\{sk_{A_i, S_\iota}\}_{\iota \in [n]}$ for $i \in [N]$, combines them, i.e. $(sk_{A_1, S_\iota}, \ldots, sk_{A_t, S_\iota})$ for each $\iota \in [n]$ in order to compute secret shares for the decryption servers. The oracle takes these keys to compute decryption shares upon running the ABE.ShareDec algorithm. Finally it runs the ABE.Decrypt algorithm and outputs a message m. Otherwise if $(policy_\kappa, \cdot) \in K$ it invokes the ABE.Extract algorithm to compute a secret key sk_{A_i} and on this input it invokes the ABE.KeyDistr algorithm to generate secret shares of the given secret key. Taking the values $\{sk_{A_i, S_\iota}\}_{\iota \in [n]}$ as input it runs the ABE.ShareDec algorithm to compute the decryption shares. Upon running the ABE.Decrypt algorithm it returns either a message m or $\perp$ to the adversary.*

Challenge: *The adversary outputs two messages m_0, m_1 and two attribute universes $\{\mathbb{A}_i^{(0)}\}_{i \in [N]}$ and $\{\mathbb{A}_i^{(1)}\}_{i \in [N]}$. The challenger chooses random bits b, b' and encrypts m_b for an attribute set $\{\mathbb{A}_i^{(b')}\}_{i \in [N]}$ and sends the encryption to $\mathcal{A}_{ind/ano}$.*

Queries: *The adversary continues to issue more queries in the same manner as earlier.*

Guess: *$\mathcal{A}_{ind/ano}$ outputs a guess β, β' that a message m_β has been encrypted by challenger using $\{\mathbb{A}_i^{(\beta')}\}_{i \in [N]}$.*

The advantage of $\mathcal{A}_{ind/ano}$ is given by
Adv$_{MABE, \mathcal{A}_{ind/ano}}(1^\lambda) = |Pr[\beta = b \wedge \beta' = b'] - 1/4|$.
A KP-MABE scheme is secure against key policy attacks if this advantage is negligible.

In the following definition we formulate another important property of our scheme, called robustness, which prevents to decrypt to a valid ciphertext under different decryption keys for two different access policies. The property was initially defined by Abdalla et al. [1] and discussed in later work [35].

Definition 8 (Robustness). *Let $\mathcal{A}_{rob}$ be a probabilistic polynomial-time adversary against the ROB–CCA security of the KP-MABE scheme.*

Init: *The adversary chooses a set J of $t-1$ decryption servers and a set I of $t-1$ attribute authorities that it wants to corrupt. Let $J = \{\iota_1, \ldots, \iota_{t-1}\} \subset \{1, \ldots, n\}$, $I = \{i_1, \ldots, i_{t-1}\} \subset \{1, \ldots, N\}$. It outputs a target policy*.*

Setup: *The challenger runs $\mathtt{Setup}(1^\lambda, 1^l, N, n, t)$, and outputs authorities master public key mpk_i and a master secret key msk_i.*

Secret Key Queries: *On input $(policy'_\kappa, \cdot)$ the adversary $\mathcal{A}_{rob}$ can issue as many secret key queries as he wants to the attribute authorities. We assume that there is at least one honest attribute authority, such that the adversary cannot recover all secret keys and combine them into one, which he can use during the decryption process. Furthermore we assume, that the secret key queries corresponding to certain $policy_\kappa$ do not satisfy another $policy'_\kappa$. We also restrict the adversary to query the same authority twice.*

Secret Share Queries: *On input $(policy_\kappa, \iota)$. We note that $policy_\kappa$ has not been the input of secret key query. We assume existence of a list K consisting of the entries $(policy_\kappa, \{sk_{A_i, S_\iota}\}_{i \in [N], \iota \in [n]})$. The oracle checks if $(policy_\kappa, \cdot) \in K$, if so it returns the corresponding secret share tuple $\{sk_{A_i, S_\iota}\}_{\iota \in [n]}$ to the adversary.*
Otherwise it invokes the $\mathtt{ABE.Extract}$ algorithm and outputs $\{sk_{A_i, S_\iota}\}_{\iota \in [n]}$ to $\mathcal{A}_{rob}$. Let I and J denote the lists of queried authorities and servers. Initially the lists K, I, J are empty.

Decryption Queries: *On input $(C, \{\mathbb{A}_{\kappa, i}\}_{i \in [N]})$, the decryption oracle checks if $(policy_\kappa, \cdot) \in K$. If so, the oracle takes t-out-of-N secret share tuples $\{sk_{A_i, S_\iota}\}_{\iota \in [n]}$ for $i \in [N]$, combines them, i.e. $(sk_{A_1, S_\iota}, \ldots, sk_{A_t, S_\iota})$ for each $\iota \in [n]$ in order to compute secret shares for the decryption servers. The oracle* *takes these keys to compute decryption shares upon running the* ABE.ShareDec *algorithm. Finally it runs the* ABE.Decrypt *algorithm and outputs a message m.*

Otherwise if $(policy_\kappa, \cdot) \in K$ *it invokes the* ABE.Extract *algorithm to compute a secret key* $sk_{\mathcal{A}_i}$ *and on this input it invokes the* ABE.KeyDistr *algorithm to generate secret shares of the given secret key. Taking the values* $\{sk_{\mathcal{A}_i, S_\iota}\}_{\iota \in [n]}$ *as input it runs the* ABE.ShareDec *algorithm to compute the decryption shares. Upon running the* ABE.Decrypt *algorithm it returns either* m *or* $\perp$ *to adversary.*

Challenge: *The adversary* $\mathcal{A}_{rob}$ *outputs* $(policy_0, policy_1, C^*)$. *The challenger takes* $\{mpk_i\}_{i \in [N]}$ *and computes* $C \leftarrow$ Encrypt$(\{mpk_i\}_{i \in [N]}, \{\mathbb{A}_i\}_{i \in [N]}, m)$. $\mathcal{A}_{rob}$ *outputs* $policy_0, policy_1$ *and a ciphertext* C^* *on challenge policy**.

Output:

(i) *If* $policy_0 = policy_1$, *challenger returns* 0.

(ii) *If* $(policy_0, \cdot) \notin K$ *or* $(policy_1, \cdot) \notin K$, *it returns* 0.

(iii) *If* $|I_0| \geq t$ *or* $|I_1| \geq t$ *and* $|J_0| \geq t$ *or* $|J_0| \geq t$, *return* 0. *Else compute decryption shares* $\delta_{0,\iota} \xleftarrow{r}$ ABE.ShareDec$(\{mpk_i\}_{i \in [N]},$ $sk^{(0)}_{\mathcal{A}_i, \iota}, C)$, *and decrypt* $m_0 \xleftarrow{r}$ ABE.Decrypt$(\{mpk_i\}_{i \in [N]}, \{\delta_{0,\iota}\}_{i \in \Omega})$; $\delta_{1,\iota} SDe(\{mpk_i\}_{i \in [N]}, (\iota, sk^{(1)}_{\mathcal{A}_i, S_\iota}), C)$, $m_1 \xleftarrow{r}$ ABE.Decrypt$(\{mpk_i\}_{i \in [N]},$ $\{\delta_{1,\iota}\}_{\iota \in \Omega})$.

If $m_0 \neq \perp$ *and* $m_1 \neq \perp$ *returns* 1.

An KP-MABE scheme is ROB–CCA *secure if* $\mathbf{Adv}^{\text{ROB–CCA}}_{KP-MABE, \mathcal{A}_{rob}}(1^\lambda)$ *is negligible.*

3.1   Construction

ABE.Setup$(1^\lambda, 1^l, N, n, t)$: On input a security parameter 1^λ and attribute bound 1^λ, number of attribute authorities N and the number of servers n with the corresponding threshold t, the algorithm generates master public key and master secret key for each of the N attribute authorities AA. Each attribute authority is in possession of an attribute universe $\{\mathbb{A}_i\}_{i \in [N]}$, where $\mathbb{A}_i = \{a_1, \ldots, a_{l_i}\}$. Considering a distribution $\mathcal{D}$ over $\mathbb{Z}_q^{\bar{m} \times w}$ chooses a matrix for each attribute authority i: $\bar{A}_i \in \mathbb{Z}_q^{\nu \times \bar{m}}$, where $m = \bar{m} + w$. For each i choose ℓ uniformly random $\boldsymbol{u}_{i,j} \in \mathbb{Z}_q^\nu$, where $j \in [\ell]$ and set $mpk_i = (\{A_{i,j}\}_{j \in [\ell]}, \bar{A}_i, \boldsymbol{u}_{i,j})$ where $A_{i,j} = [\bar{A}_i | - \bar{A}_i R_{i,j}] \in \mathbb{Z}_q^{\nu \times m}$, with corresponding master secret key $msk_i = \{R_{i,j}\}_{j \in [l]} \in \mathbb{Z}_q^{\bar{m} \times w}$ which is generated according to the TrapGen algorithm from Definition 2.

ABE.Extract$(mpk_i, msk_i, policy_\kappa)$: On input a master public key $mpk_i = (\{A_{i,j}\}, \{\bar{A}_i\}, \boldsymbol{u}_{i,j})$, master secret key msk_i of authority i, and the corresponding policy $policy_\kappa$ of user κ it generates in the first stage the secret key share for each attribute authority. In the second stage it applies secret sharing technique.

1. According to an access policy of an user k it generates a matrix $L \in \mathbb{Z}_q^{\ell \times (1+\theta)}$, where κ–th row is assigned to the binary attribute of index $\kappa \in [\ell]$. The columns $\mu \in [0, \theta]$ describe a function of $policy_\kappa$, where $\theta \leq \ell$ and κ denote the user.

2. For each authority $i \in [N]$, select θ ephemeral uniform random matrices $Z_{i,\mu} \in \mathbb{Z}_q^{\nu \times w}$.

   Let $L = (l_{rs})_{r \in [\ell], s \in [1+\theta]}$. The attribute list is given in binary mode, i.e. $l_{rs} \in \{0,1\}^\ell$. It holds that the access policy is satisfied if the rows of matrix L contain in their span the row-vector $(1, 0, \ldots, 0) \in \mathbb{Z}_q^{1+\theta}$. In order to simplify the notations of the following computations we set for simplification $\xi := (m(\ell + 1) + w(\theta - 1))$. Take the master public key mpk_i and construct a virtual encryption matrix:

$$M_i = \begin{bmatrix} A_{i,1} & & & \begin{array}{c|cccc} l_{1,0}\bar{A}_i & l_{1,1}Z_{i,1} & \cdots & l_{1,\theta}Z_{i,\theta} \\ l_{2,0}\bar{A}_i & l_{2,1}Z_{i,1} & \cdots & l_{2,\theta}Z_{i,\theta} \\ \vdots & \vdots & & \vdots \\ \end{array} \\ & A_{i,2} & & \\ & & \ddots & \\ & & & A_{i,l} \quad l_{\ell,0}\bar{A}_i \quad l_{\ell,1}Z_{i,1} \quad \cdots \quad l_{\ell,\theta}Z_{i,\theta} \end{bmatrix} \mod q \in \mathbb{Z}_q^{\nu\ell \times \xi},$$

3. Using the new trapdoor generation technique from [7,30] and Definition 2. the well known mathematical constructs - tensor product and direct sum (see Definitions 11 and 12) - we take as input $M_i \in \mathbb{Z}_q^{\nu l \times \xi}$, which can be written as

$$M_i = \left[\bigoplus_{j=1}^{\ell} A_j | l_0 \otimes \bar{A} | l_1 \otimes Z_1 | \ldots | l_\theta \otimes Z_\theta | \right], \text{ where } l_j = (l_{1,j}, \ldots, l_{\ell,j}).$$

According to the new trapdoor generation technique in [30] we modify M_i by subtracting $\bar{A}R_j$ from each l_jZ_j matrix and receive $\tilde{M}_i \in \mathbb{Z}_q^{\nu l \times \xi}$ i.e.

$$\tilde{M}_i = \begin{bmatrix} A_{i,1} & & & \begin{array}{c|cccc} l_{1,0}\bar{A}_i & l_{1,1}Z_{i,1} - \bar{A}_iR_{i,1} & \cdots & l_{1,\theta}Z_{i,\theta} - \bar{A}_iR_{i,1} \\ l_{2,0}\bar{A}_i & l_{2,1}Z_{i,1} - \bar{A}_iR_{i,2} & \cdots & l_{2,\theta}Z_{i,\theta} - \bar{A}_iR_{i,2} \\ \vdots & \vdots & & \vdots \\ \end{array} \\ & A_{i,2} & & \\ & & \ddots & \\ & & & A_{i,l} \quad l_{\ell,0}\bar{A}_i \quad l_{\ell,1}Z_{i,1} - \bar{A}_iR_{i,\ell} \quad \cdots \quad l_{\ell,\theta}Z_{i,\theta} - \bar{A}_iR_{i,\ell} \end{bmatrix},$$

where all the entries belong to $\mathbb{Z}_q$. If we set $\tilde{Z}_{i,j,\mu} = l_jZ_{i,\mu} - \bar{A}_iR_{i,j}$ for $\mu \in \{0, \theta\}$, we obtain:

$$\tilde{M}_i = \left[\bigoplus_{j=1}^{l} A_{i,j} | l_0 \otimes \bar{A}_i | \tilde{Z}_{i,j,1} | \ldots | \tilde{Z}_{i,j,\theta} | \right]$$

The intermediate secret key generated by attribute authority i is given by $sk_{A_i} = f_{\tilde{M}_i}^{-1}(\hat{u}_i) = x_i \in \mathbb{Z}_q^\xi$, for an $f : \tilde{M}_i x_i = \hat{u}_i$ and f^{-1} the corresponding inverse, i.e. $f^{-1} : \tilde{M}_i^{-1}x_i = \hat{u}_i$, where $\hat{u}_i = \bigoplus_{j=1}^{l} u_{i,j} = (u_1, \ldots, u_l)_i^t$ is the vector consisting of l vectors $u_{i,j}$. We note that $(\cdot)^t$ is a transpose of a column vector $(\cdot)$ transforming a column to a row.

ABE.KeyDistr$(mpk_i, sk_{\mathcal{A}_i}, t, n)$: On input authority's master public key mpk_i, authority's secret key $sk_{\mathcal{A}_i} = \boldsymbol{x}_i$ and threshold parameters t, n it distributes the secret key using secret sharing technique from Shamir [37] among n servers for additive groups as follows: Choose a set of values $U = \{r_0 = 0, r_1, \dots, r_\xi\} \in \mathbb{Z}_q$, such that $r_i - r_j$ is invertible in $\mathbb{Z}_q$ for every $i \neq j$. Make U public. Let $\boldsymbol{x} = (x_1, \dots, x_\xi)$. Then choose ξ formal t-degree polynomials $h_k(z) = \sum_{j=0}^{t} h_{j,k} z^j$, $k \in \{1, \dots, \xi\}$ and $h_{0,k} = x_k$, are the ξ secret components of vectors $\boldsymbol{x}_i$ which was recently generated by an attribute authority $i \in [N]$. The chosen polynomials are uniformly random and independent. Server ι is publicly associated with value $r_{i,\iota} \in \mathbb{Z}_q$, for $\iota \in [n]$ and n is the number of servers. The corresponding secret share is $sk_{\mathcal{A}_i, S_\iota} = \boldsymbol{x}_{i,\iota} = \boldsymbol{h}_k(r_{i,\iota})$, where $\boldsymbol{h}_k(r_{i,\iota}) = (h_1(r_{i,\iota}), \dots, h_\xi(r_{i,\iota}))$, where $\iota \in [n], i \in [N]$ and S_ι denotes the server with index ι.

ABE.Encrypt $(\{mpk_i\}_{i \in [N]}, \{\mathbb{A}_i\}_{i \in [N]}, m)$: On input master public key $mpk_i = (\{A_{i,j}\}_{i \in [N], j \in [\ell]}, \bar{A}, \{\boldsymbol{u}_{i,j}\}_{j \in [\ell]})$, an attribute list $\{\mathbb{A}_i\}_{i \in [N]}$ and a message $m \in \{0, 1\}$, where $i \in [N], j \in [\ell]$ the algorithm performs the following computations:

1. Construct an encryption matrix as follows: It is a direct sum of $\{A_{i,j}\}_{j \in [\ell]}$ if the $j \in \mathbb{A}_i$, i.e. an attribute in the the attribute list of authority i. Otherwise it is 0 if $j \notin \mathbb{A}_i$.

$$
F_i = \begin{bmatrix} A_{i,1} & & & & \bar{A}_i & 0 \\ & A_{i,2} & & & \bar{A}_i & 0 \\ & & \ddots & & \vdots & \vdots \\ & & & A_{i,l} & \bar{A}_i & 0 \end{bmatrix} \quad \bmod q \in \mathbb{Z}_q^{\nu\ell \times \xi}.
$$

2. Select a random vector $\boldsymbol{s} \in \mathbb{Z}_q^{\nu\ell}$.
3. Select Gaussian noise vectors $\epsilon_0 \in \mathbb{Z}_q$ and $\epsilon_{1,i} \in \mathbb{Z}_q^\xi$ to some parametric distribution as described in [12]. For each $i \in [N]$ compute ciphertext: $c_{0,i} = \boldsymbol{s}^t \hat{\boldsymbol{u}}_i + \epsilon_0 + \lfloor \frac{q}{2} m \rfloor \mod q$, $\boldsymbol{c}_{1,i} = (\boldsymbol{s}^t F_i + \epsilon_{1,i}) \mod q$, where $\hat{\boldsymbol{u}}_i$ is a ℓ-multiple concatenation of $\boldsymbol{u}_i$. Output $C_i = (c_{0,i}, \boldsymbol{c}_{1,i}) \in \mathbb{Z}_q \times \mathbb{Z}_q^\xi$.

ABE.ShareDec$(\{mpk_i\}_{i \in [N]}, sk_{\mathcal{A}_i, S_\iota}, policy_\kappa, \{C_i\}_{i \in \Omega})$: On input public and secret keys $mpk_i, sk_{\mathcal{A}_i, S_\iota}, \hat{x}$, t-out-of-N ciphertexts C_i, and known $policy_\kappa$ of a user with index κ it computes: Generate a decryption share by using secret shares $sk_{\mathcal{A}_i, S_\iota}$, namely compute $c_{1,i,\iota} = \boldsymbol{c}_{1,i} \cdot \boldsymbol{x}_{i,\iota} = \boldsymbol{s}^t F_i \boldsymbol{x}_{i,\iota} + \epsilon_{1,i} \boldsymbol{x}_{i,\iota}$ and keep $c_{0,i}$ as it is. Output decryption share $\delta_{i,\iota} = (c_{0,i}, c_{1,i,\iota})$

ABE.Decrypt$(\{mpk_i\}_{i \in [N]}, \{\delta_{i,\iota}\}_{\iota \in \Omega}, \{C_i\}_{i \in [N]})$: On input $\{mpk_i\}_{i \in [N]}$, decryption shares $\{\delta_{i,\iota}\}_{\iota \in \Omega}$, where Ω is a set with indices $i_1, \dots, i_t$, and ciphertext C compute the decryption via following steps:

1. Compute Lagrange coefficients $\lambda_{i,\iota}$ for $\iota \in [n]$ as follows $\lambda_{i,\iota} = \prod_{k \in \Omega, k \neq \iota} \frac{-k}{(\iota - k)}$, where Ω is a set of indices with $|\Omega| = t$. Using these values, compute $c_{1,i} = \sum_{i=1}^{n} \lambda_i c_{1,i,\iota} = \boldsymbol{s}^t \sum_{i=1}^{n} F_i \lambda_i \boldsymbol{x}_{i,\iota} + \epsilon_{1,i}^t \boldsymbol{x}_{i,\iota} = \boldsymbol{s}^t \hat{\boldsymbol{u}}_i + \epsilon_{1,i}^t \sum_{i=1}^{n} \boldsymbol{x}_{i,\iota}$.

2. Do: $c_0 - c_{1,i} = s^t \hat{u}_i + \epsilon_0 + \lfloor \frac{q}{2} \rfloor m - s\hat{u}_i - \epsilon_{1,i}^t \sum_{i=1}^{n} \tilde{x}_{i,\iota} = \lfloor \frac{q}{2} \rfloor m \mod q + \text{`error'}$,

where $\epsilon_{1,i}^t \sum_{i=1}^{n} \tilde{x}_{i,\iota}$ is an error value limited by Gaussian error distribution parameters multiplied by a factor at most q, since the entries of $x_{i,\iota}$ are all mod q.

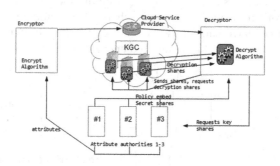

Fig. 1. Multi-authority distributed ABE scheme

3.2 Security Reduction

Theorem 1. *Our KP-MABE scheme is secure against chosen-ciphertext attack assuming that the LWE problem as given in Definition 1 is hard.*

Proof. Let $\mathcal{A}_{ind/ano}$ be a probabilistic polynomial-time adversary in a chosen-ciphertext attack against our KP-MABE scheme. We construct an adversary $\mathcal{B}$ against LWE problem which simulates the outputs for $\mathcal{A}_{ind/ano}$. The instance of LWE problem is given as a sampling oracle $\mathcal{O}$. This oracle can be either purely random $\mathcal{O}_r$ or pseudo-random $\mathcal{O}_s$ for some secret $s \in \mathbb{Z}_q^{\nu}$. $\mathcal{B}$ queries from his sampling oracle $\mathcal{O}$ and receives for each request i a fresh pair $(a_i, t_i) \in \mathbb{Z}_q^{\nu} \times \mathbb{Z}_q$. Additionally $\mathcal{B}$ request from its oracle LWE samples and obtains $(\alpha_{ij}, \beta_{ij})_{i \in \{0,...,N\}, j \in \{0,...,\ell\}}$. Before the adversary $\mathcal{A}_{ind/ano}$ can start to play the security game with a challenger, it announces a target attribute vector, it wants to be challenged on. We proceed the proof via a sequence of hybrid games. The Game0 is exactly the same game as described in the security game above. In Game1 we change the generation of public key, challenge ciphertext and the way on how secret key queries are answered. The changes should result in a minimal negligible difference with Game0. To simulate vector $u_{i,j} \in \mathbb{Z}_q^{\nu}$ it picks an LWE sample α_{00} and computes $A_{i,j} = [\bar{A}_i | A_1] = [\bar{A}_i | - \bar{A}_i R_{i,j}]$, where $R_{i,j}$ are chosen in the same way as in Game0.

Secret Key Generation Queries: KP-MABE adversary $\mathcal{A}_{ind/ano}$ can issue adaptive key queries on input $(policy_\kappa, \cdot)$, which do not satisfy the previously chosen target attribute set. We assume that $\mathbb{A}_{\kappa,i}^* = \{a_{\kappa,1}^*, \ldots, a_{\kappa,\xi}^*\}$ is the target attribute set for user κ and attribute authority $i \in [N]$, $\xi = |\mathbb{A}_{\kappa,i}|$. Simulator $\mathcal{B}$ changes the linear span program matrix L to L' which consists only of those rows of index j such that $j \in \mathbb{A}_i$ and accordingly changes L' to L'' which skips the column with index 0. Invoking TrapGen algorithm, generates ξ random matrices $Z_i \in \mathbb{Z}_q^{\nu \times m}$ for all attribute $i \in [\xi]$. It is required that the challenge attribute set does not satisfy the query on $policy_\kappa$. Build a virtual encryption matrix and generate the secret keys of attribute authorities in the same way as in the real construction. That means first to sample ξ' random matrices $Z_{i,j} \in \mathbb{Z}_q^{\nu \times m}$ using the TrapGen algorithm from Definition 2, where $i \in [N]$ and $j \in [\xi']$. We assume that those matrices have short bases $B_{i,j} \in \mathbb{Z}_q^{m \times m}$ for all attributes from the challenge attribute set $\mathbb{A}_{\kappa,i}^*$. The simulated encryption matrix is given as follows:

$$
M_i = \begin{bmatrix} A_{i,1} & & & \\ & A_{i,2} & & \\ & & \ddots & \\ & & & A_{i,\xi'} \end{bmatrix} \begin{array}{|c|ccc|} l_{1,0}\bar{A}_i & l_{1,1}Z_{i,1} & \cdots & l_{1,\theta}Z_{i,\xi'} \\ l_{2,0}\bar{A}_i & l_{2,1}Z_{i,1} & \cdots & l_{2,\theta}Z_{i,\xi'} \\ \vdots & \vdots & & \vdots \\ l_{\xi',0}\bar{A}_i & l_{\xi',1}Z_{i,1} & \cdots & l_{\xi',\xi'}Z_{i,\xi'} \end{array} \mod q \in \mathbb{Z}_q^{\nu\xi' \times \tilde{\xi}}
$$

where $\tilde{x}i := (m(\xi'+1) + w(\xi'-1))$ Let $\tilde{Z}_i$ denote the right part of the simulated M_i consisting of $l_{j,k}Z_{i,j}$, where $k \in [\xi']$ another index from 1 to ξ'. The diagonal matrix $\tilde{B}_i$ consisting of $B_{i,j}$ in the diagonal. Then holds: $\tilde{Z}_i\tilde{B}_i = 0 \mod q$, i.e. $\tilde{B}_i$ is a basis for $\Lambda_q^\perp(\tilde{Z}_i)$. Using the Inverse algorithm from Definition 3 the challenger $\mathcal{B}$ can simulate $sk_{\mathcal{A}_i} = f_{M_i}^{-1}(\hat{u})$.

Secret Share Queries: Assuming that KP-MABE adversary $\mathcal{A}_{ind/ano}$ corrupts up to $t-1$ decryption servers, he obtains $t-1$ secret shares $sk_{\mathcal{A}_i,S_\iota}$ which we give to $\mathcal{B}$. Using Lagrange coefficients $\lambda_{k,j}$ for $j \in [t-1]$ and choosing χ random polynomials $h_k(z) \in \mathbb{Z}_q[z]$ the simulator can compute $h_k(j) = \lambda_{k,0} + \sum_{i=1}^{t-1} \lambda_{k,j}j^i$, the secret share of server j. The simulator gives the generated secret key shares to the adversary $\mathcal{A}_{ind/ano}$.

Decryption Queries: $\mathcal{A}_{ind/ano}$ also makes decryption queries. To respond to such a query $\mathcal{B}$ first prepares the secret keys and secret shares as above. It then computes the decryption shares by first generating Lagrange coefficients and then taking linear combinations of t-out-of-n valid decryption shares and those Lagrange coefficients it decrypts the message m.

Challenge: $\mathcal{A}_{ind/ano}$ outputs two messages m_0, m_1 and two attribute universes $\{\mathbb{A}_i^{(0)}\}_{i\in[N]}$ and $\{\mathbb{A}_i^{(1)}\}_{i\in[N]}$. $\mathcal{B}$ simulates the ciphertext c_0, c_1 using the outputs from LWE instance. If the samples come from a purely random oracle, the ciphertext is indistinguishable from random, if it comes from a pseudo-random oracle, the ciphertext is independent of the message, since c_0 is independently distributed.

Continuation: The adversary is allowed to issue additional secret share queries and decryption queries after having seen the challenge ciphertext.

Guess: $\mathcal{A}_{ind/ano}$ finally outputs a guess, whether the ciphertext C was a valid encryption or not. Thereon, $\mathcal{B}$ decides, whether the LWE oracle was purely random not. If $\mathcal{A}$'s output is "valid", then $\mathcal{A}_{ind/ano}$ decides $random$, otherwise $\mathcal{B}$ outputs "$pseudorandom$". We also consider that $\mathcal{B}$ guesses the correct indices of corrupted servers with probability $\frac{1}{\binom{n}{t-1}}$. If the algorithm $\mathcal{A}_{ind/ano}$ succeeds to guess the correct message and policy with a minimal probability of $\frac{1}{\binom{n}{t-1}}(1/4 + \epsilon/4)$, then the algorithm $\mathcal{B}$ guesses correctly the behavior of the LWE oracle with a minimal probability of $1/4 + \epsilon/4$ such that holds $\mathbf{Adv}_{\mathcal{B},LWE} \geq \frac{1}{\binom{n}{t-1}}\mathbf{Adv}_{\mathcal{A},MABE}$. □

Theorem 2. *Our KP-MABE scheme is KP-MABE-ROB-CCA secure.*

Proof. (Sketch) For the more detailes of this proof we refer to the full version of this paper and provide only a short sketch here. Since the access policy is implemented in the encryption matrix F_i, it means that for two different policies, $policy_0, policy_1$, the encryption matrices $F_{i,0}, F_{i,1}$ are different. From this fact follows, that the ciphertexts which depend to these matrices are also different from each other.

4 Application: Distributed Attribute-Based Searchable Encryption

In this section we use our KP-MABE scheme in order to construct a distributed attribute-based searchable encryption on lattices (DABSE). We observe a cloud scenario, which allows data owner to upload encrypted data together with encrypted keywords, such that an authorized user can later perform particular search operations and download data upon a successful cloud search. Additionally we provide an improvement of the single server public key encryption with keyword search (PEKS) schemes by distributing the role of one server among n cloud servers and thus splitting the single point of failure among these multiple servers. We also assume existence of multiple attribute authorities whose role is to compute secret keys for the user using the corresponding access policy of that user. In order to enable the search operation without revealing any information of the data, the user computes trapdoor shares on certain keywords using the obtained secret keys from attribute authorities. Upon receiving a set of those shares the user sends each trapdoor share to one of the n cloud servers. In our scheme, trapdoors and their corresponding shares are encrypted according to an access control policy.

A distributed attribute-based searchable encryption scheme consists of the following six algorithms (Setup, KeyGen, DABSE, ShareTrpd, ShareTest, Test). The Setup algorithm generates a secret and a public key pair (sk_i, pk_i). The KeyGen algorithm computes authorities secret keys $sk_{\mathcal{A}_i}$ according to the corresponding access policies of an user. Via DABSE algorithm, data owner encrypts certain keywords using certain attributes which allow access control during the search process. The ShareTrpd algorithm takes as inputs the authorities secret key $sk_{\mathcal{A}_i}$ and a keyword algorithm and generates a trapdoor for some keywords.

Then it distributes the trapdoor among n servers, such that each server for $\iota \in [n]$ obtains a trapdoor share $T_{w,\iota}$. The ShareTest algorithm is run by servers where each of them takes a trapdoor share and attribute credentials, a DABSE ciphertext Φ and computes the corresponding test shares which are sent back to the user. The user takes at least t-out-of-n valid test shares and runs the Test algorithm, to check whether DABSE ciphertext encrypts the same keywords used for generating trapdoor shares. The output of Test is a message m or $\bot$.

Intuition. We present first an intuition for our construction. We note that the ABE.Setup algorithm includes the Setup algorithm of KP-MABE scheme, which first outputs master public key and master secret key mpk, msk. The attributes from KP-MABE correspond to a keyword set in a DABSE scheme. Algorithm KeyGen takes these keys and runs the ABE.Extract algorithm of KP-MABE to generate secret and public keys of each attribute authority. The data owner runs ABE.Encrypt algorithm of KP-MABE scheme to compute the ciphertext of a keyword w using an attribute set $\mathbb{A}$ of each attribute authority. This ciphertext corresponds to DABSE ciphertext Φ. The ShareTrpd algorithm takes as input secret and public key from each authority and a keyword w which is equal to an id of KP-MABE scheme and runs ABE.KeyDistr algorithm of KP-MABE scheme in order to generate trapdoor shares. Algorithm ShareTest takes as input attribute credentials and trapdoor shares and runs ABE.ShareDec algorithm to generate test shares. The Test algorithm takes as input test shares, a public key and runs Decrypt algorithm of KP-MABE and if the decryption is successful, Test algorithm outputs 1.

Definition 9 (Indistinguishability/Consistency). *We formalize the security property of our DABSE scheme, called indistinguishability and consistency against chosen ciphertext attacks (*IND/CONS − DABSE − CCA*).*

Init: *The adversary $\mathcal{A}_{ind/con}$ chooses a set I of $t-1$ decryption servers and a set J of $t-1$ attribute authorities that it wants to corrupt. Let $I = \{i_1, \ldots, i_{t-1}\} \subset \{1, \ldots, N\}$, $J = \{j_1, \ldots, j_{t-1}\}$.*

Setup: *The challenger runs $\mathsf{Setup}(1^\lambda)$ and gives the adversary the master public key mpk to the adversary. It keeps the master secret key to itself.*

Phase1: *The adversary issues queries $q_1, \ldots, q_d$, where q_i is one of the following queries:*

Secret Key Queries: *On input $(policy'_\kappa, \cdot)$ the adversary $\mathcal{A}_{ind/con}$ can issue as many secret key queries as he wants to the attribute authorities. We assume that there is at least one honest attribute authority, such that the adversary cannot recover all secret keys and combine them into one, which he can use during the decryption process. Furthermore we assume, that the secret key queries corresponding to certain $policy_j$ do not satisfy the challenge $policy^*_\kappa$ of user κ. We also restrict the adversary to query the same authority twice.*

Trapdoor Share Query on Input $(policy_\kappa, w, \iota)$: *We assume existence of a list T which consists of the entries $(policy_\kappa, \{T_{w,\iota}\}_{\iota \in [n]})$. We note that $policy_\kappa$ has not been the input of secret key query. The oracle checks if $(policy_\kappa, \cdot) \in T$, if so it returns the required trapdoor share $(\iota, T_{w,\iota})$ to the adversary.*

Otherwise if $(policy_\kappa, \cdot) \in T$ *it runs the* ShareTrpd *algorithm and generates* $\{T_{w,\iota}\}_{\iota \in [n]}$. *Finally it sends a trapdoor share* $(\iota, T_{w,\iota})$ *to the adversary. We allow an adversary to make adaptive queries, that means, he may issue a query on* q_i *with knowledge of* $q_1, \ldots, q_{i-1}$, *but with the constraint, that* $w \neq w^*$ *and the number of credential queries is at most* $t - 1$.

Test Queries on Input (Φ, w): *The challenger runs* ShareTrpd *on* w *to obtain a set of* n *trapdoor shares* $T := \{T_{w,1}, \ldots, T_{w,n}\}$. *We assume existence of a list* T *which consists of the entries* $(policy_\kappa, \{T_{w,\iota}\}_{\iota \in [n]})$. *The oracle checks if* $(policy_\kappa, \cdot) \in T$, *if so it takes the trapdoor shares* $\{T_{w,\iota}\}_{\iota \in [n]}$ *and runs* ShareTest *algorithm in order to generate the required test shares. Upon taking* t-*out-of-*n *valid test shares it runs* Test *and outputs either 1 or 0 to the adversary.*

Otherwise if $(policy_\kappa, \cdot) \in T$ *it first runs* ShareTrpd *algorithm and generates* $\{T_{w,\iota}\}_{\iota \in [n]}$. *Afterward it takes these trapdoor shares and runs* ShareTest *in order to obtain test shares* $\delta_{i_1}, \ldots, \delta_{i_t}$, *where* $i_j \in [n]$. *On input these test shares it runs* Test *to check, whether* Φ *encrypts the same keywords which were also used in trapdoor generation algorithm* ShareTrpd. *If the test process is successful, the challenger sends the result of* Test *to the* $\mathcal{A}_{ind/con}$. *Otherwise it returns* 0.

Challenge: *The adversary outputs identities* w_0, w_1 *which it wants to be challenged on. The challenger picks a random bit* $b \in \{0, 1\}$ *and computes* $\Phi \leftarrow$ DABSE($\{pk_i\}_{i \in [N]}, w_b$). *He sends* Φ^* *as challenge ciphertext to the adversary* $\mathcal{A}$.

Phase2: *The adversary and the challenger interact as in Phase1.*

Guess: *The adversary outputs a guess* $b' \in \{0, 1\}$. *It wins the game if* $b = b'$ *and trapdoor was computed on* w_{1-b}.

4.1   Construction

In this section we provide the general construction of an DABSE scheme from a secure KP-MABE scheme.

Definition 10 (KP-MABE-to-DABSE Transform).

Setup(1^λ): *On input a security parameter* 1^λ, *it runs the* Setup *algorithm of the KP-MABE scheme* $(msk_i, mpk_i) \xleftarrow{r} \text{Setup}(N, t, 1^\lambda)$ *and outputs* $sk_i = msk_i$ *and* $pk_i = mpk_i$.

KeyGen($pk_i, sk_i, policy_\kappa$): *On input public and secret key pair* pk_i, sk_i, *and* $policy_\kappa$ *of the user* κ *it runs the* ABE.Extract *algorithm of KP-MABE scheme and outputs authorities secret key* $sk_{\mathcal{A}_i}$ *which corresponds to the provided access policy.*

DABSE($\{pk_i\}_{i \in [N]}, w$): *On input authorities public keys* $\{pk_i\}_{i \in [N]}$, *keyword* w *which corresponds to an attribute set* $\bar{\mathbb{A}} = \{\mathbb{A}_1, \ldots, \mathbb{A}_N\}$ *of KP-MABE scheme, it runs the encryption algorithm of KP-MABE scheme in order to compute* $C \xleftarrow{r} \text{ABE.Encrypt}(pk, w, R)$, *where* $R \xleftarrow{r} \{0, 1\}$ *is picked randomly. It returns the DABSE ciphertext* $\Phi = (C, R)$.

ShareTrpd($pk_i, sk_{\mathcal{A}_i}, w, t, n$): *On input authority's public key pk_i and policy-related secret key $sk_{\mathcal{A}_i}$, where $i \in [N]$ is the index of corresponding authority $\mathcal{A}_i$, threshold parameters t, n, keyword w it runs the* ABE.Extract *algorithm of KP-MABE scheme to output secret key shares $\{sk_{\mathcal{A}_i, S_\iota}\}_{i \in [N], \iota \in [n]}$ of each authority i. It sets these secret shares equal to trapdoor shares $(\iota, T_{w,\iota}) = (\iota, \{sk_{\mathcal{A}_i, S_\iota}\}_{i \in [N], \iota \in [n]})$.*

ShareTest($pk_i, \{\iota, T_{w,\iota}\}_{\iota \in \Omega}, \Phi$): *On input a public key pk_i of authority $i \in [N]$, trapdoor shares $\{\iota, T_{w,\iota}\}_{\iota \in [\Omega]}$ and a DABSE ciphertext $\Phi = (C, R)$, it runs the* ShareDecrypt *algorithm of KP-MABE scheme and outputs decryption shares $\delta_\iota \xleftarrow{r}$ ShareDecrypt($pk_i, (\iota, T_{w,\iota}), \Phi$). It outputs test shares $\tau_\iota = (\delta_\iota)$.*

Test($\{pk_i\}_{i \in [N]}, \{\tau_\iota\}_{\iota \in \Omega}$): *On input a public key, at least t-out-of-n test shares τ_ι, $\iota \in \Omega$ with $|\Omega| = t$, it runs* ABE.Decrypt *algorithm of KP-MABE on input δ_ι. The algorithm outputs m if* ABE.Decrypt *algorithm was successful and $R' = R$. Otherwise outputs $\perp$.*

Note on Verifiability: Some users of cloud services require confidentiality in the search process, especially if we assume that the storage function of the cloud is corrupted or there is a malfunction in the whole storage system. Therefore there are users which would like to get guarantee about the authenticity of conducted search operations and received search results. We can adapt our scheme to those requirements by adding a verification operation to our construction. Similar to [42], the verification can be done using a digital signature scheme as an additional building block. A signature is prepared for each DABSE ciphertext and for a trapdoor where the trapdoor signature is distributed into n shares in such a way that each signature share represents a signature for a trapdoor share. After completing the search function, the cloud servers output the search result, which is either 1 or 0, and a proof which includes signatures on DABSE ciphertext and trapdoor shares. After receiving proof shares from each cloud server, the user can reconstruct the signature of a trapdoor using at least t-out-of-n signature shares from each proof share. The verification succeeds if the both signatures, one of a DABSE ciphertext and one of the trapdoor can be verified using a public key of the signature scheme.

4.2   Security Reduction

Theorem 3. *If KP-MABE scheme is* IND/ANO $-$ MABE $-$ CCA *and* ROB$-$CCA *secure, the obtained DABSE scheme in Definition 10 is* IND/CONS $-$ DABSE $-$ CCA *secure.*

Proof. We use a IND/CONS $-$ DABSE $-$ CCA adversary $\mathcal{B}_{ind/con}$ against DABSE scheme to construct a simulator of a chosen-ciphertext attack against KP-MABE scheme. That is, the simulator acts as $\mathcal{A}_{ind/ano}$ against chosen-ciphertext attack and as $\mathcal{A}_{rob}$ against ROB$-$CCA attack. The simulation of the view of $\mathcal{B}_{ind/con}$ distinguishes by two cases $C \neq C^*$ and $C = C^*$. We start to describe the simulation using game hopping technique.

Game0: The initial game is the Game0 which describes the real attack. First the challenger runs the **Setup** of the DABSE scheme on input a security parameter λ, threshold parameter t, number of servers n and N attribute authorities. The challenger gives $\mathcal{B}_{ind/con}$ the master public keys of N attribute authorities $\{mpk_i\}_{i\in[N]}$. $\mathcal{B}_{ind/con}$ issues also the required queries on authority's secret key belonging to the attribute authority $i \in [N]$. Upon receiving the secret keys $sk_{\mathcal{A}_i,S_\iota}$, challenger $\mathcal{B}_{ind/con}$ issues trapdoor share query on a keyword w and test queries on the DABSE ciphertext Φ and an attribute set $\{\mathbb{A}_i\}_{i\in[N]}$. The challenger responds with the challenge DABSE ciphertext $\Phi^*(w)$. In the challenge phase $\mathcal{B}_{ind/con}$ outputs two challenge keywords w, w'. The challenger computes the corresponding DABSE ciphertext $\Phi(pk, w)$ on w. $\mathcal{B}_{ind/con}$ issues more trapdoor share and decryption queries to which the $\mathcal{O}\mathtt{ShareTrpd}$ and $\mathcal{O}\mathtt{Test}$ respond as in the first phase. $\mathcal{B}_{ind/con}$ also issues up to $t-1$ trapdoor share queries on w and on w', where $w \neq w'$ but no decryption queries on these challenge keywords. Finally $\mathcal{B}_{ind/con}$ outputs a random bit $b' \in \{0,1\}$ which is a guess for b. We denote by $\mathcal{E}_0$ the event $b = b'$. Since Game0 is the same as the real attack, we have

$$Pr[\mathcal{E}_0] = \mathbf{Adv}_{DABSE,\mathcal{B}_{ind/con}}^{\mathrm{IND/CONS-DABSE-CCA}}(1^\lambda)$$

Game1: The first game (Game1) differs from the previous one by simulation of secret key, trapdoor share and test queries. If these shares are involved in computing the DABSE ciphertext, the simulator modifies the challenge ciphertext appropriately. $\mathcal{A}_{ind/ano}$ is given as input the attribute authorities master public key mpk_i of KP-MABE scheme which gives it to $\mathcal{B}_{ind/con}$. When $\mathcal{A}_{ind/ano}$ issues queries on the authorities secret keys, $\mathcal{B}_{ind/con}$ forwards those queries to its secret key generation oracle on input a randomly chosen access policy, which is not equal to the target $policy_\kappa$ of user κ. $\mathcal{B}'_{ind/con}$s oracle runs the **ABE.Extract** algorithm and returns the corresponding secret key to $\mathcal{A}_{ind/ano}$, which is to be forwarded to $\mathcal{B}_{ind/con}$ by the simulator. We assume that $\mathcal{B}_{ind/con}$ corrupts a set of $t-1$ servers. Therefore, challenger $\mathcal{A}_{ind/ano}$ also chooses a random set of $t-1$ servers with probability of $1/\binom{n}{t-1}$ to match the same servers as chosen by $\mathcal{B}_{ind/con}$. In the first phase the attacker $\mathcal{B}_{ind/con}$ issues queries on (w, ι), where $\iota \in [n]$. It invokes $\mathcal{A}_{ind/ano}$ which runs the simulation of received queries. It needs to simulate the corresponding trapdoor share for index ι, which is $(\iota, T_{w,\iota})$. In order to do so, it queries its own key distribution oracle $\mathcal{O}\mathtt{ABE.KeyDistr}$ on input $(policy_{kappa}, \iota)$. The oracle returns $(\iota, sk_{\mathcal{A}_i,\iota})$ to $\mathcal{A}_{ind/ano}$. The KP-MABE attacker $\mathcal{A}_{ind/ano}$ sets $sk_{\mathcal{A}_i,\iota}$ equal to a trapdoor share $T_{w,\iota}$ and returns $(\iota, T_{w,\iota})$ to $\mathcal{B}_{ind/con}$. When $\mathcal{B}_{ind/con}$ issues test queries on input (w, Φ), the KP-MABE adversary $\mathcal{A}_{ind/ano}$ chooses random $R \in \{0,1\}$, sets $w = \{\mathbb{A}_i\}_{i\in[N],b} := \bar{\mathbb{A}}_b$, such that $\bar{\mathbb{A}}_b$ satisfies $policy_{\kappa,b}$ for $b \in \{0,1\}$ and computes $C = \mathtt{Encrypt}(pk, \bar{\mathbb{A}}_b, R\|0^\lambda)$, where $R\|0^\lambda$ is the challenge message. Then it queries its decryption oracle $\mathcal{O}\mathtt{Decrypt}$ on input $(\bar{\mathbb{A}}_b, C)$ to obtain either R or 0. In case that $\mathcal{A}_{ind/ano}$ receives R, it returns 1 to $\mathcal{B}_{ind/con}$, otherwise it returns 0. The algorithm $\mathcal{A}_{ind/ano}$ returns id as a challenge and $R_0, R_1 \xleftarrow{r} \{0,1\}$ as the challenge messages. The $\mathcal{A}_{ind/ano}$ oracle forwards the queries to its decryption oracle and returns a message m to $\mathcal{B}_{ind/con}$. When $\mathcal{B}_{ind/con}$ issues test queries on Φ^*, $\mathcal{A}_{ind/ano}$ aborts the simulation.

In the second phase, $\mathcal{B}_{ind/con}$ issues additional queries to the trapdoor share and test oracle, given challenge ciphertext $C \neq C^*$ that encrypts R_b under attribute set $\bar{\mathbb{A}} = \{\mathbb{A}_i\}_{i \in [N]}$ (which is set equal to w), $\mathcal{A}_{ind/ano}$ responds to the issued queries as before. $\mathcal{B}_{ind/con}$ is allowed to issue secret key queries to his secret key generation oracle and up to $t - 1$ trapdoor share queries but no test queries on challenge keywords w, w'. Attacker $\mathcal{A}_{ind/ano}$ simulates these queries by first querying his key generation oracle which runs the ABE.Extract algorithm in order to generate the secret key for authority $i \in [N]$. Upon receiving the corresponding secret key $sk_{\mathcal{A}_i}$, the simulator forwards it to the adversary $\mathcal{B}_{ind/con}$. Then the simulator $\mathcal{A}_{ind/ano}$ issues up to $t - 1$ queries to his secret share oracle $\mathcal{O}$ABE.KeyDistr on input an authority's secret key $sk_{\mathcal{A}_i}$. Further simulation is as before. Finally $\mathcal{B}_{ind/con}$ outputs a guess $b' \in \{0, 1\}$. The simulator outputs a guess as follows: If $b = b'$ then $\mathcal{A}_{ind/ano}$ outputs 1, which means that $\mathcal{A}_{ind/ano}$ succeeded in its attack to distinguish between R_0 and R_1. Otherwise it outputs 0 meaning that $\mathcal{A}_{ind/ano}$ didn't succeed attacking the IND/ANO $-$ MABE $-$ CCA security of KP-MABE scheme. As shown above, $\mathcal{A}_{ind/ano}$ answers any secret key, trapdoor share queries and test queries using its own secret key generation, key distribution and decryption oracles. Additionally, considering the probability of correct matching corrupt servers, it is easy to see that the lower bound of success probability of $\mathcal{A}_{ind/ano}$, which always returns bit $b = \beta = 1$, is at least the same or higher as $\mathcal{B}_{ind/con}$'s success probability. The upper bound of success probability of $\mathcal{A}_{ind/ano}$, whose output is always bit 0 is limited related to the size of the message space of messages $R_0, R_1 \in \{0, 1\}^\lambda$.

$$Pr[\mathbf{Exp}_{MABE, \mathcal{A}_{ind/ano}}^{\text{IND/ANO}-\text{MABE}-\text{CCA}}(\lambda) = 1]$$
$$\geq \frac{1}{\binom{n}{t-1}} Pr[\mathbf{Exp}_{DABSE, \mathcal{B}_{ind/con}}^{\text{IND/CONS}-\text{DABSE}-\text{CCA}}(1^\lambda) = 1],$$
$$Pr[\mathbf{Exp}_{MABE, \mathcal{A}_{ind/ano}}^{\text{IND/ANO}-\text{MABE}-\text{CCA}}(\lambda) = 1] \leq 2^{-\lambda}.$$

Game2: This game differs from previous one by the simulation of queries on the challenge ciphertext Φ^*. $\mathcal{A}_{rob}$ is given as input the master public key mpk of KP-MABE scheme which it gives to $\mathcal{B}_{ind/con}$. We assume that $\mathcal{B}_{ind/con}$ corrupts a set of $t - 1$ servers. $\mathcal{A}_{rob}$ also chooses a set of $t - 1$ servers and matches the correct server with a probability of $1/\binom{n}{t-1}$. In contrast to Case 1 we have here only one phase where the attacker $\mathcal{B}_{ind/con}$ issues queries on $(policy_\kappa, \cdot)$ to the secret key generation oracle and queries on (w, ι) to the trapdoor share oracle $\mathcal{O}$ShareTrpd and to the test oracle $\mathcal{O}$Test. It invokes $\mathcal{A}_{rob}$ which runs the simulation of received queries. It needs to simulate the corresponding secret key query and the corresponding trapdoor share for index ι, which is $(\iota, T_{w, \iota})$. First $\mathcal{A}_{rob}$ approaches his secret key generation oracle on input $(policy_\kappa, \cdot)$ and receives the corresponding secret key $sk_{\mathcal{A}_i}$. In order to simulate trapdoor share queries, $\mathcal{A}_{rob}$ approaches his key distribution oracle $\mathcal{O}$ABE.KeyDistr, i.e. for a $w \neq w^*$ it takes w and queries its own key derivation oracle $\mathcal{O}$ABE.KeyDistr on input $(sk_{\mathcal{A}_i,\iota}, \iota)$. The oracle returns $(\iota, sk_{\mathcal{A}_i}, \iota)$ to $\mathcal{A}_{rob}$. The KP-MABE attacker $\mathcal{A}_{rob}$ sets $sk_{\mathcal{A}_i,\iota}$ equal to a trapdoor share $T_{w,\iota}$ and returns $(\iota, T_{w,\iota})$ to $\mathcal{B}_{ind/con}$. If $w = w^*$ the adversary $\mathcal{B}_{ind/con}$ is allowed to issue up to $t-1$ queries to the key distribution oracle. $\mathcal{A}_{rob}$ simulates the queries as before. When $\mathcal{B}_{ind/con}$ issues test queries on input (w, Φ), where $w \neq w^*$ and $\Phi = \Phi^*$ the KP-MABE adversary $\mathcal{A}_{rob}$ chooses a random $R \in \{0,1\}^\lambda$, a keyword $w = \{\mathbb{A}_i\}_{i\in[N]} := \bar{\mathbb{A}}$ and computes $C = \texttt{Encrypt}(mpk_i, w, R\|0^\lambda)$, where $R\|0^\lambda$ is the challenge message. Then it queries its decryption oracle $\mathcal{O}\texttt{Decrypt}$ on input $(\bar{\mathbb{A}}_b, C)$ in order to obtain either R or 0, such that $\bar{\mathbb{A}}_b$ satisfies either $policy_0$ or $policy_1$. In case that $\mathcal{A}_{rob}$ receives R, it returns 1 to $\mathcal{B}_{ind/con}$, otherwise it returns 0. The $\mathcal{A}_{ind/ano}$ adversary queries its own decryption oracle and returns m to $\mathcal{B}_{ind/con}$. If $w = w^*$ and we have $\Phi = \Phi^*$, the simulator $\mathcal{A}_{rob}$ sets $w = policy_0$ and $w' = policy_1$, sets $R = 0$ and $\Phi^* = C^*$. It issues then decryption queries on $policy_0, policy_1$ to its decryption oracle which returns m_0 and m_1. If $m_0 = \bot$ and $m_1 = \bot$ then $\mathcal{A}_{rob}$ returns 1 to $\mathcal{B}_{ind/con}$.

Finally $\mathcal{B}_{ind/con}$ outputs a guess $b' \in \{0, 1\}$. $\mathcal{A}_{rob}$ outputs a guess as follows: If $b = b'$ then $\mathcal{A}_{rob}$ outputs 1, which means that $\mathcal{A}_{rob}$ succeeded in its attack. Otherwise it outputs 0 meaning that $\mathcal{A}_{rob}$ didn't succeed attacking the ROB$-$CCA security of KP-MABE scheme. As shown above, $\mathcal{A}$ answers any trapdoor-share queries and test queries using its own key derivation and decryption oracles. Additionally, considering the matching probability of corrupt servers it is easy to see that for $\mathcal{B}'$s advantage holds:

$$Pr[\mathbf{Exp}^{ROB-CCA}_{MABE,\mathcal{A}_{rob}}(1^\lambda) = 1] \geq$$
$$\frac{1}{\binom{n}{t-1}}\left(Pr[\mathbf{Exp}^{IND/CONS-DABSE-CCA}_{DABSE,\mathcal{A}_{ind/con}}(1^\lambda) = 1]\right).$$

Now combining the results of the two cases follows that

$$\frac{1}{\binom{n}{t-1}}\left(\mathbf{Adv}^{IND/CONS-DABSE-CCA}_{DABSE,\mathcal{B}_{ind/con}}\right) \leq \mathbf{Adv}^{ROB-CCA}_{MABE,\mathcal{A}_{rob}}(1^\lambda) + 2^{-\lambda}$$
$$+ \mathbf{Adv}^{IND/ANO-MABE-CCA}_{MABE,\mathcal{A}_{ind/ano}}(1^\lambda)$$

5   Conclusion

In this paper we provided the first lattice-based attribute-based threshold decryption which involves multiple attribute authorities. We proved the scheme indistinguishable and anonymous against chosen ciphertext attacks. Furthermore we presented an application of our construction to the searchable encryption by providing the first lattice-based distributed attribute-based searchable encryption with multiple attribute authorities. This scheme is usually used in cloud storage schemes in order to allow data users to perform searching operations on the stored cloud data.

Appendix

Definition 11 (Tensor Product). *For vectors $v, w \in \mathbb{Z}_q^n$, where $v = (v_1, \ldots, v_n), w = (w_1, \ldots, w_n)$ the tensor product is given by*

$$\begin{pmatrix} v_1 \\ v_2 \\ \vdots \\ v_n \end{pmatrix} \otimes \begin{pmatrix} w_1 \\ w_2 \\ \vdots \\ w_n \end{pmatrix} = (v_1 w_1, \ldots, v_1 w_n, \ldots, v_n w_1, \ldots, w_n)^t$$

For a matrix $V \in \mathbb{Z}_q^{m \times m}$ and vectors $v_1, \ldots, v_n \in \mathbb{Z}_q^m$ a tensor product has the following property:

$$V(v_1 \otimes \ldots \otimes v_n) = V v_1 \otimes v_2 \otimes \ldots \otimes v_n = (V \otimes I_{m^{n-1}})(v_1 \otimes \ldots \otimes v_n).$$

Note that $v_1 \otimes \ldots \otimes v_n \in \mathbb{Z}_q^{m^n}$ and

$$V \otimes I_{m^{n-1}} = \begin{pmatrix} v_{11} & \cdots & v_{1m} \\ \vdots & \ddots & \vdots \\ v_{m1} & \cdots & v_{mm} \end{pmatrix} \otimes \begin{pmatrix} 1_{11} & 0 & \cdots & 0_{1m^{n-1}} \\ 0 & 1 & \cdots & 0 \\ \vdots & & \ddots & \vdots \\ 0_{m^{n-1}1} & \cdots & 0 & 1_{m^{n-1}m^{n-1}} \end{pmatrix}$$

$$= \begin{pmatrix} v_{11} I_{m^{n-1}} & \cdots & v_{1m} I_{m^{n-1}} \\ \vdots & \ddots & \vdots \\ v_{m1} I_{m^{n-1}} & \cdots & v_{mm} I_{m^{n-1}} \end{pmatrix} \in \mathbb{Z}_q^{m^n \times m^n}$$

We note that in our scheme the vectors $v_1, \ldots, v_n$ associates with the different public keys, which are used to decrypt an evaluated ciphertext that associates with the vector space V.

In the following definition we recall the construct of direct sums. We propose this tool in order to provide an optimization during evaluation of different ciphertexts. On the one hand it improves the dimension of evaluated ciphertexts while on the other hand it involves additional rounds of communication between the parties during the decryption process.

Definition 12 (Direct Sum). *Let $V \in \mathbb{Z}_q^n$ and $W \in \mathbb{Z}_q^m$. The vector space $V \oplus W$ which is spanned by the basis vectors of these two vector spaces, has dimension $n + m$ and is called the direct sum of V and W. The vectors from each vector space V or W can be seen as vectors of the direct sum, just by filling zeros to the full dimension $n + m$. Let $v = (v_1, \ldots, v_n) \in V$ and $w = (w_1, \ldots, w_m) \in W$. Vector v is an element of direct sum, e.g. $v = (v_1, \ldots, v_n, 0_1, \ldots, 0_m)$ and $w = (0_1, \ldots, 0_n, w_1, \ldots, w_m)$. Then the direct sum of $v \otimes w = (v_1, \ldots, v_n, w_1, \ldots, w_m)$, which is a vector of dimension $n + m$.*

The direct sum of two matrices $A \in \mathbb{Z}_q^{n \times n}, B \in \mathbb{Z}_q^{m \times m}$ is given by:

$$A \oplus B = \begin{pmatrix} A & 0 \\ 0 & B \end{pmatrix} = \begin{pmatrix} a_{11} & \dots & a_{1n} & 0 & \dots & 0 \\ \vdots & \ddots & \vdots & \vdots & \ddots & \vdots \\ a_{n1} & \dots & a_{nn} & 0 & \dots & 0 \\ 0 & \dots & 0 & b_{11} & \dots & b_{1m} \\ \vdots & \ddots & \vdots & \vdots & \ddots & \vdots \\ 0 & \dots & 0 & b_{m1} & \dots & b_{mm} \end{pmatrix}$$

In general a direct sum of n matrices of dimensions $n_1, \dots, n_n$ is given by

$$A_1 \oplus A_2 \oplus \dots \oplus A_n = \begin{pmatrix} A_1 & 0 & \dots & 0 \\ 0 & A_2 & \dots & 0 \\ \vdots & \dots & \ddots & \vdots \\ 0 & \dots & 0 & A_n \end{pmatrix}$$

The dimension of this direct sum is $n_1 + \dots + n_n$. Furthermore, $(A_1 \oplus \dots \oplus A_n)(v_1 \oplus \dots \oplus v_n) = Av_1 \oplus \dots \oplus A_n v_n$.

References

1. Abdalla, M., Bellare, M., Neven, G.: Robust encryption. In: Micciancio, D. (ed.) TCC 2010. LNCS, vol. 5978, pp. 480–497. Springer, Heidelberg (2010). doi:10.1007/978-3-642-11799-2_28

2. Agrawal, S., Boneh, D., Boyen, X.: Efficient lattice (H)IBE in the standard model. In: Gilbert, H. (ed.) EUROCRYPT 2010. LNCS, vol. 6110, pp. 553–572. Springer, Heidelberg (2010). doi:10.1007/978-3-642-13190-5_28

3. Ajtai, M.: Generating hard instances of lattice problems (extended abstract). In: STOC 1996, pp. 99–108. ACM (1996)

4. Ajtai, M., Dwork, C.: A public-key cryptosystem with worst-case/average-case equivalence. In: STOC 1997, pp. 284–293. ACM (1997)

5. Attrapadung, N., Herranz, J., Laguillaumie, F., Libert, B., de Panafieu, E., Ràfols, C.: Attribute-based encryption schemes with constant-size ciphertexts. Theor. Comput. Sci. **422**, 15–38 (2012)

6. Beimel, A.: Secure schemes for secret sharing and key distribution. Ph.D. thesis, Israel, Institute of Technology, Technion, Haifa (1996)

7. Bendlin, R., Krehbiel, S., Peikert, C.: How to share a lattice trapdoor: threshold protocols for signatures and (H)IBE. In: Jacobson, M., Locasto, M., Mohassel, P., Safavi-Naini, R. (eds.) ACNS 2013. LNCS, vol. 7954, pp. 218–236. Springer, Heidelberg (2013). doi:10.1007/978-3-642-38980-1_14

8. Bethencourt, J., Sahai, A., Waters, B.: Ciphertext-policy attribute-based encryption. In: 2007 IEEE Symposium on Security and Privacy (S&P 2007), pp. 321–334. IEEE Computer Society (2007)

9. Boneh, D., Crescenzo, G., Ostrovsky, R., Persiano, G.: Public key encryption with keyword search. In: Cachin, C., Camenisch, J.L. (eds.) EUROCRYPT 2004. LNCS, vol. 3027, pp. 506–522. Springer, Heidelberg (2004). doi:10.1007/978-3-540-24676-3_30

10. Boneh, D., Waters, B.: Conjunctive, subset, and range queries on encrypted data. In: Vadhan, S.P. (ed.) TCC 2007. LNCS, vol. 4392, pp. 535–554. Springer, Heidelberg (2007). doi:10.1007/978-3-540-70936-7_29

11. Boyen, X.: Lattice mixing and vanishing trapdoors: a framework for fully secure short signatures and more. In: Nguyen, P.Q., Pointcheval, D. (eds.) PKC 2010. LNCS, vol. 6056, pp. 499–517. Springer, Heidelberg (2010). doi:10.1007/978-3-642-13013-7_29

12. Boyen, X.: Attribute-based functional encryption on lattices. In: Sahai, A. (ed.) TCC 2013. LNCS, vol. 7785, pp. 122–142. Springer, Heidelberg (2013). doi:10.1007/978-3-642-36594-2_8

13. Cash, D., Hofheinz, D., Kiltz, E., Peikert, C.: Bonsai trees, or how to delegate a lattice basis. In: Gilbert, H. (ed.) EUROCRYPT 2010. LNCS, vol. 6110, pp. 523–552. Springer, Heidelberg (2010). doi:10.1007/978-3-642-13190-5_27

14. Chang, Y.-C., Mitzenmacher, M.: Privacy preserving keyword searches on remote encrypted data. In: Ioannidis, J., Keromytis, A., Yung, M. (eds.) ACNS 2005. LNCS, vol. 3531, pp. 442–455. Springer, Heidelberg (2005). doi:10.1007/11496137_30

15. Chase, M.: Multi-authority attribute based encryption. In: Vadhan, S.P. (ed.) TCC 2007. LNCS, vol. 4392, pp. 515–534. Springer, Heidelberg (2007). doi:10.1007/978-3-540-70936-7_28

16. Cheung, L., Newport, C.C.: Provably secure ciphertext policy ABE. In: Proceedings of the 2007 ACM Conference on Computer and Communications Security, CCS 2007, pp. 456–465. ACM (2007)

17. Curtmola, R., Garay, J.A., Kamara, S., Ostrovsky, R.: Searchable symmetric encryption: improved definitions and efficient constructions. J. Comput. Secur. 19(5), 895–934 (2011)

18. Gentry, C., Peikert, C., Vaikuntanathan, V.: Trapdoors for hard lattices and new cryptographic constructions. In: STOC 2008, pp. 197–206. ACM (2008)

19. Goh, E.: Secure indexes. IACR Cryptol. ePrint Arch. 2003, 216 (2003)

20. Golle, P., Staddon, J., Waters, B.: Secure conjunctive keyword search over encrypted data. In: Jakobsson, M., Yung, M., Zhou, J. (eds.) ACNS 2004. LNCS, vol. 3089, pp. 31–45. Springer, Heidelberg (2004). doi:10.1007/978-3-540-24852-1_3

21. Gorbunov, S., Vaikuntanathan, V., Wee, H.: Attribute-based encryption for circuits. In: Symposium on Theory of Computing Conference, STOC 2013, pp. 545–554. ACM (2013)

22. Goyal, V., Jain, A., Pandey, O., Sahai, A.: Bounded ciphertext policy attribute based encryption. In: Aceto, L., Damgård, I., Goldberg, L.A., Halldórsson, M.M., Ingólfsdóttir, A., Walukiewicz, I. (eds.) ICALP 2008. LNCS, vol. 5126, pp. 579–591. Springer, Heidelberg (2008). doi:10.1007/978-3-540-70583-3_47

23. Goyal, V., Pandey, O., Sahai, A., Waters, B.: Attribute-based encryption for fine-grained access control of encrypted data. In: CCS 2006, pp. 89–98. ACM (2006)

24. Hou, C., Liu, F., Bai, H., Ren, L.: Public-key encryption with keyword search from lattice. In: P2P, Parallel, Grid, Cloud and Internet Computing (2013)

25. Hwang, Y.H., Lee, P.J.: Public key encryption with conjunctive keyword search and its extension to a multi-user system. In: Takagi, T., Okamoto, E., Okamoto, T., Okamoto, T. (eds.) Pairing 2007. LNCS, vol. 4575, pp. 2–22. Springer, Heidelberg (2007). doi:10.1007/978-3-540-73489-5_2

26. Lewko, A., Okamoto, T., Sahai, A., Takashima, K., Waters, B.: Fully secure functional encryption: attribute-based encryption and (hierarchical) inner product encryption. In: Gilbert, H. (ed.) EUROCRYPT 2010. LNCS, vol. 6110, pp. 62–91. Springer, Heidelberg (2010). doi:10.1007/978-3-642-13190-5_4

27. Li, J., Zhang, L.: Attribute-based keyword search and data access control in cloud. In: CIS 2014, pp. 382–386. IEEE Computer Society (2014)
28. Li, M., Yu, S., Ren, K., Lou, W.: Securing personal health records in cloud computing: patient-centric and fine-grained data access control in multi-owner settings. In: Jajodia, S., Zhou, J. (eds.) SecureComm 2010. LNICSSITE, vol. 50, pp. 89–106. Springer, Heidelberg (2010). doi:10.1007/978-3-642-16161-2_6
29. Lyubashevsky, V.: Lattice signatures without trapdoors. In: Pointcheval, D., Johansson, T. (eds.) EUROCRYPT 2012. LNCS, vol. 7237, pp. 738–755. Springer, Heidelberg (2012). doi:10.1007/978-3-642-29011-4_43
30. Micciancio, D., Peikert, C.: Trapdoors for lattices: simpler, tighter, faster, smaller. In: Pointcheval, D., Johansson, T. (eds.) EUROCRYPT 2012. LNCS, vol. 7237, pp. 700–718. Springer, Heidelberg (2012). doi:10.1007/978-3-642-29011-4_41
31. Peikert, C.: Public-key cryptosystems from the worst-case shortest vector problem: extended abstract. In: STOC 2009, pp. 333–342. ACM (2009)
32. Regev, O.: New lattice based cryptographic constructions. In: STOC, 2003, pp. 407–416. ACM (2003)
33. Regev, O.: On lattices, learning with errors, random linear codes and cryptography. In: STOC 2005, pp. 84–93. ACM (2005)
34. Regev, O.: Lattice-based cryptography. In: Dwork, C. (ed.) CRYPTO 2006. LNCS, vol. 4117, pp. 131–141. Springer, Heidelberg (2006). doi:10.1007/11818175_8
35. Rouselakis, I.: Attribute-based encryption: robust and efficient constructions. In Thesis
36. Sahai, A., Waters, B.: Fuzzy identity-based encryption. In: Cramer, R. (ed.) EUROCRYPT 2005. LNCS, vol. 3494, pp. 457–473. Springer, Heidelberg (2005). doi:10.1007/11426639_27
37. Shamir, A.: How to share a secret. Commun. ACM **22**, 612–613 (1979)
38. Sun, W., Yu, S., Lou, W., Hou, Y.T., Li, H.: Protecting your right: attribute-based keyword search with fine-grained owner-enforced search authorization in the cloud. In: 2014 IEEE Conference on Computer Communications, INFOCOM, 2014, pp. 226–234 (2014)
39. van Liesdonk, P., Sedghi, S., Doumen, J., Hartel, P., Jonker, W.: Computationally efficient searchable symmetric encryption. In: Jonker, W., Petković, M. (eds.) SDM 2010. LNCS, vol. 6358, pp. 87–100. Springer, Heidelberg (2010). doi:10.1007/978-3-642-15546-8_7
40. Wang, C., Li, W., Li, Y., Xu, X.: A ciphertext-policy attribute-based encryption scheme supporting keyword search function. In: Wang, G., Ray, I., Feng, D., Rajarajan, M. (eds.) CSS 2013. LNCS, vol. 8300, pp. 377–386. Springer, Cham (2013). doi:10.1007/978-3-319-03584-0_28
41. Waters, B.: Ciphertext-policy attribute-based encryption: an expressive, efficient, and provably secure realization. In: Catalano, D., Fazio, N., Gennaro, R., Nicolosi, A. (eds.) PKC 2011. LNCS, vol. 6571, pp. 53–70. Springer, Heidelberg (2011). doi:10.1007/978-3-642-19379-8_4
42. Zheng, Q., Xu, S., Ateniese, G.: VABKS: verifiable attribute-based keyword search over outsourced encrypted data. INFOCOM **2014**, 522–530 (2014)

One-Round Exposure-Resilient Identity-Based Authenticated Key Agreement with Multiple Private Key Generators

Atsushi Fujioka[✉]

Kanagawa University, 3-27-1 Rokkakubashi, Kanagawa-ku,
Yokohama-shi, Kanagawa 221-8686, Japan
fujioka@kanagawa-u.ac.jp

Abstract. This paper considers identity-based authenticated key agreement (IBAKA) with multiple private key generators (PKGs). In conventional IBAKA scenarios, a single PKG manages all parties in a system, whereas in a multiple PKG setting, several PKGs exist in a system, and each party is given a private key by a PKG who manages the party. IBAKA is expected to maintain security against exposing private information such as static or ephemeral private keys even in a multiple PKG setting. We define a security model for IBAKA with multiple PKGs to achieve this exposure-resilience property. Based on a security notion, we propose a one-round secure protocol under the gap bilinear Diffie–Hellman assumption in the random oracle model. The protocol utilize the NAXOS approach to embed the gap bilinear Diffie–Hellman instance even when both ephemeral private keys are exposed.

Keywords: Identity-based authenticated key agreement · Multiple private key generators · Gap bilinear Diffie–Hellman assumption · Random oracle model

1 Introduction

Key establishment (KE) is an important cryptologic research area dealing with the basic problem of how to establish a shared key between two communicating parties. This key in turn allows the establishment of secure communication channels between the two parties.

Identity-based authenticated key agreement (IBAKA) enables two parties to share a key via an insecure channel and both parties are assured that only their intended peers can derive the session key, where each party is identified with information called an *identity*. Note that IBAKA does not require an authority to certify public keys, like in a *public-key infrastructure* (PKI) system, but it is necessary that an additional party exists (called a *private key generator* (PKG)) who extracts the private key of each party corresponding to the identity of that party.

© Springer International Publishing AG 2017
R.C.-W. Phan and M. Yung (Eds.): LNCS 10311, Mycrypt 2016, pp. 436–460, 2017.
DOI: 10.1007/978-3-319-61273-7_21

In an IBAKA protocol, a PKG generates a pair comprising a master public and secret keys, and extracts the private key of each party corresponding to the identity of that party. Every party has its own private key generated by the PKG and uses the identity as its own public information. Then, a party (referred to as an *initiator*) who wants to share a key with another party (referred to as a *responder*) sends ephemeral public information to the responder, and the responder sends back other ephemeral public information to the initiator. After such interactions, each party derives a session key from intermediate values computed based on the master public key, the private key of the party, private values related to the ephemeral information of the party, the identity of the peer, and the received ephemeral information. Here, the identity of the peer is used to generate the (static) public key of the peer with the master public key. An IBAKA protocol is said to be secure when no adversary can distinguish between a random key and the actual session key of the session chosen by the adversary.

1.1 Security Models

Several IBAKA security models have been investigated, and they are influenced by the security models of PKI-based authenticated key agreement (AKA). The Bellare–Rogaway (BR) model [1] was the first formal security model for AKA in a symmetric key setting followed by the work by Blake-Wilson, Johnson, and Menezes (BJM model) [3] for PKI-based AKA. Then, the Canetti–Krawczyk (CK) model [6] and the extended Canetti–Krawczyk (eCK) model [22] were proposed. The security notions defined in the above models are given as an indistinguishability game where an adversary is required to differentiate between a random key and a session key of a session. The session is called a *target session*, and is chosen by the adversary. Based on these models, the id-BR (id-BJM)[1] [4,8], id-CK [5], and id-eCK [18] models were defined, respectively. Note that in IBAKA, although the PKG has much more power than the users, no session key between users should be revealed even to the PKG. This property is called *forward secrecy against PKG* (PKG-FS). The CK security and eCK security models are stronger than the BR security model [6,9,22]; however, the CK security and eCK security models are incompatible [10,11]. These relations also hold in the security definitions for IBAKA.

Maximal exposure attacks (MEX) [20] are powerful attacks against AKE protocols, and, needless to say, against IBAKA protocols, also. Thus, IBAKA protocols should be designed to resist MEX. We say that MEX are successful if an adversary can distinguish the session key from a random value under the disclosure of the static or ephemeral private key of the initiator and the static or ephemeral private key of the responder in the session. It is clear that an adversary who obtains both the static and ephemeral private keys of the initiator (or the responder) can trivially distinguish the session key. In other words, an IBAKA protocol is secure against non-trivial exposure of private keys when it resists

---

[1] The model should be called the id-BJM model as the BR model is defined for AKA in a symmetric key setting.

MEX. Thus, we say that an IBAKA protocol is *exposure resilient* when MEX against the protocol are not successful.

The id-eCK model captures the security against MEX. It is well known that an IBAKE protocol is exposure resilient when it is id-eCK secure.

1.2 id-eCK Secure IBAKA

The first id-eCK secure IBAKA protocol was proposed by Huang and Cao [18]. They utilize pairings to construct the protocol, and prove its security under a mathematical assumption referred to as the computational bilinear Diffie–Hellman (CBDH) assumption.

Fujioka et al. [16] invented efficient protocols but these are id-eCK secure under a stronger assumption referred to as the gap bilinear Diffie–Hellman (GBDH) assumption. Later, Fujioka and Suzuki [15] analyzed sufficient conditions to construct IBAKA protocols that are id-eCK secure under the GBDH assumption. The above protocols are based on symmetric pairings and a construction on asymmetric pairings is discussed in [14].

On the other hand, Zhong and Ma proposed an id-eCK secure IBAKA protocol without pairings [29]. They proved its security under another mathematical assumption referred to as the computational Diffie–Hellman assumption.

We stress that all the above protocols assume that a single PKG exists in a system.

1.3 Multiple PKG Setting

In the conventional scenarios described above, IBAKA protocols are constructed based on the premise that a single PKG exists, i.e., the PKG manages all parties in a system. However, this requirement is somewhat rigid in a realistic scenario. Some parties belong to a group managed by a PKG but other parties belong to a different group managed by a different PKG. Therefore, this leads us to a multiple PKG setting. For example, consider a PKI system. An ideal PKI system assumes the existence of a single root CA. However, several root CAs exist in a real PKI system. Thus, it is natural to consider the multi-authority situation.

Chen and Kudla developed an IBAKA protocol with multiple PKGs (*mPKG-IBAKA* protocol) using *pairings*, and proposed a security model for IBAKA in a multiple PKG setting based on the id-BJM model to discuss its security [7].

After their proposal, many protocols have been investigated [12,21,23,24,27]; however, some of them have already been broken. McCullagh and Barreto devised an mPKG-IBAKA protocol [24] to improve the efficiency of the Chen–Kudla protocol. However, this protocol dis not stand up to the cryptanalysis by Xie [28]. Independently, Lee, Kim, Kim, and Oh proposed a new protocol [23] where each PKG generates master keys together with its own parameters although Oh, Moon, and Ma presented an attack on the protocol [26]. Following the attack, Kim, Lee, and Oh [21] modified the protocol by Lee et al., but unfortunately the protocol was analyzed considering only individual security notions.

Vallent, Yoon, and Kim proposed a pairing-free mPKG-IBAKA protocol [27] where each PKG utilizes common parameters and generates its own master keys based on the parameters. The protocol was evaluated based on heuristic security. Recently, Farash and Attari proposed a pairing-free protocol[2] [12] where each PKG generates master keys together with its own parameters; however, their protocol has vulnerabilities that were demonstrated by Mishra and Mukhopadhyay [25]. The protocols by Kim et al. and Vallent et al. have survived; however, both have only been discussed for individual security properties. Moreover, there has been no discussion on whether or not they are exposure resilient.

Therefore, we have the Chen–Kudla protocol, which has provable security but its security was only proven in the id-BJM model with multiple PKGs, meaning that it is unknown whether the protocol is exposure resilient.

On the other hand, Fujioka, Suzuki, Xagawa, and Yoneyama proposed a generic construction of IBAKA protocols based on an *identity-based key encapsulation mechanism* (IB-KEM) [17]. Because this construction is based on a modular approach, it can have a different PKG in the underlying IB-KEM scheme. Therefore, we may have an exposure-resilient IBAKA protocol in a multiple PKG setting. However, the resultant protocol is less efficient because of the following points. (1) The responder must compute a sending value from a receiving value, and thus the protocols are not one-round. Here, one-round means that the initiator and the responder can send their messages independently and simultaneously. (2) The responder needs to send two values although it is sufficient for the initiator to send a single value.

To the best of our knowledge, no one-round exposure-resilient mPKG-IBAKA protocol has yet been proposed.

1.4 Our Contributions

We define a security model for mPKG-IBAKA to acquire exposure resilience. Based on the security model, we propose a one-round exposure-resilient protocol under the GBDH assumption in a random oracle model. The proposed protocol utilizes the NAXOS approach [22] to avoid a situation where an intermediate value becomes necessary for the static keys of the initiator and the responder for different PKGs. Moreover, our protocol is achieved based on exchanging a single group element.

2 Definitions and Assumptions

2.1 Security Model for mPKG-IBAKA

We define a security model for mPKG-IBAKA based on the id-eCK security model proposed by Huang and Cao [18] since the id-eCK security model (and its predecessor, the eCK security model [22]) provides security against MEX.

---

[2] We found the same protocol in [13].

We denote a party as U_i and the identifier of U_i as $id_i^{(\iota)}$, which means that the party who has identifier id_i is managed by PKG P_ι. We outline our model for a two-pass mPKG-IBAKA protocol where parties U_A and U_B exchange ephemeral public keys X_A and X_B, i.e., U_A sends X_A to U_B and U_B sends X_B to U_A, and thereafter derive a session key. The session key depends on the exchanged ephemeral keys, identifiers of the parties, the static keys corresponding to these identifiers, and the protocol instance that is used.

In the model, each party is a probabilistic polynomial-time Turing machine in security parameter λ and obtains a static private key corresponding to its identifier string from its PKG via a secure and authenticated channel.

Session. An invocation of a protocol is called a *session*. A session is activated via an incoming message in the form of $(\Pi, \mathcal{I}, id_A^{(\alpha)}, id_B^{(\beta)})$ or $(\Pi, \mathcal{R}, id_A^{(\alpha)}, id_B^{(\beta)}, X_B)$, where Π is a protocol identifier. If U_A is activated with $(\Pi, \mathcal{I}, id_A^{(\alpha)}, id_B^{(\beta)})$, then U_A is the session *initiator*; otherwise, it is the session *responder*. After activation, U_A appends ephemeral public key X_A to the incoming message and sends it as an outgoing response. If U_A is the responder, U_A computes a session key. If U_A is the initiator, U_A that has been successfully activated via $(\Pi, \mathcal{I}, id_A^{(\alpha)}, id_B^{(\beta)})$ can be further activated via $(\Pi, \mathcal{I}, id_A^{(\alpha)}, id_B^{(\beta)}, X_A, X_B)$ to compute a session key.

If U_A is the initiator of a session, the session is identified by either $\mathsf{sid} = (\Pi, \mathcal{I}, id_A^{(\alpha)}, id_B^{(\beta)}, X_A)$ or $\mathsf{sid} = (\Pi, \mathcal{I}, id_A^{(\alpha)}, id_B^{(\beta)}, X_A, X_B)$. If U_B is the responder of a session, the session is identified by $\mathsf{sid} = (\Pi, \mathcal{R}, id_B^{(\beta)}, id_A^{(\alpha)}, X_A, X_B)$. We say that U_A is the *owner* (resp. *peer*) of session sid if the third (resp. fourth) coordinate of session sid is $id_A^{(\alpha)}$. We say that a session is *completed* if its owner computes a session key. The *matching session* of $(\Pi, \mathcal{I}, id_A^{(\alpha)}, id_B^{(\beta)}, X_A, X_B)$ is session $(\Pi, \mathcal{R}, id_B^{(\beta)}, id_A^{(\alpha)}, X_A, X_B)$ and vice versa.

Adversary. Adversary $\mathcal{A}$ is modeled as a probabilistic Turing machine that controls all communications between parties including session activation. Activation is performed via a Send(MESSAGE) query. The MESSAGE has one of the following forms: $(\Pi, \mathcal{I}, id_A^{(\alpha)}, id_B^{(\beta)})$, $(\Pi, \mathcal{R}, id_A^{(\alpha)}, id_B^{(\beta)}, X_A)$, or $(\Pi, \mathcal{I}, id_A^{(\alpha)}, id_B^{(\beta)}, X_A, X_B)$. Each party submits its responses to adversary $\mathcal{A}$, who decides the global delivery order. Note that adversary $\mathcal{A}$ does not control the communication between each party and its PKG.

The private information of a party is not accessible to adversary $\mathcal{A}$; however, leakage of private information is obtained via the following adversary queries.

- SessionKeyReveal(sid): $\mathcal{A}$ obtains the session key for session sid, provided that the session holds a session key.
- EphemeralKeyReveal(sid): $\mathcal{A}$ obtains the ephemeral private key (of the session owner) associated with session sid.
- StaticKeyReveal($id_i^{(\iota)}$): $\mathcal{A}$ learns the static private key of party U_i managed by PKG P_ι.

- MasterKeyReveal(P_ι): $\mathcal{A}$ learns the master secret key of PKG P_ι. For the sake of convenient queries, when MasterKeyReveal() is called, i.e., called with no argument, the master secret keys of all PKGs are returned.
- NewParty($id_i^{(\iota)}$): This query models malicious insiders. If a party is established by a NewParty($id_i^{(\iota)}$) query issued by $\mathcal{A}$, then we refer to the party as *dishonest*. If not, the party is referred to as *honest*.

Freshness. Our security definition requires the following "freshness" notion.

Definition 2.1 (freshness). *Let* sid* *be the session identifier of a completed session owned by honest party* U_A *with peer* U_B *who is also honest. We assume that* U_A *and* U_B *belong to the domains managed by* P_α *and* P_β, *respectively. If a matching session exists, then let* $\overline{\text{sid}^*}$ *be the session identifier of the matching session of* sid*. *We define* sid* *to be* fresh *if none of the following conditions hold.*

1. *$\mathcal{A}$ issues* SessionKeyReveal(sid*) *or* SessionKeyReveal($\overline{\text{sid}^*}$) *(if* $\overline{\text{sid}^*}$ *exists).*
2. $\overline{\text{sid}^*}$ *exists and $\mathcal{A}$ makes either of the following queries*
   - *both* StaticKeyReveal($id_A^{(\alpha)}$) *and* EphemeralKeyReveal(sid*), *or*
   - *both* StaticKeyReveal($id_B^{(\beta)}$) *and* EphemeralKeyReveal($\overline{\text{sid}^*}$).
3. $\overline{\text{sid}^*}$ *does not exist and $\mathcal{A}$ makes either of the following queries*
   - *both* StaticKeyReveal($id_A^{(\alpha)}$) *and* EphemeralKeyReveal(sid*), *or*
   - StaticKeyReveal($id_B^{(\beta)}$).

Note that if adversary $\mathcal{A}$ issues MasterKeyReveal(), *we regard $\mathcal{A}$ as having issued both* StaticKeyReveal($id_A^{(\alpha)}$) *and* StaticKeyReveal($id_B^{(\beta)}$). *In addition, if $\mathcal{A}$ issues* MasterKeyReveal(P_α) *(resp.* MasterKeyReveal(P_β)), *we regard $\mathcal{A}$ as having issued* StaticKeyReveal($id_A^{(\alpha)}$) *(resp.* StaticKeyReveal($id_B^{(\beta)}$)).

Security Experiment. Adversary $\mathcal{A}$ starts with common parameters, a set of master public keys together, and a set of honest parties for whom $\mathcal{A}$ adaptively selects identifiers. Note that each identifier is statically bound with its PKG, and thus, when $\mathcal{A}$ selects an identifier, the PKG who manages the identifier is automatically determined. The adversary makes an arbitrary sequence of the queries described above. During the experiment, $\mathcal{A}$ makes a special query, Test(sid*), and is given with equal probability either the session key held by sid* or a random key. The experiment continues until $\mathcal{A}$ makes a guess regarding whether or not the key is random. The adversary *wins* the game if the test session, sid*, is *fresh* at the end of execution and if the guess by $\mathcal{A}$ was correct.

Definition 2.2 (id(m)-eCK security). *The advantage of adversary $\mathcal{A}$ in the experiment with mPKG-IBAKA protocol Π is defined as*

$$\mathbf{Adv}_\Pi^{\text{mPKG-IBAKA}}(\mathcal{A}) = \Pr[\mathcal{A} \ wins] - \frac{1}{2}.$$

We say that Π is a secure mPKG-IBAKA protocol in the id(m)-eCK model if the following conditions hold.

1. *If two honest parties have a complete matching session, then except with negligible probability in security parameter λ, they both derive the same session key.*
2. *Advantage $\mathbf{Adv}_{\Pi}^{\mathrm{mPKG-IBAKA}}(\mathcal{A})$ is negligible in security parameter λ for any probabilistic polynomial-time adversary $\mathcal{A}$.*

Model Assumptions. We assume that there exists a *common parameter generator* (CPG) that is a probabilistic polynomial-time Turing machine in security parameter λ and the CPG generates common parameters. Note that the CPG has no secret information. Based on these common parameters, each PKG generates its master public and private keys. Note that the PKGs are also probabilistic polynomial-time Turing machines in security parameter λ.

We further assume that any identifier of a user is unique globally, i.e., every identifier managed by a PKG must differ from an identifier managed by a different PKG. In other words, there is a binding between an identifier and its PKG.

2.2   Number Theoretic Assumptions on Pairings

As a tool to actualize IBAKA protocols, we have a mathematical function called *pairing*. Symmetric pairing function[3] e is a polynomial-time computable bilinear non-degenerate map from two group elements to an element of another group where $e(g^a, g^b) = g_T^{ab}$ holds when $\mathbb{G}$ and $\mathbb{G}_T$ are two cyclic groups of order q, g is a generator of $\mathbb{G}$, $g_T = e(g,g)$ $(\in \mathbb{G}_T)$, and $a, b \in \mathbb{Z}_q$.

Roughly speaking, the computational bilinear Diffie–Hellman (CBDH) problem is to compute $e(g,g)^{uvw}$ on input (U, V, W) and the decisional bilinear Diffie–Hellman (DBDH) problem is to determine whether or not $uvw = x \bmod q$ holds on input (U, V, W, X), where $U = g^u$, $V = g^v$, $W = g^w$, and $X = e(g,g)^x$.

The gap bilinear Diffie–Hellman (GBDH) problem is described below. Assume that a CBDH solver can access a DBDH oracle. The GBDH problem is to compute $e(g,g)^{uvw}$ on input (U, V, W) with help from the DBDH oracle. Here $U = g^u$, $V = g^v$, and $W = g^w$. Let $\mathcal{A}$ be an adversary who is given inputs U, V, and W $(\in \mathbb{G})$ selected uniformly at random, accesses the DBDH oracle, and tries to compute $e(g,g)^{uvw}$. Roughly, speaking, the GBDH assumption is that the GBDH problem is hard for every polynomial-time adversary $\mathcal{A}$.

We formally state the GBDH assumption as follows. Let $\mathbb{G}$ and $\mathbb{G}_T$ be cyclic groups of order q where pairing function, $e : \mathbb{G}^2 \rightarrow \mathbb{G}_T$ exists, and $\mathbb{G}$ has generator g. CBDH function $\mathrm{CBDH} : \mathbb{G}^3 \rightarrow \mathbb{G}_T$ is a function that takes input (U, V, W) and returns $e(g,g)^{uvw}$. DBDH predicate $\mathrm{DBDH} : \mathbb{G}^3 \times \mathbb{G}_T \rightarrow \{0,1\}$ is a function that takes input (U, V, W, X) and returns bit 1 if $uvw = x \bmod q$ and bit 0 otherwise, where $U = g^u$, $V = g^v$, $W = g^w$, and $X = e(g,g)^x$. The GBDH

---

[3] Pairing $e : \mathbb{G}_1 \times \mathbb{G}_2 \rightarrow \mathbb{G}_T$ is referred to as *symmetric* when $\mathbb{G}_1 = \mathbb{G}_2$ and *asymmetric* when $\mathbb{G}_1 \neq \mathbb{G}_2$.

problem is to compute $\text{CBDH}(U, V, W)$ allowing access to oracle $\text{DBDH}(\cdot, \cdot, \cdot, \cdot)$. For adversary $\mathcal{A}$, we define advantage

$$\mathbf{Adv}^{\text{GBDH}}(\mathcal{A}) = \Pr\left[U, V, W \in_R \mathbb{G}, \mathcal{A}^{\text{DBDH}(\cdot, \cdot, \cdot, \cdot)}(U, V, W) = \text{CBDH}(U, V, W)\right],$$

where the probability is taken over the choices of U, V, W and the random tape of $\mathcal{A}$.

Definition 2.3 (GBDH assumption). *We say that $\mathbb{G}$ and $\mathbb{G}_T$ satisfy the GBDH assumption if, for any adversary $\mathcal{A}$ running in a polynomial-time, advantage $\mathbf{Adv}^{\text{GBDH}}(\mathcal{A})$ is negligible in security parameter λ.*

3 Existing Protocols

3.1 Chen–Kudla Protocol

We outline the Chen–Kudla protocol. For the precise description of the protocol, refer to [7].

PKG P_α (resp. P_β) publishes Z_α (resp. Z_β) as a master public key and keep z_α (resp. z_β) secret as a master secret key.

User U_A (resp. U_B) with identifier $id_A^{(\alpha)}$ (resp. $id_B^{(\beta)}$), which is managed by P_α (resp. P_β), is assigned static private key D_A ($= Q_A^{z_\alpha} \in \mathbb{G}$) (resp. D_B ($= Q_B^{z_\beta} \in \mathbb{G}$)) where $Q_A = H_1(id_A^{(\alpha)})$ ($\in \mathbb{G}$) (resp. $Q_B = H_1(id_B^{(\beta)})$ ($\in \mathbb{G}$)).

U_A chooses a uniformly random ephemeral private key, x_A ($\in_R \mathbb{Z}_q$), computes the ephemeral public key, $X_A = g^{x_A}$, and sends (Π, $id_A^{(\alpha)}$, $id_B^{(\beta)}$, X_A) to U_B.

Upon receiving (Π, $id_A^{(\alpha)}$, $id_B^{(\beta)}$, X_A), U_B chooses ephemeral private key x_B ($\in_R \mathbb{Z}_q$), computes the ephemeral public key, $X_B = g^{x_B}$, and responds to U_A with (Π, $id_A^{(\alpha)}$, $id_B^{(\beta)}$, X_A, X_B).

U_B also computes $Q_A = H_1(id_A^{(\alpha)})$, the shared secrets,

$$\sigma_1 = e(D_B, X_A) \cdot e(Q_A, Z_\alpha^{x_B}) \text{ and } \sigma_2 = X_A^{x_B},$$

and session key K as $K = H(\sigma_1, \sigma_2, \Pi, id_A^{(\alpha)}, id_B^{(\beta)}, X_A, X_B)$. Then, U_B completes the session with session key K.

Upon receiving (Π, $id_A^{(\alpha)}$, $id_B^{(\beta)}$, X_A, X_B), U_A computes $Q_B = H_1(id_B^{(\beta)})$, the shared secrets,

$$\sigma_1 = e(D_A, X_B) \cdot e(Q_B, Z_\beta^{x_A}) \text{ and } \sigma_2 = X_B^{x_A},$$

and session key K as $K = H(\sigma_1, \sigma_2, \Pi, id_A^{(\alpha)}, id_B^{(\beta)}, X_A, X_B)$. Then, U_A completes the session with session key K (Fig. 1).

The Chen–Kudla protocol is secure in the multiple PKG setting of the id-BJM model [7]. We discuss exposure resilience of the protocol in Subsect. 4.4.

$$\frac{Z_\alpha = g^{z_\alpha} \hspace{5cm} Z_\beta = g^{z_\beta}}{\begin{array}{ll} Q_A = H_1(id_A^{(\alpha)}) & Q_B = H_1(id_B^{(\beta)}) \\ D_A = Q_A^{z_\alpha} & D_B = Q_B^{z_\beta} \end{array}}$$

$$\begin{array}{cc} x_A \in_R \mathbb{Z}_q & \\ X_A = g^{x_A} & \xrightarrow{\;X_A\;} \quad x_B \in_R \mathbb{Z}_q \\ & \xleftarrow{\;X_B\;} \quad X_B = g^{x_B} \end{array}$$

$$\begin{array}{cc} \sigma_1 = e(D_A, X_B) \cdot e(Q_A, Z_\alpha^{x_B}) & \sigma_1 = e(D_B, X_A) \cdot e(Q_B, Z_\beta^{x_A}) \\ \sigma_2 = X_B^{x_A} & \sigma_2 = X_A^{x_B} \end{array}$$

$$K = H(\sigma_1, \sigma_2, \Pi, id_A^{(\alpha)}, id_B^{(\beta)}, X_A, X_B)$$

Fig. 1. Outline of Chen–Kudla protocol.

3.2 Protocol on FSXY Construction

Fujioka, Suzuki, Xagawa, and Yoneyama proposed a generic construction of IBAKA protocols based on an *identity-based key encapsulation mechanism* (IB-KEM) scheme [17]. We call it FSXY construction.

We outline an IBAKE protocol on the FSXY construction. For the precise description of the protocol, refer to [17].

The FSXY construction uses an IB-KEM scheme and a KEM scheme. Let $\Sigma = (\mathsf{KeyDer}, \mathsf{KeyDer}, \mathsf{EnCap}, \mathsf{DeCap})$ (resp. $\Sigma' = (\mathsf{wKeyGen}, \mathsf{wEnCap}, \mathsf{wDeCap})$) be the IB-KEM scheme (resp. the KEM scheme) where KeyDer, KeyDer, EnCap, and DeCap are the key generation algorithm, the key generation algorithm, the key derivation algorithm, the key encapsulation algorithm, and the key decapsulation algorithm in Σ, respectively, and $\mathsf{wKeyGen}$, wEnCap, and wDeCap are the key generation algorithm, the key derivation algorithm, the key encapsulation algorithm, and the key decapsulation algorithm in Σ', respectively.

PKG P_α (resp. P_β) publishes mpk_α (resp. mpk_β) as a master public key and keep msk_α (resp. msk_β) secret as a master secret key.

User U_A (resp. U_B) with identifier $id_A^{(\alpha)}$ (resp. $id_B^{(\beta)}$), which is managed by P_α (resp. P_β), is assigned static private key dk_A ($= \mathsf{KeyDer}(msk_\alpha, id_A^{(\alpha)})$) (resp. dk_B ($= \mathsf{KeyDer}(msk_\beta, id_B^{(\beta)})$)).

U_A runs EnCap with $id_B^{(\beta)}$ to obtain (ct_A, ρ_1) where ct_A and ρ_1 are a ciphertext and the (encapsulated) key in the IB-KEM scheme, respectively, runs $\mathsf{wKeyGen}$ with security parameter λ to obtain (ek_T, dk_T) where ek_T and ek_T are a encryption key and the decryption key in the KEM scheme, respectively, and sends $(\Pi, id_A^{(\alpha)}, id_B^{(\beta)}, ct_A, ek_T)$ to U_B.

Upon receiving $(\Pi, id_A^{(\alpha)}, id_B^{(\beta)}, ct_A, ek_T)$, U_B decrypts the ciphertext, ct_A, to obtain ρ_1, runs EnCap with $id_A^{(\alpha)}$ to obtain (ct_B, ρ_2) where ct_B and ρ_2 are a ciphertext and the (encapsulated) key in IB-KEM, respectively, runs wEnCap to obtain (ct_T, ρ_3) where ct_T and ρ_3 are a ciphertext and the (encapsulated) key in KEM, respectively, and responses to U_B with $(\Pi, id_A^{(\alpha)}, id_B^{(\beta)}, ct_A, ek_T, ct_B, ct_T)$.

U_B also computes the shared secrets,

$$\sigma_1 = \text{KDF}(\rho_1), \quad \sigma_2 = \text{KDF}(\rho_2), \quad \text{and } \sigma_3 = \text{KDF}(\rho_3),$$

session state information st as $st = (id_A^{(\alpha)}, id_B^{(\beta)}, ct_A, ek_T, ct_B, ct_T)$ and session key K as $K = G_{\sigma_1}(st) \oplus G_{\sigma_2}(st) \oplus G_{\sigma_3}(st)$. Then, U_B completes the session with session key K.

Upon receiving $(\Pi, id_A^{(\alpha)}, id_B^{(\beta)}, ct_A, ek_T, ct_B, ct_T)$, U_A decrypts ct_B and ct_T to obtain ρ_2 and ρ_3, respectively.

U_A also computes the shared secrets,

$$\sigma_1 = \text{KDF}(\rho_1), \quad \sigma_2 = \text{KDF}(\rho_2), \quad \text{and } \sigma_3 = \text{KDF}(\rho_3),$$

session state information st as $st = (id_A^{(\alpha)}, id_B^{(\beta)}, ct_A, ek_T, ct_B, ct_T)$ and session key K as $K = G_{\sigma_1}(st) \oplus G_{\sigma_2}(st) \oplus G_{\sigma_3}(st)$. Then, U_A completes the session with session key K (Fig. 2).

mpk_α, msk_α	mpk_β, msk_β
$dk_A = \text{KeyDer}(msk_\alpha, id_A^{(\alpha)})$	$dk_B = \text{KeyDer}(msk_\beta, id_B^{(\beta)})$
$(ct_A, \rho_1) = \text{EnCap}(mpk_\beta, id_B^{(\beta)})$	
$(ek_T, dk_T) = \text{wKeyGen}(1^\lambda)$ $\overset{ct_A, ek_T}{\longrightarrow}$ $\rho_1 = \text{DeCap}(dk_B, ct_A)$	
$(ct_B, \rho_2) = \text{EnCap}(mpk_\alpha, id_A^{(\alpha)})$	
$\rho_2 = \text{DeCap}(dk_A, ct_B)$ $\overset{ct_B, ct_T}{\longleftarrow}$ $(ct_T, \rho_3) = \text{wEnCap}(ek_T)$	
$\rho_3 = \text{wDeCap}(dk_T, ct_T)$	

$$\sigma_1 = \text{KDF}(\rho_1)$$
$$\sigma_2 = \text{KDF}(\rho_2)$$
$$\sigma_3 = \text{KDF}(\rho_3)$$
$$st = (id_A^{(\alpha)}, id_B^{(\beta)}, ct_A, ek_T, ct_B, ct_T)$$
$$K = G_{\sigma_1}(st) \oplus G_{\sigma_2}(st) \oplus G_{\sigma_3}(st)$$

Fig. 2. Outline of protocol on FSXY construction.

Note that U_A computes randomness used in EnCap as $F_{\tau_A}(r_A) \oplus F_{r'_A}(\tau'_A)$ for a security reason where F is a hash function, τ_A, τ'_A are parts of the secret key, and r_A, r'_A are ephemeral randomness. U_B computes as $F_{\tau_B}(r_B) \oplus F_{r'_B}(\tau'_B)$, also.

It is proved that a protocol on the FSXY construction is exposure resilient in the standard model [17]. Because this construction is based on a modular approach, it can have a different PKG in the underlying IB-KEM scheme. Therefore, we may have an exposure-resilient IBAKA protocol in a multiple PKG setting.

However, it is clear that the resultant protocol is not one-round as the responder's message is computed on the initiator's message, i.e., encapsulation key ek_T. Here, one-round means that the initiator and the responder can send their messages independently and simultaneously.

In addition, the resultant protocol is less efficient as the responder needs to send two ciphertexts although it is sufficient for the initiator to send a single one. We compared efficiency of the protocol on the FSXY construction with that of the proposed protocol in Subsect. 4.4.

4   Exposure-Resilient mPKG-IBAKA Protocol

4.1   Proposed Protocol

In this section, we describe actions required to execute a session.

The proposed IBAKA protocol, Π, is described as follows.

Let λ be the security parameter.

The CPG generates cyclic groups $\mathbb{G}$ and $\mathbb{G}_T$ where their orders are λ-bit prime q, g is a generator of $\mathbb{G}$, $e : \mathbb{G}^2 \to \mathbb{G}_T$ is a pairing function, and $g_T = e(g, g)$. Let $H : \{0,1\}^* \to \{0,1\}^\lambda$, $H_1 : \{0,1\}^* \longrightarrow \mathbb{G}$, and $H_2 : \{0,1\}^* \longrightarrow \mathbb{Z}_q$ be cryptographic hash functions modeled as random oracles [2]. The CPG outputs $(\mathbb{G}, \mathbb{G}_T, g, g_T, q, e, H, H_1, H_2)$ as common parameters.

Based on these common parameters, each PKG P_ι randomly selects master secret key z_ι ($\in_R \mathbb{Z}_q$), and publishes master public key Z_ι ($= g^{z_\iota} \in \mathbb{G}$).

User U_i with identifier $id_i^{(\iota)}$, which is managed by P_ι, is assigned static private key D_i ($= Q_i^{z_\iota} \in \mathbb{G}$) where $Q_i = H_1(id_i^{(\iota)})$ ($\in \mathbb{G}$). We refer to Q_i as the *static public key* of user U_i, and note that Q_i is expressed with some q_i ($\in \mathbb{Z}_q$) as $Q_i = g^{q_i}$.

Thus, the identifier and static public (resp. private) key of U_A are $id_A^{(\alpha)}$ and $Q_A = H_1(id_A^{(\alpha)}) = g^{q_A}$ (resp. $D_A = g^{z_\alpha q_A}$), and the identifier and static public (resp. private) key of U_B are $id_B^{(\beta)}$ and $Q_B = H_1(id_B^{(\beta)}) = g^{q_B}$ (resp. $D_B = g^{z_\beta q_B}$), respectively.

Key Agreement. User U_A is the session initiator and user U_B is the session responder.

1. U_A chooses a uniformly random ephemeral private key, x_A ($\in_R \mathbb{Z}_q$), computes the ephemeral public key, $X_A = g^{x'_A}$, where $x'_A = H_2(x_A, D_A)$, and sends (Π, $id_A^{(\alpha)}$, $id_B^{(\beta)}$, X_A) to U_B.
2. Upon receiving (Π, $id_A^{(\alpha)}$, $id_B^{(\beta)}$, X_A), U_B chooses ephemeral private key x_B ($\in_R \mathbb{Z}_q$), computes the ephemeral public key, $X_B = g^{x'_B}$, where $x'_B = H_2(x_B, D_B)$, and responds to U_A with (Π, $id_A^{(\alpha)}$, $id_B^{(\beta)}$, X_A, X_B). U_B also computes $Q_A = H_1(id_A^{(\alpha)})$, the shared secrets,

$$\sigma_1 = e(Q_A^{x'_B}, Z_\alpha), \quad \sigma_2 = e(D_B, X_A), \quad \text{and} \quad \sigma_3 = X_A^{x'_B},$$

and session key K as $K = H(\sigma_1, \sigma_2, \sigma_3, \Pi, id_A^{(\alpha)}, id_B^{(\beta)}, X_A, X_B)$. Then, U_B completes the session with session key K.

3. Upon receiving $(\Pi, id_A^{(\alpha)}, id_B^{(\beta)}, X_A, X_B)$, U_A computes $Q_B = H_1(id_B^{(\beta)})$, the shared secrets,

$$\sigma_1 = e(D_A, X_B), \ \sigma_2 = e(Q_B^{x'_A}, Z_\beta), \ \text{and} \ \sigma_3 = X_B^{x'_A},$$

and session key K as $K = H(\sigma_1, \sigma_2, \sigma_3, \Pi, id_A^{(\alpha)}, id_B^{(\beta)}, X_A, X_B)$. Then, U_A completes the session with session key K (Fig. 3).

$Z_\alpha = g^{z_\alpha}$	$Z_\beta = g^{z_\beta}$
$Q_A = H_1(id_A^{(\alpha)})$	$Q_B = H_1(id_B^{(\beta)})$
$D_A = Q_A^{z_\alpha}$	$D_B = Q_B^{z_\beta}$

$$x_A \in_R \mathbb{Z}_q$$
$$x'_A = H_2(x_A, D_A)$$
$$X_A = g^{x'_A} \xrightarrow{\quad X_A \quad} \quad x_B \in_R \mathbb{Z}_q$$
$$x'_B = H_2(x_B, D_B)$$
$$\xleftarrow{\quad X_B \quad} \quad X_B = g^{x'_B}$$

$\sigma_1 = e(D_A, X_B)$	$\sigma_1 = e(Q_A^{x_B}, Z_\alpha)$
$\sigma_2 = e(Q_B^{x'_A}, Z_\beta)$	$\sigma_2 = e(D_B, X_A)$
$\sigma_3 = X_B^{x'_A}$	$\sigma_3 = X_A^{x'_B}$

$$K = H(\sigma_1, \sigma_2, \sigma_3, \Pi, id_A^{(\alpha)}, id_B^{(\beta)}, X_A, X_B)$$

Fig. 3. Outline of proposed protocol.

Both parties compute the shared secrets,

$$\sigma_1 = g_T^{z_\alpha q_A x'_B}, \ \sigma_2 = g_T^{z_\beta q_B x'_A}, \ \text{and} \ \sigma_3 = g^{x'_A x'_B},$$

where $q_A = \log_g Q_A$, $q_B = \log_g Q_B$. Here, $\log_g U$ denotes the logarithm of U, i.e., $U = g^{\log_g U}$. Therefore, they can derive the same session key, K. We may omit g in $\log_g U$ as $\log U$ if obvious.

Note that our protocol is achieved based on exchanging a single group element, and thus, it is the most efficient.

An ordinary strategy to prove that an IBAKA protocol is exposure resilient under a mathematical assumption in a random oracle model [2] is to embed an instance of the problem in an intermediate value. In such a situation, the adversary is forced to query the random oracle regarding the intermediate value (or related value), and therefore, we can construct a problem solver. To accomplish this, we simply need four types of intermediate values: a value computed based on the static keys of the initiator and the responder, a value computed based on the static key of the initiator and the ephemeral key of the responder, a value computed based on the ephemeral key of the initiator and the static key of the responder, and a value computed based on the ephemeral keys of the initiator and the responder.

As discussed later (in Subsect. 4.4), it is difficult for the initiator and the responder straightforwardly to share a value computed based on the static keys of the initiator and the responder in the multiple PKG setting. Thus, we adopt the NAXOS technique [22] to avoid this situation.

The NAXOS technique, invented to construct the NAXOS protocol, is that the exponent of the ephemeral public key is generated with the ephemeral and static private keys using a hash function, i.e., H_2 in our protocol. Thus, an adversary cannot obtain any information regarding the exponent since the adversary cannot access both private keys. This implies that the solver can embed one of the GBDH instances into this ephemeral public key. We avoid the situation where the solver needs to embed the GBDH instance into the static public keys. Therefore, the protocol does not need a shared secret computed from the static private key and the peer static public key.

4.2 Security

The proposed mPKG-IBAKA protocol is secure based on our security notion under the GBDH assumption in the random oracle model [2].

Theorem 4.1. *If $\mathbb{G}$ and $\mathbb{G}_T$ are groups where the GBDH assumption holds and H, H_1, and H_2 are random oracles, the proposed IBAKA protocol, Π, is secure in the id(m)-eCK model.*

In particular, for any IBAKA adversary $\mathcal{A}$ against Π that runs in at most t time, involves at most n honest parties, activates at most s sessions, and makes at most h queries to the random oracles, there exists a GBDH solver, $\mathcal{S}$, such that

$$\mathbf{Adv}^{\mathrm{GBDH}}(\mathcal{S}) \geq \frac{1}{n^2 s^2} \mathbf{Adv}_{\Pi}^{\mathrm{mPKG-IBAKA}}(\mathcal{A}),$$

where $\mathcal{S}$ runs in time t plus time to perform $\mathcal{O}((n + s) \log q)$ group operations and makes $\mathcal{O}(h + s)$ queries to the DBDH oracle.

Proof. We need the GBDH assumption in pairing groups $\mathbb{G}$ and $\mathbb{G}_T$ of prime order q with generators g and g_T, respectively, when trying to compute the answer, $\mathrm{CBDH}(U, V, W)$, from instance (U, V, W), while accessing the DBDH oracle, $\mathrm{CBDH}(g^u, g^v, g^w) = g_T^{uvw}$, and the DBDH oracle on input (g^u, g^v, g^w, g_T^x) returns bit 1 if $uvw = x$, or bit 0 otherwise.

We show that if a polynomially bounded adversary can distinguish the session key of a fresh session from a randomly chosen session key, we can solve the GBDH problem. Let λ denote the security parameter, and let $\mathcal{A}$ be a polynomial-time adversary in security parameter λ. Here, we assume that $\lambda = \log q$. We use $\mathcal{A}$ to construct the GBDH solver, $\mathcal{S}$, that succeeds in solving a CBDH instance with non-negligible probability using the DBDH oracle. Adversary $\mathcal{A}$ is said to be successful with non-negligible probability if $\mathcal{A}$ wins the distinguishing game with probability $\frac{1}{2} + f(\lambda)$ where $f(\lambda)$ is non-negligible, and event M denotes that $\mathcal{A}$ is successful.

Let the test session be $\mathtt{sid}^*$, and $\mathtt{sid}^*$ be either $(\Pi, \mathcal{I}, id_A^{(\alpha)}, id_B^{(\beta)}, X_A, X_B)$ or $(\Pi, \mathcal{R}, id_B^{(\beta)}, id_A^{(\alpha)}, X_A, X_B)$, which is a completed session between honest users U_A and U_B where users U_A and U_B are the initiator and the responder of test session sid*, respectively. Let H^* be the event that $\mathcal{A}$ queries $(\sigma_1, \sigma_2, \sigma_3, \Pi, id_A^{(\alpha)}, id_B^{(\beta)}, X_A, X_B)$ to H. Let $\overline{H^*}$ be the complement of event H^*. Let sid be any completed session owned by an honest user such that sid $\neq$ sid* and sid does not match sid*. Since sid and sid* are distinct and non-matching, the inputs to key derivation function H are different for sid and sid*. Since H is a random oracle, adversary $\mathcal{A}$ cannot obtain any information regarding the test session key from the session keys of non-matching sessions. Hence, $\Pr[M \wedge \overline{H^*}] \leq \frac{1}{2}$ and $\Pr[M] = \Pr[M \wedge H^*] + \Pr[M \wedge \overline{H^*}] \leq \Pr[M \wedge H^*] + \frac{1}{2}$, where $f(\lambda) \leq \Pr[M \wedge H^*]$. Henceforth, M^* denotes event $M \wedge H^*$.

We denote a user as U_i. User U_i and other parties are modeled as probabilistic polynomial-time Turing machines in security parameter λ. We denote a master secret (resp. public) key of P_ι by z_ι (resp. Z_ι). For user U_i, we denote the static private (resp. public) key as D_i (resp. Q_i) and an ephemeral private (resp. public) key as x_i (resp. X_i). We also denote the session key as K. We assume that $\mathcal{A}$ succeeds in an environment with n users and at most n PKGs, and activates at most s sessions within a user.

We consider the non-exclusive classification of all possible events based on the freshness conditions in Tables 1 and 2, where users U_A and U_B are the initiator and the responder of test session sid*, respectively. For example, when the matching session of sid* does not exist, $\mathcal{A}$ is not allowed to access the session key of the test session and U_B's static private key but is allowed to access U_A's static private key. This corresponds to event E_1. When the matching session of sid* does not exist, $\mathcal{A}$ is not allowed to access the session key of the test session and U_B's static private key but is allowed to access U_A's ephemeral private key. This corresponds to event E_2. Other freshness conditions correspond to events $E_3, \ldots, E_6$, respectively. Table 1 classifies events when identifiers $id_A^{(\alpha)}$ and $id_B^{(\beta)}$

Table 1. Classification of attacks when identifiers $id_A^{(\alpha)}$ and $id_B^{(\beta)}$ are distinct and are managed by different PKGs.

	z_α	z_β	D_A	x_A	D_B	x_B	Instance embedding	Suc. Prob.
E_1	r	ok	r	ok	ok	n	$Z_\beta = U, X_A = V, Q_B = W$	$p_1/n^2 s$
E_2	ok	ok	ok	r	ok	n	$Z_\beta = U, X_A = V, Q_B = W$	$p_2/n^2 s$
E_3	r	ok	r	ok	ok	r	$Z_\beta = U, X_A = V, Q_B = W$	$p_3/n^2 s$
E_4	ok	ok	ok	r	ok	r	$Z_\beta = U, X_A = V, Q_B = W$	$p_4/n^2 s$
E_5	r	r	r	ok	r	ok	$X_A = V, X_B = W$	$p_5/n^2 s^2$
E_6	ok	r	ok	r	r	ok	$Z_\alpha = U, Q_A = V, X_B = W$	$p_6/n^2 s$

In the table, "ok" means that the secret/private key is not revealed, "r" means that the secret/private key may be revealed, and "n" means that no matching session exists. The "Instance Embedding" column shows how the simulator embeds an instance of the GBDH problem. The "Suc. Prob." column shows the probability of success of the simulator where $p_i = \Pr[E_i \wedge M^*]$ and n and s are the numbers of parties and sessions, respectively.

Table 2. Classification of attacks when identifiers $id_A^{(\alpha)}$ and $id_B^{(\beta)}$ are distinct but are managed by the same PKG, P_α $(= P_\beta)$. Here, $p_i' = \Pr[E_i' \wedge M^*]$.

	z_α	D_A	x_A	D_B	x_B	Instance embedding	Suc. Prob.
E_1'	ok	r	ok	ok	n	$Z_\alpha = U, X_A = V, Q_B = W$	$p_1'/n^2 s$
E_2'	ok	ok	r	ok	n	$Z_\alpha = U, X_A = V, Q_B = W$	$p_2'/n^2 s$
E_3'	ok	r	ok	ok	r	$Z_\alpha = U, X_A = V, Q_B = W$	$p_3'/n^2 s$
E_4'	ok	ok	r	ok	r	$Z_\alpha = U, X_A = V, Q_B = W$	$p_4'/n^2 s$
E_5'	r	r	ok	r	ok	$X_A = U, X_B = V$	$p_5'/n^2 s^2$
E_6'	ok	ok	r	r	ok	$Z_\alpha = U, Q_A = V, X_B = W$	$p_6'/n^2 s$

are distinct and are managed by different PKGs, P_α and P_β, respectively. Table 2 classifies events when identifiers $id_A^{(\alpha)}$ and $id_B^{(\beta)}$ are distinct but are managed by the same PKG, i.e., $P_\alpha = P_\beta$. Needless to say, we do not consider an event in which $\mathcal{A}$ does not query either StaticKeyReveal() or EphemeralKeyReveal() as we can embed an instance into the ephemeral and static keys in the test session.

Since the classification covers all possible events, at least one event $E_i \wedge M^*$ or $E_i' \wedge M^*$ in the tables occurs with non-negligible probability if event M^* occurs with non-negligible probability. Thus, the GBDH problem can be solved with non-negligible probability, which means that the proposed protocol is secure under the GBDH assumption.

Event E_i (and E_i'). We consider the following events that cover all cases of behavior of adversary $\mathcal{A}$.

- Let E_1 be the event for which test session $\texttt{sid}^*$ has no matching session $\overline{\texttt{sid}^*}$ and $\mathcal{A}$ queries StaticKeyReveal($id_A^{(\alpha)}$).
- Let E_2 be the event for which test session $\texttt{sid}^*$ has no matching session $\overline{\texttt{sid}^*}$ and $\mathcal{A}$ queries EphemeralKeyReveal($\texttt{sid}^*$).
- Let E_3 be the event for which test session $\texttt{sid}^*$ has matching session $\overline{\texttt{sid}^*}$ and $\mathcal{A}$ queries StaticKeyReveal($id_A^{(\alpha)}$) and EphemeralKeyReveal($\overline{\texttt{sid}^*}$).
- Let E_4 be the event for which test session $\texttt{sid}^*$ has matching session $\overline{\texttt{sid}^*}$ and $\mathcal{A}$ queries EphemeralKeyReveal($\texttt{sid}^*$) and EphemeralKeyReveal($\overline{\texttt{sid}^*}$).
- Let E_5 be the event for which test session $\texttt{sid}^*$ has matching session $\overline{\texttt{sid}^*}$ and $\mathcal{A}$ queries StaticKeyReveal($id_A^{(\alpha)}$) and StaticKeyReveal($id_B^{(\beta)}$).
- Let E_6 be the event for which test session $\texttt{sid}^*$ has matching session $\overline{\texttt{sid}^*}$ and $\mathcal{A}$ queries EphemeralKeyReveal($\texttt{sid}^*$) and StaticKeyReveal($id_B^{(\beta)}$).

Let E_i' $(i = 1, \ldots, 6)$ be similar events in the same PKG case.

To finish the proof, hereafter we investigate events $E_i \wedge M^*$ and $E_i' \wedge M^*$ $(i = 1, \ldots, 6)$ that cover all cases of event M^*.

Event $E_1 \wedge M^*$. In event E_1, test session $\texttt{sid}^*$ has no matching session $\overline{\texttt{sid}^*}$, adversary $\mathcal{A}$ obtains D_A, and adversary $\mathcal{A}$ does not obtain x_A and D_B from the condition of freshness. Thus, $\mathcal{A}$ does not obtain either x'_A or z_β where $x'_A = H_2(x_A, D_A)$ and $X_A = g^{x'_A}$. In this case, solver $\mathcal{S}$ embeds the instance as $Z_\beta = U\ (= g^u)$, $X_A = V\ (= g^v)$ and $Q_B = W\ (= g^w)$, and obtains g_T^{uvw} from shared value $\sigma_2 = e(Q_B^{x'_A}, Z_\beta)$. Note that solver $\mathcal{S}$ can perfectly simulate the StaticKeyReveal queries for other users except U_B by selecting random $q_i\ (\in_R \mathbb{Z}_q)$ and setting $Q_i = H_1(id_i^{(\iota)}) = g^{q_i}$ and $D_i = Z_\iota^{q_i}$. In addition, solver $\mathcal{S}$ can perfectly simulate the EphemeralKeyReveal queries for other sessions except sid^* and $\overline{\mathrm{sid}}^*$ by selecting random $x_i\ (\in_R \mathbb{Z}_q)$ and setting $x'_i = H_2(x_i, D_i)$ and $X_i = g^{x'_i}$. In event $E_1 \wedge M^*$, solver $\mathcal{S}$ performs the following **Setup** and **Simulation** phases.

Setup. GBDH solver $\mathcal{S}$ embeds (U, V, W), the instance of the GBDH problem where $U = g^u$, $V = g^v$, and $W = g^w$ as follows. $\mathcal{S}$ establishes n honest users, $U_1, \ldots, U_n$, and at most n honest PKGs, $P_1, \ldots, P_n$. Solver $\mathcal{S}$ randomly selects random $z_\iota\ (\in_R \mathbb{Z}_q, \iota = 1, \ldots, n)$ and computes master public keys $Z_\iota = g^{z_\iota}$. Solver $\mathcal{S}$ randomly selects two users, U_A and U_B, and integer $t \in_R [1, s]$, which is a guess of the test session with probability $1/n^2 s$. Here, we assume that U_A and U_B are managed by P_α and P_β, respectively. Solver $\mathcal{S}$ sets the master public key of PKG P_β, which manages U_B, as $Z_\beta = U$, sets the ephemeral public key of the t-th session of user U_A as $X_A = V$, and sets the static public key of $id_B^{(\beta)}$ of user U_B as $Q_B = W$. Solver $\mathcal{S}$ selects random $q_i\ (\in_R \mathbb{Z}_q)$, sets $Q_i = H_1(id_i^{(\iota)}) = g^{q_i}$ and $D_i = Z_\iota^{q_i}$ if U_i is managed by P_ι, and assigns static private key D_i to user U_i except U_B.

Solver $\mathcal{S}$ activates adversary $\mathcal{A}$ in this set of users (and PKGs), and awaits actions of $\mathcal{A}$. We next describe actions of $\mathcal{S}$ in response to user activation and oracle queries.

Simulation. Solver $\mathcal{S}$ maintains list L_H that contains queries and answers of the H oracle, list L_S that contains queries and answers of SessionKeyReveal, and list L_E that contains ephemeral private keys, ephemeral exponents, static private keys, and ephemeral public keys. For any $id_i^{(\iota)}$, $id_k^{(\kappa)}$, X_i, and X_k, solver $\mathcal{S}$ keeps L_S with consistency where $(\Pi, \mathcal{I}, id_i^{(\iota)}, id_k^{(\kappa)}, X_i, X_k)$ and $(\Pi, \mathcal{R}, id_k^{(\kappa)}, id_i^{(\iota)}, X_i, X_k)$ have the same answer. Solver $\mathcal{S}$ simulates oracle queries as follows.

1. $\mathsf{Send}(\Pi, \mathcal{I}, id_i^{(\iota)}, id_k^{(\kappa)})$: Solver $\mathcal{S}$ selects uniformly random ephemeral private key $x_i\ (\in_R \mathbb{Z}_q)$ and ephemeral exponent $x'_i\ (\in_R \mathbb{Z}_q)$, computes ephemeral public key $X_i\ (= g^{x'_i})$ honestly, records $(\Pi, \mathcal{I}, id_i^{(\iota)}, id_k^{(\kappa)}, X_i)$ in List L_S, and returns it. Solver $\mathcal{S}$ records (x_i, x'_i, D_i, X_i) in List L_E.

2. $\mathsf{Send}(\Pi, \mathcal{R}, id_k^{(\kappa)}, id_i^{(\iota)}, X_i)$: Solver $\mathcal{S}$ selects uniformly random ephemeral private key $x_k\ (\in_R \mathbb{Z}_q)$ and ephemeral exponent $x'_k\ (\in_R \mathbb{Z}_q)$, computes ephemeral public key $X_k\ (= g^{x'_k})$ honestly, records $(\Pi, \mathcal{R}, id_k^{(\kappa)}, id_i^{(\iota)}, X_i, X_k)$ in List L_S as completed, and returns it. Solver $\mathcal{S}$ records (x_k, x'_k, D_k, X_k) in List L_E.

3. $\mathsf{Send}(\Pi, \mathcal{I}, id_i^{(\iota)}, id_k^{(\kappa)}, X_i, X_k)$: If session $(\Pi, \mathcal{I}, id_i^{(\iota)}, id_k^{(\kappa)}, X_i)$ is not recorded in List L_S, Solver $\mathcal{S}$ records session $(\Pi, \mathcal{I}, id_i^{(\iota)}, id_k^{(\kappa)}, X_i, X_k)$ in List L_S as not completed. Otherwise, solver $\mathcal{S}$ records the session in List L_S as completed.

4. $H(\sigma_1, \sigma_2, \sigma_3, \Pi, id_i^{(\iota)}, id_k^{(\kappa)}, X_i, X_k)$:
   (a) If $(\sigma_1, \sigma_2, \sigma_3, \Pi, id_i^{(\iota)}, id_k^{(\kappa)}, X_i, X_k)$ is recorded in list L_H, then solver $\mathcal{S}$ returns recorded value K.
   (b) Else if session $(\Pi, \mathcal{I}, id_i^{(\iota)}, id_k^{(\kappa)}, X_i, X_k)$ or $(\Pi, \mathcal{R}, id_k^{(\kappa)}, id_i^{(\iota)}, X_i, X_k)$ is recorded in list L_S, then solver $\mathcal{S}$ checks that shared values σ_1, σ_2, and σ_3, are correctly formed, i.e., $\mathcal{S}$ checks that $\mathrm{DBDH}(Q_i, X_k, Z_\iota, \sigma_1) = 1$, $\mathrm{DBDH}(Q_k, X_i, Z_\kappa, \sigma_2) = 1$, and $e(X_i, X_k) = e(\sigma_3, g)$ hold.
       If σ_1, σ_2, and σ_3 are correctly formed, then solver $\mathcal{S}$ returns recorded value K in list L_S and records it in list L_H.
   (c) Else if $i = A$, $k = B$, and the session is t-th session of user U_A, then solver $\mathcal{S}$ checks that the shared values, σ_1, σ_2, and σ_3, are correctly formed, i.e., $\mathcal{S}$ checks that $\mathrm{DBDH}(Q_A, X_B, Z_\alpha, \sigma_1) = 1$, $\mathrm{DBDH}(Q_B, X_A, Z_\beta, \sigma_2) = 1$, and $e(X_A, X_B) = e(\sigma_3, g)$ hold.
       If σ_1, σ_2, and σ_3 are correctly formed, then solver $\mathcal{S}$ selects σ_2 $(= g_T^{uvw})$, the answer of the GBDH instance, from the shared values, and is successful by outputting the answer.
   (d) Otherwise, solver $\mathcal{S}$ returns random value K and records it in list L_H.

5. $H_1(id_Z^{(\zeta)})$: If $Z = B$, solver $\mathcal{S}$ returns $Q_B = W$. Otherwise, solver $\mathcal{S}$ computes $Z_\zeta = g^{z_\zeta}$ if Z_ζ does not exist, and responds to the query faithfully.

6. $H_2(x_i, D_i)$: If (x_i, x_i', D_i, X_i) is recorded in List L_E, then return x_i'. Otherwise, solver $\mathcal{S}$ selects uniformly random ephemeral exponent x_i' $(\in_R \mathbb{Z}_q)$, computes ephemeral public key X_i $(= g^{x_i})$ honestly, returns x_i', and records (x_i, x_i', D_i, X_i) in List L_E.

7. $\mathsf{SessionKeyReveal}((\Pi, \mathcal{I}, id_i^{(\iota)}, id_k^{(\kappa)}, X_i, X_k)$ or $(\Pi, \mathcal{R}, id_k^{(\kappa)}, id_i^{(\iota)}, X_i, X_k))$:
   (a) If session $(\Pi, \mathcal{I}, id_i^{(\iota)}, id_k^{(\kappa)}, X_i, X_k)$ or $(\Pi, \mathcal{R}, id_k^{(\kappa)}, id_i^{(\iota)}, X_i, X_k)$ $(= \mathsf{sid})$ is not completed, solver $\mathcal{S}$ returns error.
   (b) Else if sid is recorded in list L_S, then solver $\mathcal{S}$ returns recorded value K.
   (c) Else if $(\sigma_1, \sigma_2, \sigma_3, \Pi, id_i^{(\iota)}, id_k^{(\kappa)}, X_i, X_k)$ is recorded in list L_H, then solver $\mathcal{S}$ checks that σ_1, σ_2, and σ_3, are correctly formed, i.e., $\mathcal{S}$ checks that $\mathrm{DBDH}(Q_i, X_k, Z_\iota, \sigma_1) = 1$, $\mathrm{DBDH}(Q_k, X_i, Z_\kappa, \sigma_2) = 1$, and $e(X_i, X_k) = e(\sigma_3, g)$ hold.
       If σ_1, σ_2, and σ_3 are correctly formed, then solver $\mathcal{S}$ returns recorded value K in list L_H and records it in list L_S.
   (d) Otherwise, solver $\mathcal{S}$ returns random value K and records it in list L_S.

8. $\mathsf{StaticKeyReveal}(id_i^{(\iota)})$: If static public key Q_i of user U_i is W, then solver $\mathcal{S}$ aborts with failure; otherwise, solver $\mathcal{S}$ responds to the query faithfully.

9. $\mathsf{MasterKeyReveal}()$: Solver $\mathcal{S}$ aborts with failure.

10. $\mathsf{MasterKeyReveal}(P_\iota)$: $P_\iota = P_\beta$, solver $\mathcal{S}$ aborts with failure. Otherwise, solver $\mathcal{S}$ returns z_ι.

11. $\mathsf{Test}(\mathsf{sid})$: If sid is not the t-th session of U_A, then solver $\mathcal{S}$ aborts with failure. Otherwise, solver $\mathcal{S}$ responds to the query faithfully.

12. EphemeralKeyReveal(sid): If the corresponding ephemeral public key is V, then solver $\mathcal{S}$ aborts with failure. Else if the corresponding ephemeral public key is X_i, then solver $\mathcal{S}$ selects (x_i, x_i', D_i, X_i) in list L_E and returns x_i. Otherwise, solver $\mathcal{S}$ aborts with failure.

13. NewParty($id_i^{(\iota)}$): If $id_i^{(\iota')}$ is queried before, solver $\mathcal{S}$ returns error. Otherwise, solver $\mathcal{S}$ responds to the query faithfully.

14. If adversary $\mathcal{A}$ outputs guess γ, solver $\mathcal{S}$ aborts with failure.

The gap assumption is necessary to keep consistency in the oracle simulation, i.e., for the H and SessionKeyReveal oracles in Steps 4(b), 4(c), and 7(c).

Analysis. The simulation of the environment for adversary $\mathcal{A}$ is perfect except with negligible probability. The probability that adversary $\mathcal{A}$ selects the session where U_A is the initiator, U_B is the responder, ephemeral public key X_A is V, and the test session is sid^*, is at least $\frac{1}{n^2 s}$. Suppose that this is indeed the case, then solver $\mathcal{S}$ does not abort in Step 11.

Suppose that event E_1 occurs, then solver $\mathcal{S}$ does not abort in Steps 8, 9, 10, 11, and 12.

Suppose that event M^* occurs, and adversary $\mathcal{A}$ queries correctly formed σ_1, σ_2, and σ_3 to H. Therefore, solver $\mathcal{S}$ is successful as described in Step 4c, and does not abort as in Step 14.

Hence, solver $\mathcal{S}$ is successful with probability $\Pr[S_1] \geq \frac{p_1}{n^2 s}$ where p_1 is the probability that $E_1 \wedge M^*$ occurs, and S_1 is the event in which this solver is successful.

Event $E_2 \wedge M^*$. In event E_2, test session sid^* has no matching session $\overline{\text{sid}}^*$, $\mathcal{A}$ obtains x_A, and $\mathcal{A}$ does not obtain either D_A or D_B from the condition of freshness. Thus, $\mathcal{A}$ also does not obtain any of x_A', z_α, or z_β. The reduction to the GBDH assumption is similar to event $E_1 \wedge M^*$. GBDH solver $\mathcal{S}$ embeds (U, V, W), the instance of the GBDH problem, into $Z_\beta = U$, $X_A = V$, and $Q_B = W$. Hence, solver $\mathcal{S}$ is successful with probability $\Pr[S_2] \geq \frac{p_2}{n^2 s}$ where p_2 is the probability that $E_2 \wedge M^*$ occurs, and S_2 is the event in which this solver is successful.

Event $E_3 \wedge M^*$. In event E_3, test session sid^* has matching session $\overline{\text{sid}}^*$, $\mathcal{A}$ obtains D_A and x_B, and $\mathcal{A}$ does not obtain either x_A or D_B from the condition of freshness. Thus, $\mathcal{A}$ also does not obtain either x_A' or z_β. The reduction to the GBDH assumption is similar to event $E_1 \wedge M^*$. GBDH solver $\mathcal{S}$ embeds (U, V, W), the instance of the GBDH problem, into $Z_\beta = U$, $X_A = V$, and $Q_B = W$. Hence, solver $\mathcal{S}$ is successful with probability $\Pr[S_3] \geq \frac{p_3}{n^2 s}$ where p_3 is the probability that $E_3 \wedge M^*$ occurs, and S_3 is the event in which this solver is successful.

Event $E_4 \wedge M^*$. In event E_4, test session sid^* has matching session $\overline{\text{sid}}^*$, $\mathcal{A}$ obtains x_A and x_B, and $\mathcal{A}$ does not obtain either D_A or D_B from the condition of freshness. Thus, $\mathcal{A}$ also does not obtain any of x_A', z_α, or z_β. The reduction to the GBDH assumption is similar to event $E_1 \wedge M^*$. GBDH solver $\mathcal{S}$ embeds (U, V, W), the instance of the GBDH problem, into $Z_\beta = U$, $X_A = V$, and $Q_B = W$. Hence, solver $\mathcal{S}$ is successful with probability $\Pr[S_4] \geq \frac{p_4}{n^2 s}$ where p_4 is the probability that $E_4 \wedge M^*$ occurs, and S_4 is the event in which this solver is successful.

Event $E_5 \wedge M^*$. In event E_5, test session $\mathtt{sid}^*$ has matching session $\overline{\mathtt{sid}^*}$, $\mathcal{A}$ obtains z_α, z_β, D_A, and D_B, and $\mathcal{A}$ does not obtain either x_A or x_B from the condition of freshness. Thus, $\mathcal{A}$ also does not obtain either x'_A or x'_B. The reduction to the GBDH assumption is similar to event $E_1 \wedge M^*$, except for the following points.

In **Setup** and **Simulation**, $\mathcal{S}$ embeds (U, V, W), the GBDH instance, as $X_A = V$ and $X_B = W$. In **Simulation**, $\mathcal{S}$ obtains $g^{\log X_A \log X_B}$ from shared value σ_3 and can compute $e(U, g^{\log X_A \log X_B})$, the answer of the GBDH problem.

Hence, solver $\mathcal{S}$ is successful with probability $\Pr[S_5] \geq \frac{p_5}{n^2 s^2}$ where p_5 is the probability that $E_5 \wedge M^*$ occurs, and S_5 is the event in which this solver is successful.

Event $E_6 \wedge M^*$. In event E_6, test session $\mathtt{sid}^*$ has matching session $\overline{\mathtt{sid}^*}$, $\mathcal{A}$ obtains x_A and D_B, and $\mathcal{A}$ does not obtain either D_A or x_B from the condition of freshness. Thus, $\mathcal{A}$ also does not obtain either x'_B or z_α. The reduction to the GBDH assumption is similar to event $E_1 \wedge M^*$, except for the following points.

In **Setup** and **Simulation**, $\mathcal{S}$ embeds (U, V, W), the GBDH instance, as $Z_\alpha = U$, $Q_A = V$, and $X_B = W$. In **Simulation**, $\mathcal{S}$ obtains $g_T^{\log X_\alpha \log Q_A \log X_B}$, the answer of the GBDH problem, from shared value σ_1.

Hence, solver $\mathcal{S}$ is successful with probability $\Pr[S_6] \geq \frac{p_6}{n^2 s}$ where p_6 is the probability that $E_6 \wedge M^*$ occurs, and S_6 is the event in which this solver is successful.

Same PKG Case. We discuss the case of $id_A^{(\alpha)} \neq id_B^{(\beta)}$ managed by the same PKG, $P_\alpha = P_\beta$, i.e., $\alpha = \beta$.

In events E'_1, E'_2, E'_3, and E'_4, the reductions to the GBDH assumption are similar to those in events E_1, E_2, E_3, and E_4, respectively. In **Setup** and **Simulation**, $\mathcal{S}$ embeds (U, V, W), the GBDH instance, as $Z_\alpha = Z_\beta = U$, $X_A = V$, and $Q_B = W$. Note that solver $\mathcal{S}$ selects random q_A ($\in_R \mathbb{Z}_q$), and sets $Q_A = g^{q_A}$ and $D_A = U^{q_A}$ for user U_A. In **Simulation**, $\mathcal{S}$ obtains answer $g_T^{\log Z_\alpha \log X_A \log Q_B}$ for the GBDH problem from shared value σ_2.

In event E'_5, the reduction to the GBDH assumption is similar to event E_5. In **Setup** and **Simulation**, $\mathcal{S}$ embeds (U, V, W), the GBDH instance, as $X_A = U$ and $X_B = V$. In **Simulation**, $\mathcal{S}$ extracts $g^{\log X_A \log X_B}$ from the shared value, σ_3, and can compute $e(W, g^{\log X_A \log X_B})$, the answer of the GBDH problem.

In event E'_6, the reduction to the GBDH assumption is similar to event E_6. In **Setup** and **Simulation**, $\mathcal{S}$ embeds (U, V, W), the GBDH instance, as $Z_\alpha = Z_\beta = U$, $Q_A = V$, and $X_B = W$. Note that solver $\mathcal{S}$ selects random q_B ($\in_R \mathbb{Z}_q$), and sets $Q_B = g^{q_B}$, and $D_B = U^{q_B}$ for user U_B. In **Simulation**, $\mathcal{S}$ obtains $g_T^{\log Z_\alpha \log Q_A \log X_B}$, the answer of the GBDH problem, from shared value σ_1.

Total Analysis. Combining the success probabilities, we have

$$\mathbf{Adv}^{\mathrm{GBDH}}(\mathcal{S}) \geq \frac{1}{n^2 s^2} \mathbf{Adv}_{\Pi}^{\mathrm{mPKG-IBAKA}}(\mathcal{A}).$$

During the simulation, solver $\mathcal{S}$ performs $\mathcal{O}(n+s)$ exponentiations, i.e., $\mathcal{O}((n+s) \log q)$ group operations, to assign static and ephemeral keys, and make (at most) $\mathcal{O}(h+s)$ times DBDH oracle queries for simulating SessionKeyReveal and random oracle H queries. This completes the argument. □

4.3 Discussions

It is worth to note here that the proposed protocol does not guarantee *perfect forward secrecy* (PFS), where PFS holds if an adversary cannot learn the session keys of past sessions, even when the adversary learns the static private keys of all the parties. However, the protocol guarantees *weak* (PFS) [20] and *forward secrecy against PKG* (PKG-FS) [18], where weak PFS holds if an adversary cannot learn the session keys of past sessions in which the adversary did not actively interfere, even if the adversary learns the static private keys of all the parties, and PKG-FS holds if no session key between users should be revealed even to the PKG although the PKG has much more power than the users.

We show that the proposed protocol satisfies the PKG-FS even if some PKGs conspire. Assume that adversary $\mathcal{A}$ corrupts some PKGs. Then, the adversary clearly knows the static private keys of parties managed by the PKGs; however, $\mathcal{A}$ cannot obtain ephemeral private keys of the test session while still maintaining freshness. Thus, $\mathcal{A}$ does not know the value of σ_3 in the session, and this means that the adversary cannot guess the session key. This implies that the proposed protocol has the PKG-FS property, and therefore, the weak PFS property, also.

It is clear that the proposed protocol is based on symmetric pairings. When pairing function e is defined as $e : \mathbb{G}^2 \to \mathbb{G}_T$, e is referred to as *symmetric* when $\mathbb{G}$ and $\mathbb{G}_T$ are cyclic groups. When pairing function e is defined as $e : \mathbb{G}_1 \times G_2 \to \mathbb{G}_T$, e is referred to as *asymmetric* when $\mathbb{G}_1$, $\mathbb{G}_2$, and $\mathbb{G}_T$ are cyclic groups. The recent trend in cryptography tells us that asymmetric pairings have been adopted because some symmetric pairing functions have security problems. To modify the proposed protocol considering asymmetric pairings, the double-key technique seems applicable. Here the double-key technique, proposed by Fujioka *et al.* [14], is as follows. The PKG generates two static private keys for a party where one is in G_1 and the other is in G_2. An initiator and a responder exchange two ephemeral keys where one in G_1, the other in G_2, and they have the same exponent. However, we leave analysis for correctness and formal security as future work.

Note that referring standard documents such as ISO/IEC 15946-1 [19] is recommended for choosing elliptic curves and hash functions to construct the proposed protocol.

4.4   Comparison with Other Protocols

We compare the proposed protocol with others, the Chen–Kudla protocol and a protocol on FSXY construction.

Comparison with Chen–Kudla Protocol. The proposed protocol is secure in the id(m)-eCK model under the GBDH assumption and the Chen–Kudla protocol is secure in the multiple PKG setting of id-BJM model under the CBDH assumption [7]. In other words, it is proved that our protocol is exposure resilient but it is not proved that the Chen–Kudla protocol is so.

It seems difficult to prove that the Chen–Kudla protocol is exposure resilient and the rationale for this is given below. The protocol uses three types of intermediate values, a value computed based on the static key of the initiator and the ephemeral key of the responder, a value computed based on the ephemeral key of the initiator and the static key of the responder, and a value computed based on the ephemeral keys of the initiator and the responder. Thus, we cannot embed an instance of a problem to a value computed based on the static keys of the initiator and the responder.

In addition, even when we simply modify the Chen–Kudla protocol to use a value computed based on the static key of the initiator and the static key of the responder, it is still not easy to prove the security. To do so, a value computed based on the static keys of the initiator and the responder must be computed under different master public keys. Let (Z_α, z_α) and (Z_β, z_β) be pairs of master public and secret keys of PKG P_α and PKG P_β, respectively, where $Z_\alpha = g^{z_\alpha}$ and $Z_\beta = g^{z_\beta}$. Let $id_A^{(\alpha)}$ and $id_A^{(\beta)}$ be identities of parties managed by P_α and P_β, respectively. Even when private key D_A (resp. D_B) of the party is given as $D_A = Q^{z_\alpha}$ where $Q_A = H_1(id_A^{(\alpha)})$ and H_1 is a hash function, we cannot expect that $e(Z_\alpha, Q_A) = e(g, D_B)$ or $e(Z_\beta, Q_B) = e(g, D_A)$ holds.

The NAXOS technique avoids this situation, and thus, our protocol can be proved to be exposure resilient.

Comparison with Protocol on FSXY Construction. It is proved that a protocol on the FSXY construction is exposure resilient in the standard model [17]. It may be possible to have an exposure-resilient IBAKA protocol in a multiple PKG setting as the construction is based on a modular approach.

However, it is clear that the resultant protocol on the FSXY construction is not one-round, namely, two-move as the responder's message is computed on the initiator's message, i.e., encapsulation key ek_T.

In addition, the resultant protocol is less efficient as the responder needs to send two ciphertexts although it is sufficient for the initiator to send a single one.

We discuss efficiency of a protocol in communication complexity and computational complexity.

Regarding to communication complexity in a protocol on the FSXY construction, the initiator sends a ciphertext and a encapsulation key, and the responder returns two ciphertexts. Thus, communication complexity of the initiator is given as three group elements and that of the responder is given as four group elements when we assume that a ciphertext of an (IB-)KEM scheme consists of two group elements, and a encapsulation key of the KEM scheme consists of a single group element.

Regarding to computational complexity in a protocol on the FSXY construction, the initiator runs one EnCap, one wKeyGen, one DeCap, one wDeCap (1), and six hash functions including KDF, and the responder does one EnCap, one DeCap one wEnCap, and six hash functions including KDF. Thus, computational complexity of the initiator is given as four exponentiations, two pairings, and fix hashings, and that of the responder is given as three exponentiations, two pairings, and fix hashings when we assume that EnCap and DeCap of the IB-KEM scheme require one exponentiation and one pairing operation, and wEnCap and wDeCap of the KEM scheme require one exponentiation.

Table 3. Comparison with protocol on FSXY construction.

	Proposed	FSXY
Communication complexity		
Initiator	1	3
Responder	1	4
Computational complexity		
Initiator		
Exponentiation	3	4
Pairing	2	2
Hashing	3	6
Responder		
Exponentiation	3	3
Pairing	2	2
Hashing	3	6

We assume that a ciphertext of an (IB-)KEM scheme consists of two group elements, and a encapsulation key of the KEM scheme consists of a single group element. We assume that EnCap and DeCap of the IB-KEM scheme require one exponentiation and one pairing operation, and wEnCap and wDeCap of the KEM scheme require one exponentiation, also.

On the other hand, communication complexity of both initiator and responder in the proposed protocol are given as a single group element, and computational complexity of both initiator and responder in the proposed protocol are given as three exponentiations, two pairings, and three hashings.

We summarize the numbers of group elements and computations above in Table 3.

Thus, Table 3 shows that a protocol on the FSXY construction is less efficient than the proposed protocol in both communication and computational complexity even when efficient IB-KEM and KEM schemes are adopted to construct the protocol.

It is worth to note here that the security of our protocol is proved in the random oracle model however the security of the FSXY construction is done in the standard model.

4.5 Security in Other Models

As stated in **Model Assumptions**, we assume that there is a binding between an identifier and its PKG. We call this *static binding model*. Thus, it is natural to consider a stronger adversary who adaptively indicate binding between an identifier and its PKG. We call this *adaptive binding model*. We expect that the proposed protocol is secure in the adaptive binding model but the reduction ratio is worse to $\frac{1}{n^4 s^2}$ from $\frac{1}{n^2 s^2}$ as the solver needs to guess PKGs of the initiator and the responder in the test session. Here, the reduction ratio is given as the ratio between the probability that a solver, $\mathcal{S}$, breaks the assumption and the probability that an adversary, $\mathcal{A}$, breaks the protocol, i.e., $\mathbf{Adv}^{\mathrm{GBDH}}(\mathcal{S})/\mathbf{Adv}_{\Pi}^{\mathrm{mPKG-IBAKA}}(\mathcal{A})$.

We can consider another classification of the security model: an adversary can get the private key of an user only once or it is allowed to obtain several private keys from different PKGs. We call the former *separated domain model* and the latter *overlapped domain model*. In the security proof, we use the separated domain model but we expect that the proposed protocol is secure in the overlapped domain model also when we modify $Q_i = H_1(id_i^{(\iota)})$ to $Q_i = H_1(id_i^{(\iota)}, Z_\iota)$. However, we leave analysis for correctness and formal security as future work.

5 Conclusion

The id(m)-eCK model for identity-based authenticated key agreement in the multiple private key generator setting was defined to provide security against the exposure of private information.

Utilizing the NAXOS techinique, we proposed a one-round identity-based authenticated key agreement protocol with multiple private key generators and proved that the protocol is exposure resilient under the gap bilinear Diffie–Hellman assumption in a random oracle model. Moreover, the proposed protocol is achieved based on exchanging a single group element; therefore, it is the most efficient.

Acknowledgments. The author would like to thank Tsunekazu Saito, Koutarou Suzuki, and Tetsutaro Kobayashi for discussing problems in the multiple private key generator scenario. The author also would like to thank the anonymous reviewers for their comments and suggestions that helped me to improve this paper.

References

1. Bellare, M., Rogaway, P.: Entity authentication and key distribution. In: Stinson, D.R. (ed.) CRYPTO 1993. LNCS, vol. 773, pp. 232–249. Springer, Heidelberg (1994). doi:10.1007/3-540-48329-2_21

2. Bellare, M., Rogaway, P.: Random oracles are practical: a paradigm for designing efficient protocols. In: Denning, D.E., Pyle, R., Ganesan, R., Sandhu, R.S., Ashby, V. (eds.) CCS 1993, pp. 62–73. ACM, New York (1993)

3. Blake-Wilson, S., Johnson, D., Menezes, A.: Key agreement protocols and their security analysis. In: Darnell, M. (ed.) Cryptography and Coding 1997. LNCS, vol. 1355, pp. 30–45. Springer, Heidelberg (1997). doi:10.1007/BFb0024447

4. Boyd, C., Choo, K.-K.R.: Security of two-party identity-based key agreement. In: Dawson, E., Vaudenay, S. (eds.) Mycrypt 2005. LNCS, vol. 3715, pp. 229–243. Springer, Heidelberg (2005). doi:10.1007/11554868_17

5. Boyd, C., Cliff, Y., Gonzalez Nieto, J., Paterson, K.G.: Efficient one-round key exchange in the standard model. In: Mu, Y., Susilo, W., Seberry, J. (eds.) ACISP 2008. LNCS, vol. 5107, pp. 69–83. Springer, Heidelberg (2008). doi:10.1007/ 978-3-540-70500-0_6. http://eprint.iacr.org/2008/007

6. Canetti, R., Krawczyk, H.: Analysis of key-exchange protocols and their use for building secure channels. In: Pfitzmann, B. (ed.) EUROCRYPT 2001. LNCS, vol. 2045, pp. 453–474. Springer, Heidelberg (2001). doi:10.1007/3-540-44987-6_28

7. Chen, L., Kudla, C.: Identity based authenticated key agreement protocols from pairings. In: IEEE CSFW 2016, pp. 219–233. IEEE Computer Society, Washington, D.C. (2003). http://eprint.iacr.org/2002/184

8. Chen, L., Cheng, Z., Smart, N.P.: Identity-based key agreement protocols from pairings. Int. J. Inf. Secur. 6(4), 213–241 (2007)

9. Choo, K.-K.R., Boyd, C., Hitchcock, Y.: Examining indistinguishability-based proof models for key establishment protocols. In: Roy, B. (ed.) ASIACRYPT 2005. LNCS, vol. 3788, pp. 585–604. Springer, Heidelberg (2005). doi:10.1007/ 11593447_32

10. Cremers, C.J.F.: Session-state Reveal is stronger than Ephemeral Key Reveal: attacking the NAXOS authenticated key exchange protocol. In: Abdalla, M., Pointcheval, D., Fouque, P.-A., Vergnaud, D. (eds.) ACNS 2009. LNCS, vol. 5536, pp. 20–33. Springer, Heidelberg (2009). doi:10.1007/978-3-642-01957-9_2

11. Cremers, C.J.F.: Examining indistinguishability-based security models for key exchange protocols: the case of CK, CK-HMQV, and eCK. In: Chen, Y., Danezis, G., Shmatikov, V. (eds.) CCS 2011, pp. 80–91. ACM, New York (2011)

12. Farash, M.S., Attari, M.A.: Provably secure and efficient identity-based key agreement protocol for independent PKGs using ECC. ISC Int. J. Inf. Secur. 5(1), 55–70 (2013)

13. Farash, M.S., Attari, M.A.: A pairing-free ID-based key agreement protocol with different PKGs. Int. J. Netw. Secur. 16(2), 143–148 (2014)

14. Fujioka, A., Hoshino, F., Kobayashi, T., Suzuki, K., Ustaoğlu, B., Yoneyama, K.: id-eCK secure ID-based authenticated key exchange on symmetric pairing and its extension to asymmetric case. IEICE Trans. 96-A(6), 1139–1155 (2013)

15. Fujioka, A., Suzuki, K.: Sufficient condition for identity-based authenticated key exchange resilient to leakage of secret keys. In: Kim, H. (ed.) ICISC 2011. LNCS, vol. 7259, pp. 490–509. Springer, Heidelberg (2012). doi:10.1007/ 978-3-642-31912-9_32

16. Fujioka, A., Suzuki, K., Ustaoğlu, B.: Ephemeral key leakage resilient and efficient id-akes that can share identities, private and master keys. In: Joye, M., Miyaji, A., Otsuka, A. (eds.) Pairing 2010. LNCS, vol. 6487, pp. 187–205. Springer, Heidelberg (2010). doi:10.1007/978-3-642-17455-1_12

17. Fujioka, A., Suzuki, K., Xagawa, K., Yoneyama, K.: Strongly secure authenticated key exchange from factoring, codes, and lattices. In: Fischlin, M., Buchmann, J., Manulis, M. (eds.) PKC 2012. LNCS, vol. 7293, pp. 467–484. Springer, Heidelberg (2012). doi:10.1007/978-3-642-30057-8_28

18. Huang, H., Cao, Z.: An ID-based authenticated key exchange protocol based on bilinear Diffie-Hellman problem. In: Li, W., Susilo, W., Tupakula, U.K., Safavi-Naini, R., Varadharajan, V. (eds.) ASIACCS 2009, pp. 333–342. ACM, New York (2009)

19. ISO/IEC 15946-1:2016: Information technology – Security techniques – Cryptographic techniques based on elliptic curves - Part 1: General (2016)

20. Krawczyk, H.: HMQV: a high-performance secure Diffie-Hellman protocol. In: Shoup, V. (ed.) CRYPTO 2005. LNCS, vol. 3621, pp. 546–566. Springer, Heidelberg (2005). doi:10.1007/11535218_33

21. Kim, S., Lee, H., Oh, H.: Enhanced ID-based authenticated key agreement protocols for a multiple independent PKG environment. In: Qing, S., Mao, W., López, J., Wang, G. (eds.) ICICS 2005. LNCS, vol. 3783, pp. 323–335. Springer, Heidelberg (2005). doi:10.1007/11602897_28

22. LaMacchia, B., Lauter, K., Mityagin, A.: Stronger security of authenticated key exchange. In: Susilo, W., Liu, J.K., Mu, Y. (eds.) ProvSec 2007. LNCS, vol. 4784, pp. 1–16. Springer, Heidelberg (2007). doi:10.1007/978-3-540-75670-5_1

23. Lee, H., Kim, D., Kim, S., Oh, H.: Identity-based key agreement protocols in a multiple PKG environment. In: Gervasi, O., Gavrilova, M.L., Kumar, V., Laganá, A., Lee, H.P., Mun, Y., Taniar, D., Tan, C.J.K. (eds.) ICCSA 2005. LNCS, vol. 3483, pp. 877–886. Springer, Heidelberg (2005). doi:10.1007/11424925_92

24. McCullagh, N., Barreto, P.S.L.M.: A new two-party identity-based authenticated key agreement. In: Menezes, A. (ed.) CT-RSA 2005. LNCS, vol. 3376, pp. 262–274. Springer, Heidelberg (2005). doi:10.1007/978-3-540-30574-3_18

25. Mishra, D., Mukhopadhyay, S.: Cryptanalysis of pairing-free identity-based authenticated key agreement protocols. In: Bagchi, A., Ray, I. (eds.) ICISS 2013. LNCS, vol. 8303, pp. 247–254. Springer, Heidelberg (2013). doi:10.1007/978-3-642-45204-8_19

26. Oh, J., Moon, S.-J., Ma, J.: An attack on the identity-based key agreement protocols in multiple PKG environment. IEICE Trans. 89-A(3), 826–829 (2006)

27. Vallent, T.F., Yoon, E.-J., Kim, H.: An escrow-free two-party identity-based key agreement protocol without using pairings for distinct PKGs. IEEK Trans. Smart Process. Comput. 2(3), 168–175 (2013)

28. Xie, G.: Cryptanalysis of Noel McCullagh and Paulo S.L.M. Barreto's two-party identity-based key agreement. IACR Cryptology ePrint Archive. Report 2004/308 (2004). http://eprint.iacr.org/2004/308

29. Zhong, Y., Ma, J.: A highly secure identity-based authenticated key-exchange protocol for satellite communication. J. Commun. Netw. 12(6), 592–599 (2010)

Cryptanalysis Correspondence

Attacks on the Basic cMix Design: On the Necessity of Commitments and Randomized Partial Checking

Herman Galteland[1]([⊠]), Stig F. Mjølsnes[2], and Ruxandra F. Olimid[2]

[1] Department of Mathematical Sciences, NTNU,
Norwegian University of Science and Technology, Trondheim, Norway
`herman.galteland@math.ntnu.no`
[2] Department of Information Security and Communication Technology, NTNU,
Norwegian University of Science and Technology, Trondheim, Norway
`{sfm,ruxandra.olimid}@ntnu.no`

Abstract. The cMix scheme was proposed by Chaum et al. in 2016 as the first practical set of cryptographic protocols that offer sender-recipient unlinkability at scale. The claim was that the cMix is secure unless all nodes collude. We argue that their assertion does not hold for the basic description of cMix, and we sustain our statement by two different types of attacks: a tagging attack and an insider attack. For each one, we discuss the settings that make the attack feasible, and then possible countermeasures. By this, we highlight the necessity of implementing additional commitments or mechanisms that have only been mentioned as additional features.

Keywords: Cryptographic protocols · Sender-recipient unlinkability · Anonymity · Mixnets · Attacks

1 Introduction

1.1 cMix

The cMix protocol by Chaum et al. [1] is an improved mixing network [2] which aims to provide an anonymous communication tool for its users at large scales. The mixing should be such that no one is able to relate an output message to a user input message, that is, no one is able to link a sender with a recipient. An important advantage over its predecessors is that cMix performs expensive computations (like public key encryption) during a precomputation phase, keeping the real-time phase, which is in charge with actual message delivery, fast. The protocol is a part of a larger system, called Privategrity, but its authors describe cMix independently.

The authors of Ref. [1] claim that cMix is the first practical system that provides sender-recipient unlinkability, unless all nodes collude. We argue that their assertion does not hold for the basic description of the protocol (as given

© Springer International Publishing AG 2017
R.C.-W. Phan and M. Yung (Eds.): LNCS 10311, Mycrypt 2016, pp. 463–473, 2017.
DOI: 10.1007/978-3-319-61273-7_22

in [1, Sect. 4] and we sustain our statement by two different types of attacks. Each of them has its own effect on the design of the original protocol. By this, we want to highlight *the necessity* of using additional commitment mechanisms, whereas Ref. [1] mentions this as additional features.

1.2 Related Work

The cMix system is designed to be resistant to most of the usual mix network attacks. This paper focuses on the cryptanalysis of cMix, and Subsect. 2.2 introduces in detail the adversarial model from [1]. We present here a very brief survey of proposed general attacks on anonymous overlay networks.

Tagging attacks are a potential threat to all mix networks [3]. An adversary can put an identifier tag on an input message to the mix network and attempt to recognize the tag in the output messages. If successful, the adversary can break the anonymity of a specific sender. We show in Sect. 3 that cMix is vulnerable to a tagging attack.

Replay attacks are attacks in which an adversary retransmits a valid message several times, making it possible to analyze the outgoing traffic [4]. We do not analyze replay attacks against cMix system, as they are eliminated by the adversarial model (see Subsect. 2.2).

Intersection attacks and *statistical disclosure attacks* use information acquired by observing mix networks where the users can freely choose the mix node for their messages (free mix nodes) [4–6]. In such systems different batches can be distinguished since they come from different mix nodes. If a sender use the same mix nodes for every message then the adversary can separate the routes by analyzing the network flow.

Traffic analysis attacks is a family of attacks that observes the network traffic in order to deduce informational patterns in communication and targets connection-based systems. Unlike message-based systems like cMix, connection-based systems use free mix nodes that do not batch and permute messages. By *counting packets* [7] and *timing communication* [8] the adversary is able to distinguish between different paths in the (free) mix network. *Contextual attacks* [9] (or *traffic confirmation attacks* [10], or *intersection attacks* [11]) analyze the traffic when specific users and recipients use a protocol, their communication pattern, and how many messages they send and receive.

The authors of cMix recognize that their proposal is potentially vulnerable to attacks that make anonymous systems fail, like the broadband intersection attacks, contextual attacks, or DoS (Denial of Service) [1].

1.3 Results

We focus on the security analysis of the basic cMix description as described in [1, Sect. 4], and show that it is susceptible to two attacks, which differ by action type.

Tagging Attack. The cMix paper [1] introduces commitments [12] to overcome tagging attacks. The paper states that *"tagging attacks do not work before the exit node"*, and *"if a tagging attack is detected, at least the last node should be removed from the cascade"* [1, Sect. 4.3]. Therefore the authors might be aware of a possible attack that can be performed by the exit node. However, they do not consider any prevention for this. We introduce a simple tagging attack launched by the exit node. Although a prevention mechanism is immediate (by adding an extra commitment) we consider it for completeness, as an example of a possible tagging attack against the system. In personal communications, the authors of cMix acknowledged that the actual design of the system adds the additional commitment we refer to as a countermeasure [13].

Insider Attack. The cMix paper [1] claims that attacks are unsuccessful unless all nodes collude. We contradict this by showing that the last node can break the unlinkability, essentially by creating a mix network consisting of itself only. The attack will succeed by the last node deviating from the protocol rules and choose its own output. We argue that this attack remains undetected in the original version of cMix, and becomes detectable only if additional checks like *Randomized Integrity Checking* (RPC, see Subsect. 2.3) are considered (suggested by the authors of [1] as a special feature). We show the necessity of using randomized partial checking (RPC). However, an inappropriate use of RPC could allow a coalition of nodes (all except one) to link a large fraction of the senders to their recipients.

1.4 Outline

Section 2 describes the cMix scheme and presents the adversarial model. The two following sections contain our results: Sect. 3 describes a simple tag attack similar to the tag attack described in the original cMix paper [1]. Section 4 presents the insider attack where the adversary controls the last node and makes the overall mixing process independent of the preceding nodes. Section 5 concludes and indicates possible future research directions.

2 Preliminaries

2.1 cMix Description

Figure 1 describes the cMix protocol from [1]. We ignore the return steps, since they are irrelevant for our attacks. Note that this does not restrict the applicability of our results, since the same permutation is used for both forward and return paths. Once the permutation is disclosed both directions of communication are compromised.

cMix has two phases: a *precomputation phase* and a *real-time phase*. By design, the heavy public key computations are performed in the precomputation phase, which can be performed on separate hardware (for each node). Since

Precomputation Phase

Step 1 (preprocessing). Each node N_i, $1 \leq i \leq n$, selects a random $\mathbf{r}_i$, computes the encryption $\mathcal{E}(\mathbf{r}_i^{-1})$ and sends it to the network handler. The network handler computes the product of all the received values, produces $\mathcal{E}(\mathbf{R}_n^{-1}) = \prod_{i=1}^{n} \mathcal{E}(\mathbf{r}_i^{-1})$ and sends it to the first node.

Step 2 (mixing). Each node N_i, $1 \leq i \leq n$, computes $\pi_i(\mathcal{E}(\Pi_{i-1}(\mathbf{R}_n^{-1}) \times \mathbf{S}_{i-1}^{-1})) \times \mathcal{E}(\mathbf{s}_i^{-1})$, where Π_0 is the identity permutation and $\mathbf{S}_0^{-1} = \mathbf{1}$. The last node sends the vector of random components (i.e. the first component) of the ciphertext $(\mathbf{x}, \mathbf{c}) = \mathcal{E}((\Pi_n(\mathbf{R}_n) \times \mathbf{S}_n)^{-1})$ to the other nodes and stores the vector of message components (i.e. the second component) locally for the real-time phase.

Step 3 (postprocessing). Using the random component $\mathbf{x}$, each node N_i, $1 \leq i \leq n$, computes its individual decryption share for $(\mathbf{x}, \mathbf{c})$ as $\mathcal{D}_i(\mathbf{x}) = \mathbf{x}^{-e_i}$, stores it locally to use in the real-time phase and publicly commits to it.

Real-Time Phase

Step 0. Each user constructs its message MK_j^{-1} (for slot j) by multiplying the message M_j with the inverse of the key K_j and it sends it to the network handler, which collects all messages and combines them to get a vector $\mathbf{M} \times \mathbf{K}^{-1}$.

Step 1 (preprocessing). Each node N_i, $1 \leq i \leq n$, sends $\mathbf{k}_i \times \mathbf{r}_i$ to the network handler, which uses them to compute $\mathbf{M} \times \mathbf{R}_n = \mathbf{M} \times \mathbf{K}^{-1} \times \prod_{i=1}^{n} \mathbf{k}_i \times \mathbf{r}_i$ and sends the result to N_1.

Step 2 (mixing). Each node N_i, $1 \leq i \leq n$, computes $\pi_i(\Pi_{i-1}(\mathbf{M} \times \mathbf{R}_n) \times \mathbf{S}_{i-1}) \times \mathbf{s}_i$, where Π_0 is the identity permutation and $\mathbf{S}_0 = \mathbf{1}$. The last node N_n sends a commitment to its message $\Pi_n(\mathbf{M} \times \mathbf{R}_n) \times \mathbf{S}_n$ to every other node.

Step 3 (postprocessing). Each node N_i, $1 \leq i \leq n-1$, sends its precomputed decryption share for $(\mathbf{x}, \mathbf{c}) = \mathcal{E}((\Pi_n(\mathbf{R}_n) \times \mathbf{S}_n)^{-1})$ to the network handler, while the last node N_n sends its decryption share multiplied by the value in the previous step and the message component: $\Pi_n(\mathbf{M} \times \mathbf{R}_n) \times \mathbf{S}_n \times \mathcal{D}_n(\mathbf{x}) \times \mathbf{c}$. Finally, the network handler retrieves the permuted message as $\Pi_n(\mathbf{M}) = \Pi_n(\mathbf{M} \times \mathbf{R}_n) \times \mathbf{S}_n \times \prod_{i=1}^{n} \mathcal{D}_i(\mathbf{x}) \times \mathbf{c}$.

Fig. 1. The cMix protocol (forward path) [1]

the precomputation phase does not require any input from the users it can be performed offline and while a batch is being filled up with messages.

The scheme consists of a sequence of n mix nodes that process β messages at a time (a *batch* of messages); made simple, each node performs a permutation on the input and blinds the output by multiplying it with a random value. The last node N_n makes an exception, as it usually behaves differently from the other nodes (see Fig. 1).

Besides the last node there is another entity with a special role in the system - the *network handler* - that interacts both with the users and the whole set of nodes. The network handler receives messages from the users and arranges them

Table 1. Notations

U_j	User j
$\mathbf{M}$	A batch of β messages $\mathbf{M} = (M_1, \ldots, M_\beta)$, each M_i sent by a distinct user
N_i	Node i from the set of n mix nodes $\{N_1, \ldots, N_n\}$
e_i	The share of node N_i of the secret key e
d	The public key of the system, where $d = \prod_{i=1}^{n} g^{e_i}$
$\mathcal{E}(\cdot)$	A multi-party group-homomorphic encryption under the system public key d
π_i	A random permutation on a batch, applied by node N_i
Π_i	The composed permutation performed by all nodes from N_1 to N_i
$k_{i,j}$	The derived secret key shared between node N_i and the sending user of slot j
$\mathbf{k}_i$	The vector of derived secret keys shared between node N_i and all users in a batch, i.e. $\mathbf{k}_i = (k_{i,1}, \ldots, k_{i,\beta})$
K_j	The product of all shared keys for the sending user of slot j, i.e. $K_j = \prod_{i=1}^{n} k_{i,j}$
$\mathbf{r}_i, \mathbf{s}_i$	Random values of node N_i for the batch, where $\mathbf{r}_i = (r_{i,1}, \ldots, r_{i,\beta})$, respectively $\mathbf{s}_i = (s_{i,1}, \ldots, s_{i,\beta})$
$\mathbf{R}_i, \mathbf{S}_i$	The direct product of the first i values, i.e. $\mathbf{R}_i = \prod_{j=1}^{i} \mathbf{r}_j$, respectively $\mathbf{S}_i = \prod_{j=1}^{i} \mathbf{s}_j$

into batches; once a batch is full it is sent to the first node in the mix network. After the last node performs its mixing it sends the batch back to the network handler, which can then forward or broadcast the messages to the destination. The mixing should be such that no one is able to relate an output message to a user input message, that is, no one is able to link a sender with a recipient.

Before using the system each sender U_j must establish a private symmetric key with each of the nodes N_i, which they use as a seed in a pseudorandom generator to derive the secret keys $k_{i,j}$. To blind a message M_j before it is sent to the network handler, user U_j multiplies M_j with a key composed by the derived keys shared with each of the nodes $K_j = \prod_{i=1}^{n} k_{i,j}$. The network handler arranges messages into a batch and sends it through the mix network. Each node applies its permutation to the batch and the last node sends it back to the network handler. The output is a permuted batch of messages.

During the mixing step of the precomputation phase each node performs encryption under a public key of the system; the related private key is split across all nodes in the network. The encryption scheme suggested by the authors of [1] is the multi-party group-homomorphic encryption based on ElGamal [14] described by Benaloh [15]. Moreover, all computations of the protocol are performed in a prime order cyclic group G that satisfies the decisional Diffie-Hellman security assumption. We denote by G^* the set of nonidentity elements in G.

Refer to Fig. 1 for the detailed self-contained description of the cMix process, using the notation defined in Table 1.

2.2 Adversarial Model

The adversarial model in [1] assumes authenticated channels among the mix nodes and between the network handler and each mix node. This implies that the adversary can read, forward, and delete messages, but not modify, inject, or replay messages without detection. The adversary can compromise the users (up to all except two), and the mix nodes (up to all except one). Compromised nodes can behave malicious but cautious, since the attacker aims to remain undetected. Within this attacker model, the authors of cMix claim that the output is unlinkable to the input, even if the adversary knows the set of senders and the set of recipients for every batch of messages.

The security analysis in the Appendix A of the cMix paper assumes secure authenticated channels for which the adversary cannot read the content, only the length of a message. All our attacks hold under these stronger security assumptions.

2.3 Features and Extensions

The cMix paper [1] dedicates a section to special features and extensions of the system. It shortly discusses the utility of adding RPC (*Randomized Integrity Checking*) to cMix, an integrity check mechanism introduced by Jacobsson et al. [16], and further analyzed and developed in Refs. [17,18]. The usage of RPC in the cMix system is that each node commits to a randomly chosen permutation, publishes its input and output, and validates that it has followed the protocol correctly by revealing a (large) fraction of its secret input/output pairs, where these pairs are selected by the other nodes (or by a random oracle). The cMix system protects the user's privacy by putting nodes in pairs, such that each node belongs to only one pair. Nodes in a pair reveal their secret information such that none of the messages can be followed from the input of the first node to the output of the second node.

3 The Tagging Attack

Our first attack is similar to the tag attack described in the cMix paper [1], but it uses a different value to remove the tag. During the precomputation phase the nodes compute the value $(\mathbf{x}, \mathbf{c}) = \mathcal{E}((\varPi_n(\mathbf{R}_n) \times \mathbf{S}_n)^{-1})$, where the last node stores the vector of message components, $\mathbf{c}$, locally and sends the vector of random components, $\mathbf{x}$, to all other nodes. Each node computes its decryption share using $\mathbf{x}$ and commits to this value. Note that it is uncertain whether $\mathbf{c}$ is being committed to or not in the description of the basic cMix protocol.

The authors of cMix introduce commitments to detect potential tagging attacks exposing any attempt of using the decryption shares to remove the tag. However, the commitments are independent of $\mathbf{c}$, so it is possible to perform a similar attack which uses $\mathbf{c}$ instead of $\mathcal{D}_n(\mathbf{x})$ to remove the tag. The downside is that the adversary needs to corrupt the last node (which has access to $\mathbf{c}$) and the network handler (under the assumption of secure authorized channels). Figure 2 describes the tag attack.

For the tag attack to be successful we need to assume that it is possible to recognize valid messages in the output. To tag a message M_j the last node

Goal: Tag a message M_j belonging to user U_j and recognize it in the permuted batch of messages, linking the sender U_j to its recipient.

Step 1. The corrupted node N_n creates a tag vector $\mathbf{t}$ which consists of $\beta - 1$ ones and one tag $t \in G^*$ in slot j (i.e. $\mathbf{t} = (1,\ldots,1,t,1,\ldots,1)$), computes $\mathbf{k}_n \times \mathbf{r}_n \times \mathbf{t}$ and sends it to the network handler (Real-time Phase.Step 1).

Step 2. The network handler sends the set of all decryption shares $\{\mathcal{D}_i(\mathbf{x})|1 \leq i < n\}$ to the last node (Real-time Phase.Step 3). Node N_n can retrieve the permuted messages as $\Pi_n(\mathbf{M} \times \mathbf{t}) = \Pi_n(\mathbf{M} \times \mathbf{R}_n \times \mathbf{t}) \times \mathbf{S}_n \times \prod_{i=1}^{n} \mathcal{D}_i(\mathbf{x}) \times \mathbf{c}$ and recognize the tagged message in slot j'.

Step 3. The corrupted node N_n creates the inverse tag vector $\mathbf{t}^{-1}$, which consists of $\beta - 1$ ones and one tag $t^{-1} \in G^*$ in slot j', computes $\mathbf{c}' = \mathbf{c} \times \mathbf{t}^{-1}$, and sends $\Pi_n(\mathbf{M} \times \mathbf{R}_n) \times \mathbf{S}_n \times \mathcal{D}_n(\mathbf{x}) \times \mathbf{c}'$ to the network handler.

Fig. 2. The tagging attack

creates a tag vector $\mathbf{t} = (1,\ldots,t,\ldots,1)$, where t is in position j, multiply it with the keys and random values $\mathbf{k}_n \times \mathbf{r}_n \times \mathbf{t}$, and sends the result to the network handler. The tag then goes though the mixnet attached to message M_j and arrives at the last node as $\Pi_{n-1}(\mathbf{M} \times \mathbf{R}_n \times \mathbf{t}) \times \mathbf{S}_{n-1}$. Then the last node can permute and do the computations according to the protocol, and publish its commitment to the value $\Pi_n(\mathbf{M} \times \mathbf{R}_n \times \mathbf{t}) \times \mathbf{S}_n$. This triggers all other nodes to send their decryption share to the network handler, which forwards them to N_n. The last node can then retrieve the batch of permuted messages and find the invalid message $M_j t$ in slot j' of the permuted batch. The last node creates the inverse tag $\mathbf{t}^{-1}$, which has t^{-1} in slot j', and replaces the message components with the altered value $\mathbf{c}' = \mathbf{c} \times \mathbf{t}^{-1}$. The network handler then computes

$$\Pi_n(\mathbf{M} \times \mathbf{R}_n \times \mathbf{t}) \times \mathbf{S}_n \times \mathbf{c}' \times \prod_{i=1}^{n} \mathcal{D}_i(\mathbf{x}) =$$

$$\Pi_n(\mathbf{M} \times \mathbf{R}_n \times \mathbf{t}) \times \mathbf{S}_n \times (\Pi_n(\mathbf{R}_n) \times \mathbf{S}_n)^{-1} \times \mathbf{t}^{-1} = \Pi_n(\mathbf{M} \times \mathbf{t}) \times \mathbf{t}^{-1} = \Pi_n(\mathbf{M})$$

and delivers the permuted batch as normal. That is, the adversary has successfully linked a sender with a recipient without being detected.

To make this attack detectable, the last node should publish a commitment to the vector of message components $\mathbf{c}$ in the Precomputation Phase.Step 3, or the system should implement RPC as an integrity check mechanism. Although prevention can be simply achieved by natural solutions like the ones mentioned, we introduce the attack for completeness; it stands as an example of tagging attack performed by the last node, a type of attack the authors of cMix seem to be aware of (see [1], Sect. 4.2: *"tagging attacks do not work before the exit node"* and *"if a tagging attack is detected, at least the last node should be removed from the cascade"*).

At the time of writing, the authors of cMix acknowledged that the actual design of the system implements the countermeasure we refer to and commits to the vector of message components $\mathbf{c}$, as explained above [13].

4 The Insider Attack

Our second attack allows the last node to ignore all permutations introduced by the previous nodes and perform the overall mixing process by itself. Hence, the output of the real-time phase will be a batch of messages permuted with a known permutation making it easy to link all senders and recipients. To succeed, the adversary needs to corrupt the last node (which controls the output of the mixing process) and the network handler (which knows the content of the values $\mathcal{E}(\mathbf{R}_n^{-1})$ and $\mathbf{M} \times \mathbf{R}_n$, under the assumption of secure authenticated channels). Figure 3 describes the insider attack.

During Precomputation Phase.Step 1 the corrupted network handler computes and sends $\mathcal{E}(\mathbf{R}_n^{-1})$ to the first and the last nodes. The honest nodes operates as normal, where the last, dishonest, node discards the input it receives from the previous node and chooses its own output. The last node draws a random vector $\mathbf{A} = (A_1, \ldots, A_\beta)$, encrypts the inverted values, $\mathcal{E}(\mathbf{A}^{-1})$, and computes $\pi_n(\mathcal{E}(\mathbf{R}_n^{-1}) \times \mathcal{E}(\mathbf{A}^{-1})) = \pi_n(\mathcal{E}(\mathbf{R}_n^{-1} \times \mathbf{A}^{-1}))$. The last node publishes the random components, that is $\mathbf{x}$, of $\pi_n(\mathcal{E}(\mathbf{R}_n^{-1} \times \mathbf{A}^{-1})) = (\mathbf{x}, \mathbf{c})$ to the other nodes such that they can prepare their decryption shares.

In Real-Time Phase.Step 1 the network handler sends $\mathbf{M} \times \mathbf{R}_n$ to the first and the last nodes. In the mixing step the last node discards what it receives from the previous node, selects its output $\pi_n(\mathbf{M} \times \mathbf{R}_n \times \mathbf{A})$, commits to this batch of messages, and sends $\pi_n(\mathbf{M} \times \mathbf{R}_n \times \mathbf{A}) \times \mathbf{c} \times \mathcal{D}_n(\mathbf{x})$ to the network handler. As

---

Goal: Perform the mixing process with only the last node using only a known permutation to permute the batch of messages.

Step 1. The network handler computes and sends $\mathcal{E}(\mathbf{R}_n^{-1})$ to the first and last node (Precomputation Phase.Step 1). The last node discards the input it is given form the previous node and publishes the component of random elements of $\pi_n(\mathcal{E}(\mathbf{R}_n^{-1} \times \mathbf{A}^{-1}))$, for a random and invertible $\mathbf{A}$ (Precomputation Phase.Step 3).

Step 2. The network handler computes and sends $\mathbf{M} \times \mathbf{R}_n$ to the first and last node (Real-Time Phase.Step 1). The last node discards the input it is given form the previous node, publishes a commitment to $\pi_n(\mathbf{M} \times \mathbf{R}_n \times \mathbf{A})$, and sends $\pi_n(\mathbf{M} \times \mathbf{R}_n \times \mathbf{A}) \times \mathbf{c} \times \mathcal{D}_n(\mathbf{x})$ to the network handler (Real-Time Phase.Step 3).

Step 3. The network handler retrieves the permuted batch of messages as $\pi_n(\mathbf{M}) = \pi_n(\mathbf{M} \times \mathbf{R}_n \times \mathbf{A}) \times \pi_n(\mathbf{R}_n^{-1} \times \mathbf{A}^{-1})$ and publishes it. The adversary can recover $\mathbf{M}$ by applying π_n^{-1}.

Fig. 3. The insider attack

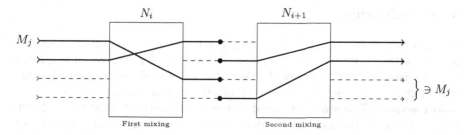

Fig. 4. RPC: two paired nodes revealing each separate half of their permutation. Continuous lines means information is revealed and dashed lines means information is not revealed

the network handler receives the decryption shares from the other nodes it can retrieve the permuted messages and forward them to the receivers:

$$\pi_n(\mathbf{M} \times \mathbf{R}_n \times \mathbf{A}) \times \mathbf{c} \times \prod_{i=1}^{n} \mathcal{D}_i(\mathbf{x}) = \pi_n(\mathbf{M} \times \mathbf{R}_n \times \mathbf{A}) \times \pi_n(\mathbf{R}_n^{-1} \times \mathbf{A}^{-1}) = \pi_n(\mathbf{M}).$$

Note that the output batch is only permuted with the permutation π_n, which is known to the last node. Hence, the adversary can easily deanonymize all of the senders by applying π_n^{-1} to the output.

The RPC mechanism ensures with high probability that each node follows its instructions, hence, this will prevent the last node from deviating from the protocol. Since our insider attack changes the entire batch, RPC will detect the attack with overwhelming probability. This shows the necessity of implementing RPC with cMix.

Notes on the RPC Mechanism. RPC makes the nodes reveal a (large) fraction of their secret information, which could break the anonymity of the users [18]. As an example, let us assume that each node performs only one permutation and proves the correctness of its output for this permutation. Further assume that an adversary corrupts all except one, honest, node and therefore only needs the permutation from this node to deanonymize the users. When using RPC, the honest node would reveal information about its permutation. Hence, the adversary can easily break the anonymity for a substantial portion of the users using the information made public by the RPC mechanism.

Even in the scenario where there are two honest nodes that are paired, the adversary can get some information about the senders and receivers [18]. Nodes in a pair reveal information such that no messages can be followed from the input of the first node to the output of the second node in a pair. This means that if a message, say M_j, is revealed by the first node, then it will not be revealed by the second node (see Fig. 4). Given enough rounds of cMix, an adversary might eventually link senders and recipients that are frequently talking with each other. Therefore, two honest nodes (a single pair) are usually not enough to protect the anonymity of all users.

5 Conclusions

We demonstrate by examples that the cMix scheme, as it was initially defined in
its basic settings, would allow linkability between senders and recipients, hence
compromising the anonymity of the users. We describe the actions an adversary
could follow to succeed for both types of attacks (the tagging attack and the
insider attack). The attacks succeed in the secure authenticated channels settings, and under the assumption that the adversary can corrupt the network
handler. This is a natural assumption that was also made by that the authors
of cMix.

By discussing the attacks, we highlight the necessity of the use of commitments and the RPC integrity mechanisms, which have only been mentioned as
additional features in cMix scheme, and where these mechanisms are not fully
included in the security proofs. However, the authors of cMix have expressed
that their demonstration software implements the commitment mechanism that
prevents our tagging attack.

This paper is restricted to a theoretical exposure of some attacks against the
cMix standalone set of cryptographic protocols. Future analysis work can include
experimental activities for practical attacks on real-world cMix implementations.
Of course, the scalability of performance, throughput, and latency are key issues.
An enterprising theoretical work would be to analyze the cMix security within
the context of the larger system Privategrity.

Remarks to the Written Rebuttal from the Authors of cMix. Both
our attacks are valid under the basic protocol description given in [1, Sect. 4].
The commitment mechanism required to overcome the first attack is not used or
referred to in the original paper. This has been acknowledged in the authors'
response [19]. Furthermore, the cMix paper describes RPC as an extension,
therefore usage of RPC can hardly be understood as *necessary* [1]. We claim
the necessity of RPC or an equivalent mechanism. RPC is not included in the
formal analysis, hence it is left outside the security theorems and performance
discussions, while we find that RPC is crucial for the security of the system, and
might introduce a significant performance penalty. The response note informs
us that proper security mechanisms protecting against the attacks we have presented, are used in their prototype and explained in a new paper, but both of
those are currently unavailable to us for inspection [19].

Acknowledgements. Herman Galteland is funded by Nasjonal sikkerhetsmyndighet
(NSM), www.nsm.stat.no.

References

1. Chaum, D., Das, D., Javani, F., Kate, A., Krasnova, A., de Ruiter, J., Sherman,
   A.T.: cMix: anonymization by high-performance scalable mixing. Cryptology
   ePrint Archive, Report 2016/008 (2016). http://eprint.iacr.org/, version 20160530:
   183553 from 30 May 2016

2. Chaum, D.: Untraceable electronic mail, return addresses, and digital pseudonyms. Commun. ACM **24**(2), 84–90 (1981)
3. Goldschlag, D.M., Reed, M.G., Syverson, P.F.: Hiding routing information. In: Anderson, R. (ed.) IH 1996. LNCS, vol. 1174, pp. 137–150. Springer, Heidelberg (1996). doi:10.1007/3-540-61996-8_37
4. Berthold, O., Pfitzmann, A., Standtke, R.: The disadvantages of free MIX routes and how to overcome them. In: Federrath, H. (ed.) Designing Privacy Enhancing Technologies. LNCS, vol. 2009, pp. 30–45. Springer, Heidelberg (2001). doi:10.1007/3-540-44702-4_3
5. Danezis, G., Diaz, C., Troncoso, C.: Two-sided statistical disclosure attack. In: Borisov, N., Golle, P. (eds.) PET 2007. LNCS, vol. 4776, pp. 30–44. Springer, Heidelberg (2007). doi:10.1007/978-3-540-75551-7_3
6. Danezis, G., Serjantov, A.: Statistical disclosure or intersection attacks on anonymity systems. In: Fridrich, J. (ed.) IH 2004. LNCS, vol. 3200, pp. 293–308. Springer, Heidelberg (2004). doi:10.1007/978-3-540-30114-1_21
7. Serjantov, A., Sewell, P.: Passive attack analysis for connection-based anonymity systems. In: Snekkenes, E., Gollmann, D. (eds.) ESORICS 2003. LNCS, vol. 2808, pp. 116–131. Springer, Heidelberg (2003). doi:10.1007/978-3-540-39650-5_7
8. Danezis, G.: The traffic analysis of continuous-time mixes. In: Martin, D., Serjantov, A. (eds.) PET 2004. LNCS, vol. 3424, pp. 35–50. Springer, Heidelberg (2005). doi:10.1007/11423409_3
9. Raymond, J.-F.: Traffic analysis: protocols, attacks, design issues, and open problems. In: Federrath, H. (ed.) Designing Privacy Enhancing Technologies. LNCS, vol. 2009, pp. 10–29. Springer, Heidelberg (2001). doi:10.1007/3-540-44702-4_2
10. Syverson, P.F., Goldschlag, D.M., Reed, M.G.: Anonymous connections and onion routing. In: Proceedings of the 1997 IEEE Symposium on Security and Privacy, SP 1997, Washington, DC, USA, pp. 44–54. IEEE Computer Society (1997)
11. Berthold, O., Langos, H.: Dummy traffic against long term intersection attacks. In: Dingledine, R., Syverson, P. (eds.) PET 2002. LNCS, vol. 2482, pp. 110–128. Springer, Heidelberg (2003). doi:10.1007/3-540-36467-6_9
12. Brassard, G., Chaum, D., Crépeau, C.: Minimum disclosure proofs of knowledge. J. Comput. Syst. Sci. **37**(2), 156–189 (1988)
13. de Ruiter, J.: Personal communication in e-mail from 28 July 2016
14. ElGamal, T.: A public key cryptosystem and a signature scheme based on discrete logarithms. In: Blakley, G.R., Chaum, D. (eds.) CRYPTO 1984. LNCS, vol. 196, pp. 10–18. Springer, Heidelberg (1985). doi:10.1007/3-540-39568-7_2
15. Benaloh, J.: Simple verifiable elections. In: Proceedings of the USENIX/Accurate Electronic Voting Technology Workshop. 2006 on Electronic Voting Technology Workshop, EVT 2006, Berkeley, CA, USA, p. 5. USENIX Association (2006)
16. Jakobsson, M., Juels, A., Rivest, R.L.: Making mix nets robust for electronic voting by randomized partial checking. In: Proceedings of the 11th USENIX Security Symposium, Berkeley, CA, USA, pp. 339–353. USENIX Association (2002)
17. Khazaei, S., Wikström, D.: Randomized partial checking revisited. In: Dawson, E. (ed.) CT-RSA 2013. LNCS, vol. 7779, pp. 115–128. Springer, Heidelberg (2013). doi:10.1007/978-3-642-36095-4_8
18. Küsters, R., Truderung, T., Vogt, A.: Formal analysis of Chaumian mix nets with randomized partial checking. In: Proceedings of the 2014 IEEE Symposium on Security and Privacy, SP 2014, Washington, DC, USA, pp. 343–358. IEEE Computer Society (2014)
19. Chaum, D., Das, D., Javani, F., Kate, A., Krasnova, A., de Ruiter, J., Sherman, A.T.: A response to... (2016)

Cryptanalysis of an Identity-Based Convertible Undeniable Signature Scheme

Rouzbeh Behnia, Syh-Yuan Tan[(✉)], and Swee-Huay Heng

Faculty of Information Science and Technology, Multimedia University,
Melaka, Malaysia
{rouzbeh,sytan,shheng}@mmu.edu.my

Abstract. In this paper, we cryptanalyze an identity-based convertible undeniable signature scheme which claimed to be secure under the random oracle model. Our result shows that the signature leaks information on signer identity and fails to provide both invisibility and anonymity under the known message attack. We propose a fix for the vulnerability by removing some information from the signature with the need for the signer to keep the record of every signed message.

Keywords: Cryptanalysis · Anonymity · Invisibility · Undeniable signature

1 Introduction

Chaum and van Antwerpen [2] introduced the notion of undeniable signature schemes to enable the signer to control the verifiability of her signature. The verification can only take place with the direct participation of the signer in the confirmation or disavowal protocol. Boyar et al. [1] introduced a new extension, namely, convertible undeniable signature (CUS) which enables the signer to selectively, or universally convert one or all of her undeniable signatures to publicly verifiable ones. If universal conversion is performed, an undeniable signature scheme turns into an ordinary signature scheme.

The ultimate goal in undeniable signatures and its extensions is to protect the privacy of the signer. Traditionally, the notion of invisibility [3] was the main requirement for an undeniable signature scheme. Invisibility implies the inability of a user to distinguish an undeniable signature from a random element in the signature space. However, as the main objective of undeniable signature is to hide the link between the signer's public key and the signature and as shown by Galbraith and Mao [4], the notion of anonymity has become the most relevant security notion for undeniable signatures and its extensions in multiuser settings. Given an undeniable signature and public keys of two or more possible signers, the notion of anonymity implies the infeasibility to determine which user has issued the signature. Galbraith and Mao highlighted that the notions of invisibility and anonymity are equivalent and proved that if an undeniable

© Springer International Publishing AG 2017
R.C.-W. Phan and M. Yung (Eds.): LNCS 10311, Mycrypt 2016, pp. 474–477, 2017.
DOI: 10.1007/978-3-319-61273-7_23

signature scheme has the property of invisibility, then it also has anonymity, and vice versa. The importance of anonymity in the context of CUS schemes was further stressed on by Huang et al. [5].

Our Contribution. In this paper, we cryptanalyze the invisibility and anonymity of the first identity-based convertible undeniable signature (IBCUS) scheme proposed by Wu et al. [7]. We find that while the scheme was claimed to be invisible, it is vulnerable to known message attack and does not provide any sense of invisibility as well as anonymity for the signer and the other involved users. Subsequently, we propose a workaround for the discovered vulnerability to resist the known message attack.

The organization of the paper is as follows. In Sect. 2, we briefly review the construction of the Wu et al.'s IBCUS [7] scheme. In Sect. 3, we demonstrate our known message attack and discuss the quick fix on the Wu et al. IBCUS scheme. Finally, we conclude the paper in Sect. 4.

2 Wu et al.'s IBCUS Scheme

In this section, we briefly recall Wu et al.'s IBCUS [7] scheme. We do not describe the confirmation, disavowal and conversion protocol due to page limit. Reader can refer to [7] for the full description.

Setup: On the input of security parameters k, generate groups $\mathbb{G}$ with the generator $g \in \mathbb{G}$ and $\mathbb{G}_1$ of prime order $q > 2^k$, and a pairing $e : \mathbb{G} \times \mathbb{G} \to \mathbb{G}_1$. Next, randomly select $s \in \mathbb{Z}_q$ as the master secret key, and compute $P_{Pub} = g^s$. Set the master public key as $mpk = (\mathbb{G}, \mathbb{G}_1, e, g, P_{pub}, H_1, H_2, H_3, H_4)$ where $H_1, H_2 : \{0,1\}^* \to \mathbb{G}$, $H_3 : \mathbb{G} \times \mathbb{G}_1 \to \mathbb{G}$, and $H_4 : \{0,1\}^* \to \mathbb{Z}_q$.

Extract: Given the user's identity ID and the master secret key s, compute the user's private keys as $SK_{ID} = H_1(ID)^s$ and $VK_{ID} = H_1(ID, undeniable)^s$. SK_{ID} is kept secret while VK_{ID} can be published as the universal conversion token at a later time.

Sign: On the input of (SK_{ID_S}, VK_{ID_S}) and a message $m \in \{0,1\}^*$ where ID_S is the signer identity, compute $U = e(VK_{ID_S}, H_2(m)), V = g^v$ and $W = SK_{ID_S} + vH_3(U, V)$ for a randomly chosen $v \in \mathbb{Z}_q$. The undeniable signature is published as $\sigma = (U, V, W)$.

Verify: Provided a message-signature pair $(m, \sigma = (U, V, W))$, check if $e(W, P) = e(H_1(ID_S), P_{Pub})e(H_3(U, V), V)$ and reject the corrupted signature if the equality does not hold. Otherwise, decide on the validity/invalidity of the pair by checking if $U = (H_2(m), VK_{ID_S})$. If the equality holds, it means that the signature is indeed generated by the signer herself and is valid.

3 The Known Message Attack

In this section, we mount known message attacks on the invisibility and anonymity of Wu et al.'s IBCUS scheme. In precise, we construct a distinguisher $\mathcal{D}_1$ who is given a challenge tuple $(m^*, \sigma^* = (U, V, W), ID_s^*)$ which says the message-signature pair (m^*, σ^*) may or may not be a valid signature of the random signers ID_s^*. $\mathcal{D}_1$ confirms the signature is valid if the equation $e(W, P) = e(H_1(ID_s^*), P_{Pub})e(H_3(U, V), V)$ holds. Otherwise, it is not a valid signature.

Next, we show that the anonymity of Wu et al.'s IBCUS is broken also by constructing a distinguisher $\mathcal{D}_2$ in a similar way. Provided a valid message-signature pair $(m^*, \sigma^* = (U, V, W))$ and public keys (in this case the mpk and identities) of two random signers ID_0^* and ID_1^*, $\mathcal{D}_2$ can decide which user has generated the signature by checking which identity (i.e. public key) satisfies the equation $e(W, P) = e(H_1(ID_b), P_{Pub})e(H_3(U, V), V)$ where $b \in \{0, 1\}$. This shows that the IBCUS completely violates the privacy that is promised to the signer.

3.1   Discussion

Although Wu et al.'s IBCUS scheme was claimed to be proven secure as the same confirmation and disavowal protocols were used in the Libert and Quisquater's provably secure IBCUS [6] scheme, the two protocols are not exactly the same. Moreover, the signing algorithm differs a lot in both schemes where the former uses two keys while the latter uses one key. Thus, it is not trivial for Wu et al.'s scheme to enjoy the security assurance from [6].

A direct yet inefficient solution is readily available for the vulnerability shown in this work. Recall that the signature is composed of three elements (U, V, W) in which U is actually the undeniable signature of Libert and Quisquater's IBCUS scheme. The elements (V, W) were added to provide a universal conversion proof but accidentally leaked information on the signing key which violates invisibility and anonymity. A workaorund is to publish (U, W) as the undeniable signature and keep V for the purpose of verification, confirmation/disavowal and conversions. However, this approach is not practical as it requires a huge storage for all signed messages and their corresponding V elements.

4   Conclusion

We mounted a known message attack on Wu et al.'s IBCUS scheme and showed that the main security properties, namely, invisibility and anonymity do not hold. This finding shows that if we extends a scheme which is provably secure, the extended scheme may not necessary inherit the provable security.

Acknowledgment. The authors would like to thank the Malaysia government's Fundamental Research Grant Scheme (FRGS/2/2014/ICT04/MMU/03/1) and (FRGS/1/2015/ICT04/MMU/03/5) for supporting this work.

References

1. Boyar, J., Chaum, D., Damgård, I., Pedersen, T.: Convertible undeniable signatures. In: Menezes, A.J., Vanstone, S.A. (eds.) CRYPTO 1990. LNCS, vol. 537, pp. 189–205. Springer, Heidelberg (1991). doi:10.1007/3-540-38424-3_14

2. Chaum, D., van Antwerpen, H.: Undeniable signatures. In: Brassard, G. (ed.) CRYPTO 1989. LNCS, vol. 435, pp. 212–216. Springer, New York (1990). doi:10.1007/0-387-34805-0_20

3. Chaum, D., Heijst, E., Pfitzmann, B.: Cryptographically strong undeniable signatures, unconditionally secure for the signer. In: Feigenbaum, J. (ed.) CRYPTO 1991. LNCS, vol. 576, pp. 470–484. Springer, Heidelberg (1992). doi:10.1007/3-540-46766-1_38

4. Galbraith, S.D., Mao, W.: Invisibility and anonymity of undeniable and confirmer signatures. In: Joye, M. (ed.) CT-RSA 2003. LNCS, vol. 2612, pp. 80–97. Springer, Heidelberg (2003). doi:10.1007/3-540-36563-X_6

5. Huang, X., Mu, Y., Susilo, W., Wu, W.: Provably secure pairing-based convertible undeniable signature with short signature length. In: Takagi, T., Okamoto, T., Okamoto, E., Okamoto, T. (eds.) Pairing 2007. LNCS, vol. 4575, pp. 367–391. Springer, Heidelberg (2007). doi:10.1007/978-3-540-73489-5_21

6. Libert, B., Quisquater, J.-J.: Identity based undeniable signatures. In: Okamoto, T. (ed.) CT-RSA 2004. LNCS, vol. 2964, pp. 112–125. Springer, Heidelberg (2004). doi:10.1007/978-3-540-24660-2_9

7. Wu, W., Mu, Y., Susilo, W., Huang, X.: Provably secure identity-based undeniable signatures with selective and universal convertibility. In: Pei, D., Yung, M., Lin, D., Wu, C. (eds.) Inscrypt 2007. LNCS, vol. 4990, pp. 25–39. Springer, Heidelberg (2008). doi:10.1007/978-3-540-79499-8_4

Invited and Insight Papers

Towards User-Friendly Cryptography

Goichiro Hanaoka[✉]

Advanced Cryptosystems Research Group,
Information Technology Research Institute,
National Institute of Advanced Industrial Science and Technology (AIST),
2-3-26 Aomi, Koto-ku, Tokyo 135-0064, Japan
hanaoka-goichiro@aist.go.jp

Abstract. In this talk, we discuss user-friendliness in cryptography and its importance. Especially, we reconsider the significance of generic constructions of cryptographic tools, using the case of proxy re-encryption as an example. We then suggest that enjoyable aspects of cryptographic tools may also be important for technology diffusion. We illustrate this using the case of card-based protocols as an example.

1 Background

Until now, there have been various proposals of cryptographic tools with additional functionalities which yield useful properties for securing complicated information and communication systems (e.g. Internet of Things). However, despite their useful properties, it seems that these cryptographic tools are not as widely used as expected. We suggest that one of the main reasons for this is that the mathematical structure of such cryptographic tools are generally very complicated (especially when providing highly functional properties), and therefore, potential users will choose other easy-to-understand solutions even though strong assumptions, such as the existence of a trusted servers or tamper-resistant hardware, are required. Hence, for technology diffusion, it is required to investigate easy-to-understand structures of cryptographic tools such that potential users can easily understand the essential mechanisms.

2 Importance of Generic Constructions

A promising approach to overcoming the above issue is the use of *generic construction* (modular construction, in other words). Namely, if it is possible to decompose the required functionality, which may be complicated, into simpler functionalities, it may also become possible to construct a cryptographic tool that yields this functionality, by a combination of simpler cryptographic tools. This may lead to the following two merits: (1) potential users can easily verify how the required functionality is realized, and (2) the security proof can also easily be verified.

© Springer International Publishing AG 2017
R.C.-W. Phan and M. Yung (Eds.): LNCS 10311, Mycrypt 2016, pp. 481–484, 2017.
DOI: 10.1007/978-3-319-61273-7_24

2.1 Group Signatures

A typical example which shows the effectiveness of this approach is the case of group signatures. Group signatures were originally proposed by Chaum and van Heyst in 1991 [9], and following [9], schemes with improved efficiency were proposed [5–7,10]. However, these improved schemes used to be complicated, and only a limited number of specialized researchers could understand their mechanisms. However, in 2003, Bellare, Micciacio, and Warinschi showed a sophisticated design principle for group signatures and presented a generic construction from ordinary digital signatures, public key encryption, and non-interactive zero knowledge proofs [2]. It is remarkable that almost all group signatures which were proposed after their work are based on this generic construction (e.g., [5,10]), and furthermore, even non-experts of this topic can easily understand functionality and security of these schemes. As a result, finally, the international standardization community accepted group signatures, and ISO/IEC 20008-2 [1], an international standard for group signatures, was published in 2012.

This example implies that even though a cryptographic tool is actually correct and secure, potential users may not be immediately convinced of this fact if the tool and the functionality it provides are complicated. For promoting better understanding, generic constructions from basic tools may be useful, and this results in technology diffusion in the real world. Note that generic constructions of cryptographic tools have already intensively been studied in the literature from the viewpoint of theoretical relationships among cryptographic primitives, but such generic constructions are usually not considered in practice due to specific (i.e. non-generic) constructions being generally more efficient. In contrast to this, we claim that we should actively use generic constructions in practice, even if this might be less efficient than specific constructions, to promote understanding among potential users.

2.2 Proxy Re-encryption

Proxy re-encryption [3,8] is another cryptographic tool whose functionality is useful but complicated, and similarly to other cryptographic tools with complicated functionality, it is not widely used so far. A proxy re-encryption scheme is basically identical to the standard public key encryption except that it is possible to change the destination (i.e. receiver) of a ciphertext without recovering the plaintext. More specifically, in addition to the sender and the receiver, another player *proxy* is set up, and it can convert a cipheretext for receiver A into that for another receiver B by using a *re-encryption key*. Note that this conversion is carried out without decrypting the original ciphertext. Proxy re-encryption technology was implemented in a commercial cloud storage system in practice, but unfortunately this service has been already finished.

In [11], Hanaoka et al. proposed a generic construction of proxy re-encryption from standard building blocks. More specifically, in this generic construction, a plaintext is encrypted by using 2-out-of-2 threshold encryption, and a re-encryption key for receiver A to another receiver B is generated as a pair of a share of A's decryption key and an encryption of the other share under public encryption key of B. When converting a ciphertext for A to that for B, the proxy carries out partial decryption of it by using the share of A's decryption key (which is not encrypted), and sets the pair of the partial decryption result and the encryption of the other share as the converted ciphertext for B. Obviously, B can recover the plaintext by decrypting the partial decryption result by using the share of decryption key which can be extracted from the converted ciphertext. We believe that this generic construction is fairly easy-to-understand and hope that it can contribute to technology diffusion of proxy re-encryption.

3   Importance of Enjoyable Aspects

For technology diffusion, usefulness alone is not always sufficient, and more attractive aspects may be required. From this viewpoint, *card-based protocols* [4,12] is a good example which shows the importance of enjoyable aspects of cryptographic tools. In this type of protocol, players can carry out secure multi-party computation only by using physical cards (like playing cards), and potential users can not only easily understand the mechanism of multi-party computation, but also enjoy using it. Due to its enjoyable aspects, card-based protocols are now considered to be a great tool for introducing multi-party computation and for encouraging potential users to utilize state-of-the-art cryptographic tools with complicated functionalities. Some universities have already introduced card-based protocols in their undergraduate programs.

References

1. Information technology – security techniques – anonymous digital signatures. ISO/IEC 20008-1
2. Bellare, M., Micciancio, D., Warinschi, B.: Foundations of group signatures: formal definitions, simplified requirements, and a construction based on general assumptions. In: Biham, E. (ed.) EUROCRYPT 2003. LNCS, vol. 2656, pp. 614–629. Springer, Heidelberg (2003). doi:10.1007/3-540-39200-9_38
3. Blaze, M., Bleumer, G., Strauss, M.: Divertible protocols and atomic proxy cryptography. In: Nyberg, K. (ed.) EUROCRYPT 1998. LNCS, vol. 1403, pp. 127–144. Springer, Heidelberg (1998). doi:10.1007/BFb0054122
4. Boer, B.: More efficient match-making and satisfiability *The Five Card Trick*. In: Quisquater, J.-J., Vandewalle, J. (eds.) EUROCRYPT 1989. LNCS, vol. 434, pp. 208–217. Springer, Heidelberg (1990). doi:10.1007/3-540-46885-4_23
5. Boneh, D., Boyen, X., Shacham, H.: Short group signatures. In: Franklin, M. (ed.) CRYPTO 2004. LNCS, vol. 3152, pp. 41–55. Springer, Heidelberg (2004). doi:10.1007/978-3-540-28628-8_3
6. Boyen, X., Waters, B.: Compact group signatures without random oracles. In: Vaudenay, S. (ed.) EUROCRYPT 2006. LNCS, vol. 4004, pp. 427–444. Springer, Heidelberg (2006). doi:10.1007/11761679_26
7. Boyen, X., Waters, B.: Full-domain subgroup hiding and constant-size group signatures. In: Okamoto, T., Wang, X. (eds.) PKC 2007, pp. 1–15. Springer, LNCS (2007)

8. Canetti, R., Hohenberger, S.: Chosen-ciphertext secure proxy re-encryption. In: Proceedings of the 14th ACM Conference on Computer and Communications Security, pp. 185–194. ACM, New York (2007)
9. Chaum, D., Heyst, E.: Group signatures. In: Davies, D.W. (ed.) EUROCRYPT 1991. LNCS, vol. 547, pp. 257–265. Springer, Heidelberg (1991). doi:10.1007/3-540-46416-6_22
10. Groth, J.: Fully anonymous group signatures without random oracles. In: Kurosawa, K. (ed.) ASIACRYPT 2007. LNCS, vol. 4833, pp. 164–180. Springer, Heidelberg (2007). doi:10.1007/978-3-540-76900-2_10
11. Hanaoka, G., Kawai, Y., Kunihiro, N., Matsuda, T., Weng, J., Zhang, R., Zhao, Y.: Generic construction of chosen ciphertext secure proxy re-encryption. In: Dunkelman, O. (ed.) CT-RSA 2012. LNCS, vol. 7178, pp. 349–364. Springer, Heidelberg (2012). doi:10.1007/978-3-642-27954-6_22
12. Mizuki, T., Kumamoto, M., Sone, H.: The five-card trick can be done with four cards. In: Wang, X., Sako, K. (eds.) ASIACRYPT 2012. LNCS, vol. 7658, pp. 598–606. Springer, Heidelberg (2012). doi:10.1007/978-3-642-34961-4_36

Multi-prover Interactive Proofs: Unsound Foundations

Claude Crépeau[1(✉)] and Nan Yang[2]

[1] McGill University, Montreal, QC, Canada
crepeau@cs.mcgill.ca
[2] Concordia University, Montreal, QC, Canada
na_yan@encs.concordia.ca

Abstract. Several Multi-Prover Interactive Proofs (MIPs) found in the literature contain proofs of soundness that are lacking. This was first observed [1] in which a notion of *Prover isolation* is defined to partly address the issue. Furthermore, some existing Zero-Knowledge MIPs suffer from a catastrophic flaw: *they outright allow the Provers to communicate via the Verifier.* Consequently, their soundness claims are now seriously in doubt, if not plain wrong. This paper outlines the lack of isolation and numerous other issues found in the (ZK)MIP literature. A follow-up paper will resolve most of these issues in detail.

1 Introduction

It has been a long-held intuition that if Alice and Bob share an inconsistent set of beliefs then, if they are questioned individually, one can expose that inconsistency. This is the idea behind the theory of Multi-Prover Interactive Proofs (MIPs): a polynomial-time Verifier who is trying to discern truth from falsity from a set of all-powerful Provers who cannot signal with each other. This theory originated from the work of Ben-Or et al. [2], and we denote the class of languages with such interactive proofs by **MIP** (and its zero-knowledge counterpart **ZKMIP**). In that paper and subsequent work of Babai, Fortnow and Lund [3], it was claimed that **ZKMIP = MIP = NEXP**.

The proof of security in [2] and many subsequent MIPs reduces the breaking of soundness to signalling. However, in the last decade, two major problems with MIPs/ZKMIPs have emerged. The first is that the Provers do not actually need to signal in order to break some MIPs, as demonstrated in the work of Cleve, Høyer, Toner and Watrous [4]; they can perform *no-signalling tasks* which do not allow communication (for example, using shared entanglement). That is, there is a fundamental and yet subtle difference between what is *local* and what is *no-signalling.* The second by Crépeau et al. [1] is that while the Provers are unable to signal between themselves, the Verifier could inadvertently perform a non-local task for them; in the extreme case, the Verifier may plainly signal *for* the Provers.

C. Crépeau—Supported in part by Québec's FRQNT and Canada's NSERC.
N. Yang—Supported in part by Prof. David Ford and by Prof. Jeremy Clark.

© Springer International Publishing AG 2017
R.C.-W. Phan and M. Yung (Eds.): Mycrypt 2016, LNCS 10311, pp. 485–493, 2017.
DOI: 10.1007/978-3-319-61273-7_25

By combining the two problems, a Verifier can perform a no-signalling task for the Provers, and thus allow them to break soundness of their protocols. In this case, not only are the Provers perfectly no-signalling, they do not even need any extra no-signalling resources (such as quantum entanglement).

The role that the Verifier must play in these MIPs was studied in [1]. It was defined and shown that a Verifier must be *isolating*, so that it will never (inadvertently or not) perform a *non-local* task (no-signalling or signalling). We show here that many existing MIPs do not satisfy isolation, even in a weak sense.

More recently, the model of Multi-Prover Interactive Proofs was extended to allow entangled Provers and the class of languages accepted under this new setting is called **MIP*** [5]. It was recently shown that **NEXP** $\subseteq$ **MIP*** [6] but we do not know whether equality holds. Similarly, the model of Multi-Prover Interactive Proofs was extended to allow No-signalling Provers and the class of languages accepted under this new setting is called **MIP**^{ns} [5]. We now know that **MIP**^{ns} = **EXP** [7]. We use some of these results to illustrate our explanation.

2   Terminology: (Non-)local, (No-)signalling and Entangled

The terms "communicating" and "signalling" are used equivalently throughout this work and should have the obvious meaning of information transfer between two or several parties. Signalling Provers are essentially the same as a single Prover because we put no restriction whatsoever on their communication (potential interesting sub-cases arise when we restrict the amount of communication they can actually use, but we do not consider them here). In the context of several parties, we consider that signalling is taking place even if no individual communicates with any other individual. In cryptographic terms, if someone uses secret-sharing to distribute a message to several parties excluding himself (even if all of them are required to communicate to reconstruct the original secret) then signalling is considered to have taken place between the sender and the secret-share-holders.

However, as soon as we restrict communication we would need to define what non-communication (or no-signalling) actually means. The initial intuition was that non-communication = locality, meaning that the Provers are allowed to share arbitrary amount of randomness before being restricted to computations involving only these local random variables. However, because of entanglement, it was later understood that certain classes of probability distributions cannot be shared in a local fashion, but they do not allow communication. The term *no-signalling* was coined to define "everything but signalling". Of course this includes locality, but also strictly more. Typical examples are the CHSH Game (on inputs a, b output x, y such that $x \oplus y = a \times b$) and the Magic Square Game [4].

This terminology mostly originates from physics. The acclaimed work from Bell [8] in the 1960s can be summarized thus, "It seems that quantum entanglement allows for non-local yet no-signalling distributions". However, it turns out that quantum physics does not allow *all* no-signalling distributions. For instance, the CHSH game cannot be achieved from quantum entanglement. An approximation that succeeds roughly 85.4% of the time can be achieved using entanglement, whereas any local strategy can only succeed up to 75% of the time [4]. Winning the CHSH game 100% of the time is impossible even using quantum entanglement.

It may seem that the only models which make sense are "local" and "entangled" because they are motivated by physical models of reality. Nevertheless, the no-signalling model turns out to be useful under certain circumstances as explained in [7] Sect. 1.2: "We show that any MIP that is sound against no-signalling cheating Provers can be converted into a 1-round delegation scheme, using a fully homomorphic encryption scheme (FHE), or alternatively, using a computational private information retrieval (PIR) scheme."

3   Issues with Existing Protocols

First, we illustrate that MIPs may be sound on their own but not when composed. It was shown in [4] that the Magic Square Game may be turned into a language which has a MIP that is sound classically but unsound when Provers share entanglement.

We present a variant of this MIP below (as Fig. 1). Given a string of six bits $r_0, r_1, r_2, c_0, c_1, c_2$, the success probability of the classical Provers is one when there exists such a matrix M and at most 8/9 when no such matrix M exists. By repeating this protocol many times, V will be able to decide with high probability whether such a matrix exists or not. However, if P_0, P_1 can win the Magic Square Game, they can systematically break the soundness of this protocol and succeed with probability one whether such a matrix exists or not.

Construction 31 *On input six bits $r_0, r_1, r_2, c_0, c_1, c_2$,*
P_0 and P_1 claim that there exists a 3×3 binary matrix

$$M := \begin{bmatrix} m_{00} & m_{01} & m_{02} \\ m_{10} & m_{11} & m_{12} \\ m_{20} & m_{21} & m_{22} \end{bmatrix}$$

such that for $0 \le t \le 2$, $m_{t0} \oplus m_{t1} \oplus m_{t2} = r_t$ and $m_{0t} \oplus m_{1t} \oplus m_{2t} = c_t$.

- *V chooses two trits a, b uniformly and sends a to P_0 and b to P_1.*
- *P_0 (P_1) replies with row $[m_{a0}, m_{a1}, m_{a2}]$ (column $[m_{0b}, m_{1b}, m_{2b}]$).*
- *V checks that $m_{a0} \oplus m_{a1} \oplus m_{a2} = r_a$, that $m_{0b} \oplus m_{1b} \oplus m_{2b} = c_b$,*
  and that m_{ab} is the same unique value from both P_0 and P_1.

Fig. 1. A MIP for a language on six bits strings.

3.1   Issues with Current Proofs of Composability

A problem with prior MIPs' proofs of soundness is that different protocols (each of which do not allow communication) can break each other.

For instance, the MIP from [7] is resistant to no-signalling strategies. Therefore if we change [7] by appending at its end an implementation of the CHSH box by the Verifier, we would still have provable soundness. However, this new MIP, when concurrently composed with any MIP vulnerable to no-signalling strategies will result in a protocol that is unsound.

The same problem exists for protocols which are vulnerable to entanglement. The MIP from [6] is resistant to entangled Provers. We can modify this protocol into one which asks the Verifier to implement some Magic Square Games [4] without affecting soundness. This new protocol, concurrently composed with the protocol of Fig. 1, breaks the soundness of the latter protocol.

In either of the above two cases, the no-communication assumption of the composed protocols is not broken. While the above examples illustrate problems with *concurrent* composition, we consider that this is indicative of the incompleteness of existing MIPs and their analyses.

3.2   Issues Specific to ZKMIPs

In this section we explain issues with the specific construction found in [2,9] which transforms an arbitrary MIP for language L into a Zero-Knowledge version of the same proof. The technique involves the Provers using commitments to show the Verifier that "if you were to see the contents of these (committed) discussions, you would accept that $x \in L$". In the case where local (or entangled) Provers are involved, it is possible to construct bit (or trit) commitment schemes that are perfectly concealing and statistically binding [1]. One of these (Construction 33) rests on the Magic Square game and is binding against classical Provers but not against entangled Provers, while a second (Construction 32) rests on the CHSH Game and is binding against classical and entangled Provers but not against No-signalling Provers.

We summarize the construction of Kilian (and BGKW) as Fig. 2. The purpose of this protocol is to convert a generic MIP $\langle P_0, P_1, V \rangle$ into a specific format given in this figure and then compile it using Bit Commitments to make it Zero-Knowledge. The issue at hand with this construction is the Steps 1 and 3 of this protocol where P_1' send messages a_0 and a_1 to V'.

In those Steps the Prover P_1' may send V' arbitrary messages as long as V' does not reject them and abort in Step 4. Imagine if we had modified the Verifier V into V^* in such a way that it ignores whatever the Provers say unless it starts with "Simon says" and the Provers P_0, P_1 accordingly. Clearly, the resulting MIP $\langle P_0^*, P_1^*, V^* \rangle$ will be as good and sound as the original $\langle P_0, P_1, V \rangle$. However, when transformed by the protocol of Fig. 2 the new MIP will allow dishonest Provers to send arbitrary messages that V^* will ignore. In Step 5, V' will imbed these arbitrary messages into Q deterministically (see Fig. 5) and feed it to P_2'. This is a communication channel P_1' may use to send arbitrary messages to

1. P_1' sends V' a message, a_0. q_1,

2. V' sends P_1' a sequence of random coin tosses, whose length depends only on $|x|$.

3. P_1' sends V' a message, a_1.

4. V' picks a random sequence of bits, q_2, whose length depends only on $|x|$.

5. Based on the values of a_0, q_1, a_1, q_2, and x, V' deterministically decides whether to abort the protocol. If not, the verifier deterministically computes Q, A_1.

6. V' sends Q to P_2'.

7. P_2' sends an answer, A_2 to V'. V' accepts iff $A_1 = A_2$.

Fig. 2. Figure from [9] Chap. 6, page 207.

P_2'. The issue here is that nowhere is it verified that *any* of these Verifiers are Isolating. The Verifier of this protocol allows P_1' to send messages to P_2' and adding Commitments will not fix that.

This issue may be used in several different ways to break soundness of the protocol. We explain only one here, but will illustrate it with several examples in the complete version of this paper.

If this protocol is composed with any other one that uses one of the Commitments of Figs. 3 and 4 where P_1' receives the string z, he can communicate it to its partner Prover. Once this has happened, the Commitments are no longer binding and whatever proof they are using loses its soundness completely.

3.3 Synchronous vs Asynchronous MIPs

In all the MIP literature, it is never clearly specified whether MIPs are *synchronous* or *asynchronous*. Can the Verifier interact with the Provers at its own (chosen) pace independently of any clock or is the whole thing very accurately clocked? Does it even matter? We argue that being *asynchronous* is much more desirable than being *synchronous*.

In the asynchronous setting, V can interact with the Provers in any order it likes, at any rate it likes. The Provers will be allowed to communicate only when both of their respective protocols are finished with V. If the Provers are not allowed to communicate and if V does not help them in that sense, we expect this asynchronous property to be satisfied. On the contrary, in the synchronous setting V must interact with the Provers in the exact order of the protocol. If the protocol has rounds, V must complete the first round before moving on to the second round and so on. If V and the Provers have a common clock they can

---

Construction 32 *All parties agree on a security parameter k.*
P_0 *and* P_1 *partition their private random tape into a $k + 1$-bit string $\langle c, w \rangle$,*
where c is a bit and $|w| = k$.

Pre-computation phase:

- V *chooses a k-bit string z uniformly at random and sends it to P_1.*
- P_1 *responds with $d = w \oplus c \times z$, where $c \times z$ is thought of as the product*
  between a scalar c and a vector z, over $\mathbb{Z}_2$.

Commit phase:

- P_0 *commits b to V as $\boxed{b} = b \oplus c$.*

Unveil phase:

- P_0 *sends w to V.*
- V *computes $c = 1$ if $d \oplus w = z$, or $c = 0$ if $d \oplus w = \mathbf{0}$ and recovers*
  $b = \boxed{b} \oplus c$. V *rejects if $d \oplus w$ does not equal to either z or $\mathbf{0}$.* □

Fig. 3. Statistically binding, perfectly concealing bit-commitment protocol.

actually have each step of the protocol happen at a very precise time and abort if any party is not ready at the expected time or if messages did not arrive by their prescribed deadline.

Clearly, a protocol that is provably sound in the asynchronous model will also be sound in the synchronous model, but certainly not the other way around. For instance, the construction of Kilian in Fig. 2 is clearly *not* sound in the *asynchronous* model because Step 6 requires Steps 1–5 to have been completed before it can be performed. It is another reason why we became suspicious of the constructions leading ZKMIPs.

We strongly believe that we should require all MIPs to be sound in the asynchronous model. Moreover, when defining Zero-Knowledge, arbitrary Verifiers should be asynchronous. This will result in a stronger notion of Zero-Knowledge as opposed to restricting the participants to be synchronous.

3.4   A Concrete Example

We give a somewhat contrived example in existing literature which is a striking example of the consequences of lacking isolation. Consider the protocol in Fig. 5. The first Prover P_1' will simulate a number of transcripts of a MIP involving a simulated Verifier and a number of simulated Provers. The actual Verifier V' will then send a random, partial transcript to the second Prover P_2'. This partial transcript contains only one of the simulated Provers' questions and answers up to a random point. P_2' must then be able to complete the transcript in an

Construction 33 *All parties agree on a security parameter k.*
P_0 and P_1 partition their private random tape into a trit and a $4k$-bit string $\langle e, w \rangle$, where e is the trit and $|w| = 4k$. P_0 and P_1 use each block of 4 bits $w_{4j}...w_{4j+3}$ to construct a 3×3 binary matrix (where $c_x = 1$ iff $x \neq e$.)

$$\begin{bmatrix} m_{00}^j & m_{01}^j & m_{02}^j \\ m_{10}^j & m_{11}^j & m_{12}^j \\ m_{20}^j & m_{21}^j & m_{22}^j \end{bmatrix} := \begin{bmatrix} w_{4j} & w_{4j+1} & w_{4j} \oplus w_{4j+1} \\ w_{4j+2} & w_{4j+3} & w_{4j+2} \oplus w_{4j+3} \\ c_0 \oplus w_{4j} \oplus w_{4j+2} & c_1 \oplus w_{4j+1} \oplus w_{4j+3} & c_2 \oplus w_{4j} \oplus w_{4j+1} \oplus w_{4j+2} \oplus w_{4j+3} \end{bmatrix}$$

Pre-computation phase:

- V chooses a k-trit string z uniformly at random and sends it to P_1.
- P_1 responds with row $\left[m_{z_j 0}^j, m_{z_j 1}^j, m_{z_j 2}^j \right]$, for $1 \le j \le k$.
- V checks that $m_{z_j 0}^j \oplus m_{z_j 1}^j \oplus m_{z_j 2}^j = 0$, for $1 \le j \le k$.

Commit phase:

- P_0 commits t to V as $\boxed{t} = t + e \bmod 3$.

Unveil phase:

- P_0 sends e to V.
- V chooses a k-bit string r uniformly at random and sends it to P_0.
- Both compute $s_j := e + r_j + 1 \bmod 3$, for $1 \le j \le k$.
- P_0 responds with column $\left[m_{0 s_j}^j, m_{1 s_j}^j, m_{2 s_j}^j \right]$, for $1 \le j \le k$.
- V checks that $m_{0 s_j}^j \oplus m_{1 s_j}^j \oplus m_{2 s_j}^j = 1$, and that $m_{z_j s_j}^j$ is the same unique value from both P_0 and P_1, for $1 \le j \le k$. $\square$

Fig. 4. Statistically binding, perfectly concealing trit-commitment protocol.

identical way, otherwise the proof is rejected. The simulated Provers and Verifiers are deterministic, which should allow this consistency check with P'_2 to succeed.

The first problem we would like to point out is that the simulated *answers* from the simulated Provers cannot be authenticated. There is no *a priori* reason why the Verifier would suspect these answers to be attempts at communication. In addition, not only is the Verifier not isolating, in this case the protocol *requires* that the Verifier actually courier some messages from one Prover to another.

Suppose that the simulated protocol has a "header" section where the simulated Provers can say anything and it will be ignored by the Verifier, but would nevertheless be part of a valid transcript. In the compositional form of Kilian's protocol, the Verifier has an auxiliary input tape (which is normally used to model prior knowledge a Verifier might have). The real Provers can use this auxiliary tape to communicate; in particular, P'_1 can send to P'_2 the value of R, the random coins V' is forcing P'_1 to use. This fixes the simulated Verifier's

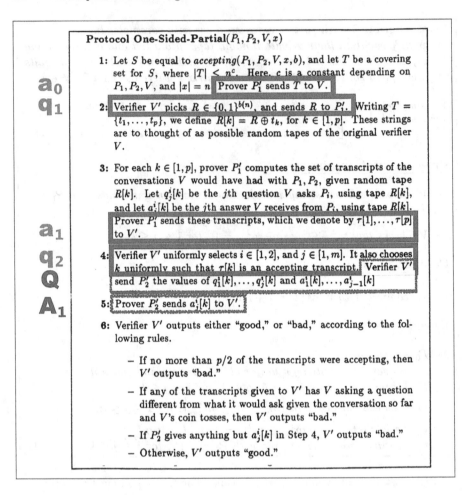

Protocol One-Sided-Partial(P_1, P_2, V, x)

1: Let S be equal to $accepting(P_1, P_2, V, x, b)$, and let T be a covering set for S, where $|T| \leq n^c$. Here, c is a constant depending on P_1, P_2, V, and $|x| = n$. Prover P_1' sends T to V'.

2: Verifier V' picks $R \in \{0,1\}^{b(n)}$, and sends R to P_1'. Writing $T = \{t_1, \ldots, t_p\}$, we define $R[k] = R \oplus t_k$, for $k \in [1, p]$. These strings are to thought of as possible random tapes of the original verifier V.

3: For each $k \in [1, p]$, prover P_1' computes the set of transcripts of the conversations V would have had with P_1, P_2, given random tape $R[k]$. Let $q_j^i[k]$ be the jth question V asks P_i, using tape $R[k]$, and let $a_j^i[k]$ be the jth answer V receives from P_i, using tape $R[k]$. Prover P_1' sends these transcripts, which we denote by $\tau[1], \ldots, \tau[p]$ to V'.

4: Verifier V' uniformly selects $i \in [1, 2]$, and $j \in [1, m]$. It also chooses k uniformly such that $\tau[k]$ is an accepting transcript. Verifier V' send P_2' the values of $q_1^i[k], \ldots, q_j^i[k]$ and $a_1^i[k], \ldots, a_{j-1}^i[k]$

5: Prover P_2' sends $a_j^i[k]$ to V'.

6: Verifier V' outputs either "good," or "bad," according to the following rules.

  – If no more than $p/2$ of the transcripts were accepting, then V' outputs "bad."

  – If any of the transcripts given to V' has V asking a question different from what it would ask given the conversation so far and V's coin tosses, then V' outputs "bad."

  – If P_2' gives anything but $a_j^i[k]$ in Step 4, V' outputs "bad."

  – Otherwise, V' outputs "good."

Fig. 5. Figure from [9] Chap. 6, page 205.

random tape, which allows the real second Prover P_2' to break any consistency checks, and therefore soundness.

Even if there is no auxiliary tape, the simulated Provers' *first* question cannot be in general authenticated by the Verifier as genuine (as the Provers are all-powerful but the Verifier is only polynomial-time). Therefore P_1' can always produce a transcript where R is disguised as the first question. Soundness breaks by the same reasoning as above. Luckily, in this case, the cheating Provers would be detected if V' asks P_2' to produce the *first* question from scratch, which it cannot since it does not know R.

Now the second problem is that in [9], the proof of soundness does not mention this near-miss. This is symptomatic of proofs of soundness of MIPs in the literature: their proofs are considerably incomplete; fixing these proofs require proper definitions of many concepts taken for granted in these papers.

4 Discussion

Considering the many issues described in the previous sections, we believe that there is a need to rethink MIPs/ZKIPs with respect to locality, synchronicity, composability and isolation. The attacks that we have demonstrated may be somewhat contrived, but they demonstrate the incompleteness of existing work.

In particular, we think that new definitions and proofs are necessary to capture the counter-intuitiveness of non-locality, including entanglement and other no-signalling tasks. Existing results must be revalidated under an upgraded model. We will explore this idea in detail in our follow-up paper.

Acknowledgements. We would like to thank Serge Fehr, Gilles Brassard, Samuel Ranellucci, Christian Schaffner, and Louis Salvail for various fruitful discussions about this work. Finally, we are grateful to Raphael C.-W. Phan and Moti Yung for inviting us to submit our work here.

References

1. Crépeau, C., Salvail, L., Simard, J.-R., Tapp, A.: Two provers in isolation. In: Lee, D.H., Wang, X. (eds.) ASIACRYPT 2011. LNCS, vol. 7073, pp. 407–430. Springer, Heidelberg (2011). doi:10.1007/978-3-642-25385-0_22
2. Ben-Or, M., Goldwasser, S., Kilian, J., Wigderson, A.: Multi-prover interactive proofs: how to remove intractability assumptions. In: Proceedings of the Twentieth Annual ACM Symposium on Theory of Computing, STOC 1988, pp. 113–131. ACM, New York (1988)
3. Babai, L., Fortnow, L., Lund, C.: Non-deterministic exponential time has two-prover interactive protocols. Comput. Complex. **2**, 374 (1992)
4. Cleve, R., Hoyer, P., Toner, B., Watrous, J.: Consequences and limits of nonlocal strategies. In: Proceedings of the 19th IEEE Annual Conference on Computational Complexity, CCC 2004, pp. 236–249, IEEE Computer Society, Washington, DC (2004)
5. Ito, T.: Polynomial-space approximation of no-signaling provers. In: Abramsky, S., Gavoille, C., Kirchner, C., Meyer auf der Heide, F., Spirakis, P.G. (eds.) ICALP 2010. LNCS, vol. 6198, pp. 140–151. Springer, Heidelberg (2010). doi:10.1007/978-3-642-14165-2_13
6. Ito, T., Vidick, T.: A multi-prover interactive proof for NEXP sound against entangled provers. In: Proceedings of the 2012 IEEE 53rd Annual Symposium on Foundations of Computer Science, FOCS 2012, pp. 243–252. IEEE Computer Society, Washington, DC (2012)
7. Kalai, Y.T., Raz, R., Rothblum, R.D.: How to delegate computations: the power of no-signaling proofs. In: Proceedings of the Forty-Sixth Annual ACM Symposium on Theory of Computing, STOC 2014, pp. 485–494. ACM, New York (2014)
8. Bell, J.S.: On the Einstein-Podolsky-Rosen paradox. Physics **1**, 195–200 (1964)
9. Kilian, J.: Uses of Randomness in Algorithms and Protocols. MIT Press, Cambridge (1990)

Human Public-Key Encryption

Houda Ferradi, Rémi Géraud[✉], and David Naccache

Information Security Group, École Normale Supérieure,
45 rue d'Ulm, 75230 Paris Cedex 05, France
{houda.ferradi,remi.geraud,david.naccache}@ens.fr

Abstract. This paper proposes a public-key cryptosystem and a short password encryption mode, where traditional hardness assumptions are replaced by specific refinements of the CAPTCHA concept called Decisional and Existential CAPTCHAs.

The public-key encryption method, achieving 128-bit security, typically requires from the sender to solve one CAPTCHA. The receiver does not need to resort to any human aid.

A second symmetric encryption method allows to encrypt messages using very short passwords shared between the sender and the receiver. Here, a simple 5-character alphanumeric password provides sufficient security for all practical purposes.

We conjecture that the automatic construction of Decisional and Existential CAPTCHAs is possible and provide candidate ideas for their implementation.

1 Introduction

CAPTCHAs[1] [1] are problems that are hard to solve by computers, while being at the reach of most untrained humans. There might be many reasons why, at a particular time, a given type of CAPTCHA is considered hard for computers. The automated solving of CAPTCHAs may either require more computational power than is available, or algorithms have yet to be invented. It might well be that computers are inherently less efficient, or even incapable, at some tasks than human beings. Whichever the cause, several candidate CAPTCHAs are widely used throughout the Internet to keep robots at bay, or at least slow them down (e.g. [2,7,8,10,16,17]).

Most CAPTCHAs are used as human-interaction proofs [3] but their full potential as cryptographic primitives has not been leveraged so far despite a few exploratory papers. Early attempts [1,4,5,9] faced the inherent difficulty of *malleability*: given a CAPTCHA Q, an adversary could generate Q', whose solution gives a solution to Q. Thus the security of such constructions could only be evaluated against unrealistic "conservative adversaries" [13]. All in all, we propose to fill the gap by providing a finer taxonomy of CAPTCHAs as well as cryptosystems based on them, which can reach real-life security standards.

---

[1] "Completely Automated Public Turing test to Tell Computers and Humans Apart".

© Springer International Publishing AG 2017
R.C.-W. Phan and M. Yung (Eds.): LNCS 10311, Mycrypt 2016, pp. 494–505, 2017.
DOI: 10.1007/978-3-319-61273-7_26

The organisation of this paper is as follows: Sect. 2 defines the classes of problems we are interested in, and estimates how many of those problems can be solved per time unit. We then refine the classical CAPTCHA concept into Decisional and Existential CAPTCHAs. Section 3 describes how to implement public-key encryption using Decisional CAPTCHAs; Sect. 4 describes a short password-based encryption mode that uses Existential CAPTCHAs to wrap high-entropy keys. Section 5 presents Decisional and Existential CAPTCHA candidates.

2 Preliminaries and Definitions

2.1 CAPTCHA Problems

Let $\mathcal{Q}$ be a class of problem instances, $\mathcal{A}$ a class of answers, and S a relation such that $S(Q, A)$ expresses the fact that "$A \in \mathcal{A}$ is a solution of $Q \in \mathcal{Q}$". Solving an instance Q of problem $\mathcal{Q}$ means exhibiting an $A \in \mathcal{A}$ such that $S(Q, A)$. We assume that for each problem there is one and only one solution, i.e. that S is bijective. This formal setting (similar to [5,13]) allows us to provide more precise definitions.

Because CAPTCHAs involve humans and considerations about the state of technology, we do not pretend to provide formal mathematical definitions but rather clarifying definitional statements.

Definition 1 (Informal). *A given problem $\mathcal{Q} \in CP$ (CAPTCHA Problem) if no known algorithm can solve a generic instance $Q \in \mathcal{Q}$ with non-negligible advantage over $1/|\mathcal{A}|$, which is the probability to answer Q correctly at random; yet most humans can provide the solution A to a random $Q \in_R \mathcal{Q}$ with very high probability in reasonable time.*

In Definition 1, it is worth pointing out that future algorithms might turn out to solve efficiently some problems that evade today's computers' reach. As such, CP is not so much a complexity class as it is a statement about technology at any given point in time.

There exist today several approaches to building CAPTCHAs, based for instance on deformed word recognition, verbal tests, logic tests or image-based tasks. We are chiefly interested in those tests that can be automatically generated.

We extend CP in two ways:

Definition 2 (Informal). *A given problem $\mathcal{Q} \in DCP$ (Decisional CP) if $\mathcal{Q} \in CP$ and, given a random instance $Q \in_R \mathcal{Q}$ and a purported solution A to Q, no known algorithm can decide whether A is a solution to Q, i.e. evaluate $S(Q, A)$, with non-negligible advantage over $1/|\mathcal{A}|$; while humans can determine with high probability $S(Q, A)$ in reasonable time.*

Finally, we introduce a further class of problems:

Definition 3 (Informal). *Let $\overline{Q} \notin CP$ be a set of "decoy data" which are not CAPTCHAs. A given problem $Q \in ECP$ (Existential CP) if $Q \in CP$ and, given a generic instance $Q \in Q$ or a decoy $Q \in \overline{Q}$, no known algorithm can decide whether $Q \in Q$ with non-negligible advantage over $|Q|/|Q \cup \overline{Q}|$; while humans can decide correctly if $Q \in Q$ or $Q \in \overline{Q}$ in reasonable time with high probability.*

Remark 1. Definition 3 depends on the set $\overline{Q}$. We silently assume that, for a given problem Q, an appropriate $\overline{Q}$ is chosen. This choice makes no difference.

When Q is not exhaustively searchable, Definition 3 means that a computer cannot decide whether a given Q is a CAPTCHA or not, let alone solve Q if Q is indeed a CAPTCHA.

Remark 2. Definition 3 can be reformulated similarly to the IND-CPA [15] security game: we pick a random bit b and provide the adversary with Q_b, where $Q_0 \in Q$ and $Q_1 \in \overline{Q}$. The adversary is expected to guess b no better than at random unless it resorts to human aid.

Remark 3. ECP, DCP $\subseteq$ CP, but there is no inclusion of ECP in DCP or *vice versa*. Informally, CP is about finding an answer, DCP is about checking an answer, and ECP is about recognizing a question.

Remark 4. Solving a problem $Q \in CP$ is either done using computers which by definition provide unreliable answers at best; or by asking a human to solve Q – effectively an oracle. However, there is a limit on the number of solutions humans can provide and on the rate at which humans can solve CAPTCHAs.

Consider a given $Q \in CP$ whose generic instances can be solved by a human in reasonable time. Let us estimate an upper bound b on the number of instances of Q that a human may solve during a lifetime. Assuming a solving rate of 10 instances per minute, and working age of 15–75 years, spent exclusively solving such problems, we get $b \sim 10^8$. Taking into account sleep and minimal life support activities, b can be brought down to $\sim 10^7$.

There should be no measurable difference between solving a problem in CP or in DCP, however it might be slightly simpler (and therefore quicker) for humans to *identify* whether a problem is a CAPTCHA without actually solving it. For simplicity we can assume that CAPTCHA recognition is ten times faster than CAPTCHA resolution.

There are various estimations on the cost of having humans solve CAPTCHAs. Some websites offer to solve 1000 CAPTCHAs for a dollar[2]. Of course, the oracle may employ more than one human, and be proportionally faster, but also proportionally more expensive.

---

[2] At a first glance, the previous figures imply that breaking a public-key (as defined in the next section) would only cost $\$10^4$. We make the economic nonlinearity conjecture there are no $\$10^4$ service suppliers allowing the scaling-up of this attack. In other words, if the solving demand d increases so will the price. We have no data allowing to quantify price(d).

3 Human Public-Key Encryption

We now describe a public-key cryptosystem using problems in DCP. Let $\mathcal{Q} \in$ DCP. We denote by $H(m)$ a cryptographic hash function (e.g. SHA-3) and by $E_k(m)$ a block cipher (e.g. AES-128). Here, m is the plaintext sent by Bob to Alice.

- *Key-pair generation*: The public key pk is a list of b instances of $\mathcal{Q}$

$$\mathsf{pk} = \{Q_1, \dots, Q_b\}$$

  The private key is the set of solutions (in the CP sense) to the Q_i:

$$\mathsf{sk} = \{A_1, \dots, A_b\}$$

  i.e. for $1 \le i \le b$, $S(Q_i, A_i)$ holds true.
- *Encryption*: Bob wants to send m to Alice. Bob picks k random problems $\{Q_{i_1}, \dots, Q_{i_k}\}$ from Alice's pk, and solves them[3]. Let $\sigma \leftarrow \{A_{i_1}, \dots, A_{i_k}\}$ and $\alpha \leftarrow \{i_1, \dots, i_k\}$. Bob computes $\kappa \leftarrow H(\alpha)$ and $c \leftarrow E_\kappa(m)$, and sends (σ, c) to Alice.
- *Decryption*: Given σ, Alice identifies the set of indices α and computes $\kappa \leftarrow H(\alpha)$. Alice then uses κ to decrypt c and retrieve m. Decryption does *not* require any human help.

The general idea of this cryptosystem is somewhat similar to Merkle's puzzles [14], however unlike Merkle's puzzle here security *is not quadratic*, thanks to problems in CP not being automatically solvable. We may assume that the A_is are pairwise different to simplify analysis.

Remark 5. Indeed if $\mathcal{Q} \in$ CP it might be the case that a machine could decide if given A, Q the relation $S(A, Q)$ holds *without* solving Q. Hence $\mathcal{Q}$ must belong to DCP.

Remark 6. A brute-force attacker will exhaust all $\binom{b}{k}$ possible values of α. Hence $\binom{b}{k}$ should be large enough. Given that b–10^7 or b–10^8, it appears that $k = 6$ provides at least 128-bit security.

Remark 7. The main drawback of the proposed protocol is the size of pk. Assuming that each Q_i can be stored in 20 bytes, a pk corresponding to b–10^8 would require 2 GB. However, given that CAPTCHAs are usually visual problems, it is reasonable to assume that pk might turn out to be compressible.

Remark 8. Instead of sending back the solutions σ in clear, Bob could hash them individually. Hashing would only make sense as long as solutions have enough entropy to resist exhaustive search.

---

[3] Here Bob must resort to human aid to solve $\{Q_{i_1}, \dots, Q_{i_k}\}$.

Remark 9. It is possible to leverage the DCP nature of the Q_is in the following way: instead of sending a random permutation of solutions, Bob could interleave into the permutation d random values (decoy answers). Alice would spot the positions of these decoy answers and both Alice and Bob would generate $\alpha = \{i_1, \ldots, i_k, j_1, \ldots, j_d\}$ where j_d are the positions of decoys. Subsequently, security will grow to $\binom{b}{k+d}/d!$. This is particularly interesting since for $b = 10^7$, $k = 1$ and $d = 6$ we exceed 128-bit security. In other words, all the sender has to do is to *solve one CAPTCHA*.

Entropy can be further increased by allowing d to vary between two small bounds. In that case the precise (per session) value of d is unknown to the attacker.

4   Short Password-Based Encryption

In the following scenario Alice and Bob share a short password w. We will show how a message m can be securely sent from Alice to Bob using *only* w. This is particularly suited to mobile devices in which storing keys is risky.

Let $\mathcal{Q} \in \mathsf{ECP} \cap \mathsf{DCP}$.

- Alice generates a full size[4] key R and uses it to encrypt m, yielding $c_0 \leftarrow E_{0|R}(m)$. She generates an instance $Q \in \mathcal{Q}$, such that $S(P, R)$. Alice computes $c_1 \leftarrow E_{1|w}(P)$ and sends (c_0, c_1) to Bob.
- Bob uses w to decrypt c_1, and solves P. He thus gets the key R that decrypts c_0.

An adversary therefore faces the choice of either "attacking Shannon" or "attacking Turing", i.e. either automatically exhaust R, or humanly exhaust w. Each candidate w yields a corresponding P that cannot be computationally identified as a CAPTCHA. The adversary must hence resort to humans to deal with every possible candidate password.

Assuming that CAPTCHA identification by humans is ten times faster than CAPTCHA resolution, it appears that w can be a 5-character alphanumeric code[5].

Remark 10. R must have enough entropy bits to provide an acceptable security level. R can be generated automatically on the user's behalf. As we write these lines we do now know if there exists $\mathcal{Q} \in \mathsf{ECP} \cap \mathsf{DCP}$ admitting 128-bits answers. If such $\mathcal{Q}$s do not exist, R could be assembled from several problem instances.

Remark 11. In the above we assume that R is generated first, and then embedded into the solution of a problem instance P. All we require from R is to provide sufficient entropy for secure block cipher encryption. Hence, it might be easier to generate P first, and collect R afterwards.

---

[4] E.g. 128-bit.
[5] There are 64 alphanumeric characters, and $64^5 > 10 \times b$.

Remark 12. The main burden resting on Bob's shoulders might not be the solving on P but the keying of the answer R. 128 bits are encoded as 22 alphanumeric characters. Inputting R is hence approximately equivalent to the typing effort required to input a credit card information into e-commerce website interfaces[6]. Alternatively, Bob may as well read the solution R to a speech-to-text interface that would convert R into digital form.

Remark 13. $Q \in \mathsf{ECP} \cap \mathsf{DCP}$ is necessary because the adversary may partially solve Q and continue using exhaustive search. Under such circumstances, c_0 serves as a clue helping the attacker to solve Q. If $Q \in \mathsf{ECP} \cap \mathsf{DCP}$, such a scenario is avoided.

5   DCP and ECP Candidate Instances

The above constructions assume that ECP and DCP instances exist and are easy to generate. Because ECP and DCP depend both on humans and on the status of technology, it is difficult to "prove" the feasibility of the proposed protocols.

We hence propose a DCP candidate an ECP candidates and submit them to public scrutiny.

5.1   DCP Candidate

As a simple way to generate DCPs, we propose to start from a standard CP (e.g. a number recognition problem) and ask a further question about the answer. The further question should be such that its answer may correspond to numerous potential contents. For instance, the further question could be whether two sequences of digits recognised in an image Q sum up to $A = 91173$ or not (see Fig. 1).

Fig. 1. A DCP candidate constructed from an existing CP.

5.2   ECP Candidates

This section proposes a few candidate Q that we conjecture to belong to ECP.

The first step is to design a task that we think is challenging for computers. Despite recent progress (see e.g. [11]), computer vision is still expensive and limited. Most computer vision algorithms have to be trained specifically to recognise

---

[6] PAN (16 characters), expiry date (4 characters) and a CVV (4 characters).

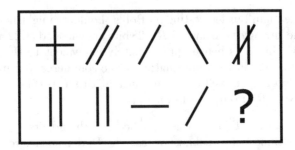

Fig. 2. An instance of a visual-logical task ECP problem. Recognizing objects in this image is insufficient to tell whether there is a solution, nor to compute the solution should there be one.

objects or features of a given kind (dog breeds, handwritten characters, etc.), and fail whenever the task at hand requires more than mere object identification. Even in that case, occlusion, distortion and noise cause drastic performance loss for most techniques. Many CAPTCHAs ideas rely on this to generate problem instances [6].

Even if image contents can be detected, we can still pose a hard challenge. Indeed, while computers excel at solving logical reasoning questions when those questions are encoded manually as logical formulae, state of the art algorithms fail at even the most basic questions when challenges are presented in visual form. Therefore, solving for instance a visual-logical task is a problem that is at least in DCP (see Fig. 2).

Good ECP candidates for cryptographic purposes should be easy to generate, they should have enough possible solutions to thwart exhaustive search attempts, and it should be hard to tell automatically whether there is a solution at all.

Temporal Sequence ECP. The intuition for this candidate is that although computer vision algorithms may reach human accuracy (and even beat it), humans can make use of external knowledge, which provides additional understanding of what is under scrutiny. Here the external knowledge is that real-life events abide by *causality*.

We provide k images (e.g. $k = 5$), each of which is a snapshot of some situation: buying goods, driving a car, dressing up, etc. The order of images is scrambled (some random images may be inserted as decoys) and the problem is to put images back in the correct order. This task, which we call *temporal sequence*, requires the contextual knowledge that some events can only happen after (or before) others. This is illustrated in Fig. 3.

We conjecture that the temporal sequence task is both in DCP and in ECP.

One drawback of this approach is that to reach an 80-bit security level we need $k = 40$ images[7] which can be unwieldy. This may be solved by using ℓ collections of κ images, and tune ℓ, κ so that $(\kappa!)^\ell > 2^{80}$.

---

[7] There are $k!$ combinations, and $40! > 2^{80}$.

Fig. 3. Three instances of the temporal sequence ECP problem. The problem consists in temporally arranging the pictures.

Temporal sequences may be automatically generated from videos, although it is not obvious how to ensure that sequences generated like this are always meaningful to humans.

Visual Letter Recognition ECP. Assume we have a CP problem Q, whose instances can successfully conceal letters (a "one-letter" CAPTCHA). We provide k instances of $Q_1, \ldots, Q_k$ corresponding to answer letters $A_1, \ldots, A_k$, and ask for the alphabetically sorted list of these A_i.

As an example, we would generate instances of Q for the letters $\{A, M, T, O, B, R\}$, and ask for the solution ABMORT. Under the assumption that $Q \in$ CP, determining whether a solution exists requires human aid. Therefore we conjecture that this problem belongs to ECP.

A further variant of this idea is illustrated in Fig. 4. Note that the visual letter recognition problem is DCP if an only if $Q \in$ DCP.

Honey Images ECP. Another candidate problem is inspired by honey encryption [12,18]. The idea is that any integer $1 \le \ell \le k$ would generate an image, but that only one value ℓ_{OK} generates a *meaningful* image[8]. All values $\ell \ne \ell_{\mathrm{OK}}$

---

[8] In the specific case of Fig. 5, translation, rotation, mirroring as well as border cropping may also generate the meaningful image corresponding to ℓ_{OK}, but the overall proportion of such images remains negligible.

Fig. 4. Visual Letter Recognition ECP: letters are concealed using an existing CP, and one digit is inserted into each sequence of letters. The ECP problem is to reorder the CAPTCHAs in increasing digit order, discarding all non-digit symbols. Here the solution consists in selecting the 4th, 5th, 2nd, 3rd, and 1st images, in that order.

Fig. 5. A honey image ECP. Left: original image; right: $Q_{\ell_{\mathrm{OK}}}$, the transformed image for ℓ_{OK}.

generate images in a way that makes them indistinguishable from meaningful images. The problem would then be to identify ℓ_{OK}, which we conjecture only humans can do reliably.

The main difficulty is that the notion of indistinguishability is tricky to define for images, and even harder to enforce: humans and computers alike use very specific visual cues to try and perform object recognition, which are hard to capture statistically. Following [18], we may try and learn from a dataset how to

Fig. 6. All values of ℓ other than ℓ_{OK} produce decoys whose statistical properties are conjectured to be indistinguishable from the correct image, with salient features but no real meaning.

properly encode images, but this is cumbersome in our context, especially when dealing with a large number of instances.

Our candidate is a simpler embodiment based on the following intuition: using biased noise (i.e. noise that is *not* random), we can elicit pareidolia in computer vision programs. Each candidate value of ℓ would then correspond to some object being recognised – but only one of those is really relevant. We conjecture that only humans are able to pick this relevant object apart.

The authors implemented this idea. We start from a black and white picture of a clearly identifiable object (Fig. 5 left, here A = "rabbit"), turn it into a collection of black dots[9] (1). The picture is then cut into blocks which are shuffled and rotated (2). Finally, noise is added, under the form of black dots whose size is distributed as the size of black dots in the original picture (3). The image is then rotated back in place (Fig. 5 right) to provide the challenge $Q_{\ell_{OK}}$.

The motivation for this approach is as follows: (1) guarantees that individual pixels contain no information on luminance, and geometric features (lines, gradients and corners) – each dot being circular destroys information about orientation; the shuffling and rotation of blocks in (2) is encoded as an integer ℓ; and (3) inserts decoy features, so that any shuffling/rotation would make geometric features appear (to lure a computer vision algorithm into detecting something).

Now, many decoys $Q_\ell \in \overline{Q}, \ell \neq \ell_{OK}$ can be generated easily from this image by shuffling and rotating blocks (Fig. 6). Each decoy shares the same statistical properties as the correct (unshuffled) image, but has no recognizable content.

Our conjecture is that the human brain can perceive structures very efficiently and assign meaning to them. Many such structures are irrelevant and inserted

---

[9] For instance using an iteratively reweighted Voronoi diagram.

so as to fool computer vision algorithms, but the familiar ones are immediately and intuitively grasped by humans. Consequently, although the original picture is severely deteriorated, we conjecture that it should still be possible for humans to tell noise and signal apart and identify correctly the contents of this image.

6   Further Applications

Beyond their cryptographic interest, DCP and ECP tasks may have interesting applications in their own right (Fig. 7).

Fig. 7. Credit card PAN and expiry date, stored as a DCP instance.

One such application is the following: users may wish to store sensitive data as a DCP instance, for instance credit card information, instead of plaintext. Indeed, attackers often browse their victims' computers looking for credit card information, which is easy to recognize automatically. By storing credentials in an ECP the attacker's task can be made harder.

References

1. von Ahn, L., Blum, M., Hopper, N.J., Langford, J.: CAPTCHA: using hard AI problems for security. In: Biham, E. (ed.) EUROCRYPT 2003. LNCS, vol. 2656, pp. 294–311. Springer, Heidelberg (2003). doi:10.1007/3-540-39200-9_18
2. von Ahn, L., Maurer, B., McMillen, C., Abraham, D., Blum, M.: Google reCAPTCHA: https://developers.google.com/recaptcha (2007)
3. Baird, H.S., Lopresti, D.P. (eds.): HIP 2005. LNCS, vol. 3517. Springer, Heidelberg (2005). doi:10.1007/b136509
4. Canetti, R., Halevi, S., Steiner, M.: Hardness amplification of weakly verifiable puzzles. In: Kilian, J. (ed.) TCC 2005. LNCS, vol. 3378, pp. 17–33. Springer, Heidelberg (2005). doi:10.1007/978-3-540-30576-7_2
5. Canetti, R., Halevi, S., Steiner, M.: Mitigating dictionary attacks on password-protected local storage. In: Dwork, C. (ed.) CRYPTO 2006. LNCS, vol. 4117, pp. 160–179. Springer, Heidelberg (2006). doi:10.1007/11818175_10
6. Chellapilla, K., Larson, K., Simard, P.Y., Czerwinski, M.: Designing human friendly human interaction proofs (HIPs). In: van der Veer, G.C., Gale, C. (eds.) Proceedings of the 2005 Conference on Human Factors in Computing Systems, CHI 2005, Portland, Oregon, USA, 2–7 April 2005, pp. 711–720. ACM (2005)
7. Chew, M., Baird, H.S.: Baffletext: a human interactive proof. In: Kanungo, T., Smith, E.H.B., Hu, J., Kantor, P.B. (eds.) SPIE Proceedings, Document Recognition and Retrieval X, Proceedings, Santa Clara, California, USA, 22–23 January 2003, vol. 5010, pp. 305–316. SPIE (2003)

8. Chow, R., Golle, P., Jakobsson, M., Wang, L., Wang, X.: Making captchas clickable. In: Spasojevic, M., Corner, M.D. (eds.) Proceedings of the 9th Workshop on Mobile Computing Systems and Applications, HotMobile 2008, Napa Valley, California, USA, 25–26 February 2008, pp. 91–94. ACM (2008)

9. Dziembowski, S.: How to pair with a human. In: Garay, J.A., Prisco, R. (eds.) SCN 2010. LNCS, vol. 6280, pp. 200–218. Springer, Heidelberg (2010). doi:10. 1007/978-3-642-15317-4_14

10. Elson, J., Douceur, J.R., Howell, J., Saul, J.: Asirra: a CAPTCHA that exploits interest-aligned manual image categorization. In: Ning, P., di Vimercati, S.D.C., Syverson, P.F. (eds.) Proceedings of the 2007 ACM Conference on Computer and Communications Security, CCS 2007, Alexandria, Virginia, USA, 28–31 October 2007, pp. 366–374. ACM (2007)

11. Goodfellow, I.J., Bulatov, Y., Ibarz, J., Arnoud, S., Shet, V.: Multi-digit number recognition from street view imagery using deep convolutional neural networks. CoRR abs/1312.6082 (2013). http://arxiv.org/abs/1312.6082

12. Juels, A., Ristenpart, T.: Honey encryption: security beyond the brute-force bound. In: Nguyen, P.Q., Oswald, E. (eds.) EUROCRYPT 2014. LNCS, vol. 8441, pp. 293–310. Springer, Heidelberg (2014). doi:10.1007/978-3-642-55220-5_17

13. Kumarasubramanian, A., Ostrovsky, R., Pandey, O., Wadia, A.: Cryptography using captcha puzzles. In: Kurosawa, K., Hanaoka, G. (eds.) PKC 2013. LNCS, vol. 7778, pp. 89–106. Springer, Heidelberg (2013). doi:10.1007/978-3-642-36362-7_7

14. Merkle, R.C.: Secure communications over insecure channels. Commun. ACM **21**(4), 294–299 (1978)

15. Naor, M., Yung, M.: Public-key cryptosystems provably secure against chosen ciphertext attacks. In: Ortiz, H. (ed.) Proceedings of the 22nd Annual ACM Symposium on Theory of Computing, Baltimore, Maryland, USA, 13–17 May 1990, pp. 427–437. ACM (1990)

16. Nayeem, M.T., Akand, M.M.R., Sakib, N., Kabir, M.W.U.: Design of a human interaction proof (HIP) using human cognition in contextual natural conversation. In: IEEE 13th International Conference on Cognitive Informatics and Cognitive Computing, ICCI*CC 2014, London, UK, 18–20 August 2014, pp. 146–154. IEEE (2014)

17. Sauer, G., Holman, J., Lazar, J., Hochheiser, H., Feng, J.: Accessible privacy and security: a universally usable human-interaction proof tool. Univ. Access Inf. Soc. **9**(3), 239–248 (2010)

18. Yoon, J.W., Kim, H., Jo, H., Lee, H., Lee, K.: Visual honey encryption: application to steganography. In: Alattar, A.M., Fridrich, J.J., Smith, N.M., Alfaro, P.C. (eds.) Proceedings of the 3rd ACM Workshop on Information Hiding and Multimedia Security, IH&MMSec 2015, Portland, OR, USA, 17–19 June 2015, pp. 65–74. ACM (2015)

Two Philosophies for Solving Non-linear Equations in Algebraic Cryptanalysis

Nicolas T. Courtois[(✉)]

Computer Science, University College London,
Room 6.18. Gower Street, London WC1E 6BT, UK
`n.courtois@ucl.ac.uk`

Abstract. Algebraic Cryptanalysis [45] is concerned with solving of particular systems of multivariate non-linear equations which occur in cryptanalysis. Many different methods for solving such problems have been proposed in cryptanalytic literature: XL and XSL method, Gröbner bases, SAT solvers, as well as many other. In this paper we survey these methods and point out that the main working principle in all of them is essentially the same. One quantity grows faster than another quantity which leads to a "phase transition" and the problem becomes efficiently solvable. We illustrate this with examples from both symmetric and asymmetric cryptanalysis.

In this paper we point out that there exists a second (more) general way of formulating algebraic attacks through dedicated **coding** techniques which involve redundancy with addition of new variables. This opens numerous new possibilities for the attackers and leads to interesting optimization problems where the existence of interesting equations may be somewhat deliberately engineered by the attacker.

Keywords: Algebraic cryptanalysis · Overdefined systems of equations · NP-hard problems · Phase transitions · XL algorithm · Gröbner bases · XSL algorithm · ElimLin · Degree falls · Error correcting codes · Algebraic codes · Elliptic curves · ECDL problem · Semaev polynomials · Block ciphers · DES · GOST · Simon

1 Two Approaches to Solving Non-linear Equations

There are two major philosophies in algebraic cryptanalysis and for the general problem of solving large system of non-linear polynomial/algebraic equations.

1. Either we expand the number of monomials.
2. Or we expand the number of variables.

Let us also recall what is the main working principle in both types of techniques: we make two values grow, yet one grows faster. We will see several examples of this in this paper. It allows one to understand why both families of techniques may and will in many cases work.

© Springer International Publishing AG 2017
R.C.-W. Phan and M. Yung (Eds.): LNCS 10311, Mycrypt 2016, pp. 506–520, 2017.
DOI: 10.1007/978-3-319-61273-7_27

1. When we add **new monomials**, we grow both the number of monomials T and the number of new equations R. For any system of equations with a certain R/T we can easily improve the R/T ratio by increasing the degree. Then R grows **faster** because there are several ways to obtain the same monomial. The number of monomials does typically NOT[1] grow as fast as the number of new equations.
2. When we add **new variables** we also grow both the number of monomials T and can generate more equations R. Here it is maybe less obvious that R can also grow **faster** and sometimes even asymptotically faster than T, well, at least and limited to[2] a certain interval. One example of this can be found in Sect. 12.3 in [29] another in [12].

This paper is organized as follows. In Sect. 2 we look briefly at the history of the tow approaches and discuss some reasons why type 2 methods have not for a long time attracted a lot of attention. In Sect. 3 we discuss various more or less standard algorithms of type 1. In Sects. 4 and 5.1 we discuss a number of existing approaches for coding cryptanalysis problems with a focus on symmetric cryptanalysis. We discuss one specific strategy for the attacker where the attacker tries to design a family of problem which becomes progressively easier to solve as the parameter grows. Finally in Sect. 6 we discuss one major "algebraic coding" problems in asymmetric cryptanalysis. Public-key cryptanalysis questions are also discussed in Sect. 4.

2 Historical Developments

Both types of methods already existed and both philosophies worked quite well in their own (somewhat disjoint) space in algebraic cryptanalysis of DES in [6]. More generally, both sorts have also been studied for solving systems of polynomial equations over finite fields at Eurocrypt 2000 [7].

It is important to see that techniques of type 1 which expand monomials are nowadays standard, well studied, fully automated by software and do not[3] require a lot of attention. The second family has not been sufficiently studied.

2.1 Difficulties with Family 2 Techniques

There have been some negative results on techniques of type 2. At Eurocrypt 2000 [7] the authors consider the general problem of solving arbitrary quadratic

---

[1] Identical monomials are generated many times, for example $x_1x_2x_3$ will be obtained 3 times, when multiplying x_1x_3 by x_2, etc. Cf. also slide 80 of [26].

[2] A situation where R grows faster than T permanently must be an illusion. Let $F \leq R$ be the number of linearly independent equations. These equations belong to the linear space of dimension T. Thus $F \leq T$ and very frequently $F \leq T - 1$, cf. [8].

[3] Such software methods are sometimes called "plug and pray" attacks, cf. [29] and the main point in this paper and in [29] is that we would like to develop a richer galaxy of attacks where the attacker plays a more active role.

or low degree equations over a finite fields. We have then the "re-linearization technique" which adds new variables (type 2) and the XL algorithm which generates new products of variables (type 1). At Eurocrypt 2000 it was concluded that re-linearization technique is highly redundant[4] and that XL works better [7]. Then researchers have discovered that at higher degrees XL is also redundant [1,4,8–10,27,28] and modern Gröbner basis techniques are precisely about removing even more redundancies in XL, cf. [1,38].

One reason why the second family has not been sufficiently studied is that gives the code breaker a **considerable degree of freedom**. Hence, it is not clear **how to even start** to design[5] an attack based on this idea. On the positive side some success was definitely achieved with methods related to hardware implementation and multiplicative complexity S-box optimizations [2,6,16,18, 19,36]. Until now there were extremely few attempts to invent new non-trivial attacks techniques based on 2nd type methods. One exception to this is [29].

It is clear that the type 1 methods always somewhat contain a type 1 method, new variables can be just monomials, which again however leads to known problems with redundancy cf. [7]. Until now cryptographic literature knows very few convincing attacks of the 2nd type. This with exception of SAT solver attacks, which very clearly greatly benefit from added variables cf. [6,16,18,19].

Combination Attacks. It is also important to note that both approaches 1 and 2 can and should be combined. To put it simply, the second approach may make the first approach work better, equations become more overdefined and the so called degree of regularity [38] is expected to decrease, i.e. system is solved by Gröbner basis software or other software at a lower degree, which implies lower running time and less memory, cf. also [12,29].

We are now going to review several classical type 1 methods in Algebraic Cryptanalysis, explain what quantities are expanded, and look at the question of how quickly these numbers grow.

3   XL Algorithm, F4, F5, and Their Variants, XSL

The XL algorithm [7] was extensively studied and there exist countless variants of this algorithm. We consider m quadratic equations with n variables over $\mathbb{F}_2$. The basic XL algorithm consists of multiplying all equations by monomials of degree $D - 2$ to create a larger number of R equations of degree D. In this and similar algorithms the number of linearly independent equations $F \leq R$ has a simple and totally predictable behavior, see [4,8]. This under some important

---

[4] Interestingly one could repair the linearization technique by some form of decimation (erasing a subset of equations) where the redundancies are removed.

[5] A related concept is the concept of "Algebraic Complexity Reduction" of [16] which has been a great success in a restricted case of a block ciphers with a lot of high-level self-similarity and which is different and stronger. In [16] the attacker also makes well chosen guesses on special combinations of variables.

technical assumptions such as that our system of equations has only one unique solution. For larger D the prediction is less accurate and also this is where more sophisticated algorithms such as F5 emerge and can make a difference. Here we show a simple example at degree 5 taken from [8].

n	24	24	24
m	16	27	32
D	5	5	5
R	37200	62775	74400
T	55455	55455	55455
F	33800	53325	55454

A ready software tool which allows to run such simulations can be found in [11]. For small D this sort of experiments can be predicted with 100% accuracy in practice, cf. [4,8]. We expect that we have always:

$$\text{For} \quad D = 5, F = \min\left(T - 1, R - (n+1)\binom{m}{2} - (n+1)m\right)$$

We obtain a curve which a collation of two closed formulas with a very neat and abrupt transition. Although no counter-examples are known, we should remain sceptical if this will be generally the case, especially at the transition boundary. The crucial object of study in this paper is precisely this "phase transition" phenomenon where we shift from one predictable curve to another equally predictable curve. The very existence of phase transitions indicates the predictions in algebraic cryptanalysis will never be an exact science and that rules can eventually be breached. However predictions are (badly) needed in order to be able to evaluate the complexity of different attacks.

3.1   XSL Algorithm

The XSL algorithm is inspired by the idea that XL algorithm is essentially a tool for dense equations in which all monomials play a similar role. This is very rarely the case in cryptanalysis. If the equations are sparse, a peculiar method was invented where a phase transition is sought by multiplying only by a selection of monomials for $D = 4$ and **only** monomials which are already used. It should be noted that two different versions of this attack exist, cf. [27,28].

One example of a data series obtained in an application of the XSL attack to a toy cipher can be found in Appendix C. of [28]. However to the best of our knowledge until now nobody has yet studied if the behavior of XSL attacks can be predicted accurately. We consider the data series from Appendix C. of [28] and used Microsoft Excel to fit a polynomial model for these data which minimizes the least square error. Let K be the number of rounds in this 6-bit toy cipher. We then observe that in this precise attack we have almost exactly (high level of accuracy $R^2 = 1$ is reported by the software in all four cases):

$$\begin{cases} R = 936K^2 - 208K + 144 \\ T = 882K^2 - 147K + 7 \\ T' = 504K + 168 \\ F = 850K^2 - 109K + 6 \end{cases}$$

In this attack, the terminating condition is $F \geq T - T'$, cf. [28] and we conclude that this attack works for up to approximately 16 rounds for this toy cipher.

4   The Algebraiz-ation Challenge

Now in order to make further progress in algebraic cryptanalysis, and both for type 1/2 methods, two interesting questions are as follows:

1. Can we efficiently generate or discover additional[6] equations the existence of which is maybe not expected or less easily predicted? This is frequently (see footnote 6) the case and it helps to solve equations with substantially lower complexity.
2. Can we solve by algebraic cryptanalysis problems which seem unsolvable or a poor fit for algebraic cryptanalysis? For example, the DES S-box do not have any strong algebraic structure. Yet algebraic coding and algebraic cryptanalysis is possible, cf. [6]. A question which is even (a lot) more difficult is a question of ECDL problem in elliptic curves, cf. [29,31,32,39,44].

These two research directions are related at more than one level. For example progress in one direction could help the other, and new type 2 methods could potentially solve both problems. The crucial question is the question "efficient algebraiz-ation", and what is the meaning of "efficient". Several fundamental definitions which we will study on the following pages will allow to understand that there is more than one interesting way to approach these problems.

4.1   I/O Equations

Definition 4.1.1 (An I/O relation, [6,22,25]). Consider a function $f : \mathbb{F}_2^n \to \mathbb{F}_2^m$, $f(x) = y$, with $x = (x_0, \ldots, x_{n-1})$, $y = (y_0, \ldots, y_{m-1})$.
    We call an I/O relation any polynomial

$$g(x_0, \ldots, x_{n-1}; y_0, \ldots, y_{m-1}) = 0$$

which hold with certainty, i.e. for every pair (x, y) such that $y = f(x)$.

---

[6] This happens for example in the cryptanalysis of the multivariate public-key cryptosystems with the discovery of so called "implicit equations" [5,37] which we call "I/O relations" in our Definition 4.1.1, see also [5,33,35]. Similarly some quite unexpected equations can be shown to always exist (worst case results) in algebraic attacks on stream ciphers [14,24,25]. We also have a closely related notion of so called "degree falls" sometimes also called "mutants" which are for example observed in ElimLin attacks [6,12,17,42].

This allows to define a very useful notion of:

Definition 4.1.2 (The I/O degree, [6,22,25]**).** Again consider a function $f : \mathbb{F}_2^n \to \mathbb{F}_2^m$. The I/O degree of f is the smallest possible degree in the linear space of existing polynomial I/O relations as defined above.

The concept of I/O can be slightly misleading as it turns out in some special cases. For example for some DES S-boxes, the I/O degree will be 2, yet these equations are too few to actually be used in an attack without adding extra equations of higher degree, cf. [6].

4.2 Describing Strategies

This difficulty leads to the notion of "Describing Degree" which is studied in [23] which amounts to saying that given the input (or the output) the system of equations should be large enough to determine the solution uniquely. This may require to mix equations of different degrees, as it was shown in [6].

4.3 Guessing Strategies

A particularly expert strategy for algebraic attacks is described in [23]. It defines a notion of [Probabilistic] "Conditional Describing Degree", see [23] which allows to split the algebraic description of a non-linear component in two parts, one is "guessed" by literally guessing some values during the attack, the other is coded by algebraic equations. This allows to develop sophisticated guess-then-determine attacks where the attacker after the initial guessing stage (which has some substantial cost) is able to substantially reduce the degree and the complexity of the algebraic problem to solve. In particular, this can be used as a "linearization strategy". For example it is shown in [23] that the "Conditional Describing Degree" of a non-linear component $\boxplus$ which is the addition modulo 2^n, can be as low as 1. This component is very frequently used inside ciphers, e.g. [16,23] and here after the initial (costly) assumption the attacker can produce a larger number of additional linear equations **simultaneously** true with high probability. This makes the last step of the attack easier. Two closely related notions are the notion of "Algebraic Complexity Reduction" [15,16] and "SAT Immunity", cf. [20].

4.4 Type 2 Techniques

All the definitions above are rather meant (not exclusively) to be used with traditional type 1 techniques. What about type 2 coding techniques? Here is another very important definition which became popular in the recent years, and is now considered as one of the four most important measures to evaluate the security of cipher components, cf. [3].

Definition 4.4.1 (Multiplicative Complexity (MC) [2,19]**).** MC is the minimum number of AND gates which are needed if we allow an unlimited number of NOT and XOR gates.

There exist in cryptanalytic literature also many other cases of efficient algebraic attacks which use coding techniques related to efficient hardware implementation, e.g. [16,18,19,36,47].

4.5   Can Algebraiz-ation Be Mandated?

All these notions of I/O relations, their advanced variants, MC and other coding methods lead to showing that some degree of algebra-ization is always possible andm in some cases **inevitable** in algebraic cryptanalysis. They lead to specific compact algebraic (or multivariate polynomial) encoding methods for various cryptanalysis problems [18,19,21,36,41,47]. How inevitable it is? In many cases, there exist "worst case" results which show that any cryptographic component of certain size can be attacked by an algebraic attack, cf. for example [14] and Theorem 1 below. Many cipher designer claim that they have verified that their cipher resists to all known attacks in symmetric cryptanalysis. Algebraic cryptanalysts will frequently insist that in absence of provable security, there is probably no way to know if a cipher resists to some already known attacks if we do not dispose on a large computing power, or/and that we do not know how well many known attacks would scale when the parameters grow, cf. [22,25,27,42].

4.6   On Small S-Boxes

If in cryptanalysis of HFE we have specific structural reasons why some algebraic polynomial "I/O relations" do exist cf. for [33,35]. In symmetric cryptanalysis, we do not have a strong internal algebraic structure, we however do have specific structural properties which are consequences of how the cipher is designed.

Any very small S-box (for example up to 4 bits) works with both definitions above: it leads to I/O relations and to relatively small MC, and therefore to a rich universe of possible algebraic attacks. For example we have a generic folklore Courtois Theorem 1 which we will find in many papers [6,14,28]:

Theorem 1 (Courtois). For any $n \times m$ S-box, $F : (x_1, \ldots, x_n) \mapsto (y_1, \ldots, y_m)$, and for any subset $\mathcal{T}$ of t out of 2^{m+n} possible monomials in the x_i and y_j, if $t > 2^n$, there are at least $t - 2^n$ linearly independent I/O equations (algebraic relations) $g(x, y)$ involving (only) monomials in $\mathcal{T}$, and that hold with probability 1, i.e. for every (x, y) such that $y = F(x)$.

Examples. Until recently the simplest S-box in cryptanalysis was the 3-bit box used in CTC2 [9,10] which was a toy cipher designed for experiments with algebraic cryptanalysis. Then the simplest "real-life" standardized ciphers were cipher such as PRESENT and GOST with 4-bit S-boxes [16]. Recent trend is to use inside block ciphers non-linear components which are yet fewer and simpler. Here the leading example is Simon, a block cipher with excessively small S-box and truly exceptionally low MC. In Simon the S-box is an AND gate, the simplest possible non-linear component one can think of. The complexity of

Simon is yet substantially lower than with CTC2, DES or GOST for which algebraic cryptanalysis were previously studied and implemented [6,10,16]. It should not therefore be surprise that Simon will be our favorite block cipher to study.

5   ElimLin Attacks on Simon

Two recent papers consider the ElimLin attacks on Simon [12,42]. ElimLin is a remarkably simple algebraic attack which to some extent break any cipher, if not too complex. The study of ElimLin is an excellent case where many interesting things happen simultaneously: degree falls, generation of extra equations the existence of which was not initially expected (like in [5]), phase transitions.

ElimLin is a curious sort of attack, cf. slide 126 in [26]. It can be described informally in 2 simple steps:

1. Find linear equations in the linear span.
2. Eliminate some variables, and iterate (try 1. again).

ElimLin is a stand-alone attack which allows one to recover the secret key of many block ciphers [9,10,13] and more recently in [30,42].

The main characteristic of ElimLin is that it quietly dissolves non-linear equations and generates linear equations. This algorithm basically makes progressively disappear the main and **the** only thing which makes cryptographic schemes not broken by simple linear algebra: non-linearity. It is not clear however why this works and how well the ElimLin attack scales for larger systems of equations. In recent 2015 work of Raddum we discover that (experimentally) ElimLin breaks up to 16 rounds of Simon cipher [42] however it is hard to know exactly what happens for 17 rounds.

5.1   The Overdefined Heuristic

Now ElimLin has something that which renders XL, XSL, T' method [26,27], and many other potentially unnecessary because a simpler attack exists and will work. Actually many methods such as XL and XSL do not work well in block cipher cryptanalysis [27]. This is primarily because they have been designed to solve problems with small quantity of data and solving such problems is difficult. We obtain systems of equations with a large value of the so called "degree of regularity" [38] which is essentially the maximum degree D of polynomials manipulated which we mentioned before for XL (however for a better algorithm this degree can be lower). Here the main observation is something which was actually known longer, at least since [7]. The fact is that **overdefined** systems of equations are substantially easier to solve. Moreover, it could be possible for the attacker, to try to design an attack which mandates or creates such systems of equations on purpose, and thus avoid or circumvent the major difficulty mentioned above. We call this the **overdefined strategy**, cf. [22,29].

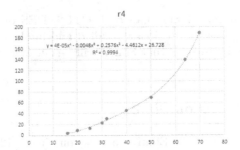

Fig. 1. Number of linearly independent equations generated at stage 4 of the ElimLin algorithm for 8 rounds of Simon 64/128 according to [12].

5.2 ElimLin or How to Make the Overdefined Strategy Work for a Block Cipher

One simple method to achieve this is to increase the data complexity in the attack which makes the "degree of regularity" decrease [9,10]. For example we consider an attack K known plaintexts and study how the complexity of the attack grows with K. Then we discover a fascinating aspect of ElimLin which has only recently attracted some attention [12].

The fact is that the number of equations generated can go through several stages cf. [12] as K grows. Initially there are no non-trivial equations whatsoever. Then we obtain a curve which grows **faster** than linear in K.

Our recent[7] paper show that super-linear growth is indeed possible [12]. Then, later on, which we do not yet see on our Fig. 1, we will eventually achieve some sort of "saturation" and the number of linear equations grows just linearly. In general, in the long run we cannot hope that the growth will be faster than linear, as the number of monomials in these equations grows linearly with K.

A recent PhD thesis [46] considers enhancements for ElimLin algorithm similar to those studied in [30] and shows that the number of equations which could be added to ElimLin can follow a curve which is a collation of 5 distinct intervals where the result is predictable with up to 100% accuracy, and yet later will switch to another curve. This suggests that we need to remain sceptical about prediction techniques however accurate they may seem.

6   Big Challenge - Coding ECC Cryptanalysis Problems

The same quest of trying to construct systems of equations which are very highly overdefined is also what motivates some recent research on coding ECC cryptanalysis problems [29,32]. We expect to achieve some sort of happy tradeoff between increasing the number of variables and lowering the regularity degree of

---

[7] One (older) example which shows that the number of equations grows faster than linear as a function of the data complexity K in ElimLin can be found at slide 153 in [26] which example is from 2006–2007 and originally comes from [13].

the equations which are then going to become more efficiently solvable. This is our recent approach which was designed as an alternative to both early and more recent attempts to design an index calculus algorithm for the ECDL problem [39,44] based on the so called Semaev Polynomials or Summation Polynomials [44]. Traditionally, ECC relations of type $P1 + P2 = P3$ will be coded by the S3 polynomial which following [43] is

$$S_3(x_1, x_2, x_3) = (x_1 - x_2)^2 x_3^2 - 2\left[(x_1 + x_2)(x_1 x_2 + A) + 2B\right] x_3 + (x_1 x_2 - A)^2 - 4B(x_1 + x_2)$$

This polynomial is of degree 6 and it is already quite complex. Can we do better?

In this paper we do not claim to provide a comprehensive treatment of this question. We will concentrate on the main idea. We want to design new type 2 techniques which work by expanding the number of variables. The principal objective for the attacker is then to code an ECC cryptanalysis problem in such a way that the "degree of regularity" and/or other measures of complexity studied earlier (e.g. MC) would decrease, i.e. system is expected to solved more easily by various techniques. We refer to [29,32,39] for a more systematic presentation of this attack strategy. In this paper we just show one concrete example on how a redundant set of ECC variables can lead to some unusually simple equations to exist.

6.1   On ECC Codes

The philosophy of adding new variables can be studied in terms of certain types of **ECC codes**.

Definition 6.1.1 (ECC Code). We call an ECC Code any injective function

$$F : E(\mathbb{F}_p)^L \to E(\mathbb{F}_p)^K$$

which is defined for all except a small number of special EC points.

Remark: We should note that error correcting codes which are defined or constructed using elliptic curves are typically defined as subsets of $\mathbb{F}^K$ where $\mathbb{F}$ is a finite field, cf. for example page 11 in [34]. In this paper and in [29] we find it more convenient to define ECC Codes as a subset of $\mathbb{E}^K$ where $\mathbb{E}$ is an elliptic curve, even though later we will just look at EC coordinates of these points which are in $\mathbb{F}^K$. We refer to [34] for additional literature pointers about error correcting codes and those which use elliptic curves.

Now we are going to exhibit one ECC property which to the best of our knowledge has not been studied before and which show that redundant sets of variables can lead to substantial simplification in the complexity of systems of polynomial equations. Our paper [29] contains more such properties and explains more in detail the process where the existence of such properties can be seen as an alternative to (or an enhancement to) some recent attempts to solve the ECDL problem in [39,40,44], which attempts so far were not a great success and better methods need precisely to be invented.

6.2 D73 - A New Family of Cubic I/O Relations

Theorem 6.2.1 (D73 Theorem). We consider the following special form of ECC Code with 3 inputs and 7 outputs for any Weierstrass elliptic curve modulo a large p. It is defined by the following application $\mathbb{E}^3 \rightarrow \mathbb{E}^7$ which transform three variables which are points on the curve into 7 "related" variables,

$$(P1, P2, P3) \mapsto (P1, P2, P3, P1 + P2, P1 + P3, P2 + P3, P1 + P2 + P3)$$

Now we consider only the x coordinates of these 7 elliptic curve points. We call them respectively in the same order:

$$(sx1, sx2, sx3, sx12, sx13, sx23, sx123)$$

If all the 7 points are distinct from the ECC neutral element ∞ we have:

```
sx1*sx2*(sx23-sx13) +sx1*sx3*(sx12-sx23) +sx2*sx3*(sx13-sx12)
+sx123[sx1*(sx13-sx12)+sx2*(sx12-sx23)+sx3*(sx23-sx13)] = 0
```

Remark. Our D73 equation is a homogenous polynomial of degree 3. We challenge the reader to discover anything comparable in terms of elegance and simplicity for an ECC Code expansion with a similar expansion factor. The main point in this paper is that having additional **redundant** variables could be a good idea. It may allow to greatly **simplify**[8] polynomial equations and effectively replace Semaev polynomials by some simpler and lower degree polynomials. This we have not demonstrated, we just demonstrate the existence of some simpler polynomials. Additional questions need to be studied in order to design a cryptanalytic attack on the ECDL problem and for the time being we refrain from making any conclusions about how our discovery might impact the complexity of such methods, as currently such methods are yet very inefficient cf. [29, 32, 39]. First of all, D73 equations will not suffice and we expect to use also polynomials of other types, cf. [29]. Then the analysis becomes increasingly complex. An interesting question is for example to construct very highly overdefined encodings of ECDL problem with properties of type $\lim_{K \to \infty} F/T = 1$ and some other "density" and "topology" properties. Some early attempts to achieve this can be found in [29] which paper also shows that there are some very substantial difficulties to make this sort of approach work.

7 Conclusion

In this paper we compare different known techniques for solving non-linear algebraic equations in algebraic cryptanalysis and show that they all can be seen as a race between two different quantities one of which grows faster. We illustrate this with examples derived from both symmetric and asymmetric cryptanalysis. The crucial question is the possibility to accurately predict the behavior of

---

[8] In terms of algebraic degree, sparsity, multiplicative complexity, etc.

such attacks and that they will later switch to another curve and a phase transition may occur. Now the question is can we do better than just contemplate these transitions? Can we explicitly engineer phase transition to happen? In this paper we point out that there exist a second somewhat more general way of formulating algebraic attacks, where the attacker plays a more active role. This is the question of **algebraic coding** which did NOT so far have great success in cryptographic literature, and was frequently ignored as a trivial first step, or rejected in some inefficient/redundant attack methods.

The main contribution of this paper is to point out that this problem of finding a "good"[9] algebraic coding was so far poorly studied. Yet it gives the attacker **a considerable degree of freedom**, especially if we allow the coding to be redundant. We need to pay more attention to specific combinatorial optimization problems such as finding non-trivial redundant representations [6,16,19,36] leading to important simplifications in algebraic description complexity. The bottom line is that we open the possibilities to invent a number of new "out-of-the-box" attacks with non-trivial à priori coding steps. For example multiplicative complexity and other S-box optimizations lead to some quite competitive attacks on block ciphers [6,16,19]. The primary challenge for the future remains how to code and re-code cryptanalysis problems in better ways. The attacker is not merely hoping that some interesting equations exist [5,37] or will be found by our attack [12], which approach to software cryptanalysis we called "plug and pray" in [29], but how to "engineer" an attack where new "interesting" equations will exist. In terms of classical Gröbner basis attacks, it will be about designing algebraic attacks where major families of **degree falls** and other interesting I/O equations are not something which happens accidentally, but something we **construct explicitly**. For example many authors have tried to develop an index calculus or a point splitting attack on the ECDL problem through the use of so called Semaev/Summation polynomials [39,44] without great success. In this paper and in [29] we suggest that the degree and complexity of ECC coding problems can be reduced with redundant coding of variables. However designing a really good working algebraic attack remains a difficult problem. Better algebraiz-ation through simpler polynomials is probably by far not enough to solve hard cryptanalysis problems. More attention needs also be paid to questions such as "densely connected equations topology" cf. Sects. 6 and 6.7 in [29], and additional "constraints coding" questions, cf. Part VI in [29].

Acknowledgments. I would like to thank the following people who have either inspired and motivated me for writing this paper, or who provided me with some valuable feedback: Moti Yung, David Naccache, Raphael Phan, Christophe Petit, Steven Galbraith, Jacques Patarin, Louis Goubin, Daniel Augot, Jonathan Bootle and Mary Maller.

---

[9] See Part 1 on slide 56 and 58 in [26].

References

1. Bardet, M., Faugère, J.-C., Salvy, B.: On the complexity of Gröbner basis computation of semi-regular overdetermined algebraic equations. In: ICPSS, Paris, France, pp. 71–75 (2004)
2. Boyar, J., Peralta, R.: A new combinational logic minimization technique with applications to cryptology. In: Festa, P. (ed.) SEA 2010. LNCS, vol. 6049, pp. 178–189. Springer, Heidelberg (2010). doi:10.1007/978-3-642-13193-6_16. An early version was published in 2009 http://eprint.iacr.org/2009/191. Accessed 13 Mar 2010
3. Boyar, J., Find, M., Peralta, R.: Four measures of nonlinearity. In: Spirakis, P.G., Serna, M. (eds.) CIAC 2013. LNCS, vol. 7878, pp. 61–72. Springer, Heidelberg (2013). doi:10.1007/978-3-642-38233-8_6
4. Yang, B.-Y., Chen, J.-M., Courtois, N.T.: On asymptotic security estimates in XL and Gröbner bases-related algebraic cryptanalysis. In: Lopez, J., Qing, S., Okamoto, E. (eds.) ICICS 2004. LNCS, vol. 3269, pp. 401–413. Springer, Heidelberg (2004). doi:10.1007/978-3-540-30191-2_31
5. Courtois, N.T.: The security of hidden field equations (HFE). In: Naccache, D. (ed.) CT-RSA 2001. LNCS, vol. 2020, pp. 266–281. Springer, Heidelberg (2001). doi:10.1007/3-540-45353-9_20
6. Courtois, N.T., Bard, G.V.: Algebraic cryptanalysis of the data encryption standard. In: Galbraith, S.D. (ed.) Cryptography and Coding 2007. LNCS, vol. 4887, pp. 152–169. Springer, Heidelberg (2007). doi:10.1007/978-3-540-77272-9_10
7. Courtois, N.T., Klimov, A., Patarin, J., Shamir, A.: Efficient algorithms for solving overdefined systems of multivariate polynomial equations. In: Preneel, B. (ed.) EUROCRYPT 2000. LNCS, vol. 1807, pp. 392–407. Springer, Heidelberg (2000). doi:10.1007/3-540-45539-6_27
8. Courtois, N.T., Patarin, J.: About the XL algorithm over $GF(2)$. In: Joye, M. (ed.) CT-RSA 2003. LNCS, vol. 2612, pp. 141–157. Springer, Heidelberg (2003). doi:10.1007/3-540-36563-X_10
9. Courtois, N.T.: How fast can be algebraic attacks on block ciphers? In: Biham, E., Handschuh, H., Lucks, S., Rijmen, V. (eds.) Online Proceedings of Dagstuhl Seminar 07021, Symmetric Cryptography 07–12 January 2007 (2007). http://drops.dagstuhl.de/portals/index.php?semnr=07021. http://eprint.iacr.org/2006/168/, ISSN 1862 - 4405
10. Courtois, N.T.: CTC2 and fast algebraic attacks on block ciphers revisited. http://eprint.iacr.org/2007/152/
11. Courtois, N.T.: Some algebraic cryptanalysis software. http://www.cryptosystem.net/aes/tools.html
12. Courtois, N.T., Papapanagiotakis-Bousy, I., Sepehrdad, P., Song, G.: Predicting outcomes of ElimLin attack on lightweight block cipher simon. In: Secrypt 2016 Proceedings (2016)
13. Courtois, N.T., Debraize, B.: Specific S-box criteria in algebraic attacks on block ciphers with several known plaintexts. In: Lucks, S., Sadeghi, A.-R., Wolf, C. (eds.) WEWoRC 2007. LNCS, vol. 4945, pp. 100–113. Springer, Heidelberg (2008). doi:10.1007/978-3-540-88353-1_9
14. Courtois, N.T.: Algebraic attacks on combiners with memory and several outputs. In: Park, C., Chee, S. (eds.) ICISC 2004. LNCS, vol. 3506, pp. 3–20. Springer, Heidelberg (2005). doi:10.1007/11496618_3. http://eprint.iacr.org/2003/125/

15. Courtois, N.T.: Security evaluation of GOST 28147-89 in view of international standardisation. Cryptologia **36**(1), 2–13 (2012)
16. Courtois, N.T.: Algebraic complexity reduction and cryptanalysis of GOST. Monograph Study of Security of GOST, 2010–2014. http://eprint.iacr.org/2011/626
17. Courtois, N.T., Sepehrdad, P., Sušil, P., Vaudenay, S.: ElimLin algorithm revisited. In: Canteaut, A. (ed.) FSE 2012. LNCS, vol. 7549, pp. 306–325. Springer, Heidelberg (2012). doi:10.1007/978-3-642-34047-5_18
18. Courtois, N.T., Hulme, D., Mourouzis, T.: Solving circuit optimisation problems in cryptography and cryptanalysis. In: SHARCS 2012, pp. 179–191 (2012). http://2012.sharcs.org/record.pdf
19. Courtois, N.T., Hulme, D., Mourouzis, T.: Multiplicative complexity and solving generalized Brent equations with SAT solvers. In: COMPUTATION TOOLS 2012, pp. 22–27 (2012)
20. Courtois, N.T., Gawinecki, J.A., Song, G.: Contradiction immunity and guess-then-determine attacks on GOST. Tatra Mt. Math. Publ. **53**(3), 65–79 (2012). http://www.sav.sk/journals/uploads/0114113604CuGaSo.pdf
21. Courtois, N.T., Mourouzis, T., Misztal, M., Quisquater, J.-J., Song, G.: Can GOST be made secure against differential cryptanalysis? Cryptologia **39**(2), 145–156 (2015)
22. Courtois, N.T.: New frontier in symmetric cryptanalysis, invited Talk at Indocrypt 2008, 14–17 December 2008 (2008). http://www.nicolascourtois.com/papers/front_indocrypt08.pdf
23. Courtois, N.T., Debraize, B.: Algebraic description and simultaneous linear approximations of addition in snow 2.0. In: Chen, L., Ryan, M.D., Wang, G. (eds.) ICICS 2008. LNCS, vol. 5308, pp. 328–344. Springer, Heidelberg (2008). doi:10.1007/978-3-540-88625-9_22
24. Courtois, N.T.: Fast algebraic attacks on stream ciphers with linear feedback. In: Boneh, D. (ed.) CRYPTO 2003. LNCS, vol. 2729, pp. 176–194. Springer, Heidelberg (2003). doi:10.1007/978-3-540-45146-4_11
25. Courtois, N.T.: General principles of algebraic attacks and new design criteria for cipher components. In: Dobbertin, H., Rijmen, V., Sowa, A. (eds.) AES 2004. LNCS, vol. 3373, pp. 67–83. Springer, Heidelberg (2005). doi:10.1007/11506447_7
26. Courtois, N.T.: Algebraic attacks vs. design of block and stream ciphers. Slides Used in GA18 Course Cryptanalysis taught at University College London, 2014–2016. http://www.nicolascourtois.com/papers/algat_all_teach_2015.pdf
27. Courtois, N.T., Pieprzyk, J.: Cryptanalysis of block ciphers with overdefined systems of equations. In: Zheng, Y. (ed.) ASIACRYPT 2002. LNCS, vol. 2501, pp. 267–287. Springer, Heidelberg (2002). doi:10.1007/3-540-36178-2_17
28. Courtois, N.T., Pieprzyk, J.: Cryptanalysis of block ciphers with overdefined systems of equations. http://eprint.iacr.org/2002/044/. Contains two different (earlier) versions of the XSL attack, see also [27]
29. Courtois, N.T.: High Saturation Complete Graph Approach for EC Point Decomposition and ECDL Problem, preprint July–September 2016 (2016). http://eprint.iacr.org/2016/704.pdf
30. Susil, P., Sepehrdad, P., Vaudenay, S., Courtois, N.: On selection of samples in algebraic attacksand a new technique to find hidden low degree equations. Int. J. Inf. Secur. **15**(1), 51–65 (2016). Springer
31. Diem, C.: On the discrete logarithm problem in elliptic curves. Compos. Math. **147**, 75–104 (2011)
32. Galbraith, S.D., Gaudry, P.: Recent progress on the elliptic curve discrete logarithm problem, preprint, 22 October 2015 (2015). https://eprint.iacr.org/2015/1022.pdf

33. Faugère, J.-C., Joux, A.: Algebraic cryptanalysis of hidden field equation (HFE) cryptosystems using Gröbner bases. In: Boneh, D. (ed.) CRYPTO 2003. LNCS, vol. 2729, pp. 44–60. Springer, Heidelberg (2003). doi:10.1007/978-3-540-45146-4_3
34. Minder, L.: Cryptography based on error correcting codes. Ph.D. thesis 3846 (2007). EPFL, 27 July 2007. http://algo.epfl.ch/_media/en/projects/lorenz_thesis.pdf
35. Huang, M.-D.A., Kosters, M., Yeo, S.L.: Last fall degree, HFE, and weil descent attacks on ECDLP. In: Gennaro, R., Robshaw, M. (eds.) CRYPTO 2015. LNCS, vol. 9215, pp. 581–600. Springer, Heidelberg (2015). doi:10.1007/978-3-662-47989-6_28
36. Mourouzis, T.: Optimizations in algebraic and differential cryptanalysis. Ph.D. thesis, under superivsion of Dr. Nicolas T. Courtois, University College London, January 2015. http://discovery.ucl.ac.uk/1462141/2/PhD_Thesis_Theodosis_Mourouzis.pdf
37. Patarin, J.: Cryptanalysis of the matsumoto and imai public key scheme of Eurocrypt'88. In: Coppersmith, D. (ed.) CRYPTO 1995. LNCS, vol. 963, pp. 248–261. Springer, Heidelberg (1995). doi:10.1007/3-540-44750-4_20
38. Perret, L.: Gröbner bases techniques in cryptography. http://web.stevens.edu/algebraic/Files/SCPQ/SCPQ-2011-03-30-talk-Perret.pdf
39. Petit, C., Kosters, M., Messeng, A.: Algebraic approaches for the elliptic curve discrete logarithm problem over prime fields. In: Cheng, C.-M., Chung, K.-M., Persiano, G., Yang, B.-Y. (eds.) PKC 2016. LNCS, vol. 9615, pp. 3–18. Springer, Heidelberg (2016). doi:10.1007/978-3-662-49387-8_1
40. Petit, C., Quisquater, J.-J.: On polynomial systems arising from a weil descent. In: Wang, X., Sako, K. (eds.) ASIACRYPT 2012. LNCS, vol. 7658, pp. 451–466. Springer, Heidelberg (2012). doi:10.1007/978-3-642-34961-4_28
41. Arabnezhad-Khanoki, H., Sadeghiyan, B., Pieprzyk, J.: Algebraic attack efficiency versus S-box representation. eprint.iacr.org/2017/007.pdf
42. Raddum, H.: Algebraic analysis of the simon block cipher family. In: Lauter, K., Rodríguez-Henríquez, F. (eds.) LATINCRYPT 2015. LNCS, vol. 9230, pp. 157–169. Springer, Cham (2015). doi:10.1007/978-3-319-22174-8_9. https://www.simula.no/file/simonpaperrevisedpdf/download
43. Semaev, I.: New algorithm for the discrete logarithm problem on elliptic curves. Preprint 10 April 2015. eprint.iacr.org/2015/310/
44. Semaev, I.: Summation polynomials and the discrete logarithm problem on elliptic curves. Preprint. eprint.iacr.org/2004/031/
45. Shannon, C.E.: Communication theory of secrecy systems. Bell Syst. Tech. J. **28**, 656–715 (1949). See in particular p. 704
46. Song, G.: Optimization and guess-then-solve attacks in cryptanalysis. Ph.D. thesis, will be presented at University College London in 2017 (2017)
47. Stoffelen, K.: Optimizing S-box implementations for several criteria using SAT solvers. In: Peyrin, T. (ed.) FSE 2016. LNCS, vol. 9783, pp. 140–160. Springer, Heidelberg (2016). doi:10.1007/978-3-662-52993-5_8. https://eprint.iacr.org/2016/198

Watermarking Cryptographic Programs

Ryo Nishimaki[✉]

Secure Platform Laboratories, NTT Corporation, Tokyo, Japan
nishimaki.ryo@lab.ntt.co.jp

Abstract. Digital watermarking embeds unremovable information called a "mark" into digital objects such as images, video, audio files, and program data without changing their functionalities. This article provides a brief overview of recent advances in watermarking for cryptographic programs and insights behind them.

Keywords: Program watermarking · Obfuscation

1 Introduction

Background. Digital watermarking enables us to embed special information called a "mark" into digital objects such as images, video, audio files, or program data. It is said that an object is *marked* if a mark is embedded into it.

There are several applications of watermarking. We explain two main applications. The first is identifying owners. If identification information is embedded as a mark into digital objects, then it is possible to identify an entity that creates the objects by verifying the embedded mark. The second application is tracing users. When a digital object is (illegally) copied, an embedded mark is also copied. It is possible to specify the owner of the original data by verifying the embedded mark. For example, consider that a software company sells software to Alice. The software is marked with Alice's identity information and she gives a copy of the software to Bob, who does not buy the software. When we verify the mark with Alice's identity in the software that Bob has, we can discover that Alice gave a copy to Bob.

There are two requirements for secure watermarking. One is preserving functionality. That is, watermarking does not change the functionality of objects. The other is unremovability. That is, it is impossible to remove a mark without destroying the functionality of the object.

There have been few studies on formal definitions and rigorous security analysis of watermarking from the cryptographic point of view despite its usefulness. Most watermarking schemes are heuristic and ad-hoc constructions.

Barak et al. provide the first theoretical treatment of program watermarking [2]. They provide formal definitions of program watermarking and basic security requirements. Unfortunately, they show that it is impossible to achieve program watermarking under a certain definition by using an impossibility result of program obfuscation. Program obfuscation converts programs into scrambled ones

© Springer International Publishing AG 2017
R.C.-W. Phan and M. Yung (Eds.): LNCS 10311, Mycrypt 2016, pp. 521–543, 2017.
DOI: 10.1007/978-3-319-61273-7_28

and hides information about the original programs while preserving their functionalities. They leave as an open problem to achieve program watermarking under some definition.

Hopper, Molnar, and Wagner provide various formal security definitions of watermarking for perceptual objects such as images and study their relationships. However, they do not give any concrete construction of secure watermarking [10].

As briefly explained above, no positive result on secure program watermarking is known. Thus, the following question is open over the past fifteen years.

Is it possible to construct secure program watermarking under some reasonable definition?

Recent Advances. Two studies have made progress in this topic.

Nishimaki provide definitions of watermarking for cryptographic functions and a concrete construction based on a number theoretic assumption [14]. His construction is secure under a standard number theoretic assumption. However, the adversary is very restricted in his model. He assumes that the adversary does not change the format of programs. That is, the adversary tries to remove a mark from a watermarked program, but it is assumed that the adversary outputs only a program based on algebraic structure (more concretely, a program consists of group elements) since his construction is based on number theory. This is a weak security guarantee and, in fact, his construction can be attacked using an indistinguishability obfuscator (iO) [2,8], which completely changes the structure of programs while preserving functionalities. Many iO candidates have been proposed since 2013.

Cohen et al. provide new formal definitions of watermarking and new negative and positive results on watermarking for cryptographic programs [6]. They avoid the impossibility result of Barak et al. by using a relaxed functionality preserving requirement. They show that it is still impossible to achieve program watermarking for some classes of programs even under the (or a more) relaxed functionality preserving requirement. They also show that we can construct watermarking schemes for (a variant of) pseudo-random functions, which are keyed functions whose outputs are indistinguishable from outputs of a truly random function. Their security definition, in which adversaries can use arbitrary strategies to remove a mark, is stronger than previous ones.

Purposes of this Article. This article is a survey and gives insights behind the recent advances in program watermarking explained above. It basically consists of three parts. First, the impossibility result by Barak et al. and how to avoid it are introduced. Second, a brief overview of the result by Nishimaki is provided and why his construction works only for restricted adversaries is explained. Finally, the idea behind the construction by Cohen et al. is explained. We start with the initial and basic ideas and explain how to extend them to satisfy a stronger security definition. We stress that all security arguments in this article are just overviews and many technical details are ignored.

2   Preliminaries

Before definitions of watermarking, we introduce standard notations and review basic cryptographic notions.

Notations. For any $n \in \mathbb{N} \backslash \{0\}$, let $[n]$ be the set $\{1, \ldots, n\}$. A bold face lower-case letter denotes a vector such as $\boldsymbol{x} = (x_1, \ldots, x_n)$. For two vectors $\boldsymbol{v}$ and $\boldsymbol{w}$, $\langle \boldsymbol{v}, \boldsymbol{w} \rangle$ denotes the inner-product $\sum_{i=1}^{n} v_i w_i$. For two strings x_1 and x_2, $x_1 \| x_2$ denotes a concatenation of x_1 and x_2. For program (or circuit) C, $C[a, b, c, \ldots]$ denotes that C contains the values $a, b, c, \ldots$ "hardwired" in its description. When D is a random variable or distribution, $y \leftarrow D$ denote that y is randomly selected from D according to its distribution. If S is a set, then $x \leftarrow S$ denotes that x is uniformly selected from S. The expression $y := z$ denotes that y is set, defined, or substituted by z. We say that function $f : \mathbb{N} \to \mathbb{R}$ is negligible in $\lambda \in \mathbb{N}$ if $f(\lambda) = \lambda^{-\omega(1)}$. Hereafter, we use $f \leq \mathsf{negl}(\lambda)$ to mean that f is negligible in λ.

Let $\mathcal{X} = \{X_\lambda\}_{\lambda \in \mathbb{N}}$ and $\mathcal{Y} = \{Y_\lambda\}_{\lambda \in \mathbb{N}}$ denote two ensembles of random variables indexed by $\lambda \in \mathbb{N}$. We define the distinguishing probability between $\mathcal{X}$ and $\mathcal{Y}$ to be $\mu(\lambda)$ if for every probabilistic polynomial-time (PPT) algorithm D,

$$\mu(\lambda) := |\Pr[D(X_\lambda) = 1] - \Pr[D(Y_\lambda) = 1]|$$

When μ is negligible, we write $\mathcal{X} \stackrel{c}{\approx} \mathcal{Y}$.

For two circuits C and D, we write $C \equiv D$ if C and D compute exactly the same function. We define the following notion of *approximating* a function.

Definition 1 (ϵ-Approximating a Function). *A circuit C is said to ϵ-approximate a function $f : \{0,1\}^n \to \{0,1\}^*$, denoted by $C \cong_\epsilon f$, if $\Pr_{x \leftarrow \{0,1\}^n} [C(x) = f(x)] \geq \epsilon$.*

Bilinear Maps (a.k.a Pairings). We consider cyclic groups $\mathbb{G}_1$, $\mathbb{G}_2$, and $\mathbb{G}_T$ of prime order p. A bilinear map is an efficient mapping $e : \mathbb{G}_1 \times \mathbb{G}_2 \to \mathbb{G}_T$ satisfying the following properties.

Bilinearity: For every $g \in \mathbb{G}_1$, $\hat{g} \in \mathbb{G}_2$ and $a, b \in \mathbb{Z}_p$, $e(g^a, \hat{g}^b) = e(g, \hat{g})^{ab}$.
Non-degeneracy: If g and $\hat{g}$ generate $\mathbb{G}_1$ and $\mathbb{G}_2$ respectively, then $e(g, \hat{g}) \neq 1$.

For simplicity, consider symmetric pairings, that is, $\mathbb{G}_1 := \mathbb{G}_2 := \mathbb{G}$, where $\mathbb{G}$ is a cyclic group of prime order p and g is a generator of $\mathbb{G}$. Let $\mathcal{G}_{\mathsf{bmp}}$ be a standard parameter generation algorithm that takes as input a security parameter λ and outputs parameters $(p, \mathbb{G}, \mathbb{G}_T, e, g)$.

Definition 2 (DLIN Assumption). *The decisional linear (DLIN) problem is to guess $\beta \in \{0, 1\}$, given $(\Gamma, g, f, h, f^x, h^y, Q_\beta) \leftarrow \mathcal{G}_\beta^{\mathsf{dlin}}(1^\lambda)$, where $\mathcal{G}_\beta^{\mathsf{dlin}}(1^\lambda)$: $\Gamma := (p, \mathbb{G}, \mathbb{G}_T, e, g) \leftarrow \mathcal{G}_{\mathsf{bmp}}(1^\lambda)$, $a, b, x, y \leftarrow \mathbb{Z}_p$, $f := g^a$, $h := g^b$, $Q_0 := g^{x+y}$, $Q_1 \leftarrow \mathbb{G}$, return $\mathcal{I} := (\Gamma, g, f, h, f^x, h^y, Q_\beta)$. This advantage $\mathsf{Adv}_\mathcal{A}^{\mathsf{dlin}}(\lambda)$ is defined as follows.*

$$\mathsf{Adv}_\mathcal{A}^{\mathsf{dlin}}(\lambda) := \left| \Pr\left[\mathcal{A}(\mathcal{I}) = 1 \mid \mathcal{I} \leftarrow \mathcal{G}_0^{\mathsf{dlin}}(1^\lambda) \right] - \Pr\left[\mathcal{A}(\mathcal{I}) = 1 \mid \mathcal{I} \leftarrow \mathcal{G}_1^{\mathsf{dlin}}(1^\lambda) \right] \right|.$$

We say that the DLIN assumption holds if for every PPT adversary $\mathcal{A}$, $\mathsf{Adv}_\mathcal{A}^{\mathsf{dlin}}(\lambda) \leq \mathsf{negl}(\lambda)$.

Definition 3 (Universal One-Way Hash Function). *A universal one-way hash function (UOWHF) family* $\mathcal{H} = \{\mathcal{H}_\lambda\}_{\lambda \in \mathbb{N}}$ *is a function family in which each function* $H \in \mathcal{H}_\lambda$ *maps a domain* D *to a range* R *and satisfies the following condition. For all PPT adversaries* $\mathcal{A} := (\mathcal{A}_1, \mathcal{A}_2)$, *it holds that*

$$\Pr\left[x \neq x^* \wedge H(x) = H(x^*) \left| \begin{array}{l} (x, s) \leftarrow \mathcal{A}_1(1^\lambda), \\ H \leftarrow \mathcal{H}_\lambda, \\ x^* \leftarrow \mathcal{A}_2(1^\lambda, H, x, s) \end{array} \right. \right] \leq \mathsf{negl}(\lambda).$$

Pseudorandom Functions. We review pseudorandom functions (PRFs).

Definition 4 (Pseudorandom Functions). *A PRF* $\mathcal{F}$ *consists of two PPT algorithms* $\mathcal{F} = (\mathsf{Key}, \mathsf{Eval})$ *and a pair of poly-time computable functions* $n(\cdot)$ *and* $m(\cdot)$ *that satisfy the following condition. For all PPT adversaries* $\mathcal{A}$ *and* $\mathsf{F} \leftarrow \mathsf{Key}(1^\lambda)$, *it holds that*

$$\left| \Pr[\mathcal{A}^{\mathsf{Eval}(\mathsf{F}, \cdot)} = 1] - \Pr[\mathcal{A}^{\mathcal{R}(\cdot)} = 1] \right| \leq \mathsf{negl}(\lambda)$$

where $\mathsf{Eval}(\mathsf{F}, \cdot)$ *computes* $\mathsf{F}(\cdot) : \{0,1\}^{n(\lambda)} \to \{0,1\}^{m(\lambda)}$, *which is a deterministic function and a function* $\mathcal{R}$ *is chosen uniformly at random from the set of all functions with the same domain/range.*

The notion of puncturable PRF (pPRF) was proposed by Sahai and Waters [17].

Definition 5 (Puncturable Pseudorandom Functions). *A pPRF* $\mathcal{F}$ *consists of three (probabilistic) algorithms* $\mathcal{F} = (\mathsf{Key}, \mathsf{Punc}, \mathsf{Eval})$ *and a pair of computable functions* $n(\cdot)$ *and* $m(\cdot)$ *that satisfy the following two conditions.*

Functionality preserving under puncturing: *For every polynomial size set* $S \subseteq \{0,1\}^{n(\lambda)}$ *and for every* $x \in \{0,1\}^{n(\lambda)} \setminus S$, *it holds that*

$$\Pr\left[\mathsf{Eval}(\mathsf{F}, x) = \mathsf{Eval}(\mathsf{F}\{S\}, x) \mid \mathsf{F} \leftarrow \mathsf{Key}(1^\lambda), \mathsf{F}\{S\} := \mathsf{Punc}(\mathsf{F}, S) \right] = 1.$$

Pseudorandom at punctured points: *For every polynomial size set* $S = \{x_1, \ldots, x_{k(\lambda)}\} \subseteq \{0,1\}^{n(\lambda)}$ *it holds that for every PPT adversary* $\mathcal{A}$,

$$\left| \Pr[\mathcal{A}(\mathsf{F}\{S\}, \{\mathsf{Eval}(\mathsf{F}, x_i)\}_{i \in [k]}) = 1] - \Pr[\mathcal{A}(\mathsf{F}\{S\}, U_{m(\lambda) \cdot |S|}) = 1] \right| \leq \mathsf{negl}(\lambda)$$

where $\mathsf{F} \leftarrow \mathsf{Key}(1^\lambda)$, $\mathsf{F}\{S\} := \mathsf{Punc}(\mathsf{F}, S)$ *and* U_ℓ *denotes the uniform distribution over* ℓ *bits.*

Hereafter, we write $\mathsf{F}(x)$ *instead of* $\mathsf{Eval}(\mathsf{F}, x)$ *for ease of notations.*

Theorem 6 [4,5,9,11]. *If one-way functions exits, then for every efficiently computable* $n(\cdot)$ *and* $m(\cdot)$, *there exists a pPRF family whose input is an* n-*bit string and output is an* m-*bit string.*

Definition 7 (Virtual Black-Box Obfuscation). *A PPT algorithm $\mathcal{O}$ is a virtual black-box (VBB) obfuscator for a collection of circuits $\mathcal{C} := \{\mathcal{C}_\lambda\}_{\lambda \in \mathbb{N}}$ if the following two conditions hold.*

Functionality: *For every security parameter $\lambda \in \mathbb{N}$, every circuit $C \in \mathcal{C}_\lambda$, and every input x, it holds that*

$$\Pr[C'(x) = C(x) \mid C' \leftarrow \mathcal{O}(1^\lambda, C)] = 1.$$

Virtual Black Box Property: *For every PPT adversary $\mathcal{A}$, there exists a PPT simulator $\mathcal{S}$, for every circuit $C \in \mathcal{C}_\lambda$,*

$$\left| \Pr[\mathcal{A}(\mathcal{O}(1^\lambda, C)) = 1] - \Pr[\mathcal{S}^C(1^\lambda) = 1] \right| \leq \mathsf{negl}(\lambda).$$

3 Definitions of Program Watermarking

We introduce definitions of program watermarking and explain the differences among them in this section.

3.1 Syntax

First, we introduce a syntax of program watermarking by Cohen et al. [6]. They consider public-key type watermarking in which everybody can extract embedded marks by using a public extraction key.

Definition 8 (Watermarking Syntax [6]). *A message-embedding watermarking scheme for a circuit class $\{\mathcal{C}_\lambda\}_{\lambda \in \mathbb{N}}$ and a message space $\mathcal{M} = \{\mathcal{M}_\lambda\}_{\lambda \in \mathbb{N}}$ consists of three probabilistic polynomial-time algorithms* (Gen, Mark, Extract).

Key Generation: Gen(1^λ) *takes as input the security parameter and outputs a pair of keys* (xk, mk), *respectively called the extraction key and mark key.*

Mark: Mark(mk, C, m) *takes as input a mark key, an arbitrary circuit C (not necessarily in $\mathcal{C}_\lambda$) and a message* m $\in \mathcal{M}$ *and outputs a marked circuit $\widetilde{C}$.*

Extract: m' $\leftarrow$ Extract(xk, C') *takes as input an extraction key and an arbitrary circuit C', and outputs* m' $\leftarrow$ Extract(xk, C') *where* m' $\in \mathcal{M} \cup \{\mathsf{unmarked}\}$.

Barak et al. consider a symmetric-key version of this syntax. That is, xk is equal to mk and it is a secret. Cohen et al. consider a message-less watermarking scheme, where a embedded message is just "marked", that is, $\mathcal{M} = \{\mathsf{marked}\}$.

3.2 Security Definitions by Barak et al.

In this section, we introduce the definitions by Barak et al. [2, Definition 8.1]. Although we can consider an asymmetric-key version, we introduce a symmetric-key version in this section since they consider a symmetric-key watermarking. We denote a symmetric key as sk in this section.

Definition 9 (Software Watermarking [2]). *A software watermarking scheme* (Gen, Mark, Extract) *is required to satisfy the following properties.*

Functionality (a.k.a Perfect Correctness): *For every circuit $C \in \mathcal{C}_\lambda$, every message $m \in \mathcal{M}_\lambda$, every input x in the domain of C, and every $\mathsf{sk} \leftarrow \mathsf{Gen}(1^\lambda)$ it holds that*

$$\mathsf{Mark}(\mathsf{sk}, C, m)(x) = C(x).$$

Extraction Correctness: *For every $C \in \mathcal{C}_\lambda$, $m \in \mathcal{M}_\lambda$ and $\mathsf{sk} \leftarrow \mathsf{Gen}(1^\lambda)$ it holds that*

$$\Pr[m' \neq m \mid m' \leftarrow \mathsf{Extract}(\mathsf{sk}, \mathsf{Mark}(\mathsf{sk}, C, m))] \leq \mathsf{negl}(\lambda).$$

Meaningfulness: *For every circuit C (not necessarily in $\mathcal{C}_\lambda$), it holds that*

$$\Pr_{\mathsf{sk} \leftarrow \mathsf{Gen}(1^\lambda)}[\mathsf{Extract}(\mathsf{sk}, C) \neq \mathsf{unmarked}] \leq \mathsf{negl}(\lambda).$$

Fragility: *For every PPT $\mathcal{A}$, there exists a PPT $\mathcal{S}$ such that for every C and m, it holds that*

$$\Pr[C^* \equiv C \wedge \mathsf{Extract}(\mathsf{sk}, C^*) \neq m \mid C^* \leftarrow \mathcal{A}(\mathsf{Mark}(\mathsf{sk}, C, m))]$$
$$\leq \Pr[C^* \equiv C \mid C^* \leftarrow \mathcal{S}^C(1^{|C|})] + \mathsf{negl}(|C|)$$

Meaningfulness means that circuits are not marked when an extraction key is randomly chosen after a circuit is fixed. That is, if a circuit is not generated via an appropriate mark algorithm, then the circuit is not marked except with negligible probability.

Fragility[1] is a simulation-based security and a strong requirement since the simulator must construct a circuit C' that is equivalent to C only with black-box access to C. The reconstructed circuit C^* by $\mathcal{S}$ is expected to be un-marked due to the meaningfulness. This means that the success probability of $\mathcal{A}$ is negligibly small. Barak et al. show that it is impossible to achieve Definition 9 if there exists one-way functions (OWFs). This is due to the impossibility of VBB obfuscation, which is a simulation-based definition of a program obfuscation [2]. Roughly speaking, VBB obfuscation means that there exists a simulator that has only a black-box access to a circuit C and it can output an indistinguishable circuit C' from a real obfuscated circuit of C (See Definition 7). We do not explain the details of this impossibility result in this article since it is not the main purpose. However, the intuition is simple. If there exists a simulator for a watermarking in Definition 9, then the simulator generates C^* such that $C^* \equiv C$ and can perfectly reconstruct $\mathcal{O}(C) \equiv \mathcal{O}(C^*)$ (i.e., a VBB obfuscated circuit) by using only the black-box access to C in Definition 7. This contradicts the impossibility of a VBB obfuscation. See the paper of Barak et al. [2] for details.

Barak et al. also consider a relaxed security requirement. They consider the following definition (from [2, Definition 8.4]) instead of the fragility in Definition 9.

---

[1] We introduce the name by Barak et al. [2] as it is.

Definition 10 (Occasional Watermarking). *For every PPT* $\mathcal{A}$, *there exists a circuit* C *and a message* m,

$$\Pr[C \equiv C^* \wedge \mathsf{Extract}(\mathsf{sk}, C^*) = \mathsf{m} \mid C^* \leftarrow \mathcal{A}(\mathsf{Mark}(\mathsf{sk}, C, \mathsf{m}))] \leq 1 - 1/\mathrm{poly}(|C|)$$

This type of definition is called a game-based definition, which is formalized as a game between a challenger and adversary. The challenger (though it is not explicitly expressed) passes a marked circuit to $\mathcal{A}$ as a target and there is no simulator. Barak et al. show that if there exists an iO then it is impossible to achieve occasional watermarking. We explain this impossibility result in Sect. 4.

3.3 Security Definitions by Cohen et al.

As we explain in Sect. 4, the impossibility result of occasional watermarking heavily relies on the *perfect correctness*. If we consider a relaxed correctness, the impossibility result does not hold. Cohen et al. consider the following definition to overcome the impossibility result of Barak et al.

Definition 11 (Watermarking Security). *A watermarking scheme* (Gen, Mark, Extract) *for circuit family* $\{\mathcal{C}_\lambda\}_{\lambda \in \mathbb{N}}$ *and with message space* $\mathcal{M} = \{\mathcal{M}_\lambda\}$ *is required to satisfy the following properties.*

Statistical Correctness: *There is a negligible function* $\nu(\lambda)$ *such that for every circuit* $C \in \mathcal{C}_\lambda$, *every message* $\mathsf{m} \in \mathcal{M}_\lambda$ *and every input* x *in the domain of* C, *it holds that*

$$\Pr\left[\widetilde{C}(x) = C(x) \;\middle|\; \begin{array}{l} (\mathsf{xk}, \mathsf{mk}) \leftarrow \mathsf{Gen}(1^\lambda) \\ \widetilde{C} \leftarrow \mathsf{Mark}(\mathsf{mk}, C, \mathsf{m}) \end{array} \right] \geq 1 - \nu(\lambda).$$

Extraction Correctness: *Same as in Definition 9 except that* xk *and* mk *are used.*

Meaningfulness: *Same as in Definition 9 except that* xk *and* mk *are used.*

ϵ**-Unremovability:** *For every PPT* $\mathcal{A}$, *we have*

$$\Pr[\mathsf{Exp}_{\mathcal{A}}^{\mathsf{nrmv}}(\lambda, \epsilon) = 1] \leq \mathsf{negl}(\lambda),$$

where ϵ *is a parameter of the scheme called the* approximation factor *and* $\mathsf{Exp}_{\mathcal{A}}^{\mathsf{nrmv}}(\lambda, \epsilon)$ *is the game defined in Definition 12.*

We say a watermarking scheme is ϵ*-secure if it satisfies these properties.*

Definition 12 (ϵ-Unremovability Security Game). *The game* $\mathsf{Exp}_{\mathcal{A}}^{\mathsf{nrmv}}(\lambda, \epsilon)$ *is defined as follows.*

1. *The challenger generates* $(\mathsf{xk}, \mathsf{mk}) \leftarrow \mathsf{Gen}(1^\lambda)$ *and gives* xk *to the adversary* $\mathcal{A}$.
2. $\mathcal{A}$ *has oracle access to the mark oracle* $\mathcal{MO}$. *If* $\mathcal{MO}$ *is queried with a circuit* C_i *(not necessarily in* $\mathcal{C}_\lambda$*) and message* m_i, *then* $\mathcal{MO}$ *answers with* $\mathsf{Mark}(\mathsf{mk}, C_i, \mathsf{m}_i)$.

3. *At some point, the adversary makes a query to the challenge oracle $\mathcal{CO}$. If $\mathcal{CO}$ is queried with a message $\mathsf{m} \in \mathcal{M}_\lambda$, it samples a circuit $C \leftarrow \mathcal{C}_\lambda$ uniformly at random and answers $\widetilde{C} \leftarrow \mathsf{Mark}(\mathsf{mk}, C, \mathsf{m})$.*
4. *Again, $\mathcal{A}$ queries many circuit and message pairs to $\mathcal{MO}$.*
5. *Finally, the adversary outputs a circuit C^*. If it holds that $C^* \cong_\epsilon C$ and $\mathsf{Extract}(\mathsf{xk}, C^*) \neq \mathsf{m}$ then the experiment outputs 1, otherwise 0.*

The definition requires the adversary to output C^* that agrees on an ϵ fraction of inputs with C. This formalizes that C^* should be similar to the original circuit C. In the symmetric key setting, i.e., $\mathsf{xk} = \mathsf{mk}$, the adversary is given access to the extraction oracle $\mathcal{XO}$, which receives a circuit and returns the embedded mark (or unmarked). In the definition, $\mathcal{MO}$ accepts a circuit outside $\mathcal{C}_\lambda$ (i.e., an arbitrary circuit). This is stronger security than that where $\mathcal{MO}$ accepts only circuits in $\mathcal{C}_\lambda$.

Lunch-Time Security. In the ϵ-unremovability game, if the adversary does not have access to $\mathcal{MO}$ after the target circuit $\widetilde{C}$ is given, then the security is called ϵ-secure against lunch-time attacks[2].

The main differences between the definitions by Barak et al. and Cohen et al. are as follows.

Correctness: Watermarked circuits do not agree on a negligible fraction of inputs with the original circuits in the definition of Cohen et al. while the definition of Barak et al. requires *perfect* correctness.

Security: There exists $\mathcal{MO}$, which gives adversaries many watermarked circuits, and a public extraction key is given to adversaries in the definition of Cohen et al.

4 Impossibility Result by Barak et al.

Barak et al. show that it is impossible to achieve secure watermarking in the sense of Definition 9 by using the impossibility of VBB [2]. It is natural to explore a more relaxed definition since Definition 9 is very strong.

Although Definition 10 is a relaxed definition, Barak et al. also show that it is impossible to achieve secure watermarking in the sense of Definition 10 if an $i\mathcal{O}$ exists and a watermarking scheme satisfies *perfect* correctness. The definition of an iO is as follows.

Indistinguishability Obfuscation. The notion of iO is proposed by Barak et al. [1, 2] and the first candidate construction is proposed by Garg, Gentry, Halevi, Raykova, Sahai, and Waters [8].

Definition 13 (Indistinguishability Obfuscation [2,8]). *A PPT algorithm $i\mathcal{O}$ is an indistinguishability obfuscator if it satisfies the following two conditions.*

---

[2] Or *non-adaptive chosen circuit attacks* (CCA1).

Functionality: *For every security parameter* $\lambda \in \mathbb{N}$, *for every circuit* C, *and every input* x, *it holds that*

$$\Pr[C'(x) = C(x) \mid C' \leftarrow i\mathcal{O}(1^\lambda, C)] = 1.$$

Indistinguishability: *For every PPT distinguisher* $\mathcal{D}$ *and every circuit ensemble* $C_0 = \{C_\lambda^{(0)}\}_{\lambda \in \mathbb{N}}$ *and* $C_1 = \{C_\lambda^{(1)}\}_{\lambda \in \mathbb{N}}$ *such that* $\forall \lambda, x : C_\lambda^{(0)}(x) = C_\lambda^{(1)}(x)$ *and* $|C_\lambda^{(0)}| = |C_\lambda^{(1)}|$ *we have:*

$$\left| \Pr[\mathcal{D}(i\mathcal{O}(1^\lambda, C_\lambda^{(0)})) = 1] - \Pr[\mathcal{D}(i\mathcal{O}(1^\lambda, C_\lambda^{(1)})) = 1] \right| \leq \mathsf{negl}(\lambda).$$

For simplicity, we write $i\mathcal{O}(C)$ instead of $i\mathcal{O}(1^\lambda, C)$ when the security parameter λ is clear from context.

Removing Marks by iO. An iO can be used as a mark-remover as follows. Consider two non-marked circuits (C_0, C_1) and a marked circuit $\widetilde{C}_0 \leftarrow \mathsf{Mark}(\mathsf{mk}, C_0)$. Here, C_0 and $\widetilde{C}_0$ are functionally equivalent due to the perfect correctness. That is, for any input x, $C_0(x) = \widetilde{C}_0(x)$. If we apply $i\mathcal{O}$ to $\widetilde{C}_0$ and C_1 (these two obfuscated circuits are functionally equivalent) and the watermarking scheme satisfies Definition 10, then the following occurs.

$$\mathsf{Extract}(\mathsf{xk}, i\mathcal{O}(\widetilde{C}_0)) \to 1$$
$$\mathsf{Extract}(\mathsf{xk}, i\mathcal{O}(C_1)) \to 0.$$

The first and second lines of the equations hold due to occasional watermarking and meaningfulness, respectively. However, this contradicts the indistinguishability of an iO. If there exists a secure iO, then it scrambles embedded marks and works as a mark-remover. The key point of this argument is that a watermarking scheme satisfies *perfect* correctness since an iO works only for two functionally equivalent circuits.

Using a relaxed correctness is a simple way to avoid the impossibility since we can no longer apply an iO to circuits that do not satisfy perfect correctness. In fact, if we use a relaxed correctness, then the iO might help to achieve secure program watermarking. Consider the following construction idea. For a negligible fraction of inputs, a watermarked circuit behaves differently from the original circuit C. Let $\mathcal{X} \subset \mathsf{D}$ be such a negligible fraction of inputs where D is the set of all inputs for C. A slightly modified circuit C' is defined as follows. For $x \in \mathsf{D} \setminus \mathcal{X}$, $C'(x)$ computes $C(x)$. For input $x \in \mathcal{X}$, circuit $C'(x)$ does not compute $C(x)$ but outputs special value y. If we can hide this small change, then (x, y) might work as a mark and extraction key for watermarking because we can verify whether a circuit is marked by checking $y = C'(x)$. Here, an iO might help hide the change and prevent adversaries from removing the mark. This was already observed by Barak et al. [2], but they do not give any provably secure construction. This observation is the starting point of the work by Cohen et al. We discuss the details of this in Sect. 6.

5   Watermarking Based on Number Theory

First, we introduce the bracket notation for ease of notation.[3] For $a, b, \alpha \in \mathbb{Z}_p$, $\boldsymbol{v} = (v_1, \ldots, v_n) \in \mathbb{Z}_p^n$, we let

$$[a] := g^a, [b]_T := e(g, g)^b,$$
$$[\boldsymbol{v}] := ([v_1], \cdots, [v_n]),$$
$$[\boldsymbol{v}] \oplus [\boldsymbol{w}] := [\boldsymbol{v} + \boldsymbol{w}] = ([v_1 + w_1], \ldots, [v_n + w_n]),$$
$$\alpha \odot [\boldsymbol{v}] := [\boldsymbol{v}]^\alpha := [\alpha \boldsymbol{v}] = ([\alpha v_1], \ldots, [\alpha v_n]),$$
$$e([\boldsymbol{v}], [\boldsymbol{w}]) := \bigoplus_{i=1}^n e([v_i], [w_i]) = [\langle \boldsymbol{v}, \boldsymbol{w} \rangle]_T.$$

In this section, we explain the number theoretic watermarking scheme for cryptographic functions by Nishimaki [14]. The scheme does not rely on strong cryptographic tools such as an iO and the construction idea itself is interesting though its security is weak.

5.1   Overview

Nishimaki proposes a watermarking scheme for (a variant of) trapdoor functions [14]. His construction idea is based on the vector decomposition (VD) problem defined over bilinear groups [7,20].

Roughly speaking, the VD problem is as follows. Let $\mathbb{V}$ be a 2-dimensional vector space and $\mathbb{V}_1$ and $\mathbb{V}_2$ two different 1-dimensional subspaces of $\mathbb{V}$. An element $[\boldsymbol{v}] \in \mathbb{V}$ can be decomposed into an element generated by a base of $\mathbb{V}_1$, $[\boldsymbol{b}_1]$, and an element generated by a base of $\mathbb{V}_2$, $[\boldsymbol{b}_2]$. That is, given $[\boldsymbol{v}] = [x\boldsymbol{b}_1] \oplus [y\boldsymbol{b}_2] \in \mathbb{V}$ where x, y are some scalars, computing $[y\boldsymbol{b}_2]$ (or $[x\boldsymbol{b}_1]$) is the VD problem. It can be generalized to the n-dimensional case. Yoshida and Fujiwara find that hidden $[y\boldsymbol{b}_2]$ is useful to achieve watermarking. If a vector $[\boldsymbol{b}^*]$ is orthogonal to $[\boldsymbol{b}_2]$, then $[\boldsymbol{b}^*]$ can be used to detect a mark since $[\boldsymbol{b}_1]$ is canceled by taking inner-product. Moreover, removing $[y\boldsymbol{b}_2]$ from $[\boldsymbol{v}]$ seems to be hard [19] since it is believed to be hard to decompose $[\boldsymbol{v}]$ into $[x\boldsymbol{b}_1]$ and $[y\boldsymbol{b}_2]$ [7,20]. This is a bright observation, but Yoshida and Fujiwara do not give a provably secure construction. The idea by Yoshida and Fujiwara is the starting point of the work by Nishimaki. The details are given below.

One way to instantiate a linear vector space defined over bilinear groups is using the concept of a dual paring vector space (DPVS) [15,16]. In a DPVS, a vector consists of group elements. By generating $n \times n$ matrices $\boldsymbol{B} := (\boldsymbol{b}_1, \ldots, \boldsymbol{b}_n)$ and $\boldsymbol{B}^* := (\boldsymbol{b}_1^*, \ldots, \boldsymbol{b}_n^*)$ such that $\boldsymbol{B}^{*\top} \cdot \boldsymbol{B} = \boldsymbol{I}$, we can generate a pair of dual orthonormal bases $[\boldsymbol{B}] := ([\boldsymbol{b}_1], \ldots, [\boldsymbol{b}_n])$ and $[\boldsymbol{B}^*] := ([\boldsymbol{b}_1^*], \ldots, [\boldsymbol{b}_n^*])$.

One of the notable features of a DPVS is the canceling property. It holds that $e([\boldsymbol{b}_i], [\boldsymbol{b}_j^*]) = [1]_T$ for $i \neq j$. Consider a hypothetical ciphertext $[x\boldsymbol{b}_1]$ that can

---

[3] This notation is used in many papers. In particular, we borrow from the work of Lin and Vaikuntanathan [13].

be decrypted by a pairing operation with $[\boldsymbol{b}_1^*]$. That is, we find x by computing $e([x\boldsymbol{b}_1], [\boldsymbol{b}_1^*]) = [x]_T$. Note that this is not a real encryption but just a hypothetical encryption for explanation. When we use a ciphertext $\widetilde{\mathsf{ct}} := [x\boldsymbol{b}_1 + y\boldsymbol{b}_2]$, it is still possible to recover x by using $\boldsymbol{b}_1^*$ since it is orthogonal to $\boldsymbol{b}_2$. Moreover, if we have $[\boldsymbol{b}_2^*]$, then we can verify whether $\widehat{\mathsf{ct}}$ includes $[\boldsymbol{b}_2]$ by checking $e(\widehat{\mathsf{ct}}, [\boldsymbol{b}_2^*]) = 1$. From the definition of $\boldsymbol{B}, \boldsymbol{B}^*$, $e(\widehat{\mathsf{ct}}, [\boldsymbol{b}_2^*]) = 1$ if $\widehat{\mathsf{ct}}$ is not spanned by $[\boldsymbol{b}_2]$. Thus, $[\boldsymbol{b}_2]$ and $[\boldsymbol{b}_2^*]$ can be used as an embedded mark and a key for detecting the mark, respectively. That is, the canceling property is useful to achieve the functionality of watermarking.

Another notable feature of a DPVS is that we can consider hidden linear subspaces spanned by subsets $[\widehat{\boldsymbol{B}}]$ and $[\widehat{\boldsymbol{B}}^*]$ of $[\boldsymbol{B}]$ and $[\boldsymbol{B}^*]$, respectively. If the DLIN assumption holds (decisional) subspace assumptions hold, which says it is hard to distinguish whether a given vector is spanned by $[\boldsymbol{B}]$ (resp. $[\boldsymbol{B}^*]$) or $[\boldsymbol{B}] \setminus [\widehat{\boldsymbol{B}}]$ (resp. $[\boldsymbol{B}^*] \setminus [\widehat{\boldsymbol{B}}^*]]$) [12,16]. Roughly speaking, an embedded mark $[\boldsymbol{b}_2]$ is hidden due to a subspace assumption. A toy example of a subspace assumption is that given $([\boldsymbol{b}_1^*], [x\boldsymbol{b}_1 + y\boldsymbol{b}_2])$, it is hard to distinguish $[v\boldsymbol{b}_1^* + w\boldsymbol{b}_2^*]$ from $[v\boldsymbol{b}_1^*]$. Subspace assumptions are useful to show that it is hard for adversaries to remove marks because if adversaries can decompose $[x\boldsymbol{b}_1 + y\boldsymbol{b}_2]$ into $[x\boldsymbol{b}_1]$ and $[y\boldsymbol{b}_1]$ then $[y\boldsymbol{b}_2]$ is used to distinguish $[v\boldsymbol{b}_1^* + w\boldsymbol{b}_2^*]$ from $[v\boldsymbol{b}_1^*]$ by just computing the pairing.

For readers that are familiar with the dual system encryption methodology, the two insights above are reminiscent of the concept of semi-functional ciphertexts and keys introduced by Waters [18]. In fact, the methodology inspires Nishimaki to come up with the application of a DPVS to watermarking.

5.2   Dual Pairing Vector Space

We review the concept of a DPVS in this section and a related assumption.

Dual Orthonormal Bases. Let $\mathsf{Dual}(\mathbb{Z}_p^n)$ be an algorithm for generating dual orthonormal bases as follows.

$\underline{\mathsf{Dual}(\mathbb{Z}_p^n)}$:   chooses $\boldsymbol{b}_i, \boldsymbol{b}_j^* \in \mathbb{Z}_p^n, \psi \leftarrow \mathbb{Z}_p^*$ such that

$$\langle \boldsymbol{b}_i, \boldsymbol{b}_j^* \rangle = 0 \bmod p \text{ for } i \neq j,$$
$$\langle \boldsymbol{b}_i, \boldsymbol{b}_i^* \rangle = \psi \bmod p \text{ for all } i \in [n]$$

outputs $\boldsymbol{B} := (\boldsymbol{b}_1, \ldots, \boldsymbol{b}_n)$ and $\boldsymbol{B}^* := (\boldsymbol{b}_1^*, \ldots, \boldsymbol{b}_n^*)$.

We now describe a parameter generation algorithm $\mathcal{G}_{\mathsf{dpvs}}(1^\lambda, n)$ for a DPVS.

$\underline{\mathcal{G}_{\mathsf{dpvs}(1^\lambda, n)}}$: $(p, \mathbb{G}, \mathbb{G}_T, e, g) \leftarrow \mathcal{G}_{\mathsf{bmp}}(1^\lambda), \mathbb{V} := \mathbb{G}^n$

$\mathsf{params}_{\mathbb{V}} := (p, \mathbb{V}, \mathbb{G}_T, e, g)$
$(\boldsymbol{B}, \boldsymbol{B}^*) \leftarrow \mathsf{Dual}(\mathbb{Z}_p^n),$
$[\boldsymbol{B}] := ([\boldsymbol{b}_1], \ldots, [\boldsymbol{b}_n]),$
$[\boldsymbol{B}^*] := ([\boldsymbol{b}_1^*], \ldots, [\boldsymbol{b}_n^*]),$
returns $(\mathsf{params}_{\mathbb{V}}, [\boldsymbol{B}], [\boldsymbol{B}^*]).$

Parameters ($\mathsf{params}_\mathbb{V}, [\boldsymbol{B}], [\boldsymbol{B}^*]$) define a DPVS over a bilinear group.

Definition 14 (Subspace Assumption). *First, we define an instance generation algorithm of the subspace problem as follow.*

$\underline{\mathcal{G}_b^{\mathsf{dss}}(1^\lambda)}:$

$\Gamma \leftarrow \mathcal{G}_{\mathsf{bmp}}(1^\lambda), (\boldsymbol{B}, \boldsymbol{B}^*) \leftarrow \mathsf{Dual}(\mathbb{Z}_p^n),$

$\eta, \beta, \tau_1, \tau_2, \tau_3, \mu_1, \mu_2, \mu_3 \leftarrow \mathbb{Z}_p,$

For $i \in [k]$

$U_i := [\mu_1 \boldsymbol{b}_i + \mu_2 \boldsymbol{b}_{k+i} + \mu_3 \boldsymbol{b}_{2k+i}],$

$V_i := [\tau_1 \eta \boldsymbol{b}_i^* + \tau_2 \beta \boldsymbol{b}_{k+i}^*],$

$W_i := [\tau_1 \eta \boldsymbol{b}_i^* + \tau_2 \beta \boldsymbol{b}_{k+i}^* + \tau_3 \boldsymbol{b}_{2k+i}^*],$

END OF For

$Q_0 := (V_1, \ldots, V_k), Q_1 := (W_1, \ldots, W_k),$

$D := ([\boldsymbol{b}_1], \ldots, [\boldsymbol{b}_{2k}], [\boldsymbol{b}_{3k+1}], \ldots, [\boldsymbol{b}_n], [\eta \boldsymbol{b}_1^*], \ldots, [\eta \boldsymbol{b}_k^*], [\beta \boldsymbol{b}_{k+1}^*], \ldots, [\beta \boldsymbol{b}_{2k}^*],$

$\qquad\quad [\boldsymbol{b}_{2k+1}^*], \ldots, [\boldsymbol{b}_n^*], U_1, \ldots, U_k, \mu_3),$

return $\mathcal{I} := (\Gamma, D, Q_b).$

The subspace problem is to guess $b \in \{0, 1\}$, given (Γ, D, Q_b). This advantage $\mathsf{Adv}_\mathcal{A}^{\mathsf{dss}}(\lambda)$ *is defined as follows.*

$$\mathsf{Adv}_\mathcal{A}^{\mathsf{dss}}(\lambda) := \left| \Pr\left[\mathcal{A}(\mathcal{I}) = 1 \mid \mathcal{I} \leftarrow \mathcal{G}_0^{\mathsf{dss}}(1^\lambda)\right] - \Pr\left[\mathcal{A}(\mathcal{I}) = 1 \mid \mathcal{I} \leftarrow \mathcal{G}_1^{\mathsf{dss}}(1^\lambda)\right] \right|$$

We say that the subspace assumption holds if for every PPT adversary $\mathcal{A}, \mathsf{Adv}_\mathcal{A}^{\mathsf{dss}}(\lambda) \leq \mathsf{negl}(\lambda).$

This assumption says that it is hard to distinguish vectors spanned by $([\boldsymbol{b}_i^*], [\boldsymbol{b}_{k+i}^*])$ from vectors spanned by $([\boldsymbol{b}_i^*], [\boldsymbol{b}_{k+i}^*], [\boldsymbol{b}_{2k+i}^*])$ if $\{[\boldsymbol{b}_{2k+i}]\}_i$, which trivially break the assumption, are not given.

Theorem 15 [12]. *The DLIN assumption implies the subspace assumption.*

5.3   Watermarking Based on DPVS

The watermarking scheme proposed by Nishimaki is a bit complicated. Thus, we introduce a hypothetical construction for ease of understanding.

Consider the following function f_C based on dual orthonormal bases $(\boldsymbol{D}, \boldsymbol{D}^*) \leftarrow \mathsf{Dual}(\mathbb{Z}_p^3)$. Given a description of a function $C := ([r_1 \boldsymbol{d}_1], [s_1 \boldsymbol{d}_1^*])$ and an input x, f_C computes $e([r_1 \boldsymbol{d}_1], x \odot [s_1 \boldsymbol{d}_1^*]) = [x \cdot r_1 s_1]_T$. This can be seen as a one-way function (assuming the hardness of the discrete logarithm problem over $\mathbb{G}_T$). Let $\mathsf{mk} := ([\boldsymbol{d}_2^*], [\boldsymbol{d}_3^*])$ and $\mathsf{xk} := ([\boldsymbol{d}_2], [\boldsymbol{d}_3])$. These keys are used to generate a vector in hidden sub-spaces. A hypothetical watermarking scheme for f_C is as follows.

$\mathsf{Gen}(1^\lambda)$: generates $(\boldsymbol{D}, \boldsymbol{D}^*) \leftarrow \mathsf{Dual}(\mathbb{Z}_p^3)$ and sets $\mathsf{mk} := ([\boldsymbol{d}_2^*], [\boldsymbol{d}_3^*])$ and $\mathsf{xk} := ([\boldsymbol{d}_2], [\boldsymbol{d}_3]).$

Mark(mk, f_C): parses $C = ([\boldsymbol{f}], [\boldsymbol{f}^*]) \in \mathbb{G}^3 \times \mathbb{G}^3$, chooses $s_2, s_3 \leftarrow \mathbb{Z}_p$, computes $[\boldsymbol{f}^*] \oplus [s_2 \boldsymbol{d}_2^* + s_3 \boldsymbol{d}_3^*]$, and output $\widetilde{C} := ([\boldsymbol{f}], [\boldsymbol{f}^* + s_2 \boldsymbol{d}_2^* + s_3 \boldsymbol{d}_3^*])$.

Extract(xk, $f_{C'}$): parses $C' = ([\boldsymbol{f}'], [\boldsymbol{f}'^*])$, chooses $r_2, r_3 \leftarrow \mathbb{Z}_p$, and verifies whether $e([r_2 \boldsymbol{d}_2 + r_3 \boldsymbol{d}_3], [\boldsymbol{f}'^*]) = [1]_T$. If the equation holds, then it outputs unmarked; otherwise, marked.

Note that we need $([\boldsymbol{d}_1], [\boldsymbol{d}_1^*])$ to generate a function. A non-marked function does not include a vector in hidden sub-spaces (spanned by $[\boldsymbol{d}_2], [\boldsymbol{d}_3], [\boldsymbol{d}_2^*]$, and $[\boldsymbol{d}_3^*]$). A marked function include a vector spanned by $([\boldsymbol{d}_1^*], [\boldsymbol{d}_2^*], [\boldsymbol{d}_3^*])$, but this does not interfere with the functionality of f_C since vectors $[\boldsymbol{d}_2^*], [\boldsymbol{d}_3^*]$ are canceled via the pairing operation with $[\boldsymbol{d}_1]$. Moreover, if we have $([\boldsymbol{d}_2], [\boldsymbol{d}_3])$, then we can verify that a vector is spanned by $[\boldsymbol{d}_2^*]$ and $[\boldsymbol{d}_3^*]$. The pairing operation yields a non-identity element in $\mathbb{G}_T$.

Intuition of Security. If an adversary can remove $[\boldsymbol{d}_2^*]$ and $[\boldsymbol{d}_3^*]$ parts from $[s_1 \boldsymbol{d}_1^* + s_2 \boldsymbol{d}_2^* + s_3 \boldsymbol{d}_3^*]$, then we can use $[s_1 \boldsymbol{d}_1^*]$ to distinguish $[r_2' \boldsymbol{d}_2 + r_3' \boldsymbol{d}_3]$ from $[r_1' \boldsymbol{d}_1 + r_2' \boldsymbol{d}_2 + r_3' \boldsymbol{d}_3]$. This contradicts the hardness of the subspace assumption.

More concretely, we can use the subspace assumption for $n = 3$ and $k = 1$ for the security. Given $([\boldsymbol{b}_1], [\boldsymbol{b}_2], [\eta \boldsymbol{b}_1^*], [\beta \boldsymbol{b}_2^*], [\boldsymbol{b}_3^*], U_1 = [\mu_1 \boldsymbol{b}_1 + \mu_2 \boldsymbol{b}_2 + \mu_3 \boldsymbol{b}_3], \mu_3, Q_{b'})$, we simulate a marked function as follows. We set

$$\boldsymbol{d}_1 := \boldsymbol{b}_3^* \qquad\qquad \boldsymbol{d}_2 := \boldsymbol{b}_1^* \qquad\qquad \boldsymbol{d}_3 := \boldsymbol{b}_2^*$$
$$\boldsymbol{d}_1^* := \boldsymbol{b}_3 \qquad\qquad \boldsymbol{d}_2^* := \boldsymbol{b}_1 \qquad\qquad \boldsymbol{d}_3^* := \boldsymbol{b}_2,$$

choose $r_1 \leftarrow \mathbb{Z}_p$, and construct a marked function $\widetilde{C} := (r_1 \odot [\boldsymbol{b}_3^*], U_1) = ([r_1 \boldsymbol{d}_1], [\mu_3 \boldsymbol{d}_1^* + \mu_1 \boldsymbol{d}_2^* + \mu_2 \boldsymbol{d}_3^*])$. If an adversary succeeds in removing $[\mu_1 \boldsymbol{d}_1^* + \mu_2 \boldsymbol{d}_2^*]$ from $\widetilde{C}$, then we can obtain $[\mu_3 \boldsymbol{d}_1^*]$. We compute $v := e(Q_{b'}, [\mu_3 \boldsymbol{d}_1^*])$. Here,

$$Q_0 = V_1 \quad = [\tau_1 \eta \boldsymbol{b}_1^* + \tau_2 \beta \boldsymbol{b}_2^*] \qquad\qquad = [\tau_1 \eta \boldsymbol{d}_2 + \tau_2 \beta \boldsymbol{d}_3]$$
$$Q_1 = W_1 \quad = [\tau_1 \eta \boldsymbol{b}_1^* + \tau_2 \beta \boldsymbol{b}_2^* + \tau_3 \boldsymbol{b}_3^*] \quad = [\tau_1 \eta \boldsymbol{d}_2 + \tau_2 \beta \boldsymbol{d}_3 + \tau_3 \boldsymbol{d}_1].$$

Thus, it should hold that $v = [0]_T$ if $b' = 0$; otherwise, $v \neq [0]_T$. This breaks the hardness of the subspace assumption. Therefore, the watermarking scheme satisfies unremovability.

On Limitation of Algebraic Construction. The security argument above works *if the adversary outputs 6 group elements as a function.* The construction above is not unremovable if adversaries take arbitrary strategies. For example, if an adversary applies an iO to the marked function and completely destroys the algebraic structure, then it succeeds in removing $[\mu_1 \boldsymbol{d}_1^* + \mu_2 \boldsymbol{d}_2^*]$ while preserving the functionality (note that the hypothetical watermarking scheme satisfies perfect correctness). As long as an adversary outputs 6 group elements, the algebraic construction is unremovable. Even if adversaries encode an obfuscated circuit into group elements [0] and [1] in a bit-by-bit manner, the number of group elements does not match that of the original algebraic function. Therefore, the watermarking scheme by Nishimaki is secure in a restricted security model [14]. In the next section, we explain how Cohen et al. go beyond the limitation.

6   Watermarking for pPRFs

In this section, we explain the construction idea of the watermarking scheme for pPRFs by Cohen et al. [6].

6.1   More Impossibility Results by Cohen et al.

Before we explain the watermarking for pPRFs, we explain more impossibility results by Cohen et al. to understand why Cohen et al. focus on pPRFs.

Cohen et al. define the notion of "waterproof" for a function family, which means it is impossible to achieve a watermarking scheme for the function family.

Definition 16 (ϵ-Waterproof). *Let $\mathcal{F} = \{\mathcal{F}_\lambda\}_{\lambda \in \mathbb{N}}$ be a circuit ensemble. We say that $\mathcal{F}$ is ϵ-waterproof if there does not exist a weak ϵ-unremovable watermarking scheme for $\mathcal{F}$, where weak ϵ-unremovable is the same as ϵ-unremovable except that the adversary is given access to neither $\mathcal{MO}$, $\mathcal{XO}$, nor xk and cannot select the message that is embedded in the target circuit.*

Cohen et al. introduce not only the statistical correctness in Definition 11 but also a more relaxed correctness called weak statistical correctness [6].

Definition 17 (Weak Statistical Correctness). *There is a negligible function $\nu(\lambda)$ such that for every circuit $C \in \mathcal{C}_\lambda$ and messages $\mathsf{m} \in \mathcal{M}_\lambda$ and it holds that*

$$\mathsf{Mark}(\mathsf{mk}, C, \mathsf{m}) \cong_\nu C.$$

Cohen et al. show that non-black-box learnable function families are ϵ-waterproof. The parameter ϵ depends on what kind of learnability is considered. Roughly speaking, non-black-box learnable means that if an efficient learner is given a circuit $\widetilde{C}$ that approximates a circuit C, then the learner can output a circuit h whose behavior is (almost) the same as that of C. The approximation notion for $\widetilde{C}$ corresponds to the statistical correctness in Definition 11 (or 17). Thus, for non-black-box learnable functions, an adversary can reconstruct the original circuit by using a learning algorithm and remove marks since the original circuit is not marked due to the meaningfulness. This is the intuition. In particular, they show that assuming one-way functions, there exists a *(standard) PRF* family $\mathcal{F}$ which is ϵ-waterproof with weak statistical correctness for any non-negligible ϵ [6]. See the paper by Cohen et al. [6] for details.

From the impossibility above, even if we consider the relaxed correctness, there still exists waterproof function families. In particular, it impossible to watermark *standard* PRFs. This is why Cohen et al. focus on *pPRFs*.

6.2   Overview of Construction

As explained in Sect. 4, the impossibility result by Barak et al. is not applicable to watermarking that satisfies only the *statistical correctness* in Definition 11.

As we see in the previous section, the impossibility still holds for some function families even if we consider the weak statistical correctness. However, an iO might be useful to achieve watermarking for *pPRFs* since the impossibility result is not applicable to a pPRF, which is an iO-friendly primitive [17]. In fact, Cohen et al. show a possibility result [6].

Consider the following toy example to understand the initial idea of Cohen et al. Let C_{F} be a circuit that computes a PRF $\mathsf{F} : \{0,1\}^n \to \{0,1\}^m$. A circuit C'_{F} is a slight modification of C_{F} as follows.

$$C'_{\mathsf{F}}(x) = \begin{cases} y_i & \text{if } x = x_i \\ C_{\mathsf{F}}(x) & \text{for } x \notin \mathcal{X}, \end{cases}$$

where $x_i \leftarrow \{0,1\}^n, y_i \leftarrow \{0,1\}^m$ for all $i \in [\ell]$, $\mathcal{X} := \{x_1, \ldots, x_\ell\}$, and $\mathcal{Y} := \{y_1, \ldots, y_\ell\}$. Although a VBB obfuscation for PRFs does not exists and PRFs are waterproof under the existence of a OWF [2,6], let us assume that it does. This is just a hypothetical argument to explain the idea of Cohen et al. If we replace the VBB obfuscation and PRFs with an iO and pPRFs, respectively, then the construction works. If we apply VBB obfuscation $\mathcal{O}$ to C'_{F}, then $\widetilde{C} := \mathcal{O}(C'_{\mathsf{F}})$ does not leak any information except the input-output behavior of C'_{F}. Here, $(\mathcal{X}, \mathcal{Y})$ are seen as mark and extraction keys and $\mathcal{X}$ is called *marked points*. If users have $(\mathcal{X}, \mathcal{Y})$, then they can detect that a mark is embedded in $\widetilde{C}$ by checking $y_i = \widetilde{C}(x_i)$ for some $i \in [\ell]$. Circuit $\widetilde{C}$ is a (message-less) watermarked circuit of C. It is easy to verify that $\widetilde{C}$ satisfies the statistical correctness since ℓ is a polynomial in λ. Values y_i for input $x_i \in \mathcal{X}$ are indistinguishable from $C_{\mathsf{F}}(x_i)$ due to the pseudorandomness of F. Intuitively speaking, the slight modification is hidden from the adversary due to the obfuscation.

A slightly formal security argument is as follows. Consider a set $\mathcal{X}' = \{x'_1, \ldots, x'_\ell\}$ where $x'_i \leftarrow \{0,1\}^n$. We stress that $\mathcal{X}'$ is *independent of* $\mathcal{X}$. First, it is shown that

$$(\mathcal{X}, \widetilde{C}) \overset{c}{\approx} (\mathcal{X}', \widetilde{C})$$

by pseudorandomness of F and the VBB property of $\mathcal{O}$. What adversaries can do is only observing the input-output behavior of $\widetilde{C}$ since C'_{F} is completely hidden due to the VBB. However, outputs for $\mathcal{X}$ and $\mathcal{X}'$ are indistinguishable due to pseudorandomness. Next, the indistinguishability above is used to show unremovability. If the adversary $\mathcal{A}$ succeeds in removing the mark in $\widetilde{C}$ and outputs un-marked C^*, then it should hold that $C^*(x_i) \neq \widetilde{C}(x_i)$ for all $i \in [\ell]$. On the other hand, for $\mathcal{X}' := \{x'_1, \ldots, x'_\ell\}$ where $x'_i \leftarrow \{0,1\}^n$ for all $i \in [\ell]$, it holds that $C^*(x'_i) = \widetilde{C}(x'_i)$ with probability ϵ due to ϵ-approximation factor and the independence of $\mathcal{X}$. That is, the probability that there exists $i \in [\ell]$ such that $C^*(x'_i) = \widetilde{C}(x'_i)$ is $1 - (1 - \epsilon)^\ell$, which is overwhelming if $\ell := \Omega(\lambda/\epsilon)$. Note that $\mathcal{X}'$ is not the marked points of $\widetilde{C}$. In summary,

with non-negligible prob. $C^*(x_i) \neq \widetilde{C}(x_i)$ for all $x_i \in \mathcal{X}$

with overwhelming prob. $C^*(x'_i) = \widetilde{C}(x'_i)$ for some $x'_i \in \mathcal{X}'$.

From this fact, it is easy to see that the adversary of ϵ-unremovability is used to distinguish $(\mathcal{X}, \widetilde{C})$ from $(\mathcal{X}', \widetilde{C})$. Thus, it is difficult to remove the mark.

Attacks on Toy Example. The toy example explained above is ϵ-unremovable in the setting where $\mathcal{A}$ is given neither $\mathcal{MO}$ nor a public extraction key (or the extraction oracle) in Definition 12. However, the toy example is insecure if $\mathcal{A}$ is given oracle access to either $\mathcal{MO}$ or $\mathcal{XO}$. Attack procedures are given below.

Attack by extraction oracle: When $\mathcal{A}$ is given $\widetilde{C}$, it generates another circuit $\widetilde{C}_P$ by using some predicate P. For input x, $\widetilde{C}_P$ outputs $\widetilde{C}(x)$ if $P(x) = 1$; otherwise, $\perp$. If $\mathcal{A}$ queries $\widetilde{C}_P$ to $\mathcal{XO}$ for various predicates P (such as $P(x) = 1$ if $x = 0\|\star$, which is all strings whose most significant bit (MSB) is 0), then it can recover marked points $\mathcal{X}$ and remove the mark in $\widetilde{C}$. More concretely, let a predicate P_s as follows. For some string $s \in \{0,1\}^{\leq |x|}$, $P_s(x) = 1$ if $x = s\|\star$; otherwise $\perp$. When $\mathcal{A}$ queries $\widetilde{C}_{P_1}$, if the extraction oracle returns marked, then it means there exists a marked point whose MSB is 1. The adversary $\mathcal{A}$ can try various predicates P_s for $s = 1\|\star$, $s = 0\|\star$, $s = 10\|\star$, $s = 01\|\star$, and so on and conduct the standard binary search. The adversary $\mathcal{A}$ can find all marked points in polynomial time since $|\mathcal{X}| = \text{poly}(\lambda)$. This attack crucially relies on the fact that the size of $\mathcal{X}$ is polynomial.

Attack by mark oracle: First, $\mathcal{A}$ queries a circuit $C_{F'}$ that computes another PRF F' to $\mathcal{MO}$ and receives a marked circuit $\widetilde{C}'$. When $\mathcal{A}$ is given a target $\widetilde{C}$, it generates another circuit $\widetilde{C}^*$ by using $\widetilde{C}$ and $\widetilde{C}'$ as follows. For input x, $\widetilde{C}^*$ outputs $\widetilde{C}(x)$ if $\widetilde{C}(x) \neq \widetilde{C}'(x)$; otherwise, $\perp$. It is easy to verify that $\text{Extract}(xk, \widetilde{C}^*) = \text{unmarked}$ since for marked points in $\mathcal{X}$ it holds that $\widetilde{C}(x) = \widetilde{C}'(x)$ and $\widetilde{C}^*(x)$ outputs $\perp$. The point is that all marked circuits use the same *fixed* $\mathcal{X}$. The hybrid circuit $\widetilde{C}^*$ also satisfies ϵ-approximation since $\mathcal{X}$ is polynomial-size. This attack works because marked points are static (all marked circuits have the same marked points $\mathcal{X}$).

Immunizing Toy Construction Against Attacks. To prevent extraction oracle attacks, we make the set of marked points $\mathcal{X} \subseteq \{0,1\}^n$ super-polynomially large, but still negligible fraction of the entire domain. The binary search by prefix predicates no longer works. Moreover, if we can set pseudorandom ciphertexts under some public-key of encryption as marked points, then we can achieve public extraction. A public extraction key enables us to sample valid ciphertext $x \leftarrow \mathcal{X}$, which is indistinguishable from uniformly random points.

To prevent mark oracle attacks, we make the set of marked points $\mathcal{X}_{F_i}$ depend on PRF F_i and different for each F_i. However, this idea yields another issue since the extraction procedure does not have the original PRF F when it is given marked circuit $\widetilde{C}$. To solve this issue, a two-step checking is used. Let z be a random point, which we call "find point". For this point, we compute $\widetilde{C}(z)$ and use the output for extraction. It is expected that $\widetilde{C}(z) = F(z)$ since z is random and unlikely a marked point. We can use $\widetilde{C}(z) = F(z)$ as a marked point. If the equation holds, the marked point depends on F.

6.3 Idealized Construction

In this section, we introduce an idealized watermarking scheme for PRF which is secure under the presence of the mark oracle and a public extraction key. The construction idea is based on the observation in the previous section.

Let $\mathsf{PKE} = (\mathsf{PKE.Gen}, \mathsf{PKE.Enc}, \mathsf{PKE.Dec})$ be a public-key encryption (PKE) that is secure against adaptive chosen ciphertext attacks (CCA-secure) and has pseudorandom ciphertext.

$\mathsf{Gen}(1^\lambda)$: generates a key pair $(\mathsf{pk}, \mathsf{sk}) \leftarrow \mathsf{PKE.Gen}(1^\lambda)$ and sets $(\mathsf{xk}, \mathsf{mk}) := (\mathsf{pk}, \mathsf{sk})$.

$\mathsf{Mark}(\mathsf{mk}, F)$: generates a circuit F_{sk}, as shown in Fig. 1, computes an obfuscated circuit $\widetilde{C} := \mathcal{O}(F_{\mathsf{sk}})$, and outputs $\widetilde{C}$ as a marked circuit. This defines the set of marked points $\mathcal{X}_F$ as follows.

$$\mathcal{X}_F := \left\{ x \in \{0,1\}^n \mid \mathsf{PKE.Dec}(\mathsf{sk}, x) = a\|b\|c \in \{0,1\}^{3\lambda} \wedge \mathsf{F}(\mathsf{PRG}(a)) = b \right\}.$$

Pre-marked Circuit F_{sk}

Hardwired: PRF F, decryption key sk
Input: $x \in \{0,1\}^n$

1. Compute $a\|b\|c \leftarrow \mathsf{PKE.Dec}(\mathsf{sk}, x)$ with $a, b, c \in \{0,1\}^\lambda$.
2. If $\mathsf{F}(\mathsf{PRG}(a)) = b$, then output c.
3. Else output $\mathsf{F}(x)$.

Fig. 1. Description of F_{sk}

$\mathsf{Extract}(\mathsf{xk}, C')$: repeats the following ℓ times
 – Choose $a, c \leftarrow \{0,1\}^\lambda$ and set $z := \mathsf{PRG}(a)$ and $b := C'(z)$.
 – Choose $x \leftarrow \mathsf{PKE.Enc}(\mathsf{pk}, a\|b\|c)$ and if $C'(x) = c$, then output marked.
If marked is not output in all iterations, then output unmarked.

A marked circuit can appropriately check that an input x is made from a marked point since sk is hard-wired in the marked circuit. Obfuscator $\mathcal{O}$ hides the information about sk. Anyone that has pk can generate an input that consists of an appropriate marked point if $b = C'(z) = \mathsf{F}(z)$ and $z = \mathsf{PRG}(a)$ hold. This condition holds with high probability since z is random due to PRG.

Intuition of Security. Let view be all variables that $\mathcal{A}$ sees in the ϵ-unremovability game. That it, view includes the public extraction key, the target marked circuit $\widetilde{C}$, and all marked circuits given by $\mathcal{MO}$. It holds that

$$(\mathsf{view}, z, x) \overset{c}{\approx} (\mathsf{view}, z', x'),$$

where $a, c \leftarrow \{0,1\}^\lambda, z := \mathsf{PRG}(a), b := F(z), x \leftarrow \mathsf{PKE.Enc}(\mathsf{pk}, a\|b\|c), z', x' \leftarrow \{0,1\}^n$. Roughly speaking, this holds by the pseudorandomness of PRG, ciphertext pseudo-randomness of the PKE, and the VBB property of $\mathcal{O}$. One notable difference from the toy example is that a simulator must simulate $\mathcal{MO}$ without the decryption key of the PKE. It is possible to simulate $\mathcal{MO}$ since the simulator has access to the decryption oracle of the PKE. The simulator heavily relies on the VBB property of $\mathcal{O}$, which enables the simulator to simulate an obfuscated circuit by using only black-box access to the circuit to be obfuscated. Again, we can use the indistinguishability above to show unremovability as in the analysis of the toy example. If there exists a successful adversary in the ϵ-unremovability game, then it holds that

$$C^*(x) \neq \widetilde{C}(x) \text{ for all } x \in \mathcal{X}_F \text{ with non-negligible probability.}$$

On the other hand, it holds that

$$C^*(x') = \widetilde{C}(x') \wedge C^*(z') = \widetilde{C}(z') \text{ for } x', z' \leftarrow \{0,1\}^n \text{ with probability at least } \epsilon^2.$$

Thus, when $\ell := \Omega(\lambda/\epsilon^2)$ pairs of random points are checked, the probability that the extraction algorithm does not output marked is at most $1 - (1 - \epsilon^2)^\ell$, which is negligible. Therefore, we can distinguish (view, z, x) from (view, z', x') by using the adversary of the ϵ-unremovability. This is a contradiction.

Removing VBB Obfuscation. A VBB obfuscation for all polynomial-sized circuits does not exist (in particular, an unobfuscatable PRF exists [2]). Instead of a VBB obfuscation, an iO is used in the real construction of Cohen et al. We can no longer use CCA-secure PKE since the security of an iO is game-based and we cannot simulate $\mathcal{MO}$ by using black-box access to the decryption oracle. Thus, to prove the security of the real construction, we need some kind of "non-black-box" version of a CCA-secure PKE. Cohen et al. introduce puncturable encryption (PE) as such a primitive.

6.4   Puncturable Encryption

Cohen et al. propose the notion of PE [6]. Roughly speaking, PE is a secure encryption even if adversaries are given a special "punctured" decryption key that can decrypt all ciphertexts except some target ciphertexts. This punctured key plays the role of the decryption oracle in a non-black-box way. Thus, PE is compatible with an iO.

Definition 18 (Puncturable Encryption Syntax). *A PE scheme* PE *for a message space* $\mathcal{M} = \{0,1\}^\ell$ *is a triple of PPT algorithms* (Gen, Puncture, Enc) *and a deterministic algorithm* Dec. *The space of ciphertexts will be* $\{0,1\}^n$ *where* $n = \mathrm{poly}(\lambda, \ell)$.

Key Generation: $(\mathsf{pk}, \mathsf{sk}) \leftarrow \mathsf{Gen}(1^\lambda)$ *takes the security parameter and outputs an encryption key* pk *and a decryption key* sk.

Puncturing: $\mathsf{sk}\{c_0, c_1\} \leftarrow \mathsf{Puncture}(\mathsf{sk}, c_0, c_1)$ *takes* sk *and a set* $\{c_0, c_1\} \subset \{0,1\}^n$. *Puncture outputs a "punctured" decryption key* $\mathsf{sk}\{c_0, c_1\}$.

Encryption: $c \leftarrow \mathsf{Enc}(\mathsf{pk}, m)$ *takes* pk *and a message* $m \in \{0,1\}^\ell$ *and outputs a ciphertext* c *in* $\{0,1\}^n$.

Decryption: m *or* $\bot \leftarrow \mathsf{Dec}(\mathsf{sk}', c')$ *takes a possibly punctured decryption key* sk' *and a string* $c' \in \{0,1\}^n$. *It outputs a message* m *or the special symbol* $\bot$.

Definition 19 (Puncturable Encryption Security). *A PE scheme* $\mathsf{PE} = (\mathsf{Gen}, \mathsf{Puncture}, \mathsf{Enc}, \mathsf{Dec})$ *with message space* $\mathcal{M}$ *is required to satisfy the following properties.*

Correctness: *We require that for all messages* m *and* $(\mathsf{pk}, \mathsf{sk}) \leftarrow \mathsf{Gen}(1^\lambda)$, *it holds that* $\mathsf{Dec}(\mathsf{sk}, \mathsf{Enc}(\mathsf{pk}, m)) = m$.

Punctured Correctness: *We require the same to hold for punctured keys. For all possible keys* $(\mathsf{pk}, \mathsf{sk}) \leftarrow \mathsf{Gen}(1^\lambda)$, *all strings* $c_0, c_1 \in \{0,1\}^n$, *all punctured keys* $\mathsf{sk}' \leftarrow \mathsf{Puncture}(\mathsf{sk}, c_0, c_1)$, *and all potential ciphertexts* $c \in \{0,1\}^n \setminus \{c_0, c_1\}$:

$$\mathsf{Dec}(\mathsf{sk}, c) = \mathsf{Dec}(\mathsf{sk}', c).$$

Ciphertext Pseudorandomness: *We define the following experiment* $\mathsf{Expt}_{\mathcal{A}}^{\mathsf{prc}}(\lambda)$.

1. $\mathcal{A}$ *sends a message* m^* *to the challenger.*
2. *The challenger does the following:*
   - *Samples* $(\mathsf{pk}, \mathsf{sk}) \leftarrow \mathsf{Gen}(1^\lambda)$
   - *Computes encryption* $c^* \leftarrow \mathsf{Enc}(\mathsf{pk}, m^*)$.
   - *Samples* $r^* \leftarrow \{0,1\}^n$.
   - *Generates the punctured key* $\mathsf{sk}' \leftarrow \mathsf{Puncture}(\mathsf{sk}, \{c^*, r^*\})$
   - *Samples* $b \leftarrow \{0,1\}$ *and sends the following to* $\mathcal{A}$:

$$(c^*, r^*, \mathsf{pk}, \mathsf{sk}') \text{ if } b = 0$$
$$(r^*, c^*, \mathsf{pk}, \mathsf{sk}') \text{ if } b = 1$$

3. $\mathcal{A}$ *outputs* b' *and the experiment outputs* 1 *if* $b = b'$; *otherwise* 0.

We say that PE *has ciphertext pseudorandomness if for every PPT adversary* $\mathcal{A}$,

$$\mathsf{Adv}_{\mathcal{A}}^{\mathsf{prc}} := |2 \cdot \Pr[\mathsf{Expt}_{\mathcal{A}}^{\mathsf{prc}} = 1] - 1| \leq \mathsf{negl}(\lambda).$$

Sparseness: *We also require that most strings are not valid ciphertexts:*

$$\Pr\left[\mathsf{Dec}(\mathsf{sk}, c) \neq \bot \mid (\mathsf{pk}, \mathsf{sk}) \leftarrow \mathsf{Gen}(1^\lambda), c \leftarrow \{0,1\}^n\right] \leq \mathsf{negl}(\lambda).$$

Theorem 20 [6]. *There exists PE if there exists an injective one-way function and iO for* P/poly.

6.5   Real Construction

In this section, a slightly simplified construction is introduced for ease of understanding. The construction of Cohen et al. can embed arbitrary length messages. See the paper of Cohen et al. [6] for full details.

Setup. The watermarking scheme by Cohen et al. works for a pPRF family $\mathcal{C}$ with domain $\{0,1\}^n$ and range $\{0,1\}^m$. Let $\mathcal{M} = \{0,1\}^m$ denote the message space. Let PE be a puncturable encryption scheme with ciphertext length n and plaintext length ℓ. Let $\mathsf{PRG} : \{0,1\}^{\ell/3} \to \{0,1\}^n$ and $\mathsf{PRG}' : \{0,1\}^{\ell/3} \to \{0,1\}^m$ be PRGs, and let $H : \{0,1\}^m \to \{0,1\}^{\ell/3}$ be a UOWHF.

Construction. For any approximation factor $\epsilon(\lambda) = \frac{1}{2} + \rho(\lambda)$ where $\rho(\lambda) = 1/\text{poly}(\lambda)$, let $Q := Q(\lambda) := \lambda/\rho(\lambda)^2$ and $R := R(\lambda) := \lambda/\rho(\lambda)^2$. The construction is as follows.

$\mathsf{Gen}(1^\lambda)$: Sample a key pair $(\mathsf{pk}, \mathsf{sk}) \leftarrow \mathsf{PE.Gen}(1^\lambda)$. Output $(\mathsf{xk}, \mathsf{mk})$ where $\mathsf{xk} = \mathsf{pk}$ and $\mathsf{mk} = \mathsf{sk}$.

$\mathsf{Mark}(\mathsf{mk}, C, \mathsf{m})$: Output an obfuscated circuit of circuit M constructed from C in Fig. 2, i.e., $i\mathcal{O}(M)$.

Constants: PE decryption key sk, circuit C, and message m
Inputs: $x \in \{0,1\}^n$

1. Try to parse $a\|b\|c \leftarrow \mathsf{PE.Dec}(\mathsf{sk}, x)$, where $|a| = |b| = |c| = \ell/3$.
2. If $a\|b\|c \neq \bot$ and $H(C(\mathsf{PRG}(a))) = b$, output $\mathsf{PRG}'(c) \oplus \mathsf{m}$.
3. Otherwise, output $C(x)$.

Fig. 2. Description of M, which is modification of C (pre-obfuscated program)

$\mathsf{Extract}(\mathsf{xk}, C')$: Let $\mathsf{m} = \mathsf{Extract}(\mathsf{xk}, C')$, where Extract is defined in Fig. 3. Extract makes use of a subroutine WeakExtract, which is defined in Fig. 4. Output m.

$\mathsf{Extract}(\mathsf{xk}, C')$:

1. For $j = 1, \ldots, Q$,
    (a) Sample uniformly random $a_j \leftarrow \{0,1\}^{\ell/3}$.
    (b) Compute $b_j = H(C'(\mathsf{PRG}(a_j)))$
    (c) Run $\mathsf{m}^{(j)} \leftarrow \mathsf{WeakExtract}(\mathsf{xk}, C', a_j, b_j)$
2. If there exists a "majority-of-majorities message" $\mathsf{m} \neq \bot$ such that $|\{j : \mathsf{m}^{(j)} = \mathsf{m}\}| > Q/2$, then output m; else output unmarked.

Fig. 3. Sub-routine algorithm $\mathsf{Extract}(\mathsf{xk}, C')$

Lunch-Time Security. Cohen et al. show the following theorem.

Theorem 21 [6]. *The real construction is $(\frac{1}{2} + \frac{1}{\text{poly}(\lambda)})$-secure against lunch-time attacks if there exists an injective one-way function and iO for P/poly.*

WeakExtract(xk, C', a, b):

1. For $k = 1, \ldots, R$,
   (a) Sample $c_k \leftarrow \{0,1\}^{\ell/3}$ and $x_k \leftarrow$ PE.Enc(pk, $a\|b\|c_k$).
   (b) Compute $\mathsf{m}^{(k)} = \mathsf{PRG}'(c_k) \oplus C'(x_k)$.
2. Define the "majority message" m such that $|\{k : \mathsf{m}^{(k)} = \mathsf{m}\}| > R/2$ if such a m exists; otherwise, define $\mathsf{m} = \bot$.

Fig. 4. Sub-routine algorithm WeakExtract(xk, C', a, b)

The formal security proof is not given here. An intuition of security proof is already given in Sect. 6.3. A PE scheme is used instead of a CCA-secure PKE. We can use the punctured programming technique [17] and hide information about M by iO since a pPRF and PE scheme are iO-friendly primitives. An UOWHF is also used to prevent adversaries querying a circuit that yields a collision with $\mathcal{MO}$. Note that the construction is secure against lunch-time attacks because adversaries might find a collision after it is given the target circuit. Even if a collision-resistant hash function is used, the full security in Definition 12 cannot be achieved because after the target circuit $\widetilde{C}$ is given, the adversary can easily generate a circuit C^* that outputs $\widetilde{C}(\mathsf{PRG}(a_i))$ for input $\mathsf{PRG}(a_i)$ where $a_i \leftarrow \{0,1\}^{\ell/3}$. See the paper of Cohen et al. [6] for details.

Message Embedding. The ideal construction in Sect. 6.3 is a message-less scheme. The extract algorithm of the real construction is more complicated than that of the ideal construction due to the message extraction. How to embed a message m is simple. The mark algorithm just outputs $\mathsf{PRG}'(c) \oplus \mathsf{m}$ instead of $\mathsf{PRG}'(c)$. However, to achieve approximation factor $\epsilon = 1/2 + \rho(\lambda)$, the following naive extraction algorithm does not work. A naive algorithm chooses a_i and c_i at the same time and computes $\mathsf{m}^{(i)} = \mathsf{PRG}'(c_i) \oplus C'(x_i)$ where $x_i \leftarrow$ PE.Enc(pk, $a_i\|b_i\|c_i$) and $b_i = H(C'(\mathsf{PRG}(a_i)))$. Consider C' ϵ-approximates C. The naive algorithm outputs a correct message when $C'(x_i)$ and $C'(\mathsf{PRG}(a_i))$ are equal to $C(x_i)$ and $C(\mathsf{PRG}(a_i))$, respectively. The probability is ϵ^2. Thus, simply taking majority works when $\epsilon \geq 1/\sqrt{2} + \rho(\lambda)$ holds[4]. To overcome this issue, the extraction algorithm of Cohen et al. uses "majority-of-majority" as in Figs. 3 and 4. Note that Cohen et al. also show that the approximation factor $\epsilon = 1/2 + \rho(\lambda)$ is optimal [6].

Related Work. Subsequent to the work of Cohen et al., Boneh et al. propose the notion of private programmable PRF and show that a symmetric-key watermarking for pPRF can be constructed from a private programmable PRF, which can be constructed from an iO [3]. However, the security of their watermarking scheme is weaker than that of Cohen et al. scheme.

---

[4] Otherwise Chernoff bound does not work.

7 Open Questions

There are several open questions about program watermarking.

Achieving full security: The construction of Cohen et al. achieves the security against lunch-time attacks (not the full security in the sense of Definition 12). In fact, if adversaries query only pPRFs (not arbitrary circuits) to $\mathcal{MO}$ and the pPRFs are "key injective"[5], then the construction of Cohen et al. satisfies the full security. Thus, it is an open question to achieve the full security without such a restriction and additional assumptions.

More functionalities: Cohen et al. show that it is possible to achieve watermarking cryptographic capabilities such as pPRF, decryption of PKE, and signing of digital signature [6]. Cohen et al. also show that it is impossible to achieve watermarking *learnable* functions [6] as we see in Sect. 6.1. It is natural to ask whether it is possible to watermark other useful functionalities.

Achieving watermarking without iO: The connection between program watermarking and program obfuscation has not been well studied. In particular, achieving program watermarking without iO is a major concern.

Acknowledgments. The author would like to thank Pooya Farshim for invaluable and constructive comments.

References

1. Barak, B., Goldreich, O., Impagliazzo, R., Rudich, S., Sahai, A., Vadhan, S., Yang, K.: On the (im)possibility of obfuscating programs. In: Kilian, J. (ed.) CRYPTO 2001. LNCS, vol. 2139, pp. 1–18. Springer, Heidelberg (2001). doi:10.1007/3-540-44647-8_1
2. Barak, B., Goldreich, O., Impagliazzo, R., Rudich, S., Sahai, A., Vadhan, S.P., Yang, K.: On the (im)possibility of obfuscating programs. J. ACM **59**(2), 6 (2012)
3. Boneh, D., Lewi, K., Wu, D.J.: Constraining pseudorandom functions privately. Cryptology ePrint Archive, Report 2015/1167 (2015). http://eprint.iacr.org/2015/1167
4. Boneh, D., Waters, B.: Constrained pseudorandom functions and their applications. In: Sako, K., Sarkar, P. (eds.) ASIACRYPT 2013. LNCS, vol. 8270, pp. 280–300. Springer, Heidelberg (2013). doi:10.1007/978-3-642-42045-0_15
5. Boyle, E., Goldwasser, S., Ivan, I.: Functional signatures and pseudorandom functions. In: Krawczyk, H. (ed.) PKC 2014. LNCS, vol. 8383, pp. 501–519. Springer, Heidelberg (2014). doi:10.1007/978-3-642-54631-0_29
6. Cohen, A., Holmgren, J., Nishimaki, R., Vaikuntanathan, V., Wichs, D.: Watermarking cryptographic capabilities. In: Wichs, D., Mansour, Y. (eds.) 48th ACM STOC, pp. 1115–1127. ACM Press, New York (2016)
7. Galbraith, S.D., Verheul, E.R.: An analysis of the vector decomposition problem. In: Cramer, R. (ed.) PKC 2008. LNCS, vol. 4939, pp. 308–327. Springer, Heidelberg (2008). doi:10.1007/978-3-540-78440-1_18

---

[5] Such pPRFs are constructed from DDH or LWE assumptions [6].

8. Garg, S., Gentry, C., Halevi, S., Raykova, M., Sahai, A., Waters, B.: Candidate indistinguishability obfuscation and functional encryption for all circuits. In: 54th FOCS, pp. 40–49. IEEE Computer Society Press, October 2013
9. Goldreich, O., Goldwasser, S., Micali, S.: How to construct random functions. J. ACM **33**(4), 792–807 (1986)
10. Hopper, N., Molnar, D., Wagner, D.: From weak to strong watermarking. In: Vadhan, S.P. (ed.) TCC 2007. LNCS, vol. 4392, pp. 362–382. Springer, Heidelberg (2007). doi:10.1007/978-3-540-70936-7_20
11. Kiayias, A., Papadopoulos, S., Triandopoulos, N., Zacharias, T.: Delegatable pseudorandom functions and applications. In: Sadeghi, A.-R., Gligor, V.D., Yung, M. (eds.) ACM CCS 2013, pp. 669–684. ACM Press, New York (2013)
12. Lewko, A.: Tools for simulating features of composite order bilinear groups in the prime order setting. In: Pointcheval, D., Johansson, T. (eds.) EUROCRYPT 2012. LNCS, vol. 7237, pp. 318–335. Springer, Heidelberg (2012). doi:10.1007/978-3-642-29011-4_20
13. Lin, H., Vaikuntanathan, V.: Indistinguishability obfuscation from DDH-like assumptions on constant-degree graded encodings. In: Dinur, I. (ed.) 57th FOCS, pp. 11–20. IEEE Computer Society Press, Washington, D.C. (2016)
14. Nishimaki, R.: How to watermark cryptographic functions. In: Johansson, T., Nguyen, P.Q. (eds.) EUROCRYPT 2013. LNCS, vol. 7881, pp. 111–125. Springer, Heidelberg (2013). doi:10.1007/978-3-642-38348-9_7
15. Okamoto, T., Takashima, K.: Homomorphic encryption and signatures from vector decomposition. In: Galbraith, S.D., Paterson, K.G. (eds.) Pairing 2008. LNCS, vol. 5209, pp. 57–74. Springer, Heidelberg (2008). doi:10.1007/978-3-540-85538-5_4
16. Okamoto, T., Takashima, K.: Fully secure functional encryption with general relations from the decisional linear assumption. In: Rabin, T. (ed.) CRYPTO 2010. LNCS, vol. 6223, pp. 191–208. Springer, Heidelberg (2010). doi:10.1007/978-3-642-14623-7_11
17. Sahai, A., Waters, B.: How to use indistinguishability obfuscation: deniable encryption, and more. In: Shmoys, D.B. (ed.) 46th ACM STOC, pp. 475–484. ACM Press, New York (2014)
18. Waters, B.: Dual system encryption: realizing fully secure IBE and HIBE under simple assumptions. In: Halevi, S. (ed.) CRYPTO 2009. LNCS, vol. 5677, pp. 619–636. Springer, Heidelberg (2009). doi:10.1007/978-3-642-03356-8_36
19. Yoshida, M., Fujiwara, T.: Toward digital watermarking for cryptographic data. IEICE Trans. **94-A**(1), 270–272 (2011)
20. Yoshida, M., Mitsunari, S., Fujiwara, T.: The vector decomposition problem. IEICE Trans. **93-A**(1), 188–193 (2010)

From Higher-Order Differentials to Polytopic Cryptyanalysis

Tyge Tiessen[✉]

DTU Compute, Technical University of Denmark,
Kgs. Lyngby, Denmark
tyti@dtu.dk

Abstract. Polytopic cryptanalysis was introduced at EUROCRYPT 2016 as a cryptanalytic technique for low-data-complexity attacks on block ciphers. In this paper, we give an account of how the technique was developed, quickly go over the basic ideas and techniques of polytopic cryptanalysis, look into how the technique differs from previously existing cryptographic techniques, and discuss whether the attack angle can be useful for developing improved cryptanalytic techniques.

1 Introduction

A few years after differential cryptanalysis [2] had been developed and successfully applied to a range of ciphers, Xuejia Lai realized that differential cryptanalysis can be generalized to higher-order differential cryptanalysis using the concept of higher-order derivatives [7]. He was not able though to give an example that demonstrates that higher-order differential attacks can be stronger than standard differential attacks. Such an example was given shortly thereafter by Knudsen [5] who broke a design proven to be secure against differential cryptanalysis.

Since and including that attack, successful applications of higher-order differential cryptanalysis have been relying on deterministic higher-order differentials, i.e., higher-order differentials of probability one. Those attacks can be put into two main categories. In the first category, upper bounds on the degree of the cipher are derived which can then be used to find higher-order derivatives that evaluate to zero. In the second category, methods of integral cryptanalysis are used to determine that for some combination of input bits, there are no terms in the polynomial representation of the output that contain all of those bits simultaneously. Such property again results in higher-order derivatives that evaluate to zero simply by taking the derivative with respect to those input bits.

The problem of working with probabilistic higher-order differentials seems to be the difficulty to estimate the probability of probabilistic differentials efficiently. While it is usually easy to derive the probability of a higher-order differential over one round, there is no straightforward method to iterate these probabilities to estimate the probability of a multiple-round higher-order differential. In particular the concept of a trail (or characteristic) does not seem to exist for higher-order differentials.

© Springer International Publishing AG 2017
R.C.-W. Phan and M. Yung (Eds.): LNCS 10311, Mycrypt 2016, pp. 544–552, 2017.
DOI: 10.1007/978-3-319-61273-7_29

In this paper, we demonstrate how attempting to construct trails for higher-order differentials leads to polytopic trails in a natural progression. We then quickly draw an outline of how the polytopic framework is a direct generalization of the differential cryptanalysis framework that includes higher-order differential cryptanalysis as a special case.

We subsequently show that attempts to apply this framework in the setting that corresponds to standard differential cryptanalysis cannot yield better results than standard differential cryptanalysis due to a large increase in the number of trails that need to be considered. We continue then by demonstrating that the upside to the increase in the number of trails is that impossible polytopic attacks can be successful where standard impossible differential attacks fail. By following this progression, we follow the development process that lead to the low-data attacks detailed in the EUROCRYPT 2016 paper [9].

The paper concludes with a comparison of polytopic cryptanalysis with existing cryptanalytic techniques, a discussion of future research directions and open problems.

2 Difficulties of Higher-Order Differentials

In differential cryptanalysis, the goal is to find a strong correlation between the difference of two input messages to a cipher and the difference of the respective output messages of the cipher. When we let E be the cipher and α and β be the input and output differences, we are hoping to find many x such that $E(x + \alpha) + E(x) = \beta$. We denote the difference operation here by a $+$ sign as we assume we are working with binary values where addition and subtraction correspond to the same operation.[1] A differential is then defined as a pair of an input difference α and an output difference β and the associated probability $\mathrm{Pr}_{\mathbf{X}}\left(E(\mathbf{X} + \alpha) + E(\mathbf{X}) = \beta\right)$ where $\mathbf{X}$ is uniformly distributed over the messages.

Lai [7] realized that by writing the output difference as a derivative, we arrive at a formalism that leads to higher-order differentials. First we define the derivative of E at x in the direction α as:

$$\Delta_\alpha E(x) := E(x + \alpha) + E(x).$$

This discrete derivative shares many of the properties of continuous derivatives over the real numbers: both are additive, commutative, and reduce the degree of the function they are applied to by at least 1. The discrete derivative furthermore features a variant of the product rule (see [7] for details).

Using this derivative, we can now define a higher-order derivative as a concatenation of discrete derivatives as defined above and evaluate it as follows:

$$\Delta_{\alpha_1,\ldots,\alpha_d} E(x) = \sum_{\alpha \in \mathcal{L}(\alpha_1,\ldots,\alpha_d)} E(x + \alpha)$$

---

[1] To be precise, we assume here that $\alpha, x \in \mathbb{F}_2^n$, $\beta \in \mathbb{F}_2^m$, and $E : \mathbb{F}_2^n \to \mathbb{F}_2^m$.

where $\mathcal{L}(\alpha_1, \ldots, \alpha_d)$ is the linear space spanned by $\alpha_1, \ldots, \alpha_d$. In this definition it is assumed we have the more interesting case where $\alpha_1, \ldots, \alpha_d$ are linearly independent as the higher-order derivative always evaluates to zero if they are not.

Similarly to the simple derivative used in standard differential cryptanalysis, we can try to find input differences and output differences that show a strong correlation in the higher-order derivative, i.e., we can try to find a set of input differences $\alpha_1, \ldots, \alpha_d$ and an output difference β such that $\Delta_{\alpha_1, \ldots, \alpha_d} E(x) = \beta$ for as many x as possible. A higher-order differential of order d is then defined as a set of input differences $\alpha_1, \ldots, \alpha_d$ and an output value β with the associated probability $\Pr_{\mathbf{X}}(\Delta_{\alpha_1, \ldots, \alpha_d} E(\mathbf{X}) = \beta)$ where $\mathbf{X}$ is again uniformly distributed over the messages.

There are two problems that we encounter when trying to find such combinations of input differences and output values, and these problems already exist in the case of standard differential cryptanalysis. Firstly, it would be computationally too expensive to evaluate the derivative at all possible values x. Secondly, in applications of differential cryptanalysis the function E is a cipher which is a function parametrized by a key unknown to the attacker, so it will not even be possible to take the exact derivative without knowledge of the key.

In standard differential cryptanalysis, both of these difficulties can be overcome by utilizing the following approach. The cipher is split into a sequence of rounds (luckily most ciphers are round-based anyhow) where each round by itself does not depend on the key and the key is only applied in between rounds. For each round it is then usually easy to determine the probability that an input pair with a given difference is mapped to a given output difference. The trick is now to approximate the probability that an input pair follows a trail of intermediate differences as the product of the probabilities of each round transition. By summing these approximation for all possible trails, we arrive at an approximation for the probability that the input difference is mapped to the output difference irregardless of the differences taken in between rounds, the probability of the differential.

While this approach is an approximation at best and can fail in certain circumstances (see for example [4]), it has proven extremely useful in the practical cryptanalysis of ciphers. It allows us often to efficiently approximate the probability of differentials (i.e., the probability that a given input difference is mapped to a given output difference) by summing only those trails that contribute significantly to its probability. For a discussion of the theoretical framework for this, we refer to [8].

Why does this approach fail when we try to apply it to higher-order differentials? Unlike a standard differential which has exactly one input difference and one output difference, a higher-order differential is taken with respect to a basis of differences (or vectors). But its output is a single value: the sum of the output messages. If we now try to iterate such a higher-order differential, we run into the problem of identifying the output value of one rounds with the set of input differences for the next rounds, or more precisely with the problem of identifying it with the space that is spanned by those basis vectors.

A first guess how to associate the output value with an input space might be to assume it is uniformly randomly chosen from all spaces of the right dimension that sum to the value. But clearly this does not work as all spaces sum to zero, so we cannot associate non-zero output values with a linear (or affine) space.

Ignoring the fact that the messages are summed at the end of a higher-order derivative and trying to work with affine subspaces as a good representation of intermediate states fails similarly: the messages will generally be in arbitrary position to each other after the first round. Thus to adequately represent the intermediate states of a higher-order differential we need to allow the messages to be in arbitrary position to each other.

Before we look into constructing trails where the states are sets of messages in arbitrary position (which is exactly what polytopic trails are), let us mention the two methods which are used to successfully evaluate higher-order differentials despite the lack of trails for them. The first one is using degree arguments. As a higher-order derivative of order d reduces the algebraic degree of function by at least d, we can determine that certain higher-order derivatives have to evaluate to zero by finding sufficient lower bounds for the degree of the cipher. The other method uses structural properties of the cipher to determine that certain terms will not be present in the algebraic representation of the cipher which again determines some higher-order derivatives to evaluate to zero (see for example [6, 10]).

3   Overview over the Polytopic Framework

To be able to accurately describe the set of values in between the rounds of the cipher which we are taking the derivative over we need to describe them as what they are, tuples of points in the state space. We will call such a tuple a polytope.

In this way, we can describe the higher-order derivative as a polytope that consists of the values in the input space which is then transformed to other polytopes while traversing the rounds and finally reduced from a polytope to a single value by summing all values contained in the polytope.

We can reduce the information that we need to describe polytopes in this usage scenario by disregarding the absolute position of the messages in the state space and only caring about their relative positions. Thus we will regard two tuples of messages that can be translated to each other by shifting in the state space as equivalent. The relative positions of the messages to each other are entirely determined by picking one message as an anchor and specifying the position of all other messages in the polytope with respect to this anchor message. We can thus reduce the number of values needed to describe the polytope by one.

Such a description of a polytope is called a d-difference where d specifies the number of differences needed to specify all relative positions, i.e., d is one lower than the number of values in the polytope. Thus for example for a pair of values, we only need one difference to describe the pair when disregarding the absolute position in state space as we know well from standard differential cryptanalysis. For a set of four messages we would then need a tuple of 3 differences, the differences of message 2, 3 and 4 with regard to the first message.

When we want to know the probability that a polytope with a given input d-difference is mapped to a polytope with a given output d-difference, we encounter the same two problems that we had in standard differential cryptanalysis. Luckily now, we can apply the same methodology to counter this. The probability of a trail of d-differences is determined as the product of the round transition probabilities, the probability of a transition over the whole cipher with fixed input and output d-differences is determined as the sum of all trails that have exactly those input and output d-differences. The analogy to a differential in standard differential cryptanalysis is a polytopic transition where we fix only the input and output d-differences but not the intermediate ones. For a rigorous description of this framework, we refer to the EUROCRYPT paper.

Let us go back to the problem of finding trails for higher-order differentials. The input to a higher-order differential uniquely corresponds to a d-difference. The output does not correspond to a single d-difference but to the set of all d-differences that sum to this output value. We can thus now describe the probability of the higher-order differential as the sum of all polytopic transitions where the input d-difference corresponds to the input to the higher-order differential and where the output d-difference sums to the output value of the higher-order differential. We can thus principally determine the probability of a higher-order differential using the same methodology that we use for standard differentials: we sum the probability of all trails that correspond to the transitions.

Can this approach be used to successfully determine the probability of higher-order differentials in practice? Unfortunately, as it turns out, this does generally not seem to be the case. The underlying problem is that the probability of polytopic trails is usually much lower than the probability of trails in differential cryptanalysis. The probabilities are so low that it is not possible to determine a good lower bound for the probability of a polytopic transition by simply summing the probabilities of trails. And thus it is not possible to practically determine a lower bound for the probability of higher-order differentials in typical cryptanalytic cases.

The fact that the probabilities of polytopic trails are so low not only makes determination of the probabilities of higher-order differentials impractical, it also make polytopic transitions uninteresting as a substitute for standard differential attacks: for any polytopic transition there exists a differential of at least the same probability. While this restricts usage of polytopic cryptanalysis in the standard setting, we will see in the next section that the framework nonetheless has practical applications.

4 Impossible Polytopic Transitions

The setting where polytopic transitions turn out to be useful is the setting where we consider transitions of probability zero: impossible transitions. The central property that determines the quality of an impossible transition attack is the ratio of impossible transitions to possible transitions, i.e., to transitions of probability strictly greater than one. While using polytopic transitions in the standard attack setting is not particularly useful due to the increased diffusion of the d-differences, in the impossible setting the diffusion is countered by an exponential increase in the total number of transitions causing the ratio of impossible to possible transitions to shift in the favor of the impossible transitions.

When we describe a polytope consisting of $d+1$ messages as a d-difference, we increase the state size, i.e., the number of possible d-differences by an exponent of d in comparison to the possible number of single differences. At the same time, the diffusion will increase by at most a constant factor. This allow us to choose the number d sufficiently large to ensure a favorable ratio of impossible to possible transitions: by increasing d we can make the ratio of possible to impossible transitions arbitrarily small.

While this principally would allow for excellent attacks, the problem now lies elsewhere: how can we efficiently tell possible from impossible transitions? The most obvious and straightforward way is to exhaustively compute a set of all reachable d-differences and use this set to distinguish possible from impossible transitions. To reduce both the memory and computational cost needed for this, a meet-in-the-middle approach can be employed. In this approach, two sets of reachable d-differences in the middle of the cipher are determined while coming from both ends of the cipher. When depending on whether or not a collision in these two sets is found, the transition is determined to be possible or impossible. This is the approach that has been used in the EUROCRYPT paper.

A more efficient way of determining the possibility of transitions is to use structural properties of the cipher to find large enough sets of impossible transitions. For standard impossible differential attacks a very successful method is the so-called miss-in-the-middle approach [1] which directly constructs a sufficiently large set of impossible differentials. An alternative approach could be to use structural properties of the cipher to determine a sufficiently small, efficiently testable super-set of all possible transitions, again indirectly giving us a large set of impossible transitions. As we demonstrated, higher-order differentials correspond to a particular collection of polytopic transitions. And indeed, those methods that use structural properties of the cipher to determine deterministic higher-order differentials can be equivalently seen as methods that efficiently determine properties of reachable output d-differences, thus specifying a super-set of all possible polytopic transitions. The standard implementation of higher-order differential attacks using structural properties of the cipher to determine non-probabilistic higher-order differentials can thus be seen as particular versions of impossible polytopic transition attacks. Apart from these attack types, it remains an open question though how structural properties of a cipher can determine such a superset when the polytopic transition does not correspond to a higher-order differential.

5   Comparison to Other Attack Vectors

There already exists a larger repertoire of attack vectors that can be wielded against ciphers. How does polyptopic cryptanalysis differ from those attack vectors? Is it just an existing attack vector in disguise?

A property that sets polytopic cryptanalysis apart from differential and linear cryptanalysis is an increased use of the correlation between different input messages and output messages. In linear cryptanalysis, correlation is only measured on the input message and the corresponding output message. In differential cryptanalysis, we measure only the correlation between two input messages and there corresponding outputs. In polytopic cryptanalysis though by considering the correlations between inputs and outputs of larger tuples, we make better use of the data that we use. This is what essentially allows the impossible polytopic attacks of [9] to have such a low data-complexity.

We already saw that polytopic cryptanalysis can be seen as an extension to the differential cryptanalysis framework sufficiently general to include higher-order differentials. To demonstrate how impossible polytopic attacks differ from differential attacks and other attack vectors, let us consider the following constructed toy cipher. Our cipher has an 8-bit block size and a round that consists of an application of the Rijndael S-box (see [3]) to the state, followed by the XOR-addition of a round key. All round keys are independent and before the first round, a whitening key is added to the input message.

Let us assume that the cipher has sixteen rounds and that our goal is a key recovery attack. Since the Rijndael S-box achieves excellent diffusion, differential and linear attack are easily be thwarted. This includes impossible differential attack as only two rounds are needed to achieve full diffusion and thus impossible differentials seize so exist after only the first few rounds. This is not true though for impossible polytopic transitions if we choose d sufficiently large.

For any d, after r rounds there are at most $2^{8 \cdot r}$ reachable d-differences out of a total of $2^{d \cdot 8}$ d-differences. Thus if we set d for example to 9, we can guarantee that after eight rounds, only a fraction of 1 over 256 d-differences is reachable. If we thus guess the last eight round keys and decrypt the last eight rounds, we can filter out 255 out of 256 key guesses using only a set of 10 texts. As we can precompute the list of reachable d-differences at a time cost of $2^{8 \cdot 8}$ and an equivalent memory cost, the total time complexity also corresponds to this value.

From the description of this attack, it is clear that the closest related attack is a meet-in-the-middle attack. And indeed impossible polytopic attacks can be seen as a meet-in-the-middle attack where the collision is not found on the state itself, but rather on the d-difference.

6   Discussion

We started out by finding a way to construct trails for higher-order differentials in the hope to be able to determine the probability of probabilistic higher-order differentials efficiently. While we succeeded with the former by constructing polytopic trails and transitions, we failed with the latter. Interestingly though the construction of the polytopic framework did not turn out to be a theoretical dead-end but proved useful when applied to probability zero transitions, impossible transitions.

The attacks that were enabled by this framework (as published in [9] were nonetheless of a somewhat restricted nature: low-data attack on few rounds. Whether or not polytopic cryptanalysis can be applied in a broader collection of attack scenarios depends on an number of open issues:

- Is it possible to use structural properties of ciphers to efficiently determine the possibility of a given polytopic transition? This excludes of course simple relabeling of the existing techniques of deterministic higher-order differentials. If such techniques exist they could be used to circumvent the restrictions of polytopic attacks to few rounds currently imposed by the strong diffusion and growth of the number of reachable d-differences.
- Are there attack scenarios where determining the probability of higher-order differentials using the representation as a collection of polytopic trails proves sufficiently efficient to be useful in an attack? Are there more effective methods of determining this probability that avoid iterating through the list of polytopic trails (potentially using structural properties)?
- Are there other efficient attack vectors that make use of the correlation between larger tuples of texts than pairs? Or are there alternatively strong theoretical arguments why efficient attacks are restricted to using single texts or pairs (such as in linear and differential cryptanalysis)?

We hope that this brief article might serve as an example of how sometimes formalizations that do not directly lead in the direction that one initially hopes for can still prove valuable when we switch the setting. Potentially some ideas may serve someone as an inspiration to derive improved, prospective attack vectors on symmetric ciphers.

References

1. Biham, E., Biryukov, A., Shamir, A.: Cryptanalysis of Skipjack reduced to 31 rounds using impossible differentials. J. Cryptol. **18**(4), 291–311 (2005)
2. Biham, E., Shamir, A.: Differential cryptanalysis of DES-like cryptosystems. J. Cryptol. **4**(1), 3–72 (1991)
3. Daemen, J., Rijmen, V.: The Design of Rijndael: AES - The Advanced Encryption Standard. Information Security and Cryptography. Springer, Heidelberg (2002)
4. Daemen, J., Rijmen, V.: Plateau characteristics. IET Inf. Secur. **1**(1), 11–17 (2007)
5. Knudsen, L.R.: Truncated and higher order differentials. In: Preneel, B. (ed.) FSE 1994. LNCS, vol. 1008, pp. 196–211. Springer, Heidelberg (1995). doi:10.1007/3-540-60590-8_16
6. Knudsen, L., Wagner, D.: Integral cryptanalysis. In: Daemen, J., Rijmen, V. (eds.) FSE 2002. LNCS, vol. 2365, pp. 112–127. Springer, Heidelberg (2002). doi:10.1007/3-540-45661-9_9
7. Lai, X.: Higher order derivatives and differential cryptanalysis. In: Blahut, R.E., Costello, D.J., Maurer, U., Mittelholzer, T. (eds.) Communications and Cryptography, Two Sides of One Tapestry, pp. 227–233. Kluwer Academic Publishers, Berlin (1994). doi:10.1007/978-1-4615-2694-0_23

8. Lai, X., Massey, J.L., Murphy, S.: Markov ciphers and differential cryptanalysis. In: Davies, D.W. (ed.) EUROCRYPT 1991. LNCS, vol. 547, pp. 17–38. Springer, Heidelberg (1991). doi:10.1007/3-540-46416-6_2
9. Tiessen, T.: Polytopic cryptanalysis. In: Fischlin, M., Coron, J.-S. (eds.) EUROCRYPT 2016. LNCS, vol. 9665, pp. 214–239. Springer, Heidelberg (2016). doi:10.1007/978-3-662-49890-3_9
10. Todo, Y.: Structural evaluation by generalized integral property. In: Oswald, E., Fischlin, M. (eds.) EUROCRYPT 2015. LNCS, vol. 9056, pp. 287–314. Springer, Heidelberg (2015). doi:10.1007/978-3-662-46800-5_12

Division Property: Efficient Method to Estimate Upper Bound of Algebraic Degree

Yosuke Todo[1,2]([✉])

[1] NTT Secure Platform Laboratories, Tokyo, Japan
`todo.yosuke@lab.ntt.co.jp`
[2] Kobe University, Kobe, Japan

Abstract. We proposed the division property, which is a new method to find integral characteristics, at EUROCRYPT2015. Then, we applied this technique to analyze the full MISTY1 at CRYPTO2015. After the proposal of the two papers, many follow-up results have been researched at major conferences. In this paper, we first expound the integral and higher-order differential cryptanalyses in detail and focus the similarities and differences. As a result, we conclude that both cryptanalyses are the same in practical. Nevertheless, both cryptanalyses use the different method to find characteristics: the propagation characteristic of integral properties is evaluated in the integral cryptanalysis and the upper bound of the algebraic degree is evaluated in the higher-order differential cryptanalysis. Our first discovery is that each of the two methods has its own advantages and disadvantages. Moreover, there are some experimental characteristics that cannot be proven by either of both methods. These observation causes significant motivation that we developed the division property. We next expound some important follow-up results, e.g., the bit-based division property at FSE2016, the parity set at CRYPTO2016, the MILP-based propagation search at ASIACRYPT2016.

Keywords: Division property · Integral cryptanalysis · Higher-order differential cryptanalysis

1 Introduction

Higher-Order Differential Cryptanalysis. After the proposal of the differential cryptanalysis [6], many extended cryptanalyses have been proposed like the impossible differential [5,20], integral [22], and meet-in-the-middle attacks [15]. The higher-order differential cryptanalysis is one of such extensions. The concept was first introduced by Lai [23] and the advantage over the conventional differential cryptanalysis was studied by Knudsen [21]. Assuming the algebraic degree of the target block cipher E_k is upper-bounded by d for all k, the dth order differential is always constant and the $(d+1)$th order differential is always zero. Then, we can distinguish the target cipher E_k as ideal block ciphers because it is unlikely that ideal block ciphers have this property, and we call this property the higher-order differential characteristics in this paper.

© Springer International Publishing AG 2017
R.C.-W. Phan and M. Yung (Eds.): LNCS 10311, Mycrypt 2016, pp. 553–571, 2017.
DOI: 10.1007/978-3-319-61273-7_30

Integral Cryptanalysis. The similar technique to the higher-order differential cryptanalysis was used as the dedicated attack against the block cipher SQUARE [13], and the dedicated attack was later referred to the square attack in several papers [14,18,39]. Then, some extensions of the square attack were proposed like the multi-set [7], saturation [25], and internal collision cryptanalyses [17]. In 2002, Knudsen and Wagner then formalized the square cryptanalysis as the integral cryptanalysis [22]. In the integral cryptanalysis, attackers first prepare N chosen plaintexts. If the XOR of all corresponding ciphertexts is 0, we say that the cipher has an integral characteristic with N chosen plaintexts. The integral characteristic is found by evaluating the propagation of four integral properties: all ($\mathcal{A}$), constant ($\mathcal{C}$), balance ($\mathcal{B}$), and unknown ($\mathcal{U}$).

Division Property. Before the introduction of the division property, it is important to understand the difference between the higher-order differential and integral cryptanalyses. Actually, we can regard both cryptanalyses as the same cryptanalysis. Nevertheless, the higher-order differential and integral characteristics are constructed by completely different methods, and either of both methods has its own advantages and disadvantages. Moreover, there are some experimental characteristics that cannot be proven by either of both methods. These observation causes significant motivation that we developed the division property.

At EUROCRYPT2015, we proposed the division property, which is a novel technique to find integral (higher-order differential) characteristics [33]. This technique is the generalization of the integral property that can also exploit the algebraic degree at the same time. As a result, the division property can find integral characteristics that cannot be found by the two conventional methods. Let $\mathbb{X}$ be a subset whose elements take n-bit values, and assume that the set fulfills the division property $\mathcal{D}_k^n$. Then, $\bigoplus_{x \in \mathbb{X}} \pi_u(x)$ is 0 when $w(u) < k$, and $\bigoplus_{x \in \mathbb{X}} \pi_u(x)$ is unknown when $w(u) \geq k$, where $w(u)$ denotes the Hamming weight of $u \in \mathbb{F}_2^n$. The division properties $\mathcal{D}_n^n$, $\mathcal{D}_2^n$, and $\mathcal{D}_1^n$ correspond to the integral properties $\mathcal{A}$, $\mathcal{B}$, and $\mathcal{U}$, respectively. Clearly, the division properties from $\mathcal{D}_3^n$ to $\mathcal{D}_{n-1}^n$ are not used in the integral property. Moreover, let us consider the set $S(\mathbb{X})$ whose elements are computed by applying the S-box S for n-bit values in $\mathbb{X}$. Then, if the algebraic degree of the S-box is at most d, the propagation of the division property is $\mathcal{D}_k^n \to \mathcal{D}_{\lceil k/d \rceil}^n$.

Follow-Up Researches. The proposal paper of the division property at EUROCRYPT2015 only shows the usefulness of generic attacks against Feistel and Substitution-Permutation networks. To insist the usefulness of the division property, we applied the new technique to the cryptanalysis on full MISTY1 at CRYPTO2015 [32]. Then, many follow-up results have been reported: application to generalized Feistel [42], LBlock [37], and TWINE [30] at INDOCRYPT2015 [41], bit-based division property at FSE2016 [34], parity set at CRYTPTO2016 [8], and MILP-based evaluation of the propagation of the division property at ASIACRYPT2016 [38]. Nowadays, new ciphers that discuss the security for the analysis using the division property in advance have been proposed like SKINNY [4], Mysterion [19], and SPARX and LAX [16].

2 Motivations of Division Property

2.1 Block Ciphers and Its Construction

Block ciphers is symmetric-key ciphers whose input and output lengths are fixed, and n-bit block ciphers denote block ciphers with n-bit input and output. The claimed security is generally κ bits when block ciphers accept κ-bit secret keys. Since block ciphers support only fixed-length message, we have to use modes of operations for block ciphers like ECB, CBC, and CTR to support variable-length message. Block ciphers are generally designed using iterating structure. First, round functions, which are high efficient but weak block ciphers, are designed, and block ciphers are constructed by iterating the round function several times with different round keys. Here, round keys are generated from the secret key using the key schedule. When round functions are iterated R times, we call it R-round block ciphers.

2.2 Higher-Order Differential Cryptanalysis

The concept of the higher-order differential cryptanalysis was first introduced by Lai [23] and the advantage over the traditional differential cryptanalysis was studied by Knudsen [21]. Let E_k be an n-bit block cipher, and the traditional differential cryptanalysis focuses on

$$\Delta_\alpha E_K(x) = E_K(x + \alpha) - E_K(x)$$

and recovers the secret key by analyzing the relationship between α and $\Delta_\alpha E_k(x)$. On the other hand, the higher-order differential cryptanalysis focuses on ith order differential as

$$\Delta^{(i)}_{\alpha_1, \alpha_2, \ldots, \alpha_i} E_K(x) = \Delta_{\alpha_i}(\Delta^{(i-1)}_{\alpha_1, \alpha_2, \ldots, \alpha_{i-1}} E_K(x))$$

and recovers the secret key by analyzing the relationship between $(\alpha_1, \alpha_2, \ldots, \alpha_i)$ and $\Delta^{(i)}_{\alpha_1, \alpha_2, \ldots, \alpha_i} E_K(x)$. For example, the 2nd order differential denotes as

$$\Delta^{(2)}_{\alpha_1, \alpha_2} E_K(x) = \Delta_{\alpha_2}(E_K(x + \alpha_1) - E_K(x))$$
$$= E_K(x + \alpha_1 + \alpha_2) - E_K(x + \alpha_2) - E_K(x + \alpha_1) + E_K(x).$$

Let $\deg(E_k)$ be the algebraic degree of E_k, and Lai showed the following relationship.

$$\deg(\Delta_\alpha E_k) \geq \deg(E_k) - 1.$$

Therefore, assuming the algebraic degree of E_k is at most d, the dth and $(d+1)$th order differentials are constant and zero, respectively. It is obvious that the algebraic degree has to be at least $\min(n - 1, \kappa - 1)$ if block ciphers accept κ-bit secret keys. Then, Knudsen showed the advantage over the conventional differential cryptanalysis by using the toy ciphers [21].

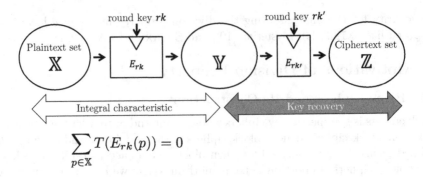

$$\sum_{p\in\mathbb{X}} T(E_{rk}(p)) = 0$$

Fig. 1. Outline of the integral cryptanalysis.

2.3   Integral Cryptanalysis

The similar technique to the higher-order differential cryptanalysis was used as the dedicated attack against the block cipher SQUARE [13]. Since this dedicated attack is powerful cryptanalysis, it is widely applied to various block ciphers, where it was referred to the square attack [14,18,39]. Then, some extensions of the square attack were proposed like the multi-set [7], saturation [25], and internal collision cryptanalyses [17]. In 2002, Knudsen and Wagner then formalized the square cryptanalysis as the integral cryptanalysis. Figure 1 shows the outline of the integral cryptanalysis. Attackers first prepare N chosen plaintexts. If the XOR of all corresponding ciphertexts is 0 for all keys, we say that the cipher has an integral characteristic with N chosen plaintexts. Namely, we prepare the plaintext set $\mathbb{X}$ satisfying

$$\sum_{p\in\mathbb{X}} T(E_{rk}(p)) = 0 \qquad (1)$$

for all round key rk, where the function T is observing function and the simple truncation function is often used. The probability that Eq. (1) holds for ideal ciphers is negligible. Therefore, we can execute a distinguishing attack by construction the integral characteristic.

The integral cryptanalysis additionally append a key recovery step to the integral characteristic. Assuming that block ciphers has an r-round integral characteristics, attackers guess round keys used in the last s rounds and attack $(r+s)$ rounds. If guessed round keys are correct, Eq. (1) always holds. Therefore, if Eq. (1) does not hold, the guessed round key is incorrect. By repeating this procedure, attackers recover the correct round keys used in the last s rounds.

2.4   What is Different Between Integral and Higher-Order Differential Cryptanalyses

The definition of the higher-order differential cryptanalysis is different from that of the integral cryptanalysis. However, they are often regarded as the same cryptanalysis. To achieve high performance under computer, almost all operations of block ciphers are defined using bit operations like XOR and bit-oriented AND. Then, both additions of the integral cryptanalysis and differences become XOR. For example, the 2nd order differential is represented as

$$\Delta^{(2)}_{\alpha_1,\alpha_2} E_K(x) = E_K(x \oplus \alpha_1 \oplus \alpha_2) \oplus E_K(x \oplus \alpha_1) \oplus E_K(x \oplus \alpha_2) \oplus E_K(x)$$
$$= \bigoplus_{p \in V[\alpha_1,\alpha_2]} E_K(p),$$

where $V[\alpha_1, \alpha_2]$ is linear subspace whose basis is (α_1, α_2). Therefore, the 2nd order differential is the same as the integral characteristic with $\mathbb{X} = V[\alpha_1, \alpha_2]$.

The higher-order differential cryptanalysis uses only a linear subspace as the input set, but the integral cryptanalysis can use arbitrary input set. However, since inputs and outputs of all components of block ciphers are represented as bit strings, almost all previous integral cryptanalyses have never used input sets that are not linear subspace. Therefore, there is no difference between the higher-order and integral characteristics once both cryptanalyses are applied to real ciphers.

2.5 How to Find Integral Characteristics

In the practical use, the distinction between the higher-order and integral crypt-analyses is not important. It is rather more important to understand how to find their characteristics. The higher-order differential cryptanalysis focuses on the algebraic degree to construct the characteristics, while the integral cryptanalysis evaluates the propagation of the integral property to construct the characteristic. Nowadays, when an analysis mainly exploits the algebraic degree, it is called "the higher-order differential cryptanalysis." On the other hand, when an analysis mainly exploits the integral property, it is called "the integral cryptanalysis."

Degree Estimation. The higher-order differential characteristics is constructed by evaluating the upper bound of the algebraic degree, but it is not easy in general. The most classical method uses the fact that the algebraic degree of an r-round block cipher is upper-bounded by d^r if the algebraic degree of each round function is at most d. However, this method is too rough evaluation, and the upper bound is generally more small. Canteaut and Videau showed tighter bound of the degree of iterated round functions [11], and Boura et al. then improved the bound in [9]. This improved bound is useful to evaluate the upper bound on Substitution-Permutation network (SPN).

Theorem 1 [9]. *Let S be a function from $\mathbb{F}_2^n$ into $\mathbb{F}_2^n$ corresponding to the concatenation of m smaller S-boxes, defined over $\mathbb{F}_2^{n_0}$. Let δ_k be the maximal degree of the product of any k bits of anyone of these S-boxes. Then, for any function G from $\mathbb{F}_2^n$ into $\mathbb{F}_2$, we have*

$$\deg(G \circ S) \leq n - \frac{n - \deg(G)}{\gamma},$$

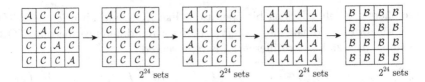

Fig. 2. Integral distinguisher on 4-round AES

where

$$\gamma = \max_{1 \leq i \leq n_0 - 1} \frac{n_0 - i}{n_0 - \delta_i}.$$

For example, let us consider SPN ciphers using four 4-bit S-boxes. Since the algebraic degree of 4-bit S-boxes is at most 3, the algebraic degree of the 2-round cipher is at most $3^2 = 9$. When we use the classical method, the algebraic degree of the 3-round cipher is at most $\min(15, 3^3 = 27) = 15$ but it is not tight. On the other hand, Theorem 1 shows that the algebraic degree is at most $\lfloor 16 - \frac{16-9}{3} \rfloor = 13$. Boura et al. showed integral distinguishers on KECCAK [12] and *Luffa* [10] by using this theorem. We cannot apply Theorem 1 to non-SPN ciphers, and real block ciphers have more complicated round function. Therefore, this method is still far from the tight lower bound.

Since the theoretical estimation of the algebraic degree is difficult, a brute-force method is often used [31], i.e., all algebraic equations are resolved. The accurate algebraic degree is evaluated by using this method, but it generally requires practically infeasible time complexity. Therefore, the application is very limited.

Integral Property. The propagation of the integral property is most widely applied to construct integral characteristics, and many integral characteristics have been constructed by evaluating the propagation of the integral property [22, 24, 37, 39, 40]. It uses four integral properties as follows:

– ALL ($\mathcal{A}$): Every value appears the same number in the multiset.
– BALANCE ($\mathcal{B}$): The XOR of all texts in the multiset is 0.
– CONSTANT ($\mathcal{C}$): The value is fixed to a constant for all texts in the multiset.
– UNKNOWN ($\mathcal{U}$): The multiset is indistinguishable from one of n-bit random values.

Knudsen and Wagner showed that AES has the 4-round integral distinguisher with 2^{32} chosen plaintexts [22] (see Fig. 2).

Unfortunately, the integral property does not find effective characteristics if block ciphers consist of non-bijective functions, e.g., DES [1] and SIMON[3] consist of non-bijection functions. Moreover, since the propagation does not clearly exploit the algebraic degree of block ciphers, it tends not to construct effective characteristics on block ciphers with low-degree round functions.

2.6   Why Division Property is Developed?

The development of the division property is motivated from following two observations.

Observation 1. *Let us consider the integral characteristic of AES. The propagation of the integral property can find 4-round integral characteristics with 2^{32} chosen plaintext, while the algebraic degree of 4-round AES is upper-bounded by 126 even if Theorem 1 is used. Therefore, the propagation of the integral property is superior to the degree estimation.*

Observation 2. *Let us consider the integral characteristic of a modified AES, where S-boxes with degree 2 are used instead of the original S-box. Since the propagation of the integral property is not affected by the algebraic degree of S-boxes, the found characteristics is the same as the original AES. On the other hand, the algebraic degree of 7-round modified AES is upper-bounded by 96 when Theorem 1 is used. Therefore, the degree estimation is superior to the propagation of the integral property.*

Observations 1 and 2 implies that each of the two methods has its own advantages and disadvantages. Specifically, S-boxes are regarded as black boxes in the propagation of the integral property, i.e., it constructs integral characteristics by mainly exploiting the diffusion part of block ciphers. On the other hand, the degree estimation can exploit the algebraic degree of S-boxes but is difficult to well exploit the diffusion part, i.e., it constructs integral characteristics by mainly exploiting the confusion part of block ciphers. Therefore, it is natural that the method that can exploit both diffusion and confusion parts of block ciphers is desirable.

3   Division Property

The division property is a novel technique to find integral (higher-order differential) characteristics and is the generalization of the integral property that can also exploit the algebraic degree at the same time. Let us consider the properties of input and output multisets for one bijective S-box with degree d. If an input multiset has $\mathcal{A}$, the output multiset also has $\mathcal{A}$. If an input multiset has $\mathcal{B}$, the output multiset has $\mathcal{U}$. Moreover, we can prepare the set of 2^{d+1} input texts that the output multiset has $\mathcal{B}$ because the algebraic degree of the S-box is d, but just using the integral property does not exploit this property. To exploit useful properties between $\mathcal{A}$ and $\mathcal{B}$, we redefine $\mathcal{A}$ and $\mathcal{B}$ by the same notation and then introduce the division property by generalizing the redefinition.

Redefinition of $\mathcal{A}$. Let $\mathbb{X}$ be a multiset whose elements take an n-bit value. We first consider features of the multiset $\mathbb{X}$ satisfying $\mathcal{A}$. If we choose one bit from n bits and calculate the XOR of the chosen bit in the multiset, the calculated value is always 0. Moreover, if we choose at most $(n-1)$ bits from n bits and calculate the XOR of the AND of chosen bits in the multiset, the calculated value is also always 0. However, if we choose all bits from n bits and calculate the XOR of the AND of n bits in the multiset, the calculated value is unknown[1]. These features are summarized as

$$\bigoplus_{x \in \mathbb{X}} \pi_u(x) = \begin{cases} 0 & w(u) < n, \\ \text{unknown} & w(u) = n, \end{cases}$$

where π_u is the bit product function defined as $\pi_u(x) := \prod_{i=1}^{n} x[i]^{u[i]}$ and $w(u)$ is the Hamming weight of $u \in \mathbb{F}_2^n$. Assuming the multiset has $\mathcal{A}$, the parity is always even for any u satisfying $w(u) < n$. On the other hand, the parity is unknown for $u = 1^n$.

Redefinition of $\mathcal{B}$. We next consider features of the multiset $\mathbb{X}$ satisfying $\mathcal{B}$. If we choose one bit from n bits and calculate the XOR of the chosen bit in the multiset, the calculated value is always 0. However, if we choose at least two bits from n bits and calculate the XOR of the AND of chosen bits in the multiset, the calculated value becomes unknown. These features are summarized as

$$\bigoplus_{x \in \mathbb{X}} \pi_u(x) = \begin{cases} 0 & w(u) < 2, \\ \text{unknown} & w(u) \geq 2, \end{cases}$$

Assuming the multiset has $\mathcal{B}$, the parity is always even for any u satisfying $w(u) < 2$. On the other hand, the parity is unknown for any u satisfying $w(u) \geq 2$.

3.1   Definition of Division Property

From the redefinition of the $\mathcal{A}$ and $\mathcal{B}$ using the bit product function, we define the division property as follows.

Definition 1 (Division Property). *Let $\mathbb{X}$ be a multiset whose elements take a value of $\mathbb{F}_2^n$, and k takes an integer value between 0 and n. When the multiset $\mathbb{X}$ has the division property $\mathcal{D}_k^n$, it fulfils the following conditions: The parity of $\pi_u(x)$ for all $x \in \mathbb{X}$ is always even if $w(u)$ is less than k. Moreover, the parity becomes unknown if $w(u)$ is greater than or equal to k.*

In summary, for the multiset $\mathbb{X}$ satisfying $\mathcal{D}_k^n$,

$$\bigoplus_{x \in \mathbb{X}} \pi_u(x) = \begin{cases} 0 & w(u) < k, \\ \text{unknown} & w(u) \geq k. \end{cases}$$

Namely, the set of $u \in \mathbb{F}_2^n$ is divided into the subset that $\bigoplus_{x \in \mathbb{X}} \pi_u(x)$ is unknown and the subset that $\bigoplus_{x \in \mathbb{X}} \pi_u(x)$ is 0.

---

[1] If all values appear the same even number in the multiset, the calculated value is always 0. If all values appear the same odd number in the multiset, the calculated value is always 1. Thus, we cannot guarantee whether the calculated value is 0 or not when we consider the multiset satisfying $\mathcal{A}$. In this case, we say the calculated value is unknown.

Example 1. Let $\mathbb{X}$ be a multiset whose elements take a value of $\mathbb{F}_2^4$. As an example, we prepare the input multiset $\mathbb{X}$ as

$$\mathbb{X} := \{0x0, 0x3, 0x3, 0x3, 0x5, 0x6, 0x8, 0xB, 0xD, 0xE\}.$$

A following table calculates the summation of $\pi_u(x)$.

u	$\sum \pi_u(x)$	$\bigoplus \pi_u(x)$	u	$\sum \pi_u(x)$	$\bigoplus \pi_u(x)$
0000	10	0	1000	4	0
0001	6	0	1001	2	0
0010	6	0	1010	2	0
0011	4	0	1011	1	1
0100	4	0	1100	2	0
0101	2	0	1101	1	1
0110	2	0	1110	1	1
0111	0	0	1111	0	0

For all u satisfying $w_u < 3$, $\bigoplus_{x \in \mathbb{X}} \pi_u(x)$ is 0. Therefore, the multiset has the division property $\mathcal{D}_3^4$.

Each definition of $\mathcal{B}$ and $\mathcal{U}$ is essentially the same as that of $\mathcal{D}_2^n$ and $\mathcal{D}_1^n$, respectively. However, the definition of $\mathcal{A}$ is different from that of $\mathcal{D}_n^n$. The multiset satisfying $\mathcal{A}$ always has the division property $\mathcal{D}_n^n$ but not vice versa. For instance, the multiset satisfying the EVEN property, which is defined that the number of occurrences is even for all values [28], does not always have $\mathcal{A}$, but it has $\mathcal{D}_n^n$. In this paper, we use only $\mathcal{D}_n^n$ instead of $\mathcal{A}$ because it is sufficient to use $\mathcal{D}_n^n$ from the viewpoint of the construction of integral characteristics.

3.2 Propagation of Division Property

Let s be an S-box whose degree is d. Let $\mathbb{X}$ be an input multiset whose elements take a value of $\mathbb{F}_2^n$. Let $S(\mathbb{X})$ be an output multiset whose elements are calculated from $s(x)$ for all $x \in \mathbb{X}$. We assume that $\mathbb{X}$ has $\mathcal{D}_k^n$, and want to evaluate the division property of $S(\mathbb{X})$. In the division property, the set of u is divided into the subset that $\bigoplus_{x \in \mathbb{X}} \pi_u(x)$ is unknown and the subset that $\bigoplus_{x \in \mathbb{X}} \pi_u(x)$ is 0. Therefore, we divide the set of v into the subset that $\bigoplus_{s(x) \in S(\mathbb{X})} \pi_v(s(x))$ is unknown and the subset that $\bigoplus_{s(x) \in S(\mathbb{X})} \pi_v(s(x))$ is 0. Since the parity of $\pi_v(s(x))$ for all $s(x) \in S(\mathbb{X})$ is equal to that of $(\pi_v \circ s)(x)$ for all $x \in \mathbb{X}$, we evaluate $\bigoplus_{x \in \mathbb{X}} (\pi_v \circ s)(x)$.

Proposition 1 (Propagation of Division Property). *Let s be an function (S-box) from n bits to n bits, and the degree is d. Assuming that an input multiset $\mathbb{X}$ has the division property $\mathcal{D}_k^n$, the output multiset $S(\mathbb{X})$ has $\mathcal{D}_{\lceil \frac{k}{d} \rceil}^n$. In addition, assuming that the S-box is a permutation, the output multiset $S(\mathbb{X})$ has $\mathcal{D}_n^n$ when the input multiset has $\mathcal{D}_n^n$.*

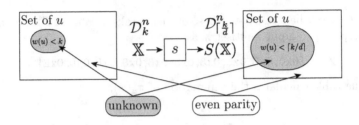

Fig. 3. Propagation of division property

Example 2. Let us consider a following 4-bit S-box.

x	0	1	2	3	4	5	6	7	8	9	A	B	C	D	E	F
$s(x)$	8	C	0	B	9	D	E	5	A	1	2	6	4	F	3	7

The S-box is bijective and the algebraic degree is 2. We now prepare the input multiset $\mathbb{X}$ as

$$\mathbb{X} := \{0\text{x}0, 0\text{x}3, 0\text{x}3, 0\text{x}3, 0\text{x}5, 0\text{x}6, 0\text{x}8, 0\text{x}B, 0\text{x}D, 0\text{x}E\},$$

which is the same as Example 1 and the division property is $\mathcal{D}_3^4$. The output multiset is calculated as

$$S(\mathbb{X}) := \{0\text{x}8, 0\text{x}B, 0\text{x}B, 0\text{x}B, 0\text{x}D, 0\text{x}E, 0\text{x}A, 0\text{x}6, 0\text{x}F, 0\text{x}3\},$$

and a following table calculates the summation of $\pi_v(y)$.

v	$\sum \pi_v(y)$	$\bigoplus \pi_v(y)$	v	$\sum \pi_v(y)$	$\bigoplus \pi_v(y)$
0000	10	0	1000	8	0
0001	6	0	1001	5	1
0010	8	0	1010	6	0
0011	5	1	1011	4	0
0100	4	0	1100	3	1
0101	2	0	1101	2	0
0110	3	1	1110	2	0
0111	1	1	1111	1	1

For all v satisfying $w_v < 2$, $\bigoplus_{y \in \mathbb{Y}} \pi_v(y)$ becomes 0. Therefore, the multiset $\mathbb{Y}$ has the division property $\mathcal{D}_2^4$.

Figure 3 shows the outline of the propagation of the division property. Let $\mathbb{X}$ and $S(\mathbb{X})$ be input and output multisets, respectively. First, the size of the set of u that $\bigoplus_{x \in \mathbb{X}} \pi_u(x)$ is unknown is small. However, the size of the set of u that $\bigoplus_{x \in \mathbb{X}} \pi_u(s(x))$ is unknown expands. If the size expands to the universal set except for 0^n, we regard that the output multiset is indistinguishable from the multiset of random texts.

4   Vectorial and Collective Division Property

Section 3 only shows the division property for one S-box. Since practical ciphers consist of several S-boxes in every round, we cannot apply the division property for practical ciphers. Therefore, we have to introduce the division property for more complicating functions.

4.1   Vectorial Division Property

First, we define vectorial division property to accept an S-box layer, where multiple S-boxes are applied in parallel.

Definition 2 (Vectorial Division Property). *Let* $\mathbb{X}$ *be the multiset whose elements take a value of* $(\mathbb{F}_2^n)^m$, *and* $\boldsymbol{k}$ *is an m-dimensional vector whose elements take a integer value between 0 and* n_i. *When the multiset* $\mathbb{X}$ *has the division property* $\mathcal{D}_{\boldsymbol{k}}^{n_1, n_2, \ldots, n_m}$, *the multiset fulfils the following conditions: The parity of* $\pi_{\boldsymbol{u}}(\boldsymbol{x})$ *for all* $\boldsymbol{x} \in \mathbb{X}$ *is unknown if* $W(\boldsymbol{u}) \succeq \boldsymbol{k}$, *Otherwise, the parity is always 0.*

Here, for any two m-dimensional vectors $\boldsymbol{k}$ and $\boldsymbol{k}'$, we define $\boldsymbol{k} \succeq \boldsymbol{k}'$ if $k_i \geq k_i'$ for all i ($1 \leq i \leq m$). Moreover, the division property for $(\mathbb{F}_2^n)^m$ is referred to as $\mathcal{D}_{\mathbb{K}}^{n^m}$ for the simplicity.

Assuming the input multiset of the S-Layer has the division property $\mathcal{D}_{\boldsymbol{k}}^{n^m}$, the output of the S-Layer is calculated as $S(\boldsymbol{x}) = (s_1(x_1), s_2(x_2), \ldots, s_m(x_m))$ for $(x_1, x_2, \ldots, x_m) \in \mathbb{X}$. We now consider the set of $\boldsymbol{v}$ that $\bigoplus_{\boldsymbol{x} \in \mathbb{X}} \pi_{\boldsymbol{v}}(S(\boldsymbol{x}))$ is unknown and the set of $\boldsymbol{v}$ that $\bigoplus_{\boldsymbol{x} \in \mathbb{X}} \pi_{\boldsymbol{v}}(S(\boldsymbol{x}))$ is 0. Since the output of each S-box is calculated independently, the propagation of the division property can also be evaluated independently. Namely, the output multiset has $\mathcal{D}_{\boldsymbol{k}'}^{n^m}$, where $k_i' = \lceil k_i / d \rceil$. Moreover, $k_i' = n$ if the S-box is bijective and $k_i = n$.

4.2   Collective Division Property

Only vectorizing the division property is still insufficient to represent the subset of $\boldsymbol{u}$ that the parity is unknown. For simplicity, we consider a multiset $\mathbb{X}$ whose elements take a value of $(\mathbb{F}_2^8)^2$. Assume that the number of elements in $\mathbb{X}$ is 256, and two elements of $\boldsymbol{x}$ take all values from 0 to 255 independently. Then, we consider the set of $\boldsymbol{u}$ that the parity of $\pi_{\boldsymbol{u}}(\boldsymbol{x})$ for all $\boldsymbol{x} \in \mathbb{X}$ is unknown.

- The parity is unknown if $W(\boldsymbol{u}) \succeq (8, 0)$.
- The parity is unknown if $W(\boldsymbol{u}) \succeq (0, 8)$.
- The parity is unknown if $W(\boldsymbol{u}) \succeq (1, 1)$.
- Otherwise, the parity is always even.

We can not express this feature by only using the vectorial division property. Therefore, we collect several vectorial division properties.

Definition 3 (Collective Division Property). *Let $\mathbb{X}$ be a multiset whose elements take a value of $(\mathbb{F}_2^{n_1} \times \mathbb{F}_2^{n_2} \times \cdots \times \mathbb{F}_2^{n_m})$. Let $\mathbb{K}$ be a set whose elements take an m-dimensional vector whose ith element takes an integer value between 0 and n_i. When the multiset $\mathbb{X}$ has the division property $\mathcal{D}_{\mathbb{K}}^{n_1,n_2,\dots,n_m}$, it fulfils the following conditions:*

$$\bigoplus_{\boldsymbol{x}\in\mathbb{X}} \pi_{\boldsymbol{u}}(\boldsymbol{x}) = \begin{cases} \text{unknown} & \text{if there exist } \boldsymbol{k} \in \mathbb{K} \text{ s.t. } W(\boldsymbol{u}) \succeq \boldsymbol{k}, \\ 0 & \text{otherwise.} \end{cases}$$

It is obvious that the collective division property with $|\mathbb{K}| = 1$ is the same as the vectorial division property.

If there are $\boldsymbol{k} \in \mathbb{K}$ and $\boldsymbol{k}' \in \mathbb{K}$ satisfying $\boldsymbol{k} \succeq \boldsymbol{k}'$, $\boldsymbol{k}$ can be removed from $\mathbb{K}$ because it is redundant. Assume that the multiset $\mathbb{X}$ has the division property $\mathcal{D}_{\mathbb{K}}^{n_1,n_2,\dots,n_m}$. If there is not an unit vector $\boldsymbol{e}_j$ in $\mathbb{K}$, where $\boldsymbol{e}_j$ is a vector whose jth element is 1 and the others are 0, $\bigoplus_{\boldsymbol{x}\in\mathbb{X}} x_j$ is 0.

4.3 Propagation Rules for Simple Operations

Practical ciphers diffuses outputs of every S-box by using linear functions. To evaluate the propagation of the division property for linear functions, four propagation rules are defined as follows.

Copy. Let F be a copy function, where the input x takes a value of $\mathbb{F}_2^n$ and the output is calculated as $[y_1, y_2] = [x, x]$. Let $\mathbb{X}$ and $S(\mathbb{X})$ be the input multiset and the output multiset, respectively. Assuming that the multiset $\mathbb{X}$ has the division property $\mathcal{D}_k^n$, the multiset $S(\mathbb{X})$ has the division property $\mathcal{D}_{\mathbb{K}'}^{n,n}$, where $\mathbb{K}'$ is calculated as follows: First, $\mathbb{K}'$ is initialized to an empty set ϕ. Then, for all i $(0 \leq i \leq k)$,

$$\mathbb{K}' = \mathbb{K}' \cup [k - i, i],$$

is calculated.

Compression by XOR. Let F be a function compressed by an XOR, where the input $[x_1, x_2]$ takes a value of $(\mathbb{F}_2^n \times \mathbb{F}_2^n)$ and the output is calculated as $y = x_1 \oplus x_2$. Let $\mathbb{X}$ and $S(\mathbb{X})$ be the input multiset and the output multiset, respectively. Assuming that the multiset $\mathbb{X}$ has the division property $\mathcal{D}_{\mathbb{K}}^{n,n}$, the division property of the multiset $S(\mathbb{X})$ is $\mathcal{D}_{k'}^n$ as

$$k' = \min_{[k_1,k_2]\in\mathbb{K}} \{k_1 + k_2\}.$$

Here, if the minimum value of k' is larger than n, the propagation characteristic of the division property is aborted. Namely, a value of $\bigoplus_{y\in S(\mathbb{X})} \pi_v(y)$ is 0 for all $v \in \mathbb{F}_2^n$.

Split. Let F be a split function, where the input x takes a value of $\mathbb{F}_2^n$ and the output is calculated as $y_1 \| y_2 = x$, where $[y_1, y_2]$ takes a value of $(\mathbb{F}_2^{n_1} \times \mathbb{F}_2^{n-n_1})$. Let $\mathbb{X}$ and $S(\mathbb{X})$ be the input multiset and the output multiset, respectively.

Assuming that the multiset $\mathbb{X}$ has the division property $\mathcal{D}_k^n$, the multiset $S(\mathbb{X})$ has the division property $\mathcal{D}_{\mathbb{K}'}^{n_1, n-n_1}$, where $\mathbb{K}'$ is calculated as follows: First, $\mathbb{K}'$ is initialized to ϕ. Then, for all i ($0 \leq i \leq k$),

$$\mathbb{K}' = \mathbb{K}' \cup [k-i, i],$$

is calculated. Here, $(k-i)$ is less than or equal to n_1, and i is less than or equal to $n - n_1$.

Concatenation. Let F be a concatenation function, where the input $[x_1, x_2]$ takes a value of $(\mathbb{F}_2^{n_1} \times \mathbb{F}_2^{n_2})$ and the output is calculated as $y = x_1 \| x_2$. Let $\mathbb{X}$ and $S(\mathbb{X})$ be the input multiset and the output multiset, respectively. Assuming that the multiset $\mathbb{X}$ has the division property $\mathcal{D}_{\mathbb{K}}^{n_1, n_2}$, the division property of the multiset $S(\mathbb{X})$ is $\mathcal{D}_{k'}^{n_1+n_2}$ as

$$k' = \min_{[k_1, k_2] \in \mathbb{K}} \{k_1 + k_2\}.$$

5 Follow-Up Works

After the proposal of the division property at EUROCRYPT2015 [33], many follow-up researches have been proposed by both ourself and third party. At CRYPTO2015, we first applied the division property to the cryptanalysis on full MISTY1 [32], where we developed an improved technique for the key-XORing S-box. Then, many follow-up results have been reported: application to generalized Feistel [42], LBlock [37], and TWINE [30] at INDOCRYPT2015 [41], application to LILLIPUT at SAC2016 [27], bit-based division property at FSE2016 [34], parity set at CRYPTO2016 [8], and MILP-based evaluation of the propagation of the division property at ASIACRYPT2016 [38]. Nowadays, new ciphers that discuss the security for the analysis using the division property in advance have been proposed like SKINNY [4], Mysterion [19], and SPARX and LAX [16].

5.1 Improving Technique

Key-XORing S-box. At CRYPTO2015, we applied the division property to cryptanalysis on MISTY1 [32], and we then proposed one extension of the division property for the application to MISTY1. The propagation of the division property, i.e., $\mathcal{D}_k^n \to \mathcal{D}_{\lceil k/d \rceil}^n$, holds for all keyed S-box whose degree is d, but many practical ciphers adopt key-XORing S-box for the efficiency. In the key-XORing S-box, the round key is first XORed with the input and the public permutation are then applied. We can sometimes improve the propagation of the division property by exploiting this structure. In the first step, note that key XORing does not affect the division property because it is linear function. In the second step, we evaluate the propagation of the division property for a public function and the focus is that we can evaluate the algebraic normal form of the public function accurately. If the algebraic degree of $\pi_v \circ s$ is smaller than $w(v) \times d$, it is possible that the propagation is improved. We analyzed two S-boxes S_9

and S_7 used in MISTY1 and found the propagation for S_7 can be improved. Thanks to this improvement, we can find the 6-round integral characteristic on MISTY1 and attack full MISTY1 by guessing round keys that are used in last two rounds. At CRYPTO2016, Bar-On and Keller improved the key recovery step of the integral cryptanalysis, and the security of full MISTY1 decreases to about 70 bits [2].

Bit-Based Division Property and Parity Set. At FSE2016, we proposed two extending variants of the division property: the conventional bit-based division property and the bit-based division property using three subsets [34]. To analyze n-bit block ciphers, the division property $\mathcal{D}_{\mathbb{K}}^{\ell_1,\ell_2,...,\ell_m}$ is used, where ℓ_i and m are chosen by attackers in the range of $n = \sum_{i=1}^{m} \ell_i$. Using small ℓ_i allows us to exploit complicating structures of target ciphers, and $\mathcal{D}_{\mathbb{K}}^{1^n}$ is the most accurate. Note that $\mathcal{D}_{\mathbb{K}}^{1^n}$ is not against the definition of the conventional division property, thus we call it the conventional bit-based division property. We applied this technique to SIMON32 [3], and 14-round integral characteristics are theoretically proven. Unfortunately, it is doubtful whether the proved 14-round integral characteristic is tight or not because 15-round integral characteristics were already reported in the experimental search by Wang et al. [36]. Therefore, there is still a gap of one round between the proof and experiment. To fill up with this gap, we then introduced the bit-based division property using three subsets. The set of $u \in \mathbb{F}_2^n$ is divided into two subsets according to whether $\bigoplus_{x \in \mathbb{X}} \pi_u(x)$ is 0 or unknown in the conventional bit-based division property. On the other hand, in the bit-based division property using three subsets, the set of $u \in \mathbb{F}_2^n$ is divided into three subsets according to whether $\bigoplus_{x \in \mathbb{X}} \pi_u(x)$ is 0, 1, or unknown. This extension allows us to find more accurate integral characteristic than the conventional one, and we theoretically proved the existence of the 15-round integral characteristic on SIMON32.

Independently of the bit-based division property, Boura and Canteaut also showed another view of the division property and introduced the concept of the parity set [8] and the link with the Reed-Muller codes. Actually, the parity set is regarded as the bit-based division property for an S-box. While the propagation of the integer-based division property focuses on the Hamming weight of $u \in \mathbb{F}_2^n$ that the parity is unknown, the parity set focus on $u \in \mathbb{F}_2^n$ itself that the parity is unknown. They evaluated the propagation for the PRESENT S-box and showed the low-data distinguishers. Moreover, the relationship between the parity set and the bit-based division property were discussed in [35].

MILP-Based Propagation Search. The most important and difficult open problem that we left is to construct efficient algorithm to evaluate the propagation of the division property. In the application to MISTY1 [33], we implemented the propagation using C++, but it requires much memory and time complexities. Moreover, at FSE2016 [34], we could not apply the bit-based division property to SIMON family except for SIMON32 because of the high complexity. This problem was solved by using the state-of-the-art technique using mixed integer linear programming by Xiang et al. at ASIACRYPT2016 [38]. The approach using MILP was first introduced by Mouha et al. for evaluating the lower bound of the number of active S-boxes on word-oriented ciphers, [26]. However, several ciphers do not have word-oriented structure. Therefore, Sun et al. then developed a method to model all possible differential propagations bit by bit even for the S-box [29]. At ASIACRYPT2016, Xiang et al. first introduced *the division trail*. To analyze r-round ciphers, the following propagation of the division property

$$\mathbb{K}_0 \to \mathbb{K}_1 \to \cdots \to \mathbb{K}_r$$

is evaluated, where $\mathcal{D}_{\mathbb{K}_{i-1}}$ is the division property for the input of ith round function. Then, for $(\boldsymbol{k}_0, \boldsymbol{k}_1, \ldots, \boldsymbol{k}_r) \in (\mathbb{K}_0 \times \mathbb{K}_1 \times \cdots \times \mathbb{K}_r)$, if $\boldsymbol{k}_{i-1}$ can propagate to $\boldsymbol{k}_i$ for all $i \in \{1, 2, \ldots, r\}$, $(\boldsymbol{k}_0, \boldsymbol{k}_1, \ldots, \boldsymbol{k}_r)$ is called r-round division trail. If there is a division trail that r-round output becomes unknown, the r-round output is unknown. They developed a method to model the propagation rules for the division property and showed that all division trails are effectively evaluated using MILP. As a result, they get new integral characteristics on SIMON family, Simeck family, PRESENT, and RECTANGLE[2].

5.2   Summary of Applications

About 1.5 years has passed since the proposal of the division property, and many applications are already reported. We summarize all results as far as we know in Table 1.

Integral characteristics on generic structure for symmetric-key ciphers were evaluated in [33], where (ℓ, d, m)-SPN and (ℓ, d, m)-AES are defined as follows.

Definition 4 $((\ell, d, m)$-SPN). *The round function consists of an S-Layer and P-Layer, where m ℓ-bit S-boxes are parallelly applied in the S-Layer, and their outputs are diffused by the P-Layer. (ℓ, d, m)-SPN is the set of ciphers that the algebraic degree of S-boxes in the S-Layer is at most d and the P-Layer is any linear function.*

Definition 5 $((\ell, d, m)$-AES). *The state is $(m \times m)$ matrix whose cell is ℓ bits. The round function has AES-like structure, i.e., m^2 ℓ-bit S-boxes are parallelly applied in SubBytes, each ℓ-bit values in the ith row is rotated $i - 1$ cells to the left in ShiftRows, and m cells in each column are diffused in MixColumns. Then, (ℓ, d, m)-AES is the set of ciphers that the algebraic degree of S-boxes in SubBytes is at most d and MixColumns is any linear function.*

Note that any cipher belonging to (ℓ, d, m)-SPN (resp. (ℓ, d, m)-AES) is always has r-round integral characteristics if (ℓ, d, m)-SPN (resp. (ℓ, d, m)-AES) has r-round integral characteristics.

---

[2] We also independently evaluated the propagation of the division property on PRESENT in [35] and get the same integral characteristics. In that paper, we introduced the compact representation for the division property to evaluate the propagation efficiently.

Table 1. Summary of integral characteristics by the division property.

Target	#Rounds	Data	Format of division property	Reference	Remarks
LED	6	2^{52}	4^{16}	[33]	$(4, 3, 4)$-AES
Joltik-BC	6	2^{52}	4^{16}		$(4, 3, 4)$-AES
PHOTON P_{100}	7	2^{97}	4^{25}		$(4, 3, 5)$-AES
PHOTON P_{144}	7	2^{132}	4^{36}		$(4, 3, 6)$-AES
PHOTON P_{196}	8	2^{192}	4^{49}		$(4, 3, 7)$-AES
PHOTON P_{256}	8	2^{249}	4^{64}		$(4, 3, 8)$-AES
Serpent	7	2^{124}	4^{32}		$(4, 3, 32)$-SPN
NOEKEON	7	2^{124}	4^{32}		$(4, 3, 32)$-SPN
KECCAK-f[800]	14	2^{798}	5^{160}		$(5, 2, 160)$-SPN
KECCAK-f[1600]	15	2^{1595}	5^{320}		$(5, 2, 320)$-SPN
MISTY1	6 w/2FL	2^{63}	$\{7, 2, 7, 7, 2, 7, 7, 2, 7, 7, 2, 7\}$	[32]	
LBlock	16	2^{63}	4^{16}	[41]	
TWINE	16	2^{63}	4^{16}		
SIMON32	14	2^{31}	bit-based	[34]	
	15	2^{31}	bit-based using 3 subsets		
PRESENT	9	2^{60}	bit-based	[35]	compact
SIMON48	16	2^{47}	bit-based	[38]	MILP
SIMON64	18	2^{63}	bit-based		
SIMON96	22	2^{95}	bit-based		
SIMON128	26	2^{127}	bit-based		
RECTANGLE	9	2^{60}	bit-based		
PRESENT	9	2^{60}	bit-based		

6   Conclusion

This paper explained the division property, which is recent technique to find integral characteristics. The motivation why we introduced the division property and its concept are explained. After the proposal of the division property at EUROCRYPT 2015, many follow-up works including technical improvement and wide applications have been proposed, and we summarized their follow-up works. We who are the developer of the division property hope that many follow-up works will be proposed in future.

References

1. Data Encryption Standard (DES). National Bureau of Standards (1977). Federal Information Processing Standards Publication 46
2. Bar-On, A., Keller, N.: A 2^{70} attack on the full MISTY1. In: Robshaw, M., Katz, J. (eds.) CRYPTO 2016. LNCS, vol. 9814, pp. 435–456. Springer, Heidelberg (2016). doi:10.1007/978-3-662-53018-4_16

3. Beaulieu, R., Shors, D., Smith, J., Treatman-Clark, S., Weeks, B., Wingers, L.: The SIMON and SPECK families of lightweight block ciphers (2013). http://eprint.iacr.org/2013/404

4. Beierle, C., et al.: The SKINNY family of block ciphers and its low-latency variant MANTIS. In: Robshaw, M., Katz, J. (eds.) CRYPTO 2016. LNCS, vol. 9815, pp. 123–153. Springer, Heidelberg (2016). doi:10.1007/978-3-662-53008-5_5

5. Biham, E., Biryukov, A., Shamir, A.: Cryptanalysis of skipjack reduced to 31 rounds using impossible differentials. In: Stern, J. (ed.) EUROCRYPT 1999. LNCS, vol. 1592, pp. 12–23. Springer, Heidelberg (1999). doi:10.1007/3-540-48910-X_2

6. Biham, E., Shamir, A.: Differential cryptanalysis of DES-like cryptosystems. In: Menezes, A.J., Vanstone, S.A. (eds.) CRYPTO 1990. LNCS, vol. 537, pp. 2–21. Springer, Heidelberg (1991). doi:10.1007/3-540-38424-3_1

7. Biryukov, A., Shamir, A.: Structural cryptanalysis of SASAS. In: Pfitzmann, B. (ed.) EUROCRYPT 2001. LNCS, vol. 2045, pp. 395–405. Springer, Heidelberg (2001). doi:10.1007/3-540-44987-6_24

8. Boura, C., Canteaut, A.: Another view of the division property. In: Robshaw, M., Katz, J. (eds.) CRYPTO 2016. LNCS, vol. 9814, pp. 654–682. Springer, Heidelberg (2016). doi:10.1007/978-3-662-53018-4_24

9. Boura, C., Canteaut, A., Cannière, C.: Higher-order differential properties of KECCAK and Luffa. In: Joux, A. (ed.) FSE 2011. LNCS, vol. 6733, pp. 252–269. Springer, Heidelberg (2011). doi:10.1007/978-3-642-21702-9_15

10. Cannière, C.D., Sato, H., Watanabe, D.: Hash function Luffa - a SHA-3 candidate (2008). http://hitachi.com/rd/yrl/crypto/luffa/round1archive/Luffa_Specification.pdf

11. Canteaut, A., Videau, M.: Degree of composition of highly nonlinear functions and applications to higher order differential cryptanalysis. In: Knudsen, L.R. (ed.) EUROCRYPT 2002. LNCS, vol. 2332, pp. 518–533. Springer, Heidelberg (2002). doi:10.1007/3-540-46035-7_34

12. Daemen, J., Bertoni, G., Peeters, M., Assche, G.V.: The Keccak reference version 3.0 (2011)

13. Daemen, J., Knudsen, L., Rijmen, V.: The block cipher Square. In: Biham, E. (ed.) FSE 1997. LNCS, vol. 1267, pp. 149–165. Springer, Heidelberg (1997). doi:10.1007/BFb0052343

14. Demirci, H.: Square-like attacks on reduced rounds of IDEA. In: Nyberg, K., Heys, H. (eds.) SAC 2002. LNCS, vol. 2595, pp. 147–159. Springer, Heidelberg (2003). doi:10.1007/3-540-36492-7_11

15. Demirci, H., Selçuk, A.A.: A meet-in-the-middle attack on 8-round AES. In: Nyberg, K. (ed.) FSE 2008. LNCS, vol. 5086, pp. 116–126. Springer, Heidelberg (2008). doi:10.1007/978-3-540-71039-4_7

16. Dinu, D., Perrin, L., Udovenko, A., Velichkov, V., Groschdl, J., Biryukov, A.: Design strategies for ARX with provable bounds: SPARX and LAX (full version) (2016). http://eprint.iacr.org/2016/984, (Accepted to ASIACRYPT 2016)

17. Gilbert, H., Minier, M.: A collision attack on 7 rounds of Rijndael. In: AES Candidate Conference, pp. 230–241 (2000)

18. He, Y., Qing, S.: Square attack on reduced camellia cipher. In: Qing, S., Okamoto, T., Zhou, J. (eds.) ICICS 2001. LNCS, vol. 2229, pp. 238–245. Springer, Heidelberg (2001). doi:10.1007/3-540-45600-7_27

19. Journault, A., Standaert, F.X., Varici, K.: Improving the security and efficiency of block ciphers based on LS-designs. Des. Codes Crypt. **82**, 1–15 (2016). http://dx.doi.org/10.1007/s10623-016-0193-8

20. Knudsen, L.: DEAL - a 128-bit block cipher. Technical report no. 151. Department of Informatics, University of Bergen, Norway, February 1998
21. Knudsen, L.R.: Truncated and higher order differentials. In: Preneel, B. (ed.) FSE 1994. LNCS, vol. 1008, pp. 196–211. Springer, Heidelberg (1995). doi:10.1007/3-540-60590-8_16
22. Knudsen, L.R., Wagner, D.: Integral cryptanalysis. In: Daemen, J., Rijmen, V. (eds.) FSE 2002. LNCS, vol. 2365, pp. 112–127. Springer, Heidelberg (2002). doi:10.1007/3-540-45661-9_9
23. Lai, X.: Higher order derivatives and differential cryptanalysis. In: Blahut, R.E., Costello Jr., D.J., Maurer, U., Mittelholzer, T. (eds.) Communications and Cryptography. The Springer International Series in Engineering and Computer Science, vol. 276, pp. 227–233. Springer, Heidelberg (1994)
24. Li, Y., Wu, W., Zhang, L.: Improved integral attacks on reduced-round CLEFIA block cipher. In: Jung, S., Yung, M. (eds.) WISA 2011. LNCS, vol. 7115, pp. 28–39. Springer, Heidelberg (2012). doi:10.1007/978-3-642-27890-7_3
25. Lucks, S.: The saturation attack - a bait for twofish. In: Matsui, M. (ed.) FSE 2001. LNCS, vol. 2355, pp. 1–15. Springer, Heidelberg (2002). doi:10.1007/3-540-45473-X_1
26. Mouha, N., Wang, Q., Gu, D., Preneel, B.: Differential and linear cryptanalysis using mixed-integer linear programming. In: Wu, C.-K., Yung, M., Lin, D. (eds.) Inscrypt 2011. LNCS, vol. 7537, pp. 57–76. Springer, Heidelberg (2012). doi:10.1007/978-3-642-34704-7_5
27. Sasaki, Y., Todo, Y.: New differential bounds and division property of lilliput: block cipher with extended generalized Feistel network. In: SAC (2016, in press)
28. Shibayama, N., Kaneko, T.: A peculiar higher order differential of CLEFIA. In: ISITA, pp. 526–530. IEEE (2012)
29. Sun, S., Hu, L., Wang, P., Qiao, K., Ma, X., Song, L.: Automatic security evaluation and (related-key) differential characteristic search: application to SIMON, PRESENT, LBlock, DES(L) and other bit-oriented block ciphers. In: Sarkar, P., Iwata, T. (eds.) ASIACRYPT 2014. LNCS, vol. 8873, pp. 158–178. Springer, Heidelberg (2014). doi:10.1007/978-3-662-45611-8_9
30. Suzaki, T., Minematsu, K., Morioka, S., Kobayashi, E.: TWINE: a lightweight block cipher for multiple platforms. In: Knudsen, L.R., Wu, H. (eds.) SAC 2012. LNCS, vol. 7707, pp. 339–354. Springer, Heidelberg (2013). doi:10.1007/978-3-642-35999-6_22
31. Tanaka, H., Hisamatsu, K., Kaneko, T.: Strength of ISTY1 without FL function for higher order differential attack. In: Fossorier, M., Imai, H., Lin, S., Poli, A. (eds.) AAECC 1999. LNCS, vol. 1719, pp. 221–230. Springer, Heidelberg (1999). doi:10.1007/3-540-46796-3_22
32. Todo, Y.: Integral cryptanalysis on full MISTY1. In: Gennaro, R., Robshaw, M. (eds.) CRYPTO 2015. LNCS, vol. 9215, pp. 413–432. Springer, Heidelberg (2015). doi:10.1007/978-3-662-47989-6_20
33. Todo, Y.: Structural evaluation by generalized integral property. In: Oswald, E., Fischlin, M. (eds.) EUROCRYPT 2015. LNCS, vol. 9056, pp. 287–314. Springer, Heidelberg (2015). doi:10.1007/978-3-662-46800-5_12
34. Todo, Y., Morii, M.: Bit-based division property and application to SIMON family. In: Peyrin, T. (ed.) FSE 2016. LNCS, vol. 9783, pp. 357–377. Springer, Heidelberg (2016). doi:10.1007/978-3-662-52993-5_18
35. Todo, Y., Morii, M.: Compact representation for division property. In: Foresti, S., Persiano, G. (eds.) CANS 2016. LNCS, vol. 10052, pp. 19–35. Springer, Cham (2016). doi:10.1007/978-3-319-48965-0_2

36. Wang, Q., Liu, Z., Varıcı, K., Sasaki, Y., Rijmen, V., Todo, Y.: Cryptanalysis of reduced-round SIMON32 and SIMON48. In: Meier, W., Mukhopadhyay, D. (eds.) INDOCRYPT 2014. LNCS, vol. 8885, pp. 143–160. Springer, Cham (2014). doi:10.1007/978-3-319-13039-2_9

37. Wu, W., Zhang, L.: LBlock: a lightweight block cipher. In: Lopez, J., Tsudik, G. (eds.) ACNS 2011. LNCS, vol. 6715, pp. 327–344. Springer, Heidelberg (2011). doi:10.1007/978-3-642-21554-4_19

38. Xiang, Z., Zhang, W., Bao, Z., Lin, D.: Applying MILP method to searching integral distinguishers based on division property for 6 lightweight block ciphers (2016). https://eprint.iacr.org/2016/857, (Accepted to ASIACRYPT 2016)

39. Yeom, Y., Park, S., Kim, I.: On the security of CAMELLIA against the square attack. In: Daemen, J., Rijmen, V. (eds.) FSE 2002. LNCS, vol. 2365, pp. 89–99. Springer, Heidelberg (2002). doi:10.1007/3-540-45661-9_7

40. Z'aba, M.R., Raddum, H., Henricksen, M., Dawson, E.: Bit-pattern based integral attack. In: Nyberg, K. (ed.) FSE 2008. LNCS, vol. 5086, pp. 363–381. Springer, Heidelberg (2008). doi:10.1007/978-3-540-71039-4_23

41. Zhang, H., Wu, W.: Structural evaluation for generalized feistel structures and applications to LBlock and TWINE. In: Biryukov, A., Goyal, V. (eds.) INDOCRYPT 2015. LNCS, vol. 9462, pp. 218–237. Springer, Cham (2015). doi:10.1007/978-3-319-26617-6_12

42. Zheng, Y., Matsumoto, T., Imai, H.: On the construction of block ciphers provably secure and not relying on any unproved hypotheses. In: Brassard, G. (ed.) CRYPTO 1989. LNCS, vol. 435, pp. 461–480. Springer, New York (1990). doi:10.1007/0-387-34805-0_42

Author Index

Aljunid, Syarifah Ruqayyah 193
Arriaga, Afonso 347

Behnia, Rouzbeh 474
Beunardeau, Marc 127
Boyd, Colin 111
Boyen, Xavier 3, 111

Carr, Christopher 111, 389
Chatterjee, Sanjit 21
Chen, Xiaofeng 252
Cheung, Henry K.F. 56
Chow, Sherman S.M. 56
Cichoń, Jacek 252
Courtois, Nicolas T. 506
Crépeau, Claude 485

Ferradi, Houda 127, 494
Francis, Danny 193
Fujioka, Atsushi 436

Galteland, Herman 233, 463
Géraud, Rémi 127, 494
Gjøsteen, Kristian 233
Gligoroski, Danilo 389

Haines, Thomas 111
Hale, Britta 389
Hanaoka, Goichiro 481
Hanzlik, Lucjan 215, 252
Hayashi, Yu-ichi 193
Heng, Swee-Huay 474

Iovino, Vincenzo 347

Kluczniak, Kamil 215, 252
Koblitz, Neal 11, 21
Kuchta, Veronika 409
Kulis, Michal 145
Kutyłowski, Mirosław 215, 252

Lai, Russell W.F. 56
Lin, Fuchun 171
Lorek, Pawel 145

Markowitch, Olivier 409
Marrière, Nicolas 321
Menezes, Alfred 21, 83
Minier, Marine 294
Mizuki, Takaaki 193
Mjølsnes, Stig F. 463

Naccache, David 127, 494
Nachef, Valérie 321
Nishida, Takuya 193
Nishimaki, Ryo 521

Olimid, Ruxandra F. 463

Phan, Raphaël C.-W. 294

Safavi-Naini, Reihaneh 171
Samajder, Subhabrata 277
Sarkar, Palash 21, 83, 277
Sharifian, Setareh 171
Singh, Shashank 83
Slamanig, Daniel 364
So, Anthony Man-Cho 56
Sone, Hideaki 193
Spreitzer, Raphael 364

Tan, Syh-Yuan 474
Tang, Qiang 347
Tiessen, Tyge 544
Todo, Yosuke 553

Unterluggauer, Thomas 364

Volte, Emmanuel 321

Wang, Jianfeng 252

Yang, Nan 485

Zagorski, Filip 145